Benjamin Peirce

Tables of the Moon

Benjamin Peirce

Tables of the Moon

ISBN/EAN: 9783337396220

Printed in Europe, USA, Canada, Australia, Japan

Cover: Foto ©berggeist007 / pixelio.de

More available books at **www.hansebooks.com**

TABLES

OF THE MOON;

CONSTRUCTED FOR THE USE OF

THE AMERICAN EPHEMERIS AND NAUTICAL ALMANAC.

BY

PROFESSOR BENJAMIN PEIRCE, LL. D.

PUBLISHED BY AUTHORITY OF THE SECRETARY OF THE NAVY.

BUREAU OF NAVIGATION,
WASHINGTON.
1865.

PREFACE

TO THE SECOND EDITION.

One of the first steps in the preparation of the American Ephemeris and Nautical Almanac, founded by Congress in March, 1849, was the collection of materials for new Tables of the Moon. The work was completed, and the first edition of the American Tables was published in 1853.

Since this preliminary step, however indispensable, unavoidably delayed the appearance of the Almanac, I am very solicitous to make known, here and elsewhere, the great obligations which American astronomical science is under to Hon. John P. Kennedy, then the Secretary of the Navy, for his just estimate of the magnitude and importance of this work. Although the processes by which it was conducted had formed no part of his study, its objects and value were perfectly understood by a scholar who had already adorned the literature of his own country by his writings.

The American Tables, as is stated in the title-page of the first edition, are "constructed from Plana's theory, with Airy's and Longstreth's corrections, Hansen's two inequalities of long period arising from the action of Venus, and Hansen's values of the secular variations of the mean motion and the motion of the perigee."

The formulas of Plana's theory were taken, not from Plana's own work, but from pages lxxiv.-lxxx. of the Introduction to the Reduction of the Observations of the Moon, made at the Royal Observatory, at Greenwich, from 1750 to 1830.

These formulas were adopted because they claimed to be corrected by Plana himself, and, what is of more importance, because they had been compared with observation, and formed the basis of Airy's corrections of the elements of the Moon.

The term 0″.8 sin [23], which occurs in that formula, is very different from the corresponding term 3″.3 sin [23], which is originally given by Plana; but it appears from the remarks given on page lxxxi. of the Introduction to the Lunar Observations, that this correction was adopted with the assent of Plana, and it was, therefore, attributed to Plana in the American Tables.

Airy's corrections of the Lunar elements were adopted from his Memoir in Vol. XVII. of the Memoirs of the Royal Astronomical Society. In making these corrections, the two inequalities given at that time by Hansen, and produced by the disturbing force of Venus, were introduced in the comparison. Therefore it was necessary to introduce these terms also into the American Tables, in order to have the theory which had been verified by observation. Very different values of these terms have since been given by Hansen, and are introduced into his Tables. A new set of corrections of the Lunar elements has also been made by Airy; but these corrections were published several years after the appearance of the American Tables.

Just before the final preparation of the American Tables, and when the first volume of the American Ephemeris was on the eve of completion, Longstreth's empirical corrections of Plana's formula for the Moon's longitude were communicated to me by their author; they were accompanied by a table of comparisons (published in the Memoirs of the American Philosophical Society for 1853) which sufficiently attest their value and propriety, and the forthcoming volume of the Ephemeris was delayed until the corresponding changes were computed and applied.

The present edition of the American Tables of the Moon is the same as the first, with the exception of the correction of typographical errors; the substitution of the Tables of the Moon's Parallax constructed from Walker's and Adams's formulas, in the place of the original Parallax Tables; and the addition of a Table adapted to a convenient modification of the method of computing the latitude, by Professor J. D. Runkle.

A third edition will shortly be issued, of which the basis will still be Plana's theory, while the Tables will be corrected to conform to the new Solar Parallax, and the corrected elements of the Moon's orbit.

WASHINGTON, Navy Department,
March, 1865.

C. H. DAVIS,
Rear-Admiral, and Chief of the Bureau of Navigation.

INTRODUCTION.

I. CONSTRUCTION OF THE TABLES.

THESE tables are constructed from the theory of PLANA, modified by the theoretical investigations of HANSEN, and the empirical corrections of AIRY and LONGSTRETH.

The moon's mean longitude is also to be corrected by the two terms of long period, arising from the action of Venus, which were discovered by HANSEN, and printed in the *Astronomische Nachrichten*, No. 597, and which are expressed by

$\delta_{♀} \theta =$ the correction of the mean longitude of the moon arising from the action of Venus.

The following notation is adopted,

$\theta =$ the mean longitude of the moon,
$\varpi =$ the longitude of the moon's perigee,
$\eta =$ the longitude of the moon's node,
$\theta_{\odot} =$ the mean longitude of the sun,
$\varpi_{\odot} =$ the longitude of the sun's perigee.
$\theta', \theta'_{\odot}, \varpi'_{\odot}$, and $\eta' =$ the respective daily motions of $\theta, \varpi, \theta_{\odot}, \varpi_{\odot}$, and η.
$\theta_0, \varpi_0, \theta_{\odot 0}, \varpi_{\odot 0}, \eta_0 =$ the values of $\theta, \varpi, \theta_{\odot}, \varpi_{\odot}$, and η at any assumed epoch, as the beginning of the nineteenth century.

The values of θ, ϖ, &c., at any number of days i from this epoch are, from the formulæ of PLANA, with the corrections given by HANSEN in the *Astronomische Nachrichten*, No. 597,

$$\theta = \theta_0 + \theta' i + 8''.598\,(10)^{-9}\, i^2 + 3''.6483\,(10)^{-16}\, i^3 + \delta_{♀}\,\theta,$$
$$\varpi = \varpi_0 + \varpi' i - 27''.217\,(10)^{-9}\, i^2 - 13''.750\,(10)^{-16}\, i^3,$$
$$\theta_{\odot} = \theta_{\odot 0} + \theta'_{\odot}\, i + 0''.91584\,(10)^{-9}\, i^2 + 0''.03879\,(10)^{-16}\, i^3,$$
$$\varpi_{\odot} = \varpi_{\odot 0} + \varpi'_{\odot}\, i + 0''.91584\,(10)^{-9}\, i^2 + 0''.03879\,(10)^{-16}\, i^3,$$
$$\eta = \eta_0 + \eta' i + 6''.0355\; 10^{-9}\, i^2 + 2''.3744\,(10)^{-16}\, i^3.$$

The values of θ_0, ϖ_0, &c., and of θ'_0, ϖ'_0, &c., derived, for the mean noon of Washington of the date 1801, Jan. 0, from the values obtained by AIRY in his Memoir upon the *Corrections of the Elements of the Moon's Orbit*, published in the *Memoirs of the Royal Astronomical Society*, Vol. XVII., and from BESSEL'S Tables of the Sun, are

$\theta_0 = 107^\circ\ 55'\ 40''.5,$		$\theta' =$	$13^\circ\ 635''.02806697,$
$\varpi_0 = 266\ \ 4\ 51.3,$		$\varpi' =$	$0\ \ 401.05788886,$
$\theta_{\odot 0} = 279\ 52\ 44.3,$		$\theta'_{\odot} =$	$0\ \ 3548.32999304,$
$\varpi_{\odot 0} = 279\ 31\ 10.4,$		$\varpi'_{\odot} =$	$0\ \ \ \ 0.16947241,$
$\eta_0 = 13\ 55\ 52.6,$		$\eta' =$	$-0\ \ 190.63366070.$

INTRODUCTION.

The assumed Western Longitude of Greenwich from Washington is $18^{h}\ 51^{m}\ 48^{s}$.

The notation

g'' = the mean anomaly of Venus,
z = the mean anomaly of the sun $= \theta_\odot - \varpi_{\odot}$,
u = the uncorrected mean longitude of the moon $= \theta - \delta_h \theta$,
x = the mean anomaly of the moon $= \theta - \varpi$,
y = the mean argument of the latitude $= \theta - \eta$,
t = the argument of variation $= \theta - \theta_\odot$,
$\bar{u}$ = the true longitude of the moon,
$\bar{y}$ = the true argument of the latitude,
♀ = the mean heliocentric longitude of Venus,
⊕ = the mean heliocentric longitude of the earth $= 180^\circ + \theta_\odot$,
♃ = the mean heliocentric longitude of Jupiter,
$H = 8g'' - 13z + 315^\circ\ 30'$,
$H' = 18g'' - 16z - x + 35^\circ\ 20'.2$,

gives the following formulæ for the longitude, latitude, and horizontal parallax of the moon.

In writing these formulæ, most of the arguments are expressed by numbers, in which the digit occupying the place of units is the coefficient of x, that in the place of tens is the coefficient of t, that in the place of hundreds is the coefficient of z, that in the place of thousands is the coefficient of y, and that in the place of tens of thousands is the coefficient of u. The accents upon the numbers indicate that the coefficient is negative. Thus the number of the argument [14′ 253′] denotes

$$u - 4y + 2z + 5t - 3x.$$

The form of the formula of the longitude is modified simply by substituting the term dependent upon the true arguments of the latitude, instead of the equivalent terms. This is accomplished by adding to the formula of longitude the equation,

$$
\begin{aligned}
0 = & -416''.7 \sin 2\bar{y} + 411''.7 \sin 2y + 4''.1 \sin (2y + 2x), \\
& + \ 0''.9 \sin (2y - 2x) - 46''.0 \sin (2y - x) + 45''.2 \sin (2y + x), \\
& + \ 0''.3 \sin (2y + 3x) - 9''.3 \sin (2y + x - 2t) + 8''.8 \sin (2y - x + 2t), \\
& + \ 0''.9 \sin (2y + x + 2t) + 0''.1 \sin (2y - x - 2t) - 0''.4 \sin (2y + 2x - 2t), \\
& - \ 0''.5 \sin (2y - 2x + 2t) + 5''.7 \sin (2y + 2t) - 3''.8 \sin (2y - 2t), \\
& - \ 0''.4 \sin (2y - x + z) - 0''.4 \sin (x + z - 2y) - 0''.2 \sin (2y - 4t + x), \\
& + \ 1''.0 \sin (2y - t) - 0''.7 \sin (2y + z) + 1''.3 \sin (2y - z), \\
& - \ 0''.4 \sin (x - 2t + z + 2y) + 0''.3 \sin (2t + 2y - z) + 0''.3 \sin (2t - z - 2y).
\end{aligned}
$$

The form of the formula for latitude is modified in a way which is evident from inspection of the formula, and which is designed to obviate the necessity of correcting the arguments for the change from mean to true longitude. The formula of parallax is retained in PLANA's form.*

The terms of the various equations are given in the *first column* of the following tables of the formulæ united with the constants which are employed in the construction of the tables. The *second column* contains the initial letter of the authority for the corresponding coefficients of the equation, whether PLANA, AIRY, or LONGSTRETH. The letter is accented when the value of the coefficient is modified by the present form of the tables. The *third column* contains the period of the argument expressed in mean solar days. The *fourth column* contains the number of the argument. The *fifth column* contains the number of the table which gives the values of the equation.

* See Formula, page 305.

EQUATION. ϑ = the moon's true longitude. = nutation	AUTHORITY.	PERIOD.	ARGUMENT.	TABLE.
+ u — 8° 3135″.2	A.			2
+ 22655.226		d.		6
+ 22639.2 sin [1]	A.	27.55455245	1	6
+ 769.5 sin [2]	P.			6
+ 36.7 sin [3]	P.			6
+ 2.0 sin [4]	P.			6
+ 0.1 sin [5]	P.			6
+ 4587.4				7
+ 4586.9 sin [21′]	A.	31.81193574	2	7
+ 31.2 sin [42′]	L.			7
+ 2457.02				8
— 122.1 sin [10]	P.	29.53058800	3	8
+ 2371.0 sin [20]	A.			8
+ 0.9 sin [30]	P.			8
+ 14.4 sin [40]	P.			8
+ 670.5				9
— 670.3 sin [100]	A.	365.259687	4	9
— 7.9 sin [200]	P.			9
+ 225.3				10
+ 18.0 sin [1′1]	P.	411.78517	5	10
— 212.4 sin [2′2]	P.			10
+ 206.9				11
+ 206.9 sin [1′21′]	L.	34.84689	6	11
+ 192.1				12
+ 192.1 sin [21]	P.	9.61372	7	12
+ 165.9				13
+ 165.9 sin [1′20]	P.	15.38731	8	13
+ 148.1				14
+ 148.1 sin [1′01]	P.	29.80283	9	14
+ 110.0				15
— 110.0 sin [101]	L.	25.62170	10	15
+ 85.0				16
— 85.0 sin [2001′]	L′.	26.87829	11	16
+ 58.7				17
— 58.7 sin [202′0]	P′.	173.31006	12′	17
+ 38.0				18
+ 38.0 sin [41′]	P.	10.08160	13	18
+ 28.8				19
— 28.8 sin [121′]	P.	29.26328	14	19
+ 25.0				20
— 25.0 sin [120]	L.	14.19161	15	20
+ 21.0				21
— 8.2 sin [11]	P.	14.25418	16	21
+ 14.1 sin [22]	P.			21
+ 17.2	P.			22
+ 17.2 sin [110]	P.	27.32168	17	22
+ 14.0				23
+ 14.0 sin [1′21]	P.	9.87359	18	23
+ 12.8				24
— 12.8 sin [2′3]	P.	24.30220	19	24
+ 9.6				25
+ 9.6 sin [1′02]	P.	14.31731	20	25
+ 9.2				26
— 9.2 sin [12′2]	L.	131.67112	21	26
+ 8.1				27
— 0.4 sin [1′10]	P.	32.12809	22	27
+ 7.8 sin [2′20]	P.			27
+ 9.8				28
— 9.8 sin [202′1]	L′.	23.77463	23	28
+ 7.3				29
— 7.3 sin [102]	P.	13.27650	24	29
+ 2.9				30
— 2.9 sin [121]	P.	9.36717	25	30
+ 1.9				31
+ 1.9 sin [41]	L.	5.82261	26	31
+ 1.6				32

EQUATION. Equations of Longitude continued.	AUTHORITY.	PERIOD.	ARGUMENT.	TABLE.
+ ″0.8 sin [1′2]	P.	d. 25.82638	27	32
— 0.9 sin [2′4]	P.			32
+ 7.5				33
+ 7.5 sin [2′21′]	P.	38.52204	28	33
+ 6.3				34
— 6.3 sin [2′021]	P′.	32.76364	29	34
+ 3.8				35
+ 3.8 sin [1′41′]	P.	10.37093	30	35
+ 3.0				36
— 3.0 sin [31′]	P.	15.31442	31	36
+ 3.0				37
+ 3.0 sin [1′42′]	L.	16.63016	32	37
+ 2.1				38
+ 2.1 sin [2′01]	P.	32.45058	33	38
+ 2.0				39
— 2.0 sin [221′]	P.	27.09271	34	39
+ 1.2				40
— 1.2 sin [201]	P.	23.94223	35	40
+ 1.0				41
+ 1.0 sin [111]	P.	13.71881	36	41
+ 0.5				42
+ 0.5 sin [43′]	P.	37.62533	37	42
+ 0.4				43
— 0.4 sin [12]	P.	9.39439	38	43
+ 0.6				44
— 0.6 sin [32′]	P.	34.47528	39	44
+ 0.4				45
+ 0.4 sin [1′03]	P.	9.42177	40	45
+ 0.4				46
— 0.4 sin [103]	P.	8.95955	41	46
+ 0.3				47
+ 0.3 sin [2′21]	P.	10.14791	42	47
+ 0.6				48
+ 0.6 sin [1′22]	P.	7.26893	43	48
+ 0.3				49
— 0.3 sin [12′3]	P.	22.78614	44	49
+ 0.2				50
— 0.2 sin [2021]	P′.	5.63335	45	50
+ 0.4				51
— 0.4 sin [2′101]	P′.	29.01328	46	51
+ 1.3				52
+ 1.3 sin [21′00]	P′.	14.13256	47	52
+ 0.5				53
— 0.5 sin [2′022]	P.	14.96709	48	53
+ 0.2				54
— 0.2 sin [202′2]	P′.	12.76271	49	54
+ 0.4				55
— 0.4 sin [2101′]	P′.	25.03597	50	55
+ 0.2				56
— 0.2 sin [2001]	P′.	9.10846	51	56
+ 1.0				57
+ 1.0 sin [201′0]	P′.	25.23137	52	57
+ 2.6				58
— 2.6 sin [212′0]	P′.	117.53942	53	58
+ 1.2				59
+ 1.2 sin [1′40]	P.	7.53494	54	59
+ 0.9				60
— 0.9 sin [141′]	P.	9.81365	55	60
+ 0.7				61
— 0.7 sin [2100]	P′.	13.11748	56	61
+ 0.8				62
+ 0.8 sin [23]	P.	5.66247	57	62
+ 0.5				63
— 0.5 sin [142′]	P.	15.24221	58	63
+ 0.6				64
— 0.6 sin [204′1]	P′.	38.96397	59	64

EQUATION. Equations of Longitude continued.	AUTHORITY.	PERIOD.	ARGUMENT.	TABLE.
+ 0″.4				65
− 0.4 sin [212'1]	P'.	d. 22.32171	60	65
+ 0.4				66
+ 0.4 sin [4000]	P'.	6.80305	61	66
+ 0.3				67
+ 0.3 sin [21'20]	P'.	7.22100	62	67
+ 0.3				68
+ 0.3 sin [11000]	P.	13.63340	63	68
+ 0.3				69
− 0.3 sin [2'1'21]	P.	35.99212	64	69
+ 0.2				70
− 0.2 sin [140]	P.	7.23638	65	70
+ 0.2				71
− 0.2 sin [2'030]	P'.	35.59582	66	71
+ 1.0				72
− 1.0 sin [1'1001]	A.	27.44332	67	72
+ 0.7				73
+ 0.7 sin [11'001]	A.	27.66669	68	73
+ 2.2				74
− 2.2 sin [1'2'2]	L.	471.89326	69	74
+ 1.5				75
+ 1.5 sin [21'2'0]	P.	329.79056	70	75
+ 1.3				76
− 1.1 sin (♀−⊕)		583.921	71	76
+ 0.4 sin 2 (♀−⊕)				76
+ 0.8				77
+ 0.7 sin (⊕−♃)		398.884	72	77
− 0.2 sin 2 (⊕−♃)				77
+ 2.0				78
+ 2.0 sin [2002']	P'.	1095.1653	73'	78
+ 0.5				79
+ 0.5 sin [111']	P.	3232.8202	74	79
+ 53.2				80
+ 23.2 sin *H*.		84753.24	75	80
+ 65 65				81
+ 27.4 sin *H'*.		95489.94	76	81
−416.9 sin 2 *g*.			77	82
+ 6.38 sin (*u* − *y*)				81A.
− 0.97 cos (*u* − *y*)				81A.

*The Moon's Latitude $= + A \sin g + B \cos g$ *

EQUATION.	AUTHORITY.	PERIOD.	ARGUMENT.	TABLE.
− 60″.				
+ 15.8				92
+ 15.8 sin [102'1]	P.	d. 188.2015	78	92
+ 22.2				93
+ 14.4 sin [1001']	P.	2190.3306	73	93
+ 0.6				94
− 0.6 sin [101'0]	P.	346.6201	12'''	94
+ 0.2				95
− 0.2 sin [1'101]	P.	438.3608	79	95
+ 0.1				96
− 0.1 sin [1101']	P.	313.0547	80	96
+ 0.7				97
+ 0.7 sin [112'1]	P.	124.2046	81	97
+ 1.8				98
+ 1.8 sin [1'021]	P.	14.8655	82	98
+ 0.7				99
+ 0.7 sin [1'2'20]	P.	39.2116	83	99
+ 0.6				100
− 0.6 sin [1002]	P'.	9.5717	84	100
+ 0.5				101
− 0.5 sin [1'011]	P.	29.9342	48	101
+ 0.4				102
+ 0.4 sin [302'0]	P'.	23.5193	85	102
+ 0.3				103
+ 0.3 sin [1'042']	P.	38.2830	86	103

* See Table cxxxvii.

EQUATION. Equations of Latitude continued.	AUTHORITY.	PERIOD.	ARGUMENT.	TABLE.
+ 0″.2				104
− 0.2 sin [1021]	P.	d. 7.1041	87	104
+ 0.2				105
+ 0.2 sin [1021']	P.	14.6664	88	105
+ 0.2				106
+ 0.2 sin [1'022]	P.	9.6561	89	106
+ 0.2				107
− 0.2 sin [1'021]	P.	32.2808	90	107
+ 6.2				108
− 6.18 sin 3 *g*	P.		77	108
+ 9.1				109
+ 2.17 cos *ū*	A.			109
− 8.80 sin *ū*	A.			109

In which the values of *A*. and *B*. are given by the following formulæ:

EQUATION.	AUTHORITY.	PERIOD.	ARGUMENT.	TABLE.
$A =$ 17600″.0	A'.			
+ 846.0				
+ 527.5 cos [20'20]	P.	d. 173.31006	12''	83
+ 25.7				
+ 25.7 cos [2002']	P.	1095.1653	73'''	84
+ 22.1				
− 22.1 cos [212'0]	P.	117.5394	53'	85
+ 1.3				
− 1.3 cos [1]	P.	27.5546	1'	86
+ 10.3				
− 10.3 cos [21'2'0]	P.	329.7906	70'	87
+ 6.63				
− 1.3 cos [100]	P'.	365.2597	4	89
+ 4.7				
+ 4.7 cos [2'2]	P.	205.8926	5'	90
$B =$ −1000.				
+ 816.0				
− 527.5 sin [202'0]	P.	173.31006	12'	83
+ 25.7				
− 25.7 sin [2002']	P.	1095.1653	73''	84
+ 22.1				
− 22.1 sin [212'0]	P.	117.5394	53	85
+ 1.3				
− 1.3 sin [1]	P.	27.5546	1	86
+ 10.3				
+ 10.3 sin [21'2'0]	P.	329.7906	70	87
+ 88.9				
+ 48.9 sin [100]	P'.	365.2597	4	88
+ 4.7				
+ 4.7 sin [2'2]	P.	205.8926	5	90
+ 1.0				
− 1.0 sin [1'1]	P'.	411.7852	5	91

The Moon's Equatorial Horizontal Parallax =

EQUATION.	AUTHORITY.	PERIOD.	ARGUMENT.	TABLE.
3000″.				
+ 348.0				110
+ 186.8 cos [1]	P.	d. 27.55455	1''	110
+ 10.3 cos [2]	P.			110
+ 0.6 cos [3]	P.			110
+ 28.6				111
− 0.9 cos [10]	P.	29.53059	3	111
+ 27.6 cos [20]	P.			111
+ 0.1 cos [40]	P.			111
+ 34.2				112
+ 33.9 cos [21']	P.	31.81194	2	112
+ 0.3 cos [42']	P.			112

* See Formula, page 305.

EQUATION.	AUTHORITY.	PERIOD.	ARGUMENT.	TABLE.
The Moon's Equatorial Horizontal Parallax continued.				
+ 1.4″		d.		113
+ 1.4 cos [1′21′]	P.	36.8469	6	113
+ 3.1				114
+ 3.1 cos [21]	P.	9.6137	7	114
+ 2.2				115
+ 2.2 cos [1′20]	P.	15.3873	8	115
+ 1.2				116
+ 1.2 cos [101]	P.	29.8028	9	116
+ 0.9				117
− 0.9 cos [1′01]	P.	25.6217	10	117
+ 1.2				118
+ 1.2 cos [2001′]	P.	26.8783	11	118
+ 0.5				119
+ 0.5 cos [41′]	P.	10.0846	13	119
+ 0.2				120
− 0.2 cos [121′]	P.	29.2633	14	120
+ 0.6				121

EQUATION.	AUTHORITY.	PERIOD.	ARGUMENT.	TABLE.
The Moon's Equatorial Horizontal Parallax continued.		d.		
− 0.6″ cos [120]	P.	14.1916	15	121
+ 0.3				122
+ 0.3 cos [22]	P.	7.1271	16	122
+ 0.2				123
+ 0.2 cos [1′21]	P.	9.8736	18	123
+ 0.1				124
+ 0.1 cos [2′2]	P.	24.3022	19	124
+ 0.4				125
− 0.4 cos [100]	P.	365.2597	4	125
+ 0.2				126
+ 0.2 cos [22′]	P.	205.8926	5	126
+ 0.2				127
− 0.2 cos [20′20]	P.	173.3101	12	127
The Moon's Equatorial Semidiameter = The Horizontal Parallax.				
× 0.272274	B.			128

This value of the Moon's Semidiameter is that deduced by BURCKHARDT from eclipses and occultations, and should be used for the computation of these phenomena. But for Meridian Observations it seems to vary with the telescope ranging from 2″ to 3″ greater than the value here given.

The Principal Terms of the Second and Fourth Differences of the Longitude for a Quarter of a Day are given in the following formulæ.

EQUATION.	ARGUMENT.	TABLE.
Second Difference of the Longitude for a Quarter of a Day =		
−130.″		
+ 76.0		6″
− 73.563 sin [1]	1	6″
− 9.993 sin [2]		6″
− 1.071 sin [3]		6″
− 0.104 sin [4]		6″
− 0.008 sin [5]		6″
+ 11.2		7″
− 11.180 sin [21′]	2	7″
− 0.304 sin [42′]		7″
+ 31.16		8″
+ 0.345 sin [10]	3	8″
− 20.808 sin [20]		8″
− 0.023 sin [30]		8″
− 0.650 sin [40]		8″
+ 0.42		11″
− 0.42 sin [1′21′]	6	11″
+ 5.12		12″
− 5.117 sin [21]	7	12″
+ 1.73		13″
− 1.727 sin [21′]	8	13″
+ 0.41		14″
− 0.411 sin [1′01]	9	14″
+ 0.41		15″
+ 0.413 sin [101]	10	15″
+ 0.29		16″
+ 0.288 sin [2001′]	11	16″
+ 0.92		18″
− 0.920 sin [41′]	13	18″
+ 0.08		19″
+ 0.083 sin [121′]	14	19″
+ 0.31		20″
+ 0.306 sin [120]	15	20″
+ 0.80		21″
+ 0.100 sin [11]	16	21″
− 0.680 sin [22]		21″
+ 0.06		22″
− 0.057 sin [110]	17	22″

EQUATION.	ARGUMENT.	TABLE.
Second Difference, continued.		
+ 0.35″		23″
− 0.354 sin [1′21]	18	23″
+ 0.05		24″
+ 0.051 sin [2′3]	19	24″
+ 0.11		25″
− 0.115 sin [1′02]	20	25″
+ 0.07		27″
− 0.074 sin [2′20]	22	27″
+ 0.10		29″
+ 0.102 sin [102]	24	29″
+ 0.08		30″
− 0.081 sin [121]	25	30″
+ 0.14		31″
− 0.137 sin [41]	26	31″
+ 0.19		35″
− 0.087 sin [1′41′]	30	35″
Fourth Difference of the Longitude for a Quarter of a Day =		
− 1.00″		
+ 0.45		6$^{iv.}$
+ 0.239 sin [1]	1	6$^{iv.}$
+ 0.130 sin [2]		6$^{iv.}$
+ 0.031 sin [3]		6$^{iv.}$
+ 0.005 sin [4]		6$^{iv.}$
+ 0.03		7$^{iv.}$
+ 0.027 sin [21′]	2	7$^{iv.}$
+ 0.003 sin [42′]		7$^{iv.}$
+ 0.31		8$^{iv.}$
+ 0.303 sin [20]	3	8$^{iv.}$
+ 0.006 sin [30]		8$^{iv.}$
+ 0.029 sin [40]		8$^{iv.}$
+ 0.14		12$^{iv.}$
+ 0.136 sin [21]	7	12$^{iv.}$
+ 0.02		13$^{iv.}$
+ 0.018 sin [21′]	8	13$^{iv.}$
+ 0.02		18$^{iv.}$
+ 0.022 sin [41′]	13	18$^{iv.}$
+ 0.03		21$^{iv.}$
+ 0.033 sin [22]	16	21$^{iv.}$

Since HANSEN's inequalities of long period, arising from the action of Venus, affect the mean longitude of the moon, they must be applied to the several arguments contained in Tables V*a*. or V*b*., and to render these corrections always additive the arguments in Tables II*a*. and II*b*. have been diminished by the following constants: —

Arg. 1 by 0.001859,	Arg. 8 by 0.0012,
Arg. 2 by 0.001410,	Arg. 9 by 0.0012,
Arg. 3 by 0.001214,	Arg. 10 by 0.0010,
Arg. 6 by 0.0013,	Arg. 11 by 0.0011,
Arg. 7 by 0.0011,	Arg. 13 by 0.0012.

And Tables V*a*. and V*b*. have been constructed by the formulæ, —

	TABLE V*a*.	TABLE V*b*.
Corr. Arg. 1	$0.00050 + 0.0004933 \sin H$	$+ 0.001359 + 0.0005826 \sin H'$,
Corr. Arg. 2	$0.00060 + 0.000569 \sin H$	$+ 0.00081 + 0.000673 \sin H'$,
Corr. Arg. 3	$0.00053 + 0.000529 \sin H$	$+ 0.000684 + 0.000624 \sin H'$,
Corr. Arg. 6	$0.0006 + 0.00062 \sin H$	$+ 0.0007 + 0.00074 \sin H'$,
Corr. Arg. 7	$0.0002 + 0.00017 \sin H$	$+ 0.0009 + 0.00021 \sin H'$,
Corr. Arg. 8	$0.0003 + 0.00028 \sin H$	$+ 0.0009 + 0.00033 \sin H'$
Corr. Arg. 9	$0.0005 + 0.00053 \sin H$	$+ 0.0007 + 0.00053 \sin H'$,
Corr. Arg. 10	$0.0005 + 0.00046 \sin H$	$+ 0.0005 + 0.00054 \sin H'$,
Corr. Arg. 11	$0.0005 + 0.00048 \sin H$	$+ 0.0006 + 0.00056 \sin H'$,
Corr. Arg. 13	$0.0002 + 0.00018 \sin H$	$+ 0.0010 + 0.00021 \sin H'$.

II. DESCRIPTION OF THE TABLES.

TABLES I*a*., I*b*., and I*c*. are Tables of Astronomical Dates in mean solar days reckoned from the beginning of the Julian Period, from which the date of any event can be reduced to days and decimals of a day when it is originally given in the usual form with reference to the Christian Era.

Tables II*a*. and II*b*. contain the values of the mean longitude, the negative of the longitude of the node, and all the arguments for the mean Washington noon of every thousandth day from 2300000 to 2500000.

Tables III*a*. and III*b*. contain the motions of the mean longitude, and of the longitude of the node for every interval of days and fractions of a day less than 1000 days, and the first ten multiples of all the periods.

Table IV. contains the constants by which the mean longitude, and the longitude of the node, and the various arguments, can be reduced, from the interval from 2400000 to 2500000 days to any previous time after the commencement of the Julian Period.

Tables V*a*. and V*b*. contain the corrections of the various arguments for HANSEN's term of long period in the action of Venus.

Tables VI. to LXXXII. inclusive contain the different equations of the moon's longitude.

Tables VI″. to XXXVII″. inclusive contain the second differences for a quarter of a day of the moon's longitude.

Tables VI.$^{IV.}$ to XVI.$^{IV.}$ inclusive contain the fourth differences for a quarter of a day of the moon's longitude.

Tables LXXXIII. to CIX. inclusive contain the terms of the moon's latitude.

Tables CX. to CXXVII. inclusive contain the terms of the moon's horizontal parallax.

Table CXXVIII. contains the table for the moon's semidiameter.

III. USE OF THE TABLES.

1. Reduce the instant for which the moon's place is to be computed to its date in days and decimals of a day from Washington mean noon, by means of Tables I*a*., I*b*., and I*c*. The number of days for the beginning of the year is to be found in Table I*a*., that for the day of the year in Table I*b*., and the decimal of the day in Table I*c*.

2. Find the longitudes and arguments for the date from Tables II*a*., II*b*., III*a*., III*b*., V*a*., and V*b*.

If the date is contained within the interval 2300000 *to* 2500000 *days*,

Find the value of the longitude or argument for the preceding thousandth day in Table II*a*. or II*b*., in the column headed with the name of the longitude or argument; multiply the number which is contained in the column headed with the accented name of the longitude or argument by one thousandth part of the excess of the date over that of the preceding thousandth day. Find in Table III*a*. the motion of the longitude for this excess of days.

The sum of the three numbers thus found for each longitude is the mean longitude for the date, and the sum of the two numbers for each argument and of the excess of the date over that of the preceding thousandth day is the argument for the date. The argument thus found must be reduced by substituting as many entire multiples of the period of the argument as it contains, and Table III*b*. will facilitate this process of reduction. Finally, correct the argument by the addition of the corrections contained in Tables V*a*. and V*b*.

If the date is less than 2300000 *days*,

Change the hundred thousands of the date to 24; find the longitude and arguments by the preceding process for the changed date, with the exception of the corrections contained in Tables V*a*. and V*b*. Take from Table IV. the constants of longitude and argument which correspond to the hundred-thousandth day next preceding the given date. Multiply the excess of the date over the next preceding hundred-thousandth day by the accented number of which the logarithm is given in the column of Table IV. next following the longitude or argument to which it corresponds when it is of sufficient magnitude to be noticed. Multiply the square of this excess by the doubly accented number of which the logarithm is given in the column of Table IV. next following that of the singly accented number, when it is of sufficient magnitude to be noticed. The sum of these two products, added to the constant, and corrected by the addition of the correction of Tables V*a*. and V*b*. is the total addition to be made to the longitude or argument obtained by the preceding process.

3. Find, by means of the several arguments, the equations of longitude contained in the Tables VI. – LXXXI*a*. inclusive, and add their sum to the value of u already obtained.

It is to be observed, in entering Tables XVII., LIII., and LXXVIII., that

Arg. 12′ = Arg. 12, but that the period is half as long.
Arg. 48′ = Arg. 48 — $6^{d}.26$, and it has a half-period.
Arg. 73′ = Arg. 73 + $173^{d}.792$, and it has a half-period.

4. To the sum of u and the equations of longitude add the value of $y - u$ already obtained, and this sum is $\bar{y}$, which is the argument with which Table LXXXII. is to be entered for the final equation of longitude, which is to be added to the sum of u and the other equations in order to obtain the true longitude, $\bar{u}$, of the moon.

5. Find the value of A by entering

Table LXXXIII.	with the Arg.	$12'' =$	Arg. $12 -$	d. 43.3275,
Table LXXXIV.	with the Arg.	$73''' =$	Arg. $73 +$	47.5826,
Table LXXXV.	with the Arg.	$53' =$	Arg. $53 -$	29.38,
Table LXXXVI.	with the Arg.	$1' =$	Arg. $1 +$	6.89,
Table LXXXVII.	with the Arg.	$70' =$	Arg. $70 -$	82.45,
Table LXXXIX.	with the Arg.	4		
Table XC.	with the Arg.	$5' =$	Arg. $5 +$	51.47.

The value of A is the sum of the equations thus obtained increased by 17600″.

6. Find the value of B by entering Tables LXXXIII. to XCI. inclusive (excepting Table LXXXIX.) with the arguments respectively written above them.

The sum of the equations thus obtained diminished by 1000″ is the value of B.

7. Find the remaining equations of the moon's latitude by entering Tables XCII. to CIX. inclusive with the arguments respectively written above them.

The sum of the equations thus obtained, diminished by 60″, and added to the sum of the products of A by sin $\bar{y}$, and of B by cos. $\bar{y}$, is the moon's latitude.

8. The moon's equatorial horizontal parallax is found by entering tables CX. to CXXVII. inclusive with their respective arguments, and adding 3000″ to the sum of the equations thus obtained.

9. The moon's equatorial semidiameter is found by multiplying the equatorial horizontal parallax by 0.272274, which product may be readily obtained by means of Table CXXVIII.

This value is to be used for the computation of eclipses, but in the reduction of meridian observations it should be increased by 2″.5 in the case of a telescope for which no special investigation has been made, or by whatever constant may be found suitable by a special investigation for each telescope.

10. *Directions for the Computation of an Ephemeris.* The longitude, exclusive of nutation and the equation of Table LXXXII., should be found for each successive noon; and also the second and fourth differences of the equations for a quarter of a day from the Tables VI.[II] to XXI.[IV.] inclusive.

If then the following notation is adopted,

u_t = the longitude for the time t in days, exclusive of nutation and the equation of Table LXXXII.

A_t = the sum of all those equations for which tables of second differences are given.

A'_t = the sum of all those equations for which tables of second differences are not given.

B_t = the sum of all the tabular second differences from those tables for which tables of fourth differences are given.

B'_t = the sum of all the tabular second differences from those tables for which tables of fourth differences are not given.

$B''_t = B_t + B'_t$.

C_t = the sum of all the tabular fourth differences for all the equations for which they are given.

$\Delta^n V_t$ = the nth differences of the series of $n + 1$ numbers

$$V_{t-\frac{1}{8}n},\quad V_{t-\frac{1}{8}n+\frac{1}{4}},\ V_{t-\frac{1}{8}n+\frac{1}{2}},\ V_{t-\frac{1}{8}n+\frac{3}{4}},\ \&c.,$$
$$V_{t+\frac{1}{8}n-\frac{3}{4}},\ V_{t+\frac{1}{8}n-\frac{1}{2}},\ V_{t+\frac{1}{8}n-\frac{1}{4}},\ V_{t+\frac{1}{8}n},$$

of which the first corresponds to the date preceding the date t by $\frac{1}{8}n$ days, and the last corresponds to the date which follows the date t by $\frac{1}{8}n$ days, and the intermediate numbers are at intervals of a quarter of a day apart.

$\Delta'^n V_t$ = the nth quarter difference of the values of the $n + 1$ numbers

$$V_{t-\frac{1}{2}n},\quad V_{t-\frac{1}{2}n+1},\ V_{t-\frac{1}{2}n+2},\ V_{t-\frac{1}{2}n+3},\ \&c.,$$
$$V_{t+\frac{1}{2}n-3},\ V_{t+\frac{1}{2}n-2},\ V_{t+\frac{1}{2}n-1},\ V_{t+\frac{1}{2}n},$$

of which the first corresponds to the date which precedes the date t by $\frac{1}{2}n$ days, and the last corresponds to the date which follows the date t by $\frac{1}{2}n$ days, the intermediate numbers occurring at intervals of a day. In taking these quarter differences, all the terms of each order of difference are to be divided by four, before their quarter differences are obtained in the same way.

In the following computations it will be sufficiently accurate to put

$$\Delta^4\ V_t = \Delta'^4\ V_t,$$
$$\Delta^3\ V_t = \Delta'^3\ V_t - \tfrac{3}{4}\,\Delta'^4\ V_t.$$

Find for each noon and interpolate by differences for every six hours

$$\Delta^4 u_t = C_t + \Delta^3 B_t + \Delta^4 A'_t - 1''.$$

Test the accuracy of B_t by the formula

$$\Delta^3 B_t = C_t + \tfrac{3}{4}\,\Delta^2 C_t - 1''.$$

Find for each noon

$$\Delta^2 u_t = B''_t + \Delta^2 A'_t - 130''.$$

Find for each day and for the instant $\frac{3}{8}$ths of a day after the noon

$$\Delta^3 u_{t+\frac{3}{8}} = \Delta B''_{t+\frac{3}{8}} + \Delta^3 A'_{t+\frac{3}{8}} + \tfrac{1}{2}\,\Delta^4 u_{t+\frac{3}{8}}.$$

Interpolate the remaining third differences of u_t by means of the fourth differences, being careful to adopt a judicious distribution of the small changes which are required in the fourth differences to accomplish this interpolation.

Interpolate the second differences of u_t with a similar regard to changes of the third differences.

Test the accuracy of A_t by the formula

$$\Delta^2 A_t = B''_t + \tfrac{3}{4}\ C_t + \tfrac{3}{4}\,\Delta^2 B'_t - 130'',$$

Find for each day and the instant $\frac{3}{8}$ths of a day after the noon

$$\Delta u_{t+\frac{3}{8}} = 3^{\circ}\ 1058''.757 + \Delta A_{t+\frac{3}{8}} + \Delta A'_{t+\frac{3}{8}} + \tfrac{1}{2}\,\Delta^2 u_{t+\frac{3}{8}} - \tfrac{1}{4}\,\Delta^4 u_{t+\frac{3}{8}}.$$

Interpolate the first differences of u_t by the aid of the second differences, and by them interpolate the values of u_t.

The values of $\bar{y}$ should be obtained for each quarter of a day, and thence the equations from Table LXXXII.

The various terms of the latitude and horizontal parallax need only be found for each noon, except A. sin $\bar{y}$, which, after the interpolation of A., should be computed for each quarter of a day.

11. *Example.*

Find the moon's longitude, latitude, equatorial horizontal parallax for the date B. C. 413, August 27, at 6 o'clock in the afternoon, mean solar time at Athens.

Solution. The western longitude of Athens from Washington being $17^{h}\ 16^{m}\ 53^{s}$, the given date is in Washington time —413 August $26^{d}\ 23^{h}\ 16^{m}\ 53^{s}$. This is reduced to the Julian date in days by Tables *Ia.*, *Ib.*, and *Ic.*

—500 years is in Julian date in days	1536798.
13 *B.*	31777.
Aug. 26	238.
23^{h}	0.958333333
16^{m}	0.011111111
53^{s}	0.000613426
—413 Aug. $26^{d}\ 23^{h}\ 16^{m}\ 53^{s}$	1570813.970057870

By substituting 2400000 for the 1500000, the date becomes 2470813.970057870, whence from Tables II. to V. inclusive the arguments are obtained as follows:

	u	y-m	Argument 1.	2.	3.
From Table II*b*., for 2470000,	22° 2502″.58	132° 758″.5	10.9057714	21.243406	8.500660
Correction × .81397,	1.29	—0.9	1142	—58	263
From Table III*a*., or the excess of date,	285 692.60	43 370.1	813.9700579	813.970058	813.970058
From Table IV. for 1500000,	4 2708.83	220 1207.8	14.9092506	22.915317	3.859920
Accented No. × 70813.97,	—1035.81	730.4	—909656	57954	—20960
Doubly accented No. × (70813.97)²,	—4.94	3.2	—5031	340	—112
Correction from Table V*a*.,			335	61	30
Correction from Table V*b*.,			16989	1202	1048
Greatest Multiple of Period for Subtraction,			826.0365734	827.110329	797.325876
Mean Longitude or Argument for date,	312 1264.55	35 3069.1	13.0588844	31.077951	28.985031

Argument	4.	5.	6.	7.	8.	9.	10.
From Table II*b*., for 2470000,	281.3878	316.3339	5.6456	9.3725	4.7758	26.1947	23.6165
Correction × .81397,	0	13	0	0	0	0	0
The excess of date,	813.9701	813.9701	813.9701	813.9701	813.9701	813.9701	813.9701
From Table IV.,	365.1276	168.9847	25.1207	7.7149	4.0279	16.1365	13.8541
Accented No. × 70813.97,	0	—1.0665	656	—454	—219	—984	—846
Doubly acc. No. × (70813.97)²,	0	—60	8	—2	—1	—5	—5
Correction from Table V*a*.,	0	0	0	0	0	0	1
Correction from Table V*b*.,	0	0	11	10	11	10	8
Gr. Mult. of Period for Subtr.,	1095.7791	1235.3555	836.3251	826.7797	815.5275	834.4791	845.5160
Argument for date,	364.7064	62.8620	8.4785	4.2332	7.2254	21.7243	5.8405

Argument	11.	12.	13.	14.	15.	16.	17.
From Table II*b*., for 2470000,	0.9587	20.6699	3.2232	5.760	5.577	4.931	15.347
The excess of date,	813.9701	813.9701	813.9701	813.970	813.970	813.970	813.970
From Table IV.,	19.0676	171.4161	9.9006	21.068	3.704	9.575	3.560
Accented No. × 70813.97,	761	1642	51	59	—21	—61	—21
Doubly acc. No. × (70813.97)²,	4	9	0	0	0	0	0
Correction from Table V*a*.,	0	0	0	0	0	0	0
Correction from Table V*b*.,	9	0	11	0	0	0	0
Gr. Mult. of Period for Subtr.,	833.2270	693.2402	826.9369	819.372	823.113	826.743	819.650
Argument for date,	0.8468	312.9810	0.1632	21.485	0.117	1.672	13.206

Argument	18.	19.	20.	21.	22.	23.	24.	25.
From Table II*b*., for 2470000,	4.364	3.891	1.139	69.94	28.372	17.824	1.155	4.760
The excess of date,	813.970	813.970	813.970	813.97	813.970	813.970	813.970	813.970
From Table IV.,	7.935	8.791	1.178	108.02	4.212	12.603	1.086	7.516
Accented No. × 70813.97,	—50	—207	—101	—73	—24	—60	—94	—47
Gr. Mult. of Period for Subtr.,	819.508	826.275	816.087	921.70	835.330	832.112	809.866	824.311
Argument for date,	6.711	0.170	0.099	69.50	11.200	12.225	6.251	1.888

Argument	26.	27.	28.	29.	30.	31.	32.	33.
From Table II*b*., for 2470000,	3.087	0.34	24.849	21.477	1.760	9.15	11.01	15.34
The excess of date,	813.970	813.97	813.970	813.970	813.970	813.97	813.97	813.97
From Table IV.,	0.368	24.56	27.803	18.084	10.182	13.03	7.35	17.58
Accented No. × 70813.97,	—38	—16	—78	—149	5	2	7	—12
Gr. Mult. of Period for Subtr.,	815.165	826.44	847.484	851.855	819.304	826.98	831.51	843.72
Argument for date,	2.222	12.27	19.120	1.527	6.613	9.19	0.89	3.05

Argument	34.	35.	36.	37.	38.	39.	40.	41.
From Table II*b*., for 2470000,	25.51	18.42	12.65	36.4	17.71	13.1	16.26	7.35
The excess of date,	813 97	813.97	813.97	814.0	813.97	814.0	813.97	813.97
From Table IV.,	19.51	12.93	9.22	33.9	2.17	10.7	4.65	4.39
Accented No. × 70813.97,	5	—8	—5	3	—10	—2	11	11
Gr. Mult. of Period for Subtr.,	839.87	837.98	823.13	865.4	826.71	827.4	829.12	824.28
Argument for date,	19.17	7.26	12.66	19.2	7.04	10.2	5.87	1.54

Argument	42.	43.	44.	45.	46.	47.	48.	49.
From Table II*b*., for 2470000,	4.08	3.31	0.5	6.42	5.3	11.98	26.58	2.2
The excess of date,	813.97	813.97	814.0	813.97	814.0	813.97	813.97	814.0
From Table IV.,	8.11	9.77	8.2	11.54	20.4	3.54	1.51	1.2
Accented No. × 70813.97,	—5	—6	0	0	0	0	—12	0
Gr. Mult. of Period for Subtr.,	821.98	814.12	820.3	828.10	812.4	819.69	838.16	816.8
Argument for date,	4.13	12.87	2.4	3.83	27.3	9.80	3.78	0.6

Argument	50.	51.	52.	53.	54.	55.	56.	57.
From Table II*b*., for 2470000,	20.9	8.23	16.97	15.27	5.10	12.40	12.36	2.93
The excess of date,	814.0	813.97	813.97	813.97	813.97	813.97	813.97	813.97
From Table IV.,	17.6	7.20	3.04	116.22	3.93	9.63	3.25	10.20
Accented No. × 70813.97,	0	—4	0	12	0	0	0	0
Gr. Mult. of Period for Subtr.,	851.2	819.76	832.64	940.32	813.77	824.35	826.40	815.40
Argument for date,	1.3	9.60	1.34	5.26	9.23	11.65	3.18	11.70

Argument	58.	59.	60.	61.	62.	63.	64.	65.
From Table II*b*., for 2470000,	11.51	32.28	14.74	0.26	7.82	0.35	20.93	12.82
The excess of date,	813.97	813.97	813.97	813.97	813.97	813.97	813.97	813.97
From Table IV.,	6.71	28.49	11.84	3.40	11.09	8.73	19.88	3.83
Accented No. × 70813.97,	0	0	0	0	0	0	0	0
Gr. Mult. of Period for Subtr.,	823.08	857.21	825.90	816.37	823.19	818.00	827.82	824.95
Argument for date,	9.11	17.53	14.65	1.26	9.69	5.05	26.96	5.67

Argument	66.	67.	68.	69.	70.	71.	72.	73.
From Table II*b*., for 2470000,	26.4	27.20	6.94	281.5	61.7	201.5	179.3	1231.1
The excess of date,	814.0	813.97	813.97	814.0	814.0	814.0	814.0	814.0
From Table IV.,	5.1	4.21	25.69	387.5	0.4	406.6	281.4	184.3
Accented No. × 70813.97,	0	—8	—11	—26	—3	0	0	7.2
Gr. Mult. of Period for Subtr.,	818.7	823.30	830.00	1415.7	659.6	1167.8	1196.7	2190.2
Argument for date,	26.8	22.00	16.49	64.7	216.2	254.3	78.0	46.4

Argument	74.	75.	76.	78.	79.	80.	81.	82.
From Table II*b*., for 2470000,	1691	54809	77263	60.489	282.2	68.3	14.23	10.000
The excess of date,	814	814	814	813.970	814.0	814.0	813.97	813.970
From Table IV.,	1905	32286	52991	170.305	401.3	26.2	112.35	2.633
Accented No. × 70813.97,	9	0	0	—.427	0	0	—28	0
Gr. Mult. of Period for Subtr.,	3233	84753	95490	941.008	1315.1	626.1	869.43	817.603
Argument for date,	1186	3156	35578	103.329	182.4	282.4	70.84	9.000

Argument	83.	84.	85.	86.	87.	88	89.	90.
From Table II*b*., for 2470000,	12.80	17.33	9.53	24.78	1.01	4.42	1.92	15.3
The excess of date,	813.97	813.97	813.97	813.97	813.97	813.97	813.97	814.0
From Table IV.,	24.97	8.28	14.44	31.21	4.82	5.08	4.68	20.4
Accented No. × 70813.97,	0	0	0	0	0	0	0	0
Gr. Mult. of Period for Subtr.,	823.44	823.17	823.18	842.22	809.87	821.32	811.11	839.3
Argument for date,	28.30	16.41	14.76	27.74	9.93	2.15	9.46	10.4

HENCE:

Arg. 12′ = 139.6709
Arg. 12″ = 96.3434
Arg. 48′ = 12.49
Arg. 12‴ = 292.9810

Arg. 73′ = 220.2
Arg. 73″ = 367.8
Arg. 73‴ = 94.0
Arg. 53′ = 93.42

Arg. 1′ = 19.95
Arg. 70′ = 133.7
Arg. 5′ = 114.33

TABLE.	EQUATION.	TABLE.	EQUATION.
VI.	45193.81	XLV.	0.04
VII.	71.73	XLVI.	0.02
VIII.	1655.54	XLVII.	0.04
IX.	1340.46	XLVIII.	0.13
X.	206.06	XLIX.	0.26
XI.	176.80	L.	0.04
XII.	360.58	LI.	0.40
XIII.	323.57	LII.	2.59
XIV.	170.69	LIII.	0.25
XV.	130.22	LIV.	0.22
XVI.	169.81	LV.	0.41
XVII.	45.31	LVI.	0.01
XVIII.	4.56	LVII.	0.92
XIX.	23.67	LVIII.	5.14
XX.	48.53	LIX.	2.16
XXI.	37.54	LX.	0.45
XXII.	34.12	LXI.	1.39
XXIII.	20.49	LXII.	0.55
XXIV.	25.40	LXIII.	0.03
XXV.	0.51	LXIV.	0.04
XXVI.	0.00	LXV.	0.38
XXVII.	10.81	LXVI.	0.30
XXVIII.	0.02	LXVII.	0.59
XXIX.	0.45	LXVIII.	0.10
XXX.	4.33	LXIX.	0.20
XXXI.	3.16	LXX.	0.36
XXXII.	1.78	LXXI.	0.13
XXXIII.	14.92	LXXII.	0.10
XXXIV.	0.49	LXXIII.	1.19
XXXV.	2.05	LXXIV.	4.40
XXXVI.	0.16	LXXV.	2.93
XXXVII.	0.13	LXXVI.	1.49
XXXVIII.	0.02	LXXVII.	1.33
XXXIX.	0.03	LXXVIII.	1.55
XL.	2.38	LXXIX.	0.50
XLI.	1.35	LXXX.	31.26
XLII.	0.99	LXXXI.	81.62
XLIII.	0.56	LXXXIA.	—4.52
XLIV.	0.73		

Sum of Equations of Longitude,	0	50222.76
or,	13	3422.76
Mean Longitude,	312	1264.55
Longitude in Orbit,	326	1087.31
$y - u$,	35	3069.1
$\bar{y}$,	2	556.4
Table LXXXII.,		—31.33
Nutation of Equinox,		10.54
$\bar{u}$ = true Longitude =	326	1066.52

TABLE.	A.	B.
LXXXIII.	1359.62	725.75
LXXXIV.	0.72	31.78
LXXXV.	26.83	43.69
LXXXVI.	1.07	0.01
LXXXVII.	13.42	20.11
LXXXVIII.		40.00
LXXXIX.	6.66	
XC.	0.00	4.73
XCI.		2.00
Constant,	17600.	—1000.
	A = 19008.32	B = — 131.93

TABLE.	EQUATION.	TABLE.	EQUATION.
XCII.	31.50	CI.	0.98
XCIII.	7.90	CII.	0.47
XCIV.	0.67	CIII.	0.39
XCV.	0.23	CIV.	0.00
XCVI.	0.11	CV.	0.00
XCVII.	0.77	CVI.	0.39
XCVIII.	3.44	CVII.	0.26
XCIX.	0.87	CVIII.	5.50
C.	0.90	CIX.	15.79
Constant,			—60.
Sum,			10.17
$A \sin \bar{y}$,			714.62
$B \cos \bar{y}$,			—131.83
☽'s Latitude =			592.96
=		9′	52.96

TABLE.	EQUATION.	TABLE.	EQUATION.
CX	367.08	CXXIII.	0.17
CXI.	57.27	CXXIV.	0.03
CXII.	28.73	CXXV.	0.14
CXIII.	2.88	CXXVI.	0.08
CXIV.	4.57	CXXVII.	0.01
CXV.	2.51	CXXVIII.	0.09
CXVI.	0.02	CXXIX.	0.07
CXVII.	0.02	CXXX.	0.06
CXVIII.	0.67	CXXXI.	0.11
CXIX.	0.31	CXXXII.	0.40
CXX.	0.46	CXXXIII.	0.63
CXXI.	0.42	CXXXIV.	0.01
CXXII.	0.21		
Constant,			3000.
sin ☽'s Horizontal Parallax =			3466.95
Table CXXXV.,			0.16
☽'s Horizontal Parallax,			3467.11

TABLE Ia.

Date of Jan. 0 in Common Years, and of Jan. 1 in Bissextile Years.

Year.	Date in Mean Solar Days.	Year.	Date in Mean Solar Days.	YEAR IN THE CENTURY.		Days from previous Centennial Date.	YEAR IN THE CENTURY.		Days from previous Centennial Date.
				If Negative.	If Positive.		If Negative.	If Positive.	
—4713 *B.*	0	—1000	1356173	100	1	0	50	51	18262
—4712	365	— 900	1392698	99	2	365	49 *B.*	52 *B.*	18628
—4711	730	— 800	1429223	98	3	730	48	53	18993
—4710	1095	— 700	1465748	97 *B.*	4 *B.*	1096	47	54	19358
—4709 *B.*	1461	— 600	1502273	96	5	1461	46	55	19723
—4708	1826	— 500	1538798	95	6	1826	45 *B.*	56 *B.*	20089
—4707	2191	— 400	1575323	94	7	2191	44	57	20454
—4706	2556	— 300	1611848	93 *B.*	8 *B.*	2557	43	58	20819
—4705 *B*	2922	— 200	1648373	92	9	2922	42	59	21184
—4704	3287	— 100	1684898	91	10	3287	41 *B.*	60 *B.*	21550
—4703	3652	1	1721423	90	11	3652	40	61	21915
—4702	4017	101	1757948	89 *B.*	12 *B.*	4018	39	62	22280
—4701 *B.*	4383	201	1794473	88	13	4383	38	63	22645
—4700	4748	301	1830998	87	14	4748	37 *B.*	64 *B.*	23011
—4600	41273	401	1867523	86	15	5113	36	65	23376
—4500	77798	501	1904048	85 *B.*	16 *B.*	5479	35	66	23741
—4400	114323	601	1940573	84	17	5844	34	67	24106
—4300	150848	701	1977098	83	18	6209	33 *B.*	68 *B.*	24472
—4200	187373	801	2013623	82	19	6574	32	69	24837
—4100	223898	901	2050148	81 *B.*	20 *B.*	6940	31	70	25202
—4000	260423	1001	2086673	80	21	7305	30	71	25567
—3900	296948	1101	2123198	79	22	7670	29 *B.*	72 *B.*	25933
—3800	333473	1201	2159723	78	23	8035	28	73	26298
—3700	369998	1301	2196248	77 *B.*	24 *B.*	8401	27	74	26663
—3600	406523	1401	2232773	76	25	8766	26	75	27028
—3500	443048	1501	2269298	75	26	9131	25 *B.*	76 *B.*	27394
—3400	479573	1583	2299238	74	27	9496	24	77	27759
—3300	516098	1584 *B.*	2299604	73 *B.*	28 *B.*	9862	23	78	28124
—3200	552623	1585	2299969	72	29	10227	22	79	28489
—3100	589148	1586	2300334	71	30	10592	21 *B.*	80 *B.*	28855
—3000	625673	1587	2300699	70	31	10957	20	81	29220
—2900	662198	1588 *B.*	2301065	69 *B.*	32 *B.*	11323	19	82	29585
—2800	698723	1589	2301430	68	33	11688	18	83	29950
—2700	735248	1590	2301795	67	34	12053	17 *B.*	84 *B.*	30316
—2600	771773	1591	2302160	66	35	12418	16	85	30681
—2500	808298	1592 *B.*	2302526	65 *B.*	36 *B.*	12784	15	86	31046
—2400	844823	1593	2302891	64	37	13149	14	87	31411
—2300	881348	1594	2303256	63	38	13514	13 *B.*	88 *B.*	31777
—2200	917873	1595	2303621	62	39	13879	12	89	32142
—2100	954398	1596 *B.*	2303987	61 *B.*	40 *B.*	14245	11	90	32507
—2000	1090923	1597	2304352	60	41	14610	10	91	32872
—1900	1127448	1598	2304717	59	42	14975	9 *B.*	92 *B.*	33238
—1800	1163973	1599	2305082	58	43	15340	8	93	33603
—1700	1200498	1600 *B.*	2305448	57 *B.*	44 *B.*	15706	7	94	33968
—1600	1237023	1601	2305813	56	45	16071	6	95	34333
—1500	1273548	1701	2312337	55	46	16436	5 *B.*	96 *B.*	34699
—1400	1210073	1801	2378861	54	47	16801	4	97	35064
—1300	1246598	1901	2415385	53 *B.*	48 *B.*	17167	3	98	35429
—1200	1283123			52	49	17532	2	99	35794
—1100	1319648			51	50	17897	1 *B.*	100 *B.*	36160
—1000	1356173			50	51	18262		100	36159

TABLE I*b*.

Number of Days from Jan. 0 in Common Years, and Jan. 1 in Bissextile Years.

Day of Month.	JANUARY. Common Year.	JANUARY. Bissextile Year.	FEBRUARY. Common Year.	FEBRUARY. Bissextile Year.	MARCH.	APRIL.	MAY.	JUNE.	JULY.	AUGUST.	SEPTEMBER.	OCTOBER.	NOVEMBER.	DECEMBER.	1582. OCTOBER.	1582. NOVEMBER.	1582. DECEMBER.
1	1	0	32	31	60	91	121	152	182	213	244	274	305	335	274	295	325
2	2	1	33	32	61	92	122	153	183	214	245	275	306	336	275	296	326
3	3	2	34	33	62	93	123	154	184	215	246	276	307	337	276	297	327
4	4	3	35	34	63	94	124	155	185	216	247	277	308	338	277	298	328
5	5	4	36	35	64	95	125	156	186	217	248	278	309	339		299	329
6	6	5	37	36	65	96	126	157	187	218	249	279	310	340		300	330
7	7	6	38	37	66	97	127	158	188	219	250	280	311	341		301	331
8	8	7	39	38	67	98	128	159	189	220	251	281	312	342		302	332
9	9	8	40	39	68	99	129	160	190	221	252	282	313	343		303	333
10	10	9	41	40	69	100	130	161	191	222	253	283	314	344		304	334
11	11	10	42	41	70	101	131	162	192	223	254	284	315	345		305	335
12	12	11	43	42	71	102	132	163	193	224	255	285	316	346		306	336
13	13	12	44	43	72	103	133	164	194	225	256	286	317	347		307	337
14	14	13	45	44	73	104	134	165	195	226	257	287	318	348		308	338
15	15	14	46	45	74	105	135	166	196	227	258	288	319	349	278	309	339
16	16	15	47	46	75	106	136	167	197	228	259	289	320	350	279	310	340
17	17	16	48	47	76	107	137	168	198	229	260	290	321	351	280	311	341
18	18	17	49	48	77	108	138	169	199	230	261	291	322	352	281	312	342
19	19	18	50	49	78	109	139	170	200	231	262	292	323	353	282	313	343
20	20	19	51	50	79	110	140	171	201	232	263	293	324	354	283	314	344
21	21	20	52	51	80	111	141	172	202	233	264	294	325	355	284	315	345
22	22	21	53	52	81	112	142	173	203	234	265	295	326	356	285	316	346
23	23	22	54	53	82	113	143	174	204	235	266	296	327	357	286	317	347
24	24	23	55	54	83	114	144	175	205	236	267	297	328	358	287	318	348
25	25	24	56	55	84	115	145	176	206	237	268	298	329	359	288	319	349
26	26	25	57	56	85	116	146	177	207	238	269	299	330	360	289	320	350
27	27	26	58	57	86	117	147	178	208	239	270	300	331	361	290	321	351
28	28	27	59	58	87	118	148	179	209	240	271	301	332	362	291	322	352
29	29	28		59	88	119	149	180	210	241	272	302	333	363	292	323	353
30	30	29			89	120	150	181	211	242	273	303	334	364	293	324	354
31	31	30			90		151		212	243		304		365	294		355

TABLE I*c*.

To Reduce Hours, Minutes, and Seconds of Time to Decimals of a Day.

Hours and Minutes.	Decimal of a Day.	Minutes.	Decimal of a Day.	Minutes and Seconds.	Decimal of a Day.	Seconds.	Decimal of a Day.
h.		m		m.		s	
1	.041666667	13	.009027778	50	.034722222	23	.000266204
2	.083333333	14	.009722222	51	.035416667	24	.000277778
3	.125000000	15	.010416667	52	.036111111	25	.000289352
4	.166666667	16	.011111111	53	.036805556	26	.000300926
5	.208333333	17	.011805556	54	.037500000	27	.000312500
6	.250000000	18	.012500000	55	.038194444	28	.000324074
7	.291666667	19	.013194444	56	.038888889	29	.000335648
8	.333333333	20	.013888889	57	.039583333	30	.000347222
9	.375000000	21	.014583333	58	.040277778	31	.000358796
10	.416666667	22	.015277778	59	.040972222	32	.000370370
11	.458333333	23	.015972222	60	.041666667	33	.000381944
12	.500000000	24	.016666667			34	.000393519
13	.541666667	25	.017361111			35	.000405093
14	.583333333	26	.018055556			36	.000416667
15	.625000000	27	.018750000			37	.000428241
16	.666666667	28	.019444444	s. 1	.000011574	38	.000439815
17	.708333333	29	.020138889	2	.000023148	39	.000451389
18	.750000000	30	.020833333	3	.000034722	40	.000462963
19	.791666667	31	.021527778	4	.000046296	41	.000474537
20	.833333333	32	.022222222	5	.000057870	42	.000486111
21	.875000000	33	.022916667	6	.000069444	43	.000497685
22	.916666666	34	.023611111	7	.000081019	44	.000509259
23	.958333333	35	.024305556	8	.000092593	45	.000520833
24	1.000000000	36	.025000000	9	.000104167	46	.000532407
		37	.025694444	10	.000115741	47	.000543981
m. 1	.000694444	38	.026388889	11	.000127315	48	.000555556
2	.001388889	39	.027083333	12	.000138889	49	.000567130
3	.002083333	40	.027777778	13	.000150463	50	.000578704
4	.002777778	41	.028472222	14	.000162037	51	.000590278
5	.003472222	42	.029166667	15	.000173611	52	.000601852
6	.004166667	43	.029861111	16	.000185185	53	.000613426
7	.004861111	44	.030555556	17	.000196759	54	.000625000
8	.005555556	45	.031250000	18	.000208333	55	.000636574
9	.006250000	46	.031944444	19	.000219907	56	.000648148
10	.006944444	47	.032638889	20	.000231481	57	.000659722
11	.007638889	48	.033333333	21	.000243056	58	.000671296
12	.008333333	49	.034027778	22	.000254630	59	.000682870
13	.009027778	50	.034722222	23	.000266204	60	.000694444

TABLE IIa.

Epochs and Arguments for each Thousandth Day, from 2300000 to 2400000.

For Washington Mean Noon.

Day of Julian Period.	u	u′	y-u	(y-u′)	Arg. 1.	1′.	2.	2′.
	° ″	″	° ″	″	d.	0.000	d.	0.000
2300000	315 909.18	−1.34	130 248.9	+0.9	22.4927509	−1188	24.229050	+079
2301000	171 2335.93	1.32	183 83.5	0.9	2.9741917	1173	6.247185	078
2302000	28 162.69	1.30	235 3518.1	0.9	11.0101863	1158	20.077254	077
2303000	244 1589.47	1.29	288 3352.7	0.9	19.0461825	1143	2.095387	076
2304000	100 3016.27	1.27	341 3187.2	0.9	27.0821802	1128	15.925454	075
2305000	317 843.09	1.26	34 3021.8	0.9	7.5636269	1113	29.755521	074
2306000	173 2269.92	1.24	87 2856.4	0.8	15.5996276	1097	11.773650	073
2307000	30 96.77	1.22	140 2690.9	0.8	23.6356298	1082	25.603715	072
2308000	246 1523.64	1.21	193 2525.4	0.8	4.1170810	1067	7.621843	071
2309000	102 2950.52	1.19	246 2359.9	0.8	12.1530862	1052	21.451905	070
2310000	319 777.42	1.17	299 2194.4	0.8	20.1890929	1037	3.470031	069
2311000	175 2204.34	1.15	352 2028.9	0.8	0.6705187	1022	17.300092	068
2312000	32 31.27	1.13	45 1863.4	0.8	8.7065584	1007	31.130151	067
2313000	248 1458.23	1.12	98 1697.8	0.8	16.7425697	0992	13.148273	066
2314000	104 2885.19	1.10	151 1532.3	0.8	24.7785824	0977	26.978331	065
2315000	321 712.18	1.09	204 1366.7	0.8	5.2600442	0962	8.996451	064
2316000	177 2139.18	1.07	257 1201.1	0.7	13.2960600	0946	22.826507	063
2317000	33 3566.20	1.05	310 1035.6	0.7	21.3320773	0931	4.814626	062
2318000	250 1393.24	1.04	3 870.0	0.7	1.8135436	0916	18.671680	061
2319000	106 2820.29	1.02	56 704.4	0.7	9.8195639	0901	0.692797	060
2320000	323 647.37	1.00	109 538.6	0.7	17.8855856	0886	14.522848	059
2321000	179 2074.15	0.98	162 373.0	0.7	25.9216089	0871	28.352899	058
2322000	35 3501.56	0.96	215 207.1	0.7	6.4030813	0856	10.371013	057
2323000	252 1328.68	0.95	268 41.7	0.7	14.4391077	0841	24.201061	056
2324000	108 2755.82	0.93	320 3176.0	0.7	22.4751355	0826	6.219173	055
2325000	325 582.98	0.92	13 3310.4	0.7	2.9566124	0811	20.049220	054
2326000	181 2010.15	0.90	66 3144.7	0.6	10.9926133	0795	2.067329	053
2327000	37 3437.34	0.88	119 2979.0	0.6	19.0286756	0780	15.897374	052
2328000	254 1264.55	0.87	172 2813.2	0.6	27.0617095	0765	29.727418	051
2329000	110 2691.77	0.85	225 2647.5	0.6	7.5161925	0750	11.745524	050
2330000	327 519.02	0.83	278 2481.8	0.6	15.5822294	0735	25.575566	049
2331000	183 1946.28	0.81	331 2316.0	0.6	23.6182678	0720	7.593671	048
2332000	39 3373.55	0.79	21 2150.3	0.6	4.0997553	0705	21.423711	047
2333000	256 1200.85	0.78	77 1984.5	0.6	12.1357967	0690	3.441814	046
2334000	112 2628.16	0.76	130 1818.7	0.6	20.1718397	0675	17.271851	045
2335000	329 455.48	0.75	183 1652.9	0.6	0.6533317	0660	31.101888	044
2336000	165 1882.83	0.73	236 1487.1	0.5	8.6893777	0644	13.119988	043
2337000	41 3310.19	0.71	289 1321.3	0.5	16.7254252	0629	26.950023	042
2338000	258 1137.57	0.70	342 1155.4	0.5	24.7614742	0614	8.968121	041
2339000	114 2564.96	0.68	35 989.6	0.5	5.2429723	0599	22.798153	040
2340000	331 392.38	0.66	88 823.7	0.5	13.2790243	0584	4.816249	039
2341000	187 1819.81	0.64	141 657.8	0.5	21.3150778	0569	18.646280	038
2342000	43 3247.25	0.62	194 491.9	0.5	1.7965804	0554	0.664373	037
2343000	260 1074.72	0.61	247 326.0	0.5	9.8326370	0538	14.494402	036
2344000	116 2502.20	0.59	300 160.1	0.4	17.8686951	0523	28.324430	035
2345000	333 329.70	0.58	352 3594.2	0.4	25.9047547	0508	10.342521	034
2346000	189 1757.21	0.56	45 3428.3	0.4	6.3862634	0493	24.172546	033
2347000	45 3184.75	0.54	98 3262.3	0.4	14.4223261	0478	6.190635	032
2348000	262 1012.30	0.53	151 3096.3	0.3	22.4583902	0462	20.020659	031
2349000	118 2439.86	0.51	204 2930.4	0.3	2.9399035	0447	2.038746	030
2350000	335 267.45	−0.49	257 2764.4	+0.3	10.9759706	−0432	15.868767	+029

TABLE IIa.

Epochs and Arguments for each Thousandth Day, from 2300000 to 2400000.

For Washington Mean Noon.

Day of Julian Period.	u	u′	y-u	(y-u′)	Arg. 1.	1′.	2.	2′.
	° ″	″	° ″	″	d.	0.000	d.	0.000
2350000	335 267.45	−0.49	257 2764.4	+0.3	10.9759706	−0132	15.868767	+029
2351000	191 1695.05	0.17	310 2598.4	0.3	19.0120393	0117	29.698788	028
2352000	47 3122.67	0.45	3 2432.4	0.3	27.0481095	0102	11.716872	027
2353000	264 950.30	0.44	56 2266.4	0.3	7.5296288	0387	25.546891	026
2354000	120 2377.96	0.42	109 2100.3	0.3	15.5657021	0372	7.564973	025
2355000	337 205.63	0.41	162 1934.3	0.3	23.6017769	0357	21.394990	024
2356000	193 1633.31	0.39	215 1768.2	0.2	4.0833007	0341	3.413069	023
2357000	49 3061.02	0.37	268 1602.2	0.2	12.1193785	0326	17.243084	022
2358000	266 888.74	0.36	321 1436.1	0.2	20.1554579	0311	31.073098	021
2359000	122 2316.48	0.34	14 1270.0	0.2	0.6369863	0296	13.091175	020
2360000	339 144.23	0.32	67 1103.9	0.2	8.6730686	0281	26.921186	019
2361000	195 1572.01	0.30	120 937.8	0.2	16.7091525	0266	8.939261	018
2362000	51 2999.80	0.28	173 771.6	0.2	24.7452379	0251	22.769271	017
2363000	268 827.60	0.27	226 605.5	0.2	5.2267724	0235	4.787344	016
2364000	124 2255.43	0.25	279 439.3	0.2	13.2628608	0220	18.617351	015
2365000	341 83.27	0.24	332 273.2	0.2	21.2989508	0205	0.635422	014
2366000	197 1511.13	0.22	25 107.0	0.1	1.7804898	0190	14.165428	013
2367000	53 2939.00	0.20	77 3540.8	0.1	9.8165828	0175	28.295432	012
2368000	270 766.90	0.19	130 3374.6	0.1	17.8526773	0159	10.313500	011
2369000	126 2194.81	0.17	183 3208.4	0.1	25.8887731	0144	24.143503	010
2370000	343 22.73	0.15	236 3042.2	0.1	6.3703184	0129	6.161539	009
2371000	199 1450.68	0.13	289 2875.9	0.1	14.4061175	0114	19.991569	008
2372000	55 2878.64	0.11	342 2709.7	0.1	22.4425181	0099	2.009633	007
2373000	272 706.62	0.10	35 2543.4	0.1	2.9240678	0083	15.839632	006
2374000	128 2134.62	0.08	88 2377.2	0.1	10.9601715	0068	29.669629	005
2375000	344 3562.63	0.06	141 2210.9	+0.1	18.9962766	0053	11.6876[illegible]0	004
2376000	201 1390.66	0.04	194 2044.6	0.0	27.0323833	0037	25.517686	003
2377000	57 2818.71	0.02	247 1878.3	0.0	7.5139391	0022	7.535745	002
2378000	274 646.77	−0.01	300 1711.9	0.0	15.5500188	−0007	21.365738	+001
2379000	130 2074.85	+0.01	353 1545.6	0.0	23.5861601	+0008	3.383795	000
2380000	346 3502.95	0.03	46 1379.2	0.0	4.0677204	0024	17.213787	−001
2381000	203 1331.07	0.05	99 1212.9	0.0	12.1038347	0039	31.043777	002
2382000	59 2759.20	0.07	152 1046.5	0.0	20.1399505	0054	13.061831	003
2383000	276 587.35	0.08	205 880.1	0.0	0.6215154	0070	26.891820	004
2384000	132 2015.52	0.10	258 713.7	0.0	8.6576343	0085	8.909872	005
2385000	348 3443.71	0.11	311 547.3	0.0	16.6937547	0100	22.739858	006
2386000	205 1271.91	0.13	4 380.9	−0.1	24.7298766	0115	4.757908	007
2387000	61 2700.13	0.15	57 214.5	0.1	5.2114176	0130	18.587893	008
2388000	278 528.37	0.16	110 48.0	0.1	13.2475726	0146	0.605910	009
2389000	134 1956.62	0.18	162 3481.6	0.1	21.2836990	0161	14.435923	010
2390000	350 3384.89	0.20	215 3315.1	0.1	1.7652747	0176	28.265904	011
2391000	207 1213.18	0.22	268 3148.6	0.1	9.8014042	0191	10.283949	012
2392000	63 2641.49	0.24	321 2982.1	0.1	17.8375353	0206	24.113929	013
2393000	280 469.81	0.25	14 2815.6	0.1	25.8736679	0222	6.131972	014
2394000	136 1898.15	0.27	67 2649.1	0.2	6.3552496	0237	19.961949	015
2395000	352 3326.51	0.28	120 2482.6	0.2	14.3913852	0252	1.979990	016
2396000	209 1154.88	0.30	173 2316.1	0.2	22.4275221	0268	15.809965	018
2397000	65 2583.27	0.32	226 2149.5	0.2	2.9091087	0283	29.639940	019
2398000	282 411.68	0.33	279 1983.0	0.3	10.9452489	0299	11.657978	020
2399000	138 1840.11	0.35	332 1816.4	0.3	18.9813907	0314	25.487950	021
2400000	354 3268.55	+0.37	25 1649.8	−0.3	27.0175339	+0329	7.505986	−022

TABLE IIa.

Epochs and Arguments for each Thousandth Day, from 2300000 to 2400000.

For Washington Mean Noon.

Day of Julian Period.	ARG. 3.	3′.	4.	5.	5′.	6.	7.	8.	9.
	d.	0.0000	d.	d.	0.00	d.	d.	d.	d.
2300000	16.095387	−27	127.1421	383.5901	−14	23.6328	8.7464	3.8010	21.5098
2301000	12.055368	27	31.3631	148.2332	14	13.0730	8.9197	3.6257	8.2136
2302000	8.015349	27	300.8437	324.6615	14	2.5132	9.0931	3.4504	24.7202
2303000	3.975330	26	205.0617	89.3046	14	26.8004	9.2663	3.2750	11.4240
2304000	29.465900	26	109.2856	265.7329	14	16.2106	9.4395	3.0997	27.9306
2305000	25.425882	26	13.5065	30.3761	14	5.6808	9.6128	2.9244	14.6345
2306000	21.385865	25	282.9872	206.8011	13	29.9679	0.1724	2.7491	1.3383
2307000	17.345848	25	187.2081	383.2328	13	19.4081	0.3457	2.5738	17.8449
2308000	13.305831	24	91.4290	147.8760	13	8.8183	0.5190	2.3984	4.5487
2309000	9.265815	24	360.9097	324.3044	13	33.1351	0.6922	2.2231	21.0554
2310000	5.225799	24	265.1306	88.9176	13	22.5756	0.8655	2.0178	7.7592
2311000	1.185783	23	169.3515	265.3761	12	12.0158	1.0388	1.8725	24.2659
2312000	26.676355	23	73.5725	30.0193	12	1.4560	1.2121	1.6972	10.9697
2313000	22.636340	23	343.0531	206.4178	12	25.7431	1.3853	1.5218	27.4763
2314000	18.596325	22	247.2711	382.8763	12	15.1833	1.5586	1.3165	14.1801
2315000	14.556311	22	151.4950	147.5196	12	4.6235	1.7319	1.1712	0.8840
2316000	10.516297	22	55.7159	323.9482	12	28.9106	1.9052	0.9959	17.3906
2317000	6.476284	21	325.1966	88.5915	11	18.3508	2.0785	0.8206	4.0945
2318000	2.436270	21	229.4175	265.0201	11	7.7910	2.2517	0.6452	20.6011
2319000	27.926845	21	133.6384	29.6635	11	32.0781	2.4250	0.4699	7.3050
2320000	23.886833	20	37.8591	206.0923	11	21.5183	2.5983	0.2946	23.8116
2321000	19.846820	20	307.3100	382.5209	11	10.9585	2.7716	15.5066	10.5151
2322000	15.806809	20	211.5609	117.1644	10	0.3987	2.9449	15.3313	27.0221
2323000	11.766797	19	115.7819	323.5930	10	24.6858	3.1182	15.1560	13.7260
2324000	7.726786	19	20.0028	88.2365	10	14.1260	3.2915	14.9807	0.4298
2325000	3.686775	19	289.4835	264.6652	10	3.5662	3.4648	14.8053	16.9365
2326000	29.177352	18	193.7044	29.3087	10	27.8533	3.6380	14.6300	3.6403
2327000	25.137342	18	97.9253	205.7374	10	17.2934	3.8113	14.4547	20.1470
2328000	21.097332	18	2.1463	382.1662	09	6.7336	3.9846	14.2794	6.8508
2329000	17.057323	17	271.6269	146.8098	09	31.0207	4.1579	14.1041	23.3575
2330000	13.017314	17	175.8478	323.2386	09	20.4609	4.3312	13.9288	10.0613
2331000	8.977305	16	80.0688	87.8822	09	9.9011	4.5045	13.7535	26.5680
2332000	4.937296	16	349.5494	264.3110	09	34.1882	4.6778	13.5782	13.2719
2333000	0.897288	16	253.7703	28.9547	09	23.6283	4.8511	13.4029	29.7785
2334000	26.387869	15	157.9913	205.3835	08	13.0685	5.0244	13.2276	16.4824
2335000	22.347861	15	62.2122	381.8124	08	2.5087	5.1977	13.0522	3.1863
2336000	18.307854	15	331.6929	146.4561	08	26.7958	5.3710	12.8769	19.6930
2337000	14.267847	14	235.9138	322.8850	08	16.2360	5.5443	12.7016	6.3968
2338000	10.227841	14	140.1347	87.5288	08	5.6761	5.7176	12.5263	22.9035
2339000	6.187835	14	44.3557	263.9577	07	29.9632	5.8909	12.3510	9.6074
2340000	2.147829	13	313.8363	28.6015	07	19.4034	6.0642	12.1757	26.1141
2341000	27.638412	13	218.0572	205.0305	07	8.8436	6.2375	12.0004	12.8179
2342000	23.598407	13	122.2782	381.4595	07	33.1306	6.4108	11.8251	29.3246
2343000	19.558402	12	26.4991	146.1033	07	22.5708	6.5841	11.6498	16.0285
2344000	15.518398	12	295.9797	322.5323	07	12.0110	6.7574	11.4745	2.7324
2345000	11.478394	12	200.2007	87.1762	06	1.4511	6.9307	11.2991	19.2391
2346000	7.438391	11	104.4216	263.6053	06	25.7382	7.1041	11.1238	5.9429
2347000	3.398387	11	8.6426	28.2492	06	15.1784	7.2774	10.9485	22.4496
2348000	28.888973	11	278.1232	204.6783	06	4.6185	7.4507	10.7732	9.1535
2349000	24.848970	10	182.3441	381.1074	06	28.9056	7.6240	10.5979	25.6602
2350000	20.808968	−10	86.5651	145.7514	−05	18.3457	7.7973	10.4226	12.3641

TABLE IIa.

Epochs and Arguments for each Thousandth Day, from 2300000 to 2400000.

For Washington Mean Noon.

Day of Julian Period.	Arg. 3.	3'.	4.	5.	5'.	6.	7.	8.	9.
	d.	0.0000	d.	d.	0.00	d.	d.	d.	d.
2350000	20.808968	−10	86.5651	145.7514	−05	18.3157	7.7973	10.4226	12.3641
2351000	16.768966	10	356.0457	322.1805	05	7.7859	7.9706	10.2473	28.8708
2352000	12.728965	09	260.2666	86.8245	05	32.0730	8.1439	10.0720	15.5747
2353000	8.688963	09	164.4876	263.2537	05	21.5131	8.3172	9.8967	2.2786
2354000	4.648963	08	68.7085	27.8977	05	10.9533	8.4905	9.7214	18.7853
2355000	0.608962	08	338.1891	204.3270	05	0.3931	8.6638	9.5460	5.4892
2356000	26.099550	08	242.4101	380.7562	04	24.6805	8.8372	9.3707	21.9959
2357000	22.059550	07	146.6310	145.4003	04	14.1207	9.0105	9.1954	8.6998
2358000	18.019551	07	50.8520	321.8296	04	3.5608	9.1838	9.0201	25.2065
2359000	13.979552	07	320.3326	86.4737	04	27.8179	9.3571	8.8448	11.9101
2360000	9.939553	06	224.5536	262.9030	04	17.2880	9.5304	8.6695	28.4172
2361000	5.899555	06	128.7745	27.5472	04	6.7282	0.0900	8.4942	15.1211
2362000	1.859557	06	32.9954	203.9765	03	31.0152	0.2633	8.3189	1.8250
2363000	27.350147	05	302.4760	380.4059	03	20.4554	0.4366	8.1436	18.3317
2364000	23.310150	05	206.6970	145.0501	03	9.8955	0.6100	7.9683	5.0356
2365000	19.270153	05	110.9179	321.4795	03	31.1825	0.7833	7.7930	21.5421
2366000	15.230156	04	15.1389	86.1238	03	23.6227	0.9566	7.6177	8.2463
2367000	11.190160	04	284.6195	262.5532	02	13.0628	1.1300	7.1124	24.7530
2368000	7.150164	04	188.8401	27.1975	02	2.5030	1.3033	7.2671	11.1569
2369000	3.110169	03	93.0614	203.6270	02	26.7900	1.4766	7.0918	27.9636
2370000	28.600762	03	362.5420	380.0565	02	16.2302	1.6499	6.9165	14.6676
2371000	24.560767	03	266.7629	144.7008	02	5.6703	1.8232	6.7412	1.3715
2372000	20.520772	02	170.9839	321.1303	02	29.9573	1.9965	6.5659	17.8783
2373000	16.480778	02	75.2048	85.7747	01	19.3975	2.1699	6.3906	4.5822
2374000	12.440784	01	344.6854	262.2043	01	8.8376	2.3432	6.2153	21.0889
2375000	8.400791	01	248.9061	26.8487	01	33.1246	2.5165	6.0400	7.7929
2376000	4.360798	01	153.1273	203.2783	01	22.5648	2.6899	5.8648	24.2996
2377000	0.320805	00	57.3483	379.7079	−01	12.0049	2.8632	5.6895	11.0035
2378000	25.811401	−00	326.8289	141.3524	00	1.4450	3.0366	5.5142	27.5103
2379000	21.771409	+00	231.0498	320.7820	00	25.7321	3.2099	5.3389	14.2142
2380000	17.731417	01	135.2708	85.4265	00	15.1722	3.3832	5.1636	0.9181
2381000	13.691425	01	39.4917	261.8562	00	4.6123	3.5565	4.9883	17.4219
2382000	9.651434	01	308.9723	26.5007	00	28.8993	3.7299	4.8130	4.1288
2383000	5.611444	02	213.1933	202.9305	00	18.3395	3.9032	4.6377	20.6356
2384000	1.571454	02	117.4142	379.3602	+01	7.7796	4.0766	4.4624	7.3395
2385000	27.062052	02	21.6352	144.0048	01	32.0666	4.2499	4.2871	23.8463
2386000	23.022062	03	291.1158	320.4346	01	21.5067	4.4232	4.1118	10.5502
2387000	18.982073	03	195.3367	85.0792	01	10.9469	4.5966	3.9365	27.0570
2388000	14.942084	03	99.5577	261.5090	01	0.3870	4.7699	3.7612	13.7610
2389000	10.902095	04	3.7786	26.1537	01	24.6740	4.9433	3.5859	0.4649
2390000	6.862107	04	273.2592	202.5835	02	14.1141	5.1166	3.4106	16.9717
2391000	2.822119	04	177.4802	379.0131	02	3.5542	5.2899	3.2353	3.6757
2392000	28.312720	05	81.7011	143.6581	02	27.8413	5.4633	3.0600	20.1824
2393000	24.272733	05	351.1817	320.0880	02	17.2814	5.6366	2.8847	6.8864
2394000	20.232746	05	255.4027	84.7327	02	6.7215	5.8100	2.7094	23.3932
2395000	16.192759	06	159.6236	261.1627	03	31.0085	5.9833	2.5341	10.0971
2396000	12.152773	06	63.8446	25.8075	03	20.4486	6.1566	2.3589	26.6039
2397000	8.112788	06	333.3252	202.2374	03	9.8887	6.3300	2.1836	13.3078
2398000	4.072802	07	237.5461	378.6674	03	34.1757	6.5033	2.0083	0.0118
2399000	0.032817	07	141.7671	143.3123	03	23.6158	6.6767	1.8330	16.5186
2400000	25.523420	+08	45.9880	319.7423	+03	13.0559	6.8500	1.6577	3.2226

TABLE IIa.

Epochs and Arguments for each Thousandth Day, from 2300000 to 2400000.

For Washington Mean Noon.

Day of Julian Period.	Arg. 10.	11.	12.	13.	14.	15.	16.	17.
	d.	d.	d.	d.	d.	d.	d.	d.
2300000	23.5709	6.1437	211.1477	9.3566	25.412	6.884	0.337	10.837
2301000	24.3246	11.6470	171.2876	0.8971	1.197	13.471	2.544	27.256
2302000	25.0784	17.1501	131.4274	2.5220	6.246	5.866	4.751	16.355
2303000	0.2101	22.6538	91.5673	4.1470	11.295	12.451	6.958	5.451
2304000	0.9642	1.2788	51.7072	5.7719	16.343	4.849	9.165	21.871
2305000	1.7179	6.7822	11.8470	7.3968	21.392	11.437	11.371	10.969
2306000	2.4716	12.2856	318.6070	9.0217	26.440	3.832	13.578	0.067
2307000	3.2251	17.7890	278.7169	0.5620	2.226	10.419	1.531	16.486
2308000	3.9791	23.2923	238.8867	2.1870	7.274	2.815	3.738	5.584
2309000	4.7329	1.9174	199.0266	3.8119	12.323	9.402	5.945	22.005
2310000	5.4866	7.4208	159.1665	5.4368	17.371	1.798	8.152	11.101
2311000	6.2403	12.9242	119.3063	7.0617	22.420	8.385	10.359	0.199
2312000	6.9941	18.4276	79.4462	8.6866	27.468	0.781	12.566	16.619
2313000	7.7478	23.9309	39.5860	0.2269	3.254	7.368	0.519	5.717
2314000	8.5016	2.5560	346.3160	1.8519	8.303	13.955	2.726	22.136
2315000	9.2553	8.0593	306.4858	3.4768	13.351	6.351	4.933	11.234
2316000	10.0091	13.5627	266.6257	5.1017	18.400	12.938	7.140	0.332
2317000	10.7629	19.0661	226.7655	6.7267	23.448	5.334	9.347	16.752
2318000	11.5166	21.5694	186.9053	8.3516	28.497	11.921	11.554	5.849
2319000	12.2704	3.1945	117.0451	9.9765	4.282	4.316	13.761	22.269
2320000	13.0241	8.6978	107.1850	1.5168	9.330	10.905	1.714	11.366
2321000	13.7779	14.2011	67.3248	3.1417	14.379	3.299	3.921	0.464
2322000	11.5316	19.7045	27.4646	4.7666	19.428	9.886	6.128	16.884
2323000	15.2854	25.2078	334.2246	6.3915	24.476	2.282	8.335	5.982
2324000	16.0391	3.8329	294.3611	8.0165	0.262	8.869	10.542	22.401
2325000	16.7929	9.3362	254.5042	9.6414	5.310	1.265	12.749	11.499
2326000	17.5167	14.8396	214.6410	1.1817	10.359	7.852	0.702	0.597
2327000	18.3004	20.3429	174.7838	2.8067	15.407	0.248	2.909	17.016
2328000	19.0542	25.8463	134.9236	4.4316	20.456	6.835	5.116	6.114
2329000	19.8079	4.4713	95.0631	6.0565	25.505	13.422	7.323	22.534
2330000	20.5617	9.9746	55.2032	7.6814	1.290	5.817	9.530	11.631
2331000	21.3155	15.4780	15.3430	9.3063	6.338	12.405	11.737	0.729
2332000	22.0693	20.9813	322.1029	0.8466	11.387	4.800	13.944	17.149
2333000	22.8231	26.4847	282.2427	2.4715	16.435	11.388	1.897	6.247
2334000	23.5768	5.1097	242.3825	4.0965	21.484	3.785	4.104	22.666
2335000	24.3306	10.6130	202.5223	5.7214	26.532	10.370	6.311	11.764
2336000	25.0844	16.1164	162.6621	7.3163	2.318	2.766	8.518	0.862
2337000	0.2164	21.6197	122.8018	8.9713	7.366	9.355	10.725	17.282
2338000	0.9702	0.2417	82.9416	0.5116	12.415	1.749	12.932	6.379
2339000	1.7240	5.7481	43.0814	2.1365	17.464	8.336	0.885	22.799
2340000	2.4778	11.2513	3.2212	3.7614	22.512	0.732	3.092	11.897
2341000	3.2316	16.7547	309.9811	5.3863	27.560	7.319	5.299	0.995
2342000	3.9854	22.2580	270.1209	7.0112	3.316	13.906	7.506	17.414
2343000	4.7392	0.8830	230.2606	8.6361	8.394	6.302	9.713	6.512
2344000	5.4930	6.3863	190.4004	0.1764	13.443	12.889	11.920	22.932
2345000	6.2467	11.8897	150.5402	1.8013	18.491	5.265	14.127	12.029
2346000	7.0005	17.3930	110.6799	3.4263	23.540	11.872	2.080	1.127
2347000	7.7543	22.8963	70.8197	5.0512	28.588	4.267	4.287	17.547
2348000	8.5081	1.5213	30.9594	6.6761	4.375	10.855	6.494	6.645
2849000	9.2619	7.0246	337.7193	8.3010	9.422	3.250	8.701	23.064
2350000	10.0157	12.5279	297.8590	9.9259	14.471	9.837	10.908	12.162

TABLE II*a*.

Epochs and Arguments for each Thousandth Day, from 2300000 to 2400000.

For Washington Mean Noon.

Day of Julian Period.	Arg. 10.	11.	12.	13.	14.	15.	16.	17.
	d.	d.	d.	d.	d.	d.	d.	d.
2350000	10.0157	12.5279	297.8590	9.9259	14.471	9.837	10.908	12.162
2351000	10.7695	18.0313	257.9988	1.4662	19.519	2.233	13.115	1.260
2352000	11.5233	23.5346	218.1385	3.0911	24.568	8.820	1.068	17.679
2353000	12.2771	2.1596	178.2782	4.7160	0.353	1.216	3.275	6.777
2354000	13.0309	7.6629	138.4180	6.3410	5.401	7.805	5.482	23.197
2355000	13.7847	13.1662	98.5577	7.9659	10.450	0.199	7.690	12.295
2356000	14.5385	18.6695	58.6974	9.5908	15.498	6.786	9.897	1.393
2357000	15.2923	24.1728	18.8372	1.1311	20.547	13.374	12.104	17.812
2358000	16.0461	2.7978	325.5970	2.7561	25.595	5.769	0.057	6.910
2359000	16.7999	8.3011	285.7367	4.3810	1.381	12.357	2.264	23.329
2360000	17.5537	13.8044	245.8764	6.0059	6.429	4.752	4.470	12.427
2361000	18.3075	19.3077	206.0161	7.6308	11.478	11.339	6.677	1.525
2362000	19.0613	24.8110	166.1558	9.2557	16.526	3.731	8.885	17.945
2363000	19.8151	3.4360	126.2956	0.7960	21.575	10.322	11.091	7.043
2364000	20.5690	8.9393	86.4353	2.4209	26.623	2.717	13.299	23.462
2365000	21.3228	14.4126	46.5750	4.0458	2.409	9.304	1.252	12.560
2366000	22.0766	19.9459	6.7147	5.6708	7.457	1.700	3.459	1.658
2367000	22.8305	25.4492	313.4745	7.2957	12.506	8.287	5.666	18.077
2368000	23.5843	4.0742	273.6142	8.9206	17.554	0.683	7.873	7.175
2369000	24.3381	9.5774	233.7539	0.4609	22.603	7.270	10.080	23.595
2370000	25.0919	15.0807	193.8936	2.0858	27.651	13.858	12.287	12.693
2371000	0.2240	20.5840	154.0333	3.7107	3.437	6.253	0.240	1.790
2372000	0.9778	26.0873	114.1729	5.3356	8.485	12.841	2.447	18.210
2373000	1.7317	4.7123	74.3126	6.9605	13.534	5.236	4.654	7.308
2374000	2.4855	10.2156	34.4523	8.5855	18.582	11.825	6.861	23.727
2375000	3.2393	15.7188	341.2121	0.1258	23.631	4.219	9.068	12.825
2376000	3.9932	21.2221	301.3518	1.7507	28.679	10.806	11.275	1.923
2377000	4.7470	26.7254	261.4914	3.3757	4.464	3.201	13.482	18.343
2378000	5.5009	5.3504	221.6311	5.0006	9.513	9.789	1.435	7.440
2379000	6.2547	10.8536	181.7708	6.6255	14.561	2.184	3.642	23.860
2380000	7.0085	16.3569	141.9104	8.2504	19.610	8.772	5.849	12.958
2381000	7.7623	21.8602	102.0501	9.8753	24.659	1.167	8.056	2.056
2382000	8.5162	0.4852	62.1897	1.4156	0.444	7.755	10.263	18.475
2383000	9.2700	5.9884	22.3294	3.0405	5.492	0.150	12.470	7.573
2384000	10.0239	11.4917	329.0892	4.6654	10.541	6.737	0.423	23.993
2385000	10.7777	16.9950	289.2288	6.2903	15.589	13.325	2.630	13.091
2386000	11.5316	22.4983	249.3685	7.9153	20.638	5.720	4.837	2.188
2387000	12.2855	1.1232	209.5081	9.5402	25.686	12.308	7.044	18.608
2388000	13.0393	6.6265	169.6477	1.0805	1.472	4.703	9.252	7.706
2389000	13.7932	12.1298	129.7874	2.7054	6.520	11.291	11.458	24.125
2390000	14.5470	17.6330	89.9270	4.3303	11.569	3.686	13.665	13.223
2391000	15.3008	23.1363	50.0666	5.9552	16.617	10.273	1.618	2.321
2392000	16.0547	1.7612	10.2063	7.5801	21.666	2.669	3.826	18.741
2393000	16.8086	7.2645	316.9660	9.2050	26.714	9.256	6.033	7.839
2394000	17.5624	12.7677	277.1056	0.7453	2.499	1.652	8.240	24.258
2395000	18.3163	18.2710	237.2453	2.3702	7.548	8.239	10.447	13.356
2396000	19.0702	23.7743	197.3849	3.9952	12.596	0.635	12.654	2.454
2397000	19.8240	2.3992	157.5245	5.6201	17.645	7.222	0.607	18.873
2398000	20.5779	7.9025	117.6641	7.2450	22.693	13.810	2.814	7.971
2399000	21.3317	13.4057	77.8037	8.8699	27.742	6.205	5.021	24.391
2400000	22.0856	18.9090	37.9133	0.4102	3.527	12.792	7.228	13.489

TABLE IIa.

Epochs and Arguments for each Thousandth Day, from 2300000 to 2400000.

For Washington Mean Noon.

Day of Julian Period.	Arg. 18.	19.	20.	21.	22.	23.	24.	25.	26.
	d.	d.	d.	d.	d.	d.	d.	d.	d.
2300000	7.892	22.048	4.908	57.34	18.074	6.402	6.714	0.194	5.702
2301000	0.786	1.356	2.696	3.97	22.101	7.868	10.976	7.274	4.214
2302000	3.553	4.966	0.484	82.27	26.133	9.333	1.963	4.987	2.725
2303000	6.320	8.576	12.589	28.90	30.162	10.799	6.225	2.699	1.237
2304000	9.086	12.185	10.378	107.20	2.063	12.264	10.487	0.412	5.572
2305000	1.980	15.795	8.166	53.83	6.093	13.730	1.474	7.491	4.083
2306000	4.747	19.405	5.954	0.46	10.122	15.196	5.736	5.204	2.595
2307000	7.513	23.014	3.742	78.76	14.152	16.661	9.999	2.916	1.107
2308000	0.407	2.322	1.530	25.39	18.181	18.127	0.985	0.629	5.441
2309000	3.174	5.932	13.635	103.70	22.210	19.592	5.247	7.708	3.953
2310000	5.942	9.542	11.423	50.33	26.239	21.058	9.510	5.422	2.465
2311000	8.708	13.151	9.211	128.63	30.268	22.524	0.495	3.134	0.977
2312000	1.602	16.761	6.999	75.26	2.170	0.215	4.758	0.847	5.311
2313000	4.369	20.371	4.787	21.89	6.199	1.680	9.021	7.926	3.823
2314000	7.136	23.981	2.575	100.19	10.228	3.146	0.007	5.639	2.335
2315000	0.029	3.288	0.363	46.82	14.258	4.612	4.269	3.352	0.847
2316000	2.796	6.898	12.468	125.12	18.287	6.077	8.532	1.064	5.181
2317000	5.563	10.508	10.256	71.75	22.316	7.543	12.794	8.144	3.693
2318000	8.330	14.117	8.044	18.38	26.346	9.008	3.780	5.857	2.204
2319000	1.224	17.727	5.832	96.69	30.375	10.474	8.042	3.569	0.716
2320000	3.991	21.337	3.620	43.32	2.277	11.940	12.305	1.282	5.051
2321000	6.758	0.644	1.408	121.62	6.305	13.406	3.292	8.361	3.563
2322000	9.525	4.254	13.513	68.25	10.335	14.871	7.554	6.074	2.074
2323000	2.419	7.861	11.301	14.88	14.364	16.337	11.816	3.786	0.586
2324000	5.186	11.474	9.090	93.18	18.393	17.802	2.803	1.499	4.921
2325000	7.952	15.084	6.878	39.81	22.423	19.268	7.065	8.579	3.432
2326000	0.846	18.693	4.666	118.11	26.452	20.734	11.327	6.291	1.944
2327000	3.613	22.305	2.454	64.74	30.481	22.199	2.314	4.004	6.279
2328000	6.380	1.611	0.242	11.37	2.382	23.665	6.576	1.717	4.790
2329000	9.146	5.221	12.347	89.67	6.412	1.355	10.838	8.796	3.302
2330000	2.041	8.830	10.135	36.30	10.441	2.821	1.825	6.509	1.814
2331000	4.808	12.440	7.923	114.61	14.471	4.287	6.088	4.221	0.326
2332000	7.574	16.050	5.711	61.24	18.500	5.753	10.350	1.934	4.660
2333000	0.468	19.660	3.499	7.87	22.529	7.218	1.336	9.014	3.172
2334000	3.235	23.269	1.287	86.17	26.558	8.684	5.599	6.727	1.684
2335000	6.002	2.577	13.392	32.80	30.588	10.150	9.861	4.439	0.196
2336000	8.769	6.187	11.180	111.10	2.489	11.615	0.847	2.151	4.530
2337000	1.663	9.797	8.969	57.73	6.518	13.081	5.110	9.231	3.041
2338000	4.430	13.407	6.757	4.36	10.547	14.546	9.372	6.944	1.554
2339000	7.196	17.016	4.545	82.66	14.577	16.012	0.358	4.656	0.065
2340000	0.090	20.626	2.333	29.29	18.606	17.478	4.621	2.369	4.400
2341000	2.857	24.236	0.121	107.60	22.636	18.944	8.884	0.082	2.912
2342000	5.624	3.544	12.226	54.23	26.665	20.410	13.146	7.161	1.423
2343000	8.391	7.151	10.014	0.86	30.694	21.875	4.132	4.874	5.758
2344000	1.285	10.765	7.803	79.16	2.595	23.341	8.395	2.586	4.270
2345000	4.052	14.375	5.591	25.79	6.625	1.032	12.657	0.299	2.781
2346000	6.818	17.985	3.379	104.09	10.654	2.497	3.644	7.379	1.293
2347000	9.585	21.595	1.167	50.72	14.683	3.963	7.906	5.091	5.628
2348000	2.479	0.900	13.272	129.03	18.713	5.428	12.168	2.804	4.139
2349000	5.246	4.510	11.061	75.66	22.742	6.894	3.155	0.517	2.651
2350000	8.013	8.120	8.848	22.29	26.771	8.360	7.417	7.596	1.163

TABLE II*a*.

Epochs and Arguments for each Thousandth Day, from 2300000 to 2400000.

For Washington Mean Noon.

Day of Julian Period.	Arg. 18.	19.	20.	21.	22.	23.	24.	25.	26.
	d	d.	d	d.	d.	d	d	d	d
2350000	8.013	8.120	8.848	22.29	26.771	8.360	7.417	7.596	1.163
2351000	0.907	11.730	6.636	100.59	30.801	9.826	11.679	5.309	5.198
2352000	3.674	15.310	4.424	47.22	2.702	11.292	2.666	3.021	4.010
2353000	6.441	18.950	2.212	125.52	6.731	12.757	6.929	0.731	2.521
2354000	9.208	22.560	0.000	72.15	10.761	14.223	11.191	7.811	1.033
2355000	2.102	1.868	12.106	18.78	14.790	15.689	2.177	5.526	5.368
2356000	4.869	5.478	9.894	97.09	18.819	17.154	6.440	3.239	3.879
2357000	7.636	9.088	7.682	43.72	22.849	18.620	10.702	0.951	2.391
2358000	0.530	12.697	5.470	122.02	26.878	20.085	1.688	8.031	0.903
2359000	3.297	16.307	3.258	68.65	30.907	21.551	5.951	5.744	5.238
2360000	6.064	19.917	1.046	15.28	2.809	23.017	10.214	3.457	3.749
2361000	8.830	23.527	13.151	93.58	6.838	0.708	1.200	1.169	2.261
2362000	1.724	2.835	10.940	40.20	10.867	2.174	5.163	8.250	0.773
2363000	4.491	6.445	8.728	118.51	14.897	3.639	9.725	5.961	5.107
2364000	7.258	10.055	6.516	65.11	18.926	5.105	0.711	3.674	3.619
2365000	0.152	13.665	4.301	11.78	22.955	6.571	4.974	1.387	2.131
2366000	2.919	17.274	2.092	90.08	26.984	8.036	9.236	8.466	0.643
2367000	5.686	20.884	14.197	36.71	31.014	9.502	0.223	6.179	4.977
2368000	8.452	0.192	11.985	115.01	2.915	10.967	4.485	3.892	3.489
2369000	1.347	3.803	9.774	61.64	6.944	12.433	8.748	1.604	2.001
2370000	4.113	7.412	7.562	8.27	10.974	13.899	13.010	8.684	0.513
2371000	6.880	11.022	5.350	86.57	15.003	15.365	3.997	6.396	4.817
2372000	9.647	14.632	3.138	33.20	19.032	16.831	8.259	4.109	3.359
2373000	2.541	18.242	0.926	111.51	23.062	18.296	12.521	1.822	1.871
2374000	5.308	21.852	13.031	58.14	27.091	19.762	3.508	8.901	0.383
2375000	8.075	1.160	10.819	4.77	31.120	21.228	7.770	6.614	4.717
2376000	0.962	4.770	8.607	83.07	3.022	22.693	12.033	4.327	3.229
2377000	3.736	8.379	6.396	29.70	7.051	0.384	3.019	2.039	1.741
2378000	6.503	11.989	4.184	108.00	11.080	1.849	7.282	9.119	0.253
2379000	9.270	15.599	1.972	51.63	15.110	3.315	11.544	6.831	4.587
2380000	2.164	19.209	14.077	1.27	19.139	4.781	2.531	4.544	3.099
2381000	4.931	22.819	11.865	79.57	23.168	6.247	6.793	2.257	1.611
2382000	7.697	2.127	9.653	26.20	27.198	7.713	11.055	9.337	0.123
2383000	0.591	5.737	7.442	104.50	31.227	9.178	2.042	7.019	4.457
2384000	3.358	9.347	5.230	51.13	3.128	10.644	6.305	4.762	2.969
2385000	6.125	12.957	3.018	129.43	7.158	12.110	10.568	2.475	1.481
2386000	8.892	16.567	0.806	76.06	11.187	13.575	1.555	0.187	5.815
2387000	1.786	20.177	12.911	22.70	15.216	15.041	5.816	7.267	4.327
2388000	4.553	23.787	10.699	101.00	19.245	16.506	10.078	4.980	2.839
2389000	7.319	3.095	8.487	47.63	23.275	17.972	1.065	2.692	1.351
2390000	0.214	6.705	6.276	125,93	27.304	19.438	5.327	0.405	5.685
2391000	2.981	10.315	4.064	72.56	31.331	20.901	9.590	7.481	4.197
2392000	5.748	13.925	1.852	19.19	3.235	22.370	0.576	5.197	2.709
2393000	8.514	17.535	13.957	97.50	7.264	0.060	4.839	2.910	1.221
2394000	1.408	21.145	11.745	41.13	11.294	1.526	9.102	0.622	5.555
2395000	4.175	0.453	9.531	122.43	15.323	2.992	0.088	7.702	4.067
2396000	6.942	4.063	7.322	69.06	19.352	4.457	4.350	5.415	2.579
2397000	9.709	7.673	5.110	15.69	23.381	5.923	8.613	3.127	1.090
2398000	2.603	11.283	2.898	93.99	27.411	7.388	12.875	0.840	5.425
2399000	5.370	14.893	0.686	40.62	31.440	8.854	3.862	7.920	3.937
2100000	8.137	18.503	12.791	118.93	3.342	10.320	8.124	5.633	2.449

TABLE IIa.

Epochs and Arguments for each Thousandth Day, from 2300000 to 2400000.

For Washington Mean Noon.

Day of Julian Period.	Arg. 27.	28.	29.	30.	31.	32.	33.	34.	35.
	d.	d.	d.	d.	d.	d.	d.	d.	d.
2300000	15.42	22.494	32.009	2.076	14.53	4.53	23.91	5.18	8.26
2301000	8.19	20.922	16.336	6.466	3.78	6.72	17.97	2.75	2.69
2302000	0.96	19.350	0.662	0.486	8.31	8.91	12.00	0.32	21.06
2303000	19.56	17.778	17.753	4.877	12.91	11.10	6.01	24.98	15.48
2304000	12.33	16.205	2.080	9.267	2.15	13.29	0.07	22.55	9.91
2305000	5.10	14.633	19.171	3.287	6.71	15.48	26.55	20.12	4.34
2306000	23.70	13.061	3.497	7.678	11.28	1.01	20.58	17.69	22.70
2307000	16.47	11.488	20.588	1.697	0.52	3.23	14.61	15.26	17.13
2308000	9.24	9.916	4.915	6.087	5.09	5.42	8.65	12.83	11.56
2309000	2.01	8.344	22.006	0.107	9.65	7.61	2.68	10.40	5.98
2310000	20.61	6.772	6.332	4.498	14.22	9.80	29.16	7.97	0.41
2311000	13.38	5.200	23.424	8.888	3.46	11.99	23.19	5.54	18.78
2312000	6.15	3.628	7.750	2.908	8.02	14.18	17.22	3.11	13.21
2313000	24.75	2.055	24.841	7.299	12.59	16.37	11.25	0.68	7.63
2314000	17.52	0.483	9.168	1.318	1.83	1.93	5.29	25.34	2.06
2315000	10.29	37.433	26.259	5.709	6.40	4.12	31.77	22.91	20.43
2316000	3.06	35.860	10.585	10.099	10.96	6.31	25.78	20.48	14.85
2317000	21.66	34.288	27.676	1.119	0.24	8.50	19.81	18.05	9.28
2318000	14.43	32.715	12.003	8.509	4.77	10.69	13.84	15.62	3.70
2319000	7.20	31.143	29.094	2.529	9.33	12.88	7.88	13.19	22.07
2320000	25.80	29.571	13.421	6.919	13.90	15.07	1.91	10.76	16.50
2321000	18.57	27.999	30.511	0.939	3.14	0.63	28.41	8.33	10.93
2322000	11.34	26.427	14.838	5.329	7.71	2.82	22.44	5.90	5.35
2323000	4.11	24.851	31.929	9.720	12.28	5.01	16.47	3.47	23.72
2324000	22.71	23.282	16.256	3.740	1.52	7.20	10.51	1.04	18.15
2325000	15.48	21.710	0.582	8.130	6.08	9.39	4.54	25.70	12.57
2326000	8.25	20.137	17.673	2.149	10.64	11.58	31.02	23.27	7.00
2327000	1.02	18.565	2.000	6.540	15.21	13.77	25.05	20.84	1.42
2328000	19.62	16.992	19.091	0.560	4.46	15.96	19.08	18.41	19.79
2329000	12.39	15.420	3.418	4.951	9.02	1.52	13.12	15.98	14.22
2330000	5.16	13.848	20.509	9.341	13.59	3.72	7.15	13.55	8.65
2331000	23.76	12.276	4.835	3.361	2.83	5.91	1.18	11.12	3.07
2332000	16.53	10.704	21.926	7.751	7.39	8.10	27.66	8.69	21.41
2333000	9.30	9.132	6.253	1.771	11.95	10.29	21.69	6.26	15.87
2334000	27.90	7.559	23.344	6.161	1.20	12.48	15.72	3.83	10.29
2335000	20.67	5.987	7.671	0.181	5.77	14.67	9.76	1.40	4.72
2336000	13.44	4.415	21.763	4.572	10.33	0.23	3.79	26.06	23.08
2337000	6.21	2.842	9.088	8.962	14.90	2.42	30.27	23.63	17.51
2338000	24.80	1.270	26.179	2.982	4.14	4.61	21.30	21.20	11.94
2339000	17.58	38.220	10.506	7.372	8.70	6.80	18.33	18.77	6.37
2340000	10.35	36.648	27.597	1.392	13.27	8.99	12.37	16.34	0.79
2341000	3.12	35.076	11.924	5.782	2.51	11.18	6.40	13.91	19.16
2342000	21.71	33.504	29.015	10.173	7.08	13.37	0.43	11.48	13.59
2343000	14.49	31.931	13.342	4.193	11.64	15.56	26.91	9.05	8.01
2344000	07.26	30.359	30.433	8.583	0.89	1.12	20.94	6.62	2.44
2345000	0.02	28.787	14.760	2.603	5.45	3.31	14.98	4.19	20.81
2346000	18.62	27.214	31.850	6.993	10.01	5.50	9.01	1.76	15.24
2347000	11.40	25.642	16.178	1.013	14.58	7.69	3.04	26.42	9.66
2348000	4.17	24 069	0.504	5.403	3.82	9.88	29.52	23.99	4.09
2349000	22.76	22.497	17.595	9.794	8.39	12.07	23.55	21.56	22.45
2350000	15.54	20.925	1.922	3.814	12.95	14.26	17.59	19.13	16.88

TABLE IIa.

Epochs and Arguments for each Thousandth Day, from 2300000 to 2400000.

For Washington Mean Noon.

Day of Julian Period.	Arg. 27.	28.	29.	30.	31.	32.	33.	34.	35.
	d	d.	d.	d.	d	d.	d	d	d.
2350000	15.54	20.925	1.922	3.814	12.95	14.26	17.59	19.13	16.88
2351000	8.31	19.353	19.013	8.201	2.20	16.45	11.62	16.70	11.31
2352000	1.08	17.780	3.310	2.224	6.76	2.01	5.65	14.27	5.73
2353000	19.67	16.208	20.431	6.614	11.32	4.20	32.13	11.84	24.10
2354000	12.45	14.635	4.757	0.634	0.57	6.39	26.16	9.41	18.53
2355000	5.23	13.063	21.848	5.024	5.13	8.58	20.20	6.98	12.96
2356000	23.81	11.491	6.175	9.415	9.70	10.77	14.23	4.55	7.38
2357000	16.58	9.918	23.266	3.435	14.26	12.96	8.26	2.12	1.81
2358000	9.36	8.346	7.593	7.825	3.51	15.15	2.29	26.78	20.18
2359000	2.13	6.773	24.684	1.845	8.07	0.71	28.77	24.35	14.60
2360000	20.72	5.201	9.011	6.235	12.64	2.90	22.81	21.92	9.03
2361000	13.50	3.629	26.102	0.255	1.88	5.09	16.84	19.49	3.46
2362000	6.27	2.057	10.429	4.645	6.41	7.28	10.87	17.06	21.82
2363000	24.86	0.481	27.520	9.086	11.01	9.47	4.90	14.63	16.25
2364000	17.63	37.434	11.846	3.056	0.25	11.66	31.38	12.20	10.68
2365000	10.41	35.862	28.938	7.446	4.82	13.85	25.42	9.77	5.10
2366000	3.18	34.289	13.265	1.466	9.38	16.01	19.45	7.34	23.47
2367000	21.77	32.717	30.355	5.856	13.95	1.60	13.48	4.91	17.90
2368000	14.54	31.144	14.682	10.247	3.19	3.79	7.51	2.48	12.32
2369000	7.32	29.572	31.773	4.266	7.75	5.99	1.54	0.05	6.75
2370000	0.08	28.000	16.100	8.657	12.32	8.18	28.03	21.71	1.18
2371000	18.68	26.428	0.427	2.676	1.56	10.37	22.06	22.28	19.55
2372000	11.46	24.856	17.518	7.067	6.13	12.56	16.09	19.85	13.97
2373000	4.23	23.283	1.845	1.087	10.69	14.75	10.12	17.42	8.40
2374000	22.82	21.711	18.936	5.477	15.26	0.31	4.15	14.99	2.82
2375000	15.59	20.139	3.263	9.868	4.50	2.50	30.64	12.56	21.19
2376000	8.37	18.566	20.354	3.887	9.06	4.69	24.67	10.13	15.62
2377000	1.14	16.994	4.682	8.278	13.63	6.88	18.70	7.70	10.04
2378000	19.73	15.421	21.772	2.298	2.87	9.07	12.73	5.27	4.47
2379000	12.50	13.849	6.099	6.688	7.44	11.26	6.76	2.84	22.84
2380000	5.28	12.277	23.190	0.708	12.00	13.45	0.80	0.41	17.27
2381000	23.87	10.705	7.516	5.098	1.25	15.64	27.28	25.07	11.69
2382000	16.64	9.132	21.608	9.489	5.81	1.20	21.31	22.61	6.12
2383000	9.42	7.560	8.934	3.508	10.37	3.39	15.34	20.21	0.54
2384000	2.19	5.987	26.026	7.899	14.94	5.58	9.37	17.78	18.91
2385000	20.78	4.415	10.352	1.918	4.19	7.77	3.41	15.35	13.34
2386000	13.55	2.813	27.414	6.309	8.75	9.96	29.89	12.92	7.77
2387000	6.33	1.270	11.770	0.329	13.31	12.15	23.92	10.49	2.19
2388000	24.92	38.220	28.861	4.719	2.56	14.34	17.95	8.06	20.56
2389000	17.69	36.647	13.188	9.110	7.12	16.53	11.98	5.63	14.99
2390000	10.47	35.075	30.279	3.129	11.68	2.09	6.02	3.20	9.41
2391000	3.24	33.503	14.606	7.520	0.93	4.28	0.05	0.77	3.84
2392000	21.83	31.931	31.697	1.539	5.49	6.47	26.53	25.43	22.21
2393000	14.61	30.358	16.024	5.930	10.06	8.66	20.56	23.00	16.63
2394000	7.38	28.786	0.351	10.321	14.62	10.85	14.59	20.57	11.06
2395000	0.14	27.214	17.442	4.340	3.87	13.04	8.63	18.14	5.49
2396000	18.74	25.641	1.769	8.731	8.43	15.23	2.66	15.71	23.85
2397000	11.52	24.069	18.860	2.750	13.00	0.79	29.14	13.28	18.28
2398000	4.29	22.496	3.187	7.141	2.24	2.98	23.17	10.85	12.71
2399000	22.88	20.924	20.278	1.161	6.80	5.17	17.20	8.42	7.13
2400000	15.66	19.352	4.605	5.551	11.37	7.36	11.24	5.99	1.56

TABLE IIa.

Epochs and Arguments for each Thousandth Day, from 2300000 to 2400000.

For Washington Mean Noon.

Day of Julian Period.	Arg. 36.	37.	38.	39.	40.	41.	42.	43.	44.
	d.	d.	d.	d.	d.	d.	d.	d.	d.
2300000	2.44	27.6	18.63	10.6	3.86	5.90	1.84	8.97	7.8
2301000	0.96	11.8	4.01	10.8	5.14	2.43	7.34	5.86	5.3
2302000	13.21	33.5	8.23	11.0	6.44	16.88	2.70	2.74	2.7
2303000	11.74	17.6	12.42	11.2	7.73	13.41	8.20	14.18	0.1
2304000	10.26	1.8	16.62	11.5	9.03	9.94	3.56	11.06	20.3
2305000	8.79	23.5	2.02	11.7	10.32	6.47	9.06	7.95	17.7
2306000	7.32	7.6	6.22	11.9	11.61	3.01	4.43	4.83	15.1
2307000	5.84	29.1	10.41	12.2	12.91	17.15	9.92	1.72	12.5
2308000	4.37	13.5	14.61	12.4	14.20	13.98	5.29	13.15	9.9
2309000	2.90	35.2	0.01	12.6	15.49	10.51	0.64	10.04	7.3
2310000	1.42	19.3	4.21	12.8	16.78	7.05	6.15	6.92	4.7
2311000	13.67	3.4	8.40	13.0	18.07	3.58	1.51	3.81	2.1
2312000	12.20	25.2	12.60	13.2	0.52	0.11	7.00	0.70	22.4
2313000	10.72	9.3	16.79	13.4	1.82	14.55	2.37	12.13	19.8
2314000	9.25	31.0	2.19	13.7	3.11	11.08	7.87	9.01	17.2
2315000	7.78	15.2	6.39	13.9	4.40	7.61	3.23	5.90	14.6
2316000	6.30	36.9	10.58	14.1	5.70	4.14	8.73	2.79	12.0
2317000	4.83	21.0	14.78	14.4	6.99	0.67	4.10	14.23	9.4
2318000	3.36	5.1	0.18	14.6	8.29	15.12	9.59	11.10	6.8
2319000	1.88	26.9	4.38	14.8	9.58	11.65	4.95	7.99	4.2
2320000	0.41	11.0	8.57	15.0	10.87	8.19	0.31	4.88	1.6
2321000	12.66	32.7	12.77	15.2	12.16	4.72	5.82	1.77	21.8
2322000	11.18	16.8	16.96	15.4	13.45	1.25	1.17	13.19	19.2
2323000	9.71	1.0	2.37	15.6	14.74	15.70	6.68	10.08	16.6
2324000	8.23	22.7	6.56	15.9	16.04	12.22	2.04	6.97	14.1
2325000	6.76	6.8	10.76	16.1	17.33	8.75	7.53	3.86	11.5
2326000	5.29	28.6	14.95	16.3	18.62	5.28	2.90	0.75	8.9
2327000	3.82	12.7	0.35	16.6	1.08	1.81	8.39	12.17	6.3
2328000	2.34	31.4	4.55	16.8	2.37	16.26	3.76	9.06	3.7
2329000	0.87	18.5	8.75	17.0	3.66	12.79	9.26	5.95	1.1
2330000	13.12	2.6	12.94	17.2	4.95	9.33	4.62	2.84	21.3
2331000	11.64	24.4	17.13	17.4	6.24	5.86	10.12	14.27	18.7
2332000	10.17	8.5	2.51	17.6	7.53	2.39	5.48	11.15	16.1
2333000	8.69	30.2	6.73	17.8	8.82	16.84	0.84	8.04	13.5
2334000	7.22	14.4	10.93	18.0	10.12	13.37	6.35	4.93	10.9
2335000	5.75	36.1	15.12	18.3	11.41	9.90	1.70	1.82	8.3
2336000	4.28	20.2	0.53	18.5	12.70	6.43	7.20	13.24	5.8
2337000	2.80	4.3	4.72	18.7	14.00	2.96	2.56	10.13	3.2
2338000	1.33	26.1	8.91	18.9	15.29	17.41	8.06	7.02	0.6
2339000	13.58	10.2	13.12	19.1	16.58	13.94	3.43	3.91	20.8
2340000	12.10	31.9	17.31	19.3	17.87	10.47	8.94	0.79	18.2
2341000	10.63	16.0	2.71	19.5	0.32	7.00	4.29	12.22	15.6
2342000	9.15	0.2	6.90	19.7	1.61	3.53	9.79	9.11	13.0
2343000	7.68	21.9	11.10	19.9	2.91	0.06	5.15	6.00	10.4
2344000	6.21	6.0	15.29	20.2	4.20	14.51	0.51	2.88	7.8
2345000	4.73	27.8	0.70	20.4	5.49	11.04	6.01	14.32	5.2
2346000	3.26	11.9	4.89	20.6	6.78	7.57	1.37	11.20	2.6
2347000	1.79	33.6	9.09	20.9	8.07	4.10	6.87	8.09	0.0
2348000	0.31	17.7	13.28	21.1	9.37	0.63	2.23	4.97	20.3
2349000	12.56	1.9	17.48	21.3	10.66	15.08	7.73	1.86	17.7
2350000	11.09	23.6	2.88	21.5	11.95	11.61	3.09	13.29	15.1

TABLE IIa.

Epochs and Arguments for each Thousandth Day, from 2300000 to 2400000.

For Washington Mean Noon.

Day of Julian Period.	Arg. 36.	37.	38.	39.	40.	41.	42.	43.	44.
	d.	d.	d.	d.	d.	d.	d.	d.	d.
2350000	11.09	23.6	2.88	21.5	11.95	11.61	3.09	13.29	15.1
2351000	9.61	7.7	7.07	21.7	13.24	8.14	8.59	10.18	12.5
2352000	8.14	29.4	11.27	21.9	14.53	4.67	3.96	7.06	9.9
2353000	6.67	13.6	15.46	22.1	15.83	1.20	9.45	3.95	7.3
2354000	5.19	35.3	0.87	22.4	17.12	15.65	4.82	0.84	4.7
2355000	3.72	19.4	5.06	22.6	18.41	12.17	0.18	12.27	2.1
2356000	2.25	3.5	9.26	22.8	0.86	8.70	5.68	9.15	22.3
2357000	0.77	25.3	13.45	23.1	2.15	5.23	1.04	6.04	19.7
2358000	13.02	9.4	17.65	23.3	3.45	1.76	6.54	2.93	17.1
2359000	11.55	31.1	3.05	23.5	4.74	16.21	1.90	14.37	14.5
2360000	10.07	15.2	7.25	23.7	6.03	12.74	7.40	11.24	12.0
2361000	8.60	37.0	11.44	23.9	7.32	9.27	2.76	8.13	9.4
2362000	7.13	21.1	15.64	24.1	8.61	5.80	8.26	5.02	6.8
2363000	5.65	5.2	1.04	24.3	9.91	2.33	3.62	1.91	4.2
2364000	4.18	27.0	5.23	24.5	11.20	16.78	9.12	13.34	1.6
2365000	2.71	11.1	9.43	24.8	12.49	13.31	4.49	10.22	21.8
2366000	1.24	32.8	13.62	25.0	13.78	9.84	9.98	7.11	19.2
2367000	13.48	17.0	17.82	25.2	15.07	6.37	5.35	4.00	16.6
2368000	12.01	1.1	3.22	25.4	16.37	2.90	0.70	0.89	14.0
2369000	10.53	22.8	7.42	25.6	17.66	17.35	6.21	12.31	11.4
2370000	9.06	6.9	11.61	25.8	0.11	13.88	1.57	9.20	8.8
2371000	7.59	28.7	15.81	26.0	1.40	10.41	7.06	6.09	6.2
2372000	6.11	12.8	1.21	26.2	2.69	6.94	2.43	2.98	3.7
2373000	4.64	34.5	5.41	26.4	3.99	3.47	7.93	14.41	1.1
2374000	3.17	18.6	9.60	26.7	5.28	0.00	3.29	11.29	21.3
2375000	1.69	2.8	13.80	26.9	6.57	14.45	8.79	8.18	18.7
2376000	0.22	24.5	17.99	27.1	7.86	10.98	4.15	5.07	16.1
2377000	12.47	8.6	3.40	27.4	9.15	7.51	9.65	1.96	13.5
2378000	11.00	30.4	7.59	27.6	10.45	4.04	5.01	13.38	10.9
2379000	9.52	14.5	11.78	27.8	11.74	0.57	0.37	10.27	8.3
2380000	8.05	36.2	15.98	28.0	13.03	15.02	5.88	7.16	5.7
2381000	6.57	20.3	1.38	28.2	14.32	11.55	1.23	4.05	3.1
2382000	5.10	4.4	5.58	28.4	15.61	8.08	6.73	0.94	0.5
2383000	3.63	26.2	9.77	28.6	16.91	4.61	2.10	12.36	20.8
2384000	2.15	10.3	13.97	28.9	18.20	1.14	7.59	9.25	18.2
2385000	0.68	32.0	18.16	29.1	0.65	15.59	2.96	6.14	15.6
2386000	12.93	16.2	3.57	29.3	1.94	12.12	8.46	3.03	13.0
2387000	11.46	0.3	7.76	29.6	3.23	8.65	3.82	14.46	10.4
2388000	9.98	22.0	11.95	29.8	4.53	5.18	9.32	11.34	7.8
2389000	8.51	6.1	16.15	30.0	5.82	1.71	4.68	8.23	5.2
2390000	7.03	27.9	1.56	30.2	7.11	16.16	0.04	5.12	2.6
2391000	5.56	12.0	5.75	30.4	8.40	12.69	5.55	2.00	0.0
2392000	4.09	33.7	9.95	30.6	9.69	9.22	0.90	13.43	20.2
2393000	2.62	17.8	14.14	30.8	10.99	5.75	6.41	10.32	17.6
2394000	1.14	2.0	18.34	31.0	12.28	2.28	1.76	7.21	15.0
2395000	13.39	23.7	3.74	31.3	13.57	16.73	7.26	4.09	12.4
2396000	11.92	7.8	7.93	31.5	14.86	13.25	2.63	0.98	9.9
2397000	10.44	29.6	12.13	31.7	16.15	9.79	8.12	12.41	7.3
2398000	8.97	13.7	16.32	31.9	17.45	6.32	3.49	9.30	4.7
2399000	7.49	35.4	1.73	32.1	18.74	2.85	8.99	6.18	2.1
2400000	6.02	19.5	5.92	32.3	1.19	17.30	4.35	3.07	22.3

TABLE IIa.

Epochs and Arguments for each Thousandth Day, from 2300000 to 2400000.

For Washington Mean Noon.

Day of Julian Period.	Arg. 45.	46.	47.	48.	49.	50.	51.	52.	53.
	d.	d.	d.	d.	d.	d.	d.	d.	d.
2300000	3.87	23.2	12.49	22.71	1.5	15.1	8.53	0.73	94.82
2301000	6.77	7.7	9.08	4.98	6.0	13.7	6.60	16.70	36.96
2302000	9.66	21.2	5.67	17.15	10.4	12.2	4.67	7.45	96.65
2303000	12.56	5.8	2.25	29.32	2.2	10.8	2.74	23.42	38.79
2304000	15.46	19.3	12.97	11.56	6.7	9.3	0.81	14.17	98.48
2305000	1.46	3.9	9.56	23.73	11.3	7.9	17.09	4.92	40.62
2306000	4.36	17.4	6.15	5.97	3.0	6.5	15.16	20.89	100.30
2307000	7.25	2.0	2.74	18.14	7.5	5.0	13.23	11.64	42.45
2308000	10.15	15.5	13.46	0.38	12.1	3.6	11.30	2.38	102.13
2309000	13.05	0.0	10.05	12.55	3.7	2.1	9.37	18.36	44.28
2310000	15.95	13.6	6.64	24.73	8.3	0.7	7.44	9.10	103.96
2311000	1.95	27.1	3.23	6.96	0.0	24.3	5.51	25.08	46.11
2312000	4.84	11.7	13.95	19.14	4.5	22.9	3.58	15.83	105.79
2313000	7.74	25.2	10.54	1.37	9.0	21.4	1.65	6.57	47.94
2314000	10.64	9.8	7.13	13.55	0.8	20.0	17.93	22.55	107.62
2315000	13.54	23.3	3.72	25.72	5.3	18.5	15.99	13.29	49.77
2316000	16.44	7.8	0.31	7.96	9.8	17.1	14.07	4.04	109.45
2317000	2.43	21.4	11.03	20.13	1.5	15.7	12.14	20.01	51.60
2318000	5.33	5.9	7.62	2.37	6.0	14.2	10.21	10.76	111.28
2319000	8.23	19.5	4.21	14.54	10.5	12.8	8.28	1.50	53.42
2320000	11.13	4.0	0.79	26.71	2.3	11.4	6.35	17.48	113.11
2321000	14.03	17.5	11.51	8.95	6.8	10.0	4.42	8.23	55.26
2322000	0.02	2.1	8.10	21.12	11.3	8.5	2.49	24.20	114.94
2323000	2.92	15.6	4.69	3.36	3.1	7.1	0.56	14.95	57.09
2324000	5.82	0.2	1.28	15.53	7.6	5.6	16.84	5.69	116.77
2325000	8.72	13.7	12.00	27.70	12.1	4.2	14.91	21.67	58.92
2326000	11.62	27.3	8.59	9.91	3.8	2.8	12.98	12.41	1.06
2327000	14.51	11.8	5.18	22.11	8.3	1.3	11.05	3.16	60.74
2328000	0.51	25.3	1.77	4.35	0.1	24.9	9.12	19.14	2.89
2329000	3.41	9.9	12.48	16.52	4.6	23.4	7.19	9.88	62.57
2330000	6.31	23.4	9.08	28.70	9.1	22.0	5.26	0.63	4.72
2331000	9.21	8.0	5.66	10.91	0.8	20.6	3.33	16.60	64.41
2332000	12.10	21.5	2.25	23.11	5.3	19.1	1.40	7.35	6.55
2333000	15.00	6.1	12.97	5.35	9.9	17.7	17.68	23.32	66.23
2334000	0.99	19.6	9.56	17.52	1.6	16.2	15.75	14.07	8.38
2335000	3.90	4.1	6.15	29.69	6.1	14.8	13.82	4.81	68.06
2336000	6.80	17.7	2.74	11.93	10.5	13.4	11.89	20.79	10.21
2337000	9.69	2.2	13.46	24.10	2.4	11.9	9.96	11.54	69.89
2338000	12.59	15.8	10.05	6.31	6.9	10.5	8.03	2.28	12.04
2339000	15.49	0.3	6.64	18.51	11.4	9.0	6.10	18.26	71.72
2340000	1.49	13.8	3.23	0.75	3.1	7.6	4.17	9.00	13.87
2341000	4.38	27.4	13.95	12.92	7.6	6.2	2.24	24.98	73.55
2342000	7.28	11.9	10.54	25.09	12.2	4.8	0.31	15.72	15.70
2343000	10.18	25.5	7.12	7.33	3.9	3.3	16.59	6.47	75.38
2344000	13.08	10.0	3.71	19.50	8.4	1.9	14.67	22.44	17.53
2345000	15.98	23.6	0.30	1.74	0.1	0.4	12.73	13.19	77.21
2346000	1.97	8.1	11.02	13.91	4.6	24.0	10.80	3.94	19.36
2347000	4.87	21.6	7.61	26.08	9.2	22.6	8.87	19.91	79.04
2348000	7.77	6.2	4.20	8.32	0.9	21.1	6.94	10.66	21.19
2349000	10.67	19.7	0.79	20.49	5.4	19.7	5.01	1.40	80.87
2350000	13.57	4.3	11.51	2.73	9.9	18.3	3.08	17.38	23.02

TABLE IIa.

Epochs and Arguments for each Thousandth Day, from 2300000 to 2400000.

For Washington Mean Noon.

Day of Julian Period.	Arg. 45.	46.	47.	48.	49.	50.	51.	52.	53.
	d.	d.	d	d	d.	d.	d	d.	d.
2350000	13.57	4.3	11.51	2.73	9.9	18.3	3.08	17.38	23.02
2351000	16.46	17.8	8.10	14.91	1.7	16.9	1.15	8.12	82.70
2352000	2.46	2.4	4.69	27.08	6.2	15.4	17.43	21.10	21.85
2353000	5.36	15.9	1.28	9.32	10.6	14.0	15.50	14.84	84.53
2354000	8.26	0.4	11.99	21.49	2.4	12.5	13.58	5.59	26.68
2355000	11.16	14.0	8.58	3.73	6.9	11.1	11.61	21.57	86.36
2356000	14.05	27.5	5.17	15.90	11.5	9.7	9.71	12.31	28.50
2357000	0.05	12.0	1.76	28.07	3.2	8.2	7.78	3.06	88.19
2358000	2.95	25.6	12.48	10.31	7.7	6.8	5.85	19.03	30.33
2359000	5.85	10.1	9.07	22.48	12.2	5.3	3.92	9.78	90.02
2360000	8.75	23.7	5.66	4.72	4.0	3.9	1.99	0.55	32.17
2361000	11.64	8.2	2.25	16.89	8.5	2.5	0.06	16.50	91.85
2362000	14.54	21.8	12.97	29.06	0.2	1.1	16.34	7.24	33.99
2363000	0.54	6.3	9.56	11.30	4.7	21.6	14.42	23.22	93.68
2364000	3.44	19.9	6.15	23.47	9.2	23.2	12.49	13.97	35.82
2365000	6.34	4.5	2.74	5.71	1.0	21.7	10.56	4.71	95.51
2366000	9.23	17.9	13.46	17.88	5.5	20.3	8.63	20.69	37.65
2367000	12.13	2.5	10.05	0.12	9.9	18.9	6.70	11.43	97.34
2368000	15.03	16.0	6.64	12.29	1.7	17.4	4.77	2.18	39.48
2369000	1.03	0.6	3.22	24.47	6.2	16.0	2.84	18.16	99.17
2370000	3.92	11.1	13.94	6.71	10.7	14.6	0.91	8.90	41.31
2371000	6.82	27.7	10.53	18.88	2.5	13.2	17.19	24.87	101.00
2372000	9.72	12.2	7.12	1.12	7.0	11.7	15.26	15.62	43.14
2373000	12.62	25.7	3.71	13.29	11.6	10.3	13.31	6.37	102.83
2374000	15.52	10.3	0.30	25.46	3.3	8.8	11.41	22.31	44.97
2375000	1.51	23.8	11.02	7.70	7.8	7.1	9.47	13.09	101.66
2376000	4.41	8.4	7.61	19.87	12.3	6.0	7.54	3.83	46.80
2377000	7.31	21.9	4.20	2.11	4.0	4.5	5.61	19.84	106.49
2378000	10.21	6.5	0.79	14.28	8.5	3.1	3.68	10.55	48.63
2379000	13.11	20.0	11.51	26.45	0.3	1.6	1.75	1.30	108.32
2380000	16.00	4.5	8.10	8.69	4.8	0.2	18.03	17.28	50.46
2381000	2.00	18.1	4.68	20.87	9.3	23.8	16.10	8.02	110.15
2382000	4.90	2.6	1.27	3.10	1.0	22.3	14.18	24.00	52.29
2383000	7.80	16.2	11.99	15.28	5.6	20.9	12.25	14.71	111.98
2384000	10.70	0.7	8.58	27.45	10.0	19.1	10.32	5.49	54.12
2385000	13.59	14.2	5.17	9.69	1.8	18.0	8.38	21.46	113.81
2386000	16.49	27.8	1.76	21.86	6.3	16.6	6.45	12.21	55.95
2387000	2.49	12.3	12.48	4.10	10.8	15.1	4.52	2.95	115.64
2388000	5.39	25.9	9.07	16.27	2.6	13.7	2.59	18.93	57.78
2389000	8.29	10.4	5.66	28.44	7.1	12.2	0.66	9.68	117.47
2390000	11.18	24.0	2.25	10.68	11.6	10.8	16.91	0.42	59.61
2391000	14.08	8.5	12.97	22.85	3.3	9.4	15.02	16.40	1.76
2392000	0.08	22.0	9.56	5.09	7.8	8.0	13.09	7.14	61.44
2393000	2.98	6.6	6.14	17.26	12.4	6.5	11.16	23.12	3.59
2394000	5.88	20.1	2.73	29.43	4.1	5.1	9.23	13.86	63.27
2395000	8.77	4.7	13.45	11.67	8.6	3.6	7.29	4.61	5.42
2396000	11.67	18.2	10.04	23.84	0.3	2.2	5.36	20.59	65.10
2397000	14.57	2.8	6.63	6.08	4.9	0.8	3.43	11.33	7.25
2398000	0.57	16.3	3.22	18.25	9.4	24.3	1.50	2.08	66.93
2399000	8.46	0.8	13.94	0.49	1.1	22.9	17.78	18.05	9.08
2400000	6.36	14.4	10.53	12.67	5.6	21.5	15.85	8.80	68.76

TABLE IIa.

Epochs and Arguments for each Thousandth Day, from 2300000 to 2400000.

For Washington Mean Noon.

Day of Julian Period.	Arg. 54.	55.	56.	57.	58.	59	60.	61.	62.
	d.	d.	d.	d.	d.	d.	d.	d.	d.
2300000	8.49	4.41	1.74	12.90	7.94	32.07	16.92	8.62	4.61
2301000	13.88	3.42	4.81	10.64	1.95	19.01	12.44	1.76	8.11
2302000	4.20	2.43	7.88	8.39	11.21	5.95	7.96	8.52	11.61
2303000	9.59	1.43	10.96	6.13	5.22	31.85	3.49	1.67	0.67
2304000	14.98	0.44	0.91	3.87	14.47	18.78	21.33	8.42	4.17
2305000	5.29	19.08	3.98	1.61	8.49	5.72	16.85	1.57	7.67
2306000	10.68	18.09	7.05	16.35	2.50	31.62	12.38	8.32	11.17
2307000	1.00	17.09	10.13	14.09	11.76	18.56	7.90	1.47	0.24
2308000	6.39	16.10	0.08	11.83	5.77	5.50	3.42	8.22	3.74
2309000	11.77	15.11	3.15	9.58	15.03	31.10	21.27	1.37	7.24
2310000	2.09	14.12	6.22	7.32	9.01	18.33	16.79	8.12	10.74
2311000	7.49	13.13	9.29	5.06	3.06	5.27	12.31	1.27	14.24
2312000	12.87	12.14	12.37	2.80	12.31	31.17	7.83	8.02	3.30
2313000	3.18	11.14	2.32	0.54	6.33	18.11	3.36	1.18	6.80
2314000	8.57	10.15	5.39	15.28	0.34	5.05	21.20	7.93	10.30
2315000	13.96	9.16	8.46	13.02	9.60	30.95	16.72	1.08	13.81
2316000	4.28	8.17	11.54	10.76	3.61	17.88	12.25	7.83	2.88
2317000	9.66	7.18	1.49	8.51	12.87	4.82	7.77	0.98	6.37
2318000	15.05	6.18	4.56	6.25	6.88	30.72	3.29	7.73	9.87
2319000	5.37	5.19	7.63	3.99	0.90	17.66	21.14	0.88	13.37
2320000	10.76	4.20	10.71	1.73	10.15	4.59	16.66	7.63	2.44
2321000	1.07	3.21	0.66	16.47	4.17	30.49	12.18	0.78	5.94
2322000	6.46	2.21	3.73	14.21	13.42	17.43	7.70	7.54	9.44
2323000	11.85	1.21	6.80	11.95	7.44	4.37	3.23	0.68	12.94
2324000	2.17	0.23	9.88	9.69	1.45	30.27	21.07	7.44	2.00
2325000	7.55	18.86	12.95	7.44	10.71	17.21	16.59	0.59	5.50
2326000	12.94	17.87	2.90	5.18	4.72	4.14	12.12	7.34	9.01
2327000	3.26	16.88	5.97	2.92	13.97	30.04	7.64	0.49	12.51
2328000	8.65	15.89	9.05	0.66	7.99	16.98	3.16	7.24	1.57
2329000	14.03	14.89	12.12	15.40	2.00	3.92	21.01	0.39	5.07
2330000	4.35	13.90	2.07	13.14	11.26	29.82	16.53	7.14	8.57
2331000	9.74	12.91	5.14	10.88	5.27	16.76	12.05	0.29	12.07
2332000	0.06	11.92	8.21	8.62	14.53	3.69	7.58	7.05	1.14
2333000	5.44	10.92	11.29	6.37	8.54	29.59	3.10	0.19	4.64
2334000	10.83	9.93	1.24	4.11	2.56	16.53	20.94	6.95	8.14
2335000	1.14	8.94	4.31	1.85	11.81	3.47	16.47	0.10	11.64
2336000	6.54	7.95	7.38	16.58	5.83	29.37	11.99	6.85	0.70
2337000	11.92	6.96	10.46	14.32	15.08	16.31	7.51	0.00	4.20
2338000	2.24	5.96	0.41	12.07	9.10	3.24	3.04	6.75	7.70
2339000	7.63	4.97	3.48	9.81	3.11	29.14	20.88	13.51	11.21
2340000	13.02	3.98	6.55	7.56	12.36	16.08	16.40	6.65	0.27
2341000	3.33	2.99	9.62	5.30	6.38	3.02	11.93	13.41	3.77
2342000	8.72	2.00	12.70	3.04	0.39	28.92	7.45	6.55	7.27
2343000	14.11	1.00	2.65	0.78	9.65	15.85	2.97	13.31	10.77
2344000	4.43	0.01	5.72	15.51	3.66	2.79	20.82	6.46	14.27
2345000	9.81	18.65	8.79	13.26	12.92	28.69	16.34	13.21	3.34
2346000	0.13	17.66	11.87	11.00	6.93	15.63	11.86	6.36	6.84
2347000	5.52	16.67	1.82	8.74	0.95	2.57	7.38	13.12	10.34
2348000	10.91	15.67	4.89	6.49	10.20	28.47	2.91	6.26	13.84
2349000	1.22	14.68	7.96	4.23	4.22	15.40	20.75	13.02	2.90
2350000	6.61	13.69	11.04	1.97	13.47	2.34	16.27	6.16	6.40

TABLE IIa.

Epochs and Arguments for each Thousandth Day, from 2300000 to 2400000.

For Washington Mean Noon.

Day of Julian Period.	Arg. 54.	55.	56.	57.	58.	59	60.	61.	62.
	d.	d.	d.	d.	d.	d	d	d.	d.
2350000	6.61	13.69	11.04	1.97	13.47	2.34	16.27	6.16	6.40
2351000	12.00	12.70	0.99	16.70	7.49	28.21	11.80	12.92	9.90
2352000	2.32	11.71	4.07	14.45	1.50	15.18	7.32	6.06	13.41
2353000	7.70	10.71	7.13	12.19	10.75	2.11	2.84	12.82	2.47
2354000	13.09	9.72	10.21	9.93	4.77	28.02	20.69	5.97	5.97
2355000	3.41	8.73	0.16	7.68	14.02	14.95	16.21	12.72	9.47
2356000	8.80	7.71	3.23	5.42	8.01	1.89	11.73	5.87	12.97
2357000	14.18	6.75	6.30	3.16	2.06	27.79	7.26	12.63	2.03
2358000	4.50	5.75	9.38	0.90	11.31	14.73	2.78	5.77	5.53
2359000	9.89	4.76	12.45	15.63	5.33	1.66	20.62	12.53	9.01
2360000	0.21	3.77	2.40	13.38	14.58	27.56	16.15	5.67	12.54
2361000	5.60	2.78	5.47	11.12	8.59	14.50	11.67	13.43	1.60
2362000	10.98	1.78	8.54	8.86	2.61	1.44	7.19	5.57	5.10
2363000	1.30	0.79	11.62	6.61	11.86	27.31	2.71	12.33	8.60
2364000	6.69	19.43	1.57	4.35	5.88	14.28	20.56	5.48	12.10
2365000	12.07	18.43	4.64	2.09	15.13	1.21	16.08	12.23	1.17
2366000	2.39	17.44	7.71	16.82	9.15	27.11	11.60	5.38	4.67
2367000	7.78	16.45	10.79	14.57	3.16	14.05	7.13	12.14	8.17
2368000	13.17	15.46	0.74	12.31	12.42	0.99	2.65	5.28	11.67
2369000	3.48	11.46	3.81	10.05	6.43	26.89	20.49	12.01	0.73
2370000	8.87	13.47	6.88	7.79	0.45	13.82	16.02	5.18	4.23
2371000	14.26	12.48	9.95	5.54	9.70	0.76	11.54	11.94	7.73
2372000	4.58	11.49	13.03	3.28	3.72	26.66	7.06	5.08	11.24
2373000	9.96	10.49	2.98	1.02	12.97	13.60	2.59	11.84	0.30
2374000	0.28	9.50	6.05	15.75	6.99	0.51	20.43	4.98	3.80
2375000	5.67	8.51	9.12	13.50	1.00	26.14	15.95	11.71	7.30
2376000	11.06	7.52	12.20	11.24	10.26	13.37	11.48	4.89	10.80
2377000	1.37	6.53	2.15	8.98	4.27	0.31	7.00	11.64	14.30
2378000	6.76	5.53	5.22	6.73	13.52	26.21	2.52	4.79	3.37
2379000	12.15	4.51	8.29	4.47	7.54	13.15	20.37	11.55	6.87
2380000	2.47	3.55	11.37	2.21	1.55	0.08	15.89	4.69	10.37
2381000	7.86	2.56	1.32	16.94	10.81	25.98	11.41	11.45	13.87
2382000	13.24	1.57	4.39	14.69	4.82	12.92	6.94	4.59	2.93
2383000	3.56	0.57	7.46	12.43	14.08	38.82	2.46	11.35	6.43
2384000	8.95	19.21	10.54	10.17	8.09	25.76	20.30	4.19	9.93
2385000	14.33	18.22	0.49	7.91	2.11	12.70	15.83	11.25	13.44
2386000	4.65	17.23	3.56	5.66	11.36	38.59	11.36	4.40	2.50
2387000	10.04	16.24	6.63	3.40	5.38	25.53	6.87	11.15	6.00
2388000	0.36	15.24	9.71	1.14	14.63	12.47	2.40	4.30	9.50
2389000	5.74	14.25	12.78	15.87	8.65	38.37	20.24	11.06	13.00
2390000	11.13	13.26	2.73	13.62	2.66	25.31	15.76	4.20	2.06
2391000	1.45	12.27	5.80	11.36	11.91	12.24	11.29	10.96	5.56
2392000	6.84	11.28	8.87	9.10	5.93	38.14	6.81	4.10	9.07
2393000	12.23	10.28	11.95	6.85	15.18	25.08	2.33	10.86	12.57
2394000	2.54	9.29	1.90	4.59	9.20	12.02	20.18	4.00	1.63
2395000	7.93	8.30	4.97	2.33	3.21	37.92	15.70	10.76	5.13
2396000	13.32	7.31	8.04	0.07	12.47	24.86	11.22	3.90	8.63
2397000	3.63	6.32	11.12	14.80	6.48	11.79	6.74	10.66	12.13
2398000	9.02	5.32	1.07	12.55	0.50	37.69	2.27	3.81	1.20
2399000	14.41	4.33	4.14	10.29	9.75	21.63	20.11	10.56	4.70
2400000	4.73	3.31	7.21	8.03	3.77	11.57	15.63	3.71	8.20

TABLE IIa.

Epochs and Arguments for each Thousandth Day, from 2300000 to 2400000.

For Washington Mean Noon.

Day of Julian Period.	Arg. 63.	64.	65.	66.	67.	68.	69.	70.	71.
	d.	d.	d.	d.	d.	d.	d.	d.	d.
2300000	8.80	11.71	9.97	32.1	11.13	18.74	163.0	233.6	122.6
2301000	13.57	3.93	11.36	35.4	23.17	22.74	219.2	244.2	538.6
2302000	4.69	32.14	12.74	3.1	7.76	26.74	275.4	254.8	370.8
2303000	9.45	24.36	14.12	6.4	19.80	3.07	331.7	265.5	203.0
2304000	0.58	16.58	1.01	9.7	4.40	7.07	387.9	276.1	35.1
2305000	5.34	8.80	2.39	13.1	16.44	11.07	444.1	286.8	451.2
2306000	10.11	1.02	3.77	16.4	1.01	15.07	28.4	297.4	283.4
2307000	1.23	29.23	5.14	19.7	13.08	19.06	84.6	308.0	115.5
2308000	6.00	21.46	6.52	23.0	25.12	23.06	140.8	318.7	531.6
2309000	10.77	13.68	7.90	26.3	9.71	27.06	197.0	329.3	363.8
2310000	1.89	5.89	9.28	29.6	21.76	3.39	253.2	10.1	195.9
2311000	6.65	34.11	10.66	32.9	6.35	7.39	309.4	20.7	28.1
2312000	11.42	26.33	12.05	0.6	18.39	11.39	365.7	31.3	444.1
2313000	2.54	18.55	13.43	4.0	2.99	15.39	421.9	42.0	276.3
2314000	7.30	10.77	0.33	7.3	15.03	19.39	6.2	52.6	108.5
2315000	12.07	2.99	1.70	10.6	27.08	23.39	62.4	63.3	524.5
2316000	3.19	31.20	3.08	13.9	11.67	27.39	118.6	73.9	356.7
2317000	7.96	23.42	4.46	17.2	23.71	3.72	174.8	84.5	188.9
2318000	12.72	15.64	5.84	20.6	8.30	7.72	231.0	95.2	21.0
2319000	3.85	7.87	7.22	23.9	20.34	11.72	287.2	105.8	437.1
2320000	8.61	0.09	8.60	27.2	4.94	15.72	343.4	116.4	269.2
2321000	13.38	28.30	9.98	30.5	16.98	19.72	399.7	127.0	101.4
2322000	4.50	20.52	11.37	33.8	1.58	23.72	455.9	137.6	517.5
2323000	9.26	12.74	12.74	1.6	13.62	0.05	40.2	148.3	349.6
2324000	0.39	4.96	14.12	4.9	25.67	4.05	96.4	158.9	181.8
2325000	5.15	33.17	1.02	8.2	10.26	8.05	152.6	169.6	13.9
2326000	9.91	25.39	2.40	11.5	22.30	12.05	208.8	180.2	430.0
2327000	1.04	17.61	3.78	14.8	6.90	16.05	265.0	190.8	262.2
2328000	5.80	9.83	5.15	18.2	18.94	20.04	321.2	201.5	94.3
2329000	10.57	2.05	6.53	21.5	3.53	24.04	377.4	212.1	510.4
2330000	1.69	30.26	7.91	24.8	15.57	0.37	433.7	222.7	342.6
2331000	6.46	22.49	9.29	28.1	0.17	4.37	18.0	233.3	174.7
2332000	11.23	14.71	10.67	31.4	12.21	8.37	74.2	243.9	6.9
2333000	2.35	6.93	12.06	34.7	24.25	12.37	130.4	254.6	423.0
2334000	7.11	35.14	13.44	2.4	8.85	16.37	186.6	265.2	255.1
2335000	11.88	27.36	0.33	5.8	20.89	20.37	242.8	275.8	87.3
2336000	3.00	19.58	1.71	9.1	5.49	21.37	299.0	286.4	503.4
2337000	7.76	11.80	3.09	12.4	17.53	0.70	355.2	297.0	335.5
2338000	12.53	4.02	4.47	15.7	2.12	4.70	411.5	307.7	167.7
2339000	3.65	32.23	5.85	19.0	14.16	8.70	467.7	318.3	583.7
2340000	8.42	24.45	7.23	22.3	26.21	12.70	52.0	328.9	415.9
2341000	13.18	16.67	8.61	25.6	10.80	16.70	108.2	9.7	248.1
2342000	4.31	8.90	9.99	28.9	22.84	20.70	164.4	20.3	80.2
2343000	9.07	1.12	11.38	32.3	7.44	24.70	220.6	31.0	496.3
2344000	0.20	29.33	12.76	0.0	19.48	1.03	276.8	41.6	328.5
2345000	4.96	21.55	14.13	3.3	4.08	5.03	333.0	52.3	160.6
2346000	9.72	13.77	1.03	6.6	16.12	9.03	389.3	62.9	576.7
2347000	0.85	5.99	2.41	9.9	0.71	13.03	445.5	73.5	408.9
2348000	5.61	34.20	3.79	13.3	12.75	17.03	29.8	84.2	241.0
2349000	10.37	26.42	5.17	16.6	24.80	21.03	86.0	94.8	73.2
2350000	1.50	18.64	6.55	19.9	9.39	25.03	142.2	105.4	489.3

TABLE IIa.

Epochs and Arguments for each Thousandth Day, from 2300000 to 2400000.

For Washington Mean Noon.

Day of Julian Period. Arg.	63.	64.	65.	66.	67.	68.	69.	70.	71.
	d.	d.	d.	d.	d.	d.	d.	d.	d.
2350000	1.50	18.64	6.55	19.9	9.39	25.03	142.2	105.4	489.3
2351000	6.26	10.86	7.93	23.2	21.43	1.36	198.4	116.0	321.4
2352000	11.03	3.09	9.31	26.5	6.04	5.36	254.6	126.6	153.6
2353000	2.16	31.30	10.68	29.9	18.08	9.35	310.9	137.3	569.6
2354000	6.92	23.52	12.07	33.2	2.67	13.36	367.1	147.9	401.8
2355000	11.69	15.74	13.45	0.9	14.71	17.36	423.3	158.6	234.0
2356000	2.81	7.96	0.35	4.2	26.76	21.36	7.6	169.2	66.1
2357000	7.57	36.17	1.73	7.5	11.35	25.35	63.8	179.8	482.2
2358000	12.34	28.39	3.10	10.9	23.39	1.68	120.0	190.5	314.4
2359000	3.46	20.61	4.48	14.2	7.99	5.68	176.2	201.1	146.5
2360000	8.22	12.83	5.86	17.5	20.02	9.68	232.5	211.7	562.6
2361000	12.99	5.05	7.24	20.8	4.62	13.68	288.7	222.3	394.8
2362000	4.11	33.27	8.62	24.1	16.66	17.68	344.9	232.9	226.9
2363000	8.88	25.49	9.99	27.5	1.26	21.68	401.1	243.6	59.1
2364000	0.01	17.71	11.38	30.8	13.30	25.68	457.3	254.2	475.1
2365000	4.77	9.93	12.76	34.1	25.35	2.01	41.6	264.9	307.3
2366000	9.53	2.15	14.14	1.8	9.94	6.01	97.8	275.5	139.5
2367000	0.66	30.36	1.04	5.1	21.98	10.01	154.0	286.1	555.5
2368000	5.42	22.58	2.41	8.4	6.57	14.01	210.3	296.8	387.7
2369000	10.18	14.80	3.79	11.8	18.61	18.01	266.5	307.4	219.9
2370000	1.31	7.02	5.17	15.1	3.21	22.01	322.7	318.0	52.0
2371000	6.07	35.23	6.55	18.4	15.25	26.01	378.9	328.6	468.1
2372000	10.84	27.46	7.93	21.7	27.30	2.34	435.1	9.4	300.2
2373000	1.96	19.68	9.31	25.0	11.89	6.34	19.4	20.1	132.4
2374000	6.73	11.90	10.69	28.3	23.93	10.34	75.7	30.7	548.5
2375000	11.50	4.12	12.07	31.7	8.53	14.34	131.9	41.4	380.6
2376000	2.62	32.33	13.45	35.0	20.57	18.34	188.1	52.0	212.8
2377000	7.38	24.55	0.35	2.7	5.16	22.34	244.3	62.6	44.9
2378000	12.15	16.77	1.73	6.0	17.20	26.34	300.5	73.3	461.0
2379000	3.27	8.99	3.11	9.3	1.80	2.67	356.7	83.9	293.2
2380000	8.03	1.21	4.49	12.6	13.84	6.67	412.9	94.5	125.4
2381000	12.80	29.43	5.87	15.9	25.89	10.67	469.1	105.1	541.4
2382000	3.92	21.65	7.25	19.2	10.48	14.67	53.5	115.7	373.6
2383000	8.68	13.87	8.63	22.6	22.52	18.67	109.7	126.4	205.7
2384000	13.45	6.09	10.01	25.9	7.12	22.67	165.9	137.0	37.9
2385000	4.58	34.30	11.39	29.3	19.16	26.67	222.1	147.7	454.0
2386000	9.34	26.52	12.77	32.6	3.75	2.99	278.3	158.3	286.1
2387000	0.47	18.74	14.15	0.3	15.79	6.99	334.5	168.9	118.3
2388000	5.23	10.96	1.05	3.6	0.39	11.00	390.7	179.6	534.4
2389000	9.99	3.19	2.43	6.9	12.43	15.00	446.9	190.2	366.5
2390000	1.12	31.40	3.81	10.2	24.48	18.99	31.3	200.8	198.7
2391000	5.88	23.62	5.19	13.5	9.07	22.99	87.5	211.4	30.9
2392000	10.65	15.84	6.57	16.8	21.13	26.99	143.7	222.0	446.9
2393000	1.77	8.06	7.95	20.2	5.71	3.32	199.9	232.7	279.1
2394000	6.53	0.28	9.33	23.5	17.75	7.32	256.1	243.3	111.2
2395000	11.30	28.49	10.70	26.8	2.35	11.32	312.3	254.0	527.3
2396000	2.43	20.71	12.09	30.1	14.39	15.32	368.5	264.6	359.5
2397000	7.19	12.94	13.47	33.4	26.43	19.32	424.8	275.2	191.6
2398000	11.96	5.16	0.37	1.2	11.02	23.32	9.1	285.9	23.8
2399000	3.08	33.37	1.75	4.5	23.06	27.32	65.3	296.5	439.8
2400000	7.84	25.59	3.13	7.8	7.66	3.65	121.5	307.1	272.0

TABLE IIa.

Epochs and Arguments for each Thousandth Day, from 2300000 to 2400000.

For Washington Mean Noon.

Day of Julian Period.	Arg. 72.	73.	74.	75.	76.	78.	79.	80.	81.
	d.	d.	d.	d.	d.	d.	d.	d.	d.
2300000	103.9	2077.0	3031	54315	2766	6.448	366.2	57.0	50.32
2301000	306.2	886.7	798	55315	3766	65.440	51.1	117.8	56.68
2302000	109.5	1886.7	1798	56315	4766	124.432	174.4	178.7	63.04
2303000	311.7	696.4	2798	57315	5767	183.424	297.7	239.5	69.41
2304000	115.1	1696.4	565	58315	6767	54.214	420.9	300.3	75.77
2305000	317.3	506.1	1565	59315	7767	113.206	105.9	48.1	82.13
2306000	120.7	1506.1	2565	60315	8768	172.198	229.1	109.0	88.49
2307000	322.9	315.8	332	61315	9768	42.988	352.1	169.8	94.85
2308000	126.2	1315.8	1332	62315	10769	101.981	37.3	230.6	101.22
2309000	328.5	125.5	2332	63315	11769	160.972	160.6	291.5	107.58
2310000	131.8	1125.5	99	64315	12769	31.763	283.9	39.2	113.94
2311000	334.1	2125.5	1099	65315	13769	90.755	407.2	100.1	120.30
2312000	137.1	935.2	2099	66315	14769	149.747	92.1	160.9	2.46
2313000	339.6	1935.2	3099	67315	15770	20.538	215.3	221.7	8.83
2314000	143.0	744.8	867	68315	16770	79.530	338.6	282.6	15.19
2315000	315.2	1744.8	1867	69315	17770	138.522	23.5	30.4	21.55
2316000	118.6	554.5	2867	70315	18771	9.312	146.8	91.2	27.92
2317000	350.8	1554.5	631	71315	19771	68.301	270.1	152.0	34.28
2318000	154.1	364.1	1631	72315	20772	127.296	393.4	212.8	40.65
2319000	356.1	1364.1	2631	73315	21772	186.288	78.3	273.7	47.01
2320000	159.7	173.9	401	74315	22772	57.079	201.6	21.5	53.37
2321000	361.9	1173.9	1401	75315	23772	116.071	324.8	82.3	59.73
2322000	165.3	2173.9	2401	76315	24773	175.061	9.8	143.2	66.09
2323000	367.5	983.6	168	77315	25773	45.854	133.0	204.0	72.46
2324000	170.9	1983.6	1168	78315	26773	104.846	256.3	264.8	78.82
2325000	373.1	793.3	2168	79315	27771	163.838	379.6	12.6	85.18
2326000	176.5	1793.3	3168	80315	28771	34.628	61.5	73.4	91.55
2327000	378.7	602.9	935	81315	29771	93.620	187.8	134.3	97.91
2328000	182.0	1602.9	1935	82315	30775	152.613	311.1	195.1	104.28
2329000	384.3	412.6	2935	83315	31775	23.103	434.3	255.9	110.64
2330000	187.6	1412.6	703	84315	32775	82.395	119.3	3.8	117.00
2331000	389.8	222.3	1703	562	33775	141.387	242.5	64.6	123.36
2332000	193.2	1222.3	2703	1562	34775	12.178	365.8	125.4	5.52
2333000	395.1	32.0	470	2562	35775	71.170	50.7	186.3	11.89
2334000	198.8	1032.0	1470	3562	36776	130.162	174.0	247.1	18.25
2335000	2.1	2032.0	2470	4562	37776	0.952	297.3	307.9	24.61
2336000	201.4	841.7	237	5562	38776	59.945	420.6	55.7	30.98
2337000	7.7	1841.7	1237	6562	39776	119.937	105.5	116.5	37.34
2338000	209.9	651.3	2237	7562	40777	177.929	228.8	177.4	43.71
2339000	13.3	1651.3	4	8562	41777	48.719	352.0	238.2	50.07
2340000	215.5	461.0	1004	9562	42777	107.712	36.9	299.1	56.43
2341000	18.9	1461.0	2004	10562	43777	166.704	160.2	46.9	62.79
2342000	221.1	270.7	3004	11562	44777	37.495	283.5	107.7	69.15
2343000	24.4	1270.7	771	12562	45777	96.487	406.8	168.6	75.52
2344000	226.7	80.4	1771	13562	46777	155.479	91.7	229.4	81.88
2345000	30.0	1080.4	2771	14562	47778	26.270	215.0	290.2	88.24
2346000	232.2	2080.4	539	15562	48778	85.262	338.3	38.0	94.61
2347000	35.6	890.1	1539	16562	49778	144.254	23.2	98.9	100.97
2348000	237.8	1890.1	2539	17562	50778	15.045	146.4	159.7	107.34
2349000	41.2	699.7	306	18562	51778	74.037	269.7	220.5	113.70
2350000	243.4	1699.7	1306	19562	52778	133.029	393.0	281.3	120.06

TABLE IIa.

Epochs and Arguments for each Thousandth Day, from 2300000 to 2400000.

For Washington Mean Noon.

Day of Julian Period.	Arg. 72.	73.	74.	75.	76.	78.	79.	80.	81.
	d.	d.	d.	d.	d.	d.	d.	d.	d.
2350000	243.4	1699.7	1306	19562	52778	133.029	393.0	281.3	120.06
2351000	46.8	509.4	2306	20562	53778	3.820	77.9	29.1	2.22
2352000	249.0	1509.4	73	21562	54778	62.812	201.2	90.0	8.58
2353000	52.3	319.1	1073	22562	55778	121.805	324.5	150.8	14.95
2354000	254.6	1319.1	2073	23562	56778	180.797	9.4	211.6	21.31
2355000	57.9	128.7	3073	24562	57778	51.588	132.7	272.5	27.67
2356000	260.1	1128.7	840	25562	58779	110.580	255.9	20.3	34.01
2357000	63.5	2128.7	1840	26562	59779	169.572	379.2	81.1	40.40
2358000	265.7	938.4	2840	27562	60779	40.363	64.1	141.9	46.77
2359000	69.1	1938.4	607	28562	61779	99.355	187.4	202.8	53.13
2360000	271.3	748.1	1607	29562	62779	158.347	310.7	263.6	59.49
2361000	74.7	1748.1	2607	30562	63779	29.138	434.0	11.4	65.85
2362000	276.9	557.8	374	31562	64779	88.131	118.9	72.2	72.21
2363000	80.2	1557.8	1374	32562	65779	147.123	242.2	133.1	78.58
2364000	282.5	367.5	2374	33562	66779	17.914	365.5	193.9	84.94
2365000	85.8	1367.5	142	34562	67779	76.906	50.4	251.7	91.30
2366000	288.0	177.1	1142	35562	68779	135.898	173.6	2.5	97.67
2367000	91.4	1177.1	2142	36562	69779	6.689	296.9	63.4	101.03
2368000	293.6	2177.1	3142	37562	70779	65.682	420.2	121.2	110.40
2369000	97.0	986.8	909	38562	71779	121.674	105.1	185.0	116.76
2370000	299.2	1986.8	1909	39562	72779	183.666	228.4	215.8	123.12
2371000	102.6	796.5	2909	40562	73779	51.457	351.7	306.7	5.28
2372000	301.8	1796.5	676	41562	74779	113.450	36.6	51.5	11.61
2373000	108.1	606.1	1676	42562	75779	172.442	159.9	115.3	18.01
2374000	310.4	1606.1	2676	43562	76779	43.233	283.2	176.1	24.37
2375000	113.7	415.8	443	44562	77779	102.225	406.4	237.0	30.73
2376000	315.9	1415.8	1443	45562	78779	161.217	91.3	297.8	37.09
2377000	119.3	225.5	2443	46562	79779	32.008	214.6	45.6	43.45
2378000	321.5	1225.5	210	47562	80779	91.000	337.9	106.4	49.82
2379000	124.9	35.1	1210	48562	81779	149.992	22.8	167.3	56.18
2380000	327.1	1035.1	2210	49562	82779	20.784	146.1	228.1	62.54
2381000	130.4	2035.1	3210	50562	83779	79.777	269.4	288.9	68.90
2382000	332.7	844.8	978	51562	84779	138.769	392.7	36.7	75.26
2383000	136.0	1844.8	1978	52562	85779	9.560	77.6	97.6	81.63
2384000	338.3	654.5	2978	53562	86779	68.552	200.9	158.4	87.99
2385000	141.6	1654.5	745	54562	87778	127.545	324.1	219.2	94.35
2386000	343.8	464.2	1745	55562	88778	186.537	9.0	280.1	100.72
2387000	147.2	1464.2	2745	56562	89778	57.328	132.3	27.9	107.08
2388000	349.4	273.8	512	57562	90778	116.321	255.6	88.7	113.45
2389000	152.8	1273.8	1512	58562	91778	175.313	378.9	149.5	119.81
2390000	355.0	83.5	2512	59562	92778	46.105	63.8	210.4	1.97
2391000	158.3	1083.5	279	60562	93778	105.097	187.1	271.2	8.33
2392000	360.6	2083.5	1279	61562	94778	164.090	310.4	19.0	14.69
2393000	163.9	893.2	2279	62562	288	34.881	433.6	79.8	21.06
2394000	366.2	1893.2	46	63562	1288	93.873	118.6	140.6	27.42
2395000	169.5	702.8	1046	64562	2288	152.866	241.8	201.5	33.78
2396000	371.7	1702.8	2046	65562	3287	23.657	365.1	262.3	40.15
2397000	175.1	512.5	3046	66562	4287	82.649	50.0	10.1	46.51
2398000	377.3	1512.5	813	67562	5287	141.642	173.3	70.9	52.88
2399000	180.7	322.1	1813	68562	6287	12.433	296.6	131.8	59.24
2100000	382.9	1322.1	2813	69562	7287	71.425	419.9	192.6	65.60

TABLE II*a*.

Epochs and Arguments for each Thousandth Day, from 2300000 to 2400000.

For Washington Mean Noon.

Day of Julian Period.	Arg. 82.	83.	84.	85.	86.	87.	88.	89.	90.
	d.	d.	d.	d.	d.	d.	d.	d.	d.
2300000	11.905	34.45	10.79	7.32	1.08	2.19	2.97	6.79	5.9
2301000	1.051	14.95	15.33	19.50	5.72	7.62	5.65	2.56	5.2
2302000	5.062	34.65	0.74	8.17	10.37	13.05	8.34	17.64	4.5
2303000	9.073	15.15	5.28	20.36	15.01	4.27	11.02	13.41	3.8
2304000	13.084	34.86	9.82	9.03	19.65	9.69	13.70	9.18	3.1
2305000	2.229	15.36	14.37	21.22	24.30	0.91	1.72	4.94	2.4
2306000	6.241	35.07	18.91	9.89	28.94	6.34	4.40	0.71	1.6
2307000	10.252	15.56	4.31	22.08	33.59	11.76	7.08	15.79	0.9
2308000	14.263	35.27	8.85	10.71	38.23	2.98	9.77	11.56	0.2
2309000	3.408	15.77	13.39	22.93	4.59	8.40	12.45	7.33	31.8
2310000	7.420	35.48	17.93	11.60	9.23	13.83	0.47	3.10	31.1
2311000	11.431	15.98	3.33	0.27	13.87	5.05	3.15	18.18	30.4
2312000	0.577	35.69	7.88	12.45	18.52	10.47	5.83	13.94	29.7
2313000	4.588	16.18	12.42	1.12	23.16	1.69	8.52	9.71	29.0
2314000	8.599	35.89	18.96	13.31	27.80	7.12	11.20	5.48	28.3
2315000	12.610	16.39	2.36	1.98	32.45	12.54	13.88	1.25	27.6
2316000	1.756	36.10	6.91	11.17	37.09	3.76	1.90	16.33	26.9
2317000	5.767	16.60	11.45	2.83	3.44	9.18	4.58	12.10	26.2
2318000	9.778	36.31	15.99	15.02	8.09	0.40	7.26	7.86	25.5
2319000	13.789	16.80	1.39	3.69	12.74	5.83	9.95	3.63	24.8
2320000	2.935	36.51	5.93	15.88	17.38	11.25	12.63	18.71	24.1
2321000	6.946	17.01	10.48	4.55	22.03	2.47	0.65	14.48	23.4
2322000	10.957	36.72	15.02	16.73	26.67	7.90	3.33	10.25	22.7
2323000	0.103	17.22	0.42	5.40	31.32	13.32	6.01	6.02	22.0
2324000	4.114	36.93	4.96	17.59	35.96	4.54	8.70	1.78	21.3
2325000	8.125	17.42	9.50	6.26	2.32	9.96	11.38	16.86	20.6
2326000	12.136	37.13	14.05	18.45	6.96	1.18	11.06	12.63	19.8
2327000	1.282	17.63	18.59	7.11	11.61	6.61	2.08	8.40	19.1
2328000	5.293	37.31	3.99	19.30	16.25	12.03	4.76	4.17	18.4
2329000	9.304	17.83	8.53	7.97	20.89	3.25	7.45	19.25	17.7
2330000	13.315	37.55	13.07	20.16	25.53	8.67	10.13	15.02	17.0
2331000	2.461	18.01	17.62	8.83	30.18	14.10	12.81	10.79	16.3
2332000	6.472	37.75	3.02	21.02	34.82	5.32	0.83	6.55	15.6
2333000	10.483	18.25	7.56	9.68	1.18	10.74	3.51	2.32	14.9
2334000	14.494	37.96	12.10	21.87	5.83	1.96	6.19	17.40	14.2
2335000	3.640	18.46	16.64	10.54	10.47	7.39	8.88	13.17	13.5
2336000	7.651	38.16	2.04	22.73	15.11	12.81	11.56	8.91	12.8
2337000	11.662	18.66	6.59	11.40	19.75	4.03	14.24	4.71	12.1
2338000	0.808	38.37	11.13	23.58	24.40	9.45	2.26	0.47	11.4
2339000	4.819	18.87	15.67	12.25	29.04	0.67	4.94	15.55	10.7
2340000	8.831	38.58	1.07	0.92	33.69	6.10	7.63	11.32	10.0
2341000	12.842	19.08	5.61	13.11	0.05	11.52	10.31	7.09	9.3
2342000	1.987	38.78	10.16	1.78	4.69	2.74	12.99	2.86	8.6
2343000	5.999	19.28	14.70	13.96	9.33	8.16	1.01	17.94	7.9
2344000	10.010	38.99	0.10	2.63	13.98	13.59	3.69	13.71	7.2
2345000	14.021	19.49	4.64	14.82	18.62	4.81	6.37	9.47	6.5
2346000	3.167	39.20	9.18	3.49	23.26	10.23	9.06	5.21	5.8
2347000	7.178	19.69	13.73	15.68	27.91	1.45	11.74	1.01	5.1
2348000	11.189	0.19	18.27	4.34	32.55	6.88	14.42	16.09	4.4
2349000	0.335	19.90	3.67	16.53	37.19	12.30	2.44	11.86	3.7
2350000	4.346	0.40	8.21	5.20	3.55	3.52	5.12	7.63	3.0

TABLE IIa.

Epochs and Arguments for each Thousandth Day, from 2300000 to 2400000.

For Washington Mean Noon.

Day of Julian Period.	Arg. 82.	83.	84.	85.	86.	87.	88.	89.	90.
	d.	d.	d.	d.	d.	d.	d.	d.	d.
2350000	4.346	0.40	8.21	5.20	3.55	3.52	5.12	7.63	3.0
2351000	8.357	20.11	12.76	17.39	8.20	8.94	7.81	3.40	2.3
2352000	12.368	0.61	17.30	6.06	12.84	0.16	10.49	18.48	1.6
2353000	1.514	20.31	2.70	18.25	17.48	5.59	13.17	14.24	0.9
2354000	5.525	0.81	7.24	6.91	22.13	11.01	1.19	10.01	0.2
2355000	9.536	20.52	11.78	19.10	26.77	2.23	3.87	5.78	31.8
2356000	13.548	1.02	16.33	7.77	31.41	7.66	6.55	1.55	31.0
2357000	2.693	20.73	1.73	19.96	36.05	13.08	9.24	16.63	30.3
2358000	6.704	1.22	6.27	8.63	2.42	4.30	11.92	12.40	29.6
2359000	10.716	20.93	10.81	20.81	7.06	9.72	14.60	8.16	28.9
2360000	14.727	1.43	15.35	9.48	11.70	0.94	2.62	3.93	28.2
2361000	3.872	21.14	0.75	21.67	16.35	6.37	5.30	19.01	27.5
2362000	7.884	1.64	5.30	10.34	20.99	11.79	7.99	11.78	26.8
2363000	11.895	21.35	9.84	22.53	25.63	3.01	10.67	10.55	26.1
2364000	1.041	1.84	14.38	11.19	30.28	8.43	13.35	6.32	25.4
2365000	5.052	21.55	18.92	23.38	34.92	13.86	1.37	2.08	24.7
2366000	9.063	2.05	4.32	12.05	1.28	5.08	4.05	17.16	24.0
2367000	13.074	21.76	8.87	0.72	5.92	10.50	6.74	12.93	23.3
2368000	2.220	2.26	13.41	12.91	10.57	1.72	9.42	8.70	22.6
2369000	6.231	21.97	17.95	1.57	15.21	7.15	12.10	4.17	21.9
2370000	10.242	2.46	3.35	13.76	19.86	12.57	0.12	0.24	21.2
2371000	14.253	22.17	7.89	2.43	24.50	3.79	2.80	15.32	20.5
2372000	3.399	2.67	12.44	14.62	29.14	9.22	5.48	11.09	19.8
2373000	7.410	22.38	16.98	3.29	33.78	0.43	8.17	6.85	19.1
2374000	11.422	2.88	2.38	15.48	0.15	5.86	10.85	2.62	18.4
2375000	0.567	22.59	6.93	4.14	4.79	11.28	13.53	17.70	17.7
2376000	4.579	3.08	11.46	16.33	9.43	2.50	1.55	13.47	16.9
2377000	8.590	22.79	16.01	5.00	14.08	7.93	4.23	9.24	16.2
2378000	12.601	3.29	1.41	17.19	18.72	13.35	6.92	5.01	15.5
2379000	1.747	23.00	5.95	5.86	23.36	4.57	9.60	0.77	14.8
2380000	5.758	3.50	10.49	18.04	28.01	10.00	12.28	15.85	14.1
2381000	9.769	23.20	15.03	6.71	32.65	1.21	0.30	11.62	13.4
2382000	13.780	3.70	0.43	18.90	37.29	6.64	2.98	7.39	12.7
2383000	2.926	23.41	4.98	7.57	3.65	12.06	5.66	3.16	12.0
2384000	6.937	3.91	9.52	19.76	8.30	3.28	8.35	18.24	11.3
2385000	10.949	23.62	14.06	8.42	12.94	8.71	11.03	14.01	10.6
2386000	0.094	4.12	18.61	20.61	17.58	14.12	13.71	9.78	9.9
2387000	4.106	23.82	4.00	9.28	22.23	5.35	1.73	5.54	9.2
2388000	8.117	4.32	8.55	21.47	26.87	10.78	4.41	1.31	8.5
2389000	12.128	24.63	13.09	10.14	31.51	2.00	7.10	16.39	7.8
2390000	1.274	4.53	17.63	22.32	36.16	7.42	9.78	12.16	7.1
2391000	5.285	24.24	3.03	10.99	2.52	12.84	12.46	7.93	6.4
2392000	9.296	4.74	7.58	23.18	7.16	4.06	0.48	3.70	5.7
2393000	13.307	24.45	12.12	11.85	11.80	9.49	3.16	18.78	5.0
2394000	2.453	4.95	16.66	0.52	16.45	0.70	5.84	14.54	4.3
2395000	6.464	24.65	2.06	12.70	21.09	6.13	8.53	10.31	3.6
2396000	10.476	5.15	6.60	1.37	25.73	11.55	11.21	6.08	2.8
2397000	14.487	24.86	11.15	13.56	30.38	2.77	13.89	1.85	2.1
2398000	3.633	5.36	15.69	2.23	35.02	8.20	1.91	16.93	1.4
2399000	7.644	25.07	1.09	14.42	1.38	13.62	4.59	12.70	0.7
2400000	11.655	5.56	5.63	3.09	6.02	4.84	7.28	8.47	0.0

TABLE IIb.

Epochs and Arguments for each Thousandth Day, from 2400000 to 2500000.

For Washington Mean Noon.

Day of Julian Period.	u	u′	y-u	(y-u′)	Arg. 1.	1′.	2.	2′.
	° ″	″	° ″	″	d.	0.000	d.	0.000
2400000	354 3268.55	+0.37	25 1649.8	−0.3	27.0175339	+0329	7.505986	−022
2401000	211 1097.01	0.39	78 1483.2	0.3	7.4991263	0344	21.335956	023
2402000	67 2525.49	0.41	131 1316.6	0.3	15.5352726	0359	3.353990	024
2403000	284 353.99	0.42	184 1149.9	0.3	23.5714205	0375	17.183959	025
2404000	140 1782.50	0.44	237 983.3	0.3	4.0530174	0390	31.013926	026
2405000	356 3211.03	0.46	290 816.7	0.3	12.0891681	0405	13.031957	027
2406000	213 1039.58	0.48	343 650.0	0.4	20.1253208	0421	26.861922	028
2407000	69 2468.14	0.50	36 483.3	0.4	0.6069224	0436	8.879951	029
2408000	286 296.72	0.51	89 316.6	0.4	8.6430779	0452	22.709914	030
2409000	142 1725.32	0.53	142 149.9	0.4	16.6792349	0467	4.727941	031
2410000	358 3153.94	0.55	194 3583.2	0.4	24.7153935	0482	18.557902	032
2411000	215 982.57	0.57	247 3416.5	0.4	5.1970011	0497	0.575927	033
2412000	71 2411.22	0.59	300 3249.8	0.4	13.2331628	0512	14.405886	034
2413000	288 239.89	0.60	353 3083.0	0.4	21.2693259	0528	28.235845	035
2414000	144 1668.57	0.62	46 2916.3	0.4	1.7509382	0543	10.253866	036
2415000	0 3097.28	0.63	99 2749.5	0.4	9.7871011	0558	24.083823	037
2416000	217 926.00	0.65	152 2582.7	0.5	17.8232722	0574	6.101842	038
2417000	73 2354.73	0.67	205 2415.9	0.5	25.8594414	0589	19.931797	039
2418000	290 183.49	0.68	258 2249.1	0.5	6.3410598	0605	1.949814	040
2419000	146 1612.26	0.70	311 2082.3	0.5	14.3772322	0620	15.779767	041
2420000	2 3041.05	0.72	4 1915.5	0.5	22.4134060	0635	29.609718	042
2421000	219 869.85	0.74	57 1748.6	0.5	2.8950290	0650	11.627732	043
2422000	75 2298.68	0.76	110 1581.8	0.5	10.9312059	0665	25.457682	044
2423000	292 127.52	0.77	163 1414.9	0.5	18.9673844	0681	7.475694	045
2424000	148 1556.38	0.79	216 1248.0	0.5	27.0035644	0696	21.305642	046
2425000	4 2985.25	0.80	269 1081.1	0.5	7.4851935	0711	3.323652	047
2426000	221 814.14	0.82	322 914.2	0.6	15.5213766	0727	17.153597	048
2427000	77 2243.05	0.84	15 747.3	0.6	23.5575612	0742	30.983542	049
2428000	294 71.98	0.85	68 580.4	0.6	4.0391919	0758	13.001549	050
2429000	150 1500.93	0.87	121 413.5	0.6	12.0753826	0773	26.831491	051
2430000	6 2929.89	0.89	174 246.5	0.6	20.1115718	0788	8.849497	052
2431000	223 758.87	0.91	227 79.5	0.6	0.5932101	0803	22.679437	053
2432000	79 2187.86	0.93	279 3512.6	0.6	8.6294024	0819	4.697440	054
2433000	296 16.88	0.94	332 3345.6	0.7	16.6655962	0834	18.527379	055
2434000	152 1445.91	0.96	25 3178.6	0.7	24.7017915	0850	0.515380	056
2435000	8 2874.96	0.97	78 3011.6	0.7	5.1834360	0865	14.375316	057
2436000	225 704.02	0.99	131 2844.5	0.7	13.2196344	0880	28.205251	058
2437000	81 2133.10	1.01	184 2677.5	0.7	21.2558343	0896	10.223250	059
2438000	297 3562.20	1.02	237 2510.5	0.8	1.7374834	0911	24.053183	060
2439000	154 1391.32	1.04	290 2343.4	0.8	9.7736864	0927	6.071179	061
2440000	10 2820.46	1.06	343 2176.3	0.8	17.8098910	0942	19.901110	062
2441000	227 649.61	1.08	36 2009.2	0.8	25.8460971	0957	1.919104	063
2442000	83 2078.78	1.10	89 1842.1	0.8	6.3277523	0972	15.749033	064
2443000	299 3507.97	1.11	142 1675.0	0.8	14.3639615	0988	29.578961	065
2444000	156 1337.17	1.13	195 1507.9	0.8	22.4001722	1003	11.596952	066
2445000	12 2766.39	1.15	248 1340.8	0.8	2.8818320	1018	25.426878	067
2446000	229 595.63	1.17	301 1173.6	0.9	10.9180457	1034	7.444867	068
2447000	85 2024.89	1.19	354 1006.5	0.9	18.9542611	1049	21.274791	069
2448000	301 3454.16	1.20	47 839.3	0.9	26.9904779	1065	3.292778	070
2449000	158 1283.45	1.22	100 672.1	0.9	7.4721439	1080	17.122700	071
2450000	14 2712.76	+1.24	153 504.9	−0.9	15.5083639	+1095	30.952622	−072

TABLE IIb.

Epochs and Arguments for each Thousandth Day, from 2400000 to 2500000.

For Washington Mean Noon.

Day of Julian Period.	n	n′	y-n	(y-n′)	Arg. 1.	1′.	2.	2′.
					d.	0.000	d.	0.000
2450000	14° 2712.76″	+1.21″	153° 504.9″	−0.9″	15.5083639	+1095	30.952622	−072
2451000	231 542.09	1.26	206 337.7	0.9	23.5115853	1110	12.970606	073
2452000	87 1971.43	1.28	259 170.5	0.9	4.0262559	1126	26.800525	074
2453000	303 3400.79	1.29	312 3.3	0.9	12.0621801	1141	8.818507	075
2454000	160 1230.17	1.31	4 3436.0	0.9	20.0987065	1157	22.648424	076
2455000	16 2659.56	1.32	57 3268.8	0.9	0.5809817	1172	4.666401	077
2456000	233 488.97	1.34	110 3101.5	1.0	8.6166109	1187	18.496319	078
2457000	89 1918.40	1.36	163 2931.2	1.0	16.6528116	1203	0.514296	079
2458000	305 3317.85	1.37	216 2766.9	1.0	24.6890738	1218	14.344209	080
2459000	162 1177.32	1.39	269 2599.6	1.0	5.1707552	1234	28.174121	081
2460000	18 2606.80	1.41	322 2432.3	1.0	13.2069906	1249	10.192097	082
2461000	235 436.30	1.43	15 2265.0	1.0	21.2432275	1264	24.022007	083
2462000	91 1865.82	1.45	68 2097.6	1.0	1.7249134	1280	6.039980	084
2463000	307 3295.35	1.46	121 1930.3	1.0	9.7611534	1295	19.869887	085
2464000	164 1124.90	1.48	174 1762.9	1.0	17.7973949	1311	1.887858	086
2465000	20 2554.17	1.49	227 1595.6	1.0	25.8336379	1326	15.717764	087
2466000	237 384.05	1.51	280 1428.2	1.1	6.3153300	1341	29.547669	088
2467000	93 1813.66	1.53	333 1260.8	1.1	14.3515761	1357	11.565637	089
2468000	309 3243.28	1.54	26 1093.4	1.1	22.3878238	1372	25.395539	090
2469000	166 1072.92	1.56	79 925.9	1.1	2.8695206	1388	7.413505	091
2470000	22 2502.58	1.58	132 758.5	1.1	10.9057714	1403	21.243406	092
2471000	239 332.25	1.60	185 591.0	1.1	18.9420237	1418	3.261370	093
2472000	95 1761.94	1.62	238 423.6	1.1	26.9782775	1434	17.091268	094
2473000	311 3191.65	1.63	291 256.1	1.1	7.4599805	1449	30.921166	095
2474000	168 1021.37	1.65	344 88.6	1.1	15.4962374	1465	12.939127	096
2475000	24 2451.11	1.67	36 3521.1	1.1	23.5324959	1480	26.769022	097
2476000	241 280.87	1.69	89 3353.6	1.2	4.0142035	1496	8.786981	099
2477000	97 1710.65	1.71	142 3186.1	1.2	12.0504650	1512	22.616874	100
2478000	313 3140.41	1.72	195 3018.6	1.2	20.0867281	1527	4.634831	101
2479000	170 970.26	1.74	248 2851.0	1.2	0.5684403	1543	18.464723	102
2480000	26 2400.09	1.76	301 2683.4	1.2	8.6047066	1558	0.482677	103
2481000	243 229.94	1.78	354 2515.9	1.2	16.6409743	1573	14.312566	104
2482000	99 1659.80	1.80	47 2348.3	1.2	24.6772436	1589	28.142454	105
2483000	315 3089.69	1.81	100 2180.7	1.3	5.1589620	1604	10.160406	106
2484000	172 919.59	1.83	153 2013.1	1.3	13.1952314	1620	23.990294	107
2485000	28 2349.50	1.84	206 1845.5	1.3	21.2315084	1635	6.008213	108
2486000	245 179.44	1.86	259 1677.8	1.3	1.7132314	1651	19.838127	109
2487000	101 1609.39	1.88	312 1510.2	1.4	9.7495084	1667	1.856075	110
2488000	317 3039.36	1.89	5 1342.5	1.4	17.7857870	1682	15.685956	111
2489000	174 869.35	1.91	58 1174.9	1.4	25.8220671	1698	29.515837	112
2490000	30 2299.35	1.93	111 1007.2	1.4	6.3037963	1713	11.533782	113
2491000	247 129.37	1.95	164 839.5	1.4	14.3400795	1728	25.363662	114
2492000	103 1559.41	1.97	217 671.8	1.4	22.3763612	1744	7.381605	115
2493000	319 2989.47	1.98	270 504.1	1.4	2.8580981	1759	21.211481	116
2494000	176 819.54	2.00	323 336.3	1.4	10.8943860	1775	3.229422	117
2495000	32 2249.63	2.02	16 168.6	1.5	18.9306754	1790	17.059297	118
2496000	219 79.74	2.04	69 0.8	1.5	26.9669663	1805	30.889171	119
2497000	105 1509.87	2.06	121 3433.1	1.5	7.4487064	1821	12.907108	120
2498000	321 2940.01	2.07	174 3265.3	1.5	15.4850004	1836	26.736981	121
2499000	178 770.17	2.09	227 3097.5	1.5	23.5212960	1852	8.754916	122
2500000	34 2220.35	+2.11	280 2929.7	−1.5	4.0030408	+1867	22.584786	−123

TABLE IIb.

Epochs and Arguments for each Thousandth Day, from 2400000 to 2500000.

For Washington Mean Noon.

Day of Julian Period.	Arg. 3.	3′.	4.	5.	5′.	6.	7.	8.	9.
	d.	0.0000	d.	d.	0.00	d.	d.	d.	d.
2400000	25.523420	+08	45.9880	319.7423	+03	13.0559	6.8500	1.6577	3.2226
2401000	21.483136	08	315.4686	84.3872	04	2.4961	7.0233	1.4824	19.7294
2402000	17.443152	08	219.6896	260.8172	04	26.7831	7.1967	1.3071	6.4331
2403000	13.403168	09	123.9105	25.4621	04	16.2232	7.3700	1.1319	22.9401
2404000	9.363185	09	28.1314	201.8922	04	5.6633	7.5434	0.9566	9.6441
2405000	5.323502	09	297.6121	378.3224	04	29.9503	7.7167	0.7813	26.1509
2406000	1.283520	10	201.8330	142.9673	05	19.3904	7.8901	0.6060	12.8549
2407000	26.774125	10	106.0540	319.3975	05	8.8305	8.0634	0.4307	29.3617
2408000	22.734144	10	10.2749	84.0425	05	33.1175	8.2368	0.2555	10.0657
2409000	18.694162	11	279.7555	260.4727	05	22.5576	8.4101	0.0802	2.7696
2410000	14.654181	11	183.9765	25.1177	05	11.9977	8.5835	15.2922	19.2765
2411000	10.614200	11	88.1974	201.5179	05	1.4377	8.7569	15.1169	5.9805
2412000	6.574220	12	357.6780	377.9782	06	25.7247	8.9302	14.9116	22.4873
2413000	2.534239	12	261.8990	142.6233	06	15.1648	9.1036	14.7664	9.1912
2414000	28.024848	12	166.1199	319.0535	06	4.6049	9.2769	14.5911	25.6981
2415000	23.984868	13	70.3108	83.6987	06	28.8919	9.4503	14.4158	12.4021
2416000	19.944889	13	339.8215	260.1290	06	18.3320	0.0100	14.2405	28.9088
2417000	15.904910	13	244.0424	21.7742	07	7.7721	0.1833	14.0652	15.6127
2418000	11.864932	14	148.2631	201.2015	07	32.0591	0.3567	13.8900	2.3168
2419000	7.824951	14	52.4843	377.6349	07	21.4992	0.5300	13.7147	18.8237
2420000	3.784977	15	321.9649	142.2801	07	10.9393	0.7034	13.5394	5.5277
2421000	29.275587	15	226.1859	318.7105	07	0.3793	0.8768	13.3641	22.0315
2422000	25.235610	15	130.4068	83.3557	07	24.6663	1.0501	13.1888	8.7385
2423000	21.195634	16	34.6277	259.7862	08	14.1061	1.2235	13.0136	25.2154
2424000	17.155657	16	304.1084	24.4315	08	3.5465	1.3969	12.8383	11.9494
2425000	13.115681	16	208.3293	200.8620	08	27.8335	1.5702	12.6630	28.4562
2426000	9.075706	17	112.5502	377.2924	08	17.2735	1.7436	12.4877	15.1602
2427000	5.035731	17	16.7712	141.9378	08	6.7136	1.9170	12.3124	1.8642
2428000	0.995756	17	286.2518	318.3683	08	31.0006	2.0903	12.1372	18.3710
2429000	26.486369	18	190.4728	83.0137	09	20.4407	2.2637	11.9619	5.0750
2430000	22.446395	18	94.6937	259.4443	09	9.8807	2.4371	11.7866	21.5819
2431000	18.406421	18	364.1743	24.0897	09	34.1677	2.6105	11.6113	8.2859
2432000	14.366448	19	268.3953	200.5203	09	23.6078	2.7838	11.4360	24.7928
2433000	10.326475	19	172.6162	376.9509	09	13.0179	2.9572	11.2608	11.4968
2434000	6.286502	19	76.8371	141.5964	10	2.4879	3.1306	11.0855	28.0036
2435000	2.246530	20	346.3178	318.0270	10	26.7749	3.3039	10.9102	14.7076
2436000	27.737146	20	250.5387	82.6725	10	16.2150	3.4773	10.7349	1.4117
2437000	23.697174	21	154.7596	259.1032	10	5.6550	3.6507	10.5596	17.9185
2438000	19.657203	21	58.9806	23.7488	10	29.9420	3.8240	10.3844	4.6225
2439000	15.617232	21	328.4612	200.1795	10	19.3821	3.9974	10.2091	21.1294
2440000	11.577261	22	232.6822	376.6103	11	8.8221	4.1708	10.0338	7.8335
2441000	7.537291	22	136.9031	141.2559	11	33.1091	4.3442	9.8585	24.3103
2442000	3.497321	22	41.1240	317.6866	11	22.5491	4.5176	9.6832	11.0443
2443000	28.987940	23	310.6047	82.3323	11	11.9892	4.6910	9.5080	27.5512
2444000	24.947970	23	214.8256	258.7631	11	1.4292	4.8643	9.3327	14.2552
2445000	20.908002	23	119.0465	23.4087	12	25.7162	5.0377	9.1574	0.9593
2446000	16.868033	24	23.2675	199.8396	12	15.1562	5.2111	8.9822	17.4661
2447000	12.828065	24	292.7481	376.2705	12	4.5963	5.3844	8.8069	4.1702
2448000	8.788097	24	196.9690	140.9162	12	28.8833	5.5578	8.6317	20.6770
2449000	4.748130	25	101.1900	317.3471	12	18.3233	5.7312	8.4564	7.3811
2450000	0.708163	+25	5.4109	81.9929	+12	7.7634	5.9046	8.2811	23.8880

TABLE IIb.

Epochs and Arguments for each Thousandth Day, from 2400000 to 2500000.

For Washington Mean Noon.

Day of Julian Period.	Arg. 3.	3′.	4.	5.	5′.	6.	7.	8.	9.
	d.	0.0000	d.	d.	0.00	d	d.	d	d.
2450000	0.708163	+25	5.4109	81.9929	+12	7.7631	5.9046	8.2811	23.8880
2451000	26.198784	25	274.8916	258.4238	13	32.0503	6.0780	8.1058	10.5921
2452000	22.158818	26	179.1125	23.0696	13	21.4904	6.2514	7.9305	27.0989
2453000	18.118852	26	83.3334	199.5006	13	10.9304	6.4248	7.7553	13.8030
2454000	14.078886	27	352.8141	375.9316	13	0.3705	6.5982	7.5800	0.5070
2455000	10.038921	27	257.0350	140.5774	13	24.6574	6.7716	7.4047	17.0139
2456000	5.998956	27	161.2559	317.0084	13	14.0975	6.9449	7.2295	3.7180
2457000	1.958991	28	65.4769	81.6513	14	3.5375	7.1183	7.0542	20.2248
2458000	27.449615	28	334.9575	258.0854	14	27.8245	7.2917	6.8790	6.9289
2459000	23.409651	28	239.1784	22.7313	14	17.2645	7.4651	6.7037	23.4358
2460000	19.369687	29	143.3994	199.1625	14	6.7045	7.6385	6.5284	10.1399
2461000	15.329724	29	47.6203	375.5936	14	30.9915	7.8119	6.3531	26.6467
2462000	11.289761	29	317.1010	140.2396	15	20.4315	7.9853	6.1779	13.3508
2463000	7.249799	30	221.3219	316.6707	15	9.8716	8.1587	6.0026	0.0549
2464000	3.209837	30	125.5428	81.3167	15	34.1585	8.3321	5.8274	16.5618
2465000	28.700463	30	29.7638	257.7479	15	23.5985	8.5055	5.6521	3.2659
2466000	24.660502	31	299.2444	22.3940	15	13.0386	8.6789	5.4768	19.7728
2467000	20.620511	31	203.4653	198.8252	15	2.4786	8.8523	5.3016	6.4768
2468000	16.580580	32	107.6863	375.2564	16	26.7655	9.0257	5.1263	22.9837
2469000	12.540620	32	11.9072	139.9025	16	16.2056	9.1991	4.9511	9.6878
2470000	8.500660	32	281.3878	316.3339	16	5.6456	9.3725	4.7758	26.1947
2471000	4.460700	33	185.6088	80.9800	16	29.9325	9.5159	4.6005	12.8988
2472000	0.420741	33	89.8297	257.4113	16	19.3725	0.1056	4.4253	29.4057
2473000	25.911370	33	359.3104	22.0575	17	8.8126	0.2790	4.2500	16.1098
2474000	21.871411	34	263.5313	198.4889	17	33.0995	0.4524	4.0748	2.8139
2475000	17.831453	34	167.7522	374.9202	17	22.5395	0.6258	3.8995	19.3208
2476000	13.791495	34	71.9732	139.5665	17	11.9796	0.7993	3.7242	6.0249
2477000	9.751538	35	341.4538	315.9979	17	1.4196	0.9727	3.5490	22.5318
2478000	5.711581	35	245.6747	80.6441	17	25.7065	1.1461	3.3737	9.2359
2479000	1.671624	35	149.8957	257.0756	18	15.1465	1.3195	3.1985	25.7428
2480000	27.162255	36	54.1166	21.7220	18	4.5865	1.4929	3.0232	12.4470
2481000	23.122299	36	323.5972	198.1534	18	28.8731	1.6663	2.8479	28.9539
2482000	19.082343	36	227.8182	374.5849	18	18.3135	1.8397	2.6727	15.6580
2483000	15.012388	37	132.0391	139.2313	18	7.7535	2.0131	2.4974	2.3621
2484000	11.002433	37	36.2601	315.6628	19	32.0404	2.1865	2.3222	18.8690
2485000	6.962478	38	305.7407	80.3092	19	21.4801	2.3599	2.1469	5.5731
2486000	2.922524	38	209.9616	256.7408	19	10.9204	2.5334	1.9716	22.0800
2487000	28.413158	38	114.1826	21.3872	19	0.3604	2.7068	1.7964	8.7841
2488000	24.373205	39	18.4035	197.8188	19	24.6473	2.8802	1.6211	25.2911
2489000	20.333251	39	287.8841	374.2504	19	14.0874	3.0536	1.4159	11.9952
2490000	16.293298	39	192.1051	138.8970	20	3.5274	3.2270	1.2706	28.5021
2491000	12.253346	40	96.3260	315.3287	20	27.8143	3.4004	1.0953	15.2063
2492000	8.213394	40	0.5470	79.9752	20	17.2543	3.5738	0.9201	1.9104
2493000	4.173442	40	270.0276	256.4069	20	6.6943	3.7472	0.7448	18.4173
2494000	0.133490	41	174.2485	21.0534	20	30.9812	3.9207	0.5696	5.1215
2495000	25.624127	41	78.4695	197.4851	20	20.4212	4.0941	0.3943	21.6284
2496000	21.584177	41	347.9501	373.9169	21	9.8612	4.2675	0.2191	8.3325
2497000	17.544226	42	252.1710	138.5635	21	34.1481	4.4410	0.0439	24.8395
2498000	13.504276	42	156.3920	314.9953	21	23.5881	4.6144	15.2559	11.5436
2499000	9.464326	43	60.6129	79.6419	21	13.0281	4.7878	15.0807	28.0505
2500000	5.424377	+43	330.0935	256.0737	+21	2.4681	4.9612	14.9054	14.7547

TABLE IIb.

Epochs and Arguments for each Thousandth Day, from 2400000 to 2500000.

For Washington Mean Noon.

Day of Julian Period.	Arg. 10.	11.	12.	13.	14.	15.	16.	17.
	d.	d.	d.	d.	d.	d.	d.	d.
2400000	22.0856	18.9090	37.9433	0.4102	3.527	12.792	7.228	13.489
2401000	22.8395	24.4122	344.7031	2.0351	8.575	5.188	9.435	2.587
2402000	23.5934	3.0372	304.8427	3.6600	13.624	11.775	11.642	19.006
2403000	24.3473	8.5401	264.9823	5.2849	18.672	4.171	13.849	8.104
2404000	25.1011	14.0436	225.1219	6.9098	23.721	10.758	1.802	24.524
2405000	0.2333	19.5469	185.2615	8.5347	28.769	3.153	4.010	13.621
2406000	0.9872	25.0501	145.4010	0.0751	4.555	9.741	6.217	2.719
2407000	1.7410	3.6751	105.5406	1.7000	9.601	2.136	8.424	19.139
2408000	2.4949	9.1783	65.6802	3.3249	11.652	8.724	10.631	8.237
2409000	3.2488	14.6815	25.8198	4.9198	19.700	1.119	12.838	24.657
2410000	4.0027	20.1848	332.5795	6.5747	24.749	7.706	0.791	13.754
2411000	4.7566	25.6880	292.7191	8.1996	0.531	0.102	2.998	2.852
2412000	5.5105	4.3129	252.8586	9.8245	5.582	6.689	5.205	19.271
2413000	6.2644	9.8162	212.9982	1.3648	10.631	13.277	7.412	8.369
2414000	7.0183	15.3194	173.1378	2.9897	15.679	5.672	9.619	24.789
2415000	7.7721	20.8226	133.2774	4.6146	20.728	12.260	11.826	13.887
2416000	8.5260	26.3259	93.4169	6.2396	25.776	4.655	14.033	2.985
2417000	9.2799	4.9508	53.5565	7.8645	1.561	11.243	1.986	19.404
2418000	10.0338	10.4540	13.6961	9.1894	6.610	3.638	4.193	8.502
2419000	10.7877	15.9572	320.4557	1.0297	11.658	10.225	6.401	24.922
2420000	11.5416	21.4605	280.5952	2.6546	16.707	2.621	8.608	14.020
2421000	12.2955	0.0854	240.7318	4.2795	21.755	9.208	10.815	3.117
2422000	13.0494	5.5886	200.8713	5.9044	26.804	1.604	13.022	19.537
2423000	13.8033	11.0918	161.0139	7.5293	2.589	8.191	0.975	8.635
2424000	14.5572	16.5950	121.1534	9.1542	7.638	0.587	3.182	25.055
2425000	15.3111	22.0983	81.2930	0.6945	12.686	7.174	5.389	14.152
2426000	16.0650	0.7232	41.4325	2.3195	17.735	13.762	7.596	3.250
2427000	16.8189	6.2261	1.5720	3.9444	22.783	6.157	9.803	19.670
2428000	17.5728	11.7296	308.3317	5.5693	27.831	12.715	12.010	8.768
2429000	18.3267	17.2328	268.4712	7.1942	3.617	5.140	14.217	25.188
2430000	19.0806	22.7360	228.6107	8.8191	8.665	11.727	2.171	14.285
2431000	19.8345	1.3609	188.7503	0.3594	13.713	4.123	4.378	3.383
2432000	20.5884	6.8641	148.8898	1.9843	18.762	10.710	6.585	19.802
2433000	21.3423	12.3673	109.0293	3.6092	23.810	3.106	8.792	8.900
2434000	22.0963	17.8706	69.1688	5.2341	28.859	9.693	10.999	25.320
2435000	22.8502	23.3738	29.3083	6.8590	4.641	2.088	13.206	14.418
2436000	23.6041	1.9987	336.0679	8.4840	9.693	8.676	1.159	3.516
2437000	24.3581	7.5019	296.2075	0.0243	14.741	1.071	3.366	19.935
2438000	25.1120	13.0051	256.3470	1.6492	19.790	7.659	5.573	9.033
2439000	0.2442	18.5083	216.4865	3.2741	24.838	0.054	7.780	25.453
2440000	0.9981	24.0115	176.6260	4.8990	0.625	6.642	9.988	14.551
2441000	1.7520	2.6363	136.7655	6.5239	5.672	13.229	12.195	3.648
2442000	2.5059	8.1395	96.9050	8.1488	10.720	5.624	0.148	20.068
2443000	3.2599	13.6427	57.0445	9.7737	15.769	12.212	2.355	9.166
2444000	4.0138	19.1459	17.1840	1.3140	20.817	4.607	4.562	25.586
2445000	4.7677	24.6491	323.9436	2.9389	25.866	11.195	6.769	14.683
2446000	5.5217	3.2740	284.0830	4.5639	1.651	3.590	8.976	3.781
2447000	6.2756	8.7772	244.2225	6.1888	6.699	10.177	11.183	20.201
2448000	7.0296	14.2804	204.3620	7.8137	11.748	2.573	13.390	9.299
2449000	7.7835	19.7836	164.5014	9.4386	16.796	9.160	1.344	25.718
2450000	8.5374	25.2868	124.6409	0.9789	21.844	1.556	3.551	14.816

TABLE IIb.

Epochs and Arguments for each Thousandth Day, from 2400000 to 2500000.

For Washington Mean Noon.

Day of Julian Period.	Arg. 10.	11.	12.	13.	14.	15.	16.	17.
	d.	d.	d.	d.	d.	d.	d.	d.
2450000	8.5374	25.2868	124.6409	0.9789	21.814	1.556	3.551	14.816
2451000	9.2913	3.9117	84.7804	2.6038	26.892	8.143	5.758	3.914
2452000	10.0453	9.4148	44.9198	4.2287	2.678	0.539	7.965	20.331
2453000	10.7992	11.9180	5.0593	5.8536	7.726	7.126	10.172	9.431
2454000	11.5532	20.4212	311.8189	7.4785	12.775	13.714	12.379	25.851
2455000	12.3071	25.9244	271.9583	9.1034	17.823	6.109	0.332	14.949
2456000	13.0611	4.5192	232.0978	0.6438	22.871	12.697	2.539	4.047
2457000	13.8150	10.0524	192.2372	2.2687	27.920	5.092	4.746	20.466
2458000	14.5690	15.5556	152.3767	3.8936	3.705	11.680	6.953	9.564
2459000	15.3229	21.0588	112.5161	5.5185	8.753	4.073	9.160	25.984
2460000	16.0769	26.5619	72.6556	7.1434	13.802	10.662	11.368	15.082
2461000	16.8309	5.1868	32.7950	8.7683	18.851	3.058	13.575	4.180
2462000	17.5848	10.6900	339.5545	0.3086	23.899	9.645	1.528	20.599
2463000	18.3388	16.1931	299.6940	1.9335	28.947	2.011	3.735	9.697
2464000	19.0927	21.6963	259.8334	3.5584	4.733	8.628	5.943	26.117
2465000	19.8467	0.3212	219.9729	5.1833	9.781	1.024	8.150	15.214
2466000	20.6007	5.8243	180.1123	6.8082	14.829	7.611	10.357	4.312
2467000	21.3546	11.3275	140.2517	8.4331	19.878	0.007	12.564	20.732
2468000	22.1086	16.8307	100.3911	10.0580	24.926	6.594	0.517	9.830
2469000	22.8625	22.3338	60.5306	1.5983	0.711	13.182	2.724	26.250
2470000	23.6165	0.9587	20.6699	3.2232	5.760	5.577	4.931	15.347
2471000	24.3705	6.4618	327.4295	4.8481	10.808	12.165	7.138	4.445
2472000	25.1245	11.9650	287.5689	6.4730	15.857	4.560	9.345	20.865
2473000	0.2568	17.4681	247.7083	8.0979	20.905	11.148	11.552	9.963
2474000	1.0107	22.9713	207.8477	9.7228	25.954	3.543	13.760	26.383
2475000	1.7647	1.5961	167.9871	1.2631	1.739	10.130	1.713	15.480
2476000	2.5187	7.0993	128.1265	2.8881	6.787	2.526	3.920	4.578
2477000	3.2727	12.6024	88.2659	4.5130	11.836	9.113	6.127	20.998
2478000	4.0266	18.1056	48.4053	6.1379	16.884	1.509	8.331	10.095
2479000	4.7806	23.6087	8.5447	7.7628	21.932	8.096	10.542	26.515
2480000	5.5346	2.2336	315.3042	9.3877	26.981	0.492	12.748	15.613
2481000	6.2886	7.7367	275.4436	0.9280	2.766	7.079	0.702	4.711
2482000	7.0426	13.2399	235.5830	2.5529	7.814	13.667	2.909	21.130
2483000	7.7965	18.7430	195.7223	4.1778	12.863	6.062	5.116	10.228
2484000	8.5506	24.2462	155.8617	5.8027	17.911	12.650	7.323	26.648
2485000	9.3046	2.8710	116.0011	7.4276	22.960	5.045	9.530	15.746
2486000	10.0585	8.3742	76.1405	9.0525	28.008	11.632	11.737	4.844
2487000	10.8125	13.8773	36.2799	0.5928	3.793	4.028	13.945	21.263
2488000	11.5665	19.3804	343.0394	2.2177	8.842	10.615	1.898	10.361
2489000	12.3205	24.8836	303.1787	3.8426	13.890	3.011	4.105	26.781
2490000	13.0745	3.5084	263.3181	5.4675	18.939	9.598	6.312	15.879
2491000	13.8285	9.0115	223.4575	7.0924	23.987	1.991	8.519	4.977
2492000	14.5825	14.5147	183.5968	8.7173	29.035	8.581	10.726	21.396
2493000	15.3365	20.0178	143.7362	0.2576	4.821	0.977	12.933	10.494
2494000	16.0905	25.5209	103.8755	1.8825	9.869	7.564	0.887	26.914
2495000	16.8445	4.1458	64.0149	3.5074	14.917	14.151	3.094	16.011
2496000	17.5985	9.6489	24.1542	5.1324	19.966	6.547	5.301	5.109
2497000	18.3525	15.1520	330.9137	6.7573	25.014	13.134	7.508	21.529
2498000	19.1065	20.6551	291.0530	8.3822	0.799	5.530	9.715	10.627
2499000	19.8605	26.1583	251.1921	10.0071	5.848	12.117	11.922	27.047
2500000	20.6145	4.7831	211.3316	1.5474	10.896	4.513	14.129	16.144

TABLE II*b*.

Epochs and Arguments for each Thousandth Day, from 2400000 to 2500000.

For Washington Mean Noon.

Day of Julian Period.	Arg. 18.	19.	20.	21.	22.	23.	24.	25.	26.
	d.	d.	d.	d.	d.	d.	d.	d.	d.
2400000	8.137	18.503	12.791	118.93	3.312	10.320	8.124	5.633	2.449
2401000	1.031	22.113	10.580	65.56	7.371	11.786	12.386	3.345	0.961
2402000	3.798	1.421	8.368	12.19	11.400	13.252	3.373	1.057	5.295
2403000	6.564	5.031	6.156	90.49	15.430	14.718	7.636	8.137	3.807
2404000	9.332	8.641	3.944	37.12	19.459	16.183	11.898	5.850	2.319
2405000	2.225	12.251	1.733	115.43	23.488	17.649	2.885	3.563	0.831
2406000	4.993	15.861	13.838	62.06	27.518	19.115	7.148	1.275	5.165
2407000	7.759	19.471	11.626	8.69	31.547	20.580	11.410	8.355	3.677
2408000	0.653	23.081	9.414	86.99	3.148	22.046	2.396	6.068	2.189
2409000	3.420	2.389	7.202	33.62	7.478	23.512	6.659	3.780	0.700
2410000	6.188	5.999	4.991	111.93	11.507	1.203	10.921	1.493	5.035
2411000	8.954	9.610	2.779	58.56	15.536	2.669	1.908	8.573	3.547
2412000	1.848	13.220	0.567	5.19	19.566	4.135	6.171	6.285	2.059
2413000	4.615	16.830	12.672	83.49	23.595	5.600	10.433	3.998	0.571
2414000	7.382	20.440	10.461	30.12	27.624	7.066	1.419	1.711	4.905
2415000	0.276	24.050	8.249	108.42	31.654	8.532	5.682	8.790	3.417
2416000	3.043	3.358	6.037	55.05	3.555	9.997	9.945	6.503	1.929
2417000	5.810	6.968	3.825	1.69	7.584	11.463	0.931	4.216	0.440
2418000	8.577	10.578	1.613	79.99	11.614	12.928	5.194	1.928	4.775
2419000	1.471	14.188	13.718	26.62	15.643	14.394	9.456	9.008	3.287
2420000	4.238	17.798	11.507	104.93	19.672	15.860	0.442	6.721	1.799
2421000	7.001	21.408	9.295	51.56	23.702	17.326	4.705	4.434	0.311
2422000	9.771	0.716	7.083	129.86	27.731	18.792	8.968	2.146	4.645
2423000	2.666	4.326	4.871	76.49	31.760	20.258	13.230	9.226	3.157
2424000	5.433	7.936	2.660	23.12	3.662	21.723	4.217	6.938	1.669
2425000	8.199	11.547	0.448	101.43	7.691	23.189	8.479	4.651	0.181
2426000	1.093	15.157	12.553	48.06	11.720	0.880	12.741	2.364	4.515
2427000	3.860	18.767	10.342	126.36	15.750	2.345	3.728	0.077	3.027
2428000	6.627	22.377	8.130	72.99	19.779	3.811	7.991	7.156	1.539
2429000	9.394	1.685	5.918	19.63	23.808	5.277	12.253	4.869	5.874
2430000	2.298	5.295	3.706	97.93	27.838	6.743	3.240	2.582	4.385
2431000	5.055	8.905	1.494	44.56	31.867	8.209	7.503	0.294	2.897
2432000	7.822	12.516	13.599	122.86	3.769	9.675	11.765	7.374	1.409
2433000	0.716	16.126	11.388	69.49	7.798	11.141	2.752	5.087	5.744
2434000	3.483	19.736	9.176	16.12	11.827	12.606	7.014	2.799	4.256
2435000	6.250	23.346	6.964	94.43	15.857	14.072	11.276	0.512	2.767
2436000	9.017	2.651	4.753	41.06	19.886	15.538	2.263	7.592	1.279
2437000	1.911	6.261	2.541	119.36	23.915	17.003	6.526	5.304	5.614
2438000	4.678	9.874	0.329	65.99	27.945	18.469	10.788	3.017	4.126
2439000	7.444	13.485	12.434	12.62	31.974	19.935	1.775	0.730	2.638
2440000	0.339	17.095	10.223	90.93	3.875	21.401	6.038	7.810	1.149
2441000	3.106	20.705	8.011	37.56	7.905	22.867	10.300	5.523	5.484
2442000	5.873	24.315	5.799	115.86	11.934	0.558	1.286	3.235	3.996
2443000	8.639	3.623	3.587	62.49	15.963	2.023	5.549	0.948	2.508
2444000	1.534	7.233	1.375	9.13	19.993	3.489	9.812	8.027	1.019
2445000	4.301	10.844	13.481	87.43	24.022	4.955	0.798	5.740	5.354
2446000	7.067	14.454	11.269	34.06	28.051	6.420	5.061	3.453	3.866
2447000	9.834	18.064	9.057	112.36	32.081	7.886	9.323	1.165	2.378
2448000	2.728	21.674	6.846	59.00	3.982	9.351	0.310	8.245	0.890
2449000	5.495	0.982	4.634	5.63	8.011	10.817	4.572	5.958	5.224
2450000	8.262	4.592	2.422	83.93	12.041	12.283	8.835	3.671	3.736

TABLE IIb.

Epochs and Arguments for each Thousandth Day, from 2400000 to 2500000.

For Washington Mean Noon.

Day of Julian Period.	Arg. 18.	19.	20.	21.	22.	23.	24.	25.	26.
	d.	d.	d.	d.	d.	d.	d.	d.	d.
2450000	8.262	4.592	2.422	83.93	12.041	12.283	8.835	3.671	3.736
2451000	1.157	8.203	0.210	30.56	16.070	13.719	13.098	1.383	2.218
2452000	3.924	11.813	12.316	108.86	20.100	15.215	4.081	8.163	0.760
2453000	6.690	15.123	10.104	55.50	24.129	16.681	8.317	6.176	5.094
2454000	9.457	19.033	7.892	2.13	28.158	18.146	12.609	3.888	3.606
2455000	2.352	22.643	5.680	80.43	0.060	19.612	3.596	1.601	2.118
2456000	5.119	1.951	3.169	27.06	4.089	21.078	7.859	8.681	0.630
2457000	7.885	5.562	1.257	105.37	8.118	22.543	12.121	6.393	4.964
2458000	0.780	9.172	13.362	52.00	12.118	0.231	3.107	4.106	3.476
2459000	3.517	12.782	11.150	130.30	16.177	1.700	7.370	1.844	1.988
2460000	6.313	16.393	8.939	76.93	20.207	3.166	11.633	8.899	0.500
2461000	9.080	20.003	6.727	23.56	24.236	4.632	2.620	6.611	4.831
2462000	1.974	23.613	4.516	101.87	28.265	6.098	6.882	4.321	3.346
2463000	4.711	2.921	2.301	48.50	0.167	7.564	11.145	2.037	1.858
2464000	7.508	6.532	0.092	126.80	4.196	9.029	2.131	9.117	0.370
2465000	0.402	10.142	12.197	73.43	8.225	10.495	6.394	6.829	4.701
2466000	3.169	13.752	9.986	20.07	12.255	11.961	10.656	4.542	3.216
2467000	5.936	17.363	7.774	98.37	16.284	13.426	1.643	2.255	1.728
2468000	8.703	20.973	5.562	45.00	20.314	14.892	5.906	9.331	0.240
2469000	1.597	0.281	3.351	123.30	24.343	16.358	10.169	7.047	4.575
2470000	4.364	3.891	1.139	69.94	28.372	17.824	1.155	4.760	3.087
2471000	7.131	7.501	13.214	16.57	0.273	19.290	5.418	2.473	1.599
2472000	0.025	11.112	11.033	94.87	4.303	20.756	9.680	0.185	0.110
2473000	2.792	14.722	8.821	41.50	8.332	22.221	0.667	7.265	4.415
2474000	5.559	18.332	6.609	119.81	12.362	23.687	4.929	4.978	2.957
2475000	8.326	21.942	4.398	66.44	16.391	1.378	9.192	2.690	1.469
2476000	1.220	1.251	2.186	13.07	20.420	2.844	0.178	0.403	5.803
2477000	3.987	4.861	14.291	91.37	24.450	4.309	4.441	7.483	4.315
2478000	6.754	8.471	12.079	38.00	28.479	5.775	8.704	5.196	2.827
2479000	9.521	12.082	9.868	116.31	0.380	7.241	12.966	2.908	1.339
2480000	2.115	15.692	7.656	62.94	4.410	8.707	3.953	0.621	5.673
2481000	5.182	19.302	5.444	9.57	8.439	10.173	8.216	7.701	4.185
2482000	7.949	22.913	3.233	87.88	12.469	11.639	12.478	5.413	2.697
2483000	0.843	2.221	1.021	34.51	16.498	13.105	3.465	3.126	1.209
2484000	3.610	5.832	13.126	112.81	20.527	14.570	7.728	0.839	5.544
2485000	6.377	9.442	10.915	59.44	24.557	16.036	11.990	7.919	4.056
2486000	9.144	13.052	8.703	6.08	28.586	17.502	2.977	5.631	2.567
2487000	2.038	16.663	6.492	84.38	0.487	18.967	7.239	3.344	1.079
2488000	4.805	20.273	4.280	31.01	4.517	20.433	11.502	1.057	5.414
2489000	7.572	23.883	2.068	109.31	8.546	21.899	2.488	8.136	3.926
2490000	0.466	3.191	14.173	55.95	12.575	23.365	6.751	5.849	2.438
2491000	3.233	6.801	11.962	2.58	16.605	1.056	11.014	3.562	0.951
2492000	6.000	10.412	9.750	80.88	20.635	2.522	2.000	1.275	5.284
2493000	8.767	14.022	7.538	27.52	24.664	3.988	6.263	8.354	3.796
2494000	1.661	17.633	5.327	105.82	28.693	5.453	10.526	6.067	2.308
2495000	4.428	21.243	3.115	52.45	0.594	6.919	1.512	3.780	0.820
2496000	7.195	0.551	0.903	130.75	4.624	8.385	5.775	1.493	5.154
2497000	0.089	4.162	13.009	77.39	8.653	9.850	10.038	8.572	3.666
2498000	2.856	7.772	10.797	24.02	12.683	11.316	1.024	6.285	2.178
2499000	5.623	11.382	8.586	102.32	16.712	12.782	5.287	3.997	0.690
2500000	8.390	14.993	6.374	48.96	20.741	14.248	9.550	1.710	5.024

TABLE II*b*.

Epochs and Arguments for each Thousandth Day, from 2400000 to 2500000.

For Washington Mean Noon.

Day of Julian Period.	Arg. 27.	28.	29.	30.	31.	32.	33.	34.	35.
	d.	d.	d.	d.	d.	d.	d.	d.	d.
2400000	15.66	19.352	4.605	5.551	11.37	7.36	11.24	5.99	1.56
2401000	8.43	17.780	21.696	9.942	0.61	9.55	5.27	3.56	19.93
2402000	1.20	16.207	6.023	3.961	5.18	11.74	31.75	1.13	14.36
2403000	19.80	14.635	23.114	8.352	9.74	13.93	25.78	25.79	8.78
2404000	12.57	13.062	7.441	2.371	14.31	16.12	19.81	23.36	3.21
2405000	5.31	11.490	24.532	6.762	3.55	1.68	13.85	20.93	21.57
2406000	23.94	9.918	8.859	0.781	8.11	3.87	7.88	18.50	16.00
2407000	16.71	8.345	25.950	5.172	12.68	6.06	1.91	16.07	10.43
2408000	9.48	6.773	10.277	9.563	1.92	8.25	28.39	13.61	4.85
2409000	2.25	5.200	27.368	3.582	6.49	10.45	22.42	11.21	23.22
2410000	20.85	3.628	11.695	7.973	11.05	12.64	16.46	8.78	17.65
2411000	13.62	2.056	28.786	1.992	0.30	14.83	10.49	6.35	12.08
2412000	6.39	0.483	13.114	6.382	4.86	0.39	4.52	3.92	6.50
2413000	24.99	37.433	30.205	0.402	9.42	2.58	31.00	1.49	0.93
2414000	17.76	35.860	14.532	4.793	13.99	4.77	25.04	26.15	19.30
2415000	10.53	34.288	31.623	9.183	3.23	6.96	19.06	23.72	13.72
2416000	3.30	32.716	15.950	3.203	7.80	9.15	13.10	21.29	8.15
2417000	21.90	31.143	0.277	7.594	12.37	11.34	7.13	18.86	2.58
2418000	14.67	29.571	17.368	1.613	1.61	13.52	1.16	16.43	20.94
2419000	7.44	27.998	1.695	6.004	6.17	15.72	27.65	14.00	15.37
2420000	0.21	26.426	18.786	0.023	10.73	1.28	21.68	11.57	9.80
2421000	18.81	24.854	3.113	4.414	15.30	3.47	15.71	9.14	4.22
2422000	11.58	23.281	20.204	8.804	4.54	5.66	9.74	6.71	22.59
2423000	4.35	21.709	4.531	2.824	9.11	7.85	3.77	4.28	17.02
2424000	22.95	20.136	21.622	7.215	13.68	10.04	30.26	1.85	11.45
2425000	15.72	18.564	5.951	1.231	2.92	12.23	24.29	26.51	5.87
2426000	8.49	17.992	23.040	5.625	7.48	14.42	18.32	24.08	0.30
2427000	1.26	15.419	7.367	10.015	12.05	16.61	12.35	21.65	18.67
2428000	19.86	13.847	24.458	4.035	1.29	2.17	6.38	19.22	13.09
2429000	12.63	12.271	8.785	8.425	5.85	4.36	0.42	16.79	7.52
2430000	5.40	10.702	25.876	2.445	10.42	6.55	26.90	14.36	1.95
2431000	24.00	9.130	10.203	6.835	14.99	8.74	20.93	11.93	20.31
2432000	16.77	7.557	27.294	0.855	4.23	10.93	14.96	9.50	14.74
2433000	9.54	5.985	11.621	5.216	8.79	13.12	9.00	7.07	9.17
2434000	2.31	4.412	28.712	9.636	13.36	15.31	3.03	4.64	3.59
2435000	20.91	2.840	13.010	3.656	2.60	0.87	29.51	2.21	21.96
2436000	13.68	1.268	30.131	8.017	7.16	3.06	23.54	26.87	16.39
2437000	6.45	38.217	14.458	2.066	11.73	5.25	17.58	24.44	10.81
2438000	25.05	36.645	31.549	6.456	0.98	7.44	11.61	22.01	5.24
2439000	17.82	35.072	15.876	0.476	5.54	9.63	5.64	19.58	23.61
2440000	10.59	33.500	0.203	4.866	10.10	11.82	32.12	17.15	18.04
2441000	3.36	31.928	17.294	9.257	14.67	14.01	26.15	14.72	12.46
2442000	21.96	30.355	1.621	3.277	3.91	16.20	20.19	12.29	6.89
2443000	14.73	28.783	18.712	7.667	8.47	1.76	14.22	9.86	1.31
2444000	7.50	27.210	3.039	1.687	13.04	3.95	8.25	7.43	19.68
2445000	0.27	25.638	20.131	6.077	2.29	6.14	2.28	5.00	14.11
2446000	18.87	24 066	4.458	0.097	6.85	8.33	28.77	2.57	8.53
2447000	11.64	22.493	21.549	4.488	11.41	10.52	22.80	0.14	2.96
2448000	4.42	20.921	5.876	8.878	0.66	12.71	16.83	24.80	21.33
2449000	23.02	19.348	22.967	2.897	5.22	14.90	10.86	22.37	15.76
2450000	15.79	17.776	7.294	7.288	9.78	0.46	4.89	19.94	10.18

TABLE IIb.

Epochs and Arguments for each Thousandth Day, from 2400000 to 2500000

For Washington Mean Noon.

Day of Julian Period.	Arg. 27.	28.	29.	30.	31.	32.	33.	34.	35.
	d	d	d.	d.	d	d.	d	d	d
2450000	15.79	17.776	7.294	7.288	9.78	0.46	4.89	19.94	10.18
2451000	8.56	16.204	21.385	1.308	14.35	2.65	31.38	17.51	4.61
2452000	1.33	14.631	8.712	5.698	3.60	4.84	25.41	15.08	22.98
2453000	19.93	13.059	25.803	10.089	8.16	7.03	19.44	12.65	17.41
2454000	12.70	11.486	10.130	4.108	12.73	9.23	13.47	10.22	11.83
2455000	5.47	9.914	27.222	8.499	1.97	11.42	7.50	7.79	6.26
2456000	24.07	8.342	11.549	2.518	6.53	13.61	1.51	5.36	0.68
2457000	16.84	6.769	28.640	6.909	11.09	15.80	28.02	2.93	19.05
2458000	9.61	5.197	12.967	0.929	0.34	1.36	22.05	0.50	13.48
2459000	2.38	3.624	30.058	5.319	4.91	3.55	16.08	25.16	7.91
2460000	20.98	2.052	14.385	9.710	9.47	5.74	10.11	22.73	2.33
2461000	13.75	0.480	31.476	3.729	14.04	7.93	4.15	20.30	20.70
2462000	6.52	37.429	15.804	8.120	3.28	10.12	30.63	17.87	15.13
2463000	25.12	35.857	0.131	2.139	7.84	12.31	24.66	15.44	9.55
2464000	17.89	34.284	17.222	6.530	12.41	14.50	18.69	13.01	3.98
2465000	10.66	32.712	1.549	0.550	1.65	0.06	12.73	10.58	22.35
2466000	3.43	31.139	18.640	4.940	6.22	2.25	6.76	8.15	16.78
2467000	22.03	29.566	2.967	9.331	10.78	4.44	0.79	5.71	11.20
2468000	14.80	27.994	20.059	3.350	15.35	6.63	27.27	3.28	5.63
2469000	7.57	26.421	4.386	7.741	4.59	8.82	21.30	0.85	0.05
2470000	0.34	24.849	21.477	1.760	9.15	11.01	15.34	25.51	18.42
2471000	18.94	23.277	5.804	6.151	13.70	13.20	9.37	23.08	12.85
2472000	11.71	21.704	22.895	0.170	2.96	15.39	3.40	20.65	7.27
2473000	4.48	20.132	7.222	4.561	7.53	0.95	29.88	18.22	1.70
2474000	23.08	18.559	24.314	8.952	12.09	3.14	23.92	15.79	20.07
2475000	15.85	16.987	8.640	2.971	1.34	5.33	17.95	13.36	14.50
2476000	8.62	15.415	25.732	7.361	5.90	7.52	11.98	10.93	8.92
2477000	1.39	13.842	10.059	1.381	10.46	9.71	6.01	8.50	3.35
2478000	19.99	12.270	27.150	5.772	15.03	11.90	0.04	6.07	21.71
2479000	12.76	10.697	11.477	10.163	4.27	14.09	26.53	3.64	16.14
2480000	5.54	9.125	28.569	4.182	8.84	16.28	20.56	1.21	10.57
2481000	24.13	7.553	12.896	8.572	13.40	1.84	14.59	25.87	5.00
2482000	16.90	5.980	29.987	2.592	2.65	4.03	8.62	23.44	23.36
2483000	9.68	4.408	14.314	6.983	7.21	6.22	2.66	21.01	17.79
2484000	2.45	2.835	31.405	1.002	11.77	8.41	29.14	18.58	12.22
2485000	21.01	1.263	15.733	5.393	1.02	10.60	23.17	16.15	6.64
2486000	13.82	38.212	0.059	9.783	5.58	12.79	17.20	13.72	1.07
2487000	6.59	36.639	17.151	3.803	10.15	14.98	11.23	11.29	19.44
2488000	25.18	35.067	1.478	8.194	14.71	0.54	5.27	8.86	13.87
2489000	17.96	33.494	18.569	2.213	3.96	2.73	31.75	6.43	8.29
2490000	10.73	31.922	2.896	6.603	8.52	4.92	25.78	4.00	2.72
2491000	3.50	30.350	19.988	0.623	13.09	7.11	19.81	1.57	21.09
2492000	22.10	28.777	4.315	5.014	2.33	9.30	13.85	26.23	15.51
2493000	14.87	27.205	21.406	9.404	6.89	11.49	7.88	23.80	9.94
2494000	7.64	25.632	5.733	3.423	11.46	13.68	1.91	21.37	4.37
2495000	0.41	24.060	22.825	7.814	0.70	15.87	28.39	18.94	22.73
2496000	19.01	22.488	7.152	1.834	5.27	1.43	22.42	16.51	17.16
2497000	11.78	20.915	24.243	6.224	9.83	3.62	16.46	14.08	11.59
2498000	4.55	19.343	8.570	0.244	14.40	5.81	10.49	11.65	6.01
2499000	23.15	17.770	25.661	4.634	3.64	8.00	4.52	9.22	0.44
2500000	15.92	16.198	9.988	9.025	8.20	10.19	31.00	6.79	18.81

TABLE IIb.

Epochs and Arguments for each Thousandth Day, from 2400000 to 2500000.

For Washington Mean Noon.

Day of Julian Period.	Arg. 36.	37.	38.	39.	40.	41.	42.	43.	44.
	d.	d.	d	d	d.	d.	d	d.	d.
2400000	6.02	19.5	5.92	32.3	1.19	17.30	4.35	3.07	22.3
2401000	4.55	3.6	10.12	32.5	2.48	13.83	9.85	14.51	19.7
2402000	3.07	25.4	14.31	32.7	3.77	10.36	5.21	11.39	17.1
2403000	1.60	9.5	18.51	32.9	5.07	6.89	0.57	8.28	14.5
2404000	0.13	31.2	3.91	33.2	6.36	3.42	6.07	5.16	11.9
2405000	12.38	15.4	8.11	33.4	7.65	17.87	1.43	2.05	9.3
2406000	10.90	37.1	12.30	33.6	8.94	14.40	6.93	13.49	6.7
2407000	9.43	21.2	16.50	33.9	10.23	10.93	2.29	10.37	4.2
2408000	7.95	5.3	1.90	34.1	11.53	7.46	7.79	7.25	1.6
2409000	6.48	27.1	6.10	34.3	12.82	3.99	3.16	4.14	21.8
2410000	5.01	11.2	10.29	0.0	11.11	0.52	8.65	1.03	19.2
2411000	3.53	32.9	14.49	0.2	15.40	14.97	4.02	12.46	16.6
2412000	2.06	17.0	18.68	0.4	16.69	11.50	9.52	9.34	14.0
2413000	0.59	1.2	4.09	0.6	17.99	8.03	4.88	6.23	11.4
2414000	12.84	22.9	8.28	0.9	0.44	4.56	0.24	3.12	8.8
2415000	11.36	7.0	12.48	1.1	1.73	1.09	5.74	0.00	6.2
2416000	9.89	28.8	16.67	1.3	3.02	15.54	1.10	11.43	3.6
2417000	8.41	12.9	2.07	1.6	4.31	12.07	6.60	8.32	1.0
2418000	6.94	34.6	6.27	1.8	5.61	8.60	1.96	5.21	21.2
2419000	5.47	18.7	10.46	2.0	6.90	5.13	7.46	2.10	18.6
2420000	3.99	2.8	14.66	2.2	8.19	1.66	2.82	13.54	16.1
2421000	2.52	24.6	0.06	2.4	9.48	16.11	8.32	10.41	13.5
2422000	1.05	8.7	4.26	2.6	10.77	12.64	3.69	7.30	10.9
2423000	13.30	30.4	8.45	2.8	12.06	9.17	9.18	4.19	8.3
2424000	11.82	14.6	12.65	3.1	13.35	5.70	4.55	1.08	5.7
2425000	10.35	36.3	16.84	3.3	14.65	2.23	10.05	12.50	3.1
2426000	8.87	20.4	2.25	3.5	15.94	16.68	5.41	9.39	0.5
2427000	7.40	4.5	6.44	3.8	17.23	13.21	0.77	6.28	20.7
2428000	5.93	26.3	10.61	4.0	18.52	9.74	6.27	3.17	18.1
2429000	4.46	10.4	14.83	4.2	0.97	6.27	1.63	0.06	15.5
2430000	2.98	32.1	0.24	4.4	2.26	2.80	7.13	11.48	12.9
2431000	1.51	16.3	4.44	4.6	3.55	17.25	2.49	8.37	10.3
2432000	0.04	0.1	8.63	4.8	4.84	13.77	7.99	5.26	7.8
2433000	12.28	22.1	12.82	5.0	6.14	10.30	3.35	2.15	5.2
2434000	10.81	6.2	17.02	5.2	7.43	6.83	8.85	13.58	2.6
2435000	9.33	28.0	2.42	5.5	8.72	3.36	4.22	10.46	0.0
2436000	7.86	12.1	6.62	5.7	10.01	17.81	9.71	7.35	20.2
2437000	6.39	33.8	10.81	5.9	11.30	14.34	5.08	4.24	17.6
2438000	4.92	18.0	15.01	6.1	12.60	10.87	0.44	1.13	15.0
2439000	3.44	2.1	0.41	6.3	13.89	7.40	5.94	12.55	12.4
2440000	1.97	23.8	4.60	6.5	15.18	3.93	1.30	9.44	9.8
2441000	0.50	7.9	8.80	6.7	16.47	0.46	6.80	6.33	7.2
2442000	12.74	29.7	12.99	6.9	17.76	14.91	2.16	3.22	4.6
2443000	11.27	13.8	17.19	7.1	0.22	11.44	7.66	0.11	2.1
2444000	9.79	35.5	2.59	7.4	1.51	7.97	3.02	11.53	22.3
2445000	8.32	19.6	6.79	7.6	2.80	4.50	8.52	8.42	19.7
2446000	6.85	3.8	10.98	7.8	4.09	1.03	3.88	5.31	17.1
2447000	5.38	25.5	15.18	8.1	5.38	15.48	9.38	2.20	14.5
2448000	3.90	9.6	0.58	8.3	6.68	12.01	4.75	13.63	11.9
2449000	2.43	31.3	4.78	8.5	7.97	8.54	0.10	10.51	9.3
2550000	0.96	15.5	8.97	8.7	9.26	5.07	5.61	7.40	6.7

TABLE IIb.

Epochs and Arguments for each Thousandth Day, from 2400000 to 2500000.

For Washington Mean Noon.

Day of Julian Period.	Arg. 36.	37.	38.	39.	40.	41.	42.	43.	44.
	d	d.	d.	d.	d	d	d	d	d
2450000	0.96	15.5	8.97	8.7	9.26	5.07	5.61	7.40	6.7
2451000	13.20	37.2	13.17	8.9	10.55	1.60	0.96	4.29	4.1
2452000	11.73	21.3	17.36	9.1	11.84	16.05	6.47	1.17	1.5
2453000	10.26	5.4	2.77	9.3	13.14	12.58	1.83	12.00	21.7
2454000	8.78	27.2	6.96	9.6	14.43	9.11	7.33	9.49	19.1
2455000	7.31	11.3	11.16	9.8	15.72	5.64	2.69	6.38	16.6
2456000	5.84	33.0	15.35	10.0	17.01	2.17	8.19	3.26	14.0
2457000	4.36	17.2	0.76	10.3	18.30	16.62	3.55	0.16	11.4
2458000	2.89	1.3	4.95	10.5	0.76	13.15	9.05	11.58	8.8
2459000	1.42	23.0	9.15	10.7	2.05	9.68	4.42	8.47	6.2
2460000	13.66	7.1	13.34	10.9	3.34	6.21	9.91	5.36	3.6
2461000	12.19	28.9	17.53	11.1	4.63	2.74	5.28	2.24	1.0
2462000	10.72	13.0	2.94	11.3	5.92	17.19	0.64	13.68	21.2
2463000	9.21	34.7	7.13	11.5	7.22	13.72	6.14	10.56	18.6
2464000	7.77	18.8	11.33	11.7	8.51	10.25	1.50	7.45	16.0
2465000	6.30	3.0	15.52	12.0	9.80	6.78	6.99	4.33	13.4
2466000	4.82	21.7	0.93	12.2	11.09	3.31	2.36	1.22	10.8
2467000	3.35	8.8	5.12	12.5	12.38	17.76	7.86	12.65	8.3
2468000	1.88	30.6	9.32	12.7	13.68	14.29	3.22	9.54	5.7
2469000	0.40	14.7	13.51	12.9	14.97	10.82	8.72	6.43	3.1
2470000	12.65	36.4	17.71	13.1	16.26	7.35	4.08	3.31	0.5
2471000	11.18	20.5	3.11	13.3	17.55	3.88	9.58	0.21	20.7
2472000	9.70	4.6	7.31	13.5	0.00	0.41	4.95	11.63	18.1
2473000	8.23	26.4	11.50	13.7	1.30	14.85	0.30	8.52	15.5
2474000	6.76	10.5	15.70	13.9	2.59	11.38	5.81	5.40	12.9
2475000	5.28	32.2	1.10	14.2	3.88	7.91	1.17	2.29	10.3
2476000	3.81	16.4	5.30	14.4	5.17	4.44	6.67	13.73	7.7
2477000	2.34	0.5	9.49	14.6	6.46	0.97	2.03	10.61	5.1
2478000	0.87	22.2	13.69	14.8	7.76	15.12	7.52	7.19	2.6
2479000	13.11	6.3	17.88	15.0	9.05	11.95	2.89	4.38	0.0
2480000	11.64	28.1	3.29	15.2	10.34	8.48	8.39	1.27	20.2
2481000	10.16	12.2	7.48	15.4	11.63	5.01	3.75	12.70	17.6
2482000	8.69	33.9	11.68	15.6	12.92	1.54	9.25	9.59	15.0
2483000	7.22	18.1	15.87	15.8	14.21	15.99	4.62	6.47	12.4
2484000	5.74	2.2	1.28	16.1	15.50	12.52	10.11	3.36	9.8
2485000	4.27	23.9	5.47	16.3	16.80	9.05	5.48	0.26	7.2
2486000	2.80	8.0	9.67	16.5	18.09	5.58	0.83	11.68	4.6
2487000	1.33	29.8	13.86	16.8	0.54	2.11	6.34	8.56	2.0
2488000	13.57	13.9	18.06	17.0	1.83	16.56	1.70	5.15	22.2
2489000	12.10	35.6	3.46	17.2	3.12	13.09	7.19	2.34	19.6
2490000	10.62	19.7	7.66	17.4	4.41	9.62	2.56	13.78	17.0
2491000	9.15	3.9	11.85	17.6	5.70	6.15	8.06	10.66	14.5
2492000	7.68	25.6	16.05	17.8	6.99	2.68	3.42	7.54	11.9
2493000	6.21	9.7	1.45	18.0	8.29	17.13	8.92	4.43	9.3
2494000	4.73	31.5	5.65	18.3	9.58	13.66	4.28	1.32	6.7
2495000	3.26	15.6	9.84	18.5	10.87	10.19	9.78	12.75	4.1
2496000	1.79	37.3	14.04	18.7	12.16	6.72	5.15	9.63	1.5
2497000	0.31	21.4	18.23	19.0	13.45	3.25	0.50	6.52	21.7
2498000	12.56	5.6	3.63	19.2	14.75	17.70	6.01	3.41	19.1
2499000	11.09	27.3	7.83	19.4	16.04	14.23	1.37	0.31	16.5
2500000	9.61	11.4	12.03	19.6	17.33	10.76	6.86	11.73	13.9

TABLE IIb.

Epochs and Arguments for each Thousandth Day, from 2400000 to 2500000.

For Washington Mean Noon.

Day of Julian Period.	Arg. 45.	46.	47.	48.	49.	50.	51.	52.	53.
	d.	d.	d.	d.	d.	d.	d.	d.	d.
2400000	6.36	11.4	10.53	12.67	5.6	21.5	15.85	8.80	68.76
2401000	9.26	27.9	7.12	24.84	10.1	20.1	13.92	21.77	10.91
2402000	12.16	12.5	3.71	7.08	1.9	18.6	11.99	15.52	70.59
2403000	15.06	26.0	0.30	19.25	6.4	17.2	10.06	6.26	12.74
2404000	1.05	10.6	11.02	1.49	10.8	15.7	8.13	22.24	72.42
2405000	3.95	24.1	7.61	13.66	2.6	14.3	6.20	12.99	14.57
2406000	6.85	8.6	4.20	25.83	7.1	12.9	4.27	3.73	74.25
2407000	9.75	22.2	0.78	8.07	11.7	11.4	2.34	19.71	16.40
2408000	12.65	6.7	11.50	20.24	3.4	10.0	0.41	10.45	76.08
2409000	15.54	20.3	8.09	2.48	7.9	8.5	16.69	1.20	18.23
2410000	1.54	4.8	4.68	14.65	12.5	7.1	14.76	17.17	77.91
2411000	4.44	18.3	1.27	26.83	4.2	5.7	12.83	7.92	20.06
2412000	7.34	2.9	11.99	9.07	8.7	4.2	10.90	23.89	79.74
2413000	10.24	16.4	8.58	21.24	0.4	2.8	8.97	14.61	21.89
2414000	13.13	1.0	5.17	3.48	4.9	1.3	7.04	5.38	81.57
2415000	16.03	11.5	1.76	15.65	9.4	24.9	5.10	21.36	23.72
2416000	2.03	28.1	12.48	27.82	1.2	23.5	3.17	12.11	83.40
2417000	4.93	12.6	9.07	10.06	5.7	22.0	1.24	2.85	25.55
2418000	7.83	26.1	5.66	22.23	10.1	20.6	17.52	18.83	85.23
2419000	10.72	10.7	2.24	4.17	1.9	19.1	15.59	9.57	27.38
2420000	13.62	24.2	12.96	16.61	6.5	17.7	13.67	0.32	87.06
2421000	16.52	8.8	9.55	28.84	10.9	16.3	11.74	16.29	29.21
2422000	2.52	22.3	6.14	11.05	2.7	14.8	9.81	7.01	88.69
2423000	5.42	6.9	2.73	23.22	7.2	13.4	7.88	23.01	31.04
2424000	8.31	20.4	13.45	5.46	11.8	11.9	5.95	13.76	90.72
2425000	11.21	4.9	10.04	17.64	3.5	10.5	4.02	4.51	32.87
2426000	14.11	18.5	6.63	29.81	8.0	9.1	2.09	20.48	92.55
2427000	0.11	3.0	3.22	12.05	12.5	7.6	0.16	11.23	34.70
2428000	3.01	16.6	13.94	24.22	4.2	6.2	16.14	1.97	94.38
2429000	5.91	1.1	10.53	6.46	8.7	4.7	11.52	17.95	36.53
2430000	8.80	14.6	7.12	18.63	0.5	3.3	12.59	8.69	96.21
2431000	11.70	28.2	3.70	0.87	5.0	1.9	10.66	21.67	38.36
2432000	14.60	12.7	0.29	13.04	9.5	0.5	8.73	15.12	98.04
2433000	0.60	26.3	11.01	25.21	1.3	21.0	6.80	6.16	40.19
2434000	3.49	10.8	7.60	7.45	5.8	22.6	4.87	22.11	99.87
2435000	6.39	21.1	4.19	19.62	10.2	21.1	2.93	12.88	32.02
2436000	9.29	8.9	0.78	1.86	2.0	19.7	1.00	3.63	91.70
2437000	12.19	22.4	11.50	14.03	6.5	18.3	17.28	19.60	33.85
2438000	15.09	7.0	8.09	26.21	11.0	16.8	15.35	10.35	93.53
2439000	1.08	20.5	4.68	8.44	2.8	15.4	13.43	1.09	35.68
2440000	3.98	5.1	1.27	20.62	7.3	14.0	11.50	17.07	105.36
2441000	6.88	18.6	11.99	2.86	11.8	12.6	9.57	7.82	47.51
2442000	9.78	3.2	8.58	15.03	3.5	11.1	7.64	23.79	107.19
2443000	12.67	16.7	5.16	27.20	8.1	9.7	5.71	14.54	49.34
2444000	15.57	1.2	1.75	9.44	12.6	8.2	3.78	5.28	109.02
2445000	1.57	14.8	12.47	21.61	4.3	6.8	1.84	21.26	51.17
2446000	4.47	28.3	9.06	3.85	8.8	5.4	18.12	12.00	110.85
2447000	7.37	12.9	5.65	16.02	0.6	3.9	16.19	2.75	53.00
2448000	10.26	26.4	2.24	28.19	5.1	2.5	14.27	18.73	112.68
2449000	13.16	11.0	12.96	10.43	9.6	1.0	12.34	9.47	54.83
2450000	16.06	24.5	9.55	22.60	1.3	24.6	10.41	0.22	114.51

TABLE IIb.

Epochs and Arguments for each Thousandth Day, from 2400000 to 2500000.

For Washington Mean Noon.

Day of Julian Period.	Arg. 45.	46.	47.	48.	49.	50.	51.	52.	53.
	d.	d	d	d	d	d.	d.	d	d
2450000	16.06	21.5	9.55	22.60	1.3	21.6	10.11	0.22	114.51
2451000	2.06	9.0	6.14	4.84	5.8	23.2	8.18	16.19	56.66
2452000	4.96	22.6	2.73	17.02	10.3	21.7	6.55	6.94	116.34
2453000	7.85	7.1	13.45	29.19	2.1	20.3	4.62	22.91	58.49
2454000	10.75	20.7	10.04	11.43	6.6	18.8	2.69	13.66	0.63
2455000	13.65	5.2	6.63	23.60	11.0	17.4	0.75	4.40	60.32
2456000	16.55	18.7	3.22	5.84	2.8	16.0	17.03	20.38	2.46
2457000	2.55	3.3	13.93	18.01	7.4	14.5	15.11	11.13	62.15
2458000	5.44	16.8	10.52	0.25	11.9	13.1	13.18	1.87	4.29
2459000	8.34	1.4	7.11	12.42	3.6	11.6	11.25	17.85	63.98
2460000	11.24	14.9	3.70	24.59	8.1	10.2	9.32	8.59	6.12
2461000	14.14	28.5	0.29	6.83	12.7	8.8	7.39	24.57	65.81
2462000	0.14	13.0	11.01	19.00	4.4	7.4	5.46	15.31	7.95
2463000	3.03	26.5	7.60	1.24	8.9	5.9	3.53	6.06	67.64
2464000	5.93	11.1	4.19	13.41	0.6	4.5	1.60	22.01	9.78
2465000	8.83	21.6	0.78	25.59	5.1	3.0	17.87	12.78	69.47
2466000	11.73	9.2	11.50	7.83	9.6	1.6	15.94	3.53	11.61
2467000	14.63	22.7	8.09	20.00	1.4	0.2	14.02	19.50	71.30
2468000	0.62	7.3	4.68	2.24	5.9	23.7	12.09	10.25	13.44
2469000	3.52	20.8	1.26	11.41	10.3	22.3	10.16	0.99	73.13
2470000	6.42	5.3	11.98	26.58	2.2	20.9	8.23	16.97	15.27
2471000	9.32	18.9	8.57	8.82	6.7	19.5	6.30	7.71	74.96
2472000	12.21	3.1	5.16	20.99	11.1	18.0	4.37	23.69	17.10
2473000	15.11	17.0	1.75	3.23	2.9	16.6	2.14	14.43	76.79
2474000	1.11	1.5	12.47	15.40	7.4	15.1	0.51	5.18	18.93
2475000	4.01	15.0	9.06	27.57	12.0	13.7	16.78	21.16	78.62
2476000	6.91	28.6	5.65	9.81	3.7	12.3	14.86	11.00	20.76
2477000	9.80	13.1	2.24	21.99	8.2	10.8	12.93	2.65	80.15
2478000	12.70	26.7	12.96	4.22	12.7	9.1	11.00	18.62	22.59
2479000	15.60	11.2	9.55	16.40	4.4	7.9	9.07	9.37	82.28
2480000	1.60	24.8	6.14	28.57	8.9	6.5	7.14	0.11	24.42
2481000	4.50	9.3	2.73	10.81	0.7	5.1	5.21	16.09	84.11
2482000	7.39	22.8	13.44	22.98	5.2	3.6	3.28	6.83	26.25
2483000	10.29	7.4	10.03	5.22	9.7	2.2	1.35	22.81	85.94
2484000	13.19	20.9	6.62	17.39	1.4	0.7	17.63	13.56	28.08
2485000	16.09	5.5	3.21	29.56	6.0	24.3	15.69	4.30	87.77
2486000	2.09	19.0	13.93	11.80	10.4	22.9	13.77	20.28	29.91
2487000	4.98	3.6	10.52	23.97	2.2	21.4	11.84	11.02	89.60
2488000	7.88	17.1	7.11	6.24	6.7	20.0	9.91	1.77	31.74
2489000	10.78	1.6	3.70	18.39	11.3	18.5	7.98	17.75	91.43
2490000	13.68	15.2	0.29	0.62	3.0	17.1	6.05	8.49	33.57
2491000	16.58	28.7	11.01	12.80	7.5	15.7	4.12	24.46	93.26
2492000	2.57	13.3	7.60	24.97	12.0	14.3	2.19	15.21	35.41
2493000	5.47	26.8	4.19	7.21	3.7	12.8	0.26	5.96	95.09
2494000	8.37	11.3	0.77	19.38	8.3	11.4	16.54	21.93	37.24
2495000	11.27	24.9	11.49	1.62	0.0	9.9	14.61	12.68	96.92
2496000	14.17	9.4	8.08	13.79	4.5	8.5	12.68	3.42	39.07
2497000	0.16	23.0	4.67	25.96	9.0	7.1	10.75	19.40	98.75
2498000	3.06	7.5	1.26	8.20	0.8	5.6	8.82	10.14	40.90
2499000	5.96	21.1	11.98	20.37	5.3	4.2	6.89	0.89	100.58
2500000	8.86	5.6	8.57	2.61	9.8	2.8	4.96	16.86	42.73

TABLE IIb.

Epochs and Arguments for each Thousandth Day, from 2400000 to 2500000.

For Washington Mean Noon.

Day of Julian Period.	Arg. 51.	55.	56.	57.	58.	59	60.	61.	62.
	d.	d	d.	d.	d	d	d	d.	d.
2400000	4.73	3.34	7.21	8.03	3.77	11.57	15.63	3.71	8.20
2401000	10.11	2.35	10.29	5.78	13.02	37.46	11.16	10.16	11.70
2402000	0.43	1.35	0.24	3.52	7.04	24.41	6.68	3.61	0.76
2403000	5.82	0.36	3.31	1.26	1.05	11.31	2.20	10.36	4.26
2404000	11.21	19.00	6.38	15.99	10.31	37.24	20.05	3.51	7.76
2405000	1.52	18.00	9.45	13.74	4.32	21.18	15.57	10.27	*11.27
2406000	6.91	17.00	12.53	11.48	13.57	11.12	11.09	3.41	0.33
2407000	12.30	16.02	2.18	9.22	7.59	37.01	6.62	10.17	3.83
2408000	2.62	15.03	5.55	6.97	1.61	23.96	2.14	3.31	7.33
2409000	8.00	14.03	8.62	4.71	10.86	10.89	19.98	10.07	10.83
2410000	13.39	13.04	11.70	2.45	4.87	36.79	15.51	3.22	14.33
2411000	3.71	12.05	1.65	0.19	14.13	23.73	11.03	9.97	3.40
2412000	9.10	11.06	4.72	14.93	8.14	10.67	6.55	3.12	6.90
2413000	14.48	10.06	7.79	12.67	2.16	36.56	2.08	9.87	10.40
2414000	4.80	9.07	10.87	10.41	11.41	23.50	19.92	3.02	13.90
2415000	10.19	8.08	0.82	8.15	5.43	10.44	15.44	9.78	2.96
2416000	0.51	7.09	3.89	5.90	14.68	36.34	10.96	2.92	6.46
2417000	5.89	6.10	6.96	3.64	8.70	23.28	6.49	9.68	9.96
2418000	11.28	5.10	10.04	1.38	2.71	10.22	2.01	2.83	13.47
2419000	1.60	4.11	13.11	16.11	11.97	36.11	19.86	9.58	2.53
2420000	6.99	3.12	3.06	13.86	5.98	23.05	15.38	2.73	6.03
2421000	12.37	2.13	6.13	11.60	15.23	9.99	10.90	9.48	9.53
2422000	2.69	1.11	9.20	9.34	9.25	35.89	6.43	2.63	13.03
2423000	8.08	0.14	12.28	7.09	3.26	22.83	1.95	9.38	2.09
2424000	13.47	18.78	2.23	4.83	12.52	9.76	19.79	2.53	5.59
2425000	3.78	17.79	5.30	2.57	6.53	35.66	15.32	9.28	9.10
2426000	9.17	16.80	8.37	0.31	0.55	22.60	10.84	2.43	12.60
2427000	14.56	15.81	11.45	15.04	9.80	9.54	6.36	9.19	1.66
2428000	4.88	14.81	1.40	12.79	3.82	35.44	1.88	2.33	5.16
2429000	10.26	13.82	4.47	10.53	13.07	22.38	19.73	9.09	8.66
2430000	0.58	12.83	7.54	8.27	7.09	9.31	15.25	2.24	12.16
2431000	5.97	11.84	10.62	6.02	1.10	35.21	10.77	8.99	1.23
2432000	11.36	10.85	0.57	3.76	10.36	22.15	6.30	2.14	4.73
2433000	1.67	9.85	3.64	1.50	4.37	9.09	1.82	8.89	8.23
2434000	7.06	8.86	6.71	16.23	13.62	34.98	19.66	2.04	11.73
2435000	12.45	7.87	9.78	13.98	7.64	21.92	15.19	8.79	0.79
2436000	2.77	6.88	12.86	11.72	1.65	8.86	10.71	1.94	4.29
2437000	8.15	5.89	2.81	9.46	10.91	34.76	6.23	8.70	7.79
2438000	13.54	4.89	5.88	7.21	4.92	21.70	1.76	1.81	11.30
2439000	3.86	3.90	8.95	4.95	14.18	8.61	19.60	8.60	0.36
2440000	9.25	2.90	12.03	2.69	8.19	34.53	15.12	1.74	3.86
2441000	14.63	1.91	1.98	0.43	2.21	21.47	10.65	8.50	7.36
2442000	4.95	0.91	5.05	15.17	11.46	8.11	6.17	1.65	10.86
2443000	10.34	19.56	8.12	12.91	5.48	34.31	1.69	8.40	14.36
2444000	0.66	18.57	11.20	10.65	14.73	21.25	19.54	1.55	3.43
2445000	6.04	17.57	1.15	8.39	8.75	8.18	15.06	8.30	6.93
2446000	11.43	16.58	4.22	6.14	2.76	34.08	10.58	1.45	10.43
2447000	1.75	15.59	7.29	3.88	12.02	21.02	6.11	8.20	13.93
2448000	7.14	14.60	10.37	1.62	6.03	7.96	1.63	1.35	2.99
2449000	12.52	13.60	0.32	16.35	0.05	33.86	19.47	8.10	6.49
2450000	2.84	12.61	3.39	14.10	9.30	20.79	15.00	1.25	9.99

TABLE II*b*.

Epochs and Arguments for each Thousandth Day, from 2100000 to 2500000.

For Washington Mean Noon.

Day of Julian Period.	Arg. 51.	55.	56.	57.	58.	59.	60.	61.	62.
	d.	d.	d.	d.	d.	d.	d.	d.	d.
2450000	2.84	12.61	3.39	14.10	9.30	20.79	15.00	1.25	9.99
2451000	8.23	11.62	6.46	11.81	3.31	7.73	10.52	8.01	13.50
2452000	13.62	10.63	9.53	9.58	12.57	33.63	6.04	1.16	2.56
2453000	3.93	9.63	12.61	7.33	6.58	20.57	1.57	7.91	6.06
2454000	9.32	8.61	2.56	5.07	0.60	7.51	19.11	1.06	9.56
2455000	11.71	7.65	5.63	2.81	9.85	33.41	14.93	7.81	13.06
2456000	5.03	6.66	8.70	0.55	3.87	20.31	10.46	0.96	2.12
2457000	10.41	5.67	11.78	15.29	13.12	7.28	5.98	7.71	5.62
2458000	0.73	4.67	1.73	13.03	7.14	33.18	1.50	0.86	9.13
2459000	6.12	3.68	4.80	10.77	1.15	20.12	19.35	7.61	12.63
2460000	11.51	2.69	7.87	8.52	10.41	7.05	14.87	0.76	1.69
2461000	1.83	1.69	10.95	6.26	4.42	32.95	10.39	7.52	5.19
2462000	7.21	0.71	0.90	4.00	13.67	19.89	5.92	0.67	8.69
2463000	12.60	19.34	3.97	1.74	7.69	6.83	1.44	7.42	12.19
2464000	2.92	18.35	7.01	16.48	1.70	32.73	19.28	0.57	1.26
2465000	8.31	17.36	10.11	14.22	10.96	19.67	14.81	7.32	4.76
2466000	13.69	16.37	0.07	11.96	4.97	6.60	10.33	0.47	8.26
2467000	4.01	15.38	3.14	9.70	14.23	32.50	5.85	7.22	11.76
2468000	9.40	14.38	6.21	7.45	8.21	19.41	1.38	0.37	0.82
2469000	14.78	13.39	9.28	5.19	2.26	6.38	19.22	7.13	4.32
2470000	5.10	12.40	12.36	2.93	11.51	32.28	14.74	0.26	7.82
2471000	10.49	11.41	2.31	0.67	5.53	19.21	10.27	7.03	11.33
2472000	0.81	10.42	5.38	15.41	14.78	6.15	5.79	0.17	0.39
2473000	6.20	9.42	8.45	13.15	8.80	32.05	1.31	6.93	3.89
2474000	11.58	8.43	11.53	10.89	2.81	18.99	19.16	0.08	7.39
2475000	1.90	7.44	1.48	8.64	12.06	5.93	14.68	6.83	10.89
2476000	7.29	6.45	4.55	6.38	6.08	31.83	10.20	13.59	14.39
2477000	12.67	5.46	7.62	4.12	0.09	18.76	5.72	6.73	3.46
2478000	2.99	4.46	10.70	1.86	9.35	5.70	1.25	13.49	6.96
2479000	8.38	3.47	0.65	16.60	3.36	31.60	19.09	6.63	10.46
2480000	13.77	2.48	3.72	14.34	12.62	18.51	14.62	13.39	13.96
2481000	4.09	1.49	6.79	12.08	6.64	5.47	10.14	6.54	3.02
2482000	9.47	0.49	9.86	9.83	0.65	31.37	5.66	13.29	6.52
2483000	14.86	19.43	12.91	7.57	9.90	18.31	1.18	6.44	10.02
2484000	5.18	18.44	2.89	5.31	3.92	5.25	19.03	13.20	13.53
2485000	10.57	17.44	5.96	3.05	13.17	31.15	14.55	6.34	2.59
2486000	0.88	16.45	9.03	0.79	7.19	18.08	10.07	13.10	6.09
2487000	6.28	15.46	12.11	15.53	1.20	5.02	5.60	6.24	9.59
2488000	11.66	14.47	2.05	13.27	10.46	30.92	1.12	13.00	13.09
2489000	1.97	13.47	5.13	11.01	4.47	17.86	18.96	6.14	2.15
2490000	7.36	12.48	8.20	8.76	13.72	4.79	14.49	12.90	5.65
2491000	12.75	11.49	11.28	6.50	7.74	30.70	10.01	6.04	9.16
2492000	3.07	10.20	1.23	4.24	1.75	17.63	5.53	12.80	12.66
2493000	8.46	9.20	4.30	1.98	11.01	4.57	1.06	5.95	1.72
2494000	13.84	8.21	7.37	16.72	5.02	30.47	18.90	12.70	5.22
2495000	4.16	7.22	10.45	14.46	14.28	17.41	14.42	5.85	8.72
2496000	9.55	6.23	0.40	12.20	8.29	4.34	9.95	12.61	12.22
2497000	14.93	5.24	3.47	9.96	2.31	30.24	5.47	5.75	1.29
2498000	5.25	4.24	6.54	7.69	11.56	17.18	0.99	12.51	4.79
2499000	10.64	3.25	9.61	5.43	5.58	4.12	18.84	5.65	8.29
2500000	0.96	2.26	12.69	3.18	14.83	30.02	14.37	12.41	11.79

TABLE IIb.

Epochs and Arguments for each Thousandth Day, from 2400000 to 2500000.

For Washington Mean Noon.

Day of Julian Period.	Arg. 63.	64.	65.	66.	67.	68.	69.	70.	71.
	d	d	d.	d.	d	d.	d.	d	d.
2400000	7.81	25.59	3.13	7.8	7.66	3.65	121.5	307.1	272.0
2401000	12.61	17.81	4.51	11.1	19.70	7.65	177.7	317.7	104.2
2402000	3.73	10.03	5.89	14.4	4.30	11.65	234.0	328.3	520.3
2403000	8.49	2.25	7.26	17.8	16.31	15.65	290.2	09.2	352.4
2404000	13.26	30.46	8.64	21.1	0.94	19.65	346.4	19.8	184.6
2405000	4.38	22.68	10.02	24.4	12.98	23.65	402.6	30.5	16.8
2406000	9.15	14.91	11.41	27.7	25.02	27.65	458.8	41.1	432.8
2407000	0.28	7.13	12.79	31.0	9.61	3.98	43.1	51.7	265.0
2408000	5.04	35.34	14.16	34.4	21.65	7.98	99.3	62.4	97.2
2409000	9.80	27.56	. 1.06	2.1	6.25	11.98	155.5	73.0	513.2
2410000	0.93	19.78	2.44	5.4	18.29	15.98	211.8	83.6	345.4
2411000	5.69	12.00	3.82	8.7	2.89	19.98	268.0	94.2	177.5
2412000	10.45	4.22	5.20	12.0	14.93	23.98	324.2	104.8	9.7
2413000	1.58	32.43	6.58	15.3	26.98	0.31	380.4	115.5	425.8
2414000	6.34	24.66	7.96	18.6	11.57	4.31	436.6	126.1	257.9
2415000	11.11	16.88	9.33	22.0	23.61	8.31	21.0	136.7	90.1
2416000	2.23	9.10	10.71	25.3	8.20	12.31	77.2	147.3	506.1
2417000	7.00	1.32	12.10	28.6	20.24	16.31	133.4	157.9	338.3
2418000	11.76	29.53	13.48	31.9	4.84	20.31	189.6	168.6	170.5
2419000	2.89	21.75	0.38	35.2	16.88	24.31	245.8	179.2	2.6
2420000	7.65	13.97	1.76	2.9	1.48	0.64	302.0	189.8	418.7
2421000	12.42	6.19	3.14	6.2	13.52	4.61	358.3	200.4	250.9
2422000	3.54	34.40	4.52	9.5	25.57	8.61	414.5	211.0	83.0
2423000	8.30	26.63	5.89	12.9	10.16	12.64	470.7	221.7	499.1
2424000	13.07	18.85	7.27	16.2	22.20	16.64	55.0	232.3	331.3
2425000	4.19	11.07	8.65	19.5	6.80	20.64	111.2	243.0	163.4
2426000	8.95	3.29	10.03	22.8	18.84	24.64	167.5	253.6	579.5
2427000	0.08	31.50	11.42	26.1	3.43	0.96	223.7	264.2	411.6
2428000	4.85	23.72	12.79	29.5	15.47	4.96	279.9	274.9	243.8
2429000	9.61	15.94	14.17	32.8	0.07	8.96	336.1	285.5	76.0
2430000	0.74	8.16	1.07	0.5	12.11	12.97	392.3	296.1	492.0
2431000	5.50	0.38	2.45	3.8	24.15	16.97	448.5	306.7	324.2
2432000	10.26	28.60	3.83	7.1	8.75	20.96	32.8	317.3	156.4
2433000	1.39	20.82	5.21	10.5	20.79	24.96	89.1	328.0	572.4
2434000	6.15	13.04	6.59	13.8	5.39	1.29	145.3	8.8	404.6
2435000	10.92	5.26	7.96	17.1	17.43	5.29	201.5	19.5	236.8
2436000	2.04	33.47	9.34	20.4	2.02	9.29	257.7	30.1	68.9
2437000	6.80	25.69	10.72	23.7	14.06	13.29	313.9	40.7	485.0
2438000	11.57	17.91	12.11	27.1	26.11	17.29	370.2	51.4	317.1
2439000	2.69	10.13	13.49	30.4	10.70	21.29	426.4	62.0	149.3
2440000	7.46	2.36	0.39	33.7	22.74	25.29	10.7	72.6	565.4
2441000	12.23	30.57	1.77	1.4	7.34	1.62	66.9	83.2	397.5
2442000	3.35	22.79	3.15	4.7	19.38	5.62	123.1	93.8	229.7
2443000	8.11	15.01	4.53	8.1	3.98	9.62	179.3	104.5	61.9
2444000	12.88	7.23	5.91	11.4	16.02	13.62	235.6	115.1	477.9
2445000	4.00	35.44	7.28	14.7	0.62	17.62	291.8	125.8	310.1
2446000	8.76	27.66	8.66	18.0	12.66	21.62	348.0	136.4	142.3
2447000	13.53	19.88	10.04	21.3	24.70	25.62	404.2	147.0	558.3
2448000	4.65	12.11	11.43	24.7	9.29	1.95	460.4	157.7	390.5
2449000	9.42	4.33	12.81	28.0	21.33	5.95	44.7	168.3	222.6
2450000	0.51	32.54	14.19	31.3	5.94	9.95	100.9	178.9	54.8

TABLE IIb.

Epochs and Arguments for each Thousandth Day, from 2400000 to 2500000.

For Washington Mean Noon.

Day of Julian Period.	Arg. 63.	64.	65.	66.	67.	68.	69.	70.	71.
	d.	d.	d.	d.	d.	d.	d.	d.	d.
2450000	0.54	32.51	14.19	31.3	5.94	9.95	100.9	178.9	51.8
2451000	5.31	24.76	1.09	34.6	17.97	13.95	157.2	189.5	470.9
2452000	10.07	16.98	2.47	2.3	2.57	17.95	213.4	200.1	303.0
2453000	1.20	9.20	3.84	5.6	14.61	21.95	269.6	210.8	135.2
2454000	5.96	1.42	5.22	8.9	26.66	25.95	325.8	221.4	551.3
2455000	10.73	29.64	6.60	12.3	11.25	2.28	382.0	232.1	383.4
2456000	1.85	21.86	7.98	15.6	23.29	6.28	438.3	242.7	215.6
2457000	6.61	14.08	9.36	18.9	7.88	10.28	22.6	253.3	47.7
2458000	11.38	6.30	10.73	22.2	19.92	14.28	78.8	264.0	463.8
2459000	2.50	34.51	12.12	25.5	4.52	18.28	135.0	274.6	296.0
2460000	7.26	26.73	13.50	28.8	16.56	22.28	191.2	285.2	128.1
2461000	12.03	18.96	0.40	32.1	1.16	26.28	247.4	295.8	544.2
2462000	3.16	11.18	1.78	35.4	13.20	2.61	303.7	306.4	376.4
2463000	7.92	3.40	3.16	3.2	25.25	6.61	359.9	317.1	208.5
2464000	12.69	31.61	4.54	6.5	9.84	10.61	416.1	327.7	40.7
2465000	3.81	23.83	5.91	9.8	21.88	14.61	0.4	8.6	456.8
2466000	8.57	16.05	7.29	13.1	6.47	18.61	56.6	19.2	288.9
2467000	13.31	8.27	8.67	16.4	18.51	22.61	112.8	29.8	121.1
2468000	4.46	0.49	10.05	19.8	3.11	26.61	169.1	40.5	537.2
2469000	9.22	28.71	11.44	23.1	15.15	2.94	225.3	51.1	369.3
2470000	0.35	20.93	12.82	26.4	27.20	6.94	281.5	61.7	201.5
2471000	5.11	13.15	14.20	29.7	11.79	10.94	337.7	72.3	33.6
2472000	9.88	5.37	1.10	33.0	23.84	14.94	393.9	82.9	449.7
2473000	1.01	33.58	2.47	0.8	8.43	18.94	450.1	93.6	281.9
2474000	5.77	25.80	3.85	4.1	20.47	22.94	34.5	104.2	114.0
2475000	10.53	18.03	5.23	7.4	5.07	26.94	90.7	114.9	530.1
2476000	1.66	10.25	6.61	10.7	17.11	3.26	146.9	125.5	362.3
2477000	6.42	2.47	7.99	14.0	1.71	7.26	203.1	136.1	194.4
2478000	11.19	30.68	9.36	17.4	13.75	11.27	259.3	146.8	26.6
2479000	2.31	22.90	10.74	20.7	25.80	15.27	315.6	157.4	442.7
2480000	7.07	15.12	12.13	24.0	10.39	19.27	371.8	168.0	274.8
2481000	11.81	7.34	13.51	27.3	22.43	23.27	428.0	178.6	107.0
2482000	2.96	35.55	0.41	30.6	7.02	27.27	12.3	189.2	523.1
2483000	7.73	27.78	1.79	34.0	19.06	3.59	68.5	199.9	355.2
2484000	12.49	20.00	3.17	1.7	3.66	7.59	124.7	210.5	187.4
2485000	3.62	12.22	4.54	5.0	15.70	11.61	181.0	221.1	19.5
2486000	8.38	4.44	5.92	8.3	0.30	15.61	237.2	231.7	435.6
2487000	13.15	32.65	7.30	11.6	12.34	19.60	293.4	242.3	267.8
2488000	4.27	24.87	8.68	15.0	24.39	23.59	349.6	253.0	99.9
2489000	9.03	17.09	10.06	18.3	8.98	27.59	405.8	263.6	516.0
2490000	0.16	9.31	11.45	21.6	21.02	3.92	462.0	274.2	348.2
2491000	4.92	1.51	12.83	24.9	5.61	7.92	46.4	284.8	180.3
2492000	9.68	29.75	14.21	28.2	17.66	11.93	102.6	295.4	12.5
2493000	0.81	21.97	1.11	31.5	2.25	15.93	158.8	306.1	428.6
2494000	5.58	14.19	2.49	34.8	14.29	19.93	215.0	316.7	260.7
2495000	10.34	6.41	3.86	2.6	26.34	23.93	271.2	327.4	92.9
2496000	1.47	34.62	5.24	5.9	10.93	0.25	327.5	8.2	508.9
2497000	6.23	26.85	6.62	9.2	22.97	4.25	383.7	18.8	341.1
2498000	11.00	19.07	8.00	12.5	7.57	8.25	439.9	29.5	173.3
2499000	2.12	11.29	9.38	15.8	19.61	12.26	24.2	40.1	5.4
2500000	6.88	3.51	10.76	19.1	4.21	16.26	80.4	50.7	421.5

TABLE IIb.

Epochs and Arguments for each Thousandth Day, from 2400000 to 2500000.

For Washington Mean Noon.

Day of Julian Period.	Arg. 72.	73.	74.	75.	76.	78.	79.	80.	81.
	d.	d.	d.	d.	d.	d.	d.	d.	d.
2400000	382.9	1322.1	2813	69562	7287	71.425	419.9	192.6	65.60
2401000	186.2	131.8	581	70562	8287	130.418	101.8	253.4	71.96
2402000	388.5	1131.8	1581	71562	9287	1.209	228.1	1.2	78.32
2403000	191.8	2131.8	2581	72562	10287	60.202	351.3	62.1	84.69
2404000	394.1	941.5	348	73562	11286	119.191	36.3	122.9	91.05
2405000	197.4	1941.5	1318	74562	12286	178.187	159.5	183.7	97.41
2406000	0.8	751.2	2318	75562	13286	48.978	282.8	214.5	103.78
2407000	203.0	1751.2	115	76562	14286	107.971	406.1	305.4	110.14
2408000	6.3	560.8	1115	77562	15285	166.963	91.0	53.2	116.51
2409000	208.6	1560.8	2115	78562	16285	37.754	214.3	114.0	122.87
2410000	11.9	370.5	3115	79562	17285	96.747	337.6	174.8	5.03
2411000	214.1	1370.5	882	80562	18285	155.710	22.5	235.7	11.39
2412000	17.5	180.1	1882	81562	19285	26.531	145.8	296.5	17.75
2413000	219.7	1180.1	2882	82562	20281	85.521	269.1	44.3	24.12
2414000	23.1	2180.1	619	83562	21284	144.516	392.3	105.1	30.49
2415000	225.3	989.8	1619	84562	22281	15.308	77.3	166.0	36.85
2416000	28.6	1989.8	2619	809	23281	74.300	200.5	226.8	43.21
2417000	230.9	799.4	416	1809	24283	133.293	323.8	287.6	49.58
2418000	34.2	1799.1	1416	2809	25283	4.084	8.7	35.4	55.91
2419000	236.5	609.1	2416	3809	26283	63.077	132.0	96.3	62.31
2420000	39.8	1609.1	183	4809	27283	122.069	255.3	157.1	68.67
2421000	242.0	418.8	1183	5809	28283	181.062	378.6	217.9	75.03
2422000	45.4	1418.8	2183	6809	29283	51.853	63.5	278.7	81.39
2423000	247.6	228.4	3183	7809	30282	110.846	186.8	26.5	87.76
2424000	51.0	1228.4	951	8809	31282	169.839	310.0	87.4	94.12
2425000	253.2	38.1	1951	9809	32282	40.630	433.3	148.2	100.48
2426000	56.5	1038.1	2951	10809	33281	99.623	118.3	209.0	106.85
2427000	258.8	2038.1	718	11809	34281	158.615	241.5	269.9	113.21
2428000	62.1	847.7	1718	12809	35280	29.407	364.8	17.7	119.58
2429000	264.4	1847.7	2718	13809	36280	88.399	49.7	78.5	1.74
2430000	67.7	657.4	485	14809	37280	147.392	173.0	139.3	8.10
2431000	269.9	1657.4	1485	15809	38280	18.183	296.3	200.2	14.46
2432000	73.3	467.1	2485	16809	39279	77.176	419.6	261.0	20.82
2433000	275.5	1467.1	252	17809	40279	136.169	104.5	8.8	27.19
2434000	78.9	276.7	1252	18809	41278	6.960	227.8	69.6	33.55
2435000	281.1	1276.7	2252	19809	42278	65.953	351.0	130.5	39.91
2436000	84.4	86.4	19	20809	43278	124.946	36.0	191.3	46.28
2437000	286.7	1086.4	1019	21809	44277	183.939	159.2	252.1	52.64
2438000	90.0	2086.4	2019	22809	45277	54.730	282.5	312.9	59.01
2439000	292.2	896.0	3019	23809	46276	113.723	405.8	60.7	65.37
2440000	95.6	1896.0	786	24809	47276	172.716	90.7	121.6	71.73
2441000	297.8	705.7	1786	25809	48276	43.507	214.0	182.4	78.09
2442000	101.2	1705.7	2786	26809	49275	102.500	337.3	243.2	84.45
2443000	303.4	515.3	551	27809	50275	161.493	22.2	304.1	90.82
2444000	106.8	1515.3	1554	28809	51274	32.284	145.5	51.9	97.18
2445000	309.0	325.0	2554	29809	52274	91.277	268.8	112.7	103.54
2446000	112.3	1325.0	321	30809	53274	150.270	392.0	173.5	109.91
2447000	314.6	134.7	1321	31809	54273	21.061	77.0	234.4	116.27
2448000	117.9	1134.7	2321	32809	55273	80.051	200.2	295.2	122.64
2449000	320.1	2134.6	88	33809	56272	139.047	323.5	43.0	4.80
2450000	123.5	944.3	1088	34809	57272	9.838	8.4	103.8	11.16

TABLE IIb.

Epochs and Arguments for each Thousandth Day, from 2400000 to 2500000.

For Washington Mean Noon.

Day of Julian Period.	Arg. 72.	73.	74.	75.	76.	78.	79.	80.	81.
	d.	d.	d.	d.	d.	d.	d.	d.	d.
2450000	123.5	911.3	1088	34809	57272	9.838	8.4	103.8	11.16
2451000	325.7	1941.3	2088	35809	58272	68.831	131.7	164.6	17.52
2452000	129.1	753.9	3088	36809	59271	127.824	255.0	225.5	23.89
2453000	331.3	1753.9	855	37809	60271	186.817	378.3	286.3	30.25
2454000	134.7	563.6	1855	38809	61270	57.609	63.2	34.1	36.62
2455000	336.9	1563.6	2855	39809	62270	116.601	186.5	94.9	42.98
2456000	140.2	373.2	622	40809	63270	175.594	309.8	155.8	49.34
2457000	342.5	1373.2	1622	41809	64269	46.386	433.0	216.6	55.71
2458000	145.8	182.9	2622	42809	65269	105.379	118.0	277.4	62.07
2459000	348.0	1182.9	389	43809	66268	164.372	241.2	25.2	68.44
2460000	151.4	2182.9	1389	44809	67268	35.163	364.5	86.0	74.80
2461000	353.6	992.5	2389	45809	68268	94.156	49.4	146.9	81.16
2462000	157.0	1992.5	157	46809	69267	153.149	172.7	207.7	87.52
2463000	359.2	802.2	1157	47809	70267	23.941	296.0	268.5	93.89
2464000	162.5	1802.2	2157	48809	71266	82.934	419.3	16.3	100.25
2465000	364.8	611.8	3157	49809	72266	141.927	101.2	77.2	106.61
2466000	168.1	1611.8	924	50809	73265	12.718	227.5	138.0	112.98
2467000	370.4	421.5	1924	51809	74264	71.711	350.8	198.8	119.34
2468000	173.7	1421.5	2924	52809	75264	130.701	35.7	259.7	1.51
2469000	375.9	231.1	691	53809	76263	1.496	159.0	7.5	7.87
2470000	179.3	1231.1	1691	54809	77263	60.489	282.2	68.3	14.23
2471000	381.5	40.8	2691	55809	78262	119.482	405.5	129.1	20.59
2472000	184.9	1040.8	458	56809	79262	178.475	90.4	189.9	26.95
2473000	387.1	2040.8	1458	57809	80261	49.267	213.7	250.8	33.32
2474000	190.4	850.1	2458	58809	81261	108.260	337.0	311.6	39.68
2475000	392.7	1850.4	225	59809	82260	167.253	21.9	59.4	46.04
2476000	196.0	660.1	1225	60809	83259	38.044	145.2	120.2	52.41
2477000	398.3	1660.1	2225	61809	84259	97.037	268.5	181.1	58.77
2478000	201.6	469.7	3225	62809	85258	156.030	391.8	241.9	65.14
2479000	5.0	1469.7	992	63809	86258	26.822	76.7	302.7	71.50
2480000	207.2	279.4	1992	64809	87257	85.815	200.0	50.5	77.86
2481000	10.5	1279.4	2992	65809	88256	144.808	323.2	111.3	84.22
2482000	212.8	89.0	759	66809	89256	15.600	8.2	172.2	90.58
2483000	16.1	1089.0	1759	67809	90255	74.593	131.4	233.0	96.95
2484000	218.3	2089.0	2759	68809	91255	133.586	254.7	293.8	103.31
2485000	21.7	898.6	527	69809	92254	4.378	378.1	41.6	109.67
2486000	223.9	1898.6	1527	70809	93253	63.371	62.8	102.5	116.04
2487000	27.3	708.3	2527	71809	94253	122.364	186.2	163.3	122.40
2488000	229.5	1708.3	294	72809	95252	181.357	309.5	224.1	4.57
2489000	32.9	517.9	1294	73809	762	52.149	432.8	284.9	10.93
2490000	235.1	1517.9	2294	74809	1761	111.142	117.7	32.7	17.29
2491000	38.4	327.6	61	75809	2760	170.135	241.0	93.6	23.65
2492000	240.7	1327.6	1061	76809	3759	40.927	364.2	154.4	30.02
2493000	44.0	137.2	2061	77809	4759	99.920	49.2	215.2	36.38
2494000	246.2	1137.2	3061	78809	5758	158.913	172.4	276.1	42.75
2495000	49.6	2137.2	828	79808	6758	29.705	295.7	23.9	49.11
2496000	251.8	946.9	1828	80808	7757	88.698	419.0	84.7	55.47
2497000	55.2	1946.8	2828	81808	8756	147.691	103.9	145.5	61.84
2498000	257.4	756.5	595	82808	9756	18.483	227.2	206.4	68.20
2499000	60.7	1756.5	1595	83808	10755	77.476	350.5	267.2	74.57
2500000	263.0	566.1	2595	55	11754	136.470	35.4	15.0	80.93

TABLE IIb.

Epochs and Arguments for each Thousandth Day, from 2400000 to 2500000.

For Washington Mean Noon.

Day of Julian Period.	Arg. 82.	83.	84.	85.	86.	87.	88.	89.	90.
	d.	d.	d.	d.	d.	d.	d.	d.	d.
2400000	11.655	5.56	5.63	3.09	6.02	4.81	7.28	8.47	0.0
2401000	0.801	25.27	10.17	15.27	10.67	10.27	9.96	4.23	31.6
2402000	4.812	5.77	14.72	3.91	15.31	1.18	12.61	0.00	30.9
2403000	8.823	25.48	0.12	16.13	19.95	6.91	0.66	15.08	30.2
2404000	12.834	5.98	4.66	4.80	24.59	12.33	3.34	10.85	29.5
2405000	1.980	25.69	9.20	16.99	29.24	3.55	6.03	6.62	28.8
2406000	5.991	6.18	13.74	5.65	33.88	8.98	8.71	2.39	28.1
2407000	10.003	25.89	18.29	17.84	0.24	0.20	11.39	17.47	27.4
2408000	14.014	6.39	3.69	6.51	4.89	5.62	14.07	13.23	26.7
2409000	3.160	26.10	8.23	18.70	9.53	11.05	2.09	9.00	26.0
2410000	7.171	6.60	12.77	7.37	14.17	2.26	4.77	4.77	25.3
2411000	11.182	26.30	17.31	19.55	18.81	7.69	7.46	0.54	24.6
2412000	0.328	6.80	2.71	8.22	23.16	13.11	10.14	15.62	23.9
2413000	4.339	26.51	7.26	20.41	28.10	4.33	12.82	11.39	23.2
2414000	8.351	7.01	11.80	9.08	32.74	9.76	0.84	7.16	22.5
2415000	12.362	26.72	16.34	21.27	37.39	0.98	3.52	2.92	21.8
2416000	1.508	7.22	1.74	9.93	3.75	6.40	6.21	18.01	21.0
2417000	5.519	26.93	6.28	22.12	8.39	11.83	8.89	13.77	20.3
2418000	9.530	7.42	10.83	10.79	13.03	3.04	11.57	9.54	19.6
2419000	13.542	27.13	15.37	22.98	17.68	8.47	14.25	5.31	18.9
2420000	2.687	7.63	0.77	11.65	22.32	13.90	2.27	1.08	18.2
2421000	6.698	27.34	5.31	0.32	26.96	5.12	4.95	16.16	17.5
2422000	10.710	7.84	9.86	12.50	31.61	10.54	7.64	11.93	16.8
2423000	14.721	27.54	14.40	1.17	36.25	1.76	10.32	7.69	16.1
2424000	3.867	8.04	18.94	13.36	2.61	7.18	13.00	3.46	15.4
2425000	7.878	27.75	4.34	2.03	7.25	12.61	1.02	18.54	14.7
2426000	11.889	8.25	8.88	14.22	11.90	3.83	3.70	14.31	14.0
2427000	1.035	27.96	13.43	2.88	16.54	9.25	6.39	10.08	13.3
2428000	5.046	8.46	17.97	15.07	21.18	0.47	9.07	5.85	12.6
2429000	9.058	28.17	3.37	3.74	25.83	5.90	11.75	1.62	11.9
2430000	13.069	8.66	7.91	15.93	30.47	11.32	14.43	16.70	11.2
2431000	2.215	28.37	12.45	4.60	35.11	2.51	2.45	12.46	10.5
2432000	6.226	8.87	17.00	16.78	1.17	7.96	5.13	8.23	9.8
2433000	10.237	28.58	2.40	5.15	6.12	13.39	7.82	4.00	9.1
2434000	14.248	9.08	6.94	17.64	10.76	4.61	10.50	19.08	8.4
2435000	3.394	28.79	11.48	6.31	15.40	10.03	13.18	14.85	7.7
2436000	7.405	9.28	16.02	18.50	20.05	1.25	1.20	10.62	7.0
2437000	11.417	28.99	1.42	7.16	24.69	6.68	3.88	6.38	6.3
2438000	0.562	9.49	5.97	19.35	29.33	12.10	6.56	2.15	5.6
2439000	4.574	29.20	10.51	8.02	33.97	3.32	9.25	17.23	4.9
2440000	8.585	9.70	15.05	20.21	0.33	8.74	11.93	13.00	4.2
2441000	12.596	29.40	0.45	8.88	4.98	14.17	14.62	8.77	3.5
2442000	1.742	9.90	4.99	21.07	9.62	5.39	2.63	4.54	2.8
2443000	5.753	29.61	9.54	9.73	14.26	10.82	5.32	0.31	2.1
2444000	9.765	10.11	14.08	21.92	18.91	2.03	8.00	15.39	1.4
2445000	13.776	29.82	18.62	10.59	23.55	7.46	10.68	11.15	0.7
2446000	2.922	10.32	4.02	22.78	28.19	12.88	13.36	6.92	32.2
2447000	6.933	30.03	8.56	11.15	32.84	4.10	1.38	2.69	31.5
2448000	10.944	10.52	13.11	0.11	37.48	9.52	4.06	17.77	30.8
2449000	0.090	30.23	17.65	12.30	3.84	0.74	6.75	13.54	30.1
2450000	4.101	10.73	3.05	0.97	8.48	6.17	9.43	9.31	29.4

TABLE IIb.

Epochs and Arguments for each Thousandth Day, from 2400000 to 2500000.

For Washington Mean Noon.

Day of Julian Period.	Arg. 82.	83.	84.	85.	86.	87.	88.	89.	90.
	d.	d.	d.	d.	d.	d.	d.	d.	d.
2450000	4.101	10.73	3.05	0.97	8.48	6.17	9.43	9.31	29.4
2451000	8.113	30.44	7.59	13.16	13.13	11.59	12.11	5.08	28.7
2452000	12.124	10.94	12.14	1.83	17.77	2.81	0.13	0.84	28.0
2453000	1.269	30.64	16.68	14.01	22.41	8.24	2.81	15.92	27.3
2454000	5.281	11.14	2.08	2.68	27.05	13.66	5.50	11.69	26.6
2455000	9.293	30.85	6.62	14.87	31.70	4.88	8.18	7.46	25.9
2456000	13.304	11.35	11.16	3.54	36.31	10.30	10.86	3.23	25.2
2457000	2.450	31.06	15.71	15.72	2.70	1.52	13.54	18.31	24.5
2458000	6.461	11.56	1.11	4.39	7.34	6.95	1.5[illegible]	14.08	23.8
2459000	10.472	31.27	5.65	16.58	11.99	12.37	4.24	9.85	23.1
2460000	14.483	11.76	10.19	5.25	16.63	3.59	6.93	5.61	22.4
2461000	3.629	31.17	14.73	17.44	21.27	9.01	9.61	1.38	21.7
2462000	7.641	11.97	0.13	6.11	25.92	0.23	12.29	16.16	21.0
2463000	11.652	31.68	4.68	18.30	30.56	5.66	0.31	12.23	20.3
2464000	0.798	12.18	9.22	6.96	35.20	11.08	2.99	8.00	19.6
2465000	4.809	31.89	13.76	19.15	1.56	2.30	5.68	3.77	18.9
2466000	8.820	12.38	18.30	7.82	6.21	7.73	8.36	18.85	18.1
2467000	12.832	32.09	3.70	20.01	10.85	13.15	11.01	11.62	17.4
2468000	1.977	12.59	8.25	8.68	15.49	4.37	13.73	10.38	16.7
2469000	5.989	32.30	12.79	20.86	20.13	9.80	1.71	6.15	16.0
2470000	10.000	12.80	17.33	9.53	21.78	1.01	4.42	1.92	15.3
2471000	14.011	32.50	2.73	21.72	29.42	6.44	7.11	17.00	14.6
2472000	3.157	13.00	7.27	10.39	31.06	11.86	9.79	12.77	13.9
2473000	7.169	32.71	11.82	22.58	0.42	3.08	12.47	8.54	13.2
2474000	11.180	13.21	16.36	11.24	5.07	8.51	0.49	4.31	12.5
2475000	0.326	32.92	1.76	23.43	9.71	13.93	3.17	0.07	11.8
2476000	4.337	13.42	6.30	12.10	14.35	5.15	5.86	15.15	11.1
2477000	8.348	33.13	10.84	0.77	19.00	10.58	8.54	10.92	10.4
2478000	12.360	13.62	15.39	12.96	23.61	1.80	11.22	6.69	9.7
2479000	1.506	33.33	0.79	1.62	28.28	7.22	13.91	2.46	9.0
2480000	5.517	13.83	5.33	13.81	32.93	12.65	1.92	17.51	8.3
2481000	9.528	33.54	9.87	2.48	37.57	3.86	4.60	13.31	7.6
2482000	13.540	14.04	14.42	14.67	3.93	9.29	7.29	9.08	6.9
2483000	2.685	33.75	18.96	3.34	8.57	11.71	9.97	4.84	6.2
2484000	6.697	14.24	4.36	15.53	13.21	5.93	12.65	0.61	5.5
2485000	10.708	33.95	8.90	4.19	17.86	11.36	0.67	15.69	4.8
2486000	14.719	14.45	13.44	16.38	22.50	2.58	3.35	11.46	4.0
2487000	3.865	34.16	17.99	5.05	27.14	8.00	6.04	7.23	3.3
2488000	7.877	14.66	3.39	17.24	31.79	13.43	8.72	3.00	2.6
2489000	11.888	34.37	7.93	5.91	36.43	4.65	11.40	18.08	1.9
2490000	1.034	14.86	12.47	18.09	2.79	10.07	14.09	13.85	1.2
2491000	5.045	34.57	17.01	6.76	7.43	1.29	2.10	9.61	0.5
2492000	9.056	15.07	2.41	18.95	12.08	6.71	4.79	5.38	32.1
2493000	13.068	34.78	6,96	7.62	16.72	12.14	7.47	1.15	31.4
2494000	2.214	15.28	11.50	19.81	21.36	3.36	10.15	16.23	30.7
2495000	6.225	34.99	16.04	8.47	26.00	8.78	12.83	12.00	30.0
2496000	10.236	15.48	1.44	20.66	30.65	0.00	0.85	7.77	29.3
2497000	14.247	35.19	5.98	9.33	35.29	5.43	3.53	3.53	28.6
2498000	3.393	15.69	10.53	21.52	1.65	10.85	6.22	18.62	27.9
2499000	7.405	35.40	15.07	10.19	6.29	2.07	8.90	14.38	27.2
2500000	11.416	15.90	0.47	22.37	10.94	7.49	11.59	10.15	26.5

TABLE IIIa.

Change of Mean Longitude, and Longitude of the Node for any Interval of Time less than 1000 Days.

Days.	u′		$\frac{1}{100}$ u′		$\frac{1}{1000}$ u′	(y-u′)		$\frac{1}{10}$ (y-u′)
0	0°	0″.00	0°	0″.00	0″.00	0°	0″.0	0″.0
10	131	2750.28	1	1143.50	47.41	0	1906.3	19.1
20	263	1900.56	2	2287.01	94.87	1	212.7	38.1
30	35	1050.84	3	3430.51	142.31	1	2119.0	57.2
40	167	201.12	5	974.01	189.74	2	425.3	76.3
50	298	2951.40	6	2117.51	237.18	2	2331.7	95.3
60	70	2101.69	7	3261.02	284.61	3	638.0	114.4
70	202	1251.97	9	804.52	332.05	3	2544.3	133.4
80	334	402.25	10	1948.02	379.48	4	850.7	152.5
90	105	3152.53	11	3091.53	426.92	4	2757.0	171.6
100	237	2302.81	13	635.03	474.35	5	1063.4	190.6
110	9	1453.09	14	1778.53	521.79	5	2969.7	209.7
120	141	603.37	15	2922.03	569.22	6	1276.1	228.8
130	272	3353.65	17	465.51	616.66	6	3182.4	247.8
140	44	2503.93	18	1609.04	664.09	7	1488.7	266.9
150	176	1654.21	19	2752.54	711.53	7	3395.1	286.0
160	308	804.50	21	296.04	758.96	8	1701.4	305.0
170	79	3554.78	22	1439.55	806.40	9	7.7	324.1
180	211	2705.06	23	2583.05	853.83	9	1914.1	343.1
190	313	1855.34	25	126.55	901.27	10	220.4	362.2
200	115	1005.62	26	1270.06	948.70	10	2126.7	381.3
210	247	155.90	27	2413.56	996.14	11	433.1	400.3
220	18	2906.18	28	3557.06	1043.57	11	2339.4	419.4
230	150	2056.46	30	1100.56	1091.01	12	645.7	438.5
240	282	1206.74	31	2244.07	1138.44	12	2552.1	457.5
250	54	357.02	32	3387.57	1185.88	13	858.4	476.6
260	185	3107.30	34	931.07	1233.31	13	2764.7	495.6
270	317	2257.59	35	2074.58	1280.75	14	1071.1	514.7
280	89	1407.87	36	3218.08	1328.18	14	2977.4	533.8
290	221	558.15	38	761.58	1375.62	15	1283.7	552.8
300	352	3308.43	39	1905.08	1423.05	15	3190.1	571.9
310	124	2458.71	40	3048.59	1470.49	16	1496.4	591.0
320	256	1608.99	42	592.09	1517.92	16	3402.7	610.0
330	28	759.27	43	1735.59	1565.36	17	1709.1	629.1
340	159	3509.55	44	2879.10	1612.79	18	15.4	648.2
350	291	2659.83	46	422.60	1660.23	18	1921.7	667.2
360	63	1810.11	47	1566.10	1707.66	19	228.1	686.3
370	195	960.39	48	2709.60	1755.10	19	2134.5	705.3
380	327	110.67	50	253.11	1802.53	20	440.7	724.4
390	98	2860.95	51	1396.61	1849.97	20	2347.1	743.5
400	230	2011.24	52	2540.11	1897.40	21	653.5	762.5
410	2	1161.52	54	83.62	1944.84	21	2559.7	781.6
420	134	311.80	55	1227.12	1992.27	22	866.1	800.7
430	265	3062.08	56	2370.62	2039.71	22	2772.5	819.7
440	37	2212.36	57	3514.12	2087.14	23	1078.7	838.8
450	169	1362.64	59	1057.63	2134.58	23	2985.1	857.9
460	301	512.92	60	2201.13	2182.01	24	1291.5	876.9
470	72	3263.20	61	3344.63	2229.45	24	3197.7	896.0
480	204	2413.48	63	688.13	2276.88	25	1504.1	915.0
490	336	1563.76	64	2031.64	2324.32	25	3410.5	934.1
500	108	714.04	65	3175.14	2371.75	26	1716.8	953.2

TABLE IIIa.

Change of Mean Longitude, and Longitude of the Node for any Interval of Time less than 1000

Days.	u′ (°)	u′ (″)	$\frac{1}{100}$ u′ (°)	$\frac{1}{100}$ u′ (″)	$\frac{1}{1000}$ u′ (″)	(y-u′) (′)	(y-u′) (″)	$\frac{1}{10}$ (y-u′) (″)
500	108°	711.04	65°	3175.14″	2371.75″	26′	1716.8″	953.2″
510	239	3464.33	67	718.64	2419.19	27	23.2	972.2
520	11	2614.61	68	1862.15	2466.62	27	1929.5	991.3
530	113	1761.89	69	3005.65	2514.06	28	235.8	1010.4
540	275	915.17	71	549.15	2561.49	28	2142.2	1029.4
550	47	65.45	72	1692.65	2608.93	29	418.6	1048.5
560	178	2815.73	73	2836.16	2656.36	29	2351.9	1067.5
570	310	1966.01	75	379.66	2703.80	30	661.2	1086.6
580	82	1116.29	76	1523.16	2751.23	30	2567.6	1105.7
590	214	266.57	77	2666.67	2798.67	31	873.9	1124.7
600	345	3016.85	79	210.17	2846.10	31	2780.2	1143.8
610	117	2167.13	80	1353.67	2893.54	32	1086.6	1162.9
620	249	1317.42	81	2497.17	2940.97	32	2993.0	1181.9
630	21	467.70	83	40.68	2988.41	33	1299.3	1201.0
640	152	3217.98	84	1184.18	3035.84	33	3205.6	1220.1
650	284	2368.26	85	2327.68	3083.28	34	1512.0	1239.1
660	56	1518.54	86	3471.19	3130.71	34	3418.3	1258.2
670	188	668.82	88	1014.69	3178.15	35	1724.6	1277.3
680	319	3419.10	89	2158.19	3225.58	36	31.0	1296.3
690	91	2569.38	90	3301.69	3273.02	36	1937.3	1315.4
700	223	1719.66	92	845.20	3320.45	37	243.6	1334.4
710	355	869.94	93	1988.70	3367.89	37	2150.0	1353.5
720	127	20.22	94	3132.20	3415.32	38	456.3	1372.6
730	258	2770.51	96	675.71	3462.76	38	2362.7	1391.6
740	30	1920.79	97	1819.21	3510.19	39	669.0	1410.7
750	162	1071.07	98	2962.71	3557.63	39	2575.3	1429.8
760	294	221.35	100	506.21	3605.06	40	881.7	1448.8
770	65	2971.63	101	1649.72	3652.50	40	2788.0	1467.9
780	197	2121.91	102	2793.22	3699.93	41	1094.3	1486.9
790	329	1272.19	104	336.72	3747.37	41	3000.6	1506.0
800	101	422.47	105	1480.22	3794.80	42	1306.9	1525.1
810	232	3172.75	106	2623.73	3842.24	42	3213.2	1544.1
820	4	2323.03	108	167.23	3889.67	43	1519.6	1563.2
830	136	1473.31	109	1310.73	3937.11	43	3425.9	1582.3
840	268	623.59	110	2454.24	3984.54	44	1732.2	1601.3
850	39	3373.88	111	3597.74	4031.98	45	38.6	1620.4
860	171	2524.16	113	1141.24	4079.41	45	1944.9	1639.4
870	303	1674.44	114	2284.74	4126.85	46	251.3	1658.5
880	75	824.72	115	3428.25	4174.28	46	2157.6	1677.6
890	206	3575.00	117	971.75	4221.72	47	463.9	1696.6
900	338	2725.28	118	2115.25	4269.15	47	2370.3	1715.7
910	110	1875.56	119	3258.76	4316.59	48	676.6	1734.8
920	242	1025.84	121	802.26	4364.02	48	2582.9	1753.8
930	14	176.12	122	1945.76	4411.46	49	889.3	1772.9
940	145	2926.40	123	3089.26	4458.89	49	2795.7	1792.0
950	277	2076.68	125	632.77	4506.33	50	1102.0	1811.0
960	49	1226.97	126	1776.27	4553.76	50	3008.3	1830.1
970	181	377.25	127	2919.77	4601.20	51	1314.7	1849.1
980	312	3127.53	129	463.28	4648.63	51	3221.0	1868.2
990	84	2277.81	130	1606.78	4696.07	52	1527.3	1887.3
1000	216	1428.09	131	2750.28	4743.50	52	3433.7	1906.3

TABLE IIIb.

Multiples of Periods of the Arguments.

No. of Periods.	Arg. 1.	2.	3.	4.	5.	6.
1	27.554552446	31.81193574	29.53058800	365.259687	411.785170	34.846892
2	55.109104892	63.62387147	59.06117599	730.519373	823.570341	69.693784
3	82.663657338	95.43580721	88.59176399	1095.779060	1235.355511	104.540676
4	110.218209784	127.24774294	118.12235199	1461.038747	1647.140681	139.387568
5	137.772762230	159.05967868	147.65293999	1826.298433	2058.925852	174.234460
6	165.327314676	190.87161442	177.18352798	2191.558120	2470.711022	209.081352
7	192.881867122	222.68355015	206.71411598	2556.817807	2882.496192	243.928244
8	220.436419568	254.49548589	236.24470398	2922.077493	3294.281362	278.775136
9	247.990972014	286.30742162	265.77529197	3287.337180	3706.066533	313.622028
10	275.545524460	318.11935736	295.30587997	3652.596867	4117.851703	348.468920

No. of Periods.	Arg. 7.	8.	9.	10.	11.	12.	13.
1	9.613718	15.387312	29.802826	25.621696	26.878290	346.620112	10.084597
2	19.227436	30.774624	59.605651	51.243393	53.756580	693.240224	20.169194
3	28.841154	46.161937	89.408477	76.865089	80.634870	1039.860336	30.253791
4	38.454872	61.549249	119.211302	102.486786	107.513160	1386.480448	40.338387
5	48.068590	76.936561	149.014128	128.108482	134.391450	1733.100559	50.422984
6	57.682308	92.323873	178.816953	153.730179	161.269740	2079.720671	60.507581
7	67.296026	107.711186	208.619779	179.351875	188.148029	2426.340783	70.592178
8	76.909744	123.098498	238.422604	204.973572	215.026319	2772.960895	80.676775
9	86.523462	138.485810	268.225430	230.595268	241.904609	3119.581007	90.761372
10	96.137180	153.873122	298.028255	256.216964	268.782899	3466.201119	100.845968

No. of Periods.	Arg. 14.	15.	16.	17.	18.	19.	20.
1	29.26328	14.19161	14.25418	27.32168	9.87359	24.30220	14.31731
2	58.52656	28.38322	28.50837	54.64336	19.74719	48.60439	28.63463
3	87.78984	42.57483	42.76255	81.96504	29.62078	72.90659	42.95194
4	117.05312	56.76644	57.01674	109.28672	39.49437	97.20878	57.26925
5	146.31640	70.95805	71.27092	136.60840	49.36797	121.51098	71.58656
6	175.57967	85.14967	85.52511	163.93008	59.24156	145.81317	85.90388
7	204.84295	99.34128	99.77929	191.25176	69.11515	170.11537	100.22119
8	234.10623	113.53289	114.03348	218.57344	78.98875	194.41757	114.53850
9	263.36951	127.72450	128.28766	245.89511	88.86234	218.71976	128.85581
10	292.63279	141.91611	142.54185	273.21679	98.73593	243.02196	143.17313

No. of Periods.	Arg. 21.	22.	23.	24.	25.	26.	27.
1	131.6711	32.12809	23.77463	13.27650	9.36717	5.82261	25.8264
2	263.3422	64.25617	47.54925	26.55300	18.73434	11.64521	51.6528
3	395.0134	96.38426	71.32388	39.82949	28.10152	17.46782	77.4791
4	526.6845	128.51234	95.09851	53.10599	37.46869	23.29042	103.3055
5	658.3556	160.64043	118.87313	66.38249	46.83586	29.11303	129.1319
6	790.0267	192.76852	142.64776	79.65899	56.20303	34.93563	154.9583
7	921.6979	224.89660	166.42238	92.93549	65.57020	40.75824	180.7847
8	1053.3690	257.02469	190.19701	106.21199	74.93738	46.58084	206.6111
9	1185.0401	289.15277	213.97164	119.48848	84.30155	52.40345	232.4374
10	1316.7112	321.28086	237.74626	132.76498	93.67172	58.22606	258.2638

TABLE IIIb.

Multiples of Periods of the Arguments.

No. of Periods.	Arg. 28.	29.	30.	31.	32.	33.	34.
1	38.52201	32.76361	10.37093	15.3144	16.6302	32.4506	27.0927
2	77.04403	65.52728	20.74186	30.6288	33.2603	64.9012	54.1854
3	115.56604	98.29092	31.11280	45.9433	49.8905	97.3517	81.2781
4	154.08805	131.05456	41.48373	61.2577	66.5206	129.8023	108.3708
5	192.61007	163.81821	51.85467	76.5721	83.1508	162.2529	135.4636
6	231.13208	196.58185	62.22559	91.8865	99.7810	194.7035	162.5563
7	269.65409	229.34549	72.59652	107.2009	116.4111	227.1541	189.6490
8	308.17610	262.10913	82.96746	122.5153	133.0413	259.6047	216.7417
9	346.69812	294.87277	93.33839	137.8298	149.6715	292.0552	243.8344
10	385.22013	327.63641	103.70932	153.1442	166.3016	324.5058	270.9271

No. of Periods.	Arg. 35.	36.	37.	38.	39.	40.	41.	42.
1	23.9422	13.7188	37.625	9.3944	34.475	18.8435	17.9191	10.1479
2	47.8845	27.4376	75.251	18.7888	68.951	37.6871	35.8382	20.2958
3	71.8267	41.1564	112.876	28.1832	103.426	56.5306	53.7573	30.4437
4	95.7689	54.8752	150.501	37.5776	137.901	75.3742	71.6764	40.5916
5	119.7112	68.5941	188.127	46.9720	172.376	94.2177	89.5955	50.7395
6	143.6534	82.3129	225.752	56.3664	206.852	113.0613	107.5146	60.8874
7	167.5956	96.0317	263.377	65.7607	241.327	131.9048	125.4337	71.0354
8	191.5379	109.7505	301.003	75.1551	275.802	150.7483	143.3528	81.1833
9	215.4801	123.4693	338.628	84.5495	310.278	169.5919	161.2720	91.3312
10	239.4223	137.1881	376.253	93.9439	344.753	188.4354	179.1911	101.4791

No. of Periods.	Arg. 43.	44.	45.	46.	47.	48.	49.	50.
1	14.5379	22.786	16.9000	29.013	14.1326	29.9342	12.763	25.036
2	29.0757	45.572	33.8001	58.027	28.2651	59.8683	25.525	50.072
3	43.6136	68.358	50.7001	87.040	42.3977	89.8025	38.288	75.108
4	58.1514	91.145	67.6001	116.053	56.5302	119.7367	51.051	100.144
5	72.6893	113.931	84.5002	145.066	70.6628	149.6709	63.814	125.180
6	87.2271	136.717	101.4002	174.080	84.7953	179.6050	76.576	150.216
7	101.7650	159.503	118.3002	203.093	98.9279	209.5392	89.339	175.252
8	116.3028	182.289	135.2003	232.106	113.0604	239.4734	102.102	200.288
9	130.8407	205.075	152.1003	261.120	127.1930	269.4075	114.864	225.324
10	145.3785	227.861	169.0004	290.133	141.3256	299.3417	127.627	250.360

No. of Periods.	Arg. 51.	52.	53.	54.	55.	56.	57.	58.
1	18.2169	25.2314	117.5394	15.070	19.6273	13.1175	16.9874	15.2422
2	36.4338	50.4627	235.0788	30.140	39.2546	26.2350	33.9748	30.4844
3	54.6508	75.6941	352.6183	45.210	58.8819	39.3524	50.9622	45.7266
4	72.8677	100.9255	470.1577	60.280	78.5092	52.4699	67.9496	60.9689
5	91.0846	126.1569	587.6971	75.349	98.1365	65.5874	84.9370	76.2111
6	109.3015	151.3882	705.2365	90.419	117.7638	78.7049	101.9244	91.4533
7	127.5184	176.6196	822.7760	105.489	137.3911	91.8223	118.9118	106.6955
8	145.7354	201.8510	940.3154	120.559	157.0184	104.9398	135.8993	121.9377
9	163.9523	227.0824	1057.8548	135.629	176.6457	118.0573	152.8867	137.1799
10	182.1692	252.3137	1175.3942	150.699	196.2730	131.1748	169.8741	152.4222

TABLE IIIb.

Multiples of Periods of the Arguments.

No. of Periods.	Arg. 59.	60.	61.	62.	63.	64.	65.
1	38.9640	22.3217	13.6061	14.4420	13.6334	35.9921	14.4728
2	77.9279	44.6431	27.2122	28.8840	27.2668	71.9842	28.9455
3	116.8919	66.9651	40.8183	43.3260	40.9002	107.9764	43.4183
4	155.8559	89.2869	54.4244	57.7680	54.5336	143.9685	57.8911
5	194.8198	111.6086	68.0306	72.2100	68.1670	179.9606	72.3639
6	233.7838	133.9303	81.6367	86.6520	81.8004	215.9527	86.8366
7	272.7478	156.2520	95.2428	101.0940	95.4338	251.9448	101.3094
8	311.7118	178.5737	108.8489	115.5360	109.0672	287.9370	115.7822
9	350.6757	200.8954	122.4550	129.9780	122.7006	323.9291	130.2549
10	389.6397	223.2171	136.0611	144.4200	136.3340	359.9212	144.7277

No. of Periods.	Arg. 66.	67.	68.	69.	70.	71.	72.	73.
1	35.596	27.4433	27.6667	471.893	329.791	583.921	398.884	2190.331
2	71.192	54.8866	55.3334	943.787	659.581	1167.843	797.768	4380.661
3	106.787	82.3300	83.0001	1415.680	989.372	1751.764	1196.652	6570.992
4	142.383	109.7733	110.6668	1887.573	1319.162	2335.685	1595.536	8761.322
5	177.979	137.2166	138.3334	2359.466	1648.953	2919.606	1994.421	10951.653
6	213.575	164.6599	166.0001	2831.360	1978.743	3503.528	2393.305	13141.984
7	249.171	192.1033	193.6668	3303.253	2308.531	4087.449	2792.189	15332.314
8	284.767	219.5466	221.3335	3775.146	2638.321	4671.370	3191.073	17522.645
9	320.362	246.9899	249.0002	4247.039	2968.115	5255.291	3589.957	19712.975
10	355.958	274.4332	276.6669	4718.933	3297.906	5839.213	3988.841	21903.306

No. of Periods.	Arg. 74.	75.	76.	78.	79.	80.	81.	82.
1	3232.82	84753	95490	188.20151	438.361	313.055	124.2046	14.86550
2	6465.64	169506	190980	376.40303	876.722	626.109	248.4092	29.73101
3	9698.46	254260	286170	564.60454	1315.082	939.164	372.6138	44.59651
4	12931.28	339013	381960	752.80605	1753.443	1252.219	496.8184	59.46202
5	16164.10	423766	477450	941.00757	2191.804	1565.273	621.0230	74.32752
6	19396.92	508519	572940	1129.20908	2630.165	1878.328	745.2276	89.19303
7	22629.74	593273	668430	1317.41059	3068.525	2191.383	869.4322	104.05853
8	25862.56	678026	763920	1505.61211	3506.886	2504.437	993.6368	118.92403
9	29095.38	762779	859409	1693.81362	3945.247	2817.492	1117.8414	133.78954
10	32328.20	847532	954899	1882.01513	4383.608	3130.547	1242.0460	148.65504

No. of Periods.	Arg. 83.	84.	85.	86.	87.	88.	89.	90.
1	39.2116	19.1434	23.5193	38.2830	14.2082	14.6664	19.3122	32.281
2	78.4233	38.2868	47.0387	76.5659	28.4164	29.3329	38.6243	64.562
3	117.6349	57.4302	70.5580	114.8489	42.6246	43.9993	57.9365	96.842
4	156.8465	76.5736	94.0774	153.1318	56.8328	58.6657	77.2486	129.123
5	196.0582	95.7170	117.5967	191.4148	71.0410	73.3321	96.5608	161.404
6	235.2698	114.8604	141.1160	229.6977	85.2492	87.9986	115.8729	193.685
7	274.4815	134.0039	164.6354	267.9807	99.4574	102.6650	135.1851	225.965
8	313.6931	153.1473	188.1547	306.2636	113.6656	117.3314	154.4972	258.246
9	352.9047	172.2907	211.6741	344.5466	127.8739	131.9978	173.8094	290.527
10	392.1164	191.4341	235.1934	382.8295	142.0821	146.6643	193.1215	322.808

TABLE. IV.

Constants of Epochs and Arguments for every Hundred Thousandth Day to 2300000.

Day.	u		Log. (-u′)	Log. (-u″)	(y-u)		Log. (y-u′)	Log. (y-u″)
0	140°	306″.18	8.514661	91.41942	312	1071″.9	8.396885	91.2329
100000	178	2905.81	8.529131	91.40094	238	1261.1	8.381369	91.2141
200000	217	2028.29	8.513363	91.38163	134	1962.1	8.365011	91.1951
300000	256	1274.81	8.496369	91.36143	30	1373.7	8.347731	91.1749
400000	295	647.57	8.478365	91.34024	286	1291.2	8.329152	91.1537
500000	334	148.78	8.459252	91.31796	182	1122.4	8.310061	91.1314
600000	12	3380.62	8.438911	91.29418	78	856.8	8.289452	91.1080
700000	51	3145.30	8.417204	91.26966	334	495.9	8.267479	91.0831
800000	90	3015.02	8.393970	91.24333	230	38.3	8.243981	91.0568
900000	129	3081.98	8.369015	91.21530	125	3082.6	8.218766	91.0288
1000000	168	3258 37	8.342103	91.18534	21	2427.3	8.191598	90.9988
1100000	207	3576.10	8.312949	91.15315	277	1671.1	8.162190	90.9666
1200000	247	438.26	8.281195	91.11839	173	812.4	8.130188	90.9319
1300000	286	1046.15	8.246395	91.08060	68	3450.0	8.095111	90.8941
1400000	325	1802 27	8.207971	91.03921	324	2382.2	8.056171	90.8527
1500000	4	2708.89	8.165161	90.99346	220	1207.8	8.013124	90.8069
1600000	44	168.01	8.116937	90.94230	115	3525.2	7.964963	90.7558
1700000	83	1382.02	8.061851	90.88131	11	2133.0	7.909646	90.6978
1800000	122	2753.06	7.997796	90.81736	267	629.9	7.845957	90.6308
1900000	162	683.32	7.921485	90.73818	162	2614.3	7.768818	90.5516
2000000	201	2375 01	7.827426	90.61127	58	884.9	7.674534	90.4547
2100000	241	630.32	7.705321	90.51633	313	2640.2	7.552206	90.3298
2200000	280	2651.16	7.532432	90.34021	209	678.7	7.378709	90.1537
2300000	320	1210.62	7.233809	90.03921	104	2199.2	7.080258	89.8526

Day.	Arg. 1.	Log. 1′.	Log. 1″.	Arg. 2.	Log. 2′.	Log. 2″.
0	5.3268824	−94.480661	−87.4274	17.176819	+94.28132	+87.2570
100000	9.5618159	94.465942	87.4089	0.643873	94.26692	87.2385
200000	13.8069578	94.450440	87.3896	15.916224	94.25166	87.2192
300000	18.0624109	94.433995	87.3694	31.181815	94.23547	87.1990
400000	22.3285183	94.416533	87.3482	14.628562	94.21824	87.1778
500000	26.6051129	94.397950	87.3260	29.880187	94.19992	87.1556
600000	3.3387953	94.378134	87.3025	13.312665	94.18034	87.1324
700000	7.6379932	94.356947	87.2776	28.549716	94.15938	87.1072
800000	11.9486771	94.334218	87.2513	11.967341	94.13689	87.0809
900000	16.2710701	94.309762	87.2233	27.189199	94.11266	87.0529
1000000	20.6053951	94.283344	87.1933	10.591328	94.08647	87.0229
1100000	24.9518750	94.254673	87.1611	25.797429	94.05802	86.9907
1200000	1.7561803	94.223400	87.1261	9.183181	94.02696	86.9560
1300000	6.1276389	94.189070	87.0886	24.373204	93.99290	86.9182
1400000	10.5119214	94.151100	87.0472	7.742576	93.95509	86.8768
1500000	14.9092506	94.108758	87.0014	22.915317	93.91295	86.8310
1600000	19.3198495	94.060984	86.9503	6.267407	93.86538	86.7799
1700000	23.7439410	94.006345	86.8923	21.422565	93.81094	86.7219
1800000	0.6271958	93.942725	86.8254	4.756769	93.74674	86.6550
1900000	5.0789415	93.866846	86.7462	19.893742	93.67183	86.5758
2000000	9.5118488	93.773213	86.6493	3.209159	93.57781	86.4789
2100000	14.0251406	93.651526	86.5243	18.327613	93.45689	86.3539
2200000	18.5200399	93.478663	86.3482	1.621271	93.28421	86.1778
2300000	23.0297695	−93.180837	−86.0172	16.723061	+92.98656	+85.8768

TABLE IV.

Constants of Epochs and Arguments for every Hundred Thousandth Day to 2300000.

Day.	Arg. 3.	Log. 3′.	Log. 3″.	Arg. 4.	Arg. 5.	Log. 5′.	Log. 5″.
0	10.8269901	−93.84505	−86.7744	121.4011	340.1155	−95.54931	−88.501
100000	20.1879472	93.83025	86.7559	40.2469	273.1718	95.53469	88.483
200000	0.0206869	93.81459	86.7366	324.3525	206.0468	95.51921	88.463
300000	9.3864354	93.79800	86.7161	243.1983	139.0429	95.50279	88.443
400000	18.7516540	93.78039	86.6952	162.0442	72.1629	95.48537	88.422
500000	28.1253925	93.76167	86.6730	80.8900	5.4095	95.46683	88.400
600000	7.9681123	93.74171	86.6495	364.9956	350.5703	95.44703	88.376
700000	17.3440390	93.72039	86.6246	283.8444	284.0778	95.42588	88.351
800000	26.7226343	93.69753	86.5983	202.6873	217.7198	95.40318	88.325
900000	6.5733596	93.67294	86.5703	121.5331	151.4988	95.37875	88.297
1000000	15.9574407	93.64640	86.5403	40.3790	85.4175	95.35236	88.267
1100000	25.3443389	93.61760	86.5081	324.4845	19.4786	95.32370	88.235
1200000	5.2035160	93.58620	86.4734	243.3304	365.4699	95.29245	88.200
1300000	14.5961974	93.55175	86.4356	162.1762	299.8237	95.25814	88.162
1400000	23.9918418	93.51367	86.3942	81.0221	231.3277	95.22020	88.121
1500000	3.8599197	93.47120	86.3484	365.1276	168.9817	95.17781	88.075
1600000	13.2616177	93.42330	86.2973	283.9735	103.7972	95.13007	88.024
1700000	22.6664901	93.36855	86.2393	202.8193	38.7679	95.07544	87.966
1800000	2.5139093	93.30482	86.1724	121.6652	385.6847	95.01177	87.899
1900000	11.9551300	93.22882	86.0932	40.5110	320.9798	94.93584	87.820
2000000	21.3696141	93.13508	85.9963	324.6166	256.1110	94.84212	87.723
2100000	1.2568232	93.01329	85.8713	243.4624	192.0710	94.72109	87.598
2200000	10.6779828	92.84032	85.6952	162.3083	127.8721	94.54826	87.422
2300000	20.1025546	−92.54238	−85.3942	81.1541	63.8178	−94.25046	−87.121

Day.	Arg. 6.	Log. 6.	Log. 6′.	Arg. 7.	Log. 7′.	Log. 7″.	Arg. 8.	Log. 8′.	Log. 8″.
0	7.2514	+94.3374	+87.597	8.9079	−94.1786	−87.121	6.1678	−93.8625	−86.792
100000	31.7296	94.3228	87.579	6.8667	94.1610	87.103	3.9547	93.8478	86.774
200000	21.3537	94.3074	87.559	4.8305	94.1483	87.083	1.7440	93.8322	86.754
300000	10.9701	94.2910	87.539	2.7996	94.1319	87.063	11.9232	93.8155	86.734
400000	0.5795	94.2737	87.518	0.7740	94.1143	87.042	12.7176	93.7980	86.713
500000	25.0277	94.2553	87.496	8.3675	94.0956	87.020	10.5116	93.7793	86.691
600000	14.6213	94.2355	87.472	6.3528	94.0757	86.996	8.3143	93.7594	86.667
700000	4.2067	94.2144	87.447	4.3138	94.0544	86.971	6.1168	93.7381	86.642
800000	28.6308	94.1918	87.421	2.3105	94.0318	86.945	3.9220	93.7152	86.616
900000	18.1996	94.1674	87.393	0.3430	94.0071	86.917	1.7300	93.6907	86.588
1000000	7.7598	94.1412	87.363	7.9653	93.9807	86.887	14.9283	93.6641	86.558
1100000	32.1583	94.1126	87.331	5.9799	93.9525	86.855	12.7421	93.6351	86.526
1200000	21.7009	94.0813	87.296	4.0006	93.9216	86.820	10.5590	93.6010	86.491
1300000	11.2345	94.0470	87.258	2.0277	93.8872	86.782	8.3788	93.5695	86.453
1400000	0.7588	94.0092	87.217	9.6749	93.8492	86.741	6.2018	93.5315	86.412
1500000	25.1207	93.9670	87.171	7.7149	93.8067	86.695	4.0279	93.4890	86.366
1600000	14.6262	93.9193	87.120	5.7616	93.7590	86.644	1.8572	93.4412	86.315
1700000	4.1219	93.8647	87.062	3.8149	93.7043	86.586	15.0770	93.3861	86.257
1800000	28.4546	93.8011	86.995	1.8751	93.6407	86.519	12.9129	93.3227	86.190
1900000	17.9305	93.7254	86.916	9.5560	93.5648	86.440	10.7521	93.2467	86.111
2000000	7.3961	93.6318	86.819	7.6302	93.4711	86.343	8.5946	93.1530	86.014
2100000	31.6982	93.5102	86.694	5.7116	93.3494	86.218	6.4407	93.0313	85.889
2200000	21.1429	93.3375	86.518	3.8003	93.1764	86.042	4.2902	92.8583	85.713
2300000	10.5769	+93.0397	+86.217	1.8964	−92.8786	−85.741	2.1433	−92.5604	−85.412

TABLE IV.

Constants of Epochs and Arguments for every Hundred Thousandth Day to 2300000.

Day.	Arg. 9.	Log. 9′.	Log. 9″.	Arg. 10.	Log. 10′.	Log. 10″.	Arg. 11.	Log. 11′.	Log. 11″.
0	25.6590	−94.5146	−87.461	13.4695	−94.4192	−87.396	13.0788	+94.4033	+87.348
100000	7.0583	94.5001	87.443	11.7147	94.4315	87.378	26.0861	94.3886	87.330
200000	18.2714	94.4846	87.423	9.9693	94.4190	87.358	12.2071	94.3732	87.310
300000	29.4957	94.4680	87.403	8.2335	94.4025	87.338	25.1975	94.3567	87.290
400000	10.9287	94.4506	87.382	6.5076	94.3850	87.317	11.3007	94.3393	87.269
500000	22.1762	94.4320	87.360	4.7917	94.3661	87.295	21.2732	94.3206	87.247
600000	3.6328	94.4123	87.336	3.0862	94.3467	87.271	10.3581	94.3009	87.223
700000	14.9044	94.3911	87.311	1.3911	94.3254	87.246	23.3119	94.2797	87.198
800000	26.1885	94.3683	87.285	25.3284	94.3027	87.220	9.3778	94.2569	87.172
900000	7.6823	94.3438	87.257	23.6518	94.2781	87.192	22.3122	94.2321	87.144
1000000	18.9920	94.3175	87.227	21.9924	94.2519	87.162	8.3583	94.2061	87.114
1100000	0.5119	94.2887	87.195	20.3413	94.2232	87.130	21.2726	94.1772	87.082
1200000	11.8181	94.2575	87.160	18.7017	94.1919	87.095	7.2982	94.1457	87.047
1300000	23.1979	94.2230	87.122	17.0738	94.1575	87.057	20.1915	94.1114	87.009
1400000	4.7587	94.1853	87.081	15.4579	94.1200	87.016	6.1959	94.0731	86.968
1500000	16.1365	94.1428	87.035	13.8541	94.0772	86.970	19.0676	94.0311	86.922
1600000	27.5286	94.0950	86.981	12.2626	94.0295	86.919	5.0499	93.9833	86.871
1700000	9.1325	94.0405	86.926	10.6837	93.9748	86.861	17.8993	93.9286	86.813
1800000	20.5511	93.9768	86.859	9.1175	93.9112	86.794	3.8589	93.8649	86.746
1900000	2.1879	93.9009	86.780	7.5643	93.8351	86.715	16.6851	93.7892	86.667
2000000	13.6398	93.8073	86.683	6.0243	93.7117	86.618	2.6212	93.6951	86.570
2100000	25.4073	93.6855	86.558	4.4976	93.6199	86.493	15.4235	93.5737	86.445
2200000	6.7878	93.5127	86.382	2.9846	93.4471	86.317	1.3351	93.4009	86.269
2300000	18.2872	−93.2119	−86.081	1.4853	−93.1492	−86.016	14.1131	+93.1030	+85.968

Day.	Arg. 12.	Log. 12′.	Log. 12″.	Arg. 13.	Log. 13′.	Arg. 14.	Log. 14.	Arg. 15.	Log. 15.
0	337.3723	+94.7409	+87.655	2.7557	+93.279	25.527	+94.346	0.927	−93.906
100000	161.6955	94.7259	87.637	3.9081	93.261	3.816	94.328	6.772	93.887
200000	338.6200	94.7101	87.617	5.0600	93.241	11.363	94.308	12.618	93.865
300000	165.9052	94.6934	87.597	6.2114	93.221	18.904	94.288	4.276	93.848
400000	339.7910	94.6757	87.576	7.3623	93.200	26.438	94.267	10.128	93.827
500000	167.0368	94.6568	87.554	8.5127	93.178	4.703	94.245	1.790	93.805
600000	340.8823	94.6368	87.530	9.6625	93.154	12.224	94.221	7.646	93.781
700000	168.0871	94.6153	87.505	0.7273	93.129	19.738	94.196	13.505	93.756
800000	341.8909	94.5921	87.479	1.8760	93.103	27.246	94.170	5.175	93.729
900000	169.0532	94.5677	87.451	3.0242	93.075	5.483	94.142	11.039	93.702
1000000	342.8138	94.5410	87.421	4.1718	93.045	12.977	94.112	2.714	93.672
1100000	169.9321	94.5122	87.389	5.3188	93.013	20.463	94.080	8.583	93.640
1200000	343.6179	94.4806	87.354	6.4652	92.978	27.942	94.045	0.263	93.605
1300000	170.7207	94.4461	87.316	7.6110	92.940	6.149	94.007	6.138	93.567
1400000	344.3903	94.4079	87.275	8.7561	92.899	13.613	93.966	12.016	93.526
1500000	171.4161	94.3653	87.229	9.9006	92.853	21.068	93.920	3.704	93.480
1600000	345.0379	94.3173	87.178	0.9598	92.802	28.515	93.869	9.588	93.429
1700000	172.0151	94.2625	87.120	2.1030	92.744	6.691	93.811	1.283	93.372
1800000	345.5877	94.1987	87.053	3.2454	92.677	14.122	93.744	7.172	93.304
1900000	172.5149	94.1225	86.974	4.3871	92.598	21.545	93.665	13.061	93.225
2000000	346.0367	94.0287	86.877	5.5281	92.501	28.959	93.568	4.768	93.128
2100000	172.9123	93.9069	86.752	6.6684	92.376	7.101	93.443	10.667	93.003
2200000	346.3818	93.7338	86.576	7.8079	92.200	14.498	93.267	2.378	92.827
2300000	173.2014	+93.4358	+86.275	8.9466	+91.899	21.885	+92.966	8.283	−92.526

TABLE IV.

Constants of Epochs and Arguments for every Hundred Thousandth Day to 2300000.

Day.	Arg. **16.**	Log. **16'.**	Arg. **17.**	Log. **17'.**	Arg. **18.**	Log. **18'.**	Arg. **19.**	Log. **19'.**
0	7.980	−94.361	19.088	−93.889	5.893	−94.274	20.567	−94.8382
100000	0.435	94.343	21.679	93.871	5.989	94.256	16.366	94.8236
200000	7.151	94.323	24.273	93.851	6.090	94.236	12.188	94.8080
300000	13.873	94.303	26.869	93.831	6.197	94.216	8.033	94.7916
400000	6.318	94.282	2.145	93.810	6.308	94.195	3.902	94.7742
500000	13.084	94.260	4.745	93.788	6.426	94.173	24.097	94.7556
600000	5.572	94.236	7.348	93.764	6.549	94.149	20.016	94.7358
700000	12.322	94.211	9.953	93.739	6.678	94.121	15.960	94.7146
800000	4.825	94.185	12.560	93.713	6.814	94.098	11.930	94.6919
900000	11.589	94.157	15.170	93.685	6.955	94.070	7.926	94.6675
1000000	4.107	94.127	17.782	93.655	7.102	94.040	3.949	94.6412
1100000	10.887	94.095	20.397	93.623	7.256	94.008	0.000	94.6123
1200000	3.420	94.060	23.014	93.588	7.415	93.973	20.381	94.5811
1300000	10.215	94.022	25.634	93.550	7.582	93.935	16.489	94.5467
1400000	2.761	93.981	0.935	93.509	7.755	93.894	12.625	94.5089
1500000	9.575	93.935	3.560	93.463	7.935	93.848	8.791	94.4665
1600000	2.411	93.884	6.189	93.412	8.121	93.797	4.987	94.4188
1700000	8.969	93.826	8.820	93.354	8.315	93.739	1.213	94.3642
1800000	1.552	93.759	11.451	93.287	8.516	93.672	21.773	94.3007
1900000	8.398	93.680	14.091	93.208	8.723	93.593	18.062	94.2243
2000000	0.998	93.583	16.731	93.111	8.938	93.496	14.384	94.1309
2100000	7.862	93.458	19.374	92.986	9.161	93.371	10.738	94.0094
2200000	0.480	93.282	22.020	92.810	9.391	93.195	7.125	93.8365
2300000	7.363	−92.981	24.669	−92.509	9.628	−92.894	3.545	−93.5387

Day.	Arg. **20.**	Log. **20'.**	Arg. **21.**	Log. **21'.**	Arg. **22.**	Log. **22'.**	Arg. **23.**	Log. **23'.**
0	0.771	−94.580	129.77	−95.438	1.111	−93.961	3.326	−94.352
100000	8.353	94.562	57.51	95.420	18.432	93.943	7.067	94.334
200000	1.628	94.542	117.00	95.400	3.628	93.923	10.814	94.314
300000	9.232	94.522	44.90	95.380	20.955	93.903	14.567	94.294
400000	2.529	94.501	104.55	95.359	6.157	93.882	18.327	94.273
500000	10.155	94.479	32.61	95.337	23.489	93.860	22.093	94.251
600000	3.475	94.455	92.42	95.313	8.696	93.836	2.092	94.227
700000	11.124	94.430	20.64	95.288	26.035	93.811	5.872	94.202
800000	4.467	94.404	80.62	95.262	11.248	93.785	9.659	94.176
900000	12.140	94.376	9.02	95.234	28.592	93.757	13.454	94.148
1000000	5.508	94.346	69.18	95.204	13.811	93.727	17.255	94.118
1100000	13.207	94.314	129.42	95.172	31.162	93.695	21.061	94.086
1200000	6.600	94.279	58.09	95.137	16.387	93.660	1.106	94.051
1300000	0.007	94.241	118.53	95.099	1.616	93.622	4.931	94.013
1400000	7.745	94.200	47.39	95.058	18.976	93.581	8.763	93.972
1500000	1.178	94.154	108.02	95.012	4.212	93.535	12.603	93.926
1600000	8.943	94.103	37.08	94.961	21.579	93.484	16.452	93.875
1700000	2.405	94.045	97.94	94.903	6.821	93.426	20.308	93.817
1800000	10.198	93.978	27.17	94.836	24.194	93.359	0.399	93.750
1900000	3.688	93.899	88.20	94.757	9.414	93.280	4.273	93.671
2000000	11.511	93.802	17.67	94.660	26.824	93.183	8.155	93.574
2100000	5.031	93.677	78.93	94.535	12.081	93.058	12.046	93.449
2200000	12.883	93.501	8.61	94.359	29.469	92.882	15.947	93.273
2300000	6.434	−93.200	70.09	−94.058	14.733	−92.581	19.856	−92.972

TABLE IV.

Constants of Epochs and Arguments for every Hundred Thousandth Day to 2300000.

Day.	Arg. 24.	Log. 24′.	Arg. 25.	Log. 25′.	Arg. 26.	Log. 26′.	Arg. 27.	Log. 27′.
0	9.555	−94.517	2.427	−94.248	3.825	−94.156	0.41	−94.788
100000	10.685	94.529	7.724	94.230	0.159	94.138	0.20	94.770
200000	11.825	94.509	3.660	94.210	2.919	94.118	25.80	94.750
300000	12.975	94.489	8.967	94.190	5.384	94.098	25.59	94.730
400000	0.859	94.468	4.913	94.169	2.030	94.077	25.40	94.709
500000	2.030	94.446	0.863	94.147	4.502	94.055	25.23	94.687
600000	3.211	94.422	6.186	94.123	1.157	94.031	25.08	94.663
700000	4.401	94.397	2.148	94.098	3.638	94.006	24.94	94.638
800000	5.607	94.371	7.482	94.072	0.302	93.980	24.82	94.612
900000	6.822	94.343	3.451	94.044	2.792	93.952	24.72	94.584
1000000	8.049	94.313	8.800	94.014	5.287	93.922	24.65	94.554
1100000	9.287	94.281	4.781	93.982	1.965	93.890	24.59	94.522
1200000	10.538	94.246	0.771	93.947	4.169	93.855	24.55	94.487
1300000	11.800	94.208	6.138	93.909	1.156	93.817	24.53	94.449
1400000	13.075	94.167	2.140	93.868	3.671	93.776	24.54	94.408
1500000	1.086	94.121	7.516	93.822	0.368	93.730	24.56	94.362
1600000	2.386	94.070	3.532	93.771	2.893	93.679	24.61	94.311
1700000	3.700	94.012	8.921	93.713	5.423	93.621	24.68	94.253
1800000	5.026	93.945	4.949	93.646	2.136	93.551	24.77	94.186
1900000	6.367	93.866	0.984	93.567	4.677	93.475	24.89	94.107
2000000	7.720	93.769	6.394	93.470	1.401	93.378	25.03	94.010
2100000	9.088	93.644	2.443	93.345	3.954	93.253	25.19	93.885
2200000	10.470	93.468	7.866	93.169	0.689	93.077	25.38	93.709
2300000	11.866	−93.167	3.929	−92.868	3.253	−92.776	25.59	−93.408

Day.	Arg. 28.	Log. 28′.	Arg. 29.	Log. 29′.	Arg. 30.	Log. 30′.	Arg. 31.	Log. 31′.
0	33.941	−94.465	8.071	−94.748	9.741	+93.294	13.80	+93.857
100000	31.030	94.447	12.988	94.730	2.861	93.276	10.70	93.839
200000	28.110	94.427	17.920	94.710	6.351	93.256	7.61	93.819
300000	25.182	94.407	22.869	94.690	9.841	93.236	4.51	93.799
400000	22.246	94.386	27.833	94.669	2.959	93.215	1.41	93.778
500000	19.301	94.364	0.051	94.647	6.417	93.193	13.62	93.756
600000	16.348	94.340	5.019	94.623	9.935	93.169	10.51	93.732
700000	13.385	94.315	10.061	94.598	3.051	93.144	7.40	93.707
800000	10.413	94.289	15.097	94.572	6.537	93.118	4.28	93.681
900000	7.432	94.261	20.148	94.544	10.023	93.090	1.17	93.653
1000000	4.442	94.231	25.217	94.514	3.138	93.060	13.36	93.623
1100000	1.442	94.199	30.304	94.482	6.622	93.028	10.24	93.591
1200000	36.951	94.164	2.617	94.447	10.106	92.993	7.11	93.556
1300000	33.934	94.126	7.773	94.409	3.218	92.955	3.98	93.518
1400000	30.904	94.085	12.919	94.368	6.700	92.914	0.85	93.477
1500000	27.863	94.039	18.084	94.322	10.182	92.868	13.03	93.431
1600000	24.812	93.988	23.270	94.271	3.292	92.817	9.90	93.380
1700000	21.750	93.930	28.476	94.213	6.772	92.759	6.75	93.322
1800000	18.677	93.863	0.940	94.146	10.252	92.692	3.61	93.255
1900000	15.593	93.784	6.189	94.067	3.360	92.613	0.46	93.176
2000000	12.498	93.687	11.459	93.970	6.839	92.516	12.63	93.079
2100000	9.391	93.562	16.751	93.845	10.315	92.391	9.48	92.954
2200000	6.272	93.386	22.066	93.669	3.420	92.215	6.32	92.778
2300000	3.142	−93.085	27.403	−93.368	6.896	+91.914	3.16	+92.477

TABLE IV.

Constants of Epochs and Arguments for every Hundred Thousandth Day to 2300000.

Day.	Arg. 32.	Log. 32′.	Arg. 33.	Log. 33′.	Arg. 34.	Log. 34′.	Arg. 35.	Log. 35′.
0	12.55	+94.405	17.07	−94.635	5.54	+94.314	20.52	−94.501
100000	15.58	94.387	4.04	94.617	6.52	94.296	13.56	94.483
200000	1.97	94.367	23.47	94.597	7.49	94.276	6.61	94.463
300000	4.99	94.347	10.46	94.577	8.45	94.256	23.62	94.443
400000	7.99	94.326	29.91	94.556	9.41	94.235	16.69	94.422
500000	11.00	94.304	16.92	94.534	10.37	94.213	9.76	94.400
600000	13.99	94.280	3.95	94.510	11.31	94.189	2.85	94.376
700000	0.35	94.255	23.43	94.485	12.25	94.164	19.89	94.351
800000	3.33	94.229	10.48	94.459	13.18	94.138	13.00	94.325
900000	6.31	94.201	30.00	94.431	14.11	94.110	6.12	94.297
1000000	9.27	94.171	17.08	94.401	15.03	94.080	23.19	94.267
1100000	12.23	94.139	4.17	94.369	15.94	94.048	16.33	94.235
1200000	15.18	94.104	23.72	94.331	16.84	94.013	9.48	94.200
1300000	1.49	94.066	10.84	94.296	17.74	93.975	2.64	94.162
1400000	4.43	94.025	30.43	94.255	18.63	93.934	19.75	94.121
1500000	7.35	93.979	17.58	94.209	19.51	93.888	12.93	94.075
1600000	10.27	93.928	4.74	94.158	20.38	93.837	6.13	94.024
1700000	13.18	93.870	24.37	94.100	21.25	93.779	23.27	93.966
1800000	16.08	93.803	11.57	94.033	22.11	93.712	16.49	93.899
1900000	2.33	93.724	31.23	93.954	22.96	93.633	9.72	93.820
2000000	5.21	93.627	18.16	93.857	23.80	93.536	2.96	93.723
2100000	8.08	93.502	5.71	93.732	24.64	93.411	20.16	93.598
2200000	10.94	93.326	25.42	93.556	25.46	93.235	13.43	93.422
2300000	13.79	+93.025	12.70	−93.255	26.28	+92.934	6.71	−93.121

Day.	Arg. 36.	Log. 36′.	Arg. 37.	Log. 37′.	Arg. 38.	Log. 38′.	Arg. 39.	Log. 39′.
0	12.32	−94.237	34.1	+95.024	8.25	−94.580	23.7	−94.79
100000	2.01	94.219	26.8	95.006	14.02	94.562	11.6	94.77
200000	5.41	94.199	19.4	94.986	1.02	94.542	33.8	94.75
300000	8.83	94.179	12.0	94.966	6.81	94.522	21.6	94.73
400000	12.25	94.158	4.6	94.945	12.62	94.501	9.3	94.71
500000	1.96	94.136	34.7	94.923	18.44	94.479	31.5	94.69
600000	5.39	94.112	27.3	94.899	5.48	94.455	19.2	94.66
700000	8.84	94.087	19.8	94.874	11.32	94.430	6.9	94.64
800000	12.29	94.061	12.2	94.848	17.18	94.404	29.0	94.61
900000	2.03	94.033	4.7	94.820	4.26	94.376	16.6	94.58
1000000	5.49	94.003	34.7	94.790	10.14	94.346	4.2	94.55
1100000	8.96	93.971	27.1	94.758	16.03	94.314	26.2	94.52
1200000	12.45	93.936	19.4	94.723	3.15	94.279	13.7	94.49
1300000	2.22	93.898	11.7	94.685	9.07	94.241	1.3	94.45
1400000	5.71	93.857	4.0	94.644	15.01	94.200	23.2	94.41
1500000	9.22	93.811	33.9	94.598	2.17	94.154	10.7	94.36
1600000	12.73	93.760	26.1	94.547	8.13	94.103	32.6	94.31
1700000	2.54	93.702	18.3	94.489	14.11	94.045	20.0	94.25
1800000	6.07	93.635	10.4	94.422	1.31	93.978	7.3	94.19
1900000	9.61	93.556	2.5	94.343	7.32	93.899	29.2	94.11
2000000	13.16	93.459	32.2	94.246	13.34	93.802	16.5	94.01
2100000	2.99	93.334	24.2	94.121	0.59	93.677	3.8	93.89
2200000	6.56	93.158	16.2	93.945	6.64	93.501	25.5	93.71
2300000	10.13	−92.857	8.1	+93.644	12.71	−93.200	12.8	−93.41

TABLE IV.

Constants of Epochs and Arguments for every Hundred Thousandth Day to 2300000.

Day.	Arg. 40.	Log. 40′.	Arg. 41.	Log. 41′.	Arg. 42.	Log. 42′.	Arg. 43.	Log. 43′.
0	3.48	+94.63	9.23	+94.61	2.57	−94.28	13.05	−94.37
100000	1.13	94.61	3.02	94.59	4.93	94.26	6.96	94.35
200000	17.62	94.59	14.71	94.57	7.29	94.24	0.88	94.33
300000	15.25	94.57	8.48	94.55	9.66	94.22	9.35	94.31
400000	12.86	94.55	2.23	94.53	1.88	94.20	3.28	94.29
500000	10.47	94.53	13.89	94.51	4.26	94.18	11.76	94.27
600000	8.06	94.51	7.62	94.49	6.65	94.16	5.71	94.25
700000	5.64	94.48	1.34	94.46	9.04	94.13	14.20	94.22
800000	3.21	94.45	12.97	94.43	1.29	94.10	8.17	94.19
900000	0.77	94.43	6.66	94.41	3.69	94.08	2.11	94.17
1000000	17.15	94.40	0.34	94.38	6.10	94.05	10.66	94.14
1100000	14.68	94.36	11.93	94.34	8.52	94.01	4.65	94.10
1200000	12.19	94.33	5.59	94.31	0.80	93.98	13.18	94.07
1300000	9.69	94.29	17.15	94.27	3.23	93.94	7.19	94.03
1400000	7.18	94.25	10.78	94.23	5.66	93.90	1.20	93.99
1500000	4.65	94.20	4.39	94.18	8.11	93.85	9.77	93.94
1600000	2.10	94.15	15.91	94.13	0.44	93.80	3.80	93.89
1700000	18.38	94.10	9.50	94.08	2.87	93.75	12.38	93.84
1800000	15.81	94.03	3.07	94.01	5.34	93.68	6.43	93.77
1900000	13.22	93.95	14.54	93.93	7.81	93.60	0.19	93.69
2000000	10.61	93.85	8.08	93.83	0.15	93.50	9.09	93.59
2100000	7.98	93.73	1.61	93.71	2.63	93.38	3.17	93.47
2200000	5.34	93.55	13.03	93.53	5.13	93.20	11.80	93.29
2300000	2.68	+93.25	6.53	+93.23	7.64	−92.90	5.89	−93.99

Day.	Arg. 44.	Arg. 45.	Arg. 46.	Arg. 47.	Arg. 48.	Log. 48′.	Arg. 49.	Arg. 50.
0	4.8	8.93	3.6	4.93	6.86	−94.66	5.5	19.8
100000	18.5	11.33	21.1	2.95	26.36	94.64	9.4	1.3
200000	9.4	13.71	15.7	0.97	15.95	94.62	0.5	7.9
300000	0.3	16.15	7.2	13.12	5.54	94.60	4.4	14.5
400000	14.1	1.67	27.7	11.13	25.09	94.58	8.3	21.1
500000	5.1	4.09	19.2	9.15	14.71	94.56	12.3	2.6
600000	18.9	6.51	10.7	7.17	4.35	94.54	3.5	9.2
700000	10.0	8.93	2.2	5.20	23.93	94.51	7.4	15.7
800000	1.1	11.36	22.6	3.22	13.59	94.48	11.4	22.3
900000	15.1	13.79	14.1	1.24	3.27	94.46	2.6	3.8
1000000	6.2	16.22	5.5	13.40	22.89	94.43	6.6	10.3
1100000	20.2	1.76	25.9	11.43	12.60	94.39	10.6	16.8
1200000	11.5	4.20	17.3	9.45	2.32	94.36	1.9	23.3
1300000	2.7	6.61	8.7	7.48	21.99	94.33	5.9	4.7
1400000	16.8	9.09	0.1	5.52	11.74	94.28	10.0	11.2
1500000	8.2	11.54	20.4	3.54	1.51	94.23	1.2	17.6
1600000	22.3	14.00	11.8	1.57	21.23	94.18	5.3	24.0
1700000	13.7	16.46	3.1	13.74	11.03	94.13	9.4	5.4
1800000	5.2	2.02	23.4	11.77	0.85	94.06	0.7	11.8
1900000	19.4	4.49	14.7	9.81	20.62	93.98	4.8	18.2
2000000	10.9	6.96	6.0	7.84	10.47	93.88	8.9	24.6
2100000	2.4	9.41	26.3	5.88	0.34	93.76	0.3	6.0
2200000	16.8	11.92	17.5	3.92	20.17	93.58	4.4	12.3
2300000	8.4	14.41	8.8	1.96	10.07	−93.28	8.6	18.7

TABLE IV.

Constants of Epochs and Arguments for every Hundred Thousandth Day to 2300000.

Day.	Arg. **51.**	Log. **51′.**	Arg. **52.**	Arg. **53.**	Log. **53′.**	Arg. **54.**	Arg. **55.**	Arg. **56.**
0	7.90	−94.13	7.92	32.88	+94.650	1.05	5.90	0.11
100000	15.10	94.11	16.01	7.18	94.632	12.27	4.85	5.56
200000	4.10	94.09	24.10	99.01	94.612	8.42	3.79	11.01
300000	11.32	94.07	6.96	73.29	94.592	4.58	2.74	3.35
400000	0.33	94.05	15.05	47.56	94.571	0.74	1.68	8.80
500000	7.56	94.03	23.14	21.82	94.549	11.97	0.62	1.13
600000	14.79	94.01	5.99	113.60	94.525	8.14	19.19	6.59
700000	3.81	93.98	14.08	87.82	94.500	4.31	18.13	12.05
800000	11.05	93.95	22.16	62.01	94.474	0.48	17.07	4.38
900000	18.29	93.93	5.02	36.24	94.446	11.72	16.01	9.84
1000000	7.32	93.90	13.10	10.43	94.416	7.90	14.95	2.18
1100000	14.58	93.86	21.18	102.11	94.384	4.09	13.89	7.64
1200000	3.62	93.83	4.03	76.30	94.349	0.27	12.82	13.10
1300000	10.88	93.79	12.11	50.14	94.311	11.53	11.76	5.45
1400000	18.15	93.75	20.19	24.57	94.270	7.73	10.69	10.91
1500000	7.20	93.70	3.04	116.22	94.224	3.93	9.63	3.25
1600000	14.48	93.65	11.11	90.31	94.173	0.13	8.56	8.72
1700000	3.54	93.60	19.19	64.39	94.115	11.40	7.49	1.07
1800000	10.83	93.53	2.03	38.45	94.048	7.61	6.43	6.53
1900000	18.12	93.45	10.11	12.50	93.969	3.82	5.36	12.00
2000000	7.20	93.35	18.18	104.07	93.872	0.04	4.29	4.35
2100000	14.50	93.23	1.02	78.08	93.747	11.32	3.22	9.82
2200000	3.58	93.05	9.09	52.07	93.571	7.55	2.15	2.17
2300000	10.90	−92.75	17.16	26.04	+93.270	3.77	1.07	7.63

Day.	Arg. **57.**	Arg. **58.**	Arg. **59.**	Arg. **60.**	Arg. **61.**	Arg. **62.**	Arg. **63.**	Arg. **64.**
0	0.44	6.20	23.11	10.63	9.27	1.09	10.05	32.81
100000	12.37	2.23	2.72	9.17	4.33	4.63	9.04	10.25
200000	7.31	13.48	21.29	7.72	13.01	8.17	8.03	23.68
300000	2.27	9.49	0.89	6.28	8.07	11.72	7.02	1.12
400000	14.21	5.49	19.45	4.84	3.11	0.82	6.01	14.58
500000	9.18	1.49	38.00	3.41	11.82	4.37	5.01	28.05
600000	4.15	12.72	17.59	1.99	6.89	7.92	4.01	5.55
700000	16.11	8.69	36.14	0.57	1.96	11.47	3.01	19.06
800000	11.10	4.66	15.72	21.49	10.64	0.58	2.01	32.59
900000	6.09	0.63	34.26	20.09	5.72	4.14	1.01	10.15
1000000	1.10	11.82	13.83	18.69	0.79	7.70	0.02	23.72
1100000	13.09	7.77	32.36	17.31	9.48	11.26	12.66	1.31
1200000	8.11	3.71	11.92	15.93	4.55	0.38	11.68	14.92
1300000	3.14	14.88	30.43	14.56	13.24	3.95	10.69	28.55
1400000	15.16	10.80	9.98	13.19	8.32	7.52	9.71	6.21
1500000	10.20	6.71	28.49	11.84	3.40	11.09	8.73	19.88
1600000	5.26	2.61	8.03	10.49	12.09	0.22	7.75	33.58
1700000	0.32	13.75	26.53	9.15	7.17	3.79	6.77	11.30
1800000	12.37	9.64	6.06	7.82	2.26	7.37	5.80	25.04
1900000	7.46	5.51	24.55	6.49	10.95	10.95	4.83	2.81
2000000	2.55	1.38	4.06	5.18	6.04	0.09	3.86	16.60
2100000	14.64	12.48	22.54	3.87	1.12	3.67	2.89	30.41
2200000	9.75	8.33	2.04	2.57	9.82	7.26	1.92	8.26
2300000	4.87	4.17	20.51	1.28	4.91	10.85	0.96	22.11

TABLE IV.

Constants of Epochs and Arguments for every Hundred Thousandth Day to 2300000.

Day.	Arg. 65.	Arg. 66.	Arg. 67.	Log. 67′.	Arg. 68.	Log. 68′.	Arg. 69.	Log. 69′.
0	6.08	15.4	3.97	−94.483	6.58	−94.630	151.6	−95.993
100000	13.62	26.5	0.26	94.465	18.84	94.612	102.3	95.975
200000	6.69	2.0	24.01	94.445	3.43	94.592	53.3	95.955
300000	14.24	13.2	20.32	94.425	15.71	94.572	4.6	95.935
400000	7.32	24.3	16.63	94.404	0.33	94.551	428.0	95.914
500000	0.41	35.5	12.97	94.382	12.63	94.529	379.8	95.892
600000	7.97	11.0	9.30	94.358	24.94	94.505	332.0	95.868
700000	1.06	22.2	5.65	94.333	9.60	94.480	284.1	95.843
800000	8.63	33.4	2.00	94.307	21.94	94.454	237.2	95.817
900000	1.73	9.0	25.81	94.279	6.62	94.426	190.2	95.789
1000000	9.30	20.2	22.19	94.249	18.99	94.396	143.6	95.759
1100000	2.41	31.4	18.57	94.217	3.70	94.364	97.3	95.727
1200000	10.00	7.0	14.96	94.182	16.09	94.329	51.1	95.692
1300000	3.11	18.2	11.37	94.144	0.83	94.291	5.8	95.654
1400000	10.70	29.5	7.78	94.103	13.25	94.250	432.4	95.613
1500000	3.83	5.1	4.21	94.057	25.69	94.204	387.5	95.567
1600000	11.43	16.4	0.65	94.006	10.47	94.153	342.9	95.516
1700000	4.56	27.6	21.51	93.948	22.91	94.095	298.7	95.458
1800000	12.16	3.3	21.00	93.881	7.75	94.028	254.9	95.391
1900000	5.30	14.6	17.47	93.802	20.25	93.949	211.4	95.312
2000000	12.92	25.9	13.95	93.705	5.10	93.852	168.4	95.215
2100000	6.06	1.6	10.44	93.580	17.63	93.727	125.7	95.090
2200000	13.69	12.9	6.95	93.404	2.52	93.551	83.4	94.914
2300000	6.84	21.3	3.47	−93.103	15.08	−93.250	41.5	−94.613

Day.	Arg. 70.	Log. 70′.	Arg. 71.	Arg. 72.	Arg. 73.	Log. 73′.	Arg. 74.	Log. 74′.
0	228.4	−95.096	500.3	85.6	320.3	+96.432	1633	+96.526
100000	300.9	95.078	65.8	364.6	1777.3	96.414	1443	96.508
200000	43.5	95.058	215.3	244.7	1043.1	96.394	1252	96.488
300000	116.2	95.038	364.7	124.8	308.2	96.374	1060	96.468
400000	188.8	95.017	514.2	4.9	1762.9	96.353	866	96.447
500000	261.4	94.995	79.8	283.8	1026.3	96.331	672	96.425
600000	4.3	94.971	229.2	163.9	289.0	96.307	477	96.401
700000	77.0	94.946	378.7	44.0	1741.2	96.282	281	96.376
800000	149.7	94.920	528.1	323.0	1002.1	96.256	84	96.350
900000	222.5	94.892	93.7	203.1	262.2	96.228	3119	96.322
1000000	295.3	94.862	243.2	83.2	1711.8	96.198	2919	96.292
1100000	38.4	94.830	392.6	362.2	970.1	96.166	2719	96.260
1200000	111.3	94.795	542.1	242.2	227.5	96.131	2517	96.225
1300000	184.2	94.757	107.6	122.3	1674.3	96.093	2314	96.187
1400000	257.2	94.716	257.1	2.4	929.8	96.052	2110	96.146
1500000	0.4	94.670	406.6	281.4	184.3	96.006	1905	96.100
1600000	73.4	94.619	556.0	161.5	1628.2	95.955	1698	96.049
1700000	146.5	94.561	121.0	41.6	880.8	95.897	1491	95.991
1800000	219.7	94.494	271.0	320.6	132.3	95.830	1282	95.924
1900000	292.9	94.415	420.5	200.7	1573.2	95.751	1071	95.845
2000000	36.3	94.318	570.0	80.7	822.6	95.654	860	95.748
2100000	109.6	94.193	135.5	359.7	71.0	95.529	647	95.623
2200000	183.0	94.017	285.0	239.8	1508.7	95.353	433	95.447
2300000	256.3	−93.716	434.5	119.9	751.9	+95.052	217	+95.146

TABLE IV.

Constants of Epochs and Arguments for every Hundred Thousandth Day to 2300000.

DAY.	ARG. 75.	ARG. 76.	ARG. 78.	LOG. 78′.	ARG. 79.	ARG. 80.	ARG. 81.	LOG. 81′.
0	57841	69646	150.587	−95.206	81.9	149.2	16.37	−95.025
100000	73091	75172	26.076	95.188	131.3	288.0	30.83	95.007
200000	3584	80662	89.831	95.168	180.7	113.6	45.32	94.987
300000	18831	86117	153.631	95.118	230.4	252.1	59.83	94.967
400000	34078	91534	29.276	95.127	280.2	77.4	74.38	94.946
500000	49325	1424	93.169	95.105	330.2	215.7	88.96	94.924
600000	64571	6766	157.111	95.081	380.3	40.8	103.57	94.900
700000	79818	12069	32.899	95.056	430.6	178.9	118.21	94.875
800000	10312	17332	96.910	95.030	42.8	3.7	8.67	94.849
900000	25558	22551	161.031	95.002	93.4	141.5	23.38	94.821
1000000	40805	27735	36.973	94.972	144.3	279.2	38.12	94.791
1100000	56052	32874	101.170	94.940	195.3	103.6	52.89	94.759
1200000	71299	37969	165.420	94.905	246.5	241.0	67.70	94.724
1300000	1792	43021	41.525	94.867	297.9	65.2	82.55	94.686
1400000	17039	48029	105.886	94.826	349.5	202.3	97.43	94.645
1500000	32286	52991	170.305	94.780	401.3	26.2	112.35	94.599
1600000	47532	57908	46.580	94.729	15.0	163.0	3.10	94.547
1700000	62779	62777	111.116	94.671	67.1	299.7	18.09	94.490
1800000	78026	67599	175.712	94.601	119.5	123.2	33.13	94.423
1900000	8519	72373	52.167	94.525	172.1	259.6	48.20	94.344
2000000	23766	77098	116.887	94.428	225.0	82.7	63.32	94.247
2100000	39013	81773	181.669	94.303	278.0	218.8	78.48	94.122
2200000	54260	86397	58.313	94.127	331.2	41.7	93.68	93.946
2300000	69506	90969	123.224	−93.826	384.7	177.4	108.92	−93.645

DAY.	ARG. 82.	ARG. 83.	ARG. 84.	ARG. 85.	ARG. 86.	ARG. 87.	ARG. 88.	ARG. 89.	ARG. 90.
0	8.720	28.58	9.80	6.63	28.32	8.97	12.82	1.48	12.4
100000	8.256	38.75	4.57	2.47	33.75	11.51	2.55	2.92	6.4
200000	7.801	9.72	18.50	21.82	0.87	11.05	6.95	4.37	0.5
300000	7.353	19.91	13.28	17.65	6.26	2.38	11.34	5.82	26.8
400000	6.913	30.10	8.06	13.48	11.61	4.93	1.06	7.29	20.9
500000	6.480	1.08	2.85	9.31	16.99	7.48	5.45	8.76	14.9
600000	6.056	11.29	16.78	5.13	22.33	10.01	9.83	10.24	9.0
700000	5.641	21.50	11.57	0.95	27.65	12.60	14.21	11.73	3.0
800000	5.233	31.71	6.37	20.28	32.95	0.95	3.91	13.23	29.4
900000	4.835	2.72	1.16	16.10	38.24	3.52	8.28	14.74	23.5
1000000	4.445	12.95	15.11	11.91	5.21	6.09	12.65	16.25	17.6
1100000	4.064	23.18	9.91	7.71	10.45	8.67	2.34	17.78	11.7
1200000	3.692	33.42	4.71	3.52	15.67	11.25	6.70	0.01	5.8
1300000	3.330	4.46	18.66	22.84	20.87	13.84	11.05	1.55	32.2
1400000	2.977	14.71	13.47	18.64	26.05	2.22	0.73	3.11	26.3
1500000	2.633	24.97	8.28	14.41	31.21	4.82	5.08	4.68	20.4
1600000	2.300	35.21	3.10	10.23	36.31	7.42	9.42	6.26	14.5
1700000	1.976	6.30	17.06	6.02	3.17	10.02	13.75	7.85	8.6
1800000	1.662	16.58	11.88	1.81	8.26	12.63	3.42	9.45	2.8
1900000	1.359	26.87	6.70	21.12	13.32	1.04	7.75	11.07	29.2
2000000	1.065	37.17	1.53	16.90	18.36	3.66	12.07	12.69	23.4
2100000	0.783	8.26	15.50	12.67	23.38	6.29	1.73	14.33	17.5
2200000	0.511	18.57	10.33	8.45	28.37	8.93	6.05	15.98	11.7
2300000	0.250	28.89	5.16	4.23	33.34	11.56	10.36	17.64	5.8

TABLE Va.

Argument 75.

Days.	Arg. 1.		2.		3.		6.	7.	8.	9.	10.	11.	13.	Days.
	Correction of Arguments													
d.	0.000	Diff.	0.00	Diff.	0.00	Diff.	0.00	0.00	0.00	0.00	0.00	0.00	0.00	d.
0	0834	186	0119	21	0084	20	01	01	01	00	01	01	00	100000
1000	0648	160	0098	18	0064	17	01	01	01	00	01	01	00	99000
2000	0488	135	0080	16	0047	15	01	01	01	00	01	01	00	98000
3000	0353	109	0064	13	0032	12	00	00	00	00	00	00	00	97000
4000	0244	83	0051	10	0020	9	00	00	00	00	00	00	00	96000
5000	0161	56	0041	7	0011	6	00	00	00	00	00	00	00	95000
6000	0105	30	0034	4	0005	3	00	00	00	00	00	00	00	94000
7000	0075	4	0030	0	0002	0	00	00	00	00	00	00	00	93000
8000	0071	22	0030	3	0002	2	00	00	00	00	00	00	00	92000
9000	0093	48	0033	6	0004	5	00	00	00	00	00	00	00	91000
10000	0141	77	0039	9	0009	8	00	00	00	00	00	00	00	90000
11000	0218	103	0048	12	0017	11	00	00	00	00	00	00	00	89000
12000	0321	129	0060	15	0028	14	00	00	00	00	00	00	00	88000
13000	0450	155	0075	18	0042	17	01	00	01	00	01	01	00	87000
14000	0605	181	0093	21	0059	19	01	00	01	00	01	01	00	86000
15000	0786	202	0114	23	0078	21	01	00	01	00	01	01	00	85000
16000	0988	223	0137	26	0099	23	01	00	01	01	01	01	00	84000
17000	1211	244	0163	28	0122	26	01	00	01	01	01	01	00	83000
18000	1455	265	0191	30	0148	29	02	01	01	01	01	02	01	82000
19000	1720	282	0224	33	0177	32	02	01	01	02	02	02	01	81000
20000	2002	298	0254	35	0209	33	02	01	01	02	02	02	01	80000
21000	2300	313	0289	36	0242	34	02	01	01	02	02	02	01	79000
22000	2613	326	0325	38	0276	35	03	01	01	03	03	03	01	78000
23000	2939	337	0363	39	0311	35	03	01	02	03	03	03	01	77000
24000	3276	345	0402	39	0346	36	03	01	02	04	03	03	01	76000
25000	3621	355	0441	40	0382	37	04	01	02	04	04	04	01	75000
26000	3979	360	0481	41	0419	38	04	01	02	04	04	04	01	74000
27000	4339	363	0522	42	0457	39	05	01	02	04	04	04	02	73000
28000	4702	365	0564	43	0496	40	05	02	03	05	05	05	02	72000
29000	5067	367	0607	43	0536	40	06	02	03	05	05	05	02	71000
30000	5431	365	0650	42	0576	39	07	02	03	05	05	05	02	70000
31000	5799	360	0692	41	0615	39	07	02	03	05	06	06	02	69000
32000	6159	352	0733	40	0654	38	08	02	03	06	06	06	02	68000
33000	6511	343	0773	40	0692	37	08	02	04	06	06	06	02	67000
34000	6854	333	0813	39	0729	35	09	02	04	07	07	07	03	66000
35000	7187	321	0852	38	0764	34	09	02	04	07	07	07	03	65000
36000	7508	307	0890	36	0798	32	09	02	04	07	07	07	03	64000
37000	7815	291	0926	34	0830	31	10	02	04	08	07	07	03	63000
38000	8106	274	0960	31	0861	30	10	03	05	08	08	08	03	62000
39000	8380	257	0991	29	0891	29	11	03	05	09	08	08	03	61000
40000	8637	238	1020	27	0920	27	11	03	05	09	08	08	03	60000
41000	8875	217	1047	24	0947	24	11	03	05	09	08	08	03	59000
42000	9092	194	1071	22	0971	21	11	03	05	09	08	08	03	58000
43000	9286	170	1093	20	0992	17	12	03	06	09	09	09	04	57000
44000	9456	144	1113	18	1009	14	12	04	06	10	09	09	04	56000
45000	9600	118	1131	15	1023	12	12	04	06	10	09	09	04	55000
46000	9718	93	1146	11	1035	9	12	04	06	10	09	09	04	54000
47000	9811	67	1157	7	1044	7	12	04	06	10	09	09	04	53000
48000	9878	41	1164	4	1051	5	12	04	06	10	10	10	04	52000
49000	9919	14	1168	1	1056	3	12	04	06	10	10	10	04	51000
50000	9933		1169		1059		12	04	06	10	10	10	04	50000

TABLE V*b*.

ARGUMENT 76.

Days.	ARG. 1.		2.		3.		6.	7.	8.	9.	10.	11.	13.	Days.
	Correction of Arguments													
d	0.00	Diff.	0.00	Diff.	0.00	Diff.	0.00	0.00	0.00	0.00	0.00	0.00	0.00	d.
0	07832	41	0145	5	0067	5	00	07	06	01	00	01	08	100000
1000	07791	16	0140	2	0062	2	00	07	06	01	00	01	08	99000
2000	07775	8	0138	1	0060	1	00	07	06	01	00	01	08	98000
3000	07783	33	0139	4	0061	3	00	07	06	01	00	01	08	97000
4000	07816	58	0143	6	0064	6	00	07	06	01	00	01	08	96000
5000	07874	82	0149	9	0070	9	00	07	06	01	00	01	08	95000
6000	07956	105	0158	12	0079	11	00	07	06	01	00	01	08	94000
7000	08061	127	0170	15	0090	14	00	07	06	01	00	01	08	93000
8000	08188	149	0185	18	0104	16	01	07	06	02	00	01	08	92000
9000	08337	171	0203	21	0120	19	01	07	06	02	00	01	08	91000
10000	08508	194	0224	23	0139	21	01	07	06	02	00	01	08	90000
11000	08702	216	0247	25	0160	24	01	07	06	02	00	01	08	89000
12000	08918	239	0272	27	0184	26	01	07	06	02	00	01	08	88000
13000	09157	261	0299	30	0210	28	02	08	07	03	01	02	09	87000
14000	09418	281	0329	32	0238	29	02	08	07	03	01	02	09	86000
15000	09699	298	0361	34	0267	31	02	08	07	03	01	02	09	85000
16000	09997	312	0395	35	0298	33	02	08	07	03	01	02	09	84000
17000	10309	322	0430	37	0331	35	02	08	07	04	02	03	09	83000
18000	10631	332	0467	39	0366	36	03	08	08	04	02	03	09	82000
19000	10963	340	0506	40	0402	37	03	08	08	05	02	03	09	81000
20000	11303	351	0546	41	0439	38	04	08	08	05	03	04	09	80000
21000	11654	362	0587	42	0477	39	04	08	08	05	03	04	09	79000
22000	12016	372	0629	43	0516	40	05	08	08	06	03	04	10	78000
23000	12388	381	0672	44	0556	41	05	09	09	06	04	05	10	77000
24000	12769	391	0716	44	0597	41	06	09	09	07	04	05	10	76000
25000	13160	390	0760	44	0638	41	06	09	09	07	04	05	10	75000
26000	13550	388	0804	44	0679	41	07	09	09	07	04	05	10	74000
27000	13938	382	0848	44	0720	41	07	09	09	08	05	06	10	73000
28000	14320	374	0892	44	0761	41	08	10	10	08	05	06	10	72000
29000	14694	366	0936	44	0802	40	08	10	10	09	06	07	11	71000
30000	15060	360	0980	43	0842	39	09	10	10	09	06	07	11	70000
31000	15420	355	1023	42	0881	38	09	10	10	09	06	07	11	69000
32000	15775	349	1065	40	0919	37	10	10	10	09	07	08	11	68000
33000	16124	343	1105	39	0956	36	10	10	11	10	07	08	11	67000
34000	16467	336	1144	37	0992	36	11	10	11	10	07	08	11	66000
35000	16803	321	1181	36	1028	34	11	10	11	10	08	09	11	65000
36000	17124	302	1217	34	1062	32	11	10	11	10	08	09	11	64000
37000	17426	281	1251	32	1094	30	12	10	11	11	08	09	11	63000
38000	17707	259	1283	30	1124	28	12	11	12	11	09	10	12	62000
39000	17966	236	1313	29	1152	26	13	11	12	11	09	10	12	61000
40000	18202	218	1342	27	1178	24	13	11	12	12	09	10	12	60000
41000	18420	200	1369	24	1202	22	13	11	12	12	09	10	12	59000
42000	18620	182	1393	21	1224	20	13	11	12	12	09	10	12	58000
43000	18802	162	1414	18	1244	17	14	11	12	13	10	11	12	57000
44000	18964	142	1432	15	1261	15	14	11	12	13	10	11	12	56000
45000	19106	117	1447	12	1276	12	14	11	12	13	10	11	12	55000
46000	19223	88	1459	10	1288	9	14	11	12	13	10	11	12	54000
47000	19311	60	1469	7	1297	6	14	11	12	13	10	11	12	53000
48000	19371	35	1476	5	1303	4	14	11	12	13	10	11	12	52000
49000	19406	10	1481	2	1307	1	14	11	12	13	10	11	12	51000
50000	19416		1483		1308		14	11	12	13	10	11	12	50000

LONGITUDE TABLES.

EQUATIONS.

TABLES VI.-LXXXII.

SECOND AND FOURTH DIFFERENCES

FOR A QUARTER OF A DAY.

TABLES VI''.-XXXV''., VI.$^{IV.}$-XXI.$^{IV.}$

TABLE VI. ARGUMENT 1.

Equation $= 22655''.226 + 22639''.2 \sin. x + 769''.5 \sin. 2x + 36''.7 \sin. 3x + 2''.0 \sin. 4x + 0''.1 \sin. 5x.$

Period, 27.55455245 days.

Days.	0		1		2		3		4	
Decimals of a Day.	Equation.	Log. Dif.	Equation.	Log. Dif.	Equation.	Log. Dif.	Equation.	Log. Dif.	Equation.	Log. Dif.
.00	60″.87	−0.5682	274″.41	+0.9117	1684″.78	+1.3038	4268″.29	+1.4981	7924″.37	+1.6172
.01	57.17	0.5539	282.60	0.9180	1704.91	1.3065	4299.78	1.4995	7965.79	1.6182
.02	53.59	0.5403	290.88	0.9243	1725.16	1.3090	4331.37	1.5011	8007.31	1.6192
.03	50.12	0.5250	299.28	0.9304	1745.53	1.3113	4363.07	1.5025	8048.92	1.6201
.04	46.77	0.5092	307.80	0.9365	1766.01	1.3139	4394.88	1.5041	8090.62	1.6209
.05	43.51	0.4942	316.44	0.9425	1786.61	1.3164	4426.80	1.5055	8132.40	1.6219
.06	40.42	0.4771	325.20	0.9484	1807.33	1.3187	4458.83	1.5069	8174.27	1.6228
.07	37.42	0.4594	334.08	0.9542	1828.16	1.3212	4490.96	1.5084	8216.23	1.6237
.08	34.54	0.4425	343.08	0.9600	1849.11	1.3236	4523.20	1.5099	8258.28	1.6247
.09	31.77	0.4232	352.20	0.9657	1870.18	1.3261	4555.55	1.5112	8300.42	1.6256
.10	29.12	0.4018	361.44	0.9713	1891.37	1.3284	4588.00	1.5127	8342.65	1.6264
.11	26.58	0.3838	370.80	0.9768	1912.67	1.3308	4620.56	1.5140	8384.96	1.6274
.12	24.16	0.3617	380.28	0.9827	1934.09	1.3332	4653.22	1.5154	8427.36	1.6283
.13	21.86	0.3385	389.89	0.9884	1955.63	1.3357	4685.99	1.5170	8469.85	1.6291
.14	19.68	0.3160	399.62	0.9930	1977.29	1.3379	4718.87	1.5183	8512.42	1.6300
.15	17.61	0.2900	409.46	0.9983	1999.06	1.3403	4751.85	1.5197	8555.08	1.6309
.16	15.66	0.2625	419.42	1.0033	2020.95	1.3426	4784.94	1.5210	8597.83	1.6318
.17	13.83	0.2355	429.50	1.0090	2042.96	1.3450	4818.13	1.5224	8640.67	1.6327
.18	12.11	0.2041	439.71	1.0140	2065.09	1.3472	4851.43	1.5238	8683.59	1.6335
.19	10.54	0.1703	450.04	1.0191	2087.33	1.3495	4884.84	1.5251	8726.59	1.6344
.20	9.03	0.1335	460.19	1.0237	2109.69	1.3516	4918.35	1.5266	8769.68	1.6353
.21	7.67	0.0969	471.05	1.0286	2132.16	1.3539	4951.97	1.5278	8812.86	1.6361
.22	6.42	0.0531	481.73	1.0338	2154.75	1.3562	4985.69	1.5292	8856.12	1.6370
.23	5.29	0.0043	492.54	1.0386	2177.46	1.3583	5019.51	1.5306	8899.47	1.6378
.24	4.28	9.9512	503.47	1.0433	2200.28	1.3606	5053.44	1.5319	8942.90	1.6386
.25	3.38	9.8921	514.52	1.0479	2223.22	1.3628	5087.47	1.5333	8986.41	1.6395
.26	2.60	9.8195	525.69	1.0526	2246.28	1.3649	5121.61	1.5345	9030.01	1.6403
.27	1.94	9.7321	536.98	1.0569	2269.45	1.3672	5155.85	1.5358	9073.69	1.6412
.28	1.40	9.6232	548.38	1.0618	2292.74	1.3692	5190.19	1.5372	9117.46	1.6420
.29	0.98	9.4911	559.91	1.0664	2316.14	1.3715	5224.64	1.5384	9161.31	1.6428
.30	0.67	9.2788	571.56	1.0708	2339.66	1.3736	5259.19	1.5399	9205.24	1.6437
.31	0.48	−8.8451	583.33	1.0749	2363.30	1.3756	5293.85	1.5411	9249.26	1.6444
.32	0.41	+8.6990	595.22	1.0795	2387.05	1.3779	5328.61	1.5424	9293.36	1.6452
.33	0.46	9.2041	607.23	1.0838	2410.92	1.3799	5363.47	1.5437	9337.54	1.6460
.34	0.62	9.4172	619.36	1.0881	2434.90	1.3820	5398.44	1.5450	9381.80	1.6468
.35	0.90	9.6021	631.61	1.0923	2459.00	1.3842	5433.51	1.5462	9426.14	1.6476
.36	1.30	9.7160	643.98	1.0965	2483.22	1.3861	5468.68	1.5474	9470.56	1.6484
.37	1.82	9.8062	656.47	1.1007	2507.55	1.3881	5503.95	1.5487	9515.06	1.6493
.38	2.46	9.8751	669.08	1.1052	2531.99	1.3903	5539.32	1.5500	9559.65	1.6500
.39	3.21	9.9395	681.82	1.1089	2556.55	1.3923	5574.80	1.5512	9604.32	1.6508
.40	4.08	9.9956	694.67	1.1130	2581.23	1.3943	5610.38	1.5524	9649.07	1.6515
.41	5.07	0.0453	707.64	1.1168	2606.02	1.3962	5646.06	1.5537	9693.89	1.6522
.42	6.18	0.0899	720.73	1.1209	2630.92	1.3982	5681.84	1.5549	9738.79	1.6530
.43	7.41	0.1303	733.94	1.1249	2655.94	1.4004	5717.72	1.5561	9783.77	1.6539
.44	8.76	0.1644	747.27	1.1287	2681.08	1.4023	5753.70	1.5574	6828.84	1.6547
.45	10.22	0.1987	760.72	1.1326	2706.33	1.4041	5789.79	1.5586	9873.99	1.6553
.46	11.80	0.2304	774.29	1.1367	2731.69	1.4062	5825.98	1.5597	9919.21	1.6560
.47	13.50	0.2601	787.99	1.1405	2757.17	1.4080	5862.26	1.5610	9964.50	1.6568
.48	15.32	0.2878	801.81	1.1410	2782.76	1.4101	5898.65	1.5622	10009.87	1.6577
.49	17.26	0.3139	815.74	1.1476	2808.47	1.4119	5935.14	1.5634	10055.33	1.6584
.50	19.32	+0.3385	829.79	+1.1513	2834.29	+1.4138	5971.73	+1.5646	10100.87	+1.6591

TABLE VI. ARGUMENT 1.

Equation $= 22655''.226 + 22639''.2 \sin. x + 769''.5 \sin. 2x + 36''.7 \sin. 3x + 2''.0 \sin. 4x + 0''.1 \sin. 5x.$

Period, 27.55455245 days.

Days.	0		1		2		3		4	
Decimals of a Day.	Equation.	Log. Dif.	Equation.	Log. Dif.	Equation.	Log. Dif.	Equation.	Log. Dif.	Equation.	Log. Dif.
.50	19″.32	+0.3385	829″.79	+1.1513	2831″.29	+1.4138	5971.73	+1.5646	10100″.87	+1.6591
.51	21.59	0.3617	843.96	1.1550	2860.22	1.4158	6008.12	1.5656	10146.48	1.6598
.52	23.80	0.3820	858.25	1.1587	2886.27	1.4176	6045.20	1.5668	10192.17	1.6606
.53	26.21	0.4031	872.66	1.1626	2912.43	1.4194	6082.08	1.5681	10237.91	1.6614
.54	28.74	0.4232	887.20	1.1662	2938.70	1.4215	6119.07	1.5692	10283.79	1.6620
.55	31.39	0.4425	901.86	1.1694	2965.09	1.4232	6156.16	1.5703	10329.71	1.6627
.56	34.16	0.4609	9[illegible]6.63	1.1730	2991.53	1.425[illegible]	6[illegible]93.34	1.5714	[illegible]5.70	1.6634
.57	37.05	0.4786	931.52	1.1764	301[illegible].21	1.426[illegible]	6230.62	1.5726	[illegible].77	1.6642
.58	40.06	0.4955	946.53	1.1801	3044.93	1.4288	6268.00	1.5738	10467.92	1.6649
.59	43.19	0.5119	961.67	1.1833	3071.77	1.4306	6305.48	1.5750	10514.15	1.6658
.60	46.44	0.5276	976.92	1.1867	3098.72	1.4325	6343.06	1.5760	10560.46	1.6663
.61	49.84	0.5446	992.29	1.1900	3125.79	1.4343	6380.73	1.5771	10606.84	1.6670
.62	53.29	0.5563	1007.78	1.1934	3152.97	1.4360	6418.50	1.5782	10653.29	1.6676
.63	56.89	0.5705	1023.39	1.1967	3180.26	1.4378	6456.37	1.5794	10699.84	1.6684
.64	60.61	0.5855	1039.12	1.2000	3207.66	1.4396	6494.34	1.5806	10746.41	1.6691
.65	64.46	0.5988	1054.97	1.2034	3235.18	1.4414	6532.41	1.5817	10793.09	1.6698
.66	68.43	0.6107	1070.94	1.2065	3262.81	1.4433	6570.58	1.5826	10839.84	1.6704
.67	72.51	0.6232	1087.03	1.20[illegible]5	3290.56	1.4[illegible]	6608.83	1.5838	10886.66	1.6711
.68	76.71	0.6355	1103.23	1.2130	3318.42	1.4467	6647.18	1.5849	10933.55	1.6718
.69	81.03	0.6474	1119.56	1.2161	3346.39	1.4484	6685.63	1.5861	10980.52	1.6725
.70	85.47	0.6599	1136.01	1.2191	3374.47	1.45[illegible]	672[illegible]	1.5872	11027.56	1.6731
.71	90.04	0.6702	1152.57	1.2225	3402.66	1.4518	6762.83	1.5882	11074.67	1.6737
.72	94.72	0.6812	1169.26	1.2256	3430.96	1.4535	6801.57	1.5892	11121.85	1.6745
.73	99.52	0.6920	1186.07	1.2284	3459.37	1.4553	6840.40	1.5903	11169.44	1.6752
.74	104.44	0.7024	1202.99	1.2314	3487.90	1.4570	6879.33	1.5914	11216.44	1.6758
.75	109.48	0.7[illegible]26	1220.03	1.2347	3516.54	1.4587	69[illegible]	1.59[illegible]	[illegible]	1.6764
.76	114.61	0.7226	1237.20	1.2375	3545.29	1.4602	6957.48	1.5935	11311.31	1.6771
.77	119.92	0.7316	1254.48	1.2405	3574.14	1.4620	6996.70	1.5945	11358.85	1.6776
.78	125.31	0.7419	1271.8[illegible]	1.243[illegible]	[illegible]	1.4637	7036.01	1.5955	11406.45	1.6783
.79	130.83	0.7513	1289.11	1.2465	3632.20	1.4652	7075.41	1.5966	11454.13	1.6790
.80	136.47	0.7597	1307.05	1.2495	3661.39	1.4669	7114.91	1.5977	11501.88	1.6796
.81	142.22	0.7[illegible]	13[illegible]4.81	1.2524	3690.69	1.4684	7154.51	1.5[illegible]	11549.70	1.6802
.82	148.10	0.7782	1342.68	1.2553	3720.10	1.4701	7194.20	1.5997	11597.59	1.6808
.83	154.10	0.7868	1360.68	1.2582	3749.62	1.4719	7233.98	1.6007	11645.54	1.6814
.84	160.22	0.7945	1378.80	1.2610	3779.26	1.4735	7273.85	1.6017	11693.56	1.6820
.85	166.45	0.8035	1397.04	1.2637	3809.04	1.4749	7313.82	1.6027	11741.65	1.6828
.86	172.81	0.8116	1415.39	1.2664	3838.86	1.4766	7353.88	1.6037	11789.82	1.6833
.87	179.29	0.8195	1433.86	1.2693	3868.82	1.4781	7394.03	1.6047	11838.05	1.6838
.88	185.89	0.8274	1452.45	1.2720	3898.89	1.4799	7434.27	1.6057	11886.34	1.6844
.89	192.61	0.8351	1471.16	1.2749	3929.08	1.4814	7474.61	1.6067	11934.70	1.6851
.90	199.45	0.8420	1489.99	1.2777	3959.38	1.4829	7515.04	1.6077	11983.13	1.6857
.91	206.40	0.8500	1508.94	1.2801	3989.78	1.4844	7555.56	1.6086	12031.63	1.6862
.92	213.48	0.8573	1528.00	1.2828	4020.29	1.4860	7596.17	1.6097	12080.19	1.6869
.93	220.68	0.8645	1547.18	1.2858	4050.91	1.4875	7636.88	1.6107	12128.82	1.6875
.94	228.00	0.8716	1566.49	1.2882	4081.64	1.4892	7677.68	1.6116	12177.52	1.6880
.95	235.44	0.8785	1585.91	1.2909	4112.48	1.4907	7718.57	1.6126	12226.28	1.6887
.96	243.00	0.8854	1605.45	1.2936	4143.43	1.4921	7759.55	1.6135	12275.11	1.6892
.97	250.68	0.8921	1625.11	1.2961	4174.48	1.4936	7800.62	1.6145	12324.00	1.6897
.98	258.48	0.8987	1644.88	1.2987	4205.64	1.4952	7841.78	1.6154	12372.95	1.6904
.99	266.40	0.9053	1664.77	1.3012	4236.94	1.4966	7883.03	1.6164	12421.97	1.6910
1.00	274.44	+0.9117	1684.78	+1.3038	4268.29	+1.4981	7924.37	+1.6172	12471.06	+1.6915

TABLE VI. ARGUMENT 1.

Equation = 22655″.226 + 22639″.2 sin. x + 769″.5 sin. $2x$ + 36″.7 sin. $3x$ + 2″.0 sin. $4x$ + 0″.1 sin. $5x$.

Period, 27.55455245 days.

Days	5		6		7		8		9	
Decimals of a Day.	Equation.	Log. Dif.	Equation.	Log. Dif.	Equation.	Log. Dif.	Equation.	Log. Dif.	Equation.	Log. Dif.
.00	12471″.06	+1.6915	17650″.56	+1.7321	23146″.27	+1.7431	28610″.81	+1.7267	33699″.84	+1.6804
.01	12520.21	1.6921	17704.52	1.7323	23201.66	1.7433	28664.11	1.7264	33747.75	1.6797
.02	12569.42	1.6927	17758.51	1.7325	23257.01	1.7433	28717.37	1.7262	33795.58	1.6791
.03	12618.70	1.6932	17812.52	1.7328	23312.42	1.7433	28770.60	1.7258	33843.35	1.6785
.04	12668.04	1.6937	17866.57	1.7331	23367.80	1.7432	28823.79	1.7255	33891.05	1.6779
.05	12717.44	1.6942	17920.66	1.7333	23423.17	1.7432	28876.94	1.7251	33938.68	1.6772
.06	12766.90	1.6949	17974.77	1.7335	23478.54	1.7432	28930.04	1.7249	33986.24	1.6766
.07	12816.43	1.6951	18028.90	1.7338	23533.90	1.7431	28983.11	1.7245	34033.73	1.6760
.08	12866.02	1.6959	18083.07	1.7340	23589.25	1.7431	29036.11	1.7242	34081.15	1.6753
.09	12915.67	1.6966	18137.27	1.7342	23644.60	1.7431	29089.13	1.7239	34128.49	1.6745
.10	12965.39	1.6970	18191.50	1.7345	23699.95	1.7430	29142.08	1.7235	34175.76	1.6739
.11	13015.16	1.6975	18245.76	1.7346	23755.29	1.7430	29194.98	1.7231	34222.96	1.6733
.12	13064.99	1.6980	18300.04	1.7349	23810.62	1.7429	29247.81	1.7228	34270.09	1.6726
.13	13114.88	1.6986	18354.35	1.7351	23865.94	1.7428	29300.66	1.7225	34317.15	1.6720
.14	13164.84	1.6991	18408.69	1.7354	23921.25	1.7428	29353.44	1.7221	34364.14	1.6713
.15	13214.85	1.6996	18463.06	1.7355	23976.56	1.7427	29406.18	1.7217	34411.05	1.6706
.16	13264.92	1.7001	18517.45	1.7358	24031.86	1.7426	29458.87	1.7214	34457.89	1.6700
.17	13315.05	1.7006	18571.87	1.7360	24087.15	1.7425	29511.52	1.7211	34504.66	1.6692
.18	13365.24	1.7010	18626.32	1.7362	24142.43	1.7425	29564.13	1.7208	34551.35	1.6686
.19	13415.48	1.7017	18680.79	1.7363	24197.70	1.7424	29616.70	1.7204	34597.97	1.6679
.20	13465.79	1.7021	18735.28	1.7366	24252.96	1.7423	29669.22	1.7199	34644.51	1.6672
.21	13516.15	1.7026	18789.80	1.7368	24308.21	1.7422	29721.69	1.7195	34690.98	1.6665
.22	13566.57	1.7031	18844.35	1.7370	24363.45	1.7421	29774.12	1.7192	34737.38	1.6658
.23	13617.05	1.7036	18898.92	1.7372	24418.68	1.7421	29826.51	1.7188	34783.70	1.6651
.24	13667.58	1.7041	18953.52	1.7374	24473.89	1.7420	29878.85	1.7185	34829.95	1.6644
.25	13718.17	1.7046	19008.14	1.7375	24529.10	1.7419	29931.15	1.7181	34876.12	1.6636
.26	13768.82	1.7050	19062.78	1.7377	24584.30	1.7418	29983.40	1.7177	34922.21	1.6630
.27	13819.52	1.7055	19117.41	1.7379	24639.48	1.7417	30035.60	1.7172	34968.23	1.6622
.28	13870.28	1.7060	19172.13	1.7381	24694.65	1.7416	30087.75	1.7169	35014.17	1.6616
.29	13921.09	1.7065	19226.84	1.7382	24749.80	1.7416	30139.86	1.7166	35060.01	1.6608
.30	13971.96	1.7069	19281.57	1.7384	24804.95	1.7414	30191.93	1.7161	35105.83	1.6600
.31	14022.88	1.7074	19336.32	1.7386	24860.08	1.7413	30243.91	1.7158	35151.54	1.6593
.32	14073.86	1.7079	19391.09	1.7388	24915.19	1.7412	30295.91	1.7153	35197.17	1.6586
.33	14124.89	1.7084	19445.89	1.7390	24970.29	1.7410	30347.82	1.7149	35242.73	1.6579
.34	14175.98	1.7087	19500.71	1.7390	25025.38	1.7409	30399.69	1.7145	35288.21	1.6572
.35	14227.12	1.7092	19555.51	1.7392	25080.45	1.7408	30451.51	1.7140	35333.62	1.6563
.36	14278.31	1.7096	19610.39	1.7394	25135.50	1.7407	30503.28	1.7137	35378.94	1.6555
.37	14329.55	1.7101	19665.27	1.7396	25190.54	1.7406	30555.00	1.7132	35424.18	1.6549
.38	14380.85	1.7105	19720.17	1.7397	25245.56	1.7405	30606.67	1.7128	35469.35	1.6541
.39	14432.20	1.7111	19775.08	1.7398	25300.57	1.7404	30658.29	1.7124	35514.41	1.6533
.40	14483.61	1.7114	19830.01	1.7400	25355.57	1.7402	30709.86	1.7120	35559.45	1.6524
.41	14535.06	1.7118	19884.96	1.7402	25410.54	1.7400	30761.38	1.7116	35604.37	1.6518
.42	14586.56	1.7123	19939.93	1.7402	25465.49	1.7398	30812.85	1.7112	35649.22	1.6510
.43	14638.12	1.7127	19994.91	1.7405	25520.42	1.7398	30864.27	1.7107	35693.99	1.6502
.44	14689.73	1.7131	20049.92	1.7405	25575.31	1.7396	30915.64	1.7102	35738.68	1.6494
.45	14741.39	1.7135	20104.94	1.7406	25630.24	1.7394	30966.95	1.7098	35783.29	1.6486
.46	14793.09	1.7140	20159.97	1.7408	25685.12	1.7393	31018.21	1.7094	35827.81	1.6479
.47	14844.85	1.7143	20215.02	1.7409	25739.98	1.7391	31069.42	1.7088	35872.26	1.6470
.48	14896.65	1.7148	20270.09	1.7410	25794.82	1.7390	31120.57	1.7084	35916.62	1.6462
.49	14948.51	1.7153	20325.17	1.7412	25849.64	1.7388	31171.67	1.7080	35960.90	1.6454
.50	15000.42	+1.7156	20380.27	+1.7413	25904.44	+1.7386	31222.72	+1.7075	36005.10	+1.6446

TABLE VI. ARGUMENT 1.

Equation = 22655″.226 + 22639″.2 sin. x + 769″.5 sin. $2x$ + 36″.7 sin. $3x$ + 2″.0 sin. $4x$ + 0″.1 sin. $5x$.

Period, 27.55455245 days.

Days.	5		6		7		8		9	
Decimals of a Day.	Equation.	Log. Dif.	Equation.	Log. Dif.	Equation.	Log. Dif.	Equation.	Log. Dif.	Equation.	Log. Dif.
.50	15000″.42	+1.7156	20380″.27	+1.7413	25904″.44	+1.7386	31222″.72	+1.7075	36005″.10	+1.64[illegible]
.51	15052.37	1.7160	20435.38	1.7413	25959.22	1.7385	31273.71	1.7070	36049.22	1.6439
.52	15104.37	1.7164	20490.50	1.7415	26013.98	1.7383	31324.65	1.7066	36093.26	1.6430
.53	15156.42	1.7167	20545.64	1.7416	26068.72	1.7382	31375.53	1.7062	36137.21	1.6422
.54	15208.51	1.7171	20600.79	1.7417	26123.44	1.7379	31426.36	1.7057	36181.08	1.6414
.55	15260.65	1.7176	20655.96	1.7418	26178.13	1.7378	31477.14	1.7052	36224.87	1.6405
.56	15312.84	1.7180	20711.14	1.7419	26232.80	1.7376	31527.86	1.7047	36268.57	1.6397
.57	15365.08	1.7184	20766.33	1.7419	26287.45	1.7374	31578.52	1.7042	36312.19	1.6389
.58	15417.36	1.7187	20821.53	1.7420	26342.07	1.7372	31629.12	1.7037	36355.73	1.6380
.59	15469.69	1.7191	20876.74	1.7421	26396.67	1.7370	31679.67	1.7032	36399.18	1.6372
.60	15522.06	1.7195	20931.96	1.7422	26451.25	1.7368	31730.16	1.7027	36442.55	1.6363
.61	15574.48	1.7198	20987.19	1.7423	26505.80	1.7366	31780.60	1.7023	36485.83	1.6355
.62	15626.94	1.7203	21042.44	1.7425	26560.33	1.7364	31830.98	1.7018	36529.03	1.6346
.63	15679.45	1.7206	21097.71	1.7425	26614.83	1.7362	31881.30	1.7012	36572.14	1.6338
.64	15732.00	1.7210	21152.98	1.7425	26669.31	1.7360	31931.56	1.7007	36615.17	1.6329
.65	15784.60	1.7213	21208.26	1.7425	26723.76	1.7358	31981.76	1.7002	36658.11	1.6320
.66	15837.24	1.7217	21263.54	1.7427	26778.19	1.7356	32031.91	1.6998	36700.97	1.6311
.67	15889.92	1.7220	21318.84	1.7427	26832.59	1.7354	32082.00	1.6992	36743.74	1.6302
.68	15942.65	1.7224	21374.14	1.7428	26886.96	1.7352	32132.03	1.6987	36786.42	1.6294
.69	15995.42	1.7227	21429.45	1.7429	26941.31	1.7350	32181.99	1.6981	36829.02	1.6285
.70	16048.23	1.7230	21484.77	1.7429	26995.63	1.7347	32231.89	1.6976	36871.53	1.6276
.71	16101.08	1.7233	21540.10	1.7430	27049.92	1.7345	32281.74	1.6972	36913.95	1.6267
.72	16153.97	1.7238	21595.44	1.7430	27104.18	1.7343	32331.53	1.6967	36956.29	1.6258
.73	16206.91	1.7241	21650.78	1.7431	27158.42	1.7341	32381.26	1.6961	36998.54	1.6249
.74	16259.89	1.7245	21706.13	1.7431	27212.63	1.7339	32430.93	1.6955	37040.70	1.6240
.75	16312.91	1.7248	21761.48	1.7432	27266.81	1.7336	32480.53	1.6950	37082.78	1.6231
.76	16365.97	1.7250	21816.84	1.7432	27320.96	1.7334	32530.07	1.6943	37124.77	1.6222
.77	16419.06	1.7254	21872.20	1.7432	27375.08	1.7330	32579.54	1.6939	37166.67	1.6213
.78	16472.20	1.7257	21927.57	1.7433	27429.16	1.7329	32628.96	1.6933	37208.48	1.6203
.79	16525.38	1.7261	21982.95	1.7433	27483.22	1.7326	32678.32	1.6928	37250.20	1.6194
.80	16578.60	1.7263	22038.33	1.7433	27537.25	1.7324	32727.61	1.6923	37291.83	1.6184
.81	16631.85	1.7266	22093.71	1.7434	27591.25	1.7321	32776.84	1.6917	37333.37	1.6175
.82	16685.14	1.7269	22149.10	1.7434	27645.21	1.7318	32826.01	1.6911	37374.82	1.6166
.83	16738.47	1.7273	22204.49	1.7434	27699.14	1.7316	32875.11	1.6906	37416.18	1.6157
.84	16791.84	1.7275	22259.88	1.7435	27753.04	1.7314	32924.15	1.6899	37457.46	1.6147
.85	16845.24	1.7279	22315.28	1.7435	27806.91	1.7311	32973.12	1.6894	37498.65	1.6137
.86	16898.68	1.7282	22370.68	1.7435	27860.75	1.7308	33022.03	1.6888	37539.74	1.6128
.87	16952.16	1.7285	22426.08	1.7435	27914.55	1.7306	33070.88	1.6882	37580.74	1.6118
.88	17005.68	1.7288	22481.48	1.7435	27968.32	1.7303	33119.66	1.6876	37621.65	1.6109
.89	17059.23	1.7290	22536.88	1.7435	28022.06	1.7300	33168.37	1.6870	37662.47	1.6099
.90	17112.81	1.7294	22592.28	1.7436	28075.76	1.7298	33217.02	1.6864	37703.20	1.6089
.91	17166.43	1.7297	22647.69	1.7435	28129.43	1.7294	33265.60	1.6859	37743.84	1.6080
.92	17220.09	1.7299	22703.09	1.7436	28183.06	1.7291	33314.12	1.6852	37784.39	1.6069
.93	17273.78	1.7302	22758.50	1.7435	28236.65	1.7289	33362.57	1.6847	37824.84	1.6061
.94	17327.51	1.7305	22813.90	1.7435	28290.21	1.7286	33410.96	1.6841	37865.21	1.6050
.95	17381.27	1.7307	22869.30	1.7435	28343.74	1.7282	33459.28	1.6834	37905.48	1.6040
.96	17435.06	1.7310	22924.70	1.7435	28397.23	1.7279	33507.52	1.6828	37945.66	1.6031
.97	17488.89	1.7313	22980.10	1.7434	28450.68	1.7276	33555.70	1.6822	37985.75	1.6020
.98	17542.75	1.7315	23035.49	1.7434	28504.09	1.7273	33603.84	1.6816	38025.74	1.6010
.99	17596.64	1.7318	23090.88	1.7434	28557.47	1.7270	33651.86	1.6811	38065.64	1.6000
1.00	17650.56	+1.7322	23146.27	+1.7434	28610.81	+1.7267	33699.84	+1.6804	38105.45	+1.5989

TABLE VI. ARGUMENT 1.

Equation $= 22655''.226 + 22639''.2 \sin. x + 769''.5 \sin. 2x + 36''.7 \sin. 3x + 2''.0 \sin. 4x + 0''.1 \sin. 5x.$

Period, 27.55455245 days.

Days.	10		11		12		13		14	
Decimals of a Day.	Equation.	Log. Dif.	Equation.	Log. Dif.	Equation.	Log. Dif.	Equation.	Log. Dif.	Equation.	Log. Dif.
.00	38105″.45	+1.5988	41582″.33	+1.4689	43962″.88	+1.2531	45160″.73	+0.7716	45164″.90	−0.7679
.01	38145.16	1.5979	41611.77	1.4673	43980.79	1.2502	45166.64	0.7627	45159.04	0.7760
.02	38184.78	1.5968	41641.10	1.4657	43998.58	1.2475	45172.43	0.7536	45153.07	0.7839
.03	38224.30	1.5958	41670.32	1.4640	44016.26	1.2442	45178.10	0.7443	45146.99	0.7917
.04	38263.73	1.5949	41699.43	1.4625	44033.81	1.2412	45183.65	0.7348	45140.80	0.8000
.05	38303.07	1.5937	41728.44	1.4608	44051.24	1.2385	45189.08	0.7259	45134.49	0.8082
.06	38342.31	1.5927	41757.33	1.4590	44068.56	1.2355	45194.40	0.7152	45128.06	0.8156
.07	38381.46	1.5916	41786.10	1.4575	44085.76	1.2322	45199.59	0.7050	45121.52	0.8228
.08	38420.51	1.5906	41814.77	1.4557	44102.83	1.2295	45204.66	0.6946	45114.87	0.8306
.09	38459.47	1.5895	41843.33	1.4539	44119.79	1.2263	45209.61	0.6839	45108.10	0.8376
.10	38498.33	1.5884	41871.77	1.4523	44136.63	1.2232	45214.44	0.6739	45101.22	0.8445
.11	38537.09	1.5871	41900.10	1.4507	44153.35	1.2201	45219.16	0.6618	45094.23	0.8519
.12	38575.76	1.5861	41928.33	1.4489	44169.95	1.2169	45223.75	0.6503	45087.12	0.8585
.13	38614.34	1.5852	41956.44	1.4470	44186.43	1.2138	45228.22	0.6385	45079.90	0.8651
.14	38652.82	1.5841	41984.43	1.4451	44202.79	1.2106	45232.57	0.6274	45072.57	0.8722
.15	38691.20	1.5830	42012.32	1.4436	44219.03	1.2076	45236.81	0.6149	45065.12	0.8785
.16	38729.48	1.5819	42040.09	1.4418	44235.16	1.2044	45240.93	0.6010	45057.56	0.8848
.17	38767.67	1.5808	42067.75	1.4401	44251.16	1.2009	45244.92	0.5888	45049.89	0.8915
.18	38805.76	1.5798	42095.30	1.4384	44267.04	1.1976	45248.80	0.5752	45042.10	0.8976
.19	38843.76	1.5786	42122.74	1.4365	44282.80	1.1945	45252.56	0.5611	45034.20	0.9036
.20	38881.66	1.5775	42150.06	1.4348	44298.45	1.1909	45256.20	0.5465	45026.19	0.9096
.21	38919.46	1.5763	42177.27	1.4330	44313.97	1.1875	45259.72	0.5315	45018.07	0.9159
.22	38957.16	1.5752	42204.37	1.4313	44329.37	1.1840	45263.12	0.5159	45009.83	0.9217
.23	38994.76	1.5741	42231.36	1.4292	44344.65	1.1810	45266.40	0.4997	45001.48	0.9274
.24	39032.27	1.5730	42258.23	1.4275	44359.82	1.1775	45269.56	0.4829	44993.02	0.9335
.25	39069.68	1.5718	42284.99	1.4257	44374.87	1.1740	45272.60	0.4654	44984.44	0.9390
.26	39106.99	1.5706	42311.64	1.4237	44389.80	1.1703	45275.52	0.4472	44975.75	0.9445
.27	39144.20	1.5695	42338.17	1.4219	44404.60	1.1668	45278.32	0.4298	44966.95	0.9499
.28	39181.31	1.5683	42364.59	1.4200	44419.28	1.1632	45281.01	0.4099	44958.04	0.9552
.29	39218.32	1.5672	42390.89	1.4181	44433.84	1.1599	45283.58	0.3874	44949.02	0.9609
.30	39255.24	1.5659	42417.08	1.4163	44448.29	1.1562	45286.02	0.3674	44939.88	0.9661
.31	39292.05	1.5648	42443.16	1.4145	44462.62	1.1526	45288.35	0.3444	44930.63	0.9713
.32	39328.76	1.5637	42469.13	1.4125	44476.83	1.1486	45290.56	0.3201	44921.27	0.9763
.33	39365.38	1.5624	42494.98	1.4106	44490.94	1.1449	45292.65	0.2945	44911.80	0.9814
.34	39401.89	1.5613	42520.72	1.4085	44504.87	1.1415	45294.62	0.2672	44902.22	0.9863
.35	39438.31	1.5600	42546.31	1.4067	44518.72	1.1377	45296.47	0.2405	44892.53	0.9917
.36	39474.62	1.5588	42571.85	1.4046	44532.45	1.1338	45298.21	0.2068	44882.72	0.9961
.37	39510.83	1.5576	42597.24	1.4028	44546.06	1.1297	45299.82	0.1761	44872.81	1.0013
.38	39546.94	1.5564	42622.52	1.4009	44559.54	1.1258	45301.32	0.1399	44862.78	1.0060
.39	39582.95	1.5552	42647.69	1.3988	44572.90	1.1222	45302.70	0.1004	44852.64	1.0107
.40	39618.86	1.5540	42672.74	1.3967	44586.15	1.1179	45303.96	0.0569	44842.39	1.0154
.41	39654.67	1.5528	42697.67	1.3949	44599.27	1.1142	45305.10	0.0086	44832.03	1.0199
.42	39690.38	1.5515	42722.49	1.3929	44612.28	1.1099	45306.12	9.9590	44821.56	1.0245
.43	39725.99	1.5502	42747.20	1.3908	44625.16	1.1059	45307.03	9.8921	44810.98	1.0290
.44	39761.49	1.5490	42771.79	1.3888	44637.92	1.1021	45307.81	9.8261	44800.29	1.0334
.45	39796.89	1.5478	42796.27	1.3867	44650.57	1.0976	45308.48	9.7401	44789.49	1.0382
.46	39832.19	1.5465	42820.63	1.3845	44663.09	1.0931	45309.03	9.6335	44778.57	1.0422
.47	39867.39	1.5452	42844.87	1.3825	44675.49	1.0895	45309.46	9.5051	44767.55	1.0464
.48	39902.48	1.5439	42869.00	1.3806	44687.78	1.0849	45309.78	9.2788	44756.42	1.0507
.49	39937.47	1.5426	42893.02	1.3784	44699.94	1.0806	45309.97	8.9031	44745.18	1.0554
.50	39972.36	+1.5414	42916.92	+1.3762	44711.98	+1.0763	45310.05	+8.6021	44733.82	−1.0592

TABLE VI. ARGUMENT 1.

Equation $= 22655''.226 + 22639''.2 \sin. x + 769''.5 \sin. 2x + 36''.7 \sin. 3x + 2''.0 \sin. 4x + 0''.1 \sin. 5x.$

Period, 27.55455245 days.

Days.	10		11		12		13		14	
Decimals of a Day.	Equation.	Log. Dif.	Equation.	Log. Dif.	Equation.	Log. Dif.	Equation.	Log. Dif.	Equation.	Log. Dif.
.50	39972.36	+1.5414	42916.92	+1.3762	44711.98	+1.0763	45310.05	+8.6021	44733.82	−1.0592
.51	40007.15	1.5402	42940.70	1.3742	44723.90	1.0723	45310.01	−9.2041	44722.36	1.0633
.52	40041.84	1.5388	42964.37	1.3720	44735.71	1.0675	45309.85	9.4314	44710.79	1.0675
.53	40076.42	1.5376	42987.92	1.3699	44747.39	1.0630	45309.58	9.6021	44699.14	1.0716
.54	40110.90	1.5362	43011.36	1.3678	44758.95	1.0584	45309.18	9.7076	44687.32	1.0755
.55	40145.27	1.5349	43034.68	1.3655	44770.39	1.0539	45308.67	9.7993	44675.42	1.0795
.56	40179.51	1.5336	43057.88	1.3634	44781.71	1.0492	45308.04	9.8692	44663.41	1.0835
.57	40213.70	1.5323	43080.97	1.3612	44792.91	1.0445	45307.30	9.9345	44651.29	1.0874
.58	40247.76	1.5310	43103.94	1.3591	44803.99	1.0399	45306.44	9.9912	44639.06	1.0913
.59	40281.72	1.5297	43126.80	1.3568	44814.95	1.0351	45305.46	0.0414	44626.72	1.0948
.60	40315.58	1.5282	43149.54	1.3545	44825.79	1.0302	45304.36	0.0864	44614.28	1.0990
.61	40349.33	1.5268	43172.16	1.3522	44836.51	1.0253	45303.14	0.1239	44601.72	1.1025
.62	40382.97	1.5255	43194.66	1.3500	44847.11	1.0203	45301.81	0.1614	44589.06	1.1066
.63	40416.51	1.5241	43217.05	1.3479	44857.59	1.0149	45300.36	0.1959	44576.28	1.1100
.64	40449.94	1.5228	43239.33	1.3456	44867.94	1.0103	45298.79	0.2279	44563.40	1.1133
.65	40483.27	1.5214	43261.49	1.3430	44878.18	1.0051	45297.10	0.2553	44550.42	1.1173
.66	40516.49	1.5201	43283.52	1.3408	44888.30	0.9996	45295.30	0.2833	44537.32	1.1206
.67	40549.61	1.5186	43305.44	1.3387	44898.29	0.9948	45293.38	0.3096	44524.12	1.1245
.68	40582.62	1.5173	43327.25	1.3363	44908.17	0.9890	45291.34	0.3324	44510.80	1.1277
.69	40615.53	1.5159	43348.94	1.3338	44917.92	0.9844	45289.19	0.3560	44497.38	1.1309
.70	40648.33	1.5145	43370.51	1.3314	44927.56	0.9786	45286.92	0.3766	44483.86	1.1348
.71	40681.03	1.5131	43391.96	1.3292	44937.08	0.9727	45284.54	0.3979	44470.22	1.1380
.72	40713.62	1.5116	43413.30	1.3267	44946.47	0.9675	45282.04	0.4183	44456.48	1.1415
.73	40746.10	1.5101	43434.52	1.3243	44955.75	0.9614	45279.42	0.4378	44442.63	1.1449
.74	40778.47	1.5088	43455.62	1.3220	44964.90	0.9562	45276.68	0.4548	44428.67	1.1479
.75	40810.74	1.5073	43476.61	1.3196	44973.94	0.9499	45273.83	0.4728	44414.61	1.1513
.76	40842.90	1.5058	43497.48	1.3171	44982.85	0.9445	45270.86	0.4886	44400.44	1.1547
.77	40874.95	1.5045	43518.23	1.3145	44991.65	0.9380	45267.78	0.5051	44386.16	1.1580
.78	40906.90	1.5030	43538.86	1.3120	45000.32	0.9325	45264.58	0.5211	44371.77	1.1611
.79	40938.74	1.5015	43559.37	1.3096	45008.88	0.9258	45261.26	0.5353	44357.28	1.1644
.80	40970.47	1.5001	43579.77	1.3071	45017.31	0.9196	45257.83	0.5502	44342.68	1.1676
.81	41002.10	1.4984	43600.05	1.3045	45025.62	0.9133	45254.28	0.5647	44327.97	1.1706
.82	41033.61	1.4970	43620.21	1.3018	45033.81	0.9069	45250.61	0.5775	44313.16	1.1738
.83	41065.02	1.4955	43640.25	1.2993	45041.88	0.9009	45246.83	0.5899	44298.24	1.1767
.84	41096.32	1.4940	43660.17	1.2969	45049.84	0.8938	45242.94	0.6042	44283.22	1.1798
.85	41127.51	1.4925	43679.98	1.2943	45057.67	0.8871	45238.92	0.6160	44268.09	1.1826
.86	41158.59	1.4911	43699.67	1.2916	45065.38	0.8802	45234.79	0.6274	44252.86	1.1858
.87	41189.57	1.4896	43719.24	1.2889	45072.97	0.8733	45230.55	0.6395	44237.52	1.1889
.88	41220.44	1.4880	43738.69	1.2863	45080.44	0.8663	45226.19	0.6503	44222.07	1.1920
.89	41251.20	1.4864	43758.02	1.2837	45087.79	0.8591	45221.72	0.6618	44206.52	1.1948
.90	41281.85	1.4849	43777.24	1.2810	45095.02	0.8525	45217.13	0.6730	44190.86	1.1976
.91	41312.39	1.4833	43796.34	1.2784	45102.14	0.8454	45212.42	0.6830	44175.10	1.2007
.92	41342.82	1.4817	43815.32	1.2756	45109.14	0.8370	45207.60	0.6937	44159.23	1.2034
.93	41373.14	1.4801	43834.18	1.2727	45116.01	0.8287	45202.66	0.7033	44143.26	1.2062
.94	41403.35	1.4787	43852.92	1.2700	45122.75	0.8215	45197.61	0.7126	44127.18	1.2089
.95	41433.46	1.4770	43871.54	1.2674	45129.38	0.8136	45192.45	0.7226	44111.00	1.2119
.96	41463.45	1.4754	43890.05	1.2644	45135.89	0.8055	45187.17	0.7324	44094.71	1.2146
.97	41493.33	1.4740	43908.43	1.2617	45142.28	0.7973	45181.77	0.7412	44078.32	1.2175
.98	41523.11	1.4722	43926.70	1.2589	45148.55	0.7889	45176.26	0.7497	44061.82	1.2201
.99	41552.77	1.4707	43944.85	1.2560	45154.70	0.7803	45170.64	0.7589	44045.22	1.2227
1.00	41582.33	+1.4689	43962.88	+1.2531	45160.73	+0.7716	45164.90	−0.7679	44028.52	−1.2256

TABLE VI. ARGUMENT 1.

Equation = 22655″.226 + 22639″.2 sin. x + 769″.5 sin. $2x$ + 36″.7 sin. $3x$ + 2″.0 sin. $4x$ + 0″.1 sin. $5x$.

Period, 27.55455245 days.

Days.	15		16		17		18		19	
Decimals of a Day	Equation.	Log. Dif.	Equation.	Log. Dif.	Equation.	Log. Dif.	Equation.	Log. Dif.	Equation.	Log. Dif.
.00	44028″.52	−1.2256	41855″.39	−1.4235	38787″.01	−1.5398	34991″.50	−1.6129	30654″.69	−1.6575
.01	44011.71	1.2282	41828.87	1.4251	38752.35	1.5406	34950.49	1.6135	30609.25	1.6579
.02	43994.80	1.2309	41802.26	1.4264	38717.63	1.5416	34909.42	1.6140	30563.77	1.6581
.03	43977.78	1.2335	41775.57	1.4278	38682.83	1.5425	34868.30	1.6145	30518.26	1.6585
.04	43960.66	1.2360	41748.79	1.4292	38647.96	1.5433	34827.13	1.6151	30472.71	1.6588
.05	43943.41	1.2385	41721.92	1.4308	38613.02	1.5443	34785.91	1.6157	30427.13	1.6591
.06	43926.12	1.2412	41694.96	1.4322	38578.00	1.5452	34744.63	1.6163	30381.52	1.6594
.07	43908.69	1.2437	41667.91	1.4336	38542.91	1.5459	34703.30	1.6168	30335.88	1.6598
.08	43891.16	1.2462	41640.77	1.4351	38507.76	1.5469	34661.91	1.6174	30290.20	1.6600
.09	43873.53	1.2490	41613.54	1.4364	38472.53	1.5478	34620.48	1.6179	30244.49	1.6603
.10	43855.79	1.2514	41586.23	1.4378	38437.23	1.5488	34578.99	1.6184	30198.75	1.6607
.11	43837.95	1.2541	41558.83	1.4393	38401.85	1.5495	34537.45	1.6189	30152.97	1.6610
.12	43820.00	1.2565	41531.33	1.4406	38366.41	1.5503	34495.86	1.6196	30107.16	1.6613
.13	43801.95	1.2589	41503.75	1.4418	38330.90	1.5512	34454.21	1.6201	30061.32	1.6616
.14	43783.80	1.2613	41476.09	1.4433	38295.32	1.5521	34412.51	1.6206	30015.45	1.6618
.15	43765.55	1.2637	41448.31	1.4446	38259.66	1.5531	34370.76	1.6212	29969.55	1.6621
.16	43747.20	1.2660	41420.50	1.4461	38223.93	1.5538	34328.96	1.6217	29923.62	1.6625
.17	43728.75	1.2686	41392.57	1.4474	38188.11	1.5548	34287.11	1.6222	29877.65	1.6628
.18	43710.19	1.2709	41364.56	1.4487	38152.27	1.5555	34245.24	1.6227	29831.65	1.6631
.19	43691.53	1.2732	41336.46	1.4501	38116.31	1.5563	34203.26	1.6232	29785.62	1.6634
.20	43672.77	1.2756	41308.27	1.4514	38080.31	1.5571	34161.26	1.6238	29739.56	1.6637
.21	43653.91	1.2779	41279.99	1.4527	38044.27	1.5581	34119.20	1.6243	29693.47	1.6639
.22	43634.95	1.2801	41251.63	1.4541	38008.12	1.5588	34077.10	1.6248	29647.35	1.6642
.23	43615.89	1.2823	41223.18	1.4553	37971.91	1.5595	34034.95	1.6253	29601.20	1.6645
.24	43596.73	1.2849	41194.65	1.4567	37935.61	1.5605	33992.75	1.6259	29555.02	1.6647
.25	43577.46	1.2872	41166.03	1.4581	37899.29	1.5613	33950.49	1.6263	29508.81	1.6650
.26	43558.09	1.2894	41137.32	1.4593	37862.87	1.5622	33908.19	1.6268	29462.57	1.6653
.27	43538.62	1.2913	41108.53	1.4605	37826.38	1.5629	33865.81	1.6274	29416.30	1.6656
.28	43519.06	1.2939	41079.66	1.4618	37789.83	1.5637	33823.44	1.6279	29370.00	1.6658
.29	43499.39	1.2961	41050.70	1.4631	37753.21	1.5645	33780.99	1.6284	29323.68	1.6661
.30	43479.62	1.2980	41021.65	1.4643	37716.53	1.5651	33738.49	1.6289	29277.33	1.6664
.31	43459.76	1.3004	40992.52	1.4657	37679.77	1.5660	33695.94	1.6293	29230.94	1.6666
.32	43439.79	1.3025	40963.30	1.4669	37642.95	1.5669	33653.35	1.6299	29184.53	1.6669
.33	43419.72	1.3047	40934.00	1.4682	37606.06	1.5677	33610.71	1.6303	29138.09	1.6672
.34	43399.55	1.3067	40904.61	1.4693	37569.10	1.5684	33568.02	1.6308	29091.62	1.6675
.35	43379.29	1.3090	40875.14	1.4706	37532.08	1.5692	33525.28	1.6312	29045.12	1.6677
.36	43358.92	1.3111	40845.59	1.4719	37494.99	1.5701	33482.50	1.6318	28998.60	1.6680
.37	43338.45	1.3131	40815.95	1.4731	37457.83	1.5707	33439.66	1.6321	28952.05	1.6682
.38	43317.89	1.3152	40786.23	1.4742	37420.61	1.5715	33396.79	1.6328	28905.47	1.6684
.39	43297.23	1.3173	40756.43	1.4756	37383.32	1.5723	33353.86	1.6332	28858.87	1.6687
.40	43276.47	1.3196	40726.54	1.4767	37345.97	1.5731	33310.89	1.6337	28812.24	1.6689
.41	43255.60	1.3214	40696.57	1.4780	37308.55	1.5739	33267.87	1.6342	28765.59	1.6691
.42	43234.64	1.3232	40666.51	1.4792	37271.06	1.5746	33224.80	1.6346	28718.91	1.6694
.43	43213.59	1.3255	40636.37	1.4803	37233.51	1.5754	33181.69	1.6351	28672.20	1.6697
.44	43192.43	1.3273	40606.15	1.4814	37195.89	1.5761	33138.53	1.6355	28625.46	1.6699
.45	43171.18	1.3296	40575.85	1.4827	37158.21	1.5768	33095.33	1.6360	28578.70	1.6700
.46	43149.82	1.3314	40545.46	1.4838	37120.47	1.5775	33052.08	1.6364	28531.92	1.6703
.47	43128.37	1.3332	40515.00	1.4850	37082.67	1.5783	33008.79	1.6369	28485.11	1.6706
.48	43106.83	1.3355	40484.45	1.4861	37044.80	1.5791	32965.45	1.6373	28438.27	1.6708
.49	43085.18	1.3373	40453.82	1.4872	37006.86	1.5798	32922.07	1.6378	28391.41	1.6709
.50	43063.44	−1.3393	40423.11	−1.4886	36968.86	−1.5806	32878.64	−1.6382	28344.53	−1.6713

TABLE VI. ARGUMENT 1.

Equation = $22655''.225 + 22639''.2 \sin. x + 769''.5 \sin. 2x + 36''.7 \sin. 3x + 2''.0 \sin. 4x + 0''.1 \sin. 5x.$

Period, 27.55455245 days.

Days.	15		16		17		18		19	
Decimals of a Day.	Equation.	Log. Dif.	Equation.	Log. Dif.	Equation.	Log. Dif.	Equation.	Log. Dif.	Equation.	Log. Dif.
.50	43063.44	−1.3393	40423.11	−1.4886	36968.86	−1.5806	32878.61	−1.6382	28311.53	−1.6713
.51	43041.60	1.3412	40392.31	1.4897	36930.79	1.5812	32835.17	1.6387	28297.62	1.6715
.52	43019.66	1.3430	40361.43	1.4909	36892.66	1.5819	32791.65	1.6391	28250.69	1.6718
.53	42997.63	1.3450	403 0.47	1.4920	36851.47	1.5827	32718.09	1.6396	28203.73	1.6719
.54	42975.50	1.3470	40299.43	1.4931	36816.22	1.5834	32704.48	1.6100	28156.75	1.6721
.55	42953.27	1.3487	40268.31	1.4942	36777.90	1.5841	32660.83	1.6404	28109.75	1.6724
.56	42930.95	1.3506	40237.11	1.4953	36739.52	1.5848	32617.11	1.6408	28062.72	1.6725
.57	42908.53	1.3526	40205.83	1.4963	36701.08	1.5856	32573.41	1.6413	28015.67	1.6727
.58	42886.01	1.3543	40174.47	1.4975	36662.57	1.5863	32529.63	1.6417	27968.60	1.6729
.59	42863.40	1.3560	40143.03	1.4985	36624.00	1.5869	32485.84	1.6421	27921.51	1.6732
.60	42840.70	1.3581	40111.51	1.4997	36585.37	1.5876	32441.95	1.6426	27874.39	1.6731
.61	42817.89	1.3598	40079.91	1.5008	36546.68	1.5883	32398.01	1.6430	27827.25	1.6735
.62	42794.99	1.3615	40048.23	1.5019	36507.93	1.5889	32354.09	1.6434	27780.09	1.6738
.63	42772.00	1.3631	40016.47	1.5030	36469.12	1.5897	32310.10	1.6438	27732.90	1.6740
.64	42748.91	1.3653	39984.63	1.5041	36430.24	1.5904	32266.07	1.6442	27685.69	1.6742
.65	42725.72	1.3670	39952.71	1.5051	36391.30	1.5911	32221.99	1.6446	27638.46	1.6744
.66	42702.44	1.3689	39920.71	1.5062	36352.30	1.5918	32177.88	1.6450	27591.21	1.6745
.67	42679.06	1.3705	39888.63	1.5073	36313.24	1.5924	32133.72	1.6454	27543.94	1.6747
.68	42655.59	1.3722	39856.47	1.5083	36274.12	1.5931	32089.52	1.6458	27496.65	1.6750
.69	42632.03	1.3740	39824.24	1.5093	36234.94	1.5937	32045.28	1.6462	27449.34	1.6752
.70	42608.37	1.3758	39791.93	1.5101	36195.70	1.5944	32001.00	1.6466	27402.01	1.6751
.71	42584.61	1.3775	39759.51	1.5114	36156.40	1.5951	31956.68	1.6470	27354.65	1.6756
.72	42560.76	1.3791	39727.07	1.5126	36117.04	1.5956	31912.32	1.6474	27307.27	1.6757
.73	42536.82	1.3807	39694.52	1.5135	36077.63	1.5964	31867.92	1.6478	27259.88	1.6759
.74	42512.79	1.3825	39661.90	1.5145	36038.15	1.5970	31823.48	1.6482	27212.47	1.6761
.75	42488.66	1.3843	39629.20	1.5156	35998.61	1.5976	31779.00	1.6486	27165.04	1.6763
.76	42464.43	1.3860	39596.42	1.5167	35959.02	1.5984	31734.48	1.6490	27117.58	1.6764
.77	42440.11	1.3876	39563.56	1.5176	35919.36	1.5990	31689.92	1.6493	27070.11	1.6766
.78	42415.70	1.3892	39530.63	1.5186	35879.64	1.5996	31645.32	1.6497	27022.62	1.6768
.79	42391.20	1.3908	39497.62	1.5197	35839.87	1.6002	31600.68	1.6500	26975.11	1.6770
.80	42366.61	1.3925	39464.53	1.5206	35800.04	1.6009	31556.01	1.6504	26927.58	1.6771
.81	42341.92	1.3943	39431.37	1.5216	35760.15	1.6015	31511.30	1.6508	26880.04	1.6773
.82	42317.13	1.3957	39398.13	1.5225	35720.20	1.6021	31466.55	1.6512	26832.47	1.6775
.83	42292.26	1.3974	39364.82	1.5236	35680.20	1.6027	31421.76	1.6516	26784.89	1.6776
.84	42267.29	1.3989	39331.43	1.5246	35640.14	1.6033	31376.93	1.6520	26737.29	1.6777
.85	42242.23	1.4006	39297.96	1.5255	35600.02	1.6040	31332.06	1.6522	26689.68	1.6780
.86	42217.08	1.4021	39264.42	1.5264	35559.84	1.6045	31287.16	1.6526	26642.04	1.6781
.87	42191.84	1.4037	39230.81	1.5275	35519.61	1.6052	31242.22	1.6530	26594.39	1.6782
.88	42166.51	1.4053	39197.12	1.5285	35479.32	1.6057	31197.24	1.6534	26546.72	1.6783
.89	42141.08	1.4068	39163.35	1.5294	35438.98	1.6061	31152.22	1.6537	26499.04	1.6785
.90	42115.56	1.4081	39129.51	1.5305	35398.58	1.6070	31107.17	1.6541	26451.34	1.6787
.91	42089.95	1.4099	39095.59	1.5314	35358.12	1.6077	31062.08	1.6544	26403.62	1.6789
.92	42064.25	1.4114	39061.60	1.5321	35317.60	1.6083	31016.96	1.6548	26355.88	1.6790
.93	42038.46	1.4130	39027.53	1.5333	35277.03	1.6087	30971.80	1.6551	26308.13	1.6791
.94	42012.58	1.4145	38993.39	1.5341	35236.41	1.6091	30926.60	1.6554	26260.37	1.6792
.95	41986.61	1.4160	38959.18	1.5351	35195.73	1.6100	30881.37	1.6558	26212.59	1.6794
.96	41960.55	1.4176	38924.89	1.5361	35154.99	1.6106	30836.10	1.6561	26164.79	1.6795
.97	41934.39	1.4191	38890.53	1.5370	35114.20	1.6112	30790.80	1.6565	26116.98	1.6796
.98	41908.11	1.4205	38856.09	1.5379	35073.35	1.6117	30745.46	1.6568	26069.16	1.6798
.99	41881.81	1.4219	38821.59	1.5388	35032.45	1.6122	30700.09	1.6571	26021.32	1.6799
1.00	41855.39	−1.4235	38787.01	−1.5398	34991.50	−1.6129	30654.69	−1.6575	25973.47	−1.6800

TABLE VI. ARGUMENT 1.

Equation = 22655″.226 + 22639″.2 sin. x + 769″.5 sin. $2x$ + 36″.7 sin. $3x$ + 2″.0 sin. $4x$ + 0″.1 sin. $5x$.

Period, 27.55155245 days.

Days.	20		21		22		23		24	
Decimals of a Day.	Equation.	Log. Dif.	Equation.	Log. Dif.	Equation.	Log. Dif.	Equation.	Log. Dif.	Equation.	Log. Dif.
.00	25973″.47	−1.6800	21150″.89	−1.6831	16392″.29	−1.6682	11901″.99	−1.6321	7879″.71	−1.5705
.01	25925.60	1.6801	21102.61	1.6831	16345.72	1.6679	11859.12	1.6316	7842.51	1.5698
.02	25877.72	1.6803	21054.40	1.6833	16299.18	1.6675	11816.30	1.6312	7805.37	1.5690
.03	25829.82	1.6804	21006.17	1.6832	16252.67	1.6673	11773.53	1.6307	7768.30	1.5682
.04	25781.91	1.6805	20957.95	1.6832	16206.18	1.6671	11730.80	1.6301	7731.30	1.5676
.05	25733.99	1.6806	20909.73	1.6832	16159.72	1.6669	11688.13	1.6297	7694.36	1.5666
.06	25686.06	1.6807	20861.51	1.6830	16113.28	1.6666	11645.50	1.6292	7657.49	1.5659
.07	25638.11	1.6808	20813.31	1.6829	16066.87	1.6663	11602.92	1.6287	7620.68	1.5652
.08	25590.15	1.6810	20765.12	1.6828	16020.49	1.6660	11560.39	1.6283	7583.94	1.5643
.09	25542.18	1.6812	20716.94	1.6828	15974.15	1.6658	11517.90	1.6277	7547.27	1.5635
.10	25494.19	1.6813	20668.76	1.6827	15927.83	1.6654	11475.47	1.6271	7510.67	1.5627
.11	25446.19	1.6813	20620.59	1.6826	15881.54	1.6652	11433.09	1.6266	7474.14	1.5619
.12	25398.19	1.6814	20572.43	1.6825	15835.28	1.6650	11390.76	1.6261	7437.67	1.5611
.13	25350.17	1.6815	20524.28	1.6825	15789.04	1.6646	11348.48	1.6257	7401.27	1.5602
.14	25302.14	1.6815	20476.14	1.6825	15742.84	1.6644	11306.24	1.6252	7364.95	1.5594
.15	25254.09	1.6816	20428.00	1.6824	15696.67	1.6641	11264.05	1.6247	7328.69	1.5586
.16	25206.04	1.6817	20379.87	1.6823	15650.53	1.6638	11221.91	1.6241	7292.50	1.5579
.17	25157.97	1.6818	20331.75	1.6821	15604.42	1.6635	11179.82	1.6235	7256.37	1.5570
.18	25109.90	1.6819	20283.65	1.6820	15558.34	1.6633	11137.79	1.6231	7220.31	1.5561
.19	25061.81	1.6820	20235.56	1.6819	15512.29	1.6630	11095.80	1.6226	7184.33	1.5552
.20	25013.71	1.6822	20187.18	1.6819	15466.27	1.6627	11053.86	1.6221	7148.42	1.5545
.21	24965.60	1.6822	20139.10	1.6817	15420.28	1.6624	11011.97	1.6216	7112.57	1.5537
.22	24917.48	1.6823	20091.34	1.6816	15374.32	1.6620	10970.13	1.6210	7076.79	1.5528
.23	24869.36	1.6823	20043.29	1.6816	15328.40	1.6618	10928.34	1.6204	7041.08	1.5518
.24	24821.22	1.6824	19995.25	1.6815	15282.51	1.6615	10886.61	1.6199	7005.45	1.5509
.25	24773.07	1.6825	19947.22	1.6814	15236.65	1.6612	10844.93	1.6194	6969.89	1.5502
.26	24724.92	1.6825	19899.20	1.6813	15190.82	1.6609	10803.30	1.6188	6934.39	1.5492
.27	24676.75	1.6826	19851.19	1.6812	15145.02	1.6606	10761.72	1.6183	6898.97	1.5484
.28	24628.58	1.6827	19803.19	1.6811	15099.25	1.6602	10720.19	1.6178	6863.62	1.5475
.29	24580.40	1.6828	19755.21	1.6810	15053.52	1.6599	10678.71	1.6173	6828.34	1.5466
.30	24532.21	1.6829	19707.24	1.6809	15007.82	1.6597	10637.28	1.6167	6793.13	1.5458
.31	24484.02	1.6830	19659.28	1.6807	14962.15	1.6593	10595.91	1.6162	6757.99	1.5450
.32	24435.81	1.6831	19611.34	1.6806	14916.52	1.6591	10554.59	1.6155	6722.92	1.5441
.33	24387.60	1.6832	19563.41	1.6805	14870.91	1.6587	10513.33	1.6150	6687.92	1.5431
.34	24339.38	1.6833	19515.49	1.6803	14825.34	1.6583	10472.12	1.6144	6653.00	1.5422
.35	24291.15	1.6833	19467.59	1.6802	14779.81	1.6580	10430.96	1.6139	6618.15	1.5413
.36	24242.92	1.6834	19419.70	1.6801	14734.31	1.6578	10389.85	1.6133	6583.37	1.5404
.37	24194.68	1.6835	19371.82	1.6801	14688.84	1.6575	10348.80	1.6127	6548.66	1.5395
.38	24146.43	1.6835	19323.95	1.6799	14643.40	1.6570	10307.81	1.6121	6514.03	1.5386
.39	24098.18	1.6836	19276.10	1.6797	14598.01	1.6568	10266.87	1.6117	6479.47	1.5377
.40	24049.92	1.6836	19228.27	1.6796	14552.64	1.6564	10225.98	1.6110	6444.98	1.5367
.41	24001.65	1.6836	19180.45	1.6795	14507.31	1.6560	10185.15	1.6105	6410.57	1.5358
.42	23953.38	1.6837	19132.64	1.6793	14462.02	1.6557	10144.37	1.6098	6376.23	1.5348
.43	23905.10	1.6837	19084.85	1.6791	14416.76	1.6553	10103.65	1.6093	6341.97	1.5339
.44	23856.82	1.6838	19037.08	1.6790	14371.54	1.6551	10062.98	1.6086	6307.78	1.5331
.45	23808.53	1.6838	18989.32	1.6789	14326.35	1.6547	10022.37	1.6080	6273.66	1.5321
.46	23760.24	1.6839	18941.57	1.6788	14281.20	1.6543	9981.82	1.6075	6239.61	1.5311
.47	23711.94	1.6839	18893.84	1.6786	14236.08	1.6540	9941.32	1.6068	6205.64	1.5302
.48	23663.64	1.6840	18846.13	1.6784	14191.00	1.6536	9900.88	1.6063	6171.74	1.5292
.49	23615.34	1.6840	18798.44	1.6783	14145.96	1.6532	9860.49	1.6056	6137.92	1.5281
.50	23567.03	−1.6841	18750.76	−1.6781	14100.96	−1.6529	9820.16	−1.6051	6104.18	−1.5271

TABLE VI. ARGUMENT 1.

Equation = 22655″.226 + 22639″.2 sin. x + 769″.5 sin. $2x$ + 36″.7 sin. $3x$ + 2″.0 sin. $4x$ + 0″.1 sin. $5x$.

Period, 27.55455245 days.

Days.	20		21		22		23		24	
Decimals of a Day	Equation.	Log. Dif.	Equation.	Log. Dif.	Equation.	Log. Dif.	Equation.	Log. Dif.	Equation.	Log. Dif.
.50	23567″.03	−1.6841	18750″.76	−1.6781	14100″.96	−1.6529	9820″.16	−1.6051	6104″.18	−1.5271
.51	23518.71	1.6841	18703.10	1.6781	14055.99	1.6525	9779.88	1.6043	6070.52	1.5263
.52	23470.40	1.6841	18655.45	1.6779	14011.06	1.6522	9739.67	1.6037	6036.92	1.5254
.53	23422.08	1.6842	18607.82	1.6777	13966.17	1.6519	9699.51	1.6031	6003.39	1.5244
.54	23373.75	1.6842	18560.21	1.6775	13921.31	1.6515	9659.41	1.6026	5969.94	1.5233
.55	23325.43	1.6842	18512.62	1.6774	13876.49	1.6511	9619.36	1.6020	5936.57	1.5223
.56	23277.10	1.6842	18465.04	1.6772	13831.71	1.6507	9579.37	1.6013	5903.28	1.5214
.57	23228.76	1.6842	18417.48	1.6771	13786.97	1.6504	9539.44	1.6007	5870.06	1.5203
.58	23180.43	1.6843	18369.94	1.6769	13742.26	1.6499	9499.57	1.6000	5836.92	1.5194
.59	23132.09	1.6843	18322.42	1.6767	13697.60	1.6495	9459.76	1.5995	5803.85	1.5184
.60	23083.75	1.6843	18274.92	1.6765	13652.98	1.6493	9420.00	1.5988	5770.86	1.5173
.61	23035.41	1.6843	18227.43	1.6764	13608.39	1.6489	9380.30	1.5981	5737.95	1.5164
.62	22987.07	1.6843	18179.96	1.6762	13563.84	1.6485	9340.66	1.5975	5705.11	1.5153
.63	22938.73	1.6843	18132.52	1.6761	13519.33	1.6481	9301.08	1.5968	5672.35	1.5143
.64	22890.38	1.6843	18085.09	1.6759	13474.86	1.6476	9261.56	1.5962	5639.67	1.5132
.65	22842.04	1.6843	18037.68	1.6757	13430.44	1.6473	9222.10	1.5955	5607.07	1.5122
.66	22793.69	1.6843	17990.29	1.6754	13386.05	1.6469	9182.70	1.5949	5574.55	1.5112
.67	22745.34	1.6843	17942.93	1.6754	13341.70	1.6465	9143.35	1.5943	5542.10	1.5101
.68	22696.99	1.6843	17895.58	1.6752	13297.39	1.6461	9104.06	1.5935	5509.73	1.5091
.69	22648.64	1.6843	17848.25	1.6750	13253.12	1.6457	9064.84	1.5929	5477.44	1.5080
.70	22600.30	1.6843	17800.94	1.6747	13208.89	1.6452	9025.68	1.5923	5445.23	1.5070
.71	22551.95	1.6843	17753.66	1.6745	13164.71	1.6448	8986.57	1.5916	5413.09	1.5059
.72	22503.60	1.6843	17706.39	1.6743	13120.57	1.6444	8947.52	1.5909	5381.03	1.5049
.73	22455.26	1.6843	17659.15	1.6742	13076.47	1.6441	8908.53	1.5901	5349.05	1.5038
.74	22406.91	1.6843	17611.92	1.6739	13032.41	1.6437	8869.61	1.5893	5317.15	1.5025
.75	22358.57	1.6843	17564.72	1.6737	12988.39	1.6433	8830.77	1.5887	5285.34	1.5017
.76	22310.22	1.6843	17517.54	1.6736	12944.41	1.6429	8791.98	1.5882	5253.60	1.5005
.77	22261.88	1.6843	17470.38	1.6733	12900.47	1.6424	8753.24	1.5875	5221.94	1.4994
.78	22213.54	1.6843	17423.25	1.6732	12856.58	1.6420	8714.56	1.5867	5190.36	1.4983
.79	22165.20	1.6843	17376.13	1.6729	12812.73	1.6416	8675.95	1.5860	5158.86	1.4973
.80	22116.86	1.6843	17329.01	1.6726	12768.92	1.6412	8637.40	1.5853	5127.44	1.4961
.81	22068.52	1.6842	17281.98	1.6726	12725.15	1.6407	8598.91	1.5846	5096.10	1.4950
.82	22020.19	1.6842	17234.93	1.6723	12681.43	1.6403	8560.48	1.5839	5064.84	1.4939
.83	21971.86	1.6842	17187.91	1.6721	12637.75	1.6399	8522.12	1.5832	5033.66	1.4926
.84	21923.53	1.6841	17140.91	1.6719	12594.11	1.6394	8483.82	1.5826	5002.57	1.4917
.85	21875.21	1.6841	17093.94	1.6717	12550.52	1.6390	8445.58	1.5817	4971.55	1.4906
.86	21826.89	1.6841	17046.99	1.6715	12506.97	1.6386	8407.41	1.5810	4940.61	1.4895
.87	21778.57	1.6840	17000.06	1.6712	12463.46	1.6381	8369.30	1.5804	4909.75	1.4881
.88	21730.27	1.6840	16953.16	1.6709	12420.00	1.6376	8331.25	1.5795	4878.98	1.4870
.89	21681.97	1.6840	16906.28	1.6707	12376.59	1.6372	8293.27	1.5788	4848.29	1.4857
.90	21633.66	1.6839	16859.43	1.6705	12333.22	1.6368	8255.35	1.5782	4817.69	1.4847
.91	21585.36	1.6839	16812.60	1.6703	12289.89	1.6364	8217.49	1.5773	4787.16	1.4836
.92	21537.06	1.6839	16765.79	1.6700	12246.60	1.6359	8179.70	1.5765	4756.71	1.4823
.93	21488.77	1.6838	16719.01	1.6698	12203.36	1.6354	8141.98	1.5759	4726.35	1.4811
.94	21440.49	1.6838	16672.26	1.6695	12160.17	1.6349	8104.32	1.5752	4696.07	1.4800
.95	21392.21	1.6837	16625.54	1.6693	12117.03	1.6345	8066.72	1.5743	4665.87	1.4787
.96	21343.93	1.6837	16578.84	1.6691	12073.93	1.6341	8029.19	1.5737	4635.76	1.4775
.97	21295.66	1.6836	16532.16	1.6689	12030.87	1.6335	7991.72	1.5729	4605.73	1.4764
.98	21247.40	1.6835	16485.51	1.6686	11987.87	1.6331	7954.32	1.5721	4575.78	1.4752
.99	21199.14	1.6834	16438.89	1.6684	11944.91	1.6327	7916.98	1.5713	4545.91	1.4741
1.00	21150.89	−1.6834	16392.29	−1.6682	11901.99	−1.6321	7879.71	−1.5705	4516.12	−1.4728

TABLE VI. ARGUMENT 1.

Equation $= 22655''.226 + 22639''.2 \sin. x + 769''.5 \sin. 2x + 36''.7 \sin. 3x + 2''.0 \sin. 4x + 0''.1 \sin. 5x.$

Period, 27.55455245 days.

Days.	25		26		27	
Decimals of a Day.	Equation.	Log. Dif.	Equation.	Log. Dif.	Equation.	Log. Dif.
.00	4516″.12	−1.4728	1986″.96	−1.3126	444″.93	−0.9996
.01	4486.42	1.4716	1966.42	1.3104	431.91	0.9948
.02	4456.80	1.4702	1945.98	1.3083	425.06	0.9903
.03	1427.27	1.4690	1925.64	1.3062	415.28	0.9854
.04	4397.82	1.4678	1905.40	1.3038	405.61	0.9800
.05	4368.46	1.4666	1885.27	1.3018	396.06	0.9750
.06	4339.18	1.4652	1865.23	1.2998	386.62	0.9699
.07	1309.99	1.4640	1845.29	1.2976	377.29	0.9647
.08	4280.88	1.4627	1825.45	1.2954	368.07	0.9590
.09	1251.86	1.4615	1805.71	1.2932	358.97	0.9542
.10	4222.92	1.4602	1786.07	1.2909	349.97	0.9481
.11	4194.07	1.4590	1766.53	1.2887	341.09	0.9430
.12	4165.30	1.4576	1747.09	1.2865	332.32	0.9370
.13	4136.62	1.4562	1727.75	1.2842	323.67	0.9320
.14	4108.03	1.4549	1708.51	1.2819	315.12	0.9258
.15	4079.52	1.4536	1689.37	1.2797	306.69	0.9201
.16	4051.10	1.4524	1670.33	1.2772	298.37	0.9138
.17	4022.76	1.4510	1651.40	1.2749	290.17	0.9085
.18	3994.51	1.4498	1632.57	1.2725	282.07	0.9020
.19	3966.34	1.4484	1613.84	1.2702	274.09	0.8960
.20	3938.26	1.4470	1595.21	1.2679	266.22	0.8899
.21	3910.27	1.4456	1576.68	1.2655	258.46	0.8831
.22	3882.36	1.4444	1558.25	1.2630	250.82	0.8768
.23	3854.51	1.4428	1539.93	1.2606	243.29	0.8698
.24	3826.82	1.4415	1521.71	1.2582	235.88	0.8639
.25	3799.18	1.4402	1503.59	1.2558	228.57	0.8567
.26	3771.62	1.4389	1485.57	1.2534	221.38	0.8494
.27	3744.15	1.4373	1467.65	1.2506	214.31	0.8426
.28	3716.78	1.4360	1419.84	1.2182	207.35	0.8357
.29	3689.49	1.4346	1432.13	1.2457	200.50	0.8287
.30	3662.29	1.4332	1414.52	1.2430	193.76	0.8209
.31	3635.18	1.4319	1397.02	1.2405	187.14	0.8136
.32	3608.15	1.4305	1379.62	1.2380	180.63	0.8055
.33	3581.21	1.4289	1362.32	1.2355	174.24	0.7980
.34	3554.36	1.4273	1345.12	1.2327	167.96	0.7903
.35	3527.61	1.4260	1328.03	1.2302	161.79	0.7818
.36	3500.94	1.4245	1311.04	1.2273	155.74	0.7738
.37	3474.36	1.4231	1294.16	1.2248	149.80	0.7649
.38	3447.87	1.4216	1277.38	1.2222	143.98	0.7566
.39	3421.47	1.4202	1260.70	1.2193	138.27	0.7474
.40	3395.16	1.4185	1244.13	1.2166	132.68	0.7388
.41	3368.94	1.4171	1227.66	1.2138	127.20	0.7292
.42	3342.81	1.4157	1211.30	1.2111	121.84	0.7202
.43	3316.77	1.4142	1195.04	1.2081	116.59	0.7110
.44	3290.82	1.4125	1178.89	1.2054	111.45	0.7007
.45	3264.97	1.4111	1162.84	1.2028	106.43	0.6902
.46	3239.20	1.4095	1146.89	1.1998	101.53	0.6803
.47	3213.52	1.4079	1131.05	1.1967	96.74	0.6702
.48	3187.94	1.4063	1115.32	1.1939	92.06	0.6590
.49	3162.45	1.4048	1099.69	1.1909	87.50	0.6474
.50	3137.05	−1.4033	1084.17	−1.1881	83.06	−0.6355

TABLE VI. ARGUMENT 1.

Equation = 22655″.226 + 22639″.2 sin. x + 769″.5 sin. $2x$ + 36″.7 sin. $3x$ + 2″.0 sin. $4x$ + 0″.1 sin. $5x$.

Period, 27.55455245 days.

Days.	25		26		27	
Decimals of a Day.	Equation.	Log. Dif.	Equation.	Log. Dif.	Equation.	Log. Dif.
.50	3137″.05	−1.4033	1081″.17	−1.1881	83″.06	−0.6355
.51	3111.74	1.4017	1068.75	1.1850	78.74	0.6243
.52	3086.52	1.4002	1053.44	1.1821	74.53	0.6128
.53	3061.39	1.3984	1038.23	1.1790	70.13	0.5999
.54	3036.36	1.3969	1023.13	1.1759	66.45	0.5877
.55	3011.42	1.3954	1008.14	1.1730	62.58	0.5740
.56	2986.57	1.3938	993.25	1.1697	58.83	0.5599
.57	2961.81	1.3920	978.47	1.1668	55.20	0.5465
.58	2937.15	1.3904	963.79	1.1635	51.68	0.5315
.59	2912.58	1.3888	949.22	1.1602	48.28	0.5159
.60	2888.10	1.3872	934.76	1.1568	45.00	0.5011
.61	2863.71	1.3854	920.41	1.1538	41.83	0.4843
.62	2839.42	1.3838	906.16	1.1507	38.78	0.4683
.63	2815.22	1.3820	892.01	1.1173	35.84	0.4502
.64	2791.12	1.3804	877.97	1.1436	33.02	0.4314
.65	2767.11	1.3788	864.05	1.1405	30.32	0.4133
.66	2743.19	1.3770	850.23	1.1373	27.73	0.3927
.67	2719.37	1.3753	836.51	1.1335	25.26	0.3711
.68	2695.64	1.3735	822.91	1.1303	22.91	0.3502
.69	2672.01	1.3718	809.41	1.1268	20.67	0.3263
.70	2618.47	1.3701	796.02	1.1232	18.55	0.3032
.71	2625.02	1.3683	782.74	1.1199	16.54	0.2765
.72	2601.67	1.3666	769.56	1.1162	14.65	0.2480
.73	2578.41	1.3647	756.49	1.1127	12.88	0.2175
.74	2555.25	1.3630	743.53	1.1089	11.23	0.1847
.75	2532.18	1.3612	730.68	1.1052	9.70	0.1523
.76	2509.21	1.3593	717.94	1.1014	8.28	0.1139
.77	2186.34	1.3575	705.31	1.0976	6.98	0.0719
.78	2463.56	1.3556	692.79	1.0941	5.80	0.0294
.79	2440.88	1.3537	680.37	1.0902	4.73	9.9777
.80	2418.30	1.3519	668.06	1.0864	3.78	9.9191
.81	2395.81	1.3502	655.86	1.0824	2.95	9.8573
.82	2373.41	1.3483	643.77	1.0785	2.23	9.7782
.83	2351.11	1.3464	631.79	1.0745	1.63	9.6812
.84	2328.91	1.3444	619.92	1.0705	1.15	9.5563
.85	2306.81	1.3426	608.16	1.0664	0.79	9.3802
.86	2284.80	1.3406	596.51	1.0626	0.55	9.1139
.87	2262.89	1.3387	584.96	1.0580	0.42	−8.0000
.88	2241.08	1.3367	573.53	1.0539	0.41	+9.0114
.89	2219.37	1.3349	562.21	1.0496	0.52	9.3617
.90	2197.75	1.3328	551.00	1.0457	0.75	9.5441
.91	2176.23	1.3308	539.89	1.0411	1.10	9.6721
.92	2154.81	1.3290	528.90	1.0367	1.57	9.7634
.93	2133.48	1.3269	518.02	1.0323	2.15	9.8451
.94	2112.25	1.3249	507.25	1.0282	2.85	9.9138
.95	2091.12	1.3228	496.58	1.0233	3.67	9.9731
.96	2070.09	1.3207	486.03	1.0187	4.61	0.0253
.97	2049.16	1.3187	475.59	1.0140	5.67	0.0682
.98	2028.33	1.3166	465.26	1.0094	6.84	0.1106
.99	2007.60	1.3147	455.01	1.0047	8.13	0.1492
1.00	1986.96	−1.3126	444.93	−0.9996	9.54	+0.1818

TABLE VII. ARGUMENT 2.

Equation $= 4587''.400 + 4586''.9 \sin. (2t - x) + 31''.2 \sin. (4t - 2x)$.

Period, 31.81193574 days.

Days.	0		1		2		3		4	
Decimals of a Day.	Equation.	Dif.	Equation.	Dif.	Equation.	Dif.	Equation.	Dif.	Equation.	Dif.
.00	2.45	−.28	62.81	+1.51	300.36	+3.25	707.06	+4.88	1267.93	+6.32
.01	2.17	.27	64.32	1.52	303.61	3.27	711.94	4.89	1274.25	6.33
.02	1.90	.25	65.84	1.54	306.88	3.28	716.83	4.91	1280.58	6.35
.03	1.65	.23	67.38	1.56	310.16	3.30	721.71	4.92	1286.93	6.36
.04	1.42	.21	68.94	1.58	313.46	3.32	726.66	4.94	1293.29	6.37
.05	1.21	.19	70.52	1.59	316.78	3.33	731.60	4.95	1299.66	6.38
.06	1.02	.17	72.11	1.62	320.11	3.35	736.55	4.97	1306.04	6.40
.07	0.85	.16	73.73	1.63	323.46	3.37	741.52	4.98	1312.44	6.42
.08	0.69	.14	75.36	1.65	326.83	3.38	746.50	5.00	1318.86	6.42
.09	0.55	.12	77.01	1.66	330.21	3.40	751.50	5.02	1325.28	6.44
.10	0.43	.10	78.67	1.68	333.61	3.42	756.52	5.03	1331.72	6.45
.11	0.33	.09	80.35	1.70	337.03	3.44	761.55	5.05	1338.17	6.46
.12	0.24	.07	82.05	1.72	340.47	3.45	766.60	5.06	1344.63	6.48
.13	0.17	.05	83.77	1.74	343.92	3.47	771.66	5.07	1351.11	6.49
.14	0.12	.03	85.51	1.75	347.39	3.48	776.73	5.10	1357.60	6.50
.15	0.09	−.01	87.26	1.78	350.87	3.51	781.83	5.10	1364.10	6.52
.16	0.08	.00	89.04	1.79	354.38	3.51	786.93	5.12	1370.62	6.53
.17	0.08	+.02	90.83	1.80	357.89	3.54	792.05	5.14	1377.15	6.55
.18	0.10	.01	92.63	1.83	361.43	3.55	797.19	5.15	1383.70	6.55
.19	0.11	.06	94.16	1.84	364.98	3.57	802.34	5.16	1390.25	6.56
.20	0.20	.07	96.30	1.86	368.55	3.58	807.50	5.19	1396.81	6.59
.21	0.27	.09	98.16	1.88	372.13	3.61	812.69	5.19	1403.40	6.59
.22	0.36	.12	100.01	1.90	375.74	3.62	817.88	5.21	1409.99	6.61
.23	0.48	.13	101.91	1.91	379.36	3.64	823.09	5.23	1416.60	6.61
.24	0.61	.14	103.85	1.93	383.00	3.65	828.32	5.24	1423.21	6.63
.25	0.75	.17	105.78	1.95	386.65	3.67	833.56	5.26	1429.84	6.65
.26	0.92	.18	107.73	1.97	390.32	3.69	838.82	5.27	1436.49	6.66
.27	1.10	.20	109.70	1.98	394.01	3.70	844.09	5.29	1443.15	6.67
.28	1.30	.22	111.68	2.00	397.71	3.72	849.38	5.30	1449.82	6.69
.29	1.52	.24	113.68	2.02	401.43	3.73	854.68	5.32	1456.51	6.69
.30	1.76	.25	115.70	2.04	405.16	3.76	860.00	5.33	1463.20	6.70
.31	2.01	.27	117.74	2.05	408.92	3.77	865.33	5.34	1469.90	6.72
.32	2.28	.30	119.79	2.07	412.69	3.78	870.67	5.36	1476.62	6.74
.33	2.58	.30	121.86	2.09	416.47	3.80	876.03	5.38	1483.36	6.74
.34	2.88	.33	123.95	2.11	420.27	3.82	881.41	5.39	1490.10	6.76
.35	3.21	.34	126.06	2.12	424.09	3.84	886.80	5.41	1496.86	6.77
.36	3.55	.36	128.18	2.14	427.93	3.85	892.21	5.42	1503.63	6.78
.37	3.91	.39	130.32	2.16	431.78	3.87	897.63	5.43	1510.41	6.79
.38	4.30	.39	132.48	2.18	435.65	3.88	903.06	5.45	1517.20	6.81
.39	4.69	.42	134.66	2.20	439.53	3.90	908.51	5.47	1524.01	6.81
.40	5.11	.43	136.86	2.21	443.43	3.92	913.98	5.47	1530.82	6.83
.41	5.54	.46	139.07	2.23	447.35	3.93	919.45	5.50	1537.65	6.85
.42	6.00	.47	141.30	2.25	451.28	3.95	924.95	5.51	1544.50	6.85
.43	6.47	.48	143.55	2.26	455.23	3.97	930.46	5.52	1551.35	6.87
.44	6.95	.51	145.81	2.29	459.20	3.98	935.98	5.54	1558.22	6.88
.45	7.46	.52	148.10	2.30	463.18	4.00	941.52	5.55	1565.10	6.90
.46	7.98	.54	150.40	2.32	467.18	4.02	947.07	5.57	1572.00	6.90
.47	8.52	.56	152.72	2.33	471.20	4.03	952.64	5.58	1578.90	6.92
.48	9.08	.58	155.05	2.35	475.23	4.05	958.22	5.60	1585.82	6.93
.49	9.66	.60	157.40	2.37	479.28	4.07	963.82	5.61	1592.75	6.93
.50	10.26	+.61	159.77	+2.39	483.35	+4.08	969.43	+5.62	1599.68	+6.96

TABLE VII. ARGUMENT 2.

Equation $= 4587''.400 + 4586''.9 \sin. (2t - x) + 31''.2 \sin. (4t - 2x)$.

Period, 31.81193574 days.

Days.	0		1		2		3		4	
Decimals of a Day.	Equation.	Dif.	Equation.	Dif.	Equation.	Dif.	Equation.	Dif.	Equation.	Dif.
.50	10.26	+ .61	159.77	+2.39	483.35	+4.08	969.43	+5.62	1599.68	+6.96
.51	10.87	.63	162.16	2.41	487.43	4.10	975.05	5.64	1606.64	6.96
.52	11.50	.65	164.57	2.42	491.53	4.11	980.69	5.65	1613.60	6.98
.53	12.15	.67	166.99	2.44	495.64	4.13	986.34	5.67	1620.58	6.98
.54	12.82	.68	169.43	2.46	499.77	4.15	992.01	5.68	1627.56	7.00
.55	13.50	.70	171.89	2.47	503.92	4.16	997.69	5.70	1634.56	7.02
.56	14.20	.72	174.36	2.49	508.08	4.18	1003.39	5.71	1641.58	7.02
.57	14.92	.74	176.85	2.51	512.26	4.19	1009.10	5.73	1648.60	7.01
.58	15.66	.76	179.36	2.53	516.45	4.21	1014.83	5.74	1655.61	7.04
.59	16.42	.77	181.89	2.54	520.66	4.23	1020.57	5.75	1662.68	7.06
.60	17.19	.79	184.43	2.57	524.89	4.25	1026.32	5.77	1669.74	7.07
.61	17.98	.81	187.00	2.57	529.14	4.26	1032.09	5.78	1676.81	7.09
.62	18.79	.83	189.57	2.60	533.40	4.27	1037.87	5.80	1683.90	7.09
.63	19.62	.85	192.17	2.62	537.67	4.30	1043.67	5.81	1690.99	7.11
.64	20.47	.86	194.79	2.63	541.97	4.31	1049.48	5.83	1698.10	7.12
.65	21.33	.88	197.42	2.65	546.28	4.32	1055.31	5.84	1705.22	7.13
.66	22.21	.90	200.07	2.66	550.60	4.34	1061.15	5.85	1712.35	7.14
.67	23.11	.92	202.73	2.68	554.94	4.36	1067.00	5.87	1719.49	7.15
.68	24.03	.93	205.41	2.71	559.30	4.37	1072.87	5.88	1726.64	7.17
.69	24.96	.96	208.12	2.71	563.67	4.38	1078.75	5.89	1733.81	7.17
.70	25.92	.97	210.83	2.74	568.05	4.41	1084.64	5.91	1740.98	7.19
.71	26.89	.99	213.57	2.75	572.46	4.42	1090.55	5.93	1748.17	7.20
.72	27.88	1.00	216.32	2.77	576.88	4.44	1096.48	5.93	1755.37	7.21
.73	28.88	1.03	219.09	2.78	581.32	4.45	1102.41	5.96	1762.58	7.22
.74	29.91	1.04	221.87	2.81	585.77	4.47	1108.37	5.96	1769.80	7.23
.75	30.95	1.06	224.68	2.82	590.24	4.48	1114.33	5.98	1777.03	7.25
.76	32.01	1.08	227.50	2.84	594.72	4.50	1120.31	6.00	1784.28	7.25
.77	33.09	1.10	230.34	2.85	599.22	4.52	1126.31	6.00	1791.53	7.27
.78	34.19	1.11	233.19	2.88	603.74	4.53	1132.31	6.02	1798.80	7.28
.79	35.30	1.13	236.07	2.89	608.27	4.54	1138.33	6.03	1806.08	7.28
.80	36.43	1.15	238.96	2.91	612.81	4.57	1144.36	6.05	1813.36	7.31
.81	37.58	1.17	241.87	2.92	617.38	4.58	1150.41	6.07	1820.67	7.31
.82	38.75	1.19	244.79	2.94	621.96	4.59	1156.48	6.07	1827.98	7.32
.83	39.94	1.20	247.73	2.96	626.55	4.61	1162.55	6.09	1835.30	7.33
.84	41.14	1.22	250.69	2.98	631.16	4.63	1168.64	6.11	1842.63	7.35
.85	42.36	1.24	253.67	2.99	635.79	4.64	1174.75	6.12	1849.98	7.35
.86	43.60	1.26	256.66	3.01	640.43	4.66	1180.87	6.13	1857.33	7.37
.87	44.86	1.27	259.67	3.03	645.09	4.67	1187.00	6.14	1864.70	7.38
.88	46.13	1.29	262.70	3.04	649.76	4.69	1193.14	6.16	1872.08	7.38
.89	47.42	1.31	265.74	3.07	654.45	4.71	1199.30	6.17	1879.46	7.40
.90	48.73	1.33	268.81	3.07	659.16	4.72	1205.47	6.18	1886.86	7.41
.91	50.06	1.34	271.88	3.10	663.88	4.73	1211.65	6.20	1894.27	7.43
.92	51.40	1.37	274.98	3.11	668.61	4.75	1217.85	6.22	1901.70	7.43
.93	52.77	1.38	278.09	3.13	673.36	4.78	1224.07	6.22	1909.13	7.44
.94	54.15	1.40	281.22	3.15	678.14	4.78	1230.29	6.24	1916.57	7.46
.95	55.55	1.42	284.37	3.16	682.92	4.80	1236.53	6.25	1924.03	7.46
.96	56.97	1.44	287.53	3.19	687.72	4.81	1242.78	6.27	1931.49	7.47
.97	58.41	1.44	290.72	3.19	692.53	4.83	1249.05	6.28	1938.96	7.49
.98	59.85	1.47	293.91	3.22	697.36	4.85	1255.33	6.29	1946.45	7.50
.99	61.32	+1.49	297.13	+3.23	702.21	+4.85	1261.62	+6.31	1953.95	+7.50
1.00	62.81		300.36		707.06		1267.93		1961.45	

TABLE VII. ARGUMENT 2.

Equation $= 4587''.400 + 4586''.9 \text{ sin. } (2t - x) + 31''.2 \text{ sin. } (4t - 2x)$.

Period, 31.81193574 days.

Days.	5		6		7		8		9	
Decimals of a Day	Equation.	Dif.	Equation.	Dif	Equation.	Dif.	Equation.	Dif	Equation.	Dif.
.00	1961″.45	+7.52	2760″.43	+8.42	3633″.06	+8.98	4544″.23	+9.19	5157″.15	+9.01
.01	1968.97	7.53	2768.85	8.43	3642.04	8.99	4553.42	9.18	5166.16	9.01
.02	1976.50	7.54	2777.28	8.43	3651.03	8.99	4562.60	9.18	5175.17	9.01
.03	1984.04	7.55	2785.71	8.44	3660.02	8.99	4571.78	9.18	5184.18	9.00
.04	1991.59	7.55	2794.15	8.45	3669.01	8.99	4580.96	9.18	5193.18	9.00
.05	1999.14	7.57	2802.60	8.45	3678.00	9.00	4590.14	9.19	5502.18	8.99
.06	2006.71	7.58	2811.05	8.47	3687.00	9.01	4599.33	9.18	5511.17	8.99
.07	2014.29	7.59	2819.52	8.46	3696.01	9.00	4608.51	9.18	5520.16	8.99
.08	2021.88	7.61	2827.98	8.48	3705.01	9.01	4617.69	9.19	5529.15	8.98
.09	2029.49	7.61	2836.46	8.48	3714.02	9.02	4626.88	9.18	5538.13	8.98
.10	2037.10	7.62	2844.94	8.49	3723.04	9.02	4636.06	9.18	5547.11	8.98
.11	2044.72	7.63	2853.43	8.50	3732.06	9.02	4645.24	9.18	5556.09	8.97
.12	2052.35	7.64	2861.93	8.50	3741.08	9.02	4654.42	9.19	5565.06	8.96
.13	2059.99	7.66	2870.43	8.51	3750.10	9.03	4663.61	9.18	5574.02	8.97
.14	2067.65	7.66	2878.94	8.52	3759.13	9.03	4672.79	9.18	5582.99	8.96
.15	2075.31	7.67	2887.46	8.53	3768.16	9.04	4681.97	9.18	5591.95	8.95
.16	2082.98	7.68	2895.99	8.53	3777.20	9.04	4691.15	9.18	5600.90	8.95
.17	2090.66	7.69	2904.52	8.54	3786.24	9.04	4700.33	9.18	5609.85	8.95
.18	2098.35	7.71	2913.06	8.54	3795.28	9.04	4709.51	9.18	5618.80	8.94
.19	2106.06	7.71	2921.60	8.55	3804.32	9.05	4718.69	9.17	5627.74	8.94
.20	2113.77	7.72	2930.15	8.56	3813.37	9.05	4727.86	9.18	5636.68	8.93
.21	2121.49	7.73	2938.71	8.57	3822.42	9.06	4737.04	9.18	5645.61	8.93
.22	2129.22	7.75	2947.28	8.57	3831.48	9.05	4746.22	9.18	5654.54	8.93
.23	2136.97	7.75	2955.85	8.58	3840.53	9.06	4755.40	9.17	5663.47	8.92
.24	2144.72	7.76	2964.43	8.59	3849.59	9.07	4764.57	9.18	5672.39	8.92
.25	2152.48	7.78	2973.02	8.59	3858.66	9.06	4773.75	9.17	5681.31	8.91
.26	2160.26	7.78	2981.61	8.60	3867.72	9.07	4782.92	9.18	5690.22	8.91
.27	2168.04	7.79	2990.21	8.60	3876.79	9.07	4792.10	9.17	5699.13	8.90
.28	2175.83	7.80	2998.81	8.61	3885.86	9.08	4801.27	9.17	5708.03	8.90
.29	2183.63	7.81	3007.42	8.62	3894.94	9.08	4810.44	9.17	5716.93	8.90
.30	2191.44	7.82	3016.04	8.62	3904.02	9.08	4819.61	9.17	5725.83	8.89
.31	2199.26	7.83	3024.66	8.63	3913.10	9.08	4828.78	9.17	5734.72	8.88
.32	2207.09	7.84	3033.29	8.64	3922.18	9.09	4837.95	9.17	5743.60	8.88
.33	2214.93	7.85	3041.93	8.64	3931.27	9.08	4847.12	9.17	5752.48	8.88
.34	2222.78	7.86	3050.57	8.65	3940.35	9.10	4856.29	9.16	5761.36	8.87
.35	2230.64	7.87	3059.22	8.65	3949.45	9.09	4865.45	9.17	5770.23	8.87
.36	2238.51	7.88	3067.87	8.67	3958.54	9.10	4874.62	9.16	5779.10	8.86
.37	2246.39	7.89	3076.54	8.66	3967.64	9.10	4883.78	9.17	5787.96	8.85
.38	2254.28	7.89	3085.20	8.68	3976.74	9.10	4892.95	9.16	5796.81	8.85
.39	2262.17	7.91	3093.88	8.67	3985.84	9.10	4902.11	9.16	5805.66	8.85
.40	2270.08	7.92	3102.55	8.69	3994.94	9.11	4911.27	9.16	5814.51	8.84
.41	2278.00	7.92	3111.24	8.69	4004.05	9.10	4920.43	9.16	5823.35	8.84
.42	2285.92	7.94	3119.93	8.70	4013.15	9.11	4929.59	9.15	5832.19	8.83
.43	2293.86	7.94	3128.63	8.70	4022.26	9.12	4938.74	9.16	5841.02	8.82
.44	2301.80	7.96	3137.33	8.71	4031.38	9.11	4947.90	9.15	5849.84	8.83
.45	2309.76	7.96	3146.04	8.72	4040.49	9.12	4957.05	9.15	5858.67	8.81
.46	2317.72	7.97	3154.76	8.72	4049.61	9.12	4966.20	9.15	5867.48	8.81
.47	2325.69	7.98	3163.48	8.73	4058.73	9.12	4975.35	9.15	5876.29	8.81
.48	2333.67	7.99	3172.21	8.73	4067.85	9.13	4984.50	9.15	5885.10	8.80
.49	2341.66	8.00	3180.94	8.73	4076.98	9.12	4993.65	9.14	5893.90	8.79
.50	2349.66	+8.00	3189.67	+8.75	4086.10	+9.13	5002.79	+9.15	5902.69	+8.79

TABLE VII. ARGUMENT 2.

Equation = 4587″.400 + 4586″.9 sin. $(2t - x)$ + 31″.2 sin. $(4t - 2x)$.

Period, 31.81193574 days.

Days.	5		6		7		8		9	
Decimals of a Day	Equation.	Dif.	Equation.	Dif.	Equation.	Dif.	Equation.	Dif.	Equation.	Dif.
.50	2349″.66	+8.00	3189″.67	+8.75	4086″.10	+9.13	5002″.79	+9.15	5902″.69	+8.79
.51	2357.66	8.02	3198.42	8.75	4095.23	9.13	5011.94	9.14	5911.48	8.79
.52	2365.68	8.03	3207.17	8.76	4104.36	9.14	5021.08	9.14	5920.27	8.78
.53	2373.71	8.03	3215.93	8.75	4113.50	9.13	5030.22	9.14	5929.05	8.77
.54	2381.74	8.04	3224.68	8.77	4122.63	9.14	5039.36	9.14	5937.82	8.77
.55	2389.78	8.05	3233.45	8.77	4131.77	9.13	5048.50	9.13	5946.59	8.77
.56	2397.83	8.07	3242.22	8.78	4140.90	9.14	5057.63	9.14	5955.36	8.75
.57	2405.90	8.07	3251.00	8.78	4150.04	9.14	5066.77	9.13	5964.11	8.76
.58	2413.97	8.08	3259.78	8.79	4159.18	9.15	5075.90	9.13	5972.87	8.75
.59	2422.05	8.08	3268.57	8.79	4168.33	9.14	5085.03	9.12	5981.62	8.73
.60	2430.13	8.10	3277.36	8.80	4177.47	9.15	5094.15	9.13	5990.35	8.73
.61	2438.23	8.11	3286.16	8.80	4186.62	9.15	5103.28	9.12	5999.08	8.73
.62	2446.34	8.11	3294.96	8.81	4195.77	9.15	5112.40	9.12	6007.81	8.73
.63	2454.45	8.12	3303.77	8.82	4204.92	9.15	5121.52	9.12	6016.54	8.72
.64	2462.57	8.13	3312.59	8.81	4214.07	9.15	5130.64	9.12	6025.26	8.71
.65	2470.70	8.11	3321.40	8.83	4223.22	9.16	5139.76	9.11	6033.97	8.70
.66	2478.81	8.15	3330.23	8.83	4232.38	9.15	5148.87	9.12	6042.67	8.71
.67	2486.99	8.16	3339.06	8.83	4241.53	9.16	5157.99	9.10	6051.38	8.69
.68	2495.15	8.17	3347.89	8.84	4250.69	9.16	5167.09	9.10	6060.07	8.69
.69	2503.32	8.16	3356.73	8.85	4259.85	9.15	5176.19	9.11	6068.76	8.67
.70	2511.48	8.18	3365.58	8.85	4269.00	9.17	5185.30	9.10	6077.43	8.68
.71	2519.66	8.20	3374.43	8.85	4278.17	9.16	5194.40	9.11	6086.11	8.67
.72	2527.86	8.20	3383.28	8.86	4287.33	9.16	5203.51	9.09	6094.78	8.66
.73	2536.06	8.20	3392.14	8.87	4296.49	9.18	5212.60	9.10	6103.44	8.66
.74	2544.26	8.22	3401.01	8.87	4305.67	9.15	5221.70	9.09	6112.10	8.65
.75	2552.48	8.22	3409.88	8.87	4314.82	9.17	5230.79	9.09	6120.75	8.65
.76	2560.70	8.23	3418.75	8.88	4323.99	9.17	5239.88	9.09	6129.40	8.64
.77	2568.93	8.24	3427.63	8.88	4333.16	9.17	5248.97	9.08	6138.04	8.63
.78	2577.17	8.25	3436.51	8.89	4342.33	9.17	5258.05	9.08	6146.67	8.63
.79	2585.42	8.25	3445.40	8.89	4351.50	9.17	5267.13	9.08	6155.30	8.62
.80	2593.67	8.27	3454.29	8.90	4360.67	9.17	5276.21	9.08	6163.92	8.61
.81	2601.94	8.27	3463.19	8.90	4369.84	9.18	5285.29	9.07	6172.53	8.60
.82	2610.21	8.28	3472.09	8.91	4379.02	9.17	5294.36	9.07	6181.13	8.61
.83	2618.49	8.29	3481.00	8.91	4388.19	9.17	5303.43	9.07	6189.74	8.59
.84	2626.78	8.29	3489.91	8.91	4397.36	9.18	5312.50	9.07	6198.33	8.59
.85	2635.07	8.31	3498.82	8.92	4406.54	9.17	5321.57	9.05	6206.92	8.58
.86	2643.38	8.31	3507.74	8.92	4415.71	9.18	5330.62	9.06	6215.50	8.57
.87	2651.69	8.32	3516.66	8.93	4424.89	9.18	5339.68	9.06	6224.07	8.57
.88	2660.01	8.32	3525.59	8.94	4434.07	9.17	5348.74	9.05	6232.64	8.57
.89	2668.33	8.34	3534.53	8.93	4443.24	9.18	5357.79	9.05	6241.21	8.55
.90	2676.67	8.34	3543.46	8.94	4452.42	9.18	5366.84	9.04	6249.76	8.55
.91	2685.01	8.35	3552.40	8.95	4461.60	9.18	5375.88	9.05	6258.31	8.54
.92	2693.36	8.36	3561.35	8.95	4470.78	9.18	5384.93	9.04	6266.85	8.53
.93	2701.72	8.37	3570.30	8.95	4479.96	9.18	5393.97	9.03	6275.38	8.53
.94	2710.09	8.37	3579.25	8.96	4489.14	9.19	5403.00	9.04	6283.91	8.52
.95	2718.46	8.38	3588.21	8.96	4498.33	9.18	5412.04	9.03	6292.43	8.52
.96	2726.84	8.39	3597.17	8.97	4507.51	9.18	5421.07	9.02	6300.95	8.50
.97	2735.23	8.39	3606.14	8.97	4516.69	9.18	5430.09	9.03	6309.45	8.50
.98	2743.62	8.41	3615.11	8.97	4525.87	9.18	5439.12	9.02	6317.95	8.50
.99	2752.03	+8.40	3624.08	+8.98	4535.05	+9.18	5448.14	+9.01	6326.45	+8.48
1.00	2760.43		3633.06		4544.23		5457.15		6334.93	

TABLE VII. ARGUMENT 2.

Equation = 4587″.400 + 4586″.9 sin. $(2t - x)$ + 31″.2 sin. $(4t - 2x)$.

Period, 31.81193574 days.

Days.	10		11		12		13		14	
Decimals of a Day	Equation.	Dif.	Equation.	Dif.	Equation.	Dif.	Equation.	Dif.	Equation.	Dif.
.00	6334″.93	+8.48	7142″.25	+7.61	7846″.93	+6.43	8421″.28	+5.00	8843″.22	+3.39
.01	6343.41	8.47	7149.86	7.60	7853.36	6.42	8426.28	4.99	8846.61	3.37
.02	6351.88	8.46	7157.46	7.58	7859.78	6.41	8431.27	4.98	8849.98	3.36
.03	6360.34	8.46	7165.04	7.57	7866.19	6.39	8436.25	4.95	8853.34	3.35
.04	6368.80	8.45	7172.61	7.57	7872.58	6.38	8441.20	4.94	8856.69	3.32
.05	6377.25	8.44	7180.18	7.55	7878.96	6.36	8446.14	4.93	8860.01	3.31
.06	6385.69	8.43	7187.73	7.54	7885.32	6.35	8451.07	4.91	8863.32	3.29
.07	6394.12	8.43	7195.27	7.53	7891.67	6.34	8455.98	4.90	8866.61	3.27
.08	6402.55	8.43	7202.80	7.52	7898.01	6.32	8460.88	4.89	8869.88	3.26
.09	6410.98	8.41	7210.32	7.52	7904.33	6.31	8465.77	4.87	8873.14	3.23
.10	6419.39	8.41	7217.84	7.50	7910.64	6.30	8470.64	4.85	8876.37	3.22
.11	6427.80	8.40	7225.34	7.49	7916.94	6.29	8475.49	4.84	8879.59	3.31
.12	6436.20	8.39	7232.83	7.48	7923.23	6.27	8480.33	4.82	8882.80	3.19
.13	6444.59	8.38	7240.31	7.46	7929.50	6.26	8485.15	4.80	8885.99	3.17
.14	6452.97	8.38	7247.77	7.46	7935.76	6.24	8489.95	4.79	8889.16	3.15
.15	6461.35	8.37	7255.23	7.45	7942.00	6.23	8494.74	4.78	8892.31	3.14
.16	6469.72	8.36	7262.68	7.43	7948.23	6.22	8499.52	4.76	8895.45	3.12
.17	6478.08	8.35	7270.11	7.43	7954.45	6.20	8504.28	4.74	8898.57	3.10
.18	6486.43	8.35	7277.54	7.41	7960.65	6.19	8509.02	4.73	8901.67	3.09
.19	6494.78	8.34	7284.95	7.41	7966.84	6.18	8513.75	4.71	8904.76	3.07
.20	6503.12	8.33	7292.36	7.39	7973.02	6.17	8518.46	4.69	8907.83	3.05
.21	6511.45	8.32	7299.75	7.39	7979.19	6.15	8523.15	4.68	8910.88	3.03
.22	6519.77	8.31	7307.14	7.37	7985.34	6.13	8527.83	4.67	8913.91	3.02
.23	6528.08	8.31	7314.51	7.36	7991.47	6.12	8532.50	4.65	8916.93	3.00
.24	6536.39	8.30	7321.87	7.35	7997.59	6.11	8537.15	4.63	8919.93	2.98
.25	6544.69	8.29	7329.22	7.33	8003.70	6.10	8541.78	4.62	8922.91	2.97
.26	6552.98	8.28	7336.55	7.33	8009.80	6.08	8546.40	4.60	8925.88	2.95
.27	6561.26	8.27	7343.88	7.31	8015.88	6.07	8551.00	4.58	8928.83	2.93
.28	6569.53	8.27	7351.19	7.31	8021.95	6.05	8555.58	4.57	8931.76	2.91
.29	6577.80	8.26	7358.50	7.30	8028.00	6.04	8560.15	4.55	8934.67	2.90
.30	6586.06	8.25	7365.80	7.28	8034.04	6.03	8564.70	4.54	8937.57	2.88
.31	6594.31	8.25	7373.08	7.27	8040.07	6.01	8569.24	4.52	8940.45	2.86
.32	6602.56	8.23	7380.35	7.26	8046.08	6.00	8573.76	4.51	8943.31	2.85
.33	6610.79	8.23	7387.61	7.24	8052.08	5.99	8578.27	4.49	8946.16	2.83
.34	6619.02	8.21	7394.85	7.24	8058.07	5.97	8582.76	4.48	8948.99	2.81
.35	6627.23	8.21	7402.09	7.23	8064.04	5.96	8587.24	4.46	8951.80	2.79
.36	6635.44	8.20	7409.32	7.21	8070.00	5.94	8591.70	4.44	8954.59	2.78
.37	6643.64	8.20	7416.53	7.20	8075.94	5.93	8596.14	4.43	8957.37	2.76
.38	6651.84	8.18	7423.73	7.20	8081.87	5.91	8600.57	4.41	8960.13	2.74
.39	6660.02	8.18	7430.93	7.18	8087.78	5.90	8604.98	4.39	8962.87	2.72
.40	6668.20	8.17	7438.11	7.17	8093.68	5.89	8609.37	4.38	8965.59	2.71
.41	6676.37	8.16	7445.28	7.15	8099.57	5.87	8613.75	4.36	8968.30	2.69
.42	6684.53	8.15	7452.43	7.15	8105.44	5.86	8618.11	4.35	8970.99	2.67
.43	6692.68	8.14	7459.58	7.13	8111.30	5.85	8622.46	4.33	8973.66	2.66
.44	6700.82	8.14	7466.71	7.13	8117.15	5.83	8626.79	4.31	8976.32	2.64
.45	6708.96	8.12	7473.84	7.11	8122.98	5.82	8631.10	4.30	8978.96	2.62
.46	6717.08	8.12	7480.95	7.09	8128.80	5.80	8635.40	4.28	8981.58	2.60
.47	6725.20	8.11	7488.04	7.09	8134.60	5.79	8639.68	4.27	8984.18	2.59
.48	6733.31	8.10	7495.13	7.08	8140.39	5.77	8643.95	4.25	8986.77	2.57
.49	6741.41	8.09	7502.21	7.06	8146.16	5.76	8648.20	4.24	8989.34	2.55
.50	6749.50	+8.08	7509.27	+7.06	8151.92	+5.75	8652.44	+4.22	8991.89	+2.53

TABLE VII. ARGUMENT 2.

Equation $= 4587''.400 + 4586''.9 \sin. (2t - x) + 31''.2 \sin. (4t - 2x)$.

Period, 31.81193574 days.

Days.	10		11		12		13		14	
Decimals of a Day.	Equation.	Dif.	Equation.	Dif.	Equation.	Dif.	Equation.	Dif.	Equation.	Dif.
.50	6749.50	+8.08	7509.27	+7.06	8151.92	+5.75	8652.44	+4.22	8991.89	+2.53
.51	6757.58	8.07	7516.33	7.04	8157.67	5.73	8656.66	4.20	8994.42	2.52
.52	6765.65	8.06	7523.37	7.03	8163.40	5.71	8660.86	4.18	8996.94	2.50
.53	6773.71	8.06	7530.40	7.01	8169.11	5.70	8665.04	4.17	8999.44	2.48
.54	6781.77	8.05	7537.41	7.01	8174.81	5.69	8669.21	4.15	9001.92	2.47
.55	6789.82	8.04	7544.42	6.99	8180.50	5.68	8673.36	4.14	9004.39	2.44
.56	6797.86	8.03	7551.41	6.98	8186.18	5.66	8677.50	4.12	9006.83	2.43
.57	6805.89	8.02	7558.39	6.97	8191.84	5.64	8681.62	4.10	9009.26	2.41
.58	6813.91	8.01	7565.36	6.96	8197.48	5.63	8685.72	4.09	9011.67	2.40
.59	6821.92	8.01	7572.32	6.95	8203.11	5.62	8689.81	4.07	9014.07	2.37
.60	6829.93	7.99	7579.27	6.93	8208.73	5.60	8693.88	4.06	9016.44	2.36
.61	6837.92	7.98	7586.20	6.92	8214.33	5.59	8697.94	4.04	9018.80	2.35
.62	6845.90	7.98	7593.12	6.91	8219.92	5.57	8701.98	4.02	9021.15	2.32
.63	6853.88	7.96	7600.03	6.90	8225.49	5.56	8706.00	4.01	9023.47	2.31
.64	6861.84	7.96	7606.93	6.89	8231.05	5.54	8710.01	3.99	9025.78	2.29
.65	6869.80	7.95	7613.82	6.86	8236.59	5.53	8714.00	3.97	9028.07	2.27
.66	6877.75	7.93	7620.68	6.86	8242.12	5.52	8717.97	3.96	9030.34	2.26
.67	6885.68	7.93	7627.54	6.85	8247.64	5.50	8721.93	3.94	9032.60	2.23
.68	6893.61	7.92	7634.39	6.84	8253.14	5.48	8725.87	3.92	9034.83	2.22
.69	6901.53	7.92	7641.23	6.83	8258.62	5.47	8729.79	3.91	9037.05	2.21
.70	6909.45	7.90	7648.06	6.81	8264.09	5.46	8733.70	3.89	9039.26	2.18
.71	6917.35	7.89	7654.87	6.80	8269.55	5.44	8737.59	3.88	9041.44	2.17
.72	6925.24	7.88	7661.67	6.78	8274.99	5.43	8741.47	3.86	9043.61	2.15
.73	6933.12	7.87	7668.45	6.78	8280.42	5.41	8745.33	3.84	9045.76	2.13
.74	6940.99	7.86	7675.23	6.76	8285.83	5.39	8749.17	3.82	9047.89	2.11
.75	6948.85	7.86	7681.99	6.75	8291.22	5.38	8752.99	3.81	9050.00	2.10
.76	6956.71	7.84	7688.74	6.73	8296.60	5.37	8756.80	3.80	9052.10	2.08
.77	6964.55	7.83	7695.47	6.73	8301.97	5.35	8760.60	3.77	9054.18	2.06
.78	6972.38	7.83	7702.20	6.71	8307.32	5.34	8764.37	3.76	9056.24	2.05
.79	6980.21	7.82	7708.91	6.70	8312.66	5.32	8768.13	3.74	9058.29	2.02
.80	6988.03	7.80	7715.61	6.69	8317.98	5.31	8771.87	3.73	9060.31	2.01
.81	6995.83	7.80	7722.30	6.68	8323.29	5.29	8775.60	3.71	9062.32	1.99
.82	7003.63	7.78	7728.98	6.66	8328.58	5.28	8779.31	3.69	9064.31	1.97
.83	7011.41	7.78	7735.64	6.65	8333.86	5.26	8783.00	3.68	9066.28	1.96
.84	7019.19	7.76	7742.29	6.63	8339.12	5.25	8786.68	3.66	9068.24	1.94
.85	7026.95	7.76	7748.92	6.63	8344.37	5.24	8790.34	3.64	9070.18	1.92
.86	7034.71	7.75	7755.55	6.61	8349.61	5.21	8793.98	3.62	9072.10	1.90
.87	7042.46	7.73	7762.16	6.59	8354.82	5.20	8797.60	3.61	9074.00	1.89
.88	7050.19	7.73	7768.75	6.59	8360.02	5.19	8801.21	3.60	9075.89	1.87
.89	7057.92	7.72	7775.34	6.58	8365.21	5.17	8804.81	3.57	9077.76	1.85
.90	7065.64	7.71	7781.92	6.56	8370.38	5.16	8808.38	3.56	9079.61	1.83
.91	7073.35	7.69	7788.48	6.54	8375.54	5.15	8811.94	3.55	9081.44	1.81
.92	7081.04	7.69	7795.02	6.54	8380.69	5.13	8815.49	3.52	9083.25	1.80
.93	7088.73	7.68	7801.56	6.52	8385.82	5.10	8819.01	3.51	9085.05	1.78
.94	7096.41	7.66	7808.08	6.51	8390.92	5.10	8822.52	3.49	9086.83	1.76
.95	7104.07	7.66	7814.59	6.49	8396.02	5.08	8826.01	3.48	9088.59	1.74
.96	7111.73	7.64	7821.08	6.48	8401.10	5.07	8829.49	3.46	9090.33	1.73
.97	7119.37	7.64	7827.56	6.48	8406.17	5.05	8832.95	3.44	9092.06	1.71
.98	7127.01	7.63	7834.04	6.45	8411.22	5.04	8836.39	3.42	9093.77	1.69
.99	7134.64	+7.61	7840.49	+6.44	8416.26	+5.02	8839.81	+3.41	9095.46	+1.67
1.00	7142.25		7846.93		8421.28		8843.22		9097.13	

TABLE VII. ARGUMENT 2.

Equation $= 4587''.400 + 4586''.9 \sin. (2t - x) + 31''.2 \sin. (4t - 2x)$.

Period, 31.81193574 days.

Days.	15		16		17		18		19	
Decimals of a Day.	Equation.	Dif.	Equation.	Dif.	Equation.	Dif.	Equation.	Dif.	Equation.	Dif.
.00	9097.13	+1.65	9174.30	− .13	9073.11	−1.90	8798.89	−3.58	8363.46	−5.11
.01	9098.78	1.64	9174.17	.15	9071.21	1.91	8795.31	3.60	8358.35	5.13
.02	9100.42	1.62	9174.02	.17	9069.30	1.91	8791.71	3.61	8353.22	5.14
.03	9102.04	1.60	9173.85	.19	9067.36	1.95	8788.10	3.63	8348.08	5.16
.04	9103.64	1.59	9173.66	.20	9065.41	1.97	8784.47	3.65	8342.92	5.17
.05	9105.23	1.57	9173.46	.22	9063.44	1.99	8780.82	3.66	8337.75	5.18
.06	9106.80	1.55	9173.21	.24	9061.45	2.00	8777.16	3.68	8332.57	5.20
.07	9108.35	1.53	9173.00	.26	9059.15	2.03	8773.48	3.69	8327.37	5.21
.08	9109.88	1.51	9172.74	.27	9057.42	2.04	8769.79	3.71	8322.16	5.23
.09	9111.39	1.49	9172.47	.30	9055.38	2.05	8766.08	3.72	8316.93	5.24
.10	9112.88	1.48	9172.17	.31	9053.33	2.07	8762.36	3.75	8311.69	5.26
.11	9114.36	1.46	9171.86	.33	9051.26	2.09	8758.61	3.76	8306.43	5.27
.12	9115.82	1.44	9171.53	.34	9049.17	2.11	8754.85	3.77	8301.16	5.28
.13	9117.26	1.42	9171.19	.37	9047.06	2.13	8751.08	3.79	8295.88	5.30
.14	9118.68	1.41	9170.82	.38	9044.93	2.14	8747.29	3.81	8290.58	5.31
.15	9120.09	1.39	9170.44	.40	9042.79	2.16	8743.48	3.82	8285.27	5.33
.16	9121.48	1.37	9170.04	.42	9040.63	2.18	8739.66	3.84	8279.94	5.34
.17	9122.85	1.35	9169.62	.44	9038.45	2.19	8735.82	3.85	8274.60	5.35
.18	9124.20	1.34	9169.18	.45	9036.26	2.21	8731.97	3.87	8269.25	5.37
.19	9125.54	1.31	9168.73	.47	9034.05	2.22	8728.10	3.88	8263.88	5.38
.20	9126.85	1.30	9168.26	.49	9031.83	2.25	8724.22	3.90	8258.50	5.40
.21	9128.15	1.29	9167.77	.51	9029.58	2.26	8720.32	3.92	8253.10	5.41
.22	9129.44	1.26	9167.26	.52	9027.32	2.28	8716.40	3.94	8247.69	5.42
.23	9130.70	1.25	9166.74	.55	9025.04	2.30	8712.46	3.95	8242.27	5.44
.24	9131.95	1.22	9166.19	.56	9022.74	2.31	8708.51	3.96	8236.83	5.45
.25	9133.17	1.21	9165.63	.58	9020.43	2.33	8704.55	3.98	8231.38	5.47
.26	9134.38	1.20	9165.05	.59	9018.10	2.35	8700.57	4.00	8225.91	5.48
.27	9135.58	1.17	9164.46	.62	9015.75	2.37	8696.57	4.01	8220.43	5.50
.28	9136.75	1.16	9163.84	.63	9013.38	2.38	8692.56	4.03	8214.93	5.50
.29	9137.91	1.14	9163.21	.65	9011.00	2.40	8688.53	4.04	8209.43	5.52
.30	9139.05	1.12	9162.56	.67	9008.60	2.42	8684.49	4.06	8203.91	5.54
.31	9140.17	1.10	9161.89	.68	9006.18	2.43	8680.43	4.08	8198.37	5.55
.32	9141.27	1.09	9161.21	.70	9003.75	2.45	8676.35	4.09	8192.82	5.56
.33	9142.36	1.06	9160.51	.73	9001.30	2.47	8672.26	4.10	8187.26	5.57
.34	9143.42	1.05	9159.78	.74	8998.83	2.48	8668.16	4.12	8181.69	5.59
.35	9144.47	1.04	9159.04	.75	8996.35	2.50	8664.04	4.14	8176.10	5.61
.36	9145.51	1.01	9158.29	.78	8993.85	2.52	8659.90	4.15	8170.49	5.62
.37	9146.52	1.00	9157.51	.79	8991.33	2.54	8655.75	4.17	8164.87	5.63
.38	9147.52	.98	9156.72	.81	8988.79	2.55	8651.58	4.19	8159.24	5.64
.39	9148.50	.96	9155.91	.82	8986.24	2.57	8647.39	4.20	8153.60	5.66
.40	9149.46	.94	9155.09	.85	8983.67	2.59	8643.19	4.21	8147.94	5.67
.41	9150.40	.92	9154.24	.86	8981.08	2.60	8638.98	4.23	8142.27	5.68
.42	9151.32	.91	9153.38	.88	8978.48	2.62	8634.75	4.24	8136.59	5.70
.43	9152.23	.89	9152.50	.90	8975.86	2.64	8630.51	4.26	8130.89	5.71
.44	9153.12	.87	9151.60	.92	8973.22	2.66	8626.25	4.28	8125.18	5.73
.45	9153.99	.85	9150.68	.94	8970.56	2.67	8621.97	4.29	8119.45	5.74
.46	9154.84	.84	9149.74	.95	8967.89	2.69	8617.68	4.31	8113.71	5.75
.47	9155.68	.81	9148.79	.97	8965.20	2.70	8613.37	4.33	8107.96	5.77
.48	9156.49	.80	9147.82	.99	8962.50	2.72	8609.04	4.34	8102.19	5.78
.49	9157.29	.78	9146.83	1.00	8959.78	2.74	8604.70	4.35	8096.41	5.79
.50	9158.07	+ .77	9145.83	−1.02	8957.04	−2.76	8600.35	−4.37	8090.62	−5.80

TABLE VII. ARGUMENT 2.

Equation $= 4587''.400 + 4586''.9 \text{ sin. } (2t - x) + 31''.2 \text{ sin. } (4t - 2x)$.

Period, 31.81193574 days.

Days.	15		16		17		18		19	
Decimals of a Day.	Equation.	Dif.	Equation.	Dif.	Equation.	Dif.	Equation.	Dif.	Equation.	Dif.
.50	9158″.07	+ .77	9145″.83	−1.02	8957″.04	−2.76	8600″.35	−4.37	8090″.62	−5.80
.51	9158.81	.74	9144.81	1.04	8954.28	2.77	8595.98	4.38	8084.82	5.82
.52	9159.58	.73	9143.77	1.06	8951.51	2.79	8591.60	4.40	8079.00	5.83
.53	9160.31	.71	9142.71	1.08	8948.72	2.80	8587.20	4.42	8073.17	5.85
.54	9161.02	.69	9141.63	1.10	8945.92	2.83	8582.78	4.43	8067.32	5.86
.55	9161.71	.68	9140.53	1.11	8943.09	2.84	8578.35	4.45	8061.46	5.87
.56	9162.39	.65	9139.42	1.12	8940.25	2.85	8573.90	4.46	8055.59	5.88
.57	9163.04	.64	9138.30	1.15	8937.40	2.88	8569.44	4.47	8049.71	5.90
.58	9163.68	.62	9137.15	1.16	8934.52	2.89	8564.97	4.49	8043.81	5.91
.59	9164.30	.60	9135.99	1.19	8931.63	2.90	8560.48	4.51	8037.90	5.93
.60	9164.90	.59	9134.80	1.20	8928.73	2.92	8555.97	4.52	8031.97	5.93
.61	9165.49	.56	9133.60	1.22	8925.81	2.94	8551.45	4.54	8026.04	5.95
.62	9166.05	.55	9132.38	1.23	8922.87	2.96	8546.91	4.55	8020.09	5.97
.63	9166.60	.53	9131.15	1.25	8919.91	2.98	8542.36	4.57	8014.12	5.97
.64	9167.13	.51	9129.90	1.27	8916.93	2.99	8537.79	4.58	8008.15	5.99
.65	9167.64	.50	9128.63	1.29	8913.94	3.00	8533.21	4.60	8002.16	6.00
.66	9168.14	.48	9127.34	1.31	8910.94	3.03	8528.61	4.62	7996.16	6.02
.67	9168.62	.46	9126.03	1.32	8907.91	3.04	8523.99	4.63	7990.14	6.03
.68	9169.08	.44	9124.71	1.34	8904.87	3.05	8519.36	4.64	7984.11	6.04
.69	9169.52	.42	9123.37	1.36	8901.82	3.07	8514.72	4.65	7978.07	6.05
.70	9169.94	.40	9122.01	1.37	8898.75	3.09	8510.07	4.67	7972.02	6.07
.71	9170.34	.39	9120.64	1.40	8895.66	3.11	8505.40	4.69	7965.95	6.08
.72	9170.73	.37	9119.24	1.41	8892.55	3.12	8500.71	4.70	7959.87	6.09
.73	9171.10	.35	9117.83	1.43	8889.43	3.14	8496.01	4.72	7953.78	6.11
.74	9171.45	.33	9116.40	1.44	8886.29	3.16	8491.29	4.73	7947.67	6.12
.75	9171.78	.32	9114.96	1.47	8883.13	3.17	8486.56	4.74	7941.55	6.13
.76	9172.10	.30	9113.49	1.48	8879.96	3.19	8481.82	4.77	7935.42	6.14
.77	9172.40	.28	9112.01	1.50	8876.77	3.21	8477.05	4.78	7929.28	6.16
.78	9172.68	.26	9110.51	1.52	8873.56	3.22	8472.27	4.79	7923.12	6.17
.79	9172.94	.24	9108.99	1.53	8870.34	3.24	8467.48	4.80	7916.95	6.18
.80	9173.18	.23	9107.46	1.55	8867.10	3.25	8462.68	4.82	7910.77	6.19
.81	9173.41	.21	9105.91	1.57	8863.85	3.27	8457.85	4.83	7904.58	6.21
.82	9173.62	.19	9104.34	1.58	8860.58	3.29	8453.02	4.85	7898.37	6.22
.83	9173.81	.17	9102.76	1.61	8857.29	3.31	8448.17	4.87	7892.15	6.23
.84	9173.98	.15	9101.15	1.62	8853.98	3.32	8443.30	4.88	7885.92	6.24
.85	9174.13	.14	9099.53	1.64	8850.66	3.33	8438.42	4.89	7879.68	6.26
.86	9174.27	.12	9097.89	1.66	8847.33	3.36	8433.53	4.91	7873.42	6.27
.87	9174.39	.10	9096.23	1.67	8843.97	3.37	8428.62	4.93	7867.15	6.28
.88	9174.49	.08	9094.56	1.69	8840.60	3.38	8423.69	4.94	7860.87	6.29
.89	9174.57	.07	9092.87	1.71	8837.22	3.40	8418.75	4.95	7854.58	6.31
.90	9174.64	.05	9091.16	1.73	8833.82	3.42	8413.80	4.97	7848.27	6.32
.91	9174.69	.03	9089.43	1.74	8830.40	3.44	8408.83	4.99	7841.95	6.33
.92	9174.72	+ .01	9087.69	1.76	8826.96	3.45	8403.84	5.00	7835.62	6.34
.93	9174.73	− .01	9085.93	1.78	8823.51	3.47	8398.84	5.01	7829.28	6.35
.94	9174.72	.03	9084.15	1.80	8820.04	3.48	8393.83	5.02	7822.93	6.37
.95	9174.69	.04	9082.35	1.81	8816.56	3.50	8388.81	5.04	7816.56	6.38
.96	9174.65	.06	9080.54	1.83	8813.06	3.52	8383.77	5.06	7810.18	6.39
.97	9174.59	.08	9078.71	1.85	8809.54	3.53	8378.71	5.07	7803.79	6.40
.98	9174.51	.10	9076.86	1.87	8806.01	3.55	8373.64	5.08	7797.39	6.42
.99	9174.41	− .11	9074.99	−1.88	8802.46	−3.57	8368.56	−5.10	7790.97	−6.43
1.00	9174.30		9073.11		8798.89		8363.46		7784.54	

TABLE VII. ARGUMENT 2.

Equation = 4587″.400 + 4586″.9 sin. $(2t - x)$ + 31″.2 sin. $(4t - 2x)$.

Period, 31.81193574 days.

Days.	20		21		22		23		21	
Decimals of a Day.	Equation.	Dif.	Equation.	Dif.	Equation.	Dif.	Equation.	Dif.	Equation.	Dif.
.00	7781.51″	−6.15	7084.88″	−7.51	6291.12″	−8.30	5134.21″	−8.79	4545.39″	−8.94
.01	7778.09	6.45	7077.37	7.53	6283.12	8.32	5125.42	8.79	4536.45	8.93
.02	7771.64	6.46	7069.81	7.54	6274.80	8.32	5116.63	8.79	4527.52	8.94
.03	7765.18	6.48	7062.30	7.54	6266.48	8.33	5107.81	8.80	4518.58	8.93
.04	7758.70	6.49	7054.76	7.56	6258.15	8.33	5399.04	8.80	4509.65	8.94
.05	7752.21	6.50	7047.20	7.56	6219.82	8.34	5390.24	8.80	4500.71	8.93
.06	7745.71	6.51	7039.64	7.58	6211.18	8.35	5381.44	8.80	4491.78	8.94
.07	7739.20	6.53	7032.06	7.58	6233.13	8.35	5372.64	8.81	4482.84	8.93
.08	7732.67	6.53	7024.48	7.59	6224.78	8.36	5363.83	8.81	4473.91	8.93
.09	7726.14	6.55	7016.89	7.60	6216.42	8.37	5355.02	8.81	4464.98	8.94
.10	7719.59	6.56	7009.29	7.62	6208.05	8.37	5346.21	8.82	4456.04	8.93
.11	7713.03	6.57	7001.67	7.62	6199.68	8.37	5337.39	8.82	4447.11	8.93
.12	7706.46	6.59	6994.05	7.62	6191.31	8.39	5328.57	8.82	4438.18	8.93
.13	7699.87	6.59	6986.43	7.64	6182.92	8.39	5319.75	8.83	4429.25	8.93
.14	7693.28	6.61	6978.79	7.65	6174.53	8.39	5310.92	8.83	4420.32	8.93
.15	7686.67	6.62	6971.14	7.65	6166.14	8.10	5302.09	8.82	4411.39	8.93
.16	7680.05	6.63	6963.49	7.67	6157.74	8.11	5293.27	8.84	4402.16	8.93
.17	7673.42	6.64	6955.82	7.67	6149.33	8.41	5284.43	8.83	4393.53	8.93
.18	7666.78	6.66	6948.15	7.68	6140.92	8.12	5275.60	8.84	4384.60	8.93
.19	7660.12	6.66	6940.47	7.69	6132.50	8.42	5266.76	8.84	4375.67	8.92
.20	7653.46	6.68	6932.78	7.71	6124.08	8.13	5257.92	8.84	4366.75	8.93
.21	7646.78	6.69	6925.07	7.70	6115.65	8.14	5249.08	8.85	4357.82	8.93
.22	7640.09	6.70	6917.37	7.72	6107.21	8.14	5240.23	8.85	4348.89	8.92
.23	7633.39	6.71	6909.65	7.73	6098.77	8.11	5231.38	8.85	4339.97	8.92
.24	7626.68	6.73	6901.92	7.73	6090.33	8.16	5222.53	8.85	4331.05	8.92
.25	7619.95	6.73	6894.19	7.75	6081.87	8.16	5213.68	8.85	4322.13	8.92
.26	7613.22	6.75	6886.44	7.75	6073.11	8.46	5204.83	8.86	4313.21	8.92
.27	7606.17	6.75	6878.69	7.76	6064.95	8.17	5195.97	8.86	4304.29	8.92
.28	7599.72	6.78	6870.93	7.77	6056.18	8.48	5187.11	8.86	4295.37	8.92
.29	7592.94	6.78	6863.16	7.78	6048.00	8.18	5178.25	8.87	4286.45	8.92
.30	7586.16	6.79	6855.38	7.79	6039.52	8.48	5169.38	8.86	4277.53	8.92
.31	7579.37	6.80	6847.59	7.79	6031.01	8.49	5160.52	8.87	4268.61	8.91
.32	7572.57	6.82	6839.80	7.81	6022.55	8.50	5151.65	8.87	4259.70	8.91
.33	7565.75	6.82	6831.99	7.81	6014.05	8.50	5142.78	8.87	4250.79	8.91
.34	7558.93	6.84	6824.18	7.82	6005.55	8.51	5133.91	8.88	4241.88	8.91
.35	7552.09	6.85	6816.36	7.83	5997.04	8.51	5125.03	8.88	4232.97	8.91
.36	7545.24	6.86	6808.53	7.83	5988.53	8.52	5116.15	8.87	4224.06	8.91
.37	7538.38	6.87	6800.70	7.85	5980.01	8.52	5107.28	8.89	4215.15	8.91
.38	7531.51	6.88	6792.85	7.85	5971.49	8.53	5098.39	8.88	4206.21	8.90
.39	7524.63	6.89	6785.00	7.87	5962.96	8.54	5089.51	8.88	4197.31	8.91
.40	7517.74	6.91	6777.13	7.87	5954.42	8.53	5080.63	8.89	4188.43	8.90
.41	7510.83	6.91	6769.26	7.88	5945.89	8.55	5071.74	8.89	4179.53	8.90
.42	7503.92	6.93	6761.38	7.88	5937.34	8.55	5062.85	8.89	4170.63	8.90
.43	7496.99	6.93	6753.50	7.90	5928.79	8.55	5053.96	8.89	4161.73	8.90
.44	7490.06	6.95	6745.60	7.90	5920.24	8.56	5045.07	8.89	4152.83	8.90
.45	7483.11	6.96	6737.70	7.92	5911.68	8.56	5036.18	8.90	4143.93	8.89
.46	7476.15	6.97	6729.78	7.91	5903.12	8.57	5027.28	8.89	4135.04	8.89
.47	7469.18	6.98	6721.87	7.93	5891.55	8.57	5018.39	8.90	4126.15	8.89
.48	7462.20	6.99	6713.94	7.93	5885.98	8.58	5009.49	8.90	4117.26	8.89
.49	7455.21	7.01	6706.01	7.95	5877.40	8.59	5000.59	8.90	4108.37	8.89
.50	7448.20	−7.01	6698.06	−7.95	5868.81	−8.58	4991.69	−8.90	4099.48	−8.88

TABLE VII. ARGUMENT 2.

Equation $= 4587''.400 + 4586''.9 \sin.(2t - x) + 31''.2 \sin.(4t - 2x)$.

Period, 31.81193574 days.

Days.	20		21		22		23		24	
Decimals of a Day	Equation.	Dif.	Equation.	Dif.	Equation.	Dif.	Equation.	Dif.	Equation.	Dif.
.50	7448.20	−7.01	6698.06	−7.95	5868.81	−8.58	4991.69	−8.90	4099.48	−8.88
.51	7441.19	7.02	6690.11	7.96	5860.23	8.60	4982.79	8.91	4090.60	8.89
.52	7434.17	7.04	6682.15	7.96	5851.63	8.59	4973.88	8.91	4081.71	8.88
.53	7427.13	7.04	6674.19	7.98	5843.04	8.61	4964.97	8.90	4072.83	8.88
.54	7420.09	7.06	6666.21	7.98	5834.43	8.60	4956.07	8.91	4063.95	8.88
.55	7413.03	7.06	6658.23	7.99	5825.83	8.61	4947.16	8.91	4055.07	8.87
.56	7405.97	7.08	6650.24	8.00	5817.22	8.62	4938.25	8.91	4046.20	8.87
.57	7398.89	7.09	6642.24	8.01	5808.60	8.62	4929.34	8.92	4037.33	8.88
.58	7391.80	7.10	6634.23	8.01	5799.98	8.63	4920.42	8.91	4028.45	8.87
.59	7384.70	7.11	6626.22	8.02	5791.35	8.63	4911.51	8.91	4019.58	8.87
.60	7377.59	7.12	6618.20	8.03	5782.72	8.63	4902.60	8.92	4010.71	8.86
.61	7370.47	7.13	6610.17	8.03	5774.09	8.64	4893.68	8.92	4001.85	8.86
.62	7363.34	7.14	6602.14	8.04	5765.45	8.64	4884.76	8.91	3992.99	8.86
.63	7356.20	7.15	6594.10	8.06	5756.81	8.65	4875.85	8.92	3984.13	8.86
.64	7349.05	7.16	6586.04	8.06	5748.16	8.65	4866.93	8.92	3975.27	8.86
.65	7341.89	7.17	6577.98	8.07	5739.51	8.66	4858.01	8.93	3966.41	8.85
.66	7334.72	7.18	6569.91	8.07	5730.85	8.66	4849.08	8.92	3957.56	8.85
.67	7327.54	7.20	6561.84	8.08	5722.19	8.66	4840.16	8.92	3948.71	8.85
.68	7320.34	7.20	6553.76	8.09	5713.53	8.67	4831.24	8.92	3939.86	8.85
.69	7313.14	7.21	6545.67	8.10	5704.86	8.67	4822.32	8.93	3931.01	8.85
.70	7305.93	7.23	6537.57	8.10	5696.19	8.68	4813.39	8.93	3922.16	8.84
.71	7298.70	7.23	6529.47	8.11	5687.51	8.68	4804.46	8.92	3913.32	8.84
.72	7291.47	7.25	6521.36	8.12	5678.83	8.69	4795.54	8.93	3904.48	8.84
.73	7284.22	7.25	6513.24	8.12	5670.14	8.69	4786.61	8.93	3895.64	8.83
.74	7276.97	7.26	6505.12	8.14	5661.45	8.69	4777.68	8.93	3886.81	8.83
.75	7269.71	7.28	6496.98	8.14	5652.76	8.70	4768.75	8.93	3877.98	8.83
.76	7262.43	7.28	6488.84	8.14	5644.06	8.70	4759.82	8.93	3869.15	8.83
.77	7255.15	7.30	6480.70	8.16	5635.36	8.70	4750.89	8.93	3860.32	8.82
.78	7247.85	7.31	6472.5[illegible]	[illegible]	[illegible]	8.71	4741.96	8.93	3851.50	8.82
.79	7240.54	7.32	6464.38	8.16	5617.95	8.72	4733.03	8.93	3842.68	8.82
.80	7233.22	7.32	6456.22	8.18	5609.23	8.72	4724.10	[illegible]	3833.86	8.82
.81	7225.90	7.34	6448.04	8.18	5600.51	8.72	4715.16	[illegible]	3825.04	8.81
.82	7218.57	7.35	6439.86	8.19	5591.79	8.72	4706.23	8.93	3816.23	8.81
.83	7211.22	7.35	6431.67	8.20	5583.07	8.73	4697.30	8.94	3807.42	8.81
.84	7203.87	7.36	6423.47	8.20	5574.34	8.73	4688.36	8.93	3798.61	8.80
.85	7196.51	7.38	6415.27	8.21	5565.61	8.74	4679.43	8.94	3789.81	8.80
.86	7189.13	7.38	6407.06	8.22	5556.87	8.74	4670.49	8.93	3781.01	8.80
.87	7181.75	7.39	6398.84	8.22	5548.13	8.74	4661.56	8.94	3772.21	8.79
.88	7174.36	7.40	6390.62	8.23	5539.39	8.75	4652.62	8.93	3763.42	8.79
.89	7166.96	7.42	6382.39	8.24	5530.64	8.75	4643.69	8.94	3754.63	8.79
.90	7159.54	7.42	6374.15	8.24	5521.89	8.75	4634.75	8.93	3745.84	8.79
.91	7152.12	7.43	6365.91	8.25	5513.14	8.76	4625.82	8.94	3737.05	8.78
.92	7144.69	7.44	6357.66	8.26	5504.38	8.76	4616.88	8.94	3728.27	8.78
.93	7137.25	7.45	6349.40	8.26	5495.62	8.76	4607.94	8.93	3719.49	8.77
.94	7129.80	7.46	6341.14	8.27	5486.86	8.77	4599.01	8.94	3710.72	8.77
.95	7122.34	7.48	6332.87	8.28	5478.09	8.77	4590.07	8.94	3701.95	8.77
.96	7114.86	7.48	6324.59	8.28	5469.32	8.77	4581.13	8.93	3693.18	8.77
.97	7107.38	7.49	6316.31	8.29	5460.55	8.78	4572.20	8.94	3684.41	8.76
.98	7099.89	7.50	6308.02	8.29	5451.77	8.78	4563.26	8.93	3675.65	8.76
.99	7092.39	−7.51	6299.73	−8.31	5442.99	−8.78	4554.33	−8.94	3666.89	−8.75
1.00	7084.88		6291.42		5434.21		4545.39		3658.14	

TABLE VII. ARGUMENT 2.

Equation $= 4587''.400 + 4586''.9 \sin. (2t - x) + 31''.2 \sin. (4t - 2x)$.

Period, 31.81193574 days.

Days.	25		26		27		28		29	
Decimals of a Day	Equation.	Dif.	Equation.	Dif.	Equation.	Dif.	Equation.	Dif.	Equation.	Dif.
.00	3658.14	−8.76	2805.57	−8.21	2019.69	−7.42	1330.30	−6.31	763.97	−4.96
.01	3649.38	8.75	2797.33	8.23	2012.27	7.41	1323.99	6.30	759.01	4.95
.02	3640.63	8.74	2789.10	8.23	2004.86	7.40	1317.69	6.29	754.06	4.93
.03	3631.89	8.74	2780.87	8.22	1997.46	7.39	1311.40	6.28	749.13	4.92
.04	3623.15	8.74	2772.65	8.21	1990.07	7.38	1305.12	6.26	744.21	4.91
.05	3614.41	8.73	2764.44	8.21	1982.69	7.36	1298.86	6.25	739.30	4.89
.06	3605.68	8.73	2756.23	8.20	1975.33	7.36	1292.61	6.24	734.41	4.87
.07	3596.95	8.73	2748.03	8.19	1967.97	7.35	1286.37	6.23	729.54	4.86
.08	3588.22	8.72	2739.84	8.19	1960.62	7.35	1280.14	6.21	724.68	4.85
.09	3579.50	8.72	2731.65	8.18	1953.27	7.33	1273.93	6.20	719.83	4.82
.10	3570.78	8.72	2723.47	8.17	1945.94	7.32	1267.73	6.19	715.01	4.82
.11	3562.06	8.71	2715.30	8.17	1938.62	7.31	1261.54	6.18	710.19	4.80
.12	3553.35	8.71	2707.13	8.16	1931.31	7.30	1255.36	6.16	705.39	4.79
.13	3544.64	8.70	2698.97	8.15	1924.01	7.29	1249.20	6.15	700.60	4.77
.14	3535.94	8.70	2690.82	8.14	1916.72	7.28	1243.05	6.14	695.83	4.76
.15	3527.24	8.69	2682.68	8.14	1909.44	7.27	1236.91	6.13	691.07	4.73
.16	3518.55	8.69	2674.54	8.12	1902.17	7.26	1230.78	6.11	686.34	4.73
.17	3509.86	8.69	2666.42	8.13	1894.91	7.25	1224.67	6.10	681.61	4.71
.18	3501.17	8.69	2658.29	8.11	1887.66	7.24	1218.57	6.09	676.90	4.70
.19	3492.48	8.68	2650.18	8.11	1880.42	7.23	1212.48	6.07	672.20	4.68
.20	3483.80	8.67	2642.07	8.10	1873.19	7.22	1206.41	6.06	667.52	4.67
.21	3475.13	8.68	2633.97	8.09	1865.97	7.21	1200.35	6.05	662.85	4.65
.22	3466.45	8.66	2625.88	8.09	1858.76	7.20	1194.30	6.04	658.20	4.63
.23	3457.79	8.67	2617.79	8.08	1851.56	7.18	1188.26	6.02	653.57	4.63
.24	3449.12	8.65	2609.71	8.07	1844.38	7.18	1182.24	6.01	648.94	4.60
.25	3440.47	8.66	2601.64	8.06	1837.20	7.17	1176.23	6.00	644.34	4.59
.26	3431.81	8.65	2593.58	8.06	1830.03	7.16	1170.23	5.98	639.75	4.58
.27	3423.16	8.64	2585.52	8.05	1822.87	7.15	1164.25	5.97	635.17	4.56
.28	3414.52	8.64	2577.47	8.04	1815.72	7.13	1158.28	5.96	630.61	4.55
.29	3405.88	8.64	2569.43	8.03	1808.59	7.13	1152.32	5.95	626.06	4.53
.30	3397.24	8.63	2561.40	8.03	1801.46	7.11	1146.37	5.93	621.53	4.51
.31	3388.61	8.63	2553.37	8.01	1794.35	7.11	1140.44	5.90	617.02	4.51
.32	3379.98	8.63	2545.36	8.01	1787.24	7.09	1134.54	5.92	612.51	4.48
.33	3371.35	8.62	2537.35	8.01	1780.15	7.09	1128.62	5.89	608.03	4.48
.34	3362.73	8.61	2529.34	7.99	1773.06	7.07	1122.73	5.88	603.55	4.45
.35	3354.12	8.61	2521.35	7.99	1765.99	7.06	1116.85	5.87	599.10	4.44
.36	3345.51	8.60	2513.36	7.98	1758.93	7.06	1110.98	5.85	594.66	4.42
.37	3336.91	8.60	2505.38	7.97	1751.87	7.04	1105.13	5.84	590.24	4.41
.38	3328.31	8.60	2497.41	7.96	1744.83	7.03	1099.29	5.83	585.83	4.39
.39	3319.71	8.59	2489.45	7.96	1737.80	7.02	1093.46	5.81	581.44	4.38
.40	3311.12	8.59	2481.49	7.95	1730.78	7.01	1087.65	5.80	577.06	4.37
.41	3302.53	8.58	2473.54	7.94	1723.77	6.99	1081.85	5.79	572.69	4.34
.42	3293.95	8.57	2465.60	7.93	1716.78	6.99	1076.06	5.77	568.35	4.34
.43	3285.38	8.57	2457.67	7.93	1709.79	6.98	1070.29	5.76	564.01	4.31
.44	3276.81	8.57	2449.74	7.91	1702.81	6.96	1064.53	5.75	559.70	4.30
.45	3268.24	8.56	2441.83	7.91	1695.85	6.96	1058.78	5.73	555.40	4.29
.46	3259.68	8.56	2433.92	7.92	1688.89	6.94	1053.05	5.72	551.11	4.27
.47	3251.12	8.55	2426.02	7.89	1681.95	6.93	1047.33	5.71	546.84	4.26
.48	3242.57	8.55	2418.13	7.89	1675.02	6.92	1041.62	5.69	542.58	4.24
.49	3234.02	8.54	2410.24	7.87	1668.10	6.91	1035.93	5.68	538.34	4.22
.50	3225.48	−8.54	2402.37	−7.87	1661.19	−6.90	1030.25	−5.67	534.12	−4.21

TABLE VII. ARGUMENT 2.

Equation $= 4587''.400 + 4586''.9 \text{ sin. } (2t - x) + 31''.2 \text{ sin. } (4t - 2x)$.

Period, 31.81193574 days.

Days.	25		26		27		28		29	
Decimals of a Day	Equation.	Dif.	Equation.	Dif.	Equation.	Dif.	Equation.	Dif.	Equation.	Dif.
.50	3225.18	−8.54	2102.37	−7.87	1661.19	−6.90	1030.25	−5.67	534.12	−4.21
.51	3216.94	8.53	2394.50	7.86	1654.29	6.89	1024.58	5.65	529.91	4.19
.52	3208.41	8.53	2386.64	7.85	1647.40	6.88	1018.93	5.64	525.72	4.18
.53	3199.88	8.52	2378.79	7.84	1640.52	6.86	1013.29	5.63	521.54	4.16
.54	3191.36	8.51	2370.95	7.83	1633.66	6.86	1007.66	5.61	517.38	4.15
.55	3182.85	8.50	2363.12	7.83	1626.80	6.84	1002.05	5.59	513.23	4.13
.56	3174.35	8.52	2355.29	7.82	1619.96	6.84	996.46	5.59	509.10	4.11
.57	3165.83	8.50	2347.47	7.80	1613.12	6.82	990.87	5.57	504.99	4.11
.58	3157.33	8.49	2339.67	7.80	1606.30	6.81	985.30	5.56	500.88	4.08
.59	3148.84	8.49	2331.87	7.80	1599.49	6.80	979.74	5.54	496.80	4.07
.60	3140.35	8.49	2324.07	7.78	1592.69	6.78	974.20	5.53	492.73	4.05
.61	3131.86	8.47	2316.29	7.78	1585.91	6.78	968.67	5.52	488.68	4.04
.62	3123.39	8.48	2308.51	7.76	1579.13	6.76	963.15	5.50	484.64	4.02
.63	3114.91	8.46	2300.75	7.76	1572.37	6.76	957.65	5.49	480.62	4.01
.64	3106.45	8.46	2292.99	7.75	1565.61	6.74	952.16	5.47	476.61	3.98
.65	3097.99	8.46	2285.24	7.74	1558.87	6.73	946.69	5.46	472.63	3.98
.66	3089.53	8.45	2277.50	7.73	1552.14	6.72	941.23	5.45	468.65	3.96
.67	3081.08	8.45	2269.77	7.72	1545.42	6.71	935.78	5.43	464.69	3.94
.68	3072.63	8.44	2262.05	7.72	1538.71	6.69	930.35	5.42	460.75	3.93
.69	3064.19	8.43	2254.33	7.70	1532.02	6.69	924.93	5.40	456.82	3.91
.70	3055.76	8.43	2246.63	7.70	1525.33	6.67	919.53	5.39	452.91	3.89
.71	3047.33	8.42	2238.93	7.69	1518.66	6.66	914.14	5.39	449.02	3.88
.72	3038.91	8.42	2231.24	7.68	1512.00	6.65	908.75	5.35	445.14	3.87
.73	3030.49	8.41	2223.56	7.67	1505.35	6.64	903.40	5.35	441.27	3.84
.74	3022.08	8.40	2215.89	7.66	1498.71	6.62	898.05	5.34	437.43	3.83
.75	3013.68	8.40	2208.23	7.65	1492.09	6.62	892.71	5.32	433.60	3.82
.76	3005.28	8.39	2200.58	7.64	1485.47	6.60	887.39	5.31	429.78	3.80
.77	2996.89	8.38	2192.94	7.64	1478.87	6.59	882.08	5.29	425.98	3.78
.78	2988.51	8.38	2185.30	7.62	1472.28	6.58	876.79	5.28	422.20	3.77
.79	2980.13	8.38	2177.68	7.62	1465.70	6.57	871.51	5.26	418.43	3.75
.80	2971.75	8.37	2170.06	7.60	1459.13	6.56	866.25	5.25	414.68	3.74
.81	2963.38	8.36	2162.46	7.60	1452.57	6.54	861.00	5.24	410.94	3.72
.82	2955.02	8.36	2154.86	7.59	1446.03	6.53	855.76	5.22	407.22	3.70
.83	2946.66	8.35	2147.27	7.58	1439.50	6.52	850.54	5.20	403.52	3.69
.84	2938.31	8.34	2139.69	7.57	1432.98	6.51	845.34	5.20	399.83	3.67
.85	2929.97	8.34	2132.12	7.56	1426.47	6.49	840.14	5.18	396.16	3.65
.86	2921.63	8.33	2124.56	7.55	1419.98	6.49	834.96	5.16	392.51	3.64
.87	2913.30	8.33	2117.01	7.55	1413.49	6.47	829.80	5.15	388.87	3.62
.88	2904.97	8.32	2109.46	7.53	1407.02	6.46	824.65	5.14	385.25	3.61
.89	2896.65	8.31	2101.93	7.52	1400.56	6.45	819.51	5.12	381.64	3.59
.90	2888.34	8.31	2094.41	7.52	1394.11	6.43	814.39	5.11	378.05	3.58
.91	2880.03	8.30	2086.89	7.50	1387.68	6.43	809.28	5.09	374.47	3.56
.92	2871.73	8.29	2079.39	7.50	1381.25	6.41	804.19	5.08	370.91	3.54
.93	2863.44	8.29	2071.89	7.48	1374.84	6.40	799.11	5.06	367.37	3.53
.94	2855.15	8.28	2064.41	7.48	1368.44	6.38	794.05	5.05	363.84	3.51
.95	2846.87	8.27	2056.93	7.47	1362.06	6.38	789.00	5.04	360.33	3.49
.96	2838.60	8.27	2049.46	7.45	1355.68	6.36	783.96	5.02	356.84	3.48
.97	2830.33	8.26	2042.01	7.45	1349.32	6.35	778.94	5.00	353.36	3.46
.98	2822.07	8.25	2034.56	7.44	1342.97	6.34	773.94	4.99	349.90	3.44
.99	2813.82	−8.25	2027.12	−7.43	1336.63	−6.33	768.95	−4.98	346.46	−3.43
1.00	2805.57		2019.69		1330.30		763.97		343.03	

TABLE VII. ARGUMENT 2.

Equation $= 4587''.400 + 4586''.9 \sin. (2t - x) + 31''.2 \sin. (4t - 2x)$.

Period, 31.81193574 days.

Days.	30		31	
Decimals of a Day.	Equation.	Difference.	Equation.	Difference.
.00	343.03	−3.41	84.67	−1.72
.01	339.62	3.40	82.95	1.70
.02	336.22	3.38	81.25	1.69
.03	332.84	3.36	79.56	1.66
.04	329.48	3.35	77.90	1.66
.05	326.13	3.33	76.24	1.63
.06	322.80	3.31	74.61	1.61
.07	319.49	3.30	73.00	1.60
.08	316.19	3.28	71.40	1.58
.09	312.91	3.27	69.82	1.56
.10	309.64	3.25	68.26	1.54
.11	306.39	3.23	66.72	1.53
.12	303.16	3.22	65.19	1.51
.13	299.94	3.19	63.68	1.49
.14	296.75	3.19	62.19	1.48
.15	293.56	3.16	60.71	1.45
.16	290.40	3.15	59.26	1.44
.17	287.25	3.14	57.82	1.42
.18	284.11	3.11	56.40	1.41
.19	281.00	3.10	54.99	1.38
.20	277.90	3.08	53.61	1.37
.21	274.82	3.07	52.24	1.35
.22	271.75	3.05	50.89	1.34
.23	268.70	3.03	49.55	1.31
.24	265.67	3.02	48.24	1.30
.25	262.65	3.00	46.94	1.28
.26	259.65	2.98	45.66	1.27
.27	256.67	2.97	44.39	1.24
.28	253.70	2.95	43.15	1.23
.29	250.75	2.93	41.92	1.21
.30	247.82	2.92	40.71	1.19
.31	244.90	2.90	39.52	1.18
.32	242.00	2.88	38.34	1.15
.33	239.12	2.87	37.19	1.14
.34	236.25	2.85	36.05	1.13
.35	233.40	2.83	34.92	1.10
.36	230.57	2.82	33.82	1.09
.37	227.75	2.80	32.73	1.07
.38	224.95	2.78	31.66	1.05
.39	222.17	2.77	30.61	1.03
.40	219.40	2.74	29.58	1.02
.41	216.66	2.74	28.56	1.00
.42	213.92	2.71	27.56	.98
.43	211.21	2.70	26.58	.96
.44	208.51	2.68	25.62	.94
.45	205.83	2.65	24.68	.93
.46	203.18	2.64	23.75	.91
.47	200.54	2.63	22.84	.89
.48	197.91	2.62	21.95	.88
.49	195.29	2.62	21.07	.85
.50	192.67	−2.58	20.22	− .84

TABLE VII. ARGUMENT 2.

Equation $= 4587''.400 + 4586''.9 \text{ sin. } (2t - x) + 31''.2 \text{ sin. } (4t - 2x)$.

Period, 31.81193574 days.

Days.	30		31	
Decimals of a Day.	Equation.	Difference.	Equation.	Difference.
.50	192.67	−2.58	20.22	−.84
.51	190.09	2.56	19.38	.82
.52	187.53	2.54	18.56	.80
.53	184.99	2.53	17.76	.79
.54	182.46	2.51	16.97	.77
.55	179.95	2.50	16.20	.75
.56	177.45	2.48	15.45	.73
.57	174.97	2.46	14.72	.71
.58	172.51	2.44	14.01	.70
.59	170.07	2.42	13.31	.67
.60	167.65	2.41	12.64	.67
.61	165.24	2.40	11.97	.64
.62	162.84	2.37	11.33	.62
.63	160.47	2.36	10.71	.61
.64	158.11	2.34	10.10	.59
.65	155.77	2.33	9.51	.57
.66	153.44	2.31	8.94	.56
.67	151.13	2.29	8.38	.53
.68	148.84	2.27	7.85	.52
.69	146.57	2.25	7.33	.50
.70	144.32	2.25	6.83	.48
.71	142.07	2.22	6.35	.47
.72	139.85	2.20	5.88	.44
.73	137.65	2.19	5.41	.43
.74	135.46	2.17	5.01	.41
.75	133.29	2.15	4.60	.40
.76	131.14	2.14	4.20	.37
.77	129.00	2.12	3.83	.36
.78	126.88	2.10	3.47	.34
.79	124.78	2.08	3.13	.32
.80	122.70	2.06	2.81	.30
.81	120.64	2.05	2.51	.29
.82	118.59	2.04	2.22	.27
.83	116.55	2.01	1.95	.25
.84	114.54	2.00	1.70	.23
.85	112.54	1.98	1.47	.21
.86	110.56	1.96	1.26	.20
.87	108.60	1.94	1.06	.18
.88	106.66	1.93	.88	.16
.89	104.73	1.91	.72	.14
.90	102.82	1.89	.58	.12
.91	100.93	1.88	.46	.11
.92	99.05	1.86	.35	.09
.93	97.19	1.84	.26	.08
.94	95.35	1.82	.18	.05
.95	93.53	1.81	.13	.03
.96	91.72	1.79	.10	−.02
.97	89.93	1.77	.08	.00
.98	88.16	1.76	.08	+.02
.99	86.40	−1.73	.10	+.03
1.00	84.67		.13	

TABLE VIII. ARGUMENT 3.

Equation = 2457″.02 — 122″.1 sin. t + 2371″.0 sin. $2t$ + 0″.9 sin. $3t$ + 14″.4 sin. $4t$.

Period, 29.530587997 days.

Days.	0		1		2		3		4		5	
Decimals of a Day.	Equation.	Diff.	Equation.	Diff.	Equation.	Diff.	Equation.	Diff.	Equation.	Diff.	Equation.	Diff.
.00	2211.80	10.42	3243.69	9.91	4133.61	7.61	4723.41	3.98	4913.02	0.28	4676.85	4.41
.01	2222.22	10.42	3253.60	9.89	4141.22	7.58	4727.39	3.94	4912.74	0.32	4672.44	4.44
.02	2232.61	10.42	3263.49	9.87	4148.80	7.55	4731.33	3.91	4912.42	0.36	4668.00	4.48
.03	2243.06	10.43	3273.36	9.85	4156.35	7.51	4735.24	3.86	4912.06	0.40	4663.52	4.52
.04	2253.49	10.43	3283.21	9.84	4163.86	7.48	4739.10	3.82	4911.66	0.45	4659.00	4.55
.05	2263.92	10.41	3293.05	9.83	4171.34	7.45	4712.92	3.78	4911.21	0.49	4654.45	4.59
.06	2274.36	10.44	3302.88	9.81	4178.79	7.42	4716.70	3.74	4910.72	0.54	4649.86	4.63
.07	2284.80	10.44	3312.69	9.80	4186.21	7.39	4750.41	3.70	4910.18	0.58	4645.23	4.67
.08	2295.24	10.45	3322.49	9.78	4193.60	7.36	4751.11	3.66	4909.60	0.62	4640.56	4.70
.09	2305.69	10.44	3332.27	9.76	4200.96	7.33	4757.80	3.62	4908.98	0.66	4635.86	4.74
.10	2316.13	10.45	3342.03	9.74	4208.29	7.30	4761.42	3.57	4908.32	0.71	4631.12	4.78
.11	2326.58	10.45	3351.77	9.73	4215.59	7.26	4764.99	3.52	4907.61	0.75	4626.34	4.82
.12	2337.03	10.46	3361.50	9.71	4222.85	7.23	4768.51	3.49	4906.86	0.79	4621.52	4.85
.13	2347.49	10.46	3371.21	9.70	4230.08	7.19	4772.00	3.15	4906.07	0.84	4616.67	4.89
.14	2357.95	10.45	3380.91	9.68	4237.27	7.16	4775.15	3.41	4905.23	0.88	4611.78	4.93
.15	2368.40	10.46	3390.59	9.66	4244.43	7.13	4778.86	3.37	4904.35	0.92	4606.85	4.97
.16	2378.86	10.47	3100.25	9.65	4251.56	7.11	4782.23	3.32	4903.43	0.96	4601.88	5.00
.17	2389.33	10.46	3409.90	9.63	4258.67	7.06	4785.55	3.29	4902.47	1.01	4596.88	5.04
.18	2399.79	10.46	3419.53	9.61	4265.73	7.04	4788.84	3.24	4901.46	1.05	4591.84	5.07
.19	2410.25	10.46	3429.14	9.59	4272.77	7.00	4792.08	3.20	4900.41	1.10	4586.77	5.11
.20	2420.71	10.46	3438.73	9.57	4279.77	6.97	4795.28	3.16	4899.31	1.13	4581.66	5.15
.21	2431.17	10.47	3448.30	9.56	4286.74	6.93	4798.44	3.12	4898.18	1.18	4576.51	5.19
.22	2441.64	10.47	3457.86	9.54	4293.67	6.91	4801.56	3.07	4897.00	1.23	4571.32	5.22
.23	2452.11	10.47	3467.40	9.52	4300.58	6.87	4804.63	3.03	4895.77	1.26	4566.10	5.25
.24	2462.58	10.46	3476.92	9.50	4307.45	6.84	4807.66	2.99	4894.51	1.32	4560.85	5.30
.25	2473.04	10.47	3486.42	9.48	4314.29	6.80	4810.65	2.95	4893.19	1.35	4555.55	5.33
.26	2483.51	10.47	3495.90	9.46	4321.09	6.77	4813.60	2.91	4891.84	1.39	4550.22	5.36
.27	2493.98	10.46	3505.36	9.44	4327.86	6.73	4816.51	2.86	4890.45	1.43	4544.86	5.40
.28	2504.44	10.46	3514.80	9.43	4334.59	6.70	4819.37	2.82	4889.02	1.48	4539.46	5.43
.29	2514.90	10.46	3524.23	9.40	4311.29	6.67	4822.19	2.79	4887.54	1.52	4534.03	5.47
.30	2525.36	10.46	3533.63	9.39	4347.96	6.63	4824.98	2.71	4886.02	1.56	4528.56	5.51
.31	2535.82	10.46	3543.02	9.37	4354.59	6.60	4827.72	2.69	4884.46	1.61	4523.05	5.54
.32	2546.28	10.46	3552.39	9.34	4361.19	6.56	4830.41	2.65	4882.85	1.65	4517.51	5.57
.33	2556.74	10.46	3561.73	9.32	4367.75	6.52	4833.06	2.62	4881.20	1.69	4511.94	5.61
.34	2567.20	10.45	3571.05	9.31	4374.27	6.49	4835.68	2.57	4879.51	1.73	4506.33	5.65
.35	2577.65	10.45	3580.36	9.28	4380.76	6.46	4838.25	2.52	4877.78	1.77	4500.68	5.68
.36	2588.10	10.45	3589.64	9.26	4387.22	6.43	4840.77	2.48	4876.01	1.82	4495.00	5.72
.37	2598.55	10.45	3598.90	9.24	4393.65	6.39	4843.25	2.44	4874.19	1.86	4489.28	5.74
.38	2609.00	10.44	3608.14	9.22	4400.04	6.35	4845.69	2.40	4872.33	1.90	4483.54	5.79
.39	2619.44	10.44	3617.36	9.20	4406.39	6.32	4848.09	2.36	4870.43	1.94	4477.75	5.82
.40	2629.88	10.44	3626.56	9.18	4412.71	6.28	4850.45	2.31	4868.49	1.98	4471.93	5.85
.41	2640.32	10.43	3635.74	9.16	4418.99	6.25	4852.76	2.27	4866.51	2.03	4466.08	5.89
.42	2650.75	10.43	3644.90	9.14	4125.24	6.21	4855.03	2.23	4864.18	2.07	4460.19	5.92
.43	2661.18	10.43	3654.04	9.11	4431.45	6.18	4857.26	2.19	4862.41	2.11	4454.27	5.95
.44	2671.61	10.43	3663.15	9.09	4437.63	6.14	4859.45	2.14	4860.30	2.15	4448.32	5.99
.45	2682.04	10.42	3672.24	9.07	4443.77	6.10	4861.59	2.10	4858.15	2.20	4442.33	6.02
.46	2692.46	10.41	3681.31	9.05	4449.87	6.07	4863.69	2.06	4855.95	2.24	4436.31	6.06
.47	2702.87	10.41	3690.36	9.03	4455.94	6.03	4865.75	2.01	4853.71	2.27	4130.25	6.09
.48	2713.28	10.40	3699.39	9.00	4461.97	5.99	4867.76	1.97	4851.44	2.32	4424.16	6.13
.49	2723.68	10.40	3708.39	8.97	4467.96	5.95	4869.73	1.92	4849.12	2.37	4418.03	6.16
.50	2734.08	10.40	3717.36	8.95	4473.91	5.93	4871.65	1.89	4846.75	2.41	4411.87	6.19

TABLE VIII. ARGUMENT 3.

Equation = 2457″.02 — 122″.1 sin. t + 2371″.0 sin. $2t$ + 0″.9 sin. $3t$ + 14″.4 sin. $4t$.

Period, 29.530587997 days.

Days.	0		1		2		3		4		5	
Decimals of a Day.	Equation.	Diff.	Equation.	Diff.	Equation.	Diff.	Equation.	Diff.	Equation.	Diff.	Equation.	Diff.
.50	2734.08	10.40	3717.36	8.95	4473.91	5.93	4871.65	1.89	4846.75	2.41	4411.87	6.19
.51	2744.48	10.39	3726.31	8.93	4479.84	5.88	4873.54	1.84	4844.34	2.44	4405.68	6.22
.52	2754.87	10.39	3735.24	8.91	4485.72	5.85	4875.38	1.80	4841.90	2.49	4399.46	6.26
.53	2765.26	10.38	3744.15	8.89	4491.57	5.81	4877.18	1.76	4839.41	2.53	4393.20	6.28
.54	2775.64	10.37	3753.04	8.86	4497.38	5.78	4878.94	1.71	4836.88	2.57	4386.92	6.32
.55	2786.01	10.37	3761.90	8.84	4503.16	5.74	4880.65	1.67	4834.31	2.61	4380.60	6.35
.56	2796.38	10.37	3770.74	8.81	4508.90	5.70	4882.32	1.63	4831.70	2.65	4374.25	6.39
.57	2806.75	10.35	3779.55	8.79	4514.60	5.67	4883.95	1.59	4829.05	2.70	4367.86	6.42
.58	2817.10	10.35	3788.34	8.77	4520.27	5.62	4885.54	1.54	4826.35	2.73	4361.44	6.45
.59	2827.45	10.34	3797.11	8.74	4525.89	5.59	4887.08	1.50	4823.62	2.77	4354.99	6.48
.60	2837.79	10.33	3805.85	8.71	4531.48	5.56	4888.58	1.45	4820.85	2.82	4348.51	6.51
.61	2848.12	10.33	3814.56	8.68	4537.04	5.50	4890.03	1.41	4818.03	2.86	4342.00	6.55
.62	2858.45	10.32	3823.24	8.67	4542.54	5.48	4891.44	1.37	4815.17	2.90	4335.45	6.58
.63	2868.77	10.31	3831.91	8.64	4548.02	5.44	4892.81	1.33	4812.27	2.93	4328.87	6.61
.64	2879.08	10.30	3840.55	8.62	4553.46	5.40	4894.14	1.28	4809.34	2.98	4322.26	6.64
.65	2889.38	10.30	3849.17	8.59	4558.86	5.36	4895.42	1.23	4806.36	3.03	4315.62	6.67
.66	2899.68	10.29	3857.76	8.57	4564.22	5.33	4896.65	1.20	4803.33	3.07	4308.95	6.71
.67	2909.97	10.28	3866.33	8.54	4569.55	5.29	4897.85	1.15	4800.26	3.10	4302.24	6.73
.68	2920.25	10.28	3874.87	8.51	4574.84	5.25	4899.00	1.11	4797.16	3.14	4295.51	6.76
.69	2930.53	10.26	3883.38	8.49	4580.09	5.21	4900.11	1.07	4794.02	3.18	4288.75	6.80
.70	2940.79	10.25	3891.87	8.46	4585.30	5.17	4901.18	1.02	4790.84	3.23	4281.95	6.83
.71	2951.04	10.25	3900.33	8.43	4590.47	5.13	4902.20	0.98	4787.61	3.26	4275.12	6.86
.72	2961.29	10.24	3908.76	8.41	4595.60	5.10	4903.18	0.94	4784.35	3.30	4268.26	6.89
.73	2971.53	10.22	3917.17	8.38	4600.70	5.06	4904.12	0.89	4781.05	3.35	4261.37	6.92
.74	2981.75	10.21	3925.55	8.35	4605.76	5.01	4905.01	0.85	4777.70	3.38	4254.45	6.96
.75	2991.96	10.20	3933.90	8.33	4610.77	4.98	4905.86	0.81	4774.32	3.43	4247.49	6.98
.76	3002.16	10.19	3942.23	8.30	4615.75	4.94	4906.67	0.76	4770.89	3.46	4240.51	7.01
.77	3012.35	10.18	3950.53	8.27	4620.69	4.90	4907.43	0.72	4767.43	3.51	4233.50	7.04
.78	3022.53	10.17	3958.80	8.25	4625.59	4.86	4908.15	0.68	4763.92	3.54	4226.46	7.07
.79	3032.70	10.16	3967.05	8.22	4630.45	4.83	4908.83	0.64	4760.38	3.58	4219.39	7.09
.80	3042.86	10.16	3975.27	8.19	4635.28	4.79	4909.47	0.59	4756.80	3.63	4212.30	7.13
.81	3053.02	10.15	3983.46	8.16	4640.07	4.74	4910.06	0.54	4753.17	3.66	4205.17	7.16
.82	3063.17	10.14	3991.62	8.13	4644.81	4.71	4910.60	0.50	4749.51	3.71	4198.01	7.19
.83	3073.31	10.13	3999.75	8.11	4649.52	4.66	4911.10	0.46	4745.80	3.75	4190.82	7.22
.84	3083.44	10.12	4007.86	8.08	4654.18	4.63	4911.56	0.41	4742.05	3.78	4183.60	7.24
.85	3093.56	10.10	4015.94	8.04	4658.81	4.59	4911.97	0.37	4738.27	3.82	4176.36	7.27
.86	3103.66	10.09	4023.98	8.02	4663.40	4.54	4912.34	0.34	4734.45	3.86	4169.09	7.30
.87	3113.75	10.07	4032.00	7.99	4667.94	4.51	4912.68	0.29	4730.59	3.90	4161.79	7.33
.88	3123.82	10.06	4039.99	7.97	4672.45	4.46	4912.97	0.24	4726.69	3.94	4154.46	7.36
.89	3133.88	10.05	4047.96	7.93	4676.91	4.43	4913.21	0.20	4722.75	3.98	4147.10	7.39
.90	3143.93	10.03	4055.89	7.91	4681.34	4.39	4913.41	0.15	4718.77	4.02	4139.71	7.42
.91	3153.96	10.03	4063.80	7.88	4685.73	4.35	4913.56	0.12	4714.75	4.06	4132.29	7.44
.92	3163.99	10.01	4071.68	7.85	4690.08	4.31	4913.68	0.07	4710.69	4.09	4124.85	7.47
.93	3174.00	10.00	4079.53	7.81	4694.39	4.26	4913.75	0.03	4706.60	4.13	4117.38	7.50
.94	3184.00	9.99	4087.34	7.79	4698.65	4.23	4913.78	0.02	4702.47	4.17	4109.88	7.52
.95	3193.99	9.97	4095.13	7.76	4702.88	4.19	4913.76	0.06	4698.30	4.21	4102.36	7.56
.96	3203.96	9.96	4102.89	7.73	4707.07	4.15	4913.70	0.10	4694.09	4.26	4094.80	7.58
.97	3213.92	9.94	4110.62	7.70	4711.22	4.10	4913.60	0.15	4689.83	4.29	4087.22	7.60
.98	3223.86	9.92	4118.32	7.66	4715.32	4.07	4913.45	0.19	4685.54	4.32	4079.62	7.65
.99	3233.78	9.91	4125.98	7.63	4719.39	4.02	4913.26	0.24	4681.22	4.37	4071.97	7.67
1.00	3243.69	9.91	4133.61	7.61	4723.41	3.98	4913.02	0.28	4676.85	4.41	4064.30	7.69

TABLE VIII. ARGUMENT 3.

Equation $= 2457''.02 - 122''.1 \sin. t + 2371''.0 \sin. 2t + 0''.9 \sin. 3t + 11''.4 \sin. 4t$.

Period, 29.530587997 days.

Days.	6		7		8		9		10		11	
Decimals of a Day.	Equation.	Diff.	Equation.	Diff.	Equation.	Diff.	Equation.	Diff.	Equation.	Diff.	Equation.	Diff.
.00	4064.30	7.69	3187.42	9.62	2199.82	9.86	1271.82	8.41	563.98	5.50	201.50	1.59
.01	4056.61	7.71	3177.80	9.62	2189.96	9.85	1263.41	8.39	558.48	5.47	199.91	1.56
.02	4048.90	7.75	3168.18	9.63	2180.11	9.85	1255.02	8.37	553.01	5.43	198.35	1.51
.03	4041.15	7.77	3158.55	9.64	2170.26	9.84	1246.65	8.35	547.58	5.40	196.84	1.47
.04	4033.38	7.79	3148.91	9.65	2160.42	9.83	1238.30	8.32	542.18	5.36	195.37	1.43
.05	4025.59	7.82	3139.26	9.66	2150.59	9.83	1229.98	8.30	536.82	5.32	193.94	1.38
.06	4017.77	7.85	3129.60	9.67	2140.76	9.82	1221.68	8.28	531.50	5.29	192.56	1.35
.07	4009.92	7.88	3119.93	9.68	2130.94	9.81	1213.40	8.25	526.21	5.25	191.21	1.30
.08	4002.04	7.90	3110.25	9.69	2121.13	9.80	1205.15	8.23	520.96	5.22	189.91	1.26
.09	3994.14	7.92	3100.56	9.70	2111.33	9.80	1196.92	8.21	515.74	5.18	188.65	1.21
.10	3986.22	7.95	3090.86	9.71	2101.53	9.79	1188.71	8.18	510.56	5.15	187.44	1.18
.11	3978.27	7.97	3081.15	9.72	2091.74	9.78	1180.53	8.16	505.41	5.11	186.26	1.13
.12	3970.30	8.00	3071.43	9.72	2081.96	9.78	1172.37	8.14	500.30	5.07	185.13	1.09
.13	3962.30	8.03	3061.71	9.74	2072.18	9.76	1164.23	8.11	495.23	5.04	184.04	1.05
.14	3954.27	8.05	3051.97	9.75	2062.42	9.76	1156.12	8.09	490.19	5.00	182.99	1.00
.15	3946.22	8.07	3042.22	9.75	2052.66	9.75	1148.03	8.06	485.19	4.97	181.99	0.95
.16	3938.15	8.10	3032.47	9.76	2042.91	9.74	1139.97	8.04	480.22	4.92	181.01	0.92
.17	3930.05	8.13	3022.71	9.77	2033.17	9.73	1131.93	8.01	475.30	4.89	180.12	0.88
.18	3921.92	8.14	3012.94	9.78	2023.44	9.72	1123.92	7.99	470.41	4.86	179.24	0.84
.19	3913.78	8.17	3003.16	9.79	2013.72	9.71	1115.93	7.96	465.55	4.82	178.40	0.79
.20	3905.61	8.20	2993.37	9.80	2004.01	9.71	1107.97	7.94	460.73	4.78	177.61	0.75
.21	3897.41	8.22	2983.57	9.80	1994.30	9.69	1100.03	7.91	455.95	4.74	176.86	0.71
.22	3889.19	8.24	2973.77	9.81	1984.61	9.68	1092.12	7.89	451.21	4.71	176.15	0.67
.23	3880.95	8.27	2963.96	9.82	1974.93	9.68	1084.23	7.86	446.50	4.67	175.48	0.62
.24	3872.68	8.29	2954.14	9.82	1965.25	9.66	1076.37	7.83	441.83	4.62	174.86	0.58
.25	3864.39	8.31	2944.32	9.83	1955.59	9.65	1068.54	7.80	437.22	4.59	174.28	0.54
.26	3856.08	8.34	2934.49	9.84	1945.94	9.65	1060.74	7.80	432.62	4.56	173.74	0.50
.27	3847.74	8.36	2924.65	9.85	1936.29	9.63	1052.94	7.75	428.06	4.52	173.24	0.45
.28	3839.38	8.38	2914.80	9.85	1926.66	9.62	1045.19	7.74	423.54	4.48	172.79	0.42
.29	3831.00	8.40	2904.95	9.85	1917.04	9.61	1037.45	7.70	419.06	4.45	172.37	0.36
.30	3822.60	8.43	2895.10	9.87	1907.43	9.60	1029.75	7.68	414.61	4.40	172.01	0.33
.31	3814.17	8.45	2885.23	9.87	1897.83	9.58	1022.07	7.65	410.21	4.37	171.68	0.28
.32	3805.72	8.47	2875.36	9.87	1888.25	9.58	1014.42	7.62	405.84	4.33	171.40	0.24
.33	3797.25	8.50	2865.49	9.88	1878.67	9.56	1006.80	7.60	401.51	4.30	171.16	0.20
.34	3788.75	8.51	2855.61	9.89	1869.11	9.56	999.20	7.57	397.21	4.25	170.96	0.15
.35	3780.24	8.53	2845.72	9.89	1859.55	9.54	991.63	7.54	392.96	4.22	170.81	0.12
.36	3771.71	8.56	2835.83	9.90	1850.01	9.53	984.09	7.52	388.74	4.18	170.69	0.07
.37	3763.15	8.58	2825.93	9.90	1840.48	9.51	976.57	7.49	384.56	4.14	170.62	0.03
.38	3754.57	8.60	2816.03	9.91	1831.97	9.50	969.08	7.46	380.42	4.10	170.59	0.02
.39	3745.97	8.62	2806.12	9.91	1821.47	9.49	961.62	7.43	376.32	4.06	170.61	0.05
.40	3737.35	8.65	2796.21	9.92	1811.98	9.48	954.19	7.40	372.26	4.02	170.66	0.10
.41	3728.70	8.66	2786.29	9.92	1802.50	9.47	946.79	7.38	368.24	3.99	170.76	0.14
.42	3720.04	8.68	2776.37	9.92	1793.03	9.45	939.41	7.35	364.25	3.95	170.90	0.19
.43	3711.36	8.70	2766.45	9.93	1783.58	9.44	932.06	7.32	360.30	3.90	171.09	0.23
.44	3702.66	8.73	2756.52	9.93	1774.14	9.43	924.74	7.29	356.40	3.87	171.32	0.27
.45	3693.93	8.75	2746.59	9.93	1764.71	9.41	917.45	7.26	352.53	3.83	171.59	0.31
.46	3685.18	8.77	2736.66	9.94	1755.30	9.40	910.19	7.23	348.70	3.80	171.90	0.35
.47	3676.41	8.78	2726.72	9.94	1745.90	9.38	902.96	7.21	344.90	3.75	172.25	0.40
.48	3667.63	8.81	2716.78	9.94	1736.52	9.36	895.75	7.17	341.15	3.70	172.65	0.45
.49	3658.82	8.83	2706.84	9.95	1727.16	9.35	888.58	7.14	337.45	3.67	173.10	0.49
.50	3649.99	8.84	2696.89	9.95	1717.81	9.35	881.44	7.12	333.78	3.64	173.59	0.52

TABLE VIII. ARGUMENT 3.

Equation = 2157″.02 — 122″.1 sin. t + 2371″.0 sin. $2t$ + 0″.9 sin. $3t$ + 14″.1 sin. $4t$.

Period, 29.530587997 days.

Days.	6		7		8		9		10		11	
Decimals of a Day.	Equation.	Diff.	Equation.	Diff.	Equation.	Diff.	Equation.	Diff.	Equation.	Diff.	Equation.	Diff.
.50	3619.99	8.81	2696.89	9.95	1717.81	9.35	881.44	7.12	333.78	3.64	173.59	0.52
.51	3641.15	8.87	2686.94	9.95	1708.46	9.32	874.32	7.09	830.14	3.59	174.11	0.57
.52	3632.28	8.88	2676.99	9.96	1699.14	9.31	867.23	7.05	326.55	3.56	174.68	0.61
.53	3623.10	8.90	2667.03	9.95	1689.83	9.30	860.18	7.03	322.99	3.51	175.29	0.65
.54	3614.50	8.92	2657.08	9.96	1680.53	9.28	853.15	7.00	319.48	3.48	175.94	0.70
.55	3605.58	8.94	2617.12	9.96	1671.25	9.27	816.15	6.97	316.00	3.44	176.64	0.74
.56	3596.64	8.96	2637.16	9.96	1661.98	9.25	839.18	6.94	312.54	3.40	177.38	0.78
.57	3587.68	8.97	2627.20	9.96	1652.73	9.23	832.24	6.91	309.15	3.35	178.16	0.82
.58	3578.71	8.99	2617.21	9.97	1613.50	9.22	825.33	6.88	305.81	3.32	178.98	0.86
.59	3569.72	9.01	2607.27	9.96	1631.28	9.21	818.45	6.85	302.49	3.28	179.81	0.92
.60	3560.71	9.03	2597.31	9.97	1625.07	9.18	811.60	6.81	299.21	3.24	180.76	0.95
.61	3551.68	9.05	2587.31	9.96	1615.89	9.17	804.79	6.79	295.97	3.20	181.71	0.99
.62	3542.63	9.06	2577.38	9.97	1606.72	9.16	798.00	6.76	292.77	3.15	182.70	1.01
.63	3533.57	9.08	2567.41	9.97	1597.56	9.14	791.21	6.72	289.62	3.12	183.74	1.08
.64	3524.49	9.10	2557.41	9.97	1588.42	9.12	784.52	6.70	286.50	3.08	184.82	1.12
.65	3515.39	9.12	2517.17	9.96	1579.30	9.10	777.82	6.66	283.12	3.03	185.94	1.17
.66	3506.27	9.13	2537.51	9.97	1570.20	9.09	771.16	6.63	280.39	2.99	187.11	1.21
.67	3197.11	9.15	2527.54	9.97	1561.11	9.07	764.53	6.60	277.40	2.96	188.32	1.25
.68	3187.99	9.16	2517.57	9.97	1552.04	9.06	757.93	6.57	274.44	2.92	189.57	1.29
.69	3178.83	9.18	2507.60	9.96	1542.98	9.03	751.36	6.51	271.52	2.87	190.86	1.33
.70	3169.65	9.20	2497.64	9.97	1533.95	9.02	741.82	6.50	268.65	2.84	192.19	1.38
.71	3160.45	9.21	2187.67	9.96	1524.93	9.00	738.32	6.18	265.81	2.78	193.57	1.42
.72	3151.24	9.23	2477.71	9.97	1515.93	8.98	731.84	6.14	263.03	2.76	194.99	1.46
.73	3142.01	9.25	2167.74	9.96	1506.95	8.97	725.40	6.11	260.27	2.71	196.45	1.50
.74	3132.76	9.26	2157.78	9.96	1197.98	8.94	718.99	6.38	257.56	2.68	197.95	1.55
.75	3123.50	9.28	2447.82	9.96	1189.01	8.93	712.61	6.35	[illegible]	2.63	[illegible]	1.58
.76	3114.22	9.29	2137.86	9.95	1480.11	8.91	706.26	6.31	252.25	2.58	201.08	1.63
.77	3104.93	9.31	2427.91	9.96	1471.20	8.89	699.95	6.27	219.67	2.54	202.71	1.69
.78	3395.62	9.32	2117.95	[illegible]	1462.31	8.87	693.68	6.24	217.13	2.52	201.10	1.70
.79	3386.30	9.33	2408.00	9.95	1453.44	8.85	687.14	6.22	241.61	2.47	206.10	1.75
.80	3376.97	9.35	2398.05	9.95	1441.59	8.83	681.22	6.18	[illegible]	[illegible]	207.85	1.80
.81	3367.62	9.37	2388.10	9.95	1435.76	8.81	675.01	6.15	[illegible]	2.37	209.65	1.84
.82	3358.25	9.38	2378.15	9.91	1426.95	8.80	668.89	6.12	237.34	2.31	211.49	1.89
.83	3348.87	9.39	2368.21	9.93	1418.15	8.77	662.77	6.08	235.00	2.34	213.38	1.93
.84	3339.48	9.41	2358.28	9.94	1409.38	8.75	656.69	6.05	232.69	2.26	215.31	1.97
.85	3330.07	9.42	2348.34	9.93	1400.63	8.74	650.64	6.02	230.43	2.22	217.28	2.00
.86	3320.65	9.43	2338.41	9.93	1391.89	8.71	644.62	5.98	228.21	2.18	219.28	2.05
.87	3311.22	9.45	2328.48	9.92	1383.18	8.69	638.64	5.95	226.03	2.13	221.33	2.10
.88	3301.77	9.46	2318.56	9.92	1374.49	8.67	632.69	5.91	223.90	2.10	223.43	2.13
.89	3292.31	9.47	2308.64	9.92	1365.82	8.65	626.78	5.88	221.80	2.05	225.56	2.18
.90	3282.84	9.49	2298.72	9.91	1357.17	8.64	620.90	5.85	219.75	2.02	227.74	2.21
.91	3273.35	9.50	2288.81	9.91	1348.53	8.61	615.05	5.82	217.73	1.97	229.95	2.26
.92	3263.85	9.51	2278.90	9.90	1339.92	8.58	609.23	5.77	215.76	1.93	232.21	2.30
.93	3254.34	9.52	2269.00	9.90	1331.34	8.57	603.46	5.75	213.83	1.88	234.51	2.34
.94	3244.82	9.54	2259.10	9.89	1322.77	8.55	597.71	5.71	211.95	1.85	236.85	2.39
.95	3235.28	9.55	2249.21	9.89	1314.22	8.52	592.00	5.68	210.10	1.81	239.24	2.42
.96	3225.73	9.56	2239.32	9.88	1305.70	8.51	586.32	5.64	208.29	1.76	241.66	2.47
.97	3216.17	9.57	2229.44	9.88	1297.19	8.48	580.68	5.60	206.53	1.72	244.13	2.51
.98	3206.60	9.58	2219.56	9.87	1288.71	8.46	575.08	5.57	204.81	1.68	246.64	2.55
.99	3197.02	9.60	2209.69	9.87	1280.25	8.43	569.51	5.53	203.13	1.63	249.19	2.60
1.00	3187.42	9.62	2199.82	9.86	1271.82	8.41	563.98	5.50	201.50	1.59	251.79	2.63

TABLE VIII. ARGUMENT 3.

Equation = 2457″.02 — 122″.1 sin. t + 2371″.0 sin. $2t$ + 0″.9 sin. $3t$ + 14″.4 sin. $4t$.

Period, 29.530587997 days.

Days.	12		13		14		15		16		17	
Decimals of a Day.	Equation.	Diff.	Equation.	Diff.	Equation.	Diff.	Equation.	Diff.	Equation.	Diff.	Equation.	Diff.
.00	251″.79	2.63	709″.32	6.41	1492″.67	9.03	2457″.02	9.96	3421″.36	9.02	4204″.72	6.39
.01	254.42	2.67	715.73	6.45	1501.70	9.05	2466.98	9.95	3430.38	9.00	4211.11	6.34
.02	257.09	2.72	722.18	6.48	1510.75	9.06	2476.93	9.96	3439.38	8.97	4217.45	6.32
.03	259.81	2.75	728.66	6.52	1519.81	9.09	2486.89	9.95	3448.35	8.96	4223.77	6.28
.04	262.56	2.80	735.18	6.54	1528.90	9.10	2496.84	9.96	3457.31	8.93	4230.05	6.25
.05	265.36	2.84	741.72	6.58	1538.00	9.12	2506.80	9.95	3466.24	8.92	4236.30	6.22
.06	268.20	2.88	748.30	6.62	1547.12	9.13	2516.75	9.95	3475.16	8.91	4242.52	6.18
.07	271.08	2.92	754.92	6.64	1556.25	9.16	2526.70	9.96	3484.07	8.86	4248.70	6.15
.08	274.00	2.96	761.56	6.68	1565.41	9.17	2536.66	9.95	3492.93	8.86	4254.85	6.11
.09	276.96	3.01	768.24	6.70	1574.58	9.18	2546.61	9.94	3501.79	8.85	4260.96	6.08
.10	279.97	3.05	774.94	6.74	1583.76	9.21	2556.55	9.95	3510.64	8.81	4267.04	6.04
.11	283.02	3.09	781.68	6.78	1592.97	9.23	2566.50	9.95	3519.45	8.80	4273.08	6.01
.12	286.11	3.12	788.46	6.81	1602.20	9.24	2576.45	9.95	3528.25	8.78	4279.09	5.98
.13	289.23	3.17	795.27	6.83	1611.44	9.25	2586.40	9.94	3537.03	8.75	4285.07	5.94
.14	292.40	3.21	802.10	6.87	1620.69	9.27	2596.34	9.93	3545.78	8.74	4291.01	5.91
.15	295.61	3.25	808.97	6.90	1629.96	9.28	2606.27	9.94	3554.52	8.71	4296.92	5.88
.16	298.86	3.28	815.87	6.93	1639.24	9.31	2616.21	9.93	3563.23	8.70	4302.80	5.84
.17	302.14	3.33	822.80	6.96	1648.55	9.31	2626.14	9.93	3571.93	8.67	4308.64	5.80
.18	305.47	3.37	829.76	6.99	1657.86	9.33	2636.07	9.93	3580.60	8.65	4314.44	5.76
.19	308.84	3.41	836.75	7.02	1667.19	9.35	2646.00	9.92	3589.25	8.64	4320.20	5.73
.20	312.25	3.46	843.77	7.05	1676.54	9.36	2655.92	9.91	3597.89	8.61	4325.93	5.70
.21	315.71	3.49	850.82	7.08	1685.90	9.38	2665.83	9.92	3606.50	8.58	4331.63	5.66
.22	319.20	3.53	857.90	7.12	1695.28	9.39	2675.75	9.91	3615.08	8.57	4337.29	5.63
.23	322.73	3.57	865.02	7.14	1704.67	9.41	2685.66	9.90	3623.65	8.53	4342.92	5.58
.24	326.30	3.62	872.16	7.18	1714.08	9.42	2695.56	9.91	3632.18	8.53	4348.50	5.56
.25	329.92	3.65	879.34	7.21	1723.50	9.44	2705.47	9.90	3640.71	8.49	4354.06	5.52
.26	333.57	3.69	886.55	7.23	1732.94	9.45	2715.37	9.89	3649.20	8.48	4359.58	5.48
.27	337.26	3.73	893.78	7.26	1742.39	9.46	2725.26	9.89	3657.68	8.45	4365.06	5.44
.28	340.99	3.77	901.04	7.30	1751.85	9.47	2735.15	9.88	3666.13	8.43	4370.50	5.41
.29	344.76	3.81	908.34	7.32	1761.32	9.49	2745.03	9.87	3674.56	8.41	4375.91	5.37
.30	348.57	3.85	915.66	7.35	1770.81	9.50	2754.90	9.87	3682.97	8.38	4381.28	5.34
.31	352.42	3.89	923.01	7.38	1780.31	9.52	2764.77	9.86	3691.35	8.36	4386.62	5.30
.32	356.31	3.93	930.39	7.41	1789.83	9.52	2774.63	9.86	3699.71	8.33	4391.92	5.27
.33	360.24	3.97	937.80	7.44	1799.35	9.54	2784.49	9.85	3708.04	8.31	4397.19	5.23
.34	364.21	4.01	945.24	7.46	1808.89	9.56	2794.34	9.84	3716.35	8.29	4402.42	5.19
.35	368.22	4.06	952.70	7.50	1818.45	9.57	2804.18	9.84	3724.64	8.26	4407.61	5.15
.36	372.28	4.09	960.20	7.53	1828.02	9.58	2814.02	9.83	3732.90	8.24	4412.76	5.12
.37	376.37	4.12	967.73	7.56	1837.60	9.58	2823.85	9.83	3741.14	8.21	4417.88	5.08
.38	380.49	4.17	975.29	7.58	1847.18	9.60	2833.68	9.82	3749.35	8.19	4422.96	5.04
.39	384.66	4.20	982.87	7.61	1856.78	9.61	2843.50	9.81	3757.54	8.17	4428.00	5.01
.40	388.86	4.24	990.48	7.63	1866.39	9.62	2853.31	9.80	3765.71	8.14	4433.01	4.97
.41	393.10	4.28	998.11	7.67	1876.01	9.64	2863.11	9.79	3773.85	8.11	4437.98	4.93
.42	397.38	4.32	1005.78	7.69	1885.65	9.64	2872.90	9.79	3781.96	8.09	4442.91	4.89
.43	401.70	4.36	1013.47	7.72	1895.29	9.65	2882.69	9.77	3790.05	8.06	4447.80	4.85
.44	406.06	4.40	1021.19	7.75	1904.94	9.67	2892.46	9.77	3798.11	8.04	4452.65	4.82
.45	410.46	4.44	1028.94	7.77	1914.61	9.67	2902.23	9.76	3806.15	8.01	4457.47	4.78
.46	414.90	4.47	1036.71	7.80	1924.28	9.69	2911.99	9.75	3814.16	7.99	4462.25	4.74
.47	419.37	4.52	1044.51	7.83	1933.97	9.70	2921.74	9.74	3822.15	7.96	4466.99	4.71
.48	423.89	4.56	1052.34	7.85	1943.67	9.71	2931.48	9.74	3830.11	7.93	4471.70	4.66
.49	428.45	4.59	1060.19	7.89	1953.38	9.71	2941.22	9.73	3838.04	7.92	4476.36	4.64
.50	433.04	4.63	1068.08	7.91	1963.09	9.73	2950.95	9.71	3845.96	7.88	4481.00	4.59

TABLE VIII. ARGUMENT 3.

Equation = 2457″.02 — 122″.1 sin. t + 2371″.0 sin. $2t$ + 0″.9 sin. $3t$ + 14″.4 sin. $4t$.

Period, 29.530587997 days.

Days.	12		13		14		15		16		17	
Decimals of a Day.	Equation.	Diff.	Equation.	Diff.	Equation.	Diff.	Equation.	Diff.	Equation.	Diff.	Equation.	Diff.
.50	433″.01	4.63	1068″.08	7.91	1963″.09	9.73	2950″.95	9.71	3845″.96	7.88	4481″.00	4.59
.51	437.67	4.66	1075.99	7.94	1972.82	9.73	2960.66	9.71	3853.84	7.85	4485.59	4.56
.52	442.33	4.71	1083.93	7.96	1982.55	9.74	2970.37	9.69	3861.69	7.83	4490.15	4.54
.53	447.04	4.74	1091.89	7.98	1992.29	9.76	2980.06	9.69	3869.52	7.80	4494.66	4.47
.54	451.78	4.78	1099.87	8.01	2002.05	9.76	2989.75	9.67	3877.33	7.77	4499.13	4.44
.55	456.56	4.82	1107.88	8.04	2011.81	9.76	2999.42	9.67	3885.10	7.74	4503.57	4.40
.56	461.38	4.85	1115.92	8.06	2021.57	9.78	3009.09	9.65	3892.84	7.72	4507.97	4.36
.57	466.23	4.89	1123.98	8.09	2031.35	9.78	3018.74	9.65	3900.56	7.69	4512.33	4.32
.58	471.12	4.93	1132.07	8.12	2041.13	9.80	3028.39	9.63	3908.25	7.67	4516.65	4.28
.59	476.05	4.97	1140.19	8.14	2050.93	9.80	3038.02	9.62	3915.92	7.63	4520.93	4.24
.60	481.02	5.01	1148.33	8.16	2060.73	9.81	3047.64	9.61	3923.55	7.61	4525.17	4.21
.61	486.03	5.04	1156.49	8.19	2070.54	9.82	3057.25	9.60	3931.16	7.59	4529.38	4.17
.62	491.07	5.08	1164.68	8.22	2080.36	9.82	3066.85	9.60	3938.75	7.56	4533.55	4.13
.63	496.15	5.10	1172.90	8.23	2090.18	9.83	3076.45	9.58	3946.31	7.52	4537.68	4.08
.64	501.25	5.17	1181.13	8.26	2100.01	9.84	3086.03	9.56	3953.83	7.50	4541.76	4.05
.65	506.42	5.20	1189.39	8.29	2109.85	9.84	3095.59	9.55	3961.33	7.47	4545.81	4.01
.66	511.62	5.23	1197.68	8.31	2119.69	9.85	3105.14	9.54	3968.80	7.43	4549.82	3.97
.67	516.85	5.26	1205.99	8.33	2129.51	9.86	3114.68	9.53	3976.23	7.41	4553.79	3.93
.68	522.11	5.30	1214.32	8.36	2139.40	9.86	3124.21	9.51	3983.64	7.38	4557.72	3.89
.69	527.45	5.32	1222.68	8.39	2149.26	9.87	3133.72	9.50	3991.02	7.35	4561.61	3.85
.70	532.77	5.35	1231.07	8.41	2159.13	9.87	3143.22	9.49	3998.37	7.33	4565.46	3.81
.71	538.12	5.41	1239.48	8.42	2169.00	9.88	3152.71	9.47	4005.70	7.29	4569.27	3.77
.72	543.53	5.44	1247.90	8.46	2178.88	9.89	3162.18	9.46	4012.99	7.26	4573.04	3.73
.73	548.97	5.49	1256.36	8.47	2188.77	9.90	3171.64	9.45	4020.25	7.23	4576.77	3.69
.74	554.46	5.51	1264.83	8.50	2198.67	9.90	3181.09	9.44	4027.48	7.21	4580.46	3.66
.75	559.97	5.56	1273.33	8.52	2208.57	9.90	3190.53	9.43	4034.69	7.18	4584.12	3.61
.76	565.53	5.59	1281.85	8.54	2218.47	9.90	3199.96	9.40	4041.87	7.14	4587.73	3.57
.77	571.12	5.62	1290.39	8.56	2228.37	9.91	3209.36	9.39	4049.01	7.12	4591.30	3.51
.78	576.74	5.66	1298.95	8.59	2238.28	9.91	3218.75	9.38	4056.13	7.08	4594.81	3.49
.79	582.40	5.70	1307.54	8.61	2248.19	9.92	3228.13	9.36	4063.21	7.05	4598.33	3.45
.80	588.10	5.73	1316.15	8.63	2258.11	9.92	3237.49	9.35	4070.26	7.02	4601.78	3.41
.81	593.83	5.77	1324.78	8.65	2268.03	9.93	3246.84	9.33	4077.28	6.99	4605.19	3.37
.82	599.60	5.80	1333.43	8.68	2277.96	9.93	3256.17	9.32	4084.27	6.96	4608.56	3.33
.83	605.40	5.84	1342.11	8.69	2287.89	9.93	3265.49	9.30	4091.23	6.93	4611.89	3.29
.84	611.24	5.87	1350.80	8.71	2297.82	9.93	3274.79	9.28	4098.16	6.90	4615.18	3.25
.85	617.11	5.91	1359.51	8.74	2307.75	9.94	3284.07	9.27	4105.06	6.87	4618.43	3.20
.86	623.02	5.94	1368.25	8.76	2317.69	9.95	3293.34	9.26	4111.93	6.84	4621.63	3.17
.87	628.96	5.98	1377.01	8.77	2327.64	9.94	3302.60	9.24	4118.77	6.81	4624.80	3.14
.88	634.94	6.01	1385.78	8.81	2337.58	9.94	3311.84	9.22	4125.58	6.77	4627.94	3.08
.89	640.95	6.04	1394.57	8.82	2347.52	9.95	3321.06	9.21	4132.35	6.74	4631.02	3.05
.90	646.99	6.08	1403.41	8.83	2357.47	9.95	3330.27	9.19	4139.09	6.71	4634.07	3.00
.91	653.07	6.12	1412.24	8.86	2367.42	9.95	3339.46	9.17	4145.80	6.67	4637.07	2.96
.92	659.19	6.14	1421.10	8.88	2377.37	9.95	3348.63	9.15	4152.47	6.65	4640.03	2.92
.93	665.33	6.19	1429.98	8.90	2387.32	9.95	3357.78	9.14	4159.12	6.61	4642.95	2.88
.94	671.52	6.21	1438.88	8.91	2397.27	9.96	3366.92	9.11	4165.73	6.58	4645.83	2.84
.95	677.73	6.25	1447.79	8.94	2407.23	9.95	3376.03	9.10	4172.31	6.55	4648.67	2.80
.96	683.98	6.28	1456.73	8.95	2417.18	9.96	3385.13	9.09	4178.86	6.51	4651.47	2.75
.97	690.26	6.32	1465.68	8.98	2427.14	9.96	3394.22	9.07	4185.37	6.48	4654.22	2.72
.98	696.58	6.35	1474.66	8.99	2437.10	9.96	3403.29	9.04	4191.85	6.45	4656.94	2.67
.99	702.93	6.39	1483.65	9.02	2447.06	9.96	3412.33	9.03	4198.30	6.42	4659.61	2.64
1.00	709.32	6.41	1492.67	9.03	2457.02	9.96	3421.36	9.02	4204.72	6.39	4662.25	2.59

TABLE VIII. ARGUMENT 3.

Equation $= 2457''.02 - 122''.1 \sin. t + 2371''.0 \sin. 2t + 0''.9 \sin. 3t + 14''.4 \sin. 4t$.

Period, 29.530587997 days.

Days.	18		19		20		21		22		23	
Decimals of a Day.	Equation.	Diff.	Equation.	Diff.	Equation.	Diff.	Equation.	Diff.	Equation.	Diff.	Equation.	Diff.
.00	4662.25	2.59	4712.51	1.61	4350.06	5.53	3642.21	8.43	2714.21	9.86	1726.64	9.60
.01	4664.84	2.53	4710.90	1.68	4344.53	5.58	3633.78	8.46	2704.35	9.87	1717.04	9.58
.02	4667.39	2.51	4709.22	1.72	4338.95	5.60	3625.32	8.48	2694.48	9.88	1707.46	9.57
.03	4669.90	2.47	4707.50	1.76	4333.35	5.65	3616.81	8.51	2684.60	9.88	1697.89	9.56
.04	4672.37	2.42	4705.74	1.81	4327.70	5.67	3608.33	8.52	2674.72	9.89	1688.33	9.55
.05	4674.79	2.39	4703.93	1.85	4322.03	5.71	3599.81	8.55	2664.83	9.89	1678.78	9.54
.06	4677.18	2.34	4702.08	1.88	4316.32	5.75	3591.26	8.57	2654.94	9.90	1669.24	9.53
.07	4679.52	2.30	4700.20	1.93	4310.57	5.77	3582.69	8.58	2645.01	9.91	1659.71	9.51
.08	4681.82	2.26	4698.27	1.98	4304.80	5.82	3574.11	8.61	2635.14	9.91	1650.20	9.50
.09	4684.08	2.21	4696.29	2.01	4298.98	5.85	3565.50	8.63	2625.23	9.91	1640.70	9.48
.10	4686.29	2.18	4694.28	2.05	4293.13	5.88	3556.87	8.66	2615.32	9.92	1631.22	9.48
.11	4688.47	2.13	4692.23	2.10	4287.25	5.91	3548.21	8.67	2605.40	9.92	1621.74	9.46
.12	4690.60	2.10	4690.13	2.14	4281.34	5.95	3539.54	8.69	2595.48	9.92	1612.28	9.45
.13	4692.70	2.05	4687.99	2.17	4275.39	5.99	3530.85	8.72	2585.56	9.93	1602.83	9.43
.14	4694.75	2.01	4685.82	2.22	4269.40	6.01	3522.13	8.73	2575.63	9.93	1593.40	9.42
.15	4696.76	1.97	4683.60	2.25	4263.39	6.05	3513.40	8.75	2565.70	9.93	1583.98	9.41
.16	4698.73	1.93	4681.35	2.31	4257.34	6.08	3504.65	8.78	2555.77	9.94	1574.57	9.39
.17	4700.66	1.88	4679.04	2.34	4251.26	6.12	3495.87	8.79	2545.83	9.95	1565.18	9.38
.18	4702.54	1.84	4676.70	2.38	4245.14	6.15	3487.08	8.81	2535.88	9.94	1555.80	9.37
.19	4704.38	1.80	4674.32	2.43	4238.99	6.18	3478.27	8.83	2525.94	9.94	1546.43	9.35
.20	4706.18	1.75	4671.89	2.47	4232.81	6.22	3469.44	8.86	2516.00	9.96	1537.08	9.33
.21	4707.93	1.72	4669.42	2.50	4226.59	6.24	3460.58	8.87	2506.04	9.95	1527.75	9.33
.22	4709.65	1.67	4666.92	2.55	4220.35	6.28	3451.71	8.89	2496.09	9.95	1518.42	9.30
.23	4711.32	1.63	4664.37	2.59	4214.07	6.32	3442.82	8.90	2486.14	9.96	1509.12	9.29
.24	4712.95	1.59	4661.78	2.63	4207.75	6.33	3433.92	8.93	2476.18	9.96	1499.83	9.28
.25	4714.54	1.54	4659.15	2.67	4201.42	6.38	3424.99	8.95	2466.22	9.96	1490.55	9.26
.26	4716.08	1.50	4656.48	2.72	4195.04	6.41	3416.04	8.96	2456.26	9.96	1481.29	9.25
.27	4717.58	1.46	4653.76	2.75	4188.63	6.44	3407.08	8.98	2446.30	9.96	1472.04	9.23
.28	4719.04	1.42	4651.01	2.79	4182.19	6.48	3398.10	9.00	2436.34	9.97	1462.81	9.21
.29	4720.46	1.38	4648.22	2.84	4175.71	6.50	3389.10	9.02	2426.37	9.96	1453.60	9.20
.30	4721.84	1.33	4645.38	2.87	4169.21	6.54	3380.08	9.03	2416.41	9.97	1444.40	9.18
.31	4723.17	1.29	4642.51	2.92	4162.67	6.57	3371.05	9.06	2406.44	9.97	1435.22	9.16
.32	4724.46	1.26	4639.59	2.95	4156.10	6.60	3361.99	9.07	2396.47	9.97	1426.06	9.15
.33	4725.72	1.21	4636.64	2.99	4149.50	6.63	3352.92	9.09	2386.50	9.96	1416.91	9.13
.34	4726.93	1.16	4633.65	3.02	4142.87	6.66	3343.83	9.10	2376.54	9.97	1407.78	9.12
.35	4728.09	1.12	4630.63	3.10	4136.21	6.70	3334.73	9.12	2366.57	9.97	1398.66	9.10
.36	4729.21	1.08	4627.53	3.12	4129.51	6.72	3325.61	9.14	2356.60	9.97	1389.56	9.08
.37	4730.29	1.04	4624.41	3.16	4122.79	6.76	3316.47	9.16	2346.63	9.97	1380.48	9.06
.38	4731.33	0.99	4621.25	3.19	4116.03	6.79	3307.31	9.17	2336.66	9.96	1371.42	9.05
.39	4732.32	0.95	4618.06	3.24	4109.24	6.81	3298.14	9.18	2326.70	9.97	1362.37	9.03
.40	4733.27	0.91	4614.82	3.28	4102.43	6.85	3288.96	9.21	2316.73	9.97	1353.34	9.01
.41	4734.18	0.87	4611.54	3.32	4095.58	6.88	3279.75	9.22	2306.76	9.96	1344.33	8.99
.42	4735.05	0.82	4608.22	3.35	4088.70	6.91	3270.53	9.23	2296.80	9.96	1335.34	8.98
.43	4735.87	0.78	4604.87	3.40	4081.79	6.94	3261.30	9.25	2286.84	9.96	1326.36	8.95
.44	4736.65	0.74	4601.47	3.44	4074.85	6.97	3252.05	9.27	2276.88	9.96	1317.41	8.94
.45	4737.39	0.70	4598.03	3.47	4067.88	7.00	3242.78	9.28	2266.92	9.96	1308.47	8.92
.46	4738.09	0.65	4594.56	3.52	4060.88	7.03	3233.50	9.30	2256.96	9.96	1299.55	8.90
.47	4738.74	0.61	4591.04	3.56	4053.85	7.06	3224.20	9.31	2247.00	9.95	1290.65	8.88
.48	4739.35	0.57	4587.48	3.59	4046.79	7.08	3214.89	9.32	2237.05	9.95	1281.77	8.87
.49	4739.92	0.54	4583.89	3.63	4039.71	7.11	3205.57	9.34	2227.10	9.95	1272.90	8.85
.50	4740.46	0.47	4580.26	3.67	4032.60	7.15	3196.23	9.35	2217.15	9.95	1264.05	8.82

TABLE VIII. ARGUMENT 3.

Equation = 2457″.02 — 122″.1 sin. t + 2371″.0 sin. $2t$ + 0″.9 sin. $3t$ + 14″.4 sin. $4t$.

Period, 29.530587997 days.

Days.	18		19		20		21		22		23	
Decimals of a Day.	Equation	Diff	Equation.	Diff	Equation.	Diff.	Equation.	Diff	Equation	Diff	Equation	Diff
.50	4740.46	0.47	4580.26	3.67	4032.60	7.15	3196.23	9.35	2217.15	9.95	1261.05	8.82
.51	4740.93	0.45	4576.59	3.71	4025.45	7.17	3186.88	9.37	2207.20	9.94	1255.23	8.81
.52	4741.38	0.40	4572.88	3.75	4018.28	7.21	3177.51	9.39	2197.26	9.94	1246.42	8.79
.53	4741.78	0.35	4569.13	3.80	4011.07	7.23	3168.12	9.40	2187.32	9.94	1237.63	8.77
.54	4742.13	0.32	4565.33	3.83	4003.84	7.26	3158.72	9.41	2177.38	9.93	1228.86	8.75
.55	4742.45	0.27	4561.50	3.86	3996.58	7.29	3149.31	9.42	2167.45	9.93	1220.11	8.72
.56	4742.72	0.21	4557.64	3.91	3989.29	7.32	3139.89	9.43	2157.52	9.93	1211.39	8.70
.57	4742.93	0.20	4553.73	3.95	3981.97	7.35	3130.46	9.46	2147.59	9.93	1202.69	8.68
.58	4743.13	0.14	4549.78	3.98	3974.62	7.38	3121.00	9.47	2137.66	9.91	1194.01	8.66
.59	4743.27	0.10	4545.80	4.03	3967.24	7.40	3111.53	9.48	2127.75	9.92	1185.35	8.65
.60	4743.37	0.05	4541.77	4.06	3959.81	7.43	3102.05	9.49	2117.83	9.91	1176.70	8.62
.61	4743.42	0.02	4537.71	4.10	3952.41	7.46	3092.56	9.50	2107.92	9.91	1168.08	8.60
.62	4743.44	0.03	4533.61	4.14	3944.95	7.49	3083.06	9.52	2098.01	9.90	1159.48	8.58
.63	4743.41	0.07	4529.47	4.18	3937.46	7.52	3073.54	9.52	2088.11	9.90	1150.90	8.56
.64	4743.34	0.12	4525.29	4.22	3929.94	7.54	3064.02	9.54	2078.21	9.89	1142.34	8.53
.65	4743.22	0.14	4521.07	4.26	3922.40	7.57	3054.48	9.56	2068.32	9.89	1133.84	8.52
.66	4743.08	0.20	4516.81	4.29	3914.83	7.60	3044.92	9.56	2058.43	9.88	1125.29	8.49
.67	4742.88	0.24	4512.52	4.33	3907.23	7.62	3035.36	9.58	2048.55	9.87	1116.80	8.47
.68	4742.64	0.29	4508.19	4.37	3899.61	7.65	3025.78	9.59	2038.68	9.87	1108.33	8.45
.69	4742.35	0.32	4503.82	4.40	3891.96	7.68	3016.19	9.59	2028.81	9.86	1099.88	8.43
.70	4742.03	0.37	4499.42	4.45	3884.28	7.70	3006.60	9.61	2018.95	9.86	1091.45	8.40
.71	4741.66	0.42	4494.97	4.48	3876.58	7.73	2996.99	9.62	2009.09	9.85	1083.05	8.39
.72	4741.24	0.45	4490.49	4.52	3868.85	7.76	2987.37	9.64	1999.24	9.85	1074.66	8.35
.73	4740.79	0.50	4485.97	4.55	3861.09	7.78	2977.73	9.64	1989.39	9.84	1066.31	8.34
.74	4740.29	0.54	4481.42	4.59	3853.31	7.81	2968.09	9.65	1979.55	9.83	1057.97	8.31
.75	4739.75	0.58	4476.83	4.63	3845.50	7.84	2958.44	9.66	1969.72	9.82	1049.66	8.29
.76	4739.17	0.62	4472.20	4.67	3837.66	7.86	2948.78	9.68	1959.90	9.82	1041.37	8.27
.77	4738.55	0.67	4467.53	4.71	3829.80	7.89	2939.10	9.68	1950.08	9.81	1033.10	8.24
.78	4737.88	0.71	4462.82	4.74	3821.91	7.91	2929.42	9.70	1940.27	9.81	1024.86	8.22
.79	4737.17	0.75	4458.08	4.78	3814.00	7.91	2919.72	9.70	1930.46	9.79	1016.64	8.20
.80	4736.42	0.79	4453.30	4.82	3806.06	7.96	2910.02	9.71	1920.67	9.79	1008.44	8.17
.81	4735.63	0.84	4448.48	4.86	3798.10	7.99	2900.31	9.72	1910.88	9.78	1000.27	8.15
.82	4734.79	0.85	4443.62	4.89	3790.11	8.01	2890.59	9.73	1901.10	9.77	992.12	8.12
.83	4733.94	0.94	4438.73	4.92	3782.10	8.01	2880.86	9.74	1891.33	9.76	984.00	8.10
.84	4733.09	0.96	4433.81	4.97	3774.06	8.06	2871.12	9.75	1881.57	9.75	975.90	8.07
.85	4732.04	1.00	4428.81	5.00	3766.00	8.09	2861.37	9.76	1871.82	9.75	967.83	8.05
.86	4731.04	1.05	4423.81	5.04	3757.91	8.11	2851.61	9.76	1862.07	9.74	959.78	8.03
.87	4729.99	1.09	4418.80	5.07	3749.80	8.14	2841.85	9.78	1852.33	9.72	951.75	8.00
.88	4728.90	1.13	4413.73	5.11	3741.66	8.15	2832.07	9.78	1842.61	9.72	943.75	7.97
.89	4727.77	1.18	4408.62	5.14	3733.51	8.19	2822.29	9.79	1832.89	9.71	935.78	7.95
.90	4726.59	1.22	4403.48	5.19	3725.32	8.20	2812.50	9.80	1823.18	9.70	927.83	7.93
.91	4725.37	1.25	4398.29	5.21	3717.12	8.24	2802.70	9.80	1813.48	9.69	919.90	7.90
.92	4724.12	1.31	4393.08	5.26	3708.88	8.25	2792.90	9.82	1803.79	9.68	912.00	7.87
.93	4722.81	1.34	4387.82	5.29	3700.63	8.28	2783.08	9.82	1794.11	9.67	904.13	7.85
.94	4721.47	1.39	4382.53	5.32	3692.35	8.30	2773.26	9.82	1784.44	9.66	896.28	7.82
.95	4720.08	1.42	4377.21	5.36	3684.05	8.32	2763.44	9.84	1774.78	9.65	888.46	7.79
.96	4718.66	1.47	4371.85	5.40	3675.73	8.35	2753.60	9.84	1765.13	9.64	880.67	7.77
.97	4717.19	1.52	4366.45	5.43	3667.38	8.37	2743.76	9.84	1755.49	9.63	872.90	7.74
.98	4715.67	1.55	4361.02	5.47	3659.01	8.39	2733.92	9.86	1745.86	9.61	865.16	7.72
.99	4714.12	1.58	4355.55	5.49	3650.62	8.41	2724.06	9.85	1736.25	9.61	857.44	7.70
1.00	4712.54	1.64	4350.06	5.53	3642.21	8.43	2714.21	9.86	1726.64	9.60	849.74	7.66

TABLE VIII. ARGUMENT 3.

Equation = 2457″.02 — 122″.1 sin. t + 2371″.0 sin. $2t$ + 0″.9 sin. $3t$ + 14″.4 sin. $4t$.

Period, 29.530587997 days.

Days.	24		25		26		27		28		29	
Decimals of a Day.	Equation.	Diff.	Equation.	Diff.	Equation.	Diff.	Equation.	Diff.	Equation.	Diff.	Equation.	Diff.
.00	819″.71	7.66	237″.19	4.36	1″.02	.23	190″.64	4.02	780″.43	7.64	1670″.35	9.91
.01	842.08	7.64	232.83	4.33	0.79	.19	194.66	4.07	788.07	7.66	1680.26	9.93
.02	834.41	7.61	228.50	4.29	0.60	.15	198.73	4.10	795.73	7.70	1690.19	9.95
.03	826.83	7.58	224.21	4.24	0.45	.10	202.83	4.15	803.43	7.73	1700.14	9.95
.04	819.25	7.55	219.97	4.21	0.35	.06	206.98	4.19	811.16	7.76	1710.09	9.98
.05	811.70	7.53	215.76	4.18	0.29	.00	211.17	4.23	818.92	7.78	1720.07	9.98
.06	801.17	7.50	211.58	4.13	0.29	.01	215.40	4.26	826.70	7.82	1730.05	10.00
.07	796.67	7.47	207.45	4.10	0.30	.07	219.66	4.31	834.52	7.85	1740.05	10.01
.08	789.20	7.44	203.35	4.05	0.37	.12	223.97	4.35	842.37	7.88	1750.06	10.03
.09	781.76	7.42	199.30	4.02	0.49	.15	228.32	4.39	850.25	7.91	1760.09	10.04
.10	774.34	7.39	195.28	3.98	0.64	.20	232.71	4.43	858.16	7.94	1770.13	10.05
.11	766.95	7.36	191.30	3.94	0.84	.25	237.14	4.46	866.10	7.97	1780.18	10.06
.12	759.59	7.33	187.36	3.90	1.09	.28	241.60	4.51	874.05	7.99	1790.21	10.07
.13	752.26	7.30	183.46	3.86	1.37	.33	246.11	4.55	882.01	8.02	1800.31	10.08
.14	741.96	7.28	179.60	3.82	1.70	.38	250.66	4.58	890.06	8.05	1810.39	10.10
.15	737.68	7.24	175.78	3.79	2.08	.41	255.24	4.63	898.11	8.08	1820.49	10.11
.16	730.44	7.21	171.99	3.74	2.49	.45	259.87	4.67	906.19	8.11	1830.60	10.12
.17	723.23	7.19	168.25	3.74	2.94	.51	264.54	4.71	914.30	8.14	1840.72	10.14
.18	716.04	7.16	164.51	3.66	3.45	.55	269.25	4.74	922.41	8.16	1850.86	10.14
.19	708.88	7.13	160.88	3.62	4.00	.59	273.99	4.79	930.60	8.19	1861.00	10.16
.20	701.75	7.10	157.26	3.59	4.59	.63	278.78	4.82	938.79	8.22	1871.16	10.16
.21	694.65	7.06	153.67	3.54	5.22	.68	283.60	4.87	947.01	8.25	1881.32	10.18
.22	687.59	7.04	150.13	3.51	5.90	.72	288.47	4.90	955.26	8.27	1891.50	10.19
.23	680.55	7.01	146.62	3.46	6.62	.76	293.37	4.94	963.53	8.30	1901.69	10.20
.24	673.54	6.99	143.16	3.43	7.38	.81	298.31	4.97	971.83	8.32	1911.89	10.20
.25	666.55	6.95	139.73	3.38	8.19	.85	303.28	5.02	980.15	8.35	1922.09	10.21
.26	659.60	6.92	136.35	3.35	9.01	.89	308.30	5.05	988.50	8.38	1932.30	10.22
.27	652.68	6.89	133.00	3.30	9.93	.94	313.35	5.10	996.88	8.41	1942.52	10.24
.28	645.79	6.86	129.70	3.27	10.87	.98	318.45	5.13	1005.29	8.44	1952.76	10.25
.29	638.93	6.82	126.43	3.22	11.85	1.02	323.58	5.17	1013.73	8.46	1963.01	10.25
.30	632.11	6.80	123.21	3.18	12.87	1.07	328.75	5.21	1022.19	8.48	1973.26	10.27
.31	625.31	6.77	120.03	3.14	13.94	1.11	333.96	5.25	1030.67	8.52	1983.53	10.27
.32	618.54	6.73	116.89	3.11	15.05	1.14	339.21	5.29	1039.19	8.54	1993.80	10.28
.33	611.81	6.71	113.78	3.06	16.19	1.20	344.50	5.33	1047.73	8.56	2004.08	10.29
.34	605.10	6.67	110.72	3.03	17.39	1.24	349.83	5.36	1056.29	8.59	2014.37	10.30
.35	598.43	6.64	107.69	2.97	18.63	1.28	355.19	5.40	1064.88	8.62	2024.67	10.30
.36	591.79	6.61	104.72	2.94	19.91	1.33	360.59	5.43	1073.50	8.64	2034.97	10.32
.37	585.18	6.58	101.78	2.90	21.24	1.36	366.02	5.48	1082.14	8.66	2045.28	10.32
.38	578.60	6.54	98.88	2.86	22.60	1.42	371.50	5.52	1090.80	8.69	2055.60	10.33
.39	572.06	6.52	96.02	2.81	24.02	1.45	377.02	5.55	1099.49	8.72	2065.93	10.33
.40	565.54	6.48	93.21	2.78	25.47	1.50	382.57	5.59	1108.21	8.74	2076.26	10.34
.41	559.06	6.45	90.43	2.73	26.97	1.54	388.16	5.63	1116.95	8.76	2086.60	10.35
.42	552.61	6.42	87.70	2.70	28.51	1.59	393.79	5.66	1125.71	8.79	2096.95	10.36
.43	546.19	6.38	85.00	2.65	30.10	1.62	399.45	5.70	1134.50	8.82	2107.31	10.36
.44	539.81	6.36	82.35	2.61	31.72	1.68	405.15	5.74	1143.32	8.83	2117.67	10.37
.45	533.45	6.32	79.74	2.57	33.40	1.71	410.89	5.78	1152.15	8.87	2128.04	10.37
.46	527.13	6.29	77.17	2.53	35.11	1.76	416.67	5.81	1161.02	8.88	2138.41	10.38
.47	520.84	6.25	74.64	2.48	36.87	1.80	422.48	5.85	1169.90	8.91	2148.79	10.39
.48	514.59	6.22	72.16	2.45	38.67	1.84	428.33	5.89	1178.81	8.93	2159.18	10.39
.49	508.37	6.20	69.71	2.41	40.51	1.88	434.22	5.91	1187.74	8.95	2169.57	10.39
.50	502.17	6.16	67.30	2.36	42.39	1.93	440.13	5.96	1196.69	8.98	2179.96	10.40

TABLE VIII. ARGUMENT 3.

Equation = 2457″.02 — 122″.1 sin. t + 2371″.0 sin. $2t$ + 0″.9 sin. $3t$ + 14″.4 sin. $4t$.

Period, 29.530587997 days.

Days.	24		25		26		27		28		29	
Decimals of a Day.	Equation	Diff.	Equation.	Diff.	Equation.	Diff.	Equation.	Diff.	Equation.	Diff.	Equation.	Diff.
.50	502″.17	6.16	67″.30	2.36	42″.39	1.93	440″.13	5.96	1196″.69	8.98	2179″.96	10.40
.51	496.01	6.12	64.91	2.32	44.32	1.97	446.09	5.99	1205.67	9.00	2190.36	10.41
.52	489.89	6.09	62.62	2.28	46.29	2.01	452.08	6.03	1214.67	9.03	2200.77	10.41
.53	483.80	6.06	60.31	2.21	48.30	2.06	458.11	6.07	1223.70	9.04	2211.18	10.41
.54	477.74	6.02	58.10	2.19	50.36	2.10	464.18	6.11	1232.74	9.07	2221.59	10.42
.55	471.72	5.99	55.91	2.16	52.46	2.14	470.29	6.14	1241.81	9.09	2232.01	10.43
.56	465.73	5.95	53.75	2.14	54.60	2.19	476.43	6.17	1250.90	9.12	2242.44	10.43
.57	459.78	5.92	51.61	2.07	56.79	2.23	482.60	6.21	1260.02	9.13	2252.87	10.43
.58	453.86	5.89	49.57	2.02	59.02	2.27	488.81	6.25	1269.15	9.16	2263.30	10.44
.59	447.97	5.85	47.55	1.99	61.29	2.32	495.06	6.28	1278.31	9.18	2273.74	10.43
.60	442.12	5.82	45.56	1.94	63.61	2.35	501.34	6.31	1287.49	9.20	2284.17	10.44
.61	436.30	5.79	43.62	1.90	65.96	2.39	507.65	6.36	1296.69	9.21	2294.61	10.44
.62	430.51	5.75	41.72	1.86	68.35	2.44	514.01	6.39	1305.90	9.24	2305.05	10.45
.63	424.76	5.71	39.86	1.82	70.79	2.49	520.40	6.42	1315.14	9.27	2315.50	10.45
.64	419.05	5.68	38.04	1.77	73.28	2.52	526.82	6.47	1324.41	9.28	2325.95	10.45
.65	413.37	5.65	36.27	1.73	75.80	2.57	533.29	6.48	1333.69	9.31	2336.40	10.46
.66	407.72	5.61	34.54	1.69	78.37	2.61	539.77	6.53	1343.00	9.32	2346.86	10.45
.67	402.11	5.57	32.85	1.65	80.98	2.66	546.30	6.56	1352.32	9.35	2357.31	10.46
.68	396.54	5.51	31.20	1.60	83.64	2.69	552.86	6.60	1361.67	9.36	2367.77	10.46
.69	391.00	5.51	29.60	1.57	86.33	2.74	559.16	6.63	1371.03	9.39	2378.23	10.46
.70	385.49	5.47	28.03	1.52	89.07	2.79	566.09	6.67	1380.42	9.40	2388.69	10.46
.71	380.02	5.43	26.51	1.47	91.86	2.82	572.76	6.70	1389.82	9.43	2399.15	10.46
.72	374.59	5.40	25.01	1.44	94.68	2.86	579.46	6.73	1399.25	9.44	2409.61	10.47
.73	369.19	5.36	23.60	1.39	97.54	2.91	586.19	6.77	1408.69	9.47	2420.08	10.47
.74	363.83	5.31	22.21	1.36	100.45	2.94	592.96	6.80	1418.16	9.48	2430.55	10.46
.75	358.40	5.29	20.85	1.31	103.39	2.99	599.73	6.84	1427.64	9.49	2441.01	10.47
.76	353.20	5.26	19.54	1.26	106.38	3.04	606.60	6.87	1437.13	9.52	2451.48	10.47
.77	347.94	5.22	18.28	1.22	109.42	3.07	613.47	6.90	1446.65	9.54	2461.95	10.46
.78	342.72	5.18	17.06	1.18	112.49	3.12	620.37	6.94	1456.19	9.56	2472.41	10.47
.79	337.54	5.15	15.88	1.14	115.61	3.16	627.31	6.97	1465.75	9.57	2482.88	10.46
.80	332.39	5.11	14.74	1.09	118.77	3.20	634.28	7.00	1475.3[illegible]	9.59	2493.33	10.47
.81	327.28	5.07	13.65	1.05	121.97	3.24	641.28	7.04	148[illegible].9[illegible]	9.61	2503.80	10.46
.82	322.21	5.04	12.60	1.01	125.21	3.29	648.32	7.06	1494.52	9.63	2514.26	10.47
.83	317.17	5.00	11.59	.97	128.50	3.32	655.38	7.10	1504.15	9.65	2524.73	10.46
.84	312.17	4.97	10.62	.92	131.82	3.37	662.48	7.13	1513.80	9.66	2535.19	10.46
.85	307.20	4.93	9.70	.88	135.19	3.41	669.61	7.17	1523.46	9.68	2545.65	10.46
.86	302.27	4.89	8.82	.83	138.60	3.44	676.78	7.19	1533.14	9.69	2556.11	10.46
.87	297.38	4.85	7.99	.80	142.01	3.49	683.97	7.23	1542.83	9.71	2566.57	10.45
.88	292.53	4.82	7.19	.75	145.53	3.54	691.20	7.27	1552.54	9.74	2577.02	10.46
.89	287.71	4.78	6.44	.71	149.07	3.57	698.47	7.29	1562.28	9.74	2587.48	10.45
.90	282.93	4.74	5.73	.66	152.64	3.62	705.76	7.33	1572.02	9.77	2597.93	10.44
.91	278.19	4.70	5.07	.62	156.26	3.65	713.09	7.36	1581.79	9.78	2608.37	10.44
.92	273.49	4.67	4.45	.58	159.91	3.70	720.45	7.39	1591.57	9.79	2618.81	10.44
.93	268.82	4.63	3.87	.53	163.61	3.74	727.84	7.42	1601.36	9.81	2629.25	10.44
.94	264.19	4.59	3.34	.50	167.35	3.78	735.26	7.45	1611.17	9.82	2639.69	10.44
.95	259.60	4.56	2.84	.45	171.13	3.82	742.71	7.48	1620.99	9.85	2650.13	10.43
.96	255.04	4.51	2.39	.40	174.95	3.87	750.19	7.52	1630.84	9.86	2660.56	10.43
.97	250.53	4.48	1.99	.36	178.82	3.90	757.71	7.55	1640.70	9.87	2670.99	10.42
.98	246.05	4.44	1.63	.32	182.72	3.94	765.26	7.57	1650.57	9.88	2681.41	10.42
.99	241.61	4.42	1.31	.29	186.66	3.98	772.83	7.60	1660.45	9.90	2691.83	10.42
1.00	237.19	4.36	1.02	.23	190.64	4.02	780.43	7.64	1670.35	9.91	2702.25	10.42

TABLE IX. ARGUMENT 4.

Equation $= 670''.500 - 670''.3 \text{ sin. } z - 7''.9 \text{ sin. } 2z.$

Period, 365.259687 days.

Days.	0		10		20		30		40	
Days.	Equation.	Difference.	Equation.	Difference.	Equation.	Difference.	Equation.	Difference.	Equation.	Difference.
d 0.0	1340.68″	+.01	1334.19″	−.16	1307.89″	−.37	1262.33″	−.55	1198.72″	−.73
0.1	1340.72	.03	1334.03	.17	1307.52	.36	1261.78	.55	1197.99	.72
0.2	1340.75	.03	1333.86	.17	1307.16	.37	1261.23	.55	1197.27	.72
0.3	1340.78	.03	1333.69	.17	1306.79	.36	1260.68	.56	1196.55	.73
0.4	1340.81	.02	1333.52	.17	1306.43	.37	1260.12	.55	1195.82	.73
0.5	1340.83	.03	1333.35	.18	1306.06	.37	1259.57	.56	1195.09	.73
0.6	1340.86	.02	1333.17	.18	1305.69	.38	1259.01	.56	1194.36	.73
0.7	1340.88	.02	1332.99	.18	1305.31	.37	1258.45	.56	1193.63	.73
0.8	1340.90	.02	1332.81	.18	1304.94	.38	1257.89	.57	1192.90	.74
0.9	1340.92	.01	1332.63	.18	1304.56	.38	1257.32	.56	1192.16	.74
1.0	1340.93	.02	1332.45	.19	1304.18	.38	1256.76	.57	1191.42	.73
1.1	1340.95	.01	1332.26	.18	1303.80	.38	1256.19	.57	1190.69	.74
1.2	1340.96	.01	1332.08	.19	1303.42	.39	1255.62	.57	1189.95	.74
1.3	1340.97	.01	1331.89	.19	1303.03	.38	1255.05	.57	1189.21	.75
1.4	1340.98	.00	1331.70	.19	1302.65	.39	1254.48	.58	1188.46	.74
1.5	1340.98	.00	1331.51	.20	1302.26	.39	1253.90	.57	1187.72	.75
1.6	1340.98	+.01	1331.31	.20	1301.87	.40	1253.33	.58	1186.97	.74
1.7	1340.99	.00	1331.11	.20	1301.47	.39	1252.75	.58	1186.23	.75
1.8	1340.99	−.01	1330.91	.20	1301.08	.40	1252.17	.58	1185.48	.75
1.9	1340.98	.00	1330.71	.21	1300.68	.40	1251.59	.59	1184.73	.75
2.0	1340.98	.01	1330.50	.20	1300.28	.40	1251.00	.58	1183.98	.76
2.1	1340.97	.00	1330.30	.20	1299.88	.40	1250.42	.59	1183.22	.76
2.2	1340.97	.01	1330.10	.21	1299.48	.40	1249.83	.59	1182.46	.75
2.3	1340.96	.01	1329.89	.21	1299.08	.41	1249.24	.59	1181.71	.76
2.4	1340.95	.02	1329.68	.22	1298.67	.41	1248.65	.59	1180.95	.76
2.5	1340.93	.01	1329.46	.21	1298.26	.41	1248.06	.59	1180.19	.76
2.6	1340.92	.02	1329.25	.22	1297.85	.41	1247.47	.60	1179.43	.77
2.7	1340.90	.02	1329.03	.22	1297.44	.41	1246.87	.60	1178.66	.76
2.8	1340.88	.03	1328.81	.22	1297.03	.42	1246.27	.60	1177.90	.77
2.9	1340.85	.02	1328.59	.22	1296.61	.41	1245.67	.60	1177.13	.77
3.0	1340.83	.03	1328.37	.23	1296.20	.42	1245.07	.60	1176.36	.77
3.1	1340.80	.03	1328.14	.22	1295.78	.42	1244.47	.61	1175.59	.77
3.2	1340.77	.03	1327.92	.23	1295.36	.42	1243.86	.60	1174.82	.77
3.3	1340.74	.03	1327.69	.23	1294.94	.43	1243.26	.61	1174.05	.78
3.4	1340.71	.03	1327.46	.24	1294.51	.43	1242.65	.61	1173.27	.77
3.5	1340.68	.04	1327.22	.23	1294.08	.43	1242.04	.61	1172.50	.78
3.6	1340.64	.04	1326.99	.24	1293.65	.43	1241.43	.62	1171.72	.78
3.7	1340.60	.04	1326.75	.24	1293.22	.43	1240.81	.61	1170.94	.78
3.8	1340.56	.04	1326.51	.24	1292.79	.43	1240.20	.62	1170.16	.78
3.9	1340.52	.04	1326.27	.24	1292.36	.44	1239.58	.62	1169.38	.79
4.0	1340.48	.05	1326.03	.24	1291.92	.44	1238.96	.62	1168.59	.78
4.1	1340.43	.05	1325.79	.25	1291.48	.44	1238.34	.62	1167.81	.79
4.2	1340.38	.05	1325.54	.25	1291.04	.44	1237.72	.62	1167.02	.79
4.3	1340.33	.05	1325.29	.25	1290.60	.44	1237.10	.63	1166.23	.79
4.4	1340.28	.05	1325.04	.25	1290.16	.45	1236.47	.63	1165.44	.79
4.5	1340.23	.06	1324.79	.26	1289.71	.45	1235.84	.63	1164.65	.79
4.6	1340.17	.06	1324.53	.25	1289.26	.45	1235.21	.63	1163.86	.80
4.7	1340.11	.06	1324.28	.26	1288.81	.45	1234.58	.63	1163.06	.79
4.8	1340.05	.06	1324.02	.26	1288.36	.45	1233.95	.63	1162.27	.80
4.9	1339.99	.06	1323.76	.27	1287.91	.46	1233.32	.64	1161.47	.80
5.0	1339.93	−.07	1323.49	−.26	1287.45	−.45	1232.68	−.64	1160.67	−.80

TABLE IX. ARGUMENT 4.

Equation = 670″.500 — 670″.3 sin. z — 7″.9 sin. 2z.

Period, 365.259687 days.

Days.	0		10		20		30		40	
Days.	Equation.	Difference.	Equation.	Difference.	Equation.	Difference.	Equation.	Difference.	Equation.	Difference.
d 5.0	″ 1339.93	−.07	″ 1323.49	−.26	″ 1287.45	−.45	″ 1232.68	−.64	″ 1160.67	−.80
5.1	1339.86	.07	1323.23	.27	1287.00	.46	1232.04	.64	1159.87	.80
5.2	1339.79	.07	1322.96	.26	1286.54	.46	1231.40	.64	1159.07	.81
5.3	1339.72	.07	1322.70	.27	1286.08	.47	1230.76	.64	1158.26	.80
5.4	1339.65	.07	1322.43	.28	1285.61	.46	1230.12	.64	1157.46	.81
5.5	1339.58	.08	1322.15	.27	1285.15	.47	1229.48	.65	1156.65	.81
5.6	1339.50	.07	1321.88	.27	1284.68	.47	1228.83	.65	1155.84	.80
5.7	1339.43	.08	1321.61	.28	1284.21	.47	1228.18	.65	1155.04	.82
5.8	1339.35	.09	1321.33	.28	1283.74	.47	1227.53	.65	1154.22	.81
5.9	1339.26	.08	1321.05	.29	1283.27	.47	1226.88	.65	1153.41	.81
6.0	1339.18	.09	1320.76	.28	1282.80	.48	1226.23	.66	1152.60	.82
6.1	1339.09	.08	1320.48	.29	1282.32	.47	1225.57	.65	1151.78	.81
6.2	1339.01	.09	1320.19	.28	1281.85	.48	1224.92	.66	1150.97	.82
6.3	1338.92	.10	1319.91	.29	1281.37	.48	1224.26	.66	1150.15	.82
6.4	1338.82	.09	1319.62	.30	1280.89	.49	1223.60	.66	1149.33	.82
6.5	1338.73	.10	1319.32	.29	1280.40	.48	1222.94	.67	1148.51	.83
6.6	1338.63	.09	1319.03	.30	1279.92	.49	1222.27	.66	1147.68	.82
6.7	1338.54	.10	1318.73	.29	1279.43	.49	1221.61	.67	1146.86	.83
6.8	1338.44	.10	1318.44	.30	1278.94	.49	1220.94	.67	1146.03	.83
6.9	1338.34	.11	1318.14	.30	1278.45	.49	1220.27	.67	1145.20	.82
7.0	1338.23	.10	1317.84	.31	1277.96	.49	1219.60	.67	1144.38	.84
7.1	1338.13	.11	1317.53	.30	1277.47	.50	1218.93	.67	1143.54	.83
7.2	1338.02	.11	1317.23	.31	1276.97	.50	1218.26	.68	1142.71	.83
7.3	1337.91	.11	1316.92	.31	1276.47	.50	1217.58	.68	1141.88	.84
7.4	1337.80	.12	1316.61	.31	1275.97	.50	1216.90	.68	1141.04	.83
7.5	1337.68	.11	1316.30	.32	1275.47	.50	1216.22	.68	1140.21	.84
7.6	1337.57	.12	1315.98	.31	1274.97	.51	1215.54	.68	1139.37	.84
7.7	1337.45	.12	1315.67	.32	1274.46	.51	1214.86	.68	1138.53	.84
7.8	1337.33	.12	1315.35	.32	1273.95	.51	1214.18	.69	1137.69	.84
7.9	1337.21	.13	1315.03	.32	1273.44	.51	1213.49	.68	1136.85	.85
8.0	1337.08	.12	1314.71	.32	1272.93	.51	1212.81	.69	1136.00	.84
8.1	1336.96	.13	1314.39	.33	1272.42	.51	1212.12	.69	1135.16	.85
8.2	1336.83	.13	1314.06	.32	1271.91	.52	1211.43	.70	1134.31	.85
8.3	1336.70	.13	1313.74	.33	1271.39	.52	1210.73	.69	1133.46	.84
8.4	1336.57	.13	1313.41	.33	1270.87	.52	1210.04	.70	1132.62	.86
8.5	1336.44	.14	1313.08	.34	1270.35	.52	1209.34	.69	1131.76	.85
8.6	1336.30	.14	1312.74	.33	1269.83	.52	1208.65	.70	1130.91	.85
8.7	1336.16	.14	1312.41	.34	1269.31	.53	1207.95	.70	1130.06	.86
8.8	1336.02	.14	1312.07	.33	1268.78	.53	1207.25	.71	1129.20	.85
8.9	1335.88	.15	1311.74	.34	1268.25	.53	1206.54	.70	1128.35	.86
9.0	1335.73	.14	1311.40	.35	1267.72	.53	1205.84	.70	1127.49	.86
9.1	1335.59	.15	1311.05	.34	1267.19	.53	1205.14	.71	1126.63	.86
9.2	1335.44	.14	1310.71	.35	1266.66	.54	1204.43	.71	1125.77	.86
9.3	1335.30	.16	1310.36	.34	1266.12	.53	1203.72	.71	1124.91	.87
9.4	1335.14	.15	1310.02	.35	1265.59	.54	1203.01	.71	1124.04	.86
9.5	1334.99	.16	1309.67	.36	1265.05	.54	1202.30	.71	1123.18	.87
9.6	1334.83	.16	1309.31	.35	1264.51	.54	1201.59	.72	1122.31	.87
9.7	1334.67	.15	1308.96	.36	1263.97	.55	1200.87	.72	1121.44	.87
9.8	1334.52	.16	1308.60	.36	1263.42	.54	1200.15	.71	1120.57	.87
9.9	1334.36	−.17	1308.24	−.35	1262.88	−.55	1199.44	−.72	1119.70	−.87
10.0	1334.19		1307.89		1262.33		1198.72		1118.83	

TABLE IX. ARGUMENT 4.

Equation = 670″.500 — 670″.3 sin. z — 7″.9 sin. 2z.

Period, 365.259687 days.

Days.	50		60		70		80		90	
Days.	Equation.	Difference.	Equation.	Difference.	Equation.	Difference.	Equation.	Difference.	Equation.	Difference.
d. 0.0	1118″.83	−.87	1025″.03	−1.00	920″.14	−1.09	807″.42	−1.16	690″.38	−1.18
0.1	1117.96	.88	1024.03	1.00	919.05	1.10	806.26	1.15	689.20	1.18
0.2	1117.08	.87	1023.03	1.00	917.95	1.09	805.11	1.16	688.02	1.18
0.3	1116.21	.88	1022.03	1.01	916.86	1.10	803.95	1.15	686.84	1.18
0.4	1115.33	.88	1021.02	1.00	915.76	1.10	802.80	1.16	685.66	1.18
0.5	1114.45	.89	1020.02	1.01	914.66	1.09	801.64	1.16	684.48	1.18
0.6	1113.56	.87	1019.01	1.00	913.57	1.10	800.48	1.15	683.30	1.17
0.7	1112.69	.89	1018.01	1.01	912.47	1.10	799.33	1.16	682.13	1.18
0.8	1111.80	.88	1017.00	1.01	911.37	1.10	798.17	1.16	680.95	1.19
0.9	1110.92	.89	1015.99	1.00	910.27	1.10	797.01	1.16	679.76	1.18
1.0	1110.03	.88	1014.99	1.01	909.17	1.11	795.85	1.16	678.58	1.17
1.1	1109.15	.89	1013.98	1.02	908.06	1.10	794.69	1.16	677.41	1.18
1.2	1108.26	.89	1012.96	1.01	906.96	1.10	793.53	1.16	676.23	1.19
1.3	1107.37	.89	1011.95	1.01	905.86	1.11	792.37	1.16	675.04	1.18
1.4	1106.48	.90	1010.94	1.01	904.75	1.10	791.21	1.16	673.86	1.18
1.5	1105.58	.89	1009.93	1.02	903.65	1.11	790.05	1.16	672.68	1.18
1.6	1104.69	.90	1008.91	1.02	902.54	1.10	788.89	1.16	671.50	1.18
1.7	1103.79	.89	1007.89	1.01	901.44	1.11	787.73	1.16	670.32	1.18
1.8	1102.90	.90	1006.88	1.02	900.33	1.11	786.57	1.17	669.14	1.18
1.9	1102.00	.90	1005.86	1.02	899.22	1.11	785.40	1.16	667.96	1.18
2.0	1101.10	.90	1004.84	1.02	898.11	1.10	784.24	1.16	666.78	1.18
2.1	1100.20	.90	1003.82	1.03	897.01	1.11	783.08	1.16	665.60	1.18
2.2	1099.30	.91	1002.79	1.02	895.90	1.12	781.92	1.17	664.42	1.18
2.3	1098.39	.90	1001.77	1.02	894.78	1.11	780.75	1.16	663.24	1.18
2.4	1097.49	.91	1000.75	1.03	893.67	1.11	779.59	1.17	662.06	1.18
2.5	1096.58	.91	999.72	1.02	892.56	1.11	778.42	1.16	660.88	1.18
2.6	1095.67	.91	998.70	1.03	891.45	1.11	777.26	1.17	659.70	1.18
2.7	1094.76	.91	997.67	1.03	890.34	1.12	776.09	1.16	658.52	1.18
2.8	1093.85	.91	996.64	1.03	889.22	1.11	774.93	1.17	657.34	1.18
2.9	1092.94	.91	995.61	1.03	888.11	1.12	773.76	1.16	656.16	1.18
3.0	1092.03	.91	994.58	1.03	886.99	1.11	772.60	1.17	654.98	1.18
3.1	1091.12	.92	993.55	1.03	885.88	1.12	771.43	1.17	653.80	1.18
3.2	1090.20	.92	992.52	1.03	884.76	1.12	770.26	1.16	652.62	1.18
3.3	1089.28	.91	991.49	1.04	883.64	1.11	769.10	1.17	651.44	1.18
3.4	1088.37	.92	990.45	1.03	882.53	1.12	767.93	1.17	650.26	1.18
3.5	1087.45	.92	989.42	1.04	881.41	1.12	766.76	1.17	649.08	1.18
3.6	1086.53	.93	988.38	1.03	880.29	1.12	765.59	1.17	647.90	1.18
3.7	1085.60	.92	987.35	1.04	879.17	1.12	764.42	1.16	646.72	1.18
3.8	1084.68	.92	986.31	1.04	878.05	1.13	763.26	1.17	645.54	1.18
3.9	1083.76	.93	985.27	1.04	876.92	1.12	762.09	1.17	644.36	1.17
4.0	1082.83	.93	984.23	1.04	875.80	1.12	760.92	1.17	643.19	1.18
4.1	1081.90	.92	983.19	1.04	874.68	1.12	759.75	1.17	642.01	1.18
4.2	1080.98	.93	982.15	1.05	873.56	1.13	758.58	1.17	640.83	1.18
4.3	1080.05	.94	981.10	1.04	872.43	1.12	757.41	1.17	639.65	1.18
4.4	1079.11	.93	980.06	1.05	871.31	1.13	756.24	1.17	638.47	1.18
4.5	1078.18	.93	979.01	1.04	870.18	1.12	755.07	1.17	637.29	1.18
4.6	1077.25	.94	977.97	1.05	869.06	1.13	753.90	1.17	636.11	1.18
4.7	1076.31	.93	976.92	1.05	867.93	1.13	752.73	1.17	634.93	1.18
4.8	1075.38	.94	975.87	1.05	866.80	1.13	751.56	1.18	633.75	1.17
4.9	1074.44	.94	974.82	1.05	865.67	1.12	750.38	1.17	632.58	1.18
5.0	1073.50	−.94	973.77	−1.05	864.55	−1.13	749.21	−1.17	631.40	−1.18

TABLE IX. ARGUMENT 4.

Equation = 670″.500 — 670″.3 sin. z — 7″.9 sin. 2z.

Period, 365.259687 days.

Days.	50		60		70		80		90	
Days.	Equation.	Difference.	Equation.	Difference.	Equation.	Difference.	Equation.	Difference.	Equation.	Difference.
d.	″		″		″		″		″	
5.0	1073.50	— .94	973.77	—1.05	864.55	—1.13	749.21	—1.17	631.40	—1.18
5.1	1072.56	.94	972.72	1.05	863.42	1.13	748.04	1.17	630.22	1.18
5.2	1071.62	.94	971.67	1.05	862.29	1.13	746.87	1.17	629.04	1.18
5.3	1070.68	.94	970.62	1.06	861.16	1.13	745.70	1.18	627.86	1.18
5.4	1069.74	.95	969.56	1.05	860.03	1.14	744.52	1.17	626.68	1.17
5.5	1068.79	.95	968.51	1.06	858.89	1.13	743.35	1.17	625.51	1.18
5.6	1067.84	.94	967.45	1.05	857.76	1.13	742.18	1.18	624.33	1.18
5.7	1066.90	.95	966.40	1.06	856.63	1.13	741.00	1.17	623.15	1.17
5.8	1065.95	.95	965.34	1.06	855.50	1.14	739.83	1.17	621.98	1.18
5.9	1065.00	.95	964.28	1.05	854.36	1.13	738.66	1.18	620.80	1.18
6.0	1064.05	.95	963.23	1.06	853.23	1.14	737.48	1.17	619.62	1.17
6.1	1063.10	.96	962.17	1.06	852.09	1.13	736.31	1.18	618.45	1.18
6.2	1062.14	.95	961.11	1.07	850.96	1.14	735.13	1.17	617.27	1.18
6.3	1061.19	.96	960.04	1.06	849.82	1.14	733.96	1.18	616.09	1.17
6.4	1060.23	.95	958.98	1.06	848.68	1.13	732.78	1.17	614.92	1.18
6.5	1059.28	.96	957.92	1.07	847.55	1.14	731.61	1.18	613.74	1.17
6.6	1058.32	.96	956.85	1.07	846.41	1.14	730.43	1.17	612.57	1.18
6.7	1057.36	.96	955.78	1.06	845.27	1.14	729.26	1.18	611.39	1.18
6.8	1056.40	.97	954.72	1.07	844.13	1.14	728.08	1.17	610.21	1.17
6.9	1055.43	.96	953.65	1.07	842.99	1.14	726.91	1.18	609.04	1.18
7.0	1054.47	.96	952.58	1.06	841.85	1.14	725.73	1.17	607.86	1.17
7.1	1053.51	.97	951.52	1.07	840.71	1.14	724.56	1.18	606.69	1.18
7.2	1052.54	.97	950.45	1.08	839.57	1.14	723.38	1.18	605.51	1.17
7.3	1051.57	.97	949.37	1.07	838.43	1.14	722.20	1.18	604.34	1.17
7.4	1050.60	.96	948.30	1.07	837.29	1.14	721.02	1.17	603.17	1.18
7.5	1049.64	.97	947.23	1.07	836.15	1.15	719.85	1.18	601.99	1.17
7.6	1048.67	.98	946.16	1.08	835.00	1.14	718.67	1.18	600.82	1.18
7.7	1047.69	.97	945.08	1.07	833.86	1.15	717.49	1.17	599.64	1.17
7.8	1046.72	.97	944.01	1.08	832.71	1.14	716.32	1.18	598.47	1.17
7.9	1045.75	.98	942.93	1.07	831.57	1.14	715.14	1.18	597.30	1.17
8.0	1044.77	.98	941.86	1.08	830.43	1.15	713.96	1.18	596.13	1.18
8.1	1043.79	.97	940.78	1.08	829.28	1.14	712.78	1.18	594.95	1.17
8.2	1042.82	.98	939.70	1.08	828.14	1.15	711.60	1.17	593.78	1.17
8.3	1041.84	.98	938.62	1.08	826.99	1.15	710.43	1.18	592.61	1.17
8.4	1040.86	.98	937.54	1.08	825.84	1.15	709.25	1.18	591.44	1.18
8.5	1039.88	.98	936.46	1.08	824.69	1.15	708.07	1.18	590.26	1.17
8.6	1038.90	.99	935.38	1.09	823.54	1.15	706.89	1.18	589.09	1.17
8.7	1037.91	.99	934.29	1.08	822.39	1.14	705.71	1.18	587.92	1.17
8.8	1036.92	.98	933.21	1.09	821.25	1.15	704.53	1.17	586.75	1.17
8.9	1035.94	.99	932.12	1.08	820.10	1.15	703.36	1.18	585.58	1.17
9.0	1034.95	.98	931.04	1.09	818.95	1.16	702.18	1.18	584.41	1.17
9.1	1033.97	.99	929.95	1.08	817.79	1.15	701.00	1.18	583.24	1.17
9.2	1032.98	.99	928.87	1.09	816.64	1.15	699.82	1.18	582.07	1.17
9.3	1031.99	.99	927.78	1.09	815.49	1.15	698.64	1.18	580.90	1.17
9.4	1031.00	.99	926.69	1.09	814.34	1.15	697.46	1.18	579.73	1.17
9.5	1030.01	1.00	925.60	1.09	813.19	1.16	696.28	1.18	578.56	1.17
9.6	1029.01	.99	924.51	1.09	812.03	1.15	695.10	1.18	577.39	1.17
9.7	1028.02	1.00	923.42	1.09	810.88	1.15	693.92	1.18	576.22	1.16
9.8	1027.02	1.00	922.33	1.10	809.73	1.16	692.74	1.18	575.06	1.17
9.9	1026.02	— .99	921.23	—1.09	808.57	—1.15	691.56	—1.18	573.89	—1.17
10.0	1025.03		920.14		807.42		690.38		572.72	

TABLE IX. ARGUMENT 4.

Equation = 670″.500 — 670″.3 sin. z — 7″.9 sin. 2z.

Period, 365.259687 days.

Days.	100		110		120		130		140	
Days.	Equation.	Difference.	Equation.	Difference.	Equation.	Difference.	Equation.	Difference.	Equation.	Difference.
d. 0.0	″ 572.72	−1.16	″ 458.14	−1.12	″ 350.24	−1.01	″ 252.36	−.92	″ 167.49	−.77
0.1	571.56	1.17	457.02	1.11	319.20	1.03	251.44	.91	166.72	.77
0.2	570.39	1.17	455.91	1.12	318.17	1.03	250.53	.92	165.95	.77
0.3	569.22	1.17	454.79	1.12	347.14	1.03	249.61	.91	165.18	.77
0.4	568.05	1.16	453.67	1.11	346.11	1.03	248.70	.92	164.41	.77
0.5	566.89	1.17	452.56	1.12	315.08	1.03	247.78	.91	163.64	.77
0.6	565.72	1.16	451.11	1.11	311.05	1.03	246.87	.91	162.87	.76
0.7	561.56	1.17	450.33	1.11	343.02	1.03	245.96	.91	162.11	.76
0.8	563 39	1.16	449.22	1.12	341.99	1.02	245.05	.90	161.35	.76
0.9	562.23	1.17	448.10	1.11	340.97	1.02	244.15	.91	160.59	.77
1.0	561.06	1.16	446.99	1.11	339.95	1.03	243.21	.90	159.82	.75
1.1	559.90	1.16	445.88	1.11	338.92	1.02	242.31	.91	159.07	.76
1.2	558.74	1.17	444.77	1.11	337.90	1.02	241.43	.90	158.31	.75
1.3	557.57	1.16	443.66	1.11	336.88	1.02	240.53	.90	157.56	.76
1.4	556.41	1.16	442.55	1.11	335.86	1.02	239.63	.90	156.80	.75
1.5	555.25	1.16	441.41	1.10	334.84	1.02	238.73	.90	156.05	.75
1.6	554.09	1.17	440.31	1.11	333.82	1.02	237.83	.89	155.30	.75
1.7	552.92	1.16	439.23	1.11	332.80	1.01	236.91	.90	154.55	.75
1.8	551.76	1.16	438.12	1.10	331.79	1.02	236.01	.89	153.80	.74
1.9	550.60	1.16	437.02	1.11	330.77	1.01	235.15	.89	153.06	.75
2.0	549.44	1.16	435.91	1.10	329.76	1.02	234.26	.89	152.31	.74
2.1	548.28	1.16	434.81	1.10	328.74	1.01	233.37	.89	151.57	.74
2.2	547.12	1.16	433.71	1.10	327.73	1.01	232.48	.89	150.83	.74
2.3	545.96	1.16	432.61	1.11	326.72	1.01	231.59	.89	150.09	.74
2.1	544.80	1.16	431.50	1.10	325.71	1.01	230.70	.88	149.35	.73
2.5	543.64	1.15	430.40	1.10	324.70	1.01	229.82	.89	148.62	.74
2.6	542.49	1.16	429.30	1.10	323.69	1.00	228.93	.88	147.88	.73
2.7	541.33	1.16	428.20	1.09	322.69	1.01	228.05	.88	147.15	.73
2.8	540.17	1.16	427.11	1.10	321.68	1.00	227.17	.88	146.42	.73
2.9	539.01	1.15	426.01	1.10	320.68	1.00	226.29	.88	145.69	.73
3.0	537.86	1.16	424.91	1.10	319.68	1.01	225.41	.88	144.96	.72
3.1	536.70	1.15	423.81	1.09	318.67	1.00	224.53	.87	144.24	.73
3.2	535.55	1.16	422.72	1.10	317.67	1.00	223.66	.88	143.51	.72
3.3	534.39	1.15	421.62	1.09	316.67	1.00	222.78	.87	142.79	.72
3.4	533.24	1.16	420.53	1.09	315.67	.99	221.91	.87	142.07	.72
3.5	532.08	1.15	419.44	1.10	314.68	1.00	221.04	.87	141.35	.72
3.6	530.93	1.16	418.34	1.09	313.68	1.00	220.17	.87	140.63	.72
3.7	529.77	1.15	417.25	1.09	312.68	.99	219.30	.87	139.91	.71
3.8	528.62	1.15	416.16	1.09	311.69	.99	218.43	.87	139.20	.71
3.9	527.47	1.15	415.07	1.09	310.70	.99	217.56	.86	138.49	.72
4.0	526.32	1.16	413.98	1.08	309.71	1.00	216.70	.86	137.77	.71
4.1	525.16	1.15	412.90	1.09	308.71	.99	215.84	.87	137.06	.70
4.2	524.01	1.15	411.81	1.09	307.72	.98	214.97	.86	136.36	.71
4.3	522.86	1.15	410.72	1.08	306.74	.99	214 11	.85	135.65	.70
4.4	521.71	1.15	409.64	1.09	305.75	.99	213.26	.86	134.95	.71
4 5	520.56	1.15	408.55	1.08	304.76	.98	212.40	.86	134.24	.70
4.6	519.41	1.15	407.47	1.09	303.78	.99	211.54	.85	133.54	.70
4.7	518.26	1.15	406.38	1.08	302.79	.98	210.69	.86	132.84	.70
4.8	517.11	1.14	405.30	1.08	301.81	.98	209.83	.85	132.14	.69
4.9	515.97	1.15	404.22	1.08	300.83	.98	208.98	.85	131.45	.70
5.0	514.82	−1.15	403.14	−1.08	299.85	− .98	208.13	−.85	130.75	−.69

TABLE IX. ARGUMENT 4.

Equation = 670″.500 — 670″.3 sin. z — 7″.9 sin. 2z.

Period, 365.259687 days.

Days.	100		110		120		130		140	
Days.	Equation.	Difference.	Equation.	Difference.	Equation.	Difference.	Equation.	Difference.	Equation.	Difference.
d 5.0	514.82″	−1.15	403.14″	−1.08	299.85″	−.98	208.13″	−.85	130.75″	−.69
5.1	513.67	1.15	402.06	1.08	298.87	.98	207.28	.85	130.06	.69
5.2	512.52	1.14	400.98	1.08	297.89	.98	206.43	.84	129.37	.69
5.3	511.38	1.15	399.90	1.08	296.91	.97	205.59	.85	128.68	.69
5.4	510.23	1.14	398.82	1.07	295.94	.98	204.74	.84	127.99	.69
5.5	509.09	1.15	397.75	1.08	294.96	.97	203.90	.84	127.30	.68
5.6	507.94	1.14	396.67	1.07	293.99	.98	203.06	.84	126.62	.69
5.7	506.80	1.15	395.60	1.08	293.01	.97	202.22	.84	125.93	.68
5.8	505.65	1.14	394.52	1.07	292.04	.97	201.38	.84	125.25	.68
5.9	504.51	1.14	393.45	1.07	291.07	.96	200.54	.83	124.57	.68
6.0	503.37	1.14	392.38	1.08	290.11	.97	199.71	.84	123.89	.67
6.1	502.23	1.14	391.30	1.07	289.14	.97	198.87	.83	123.22	.68
6.2	501.09	1.15	390.23	1.07	288.17	.96	198.04	.83	122.54	.67
6.3	499.94	1.13	389.16	1.07	287.21	.97	197.21	.83	121.87	.67
6.4	498.81	1.15	388.09	1.06	286.24	.96	196.38	.83	121.20	.67
6.5	497.66	1.13	387.03	1.07	285.28	.96	195.55	.83	120.53	.67
6.6	496.53	1.14	385.96	1.07	284.32	.96	194.72	.83	119.86	.66
6.7	495.39	1.14	384.89	1.06	283.36	.96	193.89	.82	119.20	.67
6.8	494.25	1.14	383.83	1.07	282.40	.96	193.07	.82	118.53	.66
6.9	493.11	1.14	382.76	1.06	281.44	.96	192.25	.82	117.87	.66
7.0	491.97	1.13	381.70	1.06	280.48	.95	191.43	.82	117.21	.66
7.1	490.84	1.14	380.64	1.06	279.53	.96	190.61	.82	116.55	.66
7.2	489.70	1.13	379.58	1.06	278.57	.95	189.79	.82	115.89	.66
7.3	488.57	1.14	378.52	1.06	277.62	.95	188.97	.81	115.23	.65
7.4	487.43	1.13	377.46	1.06	276.67	.95	188.16	.82	114.58	.66
7.5	486.30	1.14	376.40	1.06	275.72	.95	187.34	.81	113.92	.65
7.6	485.16	1.13	375.34	1.06	274.77	.95	186.53	.81	113.27	.65
7.7	484.03	1.13	374.28	1.05	273.82	.95	185.72	.81	112.62	.64
7.8	482.90	1.13	373.23	1.06	272.87	.94	184.91	.81	111.98	.65
7.9	481.77	1.13	372.17	1.05	271.93	.95	184.10	.80	111.33	.64
8.0	480.64	1.14	371.12	1.05	270.98	.94	183.30	.81	110.69	.65
8.1	479.50	1.12	370.07	1.06	270.04	.94	182.49	.80	110.04	.64
8.2	478.38	1.13	369.01	1.05	269.10	.94	181.69	.80	109.40	.64
8.3	477.25	1.13	367.96	1.05	268.16	.94	180.89	.80	108.76	.63
8.4	476.12	1.13	366.91	1.05	267.22	.94	180.09	.80	108.13	.64
8.5	474.99	1.13	365.86	1.05	266.28	.94	179.29	.80	107.49	.63
8.6	473.86	1.13	364.81	1.04	265.34	.93	178.49	.79	106.86	.63
8.7	472.73	1.12	363.77	1.05	264.41	.94	177.70	.80	106.23	.63
8.8	471.61	1.13	362.72	1.05	263.47	.93	176.90	.79	105.60	.63
8.9	470.48	1.12	361.67	1.04	262.54	.93	176.11	.79	104.97	.63
9.0	469.36	1.13	360.63	1.05	261.61	.93	175.32	.79	104.34	.62
9.1	468.23	1.12	359.58	1.04	260.68	.93	174.53	.79	103.72	.63
9.2	467.11	1.13	358.54	1.04	259.75	.93	173.74	.78	103.09	.62
9.3	465.98	1.12	357.50	1.04	258.82	.93	172.96	.79	102.47	.62
9.4	464.86	1.12	356.46	1.04	257.89	.92	172.17	.78	101.85	.62
9.5	463.74	1.12	355.42	1.04	256.97	.93	171.39	.78	101.23	.61
9.6	462.62	1.12	354.38	1.04	256.04	.92	170.61	.78	100.62	.62
9.7	461.50	1.12	353.34	1.04	255.12	.92	169.83	.78	100.00	.61
9.8	460.38	1.12	352.30	1.03	254.20	.92	169.05	.78	99.39	.61
9.9	459.26	−1.12	351.27	−1.03	253.28	−.92	168.27	−.78	98.78	−.61
10.0	458.14		350.24		252.36		167.49		98.17	

TABLE IX. ARGUMENT 4.

Equation = 670″.500 — 670″.3 sin. z — 7″.9 sin. 2z.

Period, 365.259687 days.

Days.	150		160		170		180		190	
Days.	Equation.	Difference.	Equation.	Difference.	Equation.	Difference.	Equation.	Difference.	Equation.	Difference.
d.	″		″		″		″		″	
0.0	98.17	—.61	46.36	—.42	13.47	—.23	0.28	—.03	6.94	+.16
0.1	97.56	.61	45.94	.42	13.24	.22	0.25	.03	7.10	.17
0.2	96.95	.60	45.52	.42	13.02	.23	0.22	.03	7.27	.17
0.3	96.35	.60	45.10	.42	12.79	.23	0.19	.02	7.44	.17
0.4	95.75	.60	44.68	.42	12.56	.22	0.17	.03	7.61	.18
0.5	95.15	.60	44.26	.41	12.34	.22	0.14	.02	7.79	.17
0.6	94.55	.60	43.85	.42	12.12	.22	0.12	.02	7.96	.18
0.7	93.95	.59	43.43	.41	11.90	.21	0.10	.02	8.14	.18
0.8	93.36	.60	43.02	.41	11.69	.22	0.08	.02	8.32	.18
0.9	92.76	.59	42.61	.41	11.47	.21	0.06	.01	8.50	.18
1.0	92.17	.59	42.20	.40	11.26	.21	0.05	.01	8.68	.19
1.1	91.58	.59	41.80	.40	11.05	.21	0.04	.01	8.87	.19
1.2	90.99	.58	41.40	.40	10.84	.20	0.03	.01	9.06	.18
1.3	90.41	.59	41.00	.40	10.64	.21	0.02	.00	9.24	.20
1.4	89.82	.58	40.60	.40	10.43	.20	0.02	—.01	9.44	.19
1.5	89.24	.58	40.20	.40	10.23	.20	0.01	.00	9.63	.19
1.6	88.66	.58	39.80	.39	10.03	.20	0.01	.00	9.82	.20
1.7	88.08	.58	39.41	.39	9.83	.20	0.01	.00	10.02	.20
1.8	87.50	.58	39.02	.39	9.63	.19	0.01	+.01	10.22	.20
1.9	86.92	.57	38.63	.39	9.44	.20	0.02	.01	10.42	.20
2.0	86.35	.57	38.24	.39	9.24	.19	0.03	.01	10.62	.21
2.1	85.78	.57	37.85	.38	9.05	.19	0.01	.01	10.83	.21
2.2	85.21	.57	37.47	.38	8.86	.18	0.05	.01	11.04	.20
2.3	84.64	.56	37.09	.38	8.68	.19	0.06	.01	11.24	.22
2.4	84.08	.57	36.71	.38	8.19	.18	0.07	.02	11.46	.21
2.5	83.51	.56	36.33	.38	8.31	.18	0.09	.01	11.67	.21
2.6	82.95	.56	35.95	.37	8.13	.18	0.10	.02	11.88	.22
2.7	82.39	.56	35.58	.38	7.95	.17	0.12	.03	12.10	.22
2.8	81.83	.56	35.20	.37	7.78	.18	0.15	.02	12.32	.22
2.9	81.27	.56	34.83	.37	7.60	.17	0.17	.03	12.54	.22
3.0	80.71	.55	34.46	.36	7.43	.17	0.20	.03	12.76	.22
3.1	80.16	.55	34.10	.37	7.26	.17	0.23	.03	12.98	.23
3.2	79.61	.55	33.73	.36	7.09	.17	0.26	.03	13.21	.23
3.3	79.06	.55	33.37	.36	6.92	.16	0.29	.03	13.44	.23
3.4	78.51	.55	33.01	.36	6.76	.17	0.32	.04	13.67	.23
3.5	77.96	.55	32.65	.36	6.59	.16	0.36	.04	13.90	.23
3.6	77.41	.54	32.29	.35	6.43	.16	0.40	.04	14.13	.24
3.7	76.87	.54	31.94	.36	6.27	.15	0.44	.04	14.37	.23
3.8	76.33	.54	31.58	.35	6.12	.16	0.48	.04	14.60	.24
3.9	75.79	.54	31.23	.35	5.96	.15	0.52	.05	14.84	.24
4.0	75.25	.54	30.88	.35	5.81	.15	0.57	.05	15.08	.25
4.1	74.71	.53	30.53	.34	5.66	.15	0.62	.05	15.33	.24
4.2	74.18	.53	30.19	.35	5.51	.15	0.67	.05	15.57	.25
4.3	73.65	.53	29.84	.34	5.36	.14	0.72	.05	15.82	.25
4.4	73.12	.53	29.50	.34	5.22	.15	0.77	.06	16.07	.25
4.5	72.59	.53	29.16	.34	5.07	.14	0.83	.06	16.32	.25
4.6	72.06	.52	28.82	.33	4.93	.14	0.89	.06	16.57	.26
4.7	71.54	.53	28.49	.34	4.79	.13	0.95	.06	16.83	.25
4.8	71.01	.52	28.15	.33	4.66	.14	1.01	.06	17.08	.26
4.9	70.49	.52	27.82	.33	4.52	.13	1.07	.07	17.34	.26
5.0	69.97	—.52	27.49	—.33	4.39	—.13	1.14	+.07	17.60	+.26

TABLE IX: ARGUMENT 4.

Equation = 670″.500 — 670″.3 sin. z — 7″.9 sin. 2z.

Period, 365.259687 days.

Days.	150		160		170		180		190	
Days.	Equation.	Difference.	Equation.	Difference.	Equation.	Difference.	Equation.	Difference.	Equation.	Difference.
d.	″		″		″		″		″	
5.0	69.97	−.52	27.49	−.33	4.39	−.13	1.14	+.07	17.60	+.26
5.1	69.45	.51	27.16	.33	4.26	.13	1.21	.07	17.86	.27
5.2	68.94	.52	26.83	.32	4.13	.13	1.28	.07	18.13	.26
5.3	68.42	.51	26.51	.32	4.00	.12	1.35	.07	18.39	.27
5.4	67.91	.51	26.19	.32	3.88	.13	1.42	.08	18.66	.27
5.5	67.40	.51	25.87	.32	3.75	.12	1.50	.08	18.93	.27
5.6	66.89	.50	25.55	.32	3.63	.12	1.58	.08	19.20	.28
5.7	66.39	.51	25.23	.31	3.51	.11	1.66	.08	19.48	.27
5.8	65.88	.50	24.92	.31	3.40	.12	1.74	.08	19.75	.28
5.9	65.38	.50	24.61	.31	3.28	.11	1.82	.09	20.03	.28
6.0	64.88	.50	24.30	.31	3.17	.11	1.91	.08	20.31	.28
6.1	64.38	.50	23.99	.31	3.06	.11	1.99	.09	20.59	.28
6.2	63.88	.49	23.68	.31	2.95	.11	2.08	.10	20.87	.29
6.3	63.39	.50	23.37	.30	2.84	.10	2.18	.09	21.16	.28
6.4	62.89	.49	23.07	.30	2.74	.11	2.27	.09	21.44	.29
6.5	62.40	.49	22.77	.30	2.63	.10	2.36	.10	21.73	.29
6.6	61.91	.49	22.47	.29	2.53	.10	2.46	.10	22.02	.30
6.7	61.42	.48	22.18	.30	2.43	.09	2.56	.10	22.32	.29
6.8	60.94	.49	21.88	.29	2.34	.10	2.66	.10	22.61	.30
6.9	60.45	.48	21.59	.29	2.24	.09	2.76	.11	22.91	.29
7.0	59.97	.48	21.30	.29	2.15	.09	2.87	.11	23.20	.30
7.1	59.49	.48	21.01	.29	2.06	.09	2.98	.11	23.50	.31
7.2	59.01	.48	20.72	.28	1.97	.09	3.09	.11	23.81	.30
7.3	58.53	.47	20.44	.29	1.88	.08	3.20	.11	24.11	.31
7.4	58.06	.48	20.15	.28	1.80	.09	3.31	.12	24.42	.30
7.5	57.58	.47	19.87	.28	1.71	.08	3.43	.11	24.72	.31
7.6	57.11	.47	19.59	.28	1.63	.08	3.54	.12	25.03	.31
7.7	56.64	.46	19.31	.27	1.55	.08	3.66	.12	25.34	.32
7.8	56.18	.47	19.04	.28	1.47	.07	3.78	.12	25.66	.31
7.9	55.71	.46	18.76	.27	1.40	.07	3.90	.13	25.97	.32
8.0	55.25	.47	18.49	.27	1.33	.08	4.03	.13	26.29	.32
8.1	54.78	.46	18.22	.27	1.25	.06	4.16	.13	26.61	.32
8.2	54.32	.45	17.95	.26	1.19	.07	4.29	.13	26.93	.32
8.3	53.87	.46	17.69	.27	1.12	.07	4.42	.13	27.25	.33
8.4	53.41	.46	17.42	.26	1.05	.06	4.55	.13	27.58	.32
8.5	52.95	.45	17.16	.26	0.99	.06	4.68	.14	27.90	.33
8.6	52.50	.45	16.90	.25	0.93	.06	4.82	.14	28.23	.33
8.7	52.05	.45	16.65	.26	0.87	.06	4.96	.14	28.56	.33
8.8	51.60	.45	16.39	.25	0.81	.05	5.10	.14	28.89	.34
8.9	51.15	.44	16.14	.26	0.76	.06	5.24	.15	29.23	.33
9.0	50.71	.44	15.88	.25	0.70	.05	5.39	.14	29.56	.34
9.1	50.27	.44	15.63	.24	0.65	.05	5.53	.15	29.90	.34
9.2	49.83	.44	15.39	.25	0.60	.05	5.68	.15	30.24	.34
9.3	49.39	.44	15.14	.24	0.55	.04	5.83	.15	30.58	.34
9.4	48.95	.44	14.90	.25	0.51	.04	5.98	.16	30.92	.34
9.5	48.51	.43	14.65	.24	0.47	.05	6.14	.15	31.26	.35
9.6	48.08	.43	14.41	.23	0.42	.04	6.29	.16	31.61	.35
9.7	47.65	.43	14.18	.24	0.38	.03	6.45	.16	31.96	.35
9.8	47.22	.43	13.94	.23	0.35	.04	6.61	.16	32.31	.35
9.9	46.79	−.43	13.71	−.24	0.31	−.03	6.77	+.17	32.66	+.36
10.0	46.36		13.47		0.28		6.94		33.02	

TABLE IX. ARGUMENT 4.

Equation = 670″.500 — 670″.3 sin. z — 7″.9 sin. 2z.

Period, 365.259687 days.

Days.	200		210		220		230		240	
Days.	Equation.	Difference.	Equation.	Difference.	Equation.	Difference.	Equation.	Difference.	Equation.	Difference.
d.	″		″		″		″		″	
0.0	33.02	+.35	77.53	+.54	139.00	+.69	215.49	+.83	304.72	+ .95
0.1	33.37	.36	78.07	.53	139.69	.70	216.32	.84	305.67	.95
0.2	33.73	.36	78.60	.54	140.39	.70	217.16	.83	306.62	.95
0.3	34.09	.36	79.14	.54	141.09	.69	217.99	.84	307.57	.95
0.4	34.45	.37	79.68	.54	141.78	.70	218.83	.84	308.52	.95
0.5	34.82	.36	80.22	.55	142.48	.71	219.67	.84	309.47	.95
0.6	35.18	.37	80.77	.55	143.19	.70	220.51	.84	310.42	.96
0.7	35.55	.37	81.32	.54	143.89	.70	221.35	.84	311.38	.95
0.8	35.92	.37	81.86	.55	144.59	.71	222.19	.85	312.33	.96
0.9	36.29	.37	82.41	.54	145.30	.71	223.04	.84	313.29	.96
1.0	36.66	.38	82.95	.55	146.01	.70	223.88	.85	314.25	.95
1.1	37.04	.37	83.50	.55	146.71	.72	224.73	.84	315.20	.96
1.2	37.41	.38	84.05	.55	147.43	.71	225.57	.85	316.16	.96
1.3	37.79	.38	84.60	.56	148.14	.71	226.42	.85	317.12	.96
1.4	38.17	.38	85.16	.56	148.85	.72	227.27	.85	318.08	.97
1.5	38.55	.39	85.72	.56	149.57	.71	228.12	.85	319.05	.96
1.6	38.94	.38	86.28	.56	150.28	.72	228.97	.86	320.01	.96
1.7	39.32	.39	86.84	.56	151.00	.72	229.83	.85	320.97	.97
1.8	39.71	.39	87.40	.56	151.72	.72	230.68	.86	321.94	.96
1.9	40.10	.39	87.96	.57	152.44	.72	231.54	.86	322.90	.97
2.0	40.49	.39	88.53	.56	153.16	.73	232.40	.85	323.87	.96
2.1	40.88	.40	89.09	.57	153.89	.72	233.25	.86	324.83	.98
2.2	41.28	.39	89.66	.57	154.61	.73	234.11	.86	325.81	.97
2.3	41.67	.40	90.23	.58	155.34	.73	234.97	.87	326.78	.97
2.4	42.07	.40	90.81	.57	156.07	.73	235.84	.86	327.75	.97
2.5	42.47	.40	91.38	.58	156.80	.73	236.70	.86	328.72	.97
2.6	42.87	.41	91.96	.57	157.53	.73	237.56	.87	329.69	.98
2.7	43.28	.40	92.53	.58	158.26	.73	238.43	.87	330.67	.97
2.8	43.68	.41	93.11	.58	158.99	.74	239.30	.86	331.64	.98
2.9	44.09	.41	93.69	.58	159.73	.74	240.16	.87	332.62	.97
3.0	44.50	.41	94.27	.59	160.47	.73	241.03	.87	333.59	.98
3.1	44.91	.41	94.86	.58	161.20	.74	241.90	.88	334.57	.98
3.2	45.32	.41	95.44	.59	161.94	.74	242.78	.87	335.55	.98
3.3	45.73	.42	96.03	.59	162.68	.75	243.65	.87	336.53	.98
3.4	46.15	.42	96.62	.59	163.43	.74	244.52	.88	337.51	.98
3.5	46.57	.42	97.21	.59	164.17	.75	245.40	.87	338.49	.98
3.6	46.99	.42	97.80	.60	164.92	.74	246.27	.88	339.47	.98
3.7	47.41	.42	98.40	.59	165.66	.75	247.15	.88	340.45	.99
3.8	47.83	.43	98.99	.60	166.41	.75	248.03	.88	341.44	.98
3.9	48.26	.43	99.59	.60	167.16	.75	248.91	.88	342.42	.99
4.0	48.69	.42	100.19	.60	167.91	.75	249.79	.88	343.41	.98
4.1	49.11	.44	100.79	.60	168.66	.76	250.67	.89	344.39	.99
4.2	49.55	.43	101.39	.60	169.42	.75	251.56	.88	345.38	.99
4.3	49.98	.43	101.99	.61	170.17	.76	252.44	.89	346.37	.99
4.4	50.41	.44	102.60	.60	170.93	.76	253.33	.88	347.36	.99
4.5	50.85	.44	103.20	.61	171.69	.76	254.21	.89	348.35	.99
4.6	51.29	.44	103.81	.61	172.45	.76	255.10	.89	349.34	.99
4.7	51.73	.44	104.42	.61	173.21	.76	255.99	.89	350.33	.99
4.8	52.17	.44	105.03	.61	173.97	.77	256.88	.89	351.32	1.00
4.9	52.61	.44	105.64	.62	174.74	.76	257.77	.90	352.32	.99
5.0	53.05	+.45	106.26	+.61	175.50	+.77	258.67	+.89	353.31	+1.00

TABLE IX. ARGUMENT 4.

Equation = 670″.500 — 670″.3 sin. z — 7″.9 sin. 2z.

Period, 365.259687 days.

Days.	200		210		220		230		240	
Days.	Equation.	Difference.	Equation.	Difference.	Equation.	Difference.	Equation.	Difference.	Equation.	Difference.
d.	″		″		″		″		″	
5.0	53.05	+.45	106.26	+.61	175.50	+.77	258.67	+.89	353.31	+1.00
5.1	53.50	.45	106.87	.62	176.27	.77	259.56	.90	354.31	1.00
5.2	53.95	.45	107.49	.62	177.04	.77	260.46	.89	355.31	.99
5.3	54.40	.45	108.11	.62	177.81	.77	261.35	.90	356.30	1.00
5.4	54.85	.46	108.73	.63	178.58	.77	262.25	.90	357.30	1.00
5.5	55.31	.45	109.36	.63	179.35	.77	263.15	.90	358.30	1.00
5.6	55.76	.46	109.99	.62	180.12	.78	264.05	.90	359.30	1.00
5.7	56.22	.46	110.61	.62	180.90	.77	264.95	.90	360.30	1.00
5.8	56.68	.46	111.23	.63	181.67	.78	265.85	.90	361.30	1.00
5.9	57.14	.46	111.86	.63	182.45	.78	266.75	.91	362.30	1.01
6.0	57.60	.46	112.49	.64	183.23	.78	267.66	.90	363.31	1.00
6.1	58.06	.47	113.13	.63	184.01	.78	268.56	.91	364.31	1.01
6.2	58.53	.47	113.76	.63	184.79	.78	269.47	.91	365.32	1.00
6.3	59.00	.47	114.39	.64	185.57	.79	270.38	.91	366.32	1.01
6.4	59.47	.47	115.03	.64	186.36	.78	271.29	.91	367.33	1.01
6.5	59.94	.47	115.67	.64	187.14	.79	272.20	.91	368.34	1.00
6.6	60.41	.48	116.31	.64	187.93	.79	273.11	.91	369.34	1.01
6.7	60.89	.47	116.95	.65	188.72	.79	274.02	.91	370.35	1.01
6.8	61.36	.48	117.60	.64	189.51	.79	274.93	.92	371.36	1.01
6.9	61.84	.48	118.24	.65	190.30	.79	275.85	.91	372.37	1.02
7.0	62.32	.49	118.89	.64	191.09	.80	276.76	.92	373.39	1.01
7.1	62.81	.48	119.53	.65	191.89	.79	277.68	.92	374.40	1.01
7.2	63.29	.49	120.18	.66	192.68	.80	278.60	.91	375.41	1.02
7.3	63.78	.48	120.84	.65	193.48	.80	279.51	.92	376.43	1.01
7.4	64.26	.49	121.49	.65	194.28	.80	280.43	.93	377.44	1.02
7.5	64.75	.49	122.14	.66	195.08	.80	281.36	.92	378.46	1.01
7.6	65.24	.49	122.80	.66	195.88	.80	282.28	.92	379.47	1.02
7.7	65.73	.50	123.46	.66	196.68	.80	283.20	.92	380.49	1.02
7.8	66.23	.49	124.12	.66	197.48	.81	284.12	.93	381.51	1.02
7.9	66.72	.50	124.78	.66	198.29	.80	285.05	.92	382.53	1.02
8.0	67.22	.50	125.44	.66	199.09	.81	285.97	.93	383.55	1.02
8.1	67.72	.50	126.10	.67	199.90	.81	286.90	.93	384.57	1.02
8.2	68.22	.50	126.77	.66	200.71	.81	287.83	.93	385.59	1.02
8.3	68.72	.51	127.43	.67	201.52	.81	288.76	.93	386.61	1.02
8.4	69.23	.51	128.10	.67	202.33	.81	289.69	.93	387.63	1.03
8.5	69.74	.50	128.77	.67	203.14	.82	290.62	.93	388.66	1.02
8.6	70.24	.51	129.44	.68	203.96	.81	291.55	.94	389.68	1.03
8.7	70.75	.52	130.12	.67	204.77	.82	292.49	.93	390.71	1.02
8.8	71.27	.51	130.79	.67	205.59	.82	293.42	.94	391.73	1.03
8.9	71.78	.51	131.46	.68	206.41	.82	294.36	.93	392.76	1.03
9.0	72.29	.52	132.14	.68	207.23	.82	295.29	.94	393.79	1.03
9.1	72.81	.52	132.82	.68	208.05	.82	296.23	.94	394.82	1.02
9.2	73.33	.52	133.50	.68	208.87	.82	297.17	.94	395.84	1.03
9.3	73.85	.52	134.18	.69	209.69	.83	298.11	.94	396.87	1.04
9.4	74.37	.52	134.87	.68	210.52	.82	299.05	.94	397.91	1.03
9.5	74.89	.53	135.55	.69	211.34	.83	299.99	.95	398.94	1.03
9.6	75.42	.52	136.24	.69	212.17	.83	300.94	.94	399.97	1.03
9.7	75.94	.53	136.93	.69	213.00	.83	301.88	.95	401.00	1.04
9.8	76.47	.53	137.62	.69	213.83	.83	302.83	.94	402.04	1.03
9.9	77.00	+.53	138.31	+.69	214.66	+.83	303.77	+.95	403.07	+1.04
10.0	77.53		139.00		215.49		304.72		404.11	

TABLE IX. ARGUMENT 4.

Equation = 670″.500 — 670″.3 sin. z — 7″.9 sin. 2z.

Period, 365.259687 days.

Days.	250		260		270		280		290	
Days.	Equation.	Difference.	Equation.	Difference.	Equation.	Difference.	Equation.	Difference.	Equation.	Difference.
d.	″		″		″		″		″	
0.0	404.11	+1.03	510.84	+1.10	621.96	+1.12	734.41	+1.12	845.11	+1.09
0.1	405.14	1.04	511.94	1.09	623.08	1.13	735.53	1.12	846.20	1.09
0.2	406.18	1.03	513.03	1.10	624.21	1.12	736.65	1.12	847.29	1.08
0.3	407.21	1.04	514.13	1.09	625.33	1.13	737.77	1.12	848.37	1.09
0.4	408.25	1.04	515.22	1.10	626.46	1.12	738.89	1.12	849.46	1.08
0.5	409.29	1.04	516.32	1.10	627.58	1.12	740.01	1.12	850.54	1.09
0.6	410.33	1.04	517.42	1.09	628.70	1.13	741.13	1.12	851.63	1.08
0.7	411.37	1.04	518.51	1.10	629.83	1.12	742.25	1.12	852.71	1.09
0.8	412.41	1.04	519.61	1.10	630.95	1.13	743.37	1.12	853.80	1.08
0.9	413.45	1.05	520.71	1.09	632.08	1.12	744.49	1.12	854.88	1.09
1.0	414.50	1.04	521.80	1.10	633.20	1.12	745.61	1.12	855.97	1.08
1.1	415.54	1.04	522.90	1.10	634.32	1.13	746.73	1.12	857.05	1.08
1.2	416.58	1.05	524.00	1.10	635.45	1.12	747.85	1.12	858.13	1.08
1.3	417.63	1.04	525.10	1.10	636.57	1.13	748.97	1.11	859.21	1.08
1.4	418.67	1.05	526.20	1.10	637.70	1.12	750.08	1.12	860.29	1.08
1.5	419.72	1.04	527.30	1.10	638.82	1.12	751.20	1.12	861.37	1.08
1.6	420.76	1.05	528.40	1.10	639.94	1.13	752.32	1.12	862.45	1.08
1.7	421.81	1.05	529.50	1.11	641.07	1.12	753.44	1.12	863.53	1.08
1.8	422.86	1.05	530.61	1.10	642.19	1.13	754.56	1.11	864.61	1.08
1.9	423.91	1.05	531.71	1.10	643.32	1.12	755.67	1.12	865.69	1.08
2.0	424.96	1.05	532.81	1.10	644.44	1.13	756.79	1.12	866.77	1.08
2.1	426.01	1.05	533.91	1.11	645.57	1.12	757.91	1.11	867.85	1.07
2.2	427.06	1.05	535.02	1.10	646.69	1.13	759.02	1.12	868.92	1.08
2.3	428.11	1.05	536.12	1.10	647.82	1.12	760.14	1.11	870.00	1.08
2.4	429.16	1.05	537.22	1.11	648.94	1.13	761.25	1.12	871.08	1.07
2.5	430.21	1.06	538.33	1.10	650.07	1.12	762.37	1.12	872.15	1.08
2.6	431.27	1.05	539.43	1.11	651.19	1.13	763.49	1.11	873.23	1.07
2.7	432.32	1.06	540.54	1.10	652.32	1.13	764.60	1.12	874.30	1.07
2.8	433.38	1.05	541.64	1.11	653.45	1.12	765.72	1.11	875.37	1.08
2.9	434.43	1.06	542.75	1.10	654.57	1.13	766.83	1.11	876.45	1.07
3.0	435.49	1.05	543.85	1.11	655.70	1.12	767.94	1.12	877.52	1.07
3.1	436.54	1.06	544.96	1.11	656.82	1.13	769.06	1.11	878.59	1.07
3.2	437.60	1.06	546.07	1.10	657.95	1.12	770.17	1.12	879.66	1.07
3.3	438.66	1.06	547.17	1.11	659.07	1.13	771.29	1.11	880.73	1.07
3.4	439.72	1.06	548.28	1.11	660.20	1.13	772.40	1.11	881.80	1.07
3.5	440.78	1.06	549.39	1.11	661.33	1.12	773.51	1.12	882.87	1.07
3.6	441.84	1.06	550.50	1.10	662.45	1.13	774.63	1.11	883.94	1.07
3.7	442.90	1.06	551.60	1.11	663.58	1.12	775.74	1.11	885.01	1.07
3.8	443.96	1.06	552.71	1.11	664.70	1.13	776.85	1.11	886.08	1.07
3.9	445.02	1.06	553.82	1.11	665.83	1.12	777.96	1.11	887.15	1.06
4.0	446.08	1.06	554.93	1.11	666.95	1.13	779.07	1.11	888.21	1.07
4.1	447.14	1.07	556.04	1.11	668.08	1.13	780.18	1.12	889.28	1.07
4.2	448.21	1.06	557.15	1.11	669.21	1.12	781.30	1.11	890.35	1.06
4.3	449.27	1.07	558.26	1.11	670.33	1.13	782.41	1.11	891.41	1.06
4.4	450.34	1.06	559.37	1.11	671.46	1.12	783.52	1.11	892.47	1.07
4.5	451.40	1.07	560.48	1.11	672.58	1.13	784.63	1.11	893.54	1.06
4.6	452.47	1.06	561.59	1.12	673.71	1.13	785.74	1.11	894.60	1.06
4.7	453.53	1.07	562.71	1.11	674.84	1.12	786.85	1.11	895.66	1.07
4.8	454.60	1.07	563.82	1.11	675.96	1.13	787.96	1.10	896.73	1.06
4.9	455.67	1.07	564.93	1.11	677.09	1.12	789.06	1.11	897.79	1.06
5.0	456.74	+1.06	566.04	+1.11	678.21	+1.13	790.17	+1.11	898.85	+1.06

TABLE IX. ARGUMENT 4.

Equation $= 670''.500 - 670''.3 \sin. z - 7''.9 \sin. 2z.$

Period, 365.259687 days.

Days.	250		260		270		280		290	
Days.	Equation.	Difference.	Equation.	Difference.	Equation.	Difference.	Equation.	Difference.	Equation.	Difference.
d.	"		"		"		"		"	
5.0	456.74	+1.06	566.04	+1.11	678.21	+1.13	790.17	+1.11	898.85	+1.06
5.1	457.80	1.07	567.15	1.12	679.34	1.12	791.28	1.11	899.91	1.06
5.2	458.87	1.07	568.27	1.11	680.46	1.13	792.39	1.11	900.97	1.06
5.3	459.94	1.08	569.38	1.11	681.59	1.13	793.50	1.10	902.03	1.05
5.4	461.02	1.07	570.49	1.12	682.72	1.12	794.60	1.11	903.08	1.06
5.5	462.09	1.07	571.61	1.11	683.84	1.13	795.71	1.11	904.14	1.06
5.6	463.16	1.07	572.72	1.12	684.97	1.12	796.82	1.10	905.20	1.05
5.7	464.23	1.07	573.84	1.11	686.09	1.13	797.92	1.11	906.25	1.06
5.8	465.30	1.08	574.95	1.12	687.22	1.12	799.03	1.10	907.31	1.05
5.9	466.38	1.07	576.07	1.11	688.34	1.11	800.13	1.11	908.36	1.06
6.0	467.45	1.07	577.18	1.11	689.45	1.14	801.24	1.10	909.42	1.05
6.1	468.52	1.08	578.29	1.12	690.59	1.13	802.34	1.11	910.47	1.05
6.2	469.60	1.08	579.41	1.12	691.72	1.12	803.45	1.10	911.52	1.06
6.3	470.68	1.07	580.53	1.11	692.84	1.13	804.55	1.10	912.58	1.05
6.4	471.75	1.08	581.64	1.12	693.97	1.12	805.65	1.11	913.63	1.05
6.5	472.83	1.08	582.76	1.12	695.09	1.13	806.76	1.10	914.68	1.05
6.6	473.91	1.07	583.88	1.11	696.22	1.13	807.86	1.10	915.73	1.05
6.7	474.98	1.08	584.99	1.12	697.35	1.12	808.96	1.10	916.78	1.05
6.8	476.06	1.08	586.11	1.12	698.47	1.13	810.06	1.11	917.83	1.04
6.9	477.14	1.08	587.23	1.12	699.60	1.12	811.17	1.10	918.87	1.05
7.0	478.22	1.08	588.35	1.11	700.72	1.12	812.27	1.10	919.92	1.05
7.1	479.30	1.08	589.46	1.12	701.84	1.13	813.37	1.10	920.97	1.04
7.2	480.38	1.08	590.58	1.12	702.97	1.12	814.47	1.10	922.01	1.05
7.3	481.46	1.09	591.70	1.12	704.09	1.13	815.57	1.10	923.06	1.04
7.4	482.55	1.08	592.82	1.12	705.22	1.12	816.67	1.10	924.10	1.05
7.5	483.63	1.08	593.94	1.11	706.34	1.13	817.77	1.10	925.15	1.04
7.6	484.71	1.08	595.05	1.12	707.47	1.12	818.87	1.09	926.19	1.04
7.7	485.79	1.09	596.17	1.12	708.59	1.12	819.96	1.10	927.23	1.05
7.8	486.88	1.08	597.29	1.12	709.71	1.13	821.06	1.10	928.28	1.04
7.9	487.96	1.09	598.41	1.12	710.84	1.12	822.16	1.10	929.32	1.04
8.0*	489.05	1.08	599.53	1.12	711.96	1.13	823.26	1.09	930.36	1.04
8.1	490.13	1.09	600.65	1.12	713.09	1.12	824.35	1.10	931.40	1.04
8.2	491.22	1.08	601.77	1.12	714.21	1.12	825.45	1.10	932.44	1.04
8.3	492.30	1.09	602.89	1.12	715.33	1.13	826.55	1.09	933.48	1.03
8.4	493.39	1.09	604.01	1.12	716.46	1.12	827.64	1.10	934.51	1.04
8.5	494.48	1.08	605.13	1.12	717.58	1.12	828.74	1.09	935.55	1.04
8.6	495.56	1.09	606.25	1.12	718.70	1.13	829.83	1.10	936.59	1.03
8.7	496.65	1.09	607.37	1.13	719.83	1.12	830.93	1.09	937.62	1.04
8.8	497.74	1.09	608.50	1.12	720.95	1.12	832.02	1.09	938.66	1.03
8.9	498.83	1.09	609.62	1.12	722.07	1.12	833.11	1.10	939.69	1.03
9.0	499.92	1.09	610.74	1.12	723.19	1.13	834.21	1.09	940.72	1.04
9.1	501.01	1.09	611.86	1.12	724.32	1.12	835.30	1.09	941.76	1.03
9.2	502.10	1.09	612.98	1.12	725.44	1.12	836.39	1.09	942.79	1.03
9.3	503.19	1.09	614.10	1.13	726.56	1.12	837.48	1.09	943.82	1.03
9.4	504.28	1.09	615.23	1.12	727.68	1.12	838.57	1.09	944.85	1.03
9.5	505.37	1.10	616.35	1.12	728.80	1.13	839.66	1.09	945.88	1.03
9.6	506.47	1.09	617.47	1.12	729.93	1.12	840.75	1.09	946.91	1.02
9.7	507.56	1.09	618.59	1.13	731.05	1.12	841.84	1.09	947.93	1.03
9.8	508.65	1.10	619.72	1.12	732.17	1.12	842.93	1.09	948.96	1.03
9.9	509.75	+1.09	620.84	+1.12	733.29	+1.12	844.02	+1.09	949.99	+1.02
10.0	510.84		621.96		734.41		845.11		951.01	

TABLE IX. ARGUMENT 4.

Equation = 670″.500 — 670″.3 sin. z — 7″.9 sin. $2z$.

Period, 365.259687 days.

Days.	300		310		320		330		310	
Days.	Equation.	Difference.	Equation.	Difference.	Equation.	Difference.	Equation.	Difference.	Equation.	Difference.
d.										
0.0	951″.01	+1.03	1049″.17	+.93	1136″.80	+.81	1211″.36	+.67	1270″.60	+.51
0.1	952.04	1.02	1050.10	.93	1137.61	.81	1212.03	.67	1271.11	.51
0.2	953.06	1.02	1051.03	.93	1138.42	.82	1212.70	.67	1271.62	.50
0.3	954.08	1.03	1051.96	.93	1139.24	.81	1213.37	.67	1272.12	.51
0.4	955.11	1.02	1052.89	.93	1140.05	.80	1214.04	.66	1272.63	.50
0.5	956.13	1.02	1053.82	.93	1140.85	.81	1214.70	.67	1273.13	.50
0.6	957.15	1.02	1054.75	.93	1141.66	.81	1215.37	.66	1273.63	.50
0.7	958.17	1.02	1055.68	.92	1142.47	.80	1216.03	.66	1274.13	.49
0.8	959.19	1.02	1056.60	.92	1143.27	.81	1216.69	.66	1274.62	.50
0.9	960.21	1.01	1057.52	.93	1144.08	.80	1217.35	.66	1275.12	.49
1.0	961.22	1.02	1058.45	.92	1144.88	.80	1218.01	.65	1275.61	.49
1.1	962.24	1.02	1059.37	.92	1145.68	.80	1218.66	.66	1276.10	.49
1.2	963.26	1.01	1060.29	.92	1146.48	.80	1219.32	.65	1276.59	.49
1.3	964.27	1.01	1061.21	.92	1147.28	.80	1219.97	.65	1277.08	.49
1.4	965.28	1.02	1062.13	.92	1148.08	.79	1220.62	.65	1277.57	.48
1.5	966.30	1.01	1063.05	.91	1148.87	.80	1221.27	.65	1278.05	.48
1.6	967.31	1.01	1063.96	.92	1149.67	.79	1221.92	.65	1278.53	.48
1.7	968.32	1.01	1064.88	.91	1150.46	.79	1222.57	.64	1279.01	.48
1.8	969.33	1.01	1065.79	.92	1151.25	.79	1223.21	.65	1279.49	.48
1.9	970.34	1.01	1066.71	.91	1152.04	.79	1223.86	.64	1279.97	.47
2.0	971.35	1.01	1067.62	.91	1152.83	.79	1224.50	.64	1280.44	.48
2.1	972.36	1.01	1068.53	.91	1153.62	.79	1225.14	.64	1280.92	.47
2.2	973.37	1.01	1069.44	.91	1154.41	.78	1225.78	.63	1281.39	.47
2.3	974.38	1.00	1070.35	.91	1155.19	.78	1226.41	.64	1281.86	.47
2.4	975.38	1.01	1071.26	.91	1155.97	.79	1227.05	.63	1282.33	.47
2.5	976.39	1.00	1072.17	.90	1156.76	.78	1227.68	.64	1282.80	.46
2.6	977.39	1.01	1073.07	.91	1157.54	.78	1228.32	.63	1283.26	.46
2.7	978.40	1.00	1073.98	.90	1158.32	.78	1228.95	.63	1283.72	.46
2.8	979.40	1.00	1074.88	.90	1159.10	.77	1229.58	.63	1284.18	.46
2.9	980.40	1.00	1075.78	.90	1159.87	.78	1230.21	.62	1284.64	.46
3.0	981.40	1.00	1076.68	.90	1160.65	.77	1230.83	.63	1285.10	.46
3.1	982.40	1.00	1077.58	.90	1161.42	.77	1231.46	.62	1285.56	.45
3.2	983.40	1.00	1078.48	.90	1162.19	.78	1232.08	.62	1286.01	.45
3.3	984.40	1.00	1079.38	.90	1162.97	.77	1232.70	.62	1286.46	.45
3.4	985.40	.99	1080.28	.89	1163.74	.76	1233.32	.62	1286.91	.45
3.5	986.39	1.00	1081.17	.90	1164.50	.77	1233.94	.62	1287.36	.45
3.6	987.39	.99	1082.07	.89	1165.27	.77	1234.56	.61	1287.81	.45
3.7	988.38	1.00	1082.96	.89	1166.04	.76	1235.17	.61	1288.26	.44
3.8	989.38	.99	1083.85	.89	1166.80	.76	1235.78	.62	1288.70	.44
3.9	990.37	.99	1084.74	.89	1167.56	.77	1236.40	.61	1289.14	.44
4.0	991.36	.99	1085.63	.89	1168.33	.76	1237.01	.61	1289.58	.44
4.1	992.35	.99	1086.52	.89	1169.09	.75	1237.62	.60	1290.02	.44
4.2	993.34	.99	1087.41	.88	1169.84	.76	1238.22	.61	1290.46	.43
4.3	994.33	.99	1088.29	.89	1170.60	.76	1238.83	.60	1290.89	.44
4.4	995.32	.99	1089.18	.88	1171.36	.75	1239.43	.60	1291.33	.43
4.5	996.31	.99	1090.06	.88	1172.11	.75	1240.03	.60	1291.76	.43
4.6	997.30	.98	1090.94	.89	1172.86	.75	1240.63	.60	1292.19	.42
4.7	998.28	.99	1091.83	.88	1173.61	.75	1241.23	.60	1292.61	.43
4.8	999.27	.98	1092.71	.88	1174.36	.75	1241.83	.60	1293.04	.42
4.9	1000.25	.99	1093.59	.87	1175.11	.75	1242.43	.59	1293.46	.43
5.0	1001.24	+ .98	1094.46	+.88	1175.86	+.75	1243.02	+.59	1293.89	+.42

TABLE IX. ARGUMENT 4.

Equation $= 670''.500 - 670''.3 \text{ sin. } z - 7''.9 \text{ sin. } 2z$.

Period, 365.259687 days.

Days.	300		310		320		330		340	
Days.	Equation.	Difference.	Equation.	Difference.	Equation.	Difference.	Equation.	Difference.	Equation.	Difference.
d.	″		″		″		″		″	
5.0	1001.24	+.98	1094.46	+.88	1175.86	+.75	1243.02	+.59	1293.89	+.42
5.1	1002.22	.98	1095.34	.88	1176.61	.74	1243.61	.59	1294.31	.41
5.2	1003.20	.98	1096.22	.87	1177.35	.74	1244.20	.59	1294.72	.42
5.3	1004.18	.98	1097.09	.87	1178.09	.74	1244.79	.59	1295.14	.42
5.4	1005.16	.98	1097.96	.88	1178.83	.74	1245.38	.59	1295.56	.41
5.5	1006.14	.97	1098.84	.87	1179.57	.74	1245.97	.58	1295.97	.41
5.6	1007.11	.98	1099.71	.87	1180.31	.74	1246.55	.58	1296.38	.41
5.7	1008.09	.98	1100.58	.87	1181.05	.74	1247.13	.58	1296.79	.41
5.8	1009.07	.97	1101.45	.86	1181.79	.73	1247.71	.58	1297.20	.40
5.9	1010.04	.98	1102.31	.87	1182.52	.73	1248.29	.58	1297.60	.41
6.0	1011.02	.97	1103.18	.86	1183.25	.73	1248.87	.58	1298.01	.40
6.1	1011.99	.97	1104.04	.87	1183.98	.73	1249.45	.57	1298.41	.40
6.2	1012.96	.97	1104.91	.86	1184.71	.73	1250.02	.57	1298.81	.40
6.3	1013.93	.97	1105.77	.86	1185.44	.73	1250.59	.58	1299.21	.40
6.4	1014.90	.97	1106.63	.86	1186.17	.73	1251.17	.56	1299.61	.39
6.5	1015.87	.97	1107.49	.86	1186.90	.72	1251.73	.57	1300.00	.39
6.6	1016.84	.97	1108.35	.86	1187.62	.72	1252.30	.57	1300.39	.39
6.7	1017.81	.96	1109.21	.85	1188.34	.72	1252.87	.56	1300.78	.39
6.8	1018.77	.97	1110.06	.86	1189.06	.72	1253.43	.57	1301.17	.39
6.9	1019.74	.96	1110.92	.85	1189.78	.72	1254.00	.56	1301.56	.39
7.0	1020.70	.97	1111.77	.85	1190.50	.72	1254.56	.56	1301.95	.38
7.1	1021.67	.96	1112.62	.85	1191.22	.71	1255.12	.55	1302.33	.38
7.2	1022.63	.96	1113.47	.85	1191.93	.72	1255.67	.56	1302.71	.39
7.3	1023.59	.96	1114.32	.85	1192.65	.71	1256.23	.55	1303.10	.37
7.4	1024.55	.96	1115.17	.85	1193.36	.71	1256.78	.56	1303.47	.38
7.5	1025.51	.96	1116.02	.85	1194.07	.71	1257.34	.55	1303.85	.38
7.6	1026.47	.96	1116.87	.84	1194.78	.71	1257.89	.55	1304.23	.37
7.7	1027.43	.95	1117.71	.85	1195.49	.70	1258.44	.54	1304.60	.37
7.8	1028.38	.96	1118.56	.84	1196.19	.71	1258.98	.55	1304.97	.37
7.9	1029.34	.95	1119.40	.84	1196.90	.70	1259.53	.55	1305.34	.37
8.0	1030.29	.96	1120.24	.84	1197.60	.71	1260.08	.54	1305.71	.36
8.1	1031.25	.95	1121.08	.84	1198.31	.70	1260.62	.54	1306.07	.37
8.2	1032.20	.95	1121.92	.84	1199.01	.69	1261.16	.54	1306.44	.36
8.3	1033.15	.95	1122.76	.83	1199.70	.70	1261.70	.54	1306.80	.36
8.4	1034.10	.95	1123.59	.84	1200.40	.70	1262.24	.53	1307.16	.36
8.5	1035.05	.95	1124.43	.83	1201.10	.69	1262.77	.54	1307.52	.35
8.6	1036.00	.94	1125.26	.84	1201.79	.69	1263.31	.53	1307.87	.36
8.7	1036.94	.95	1126.10	.83	1202.48	.70	1263.84	.53	1308.23	.35
8.8	1037.89	.94	1126.93	.83	1203.18	.69	1264.37	.53	1308.58	.35
8.9	1038.83	.95	1127.76	.82	1203.87	.69	1264.90	.53	1308.93	.35
9.0	1039.78	.94	1128.58	.83	1204.56	.68	1265.43	.52	1309.28	.35
9.1	1040.72	.95	1129.41	.83	1205.24	.69	1265.95	.52	1309.63	.34
9.2	1041.67	.94	1130.24	.82	1205.93	.68	1266.47	.53	1309.97	.35
9.3	1042.61	.94	1131.06	.82	1206.61	.68	1267.00	.52	1310.32	.34
9.4	1043.55	.94	1131.88	.83	1207.29	.69	1267.52	.52	1310.66	.34
9.5	1044.49	.93	1132.71	.82	1207.98	.67	1268.04	.51	1311.00	.34
9.6	1045.42	.94	1133.53	.82	1208.65	.68	1268.55	.52	1311.34	.33
9.7	1046.36	.94	1134.35	.81	1209.33	.68	1269.07	.51	1311.67	.34
9.8	1047.30	.93	1135.16	.82	1210.01	.67	1269.58	.51	1312.01	.33
9.9	1048.23	+.94	1135.98	+.82	1210.68	+.68	1270.09	+.51	1312.34	+.33
10.0	1049.17		1136.80		1211.36		1270.60		1312.67	

TABLE IX. ARGUMENT 4.

Equation = 670″.500 — 670″.3 sin. s — 7″.9 sin. 2s.

Period, 365.259687 days.

Days.	350		360	
Days	Equation.	Difference.	Equation.	Difference.
d. 0.0	″ 1312.67	+.33	″ 1336.14	+.13
0.1	1313.00	.33	1336.27	.14
0.2	1313.33	.32	1336.41	.13
0.3	1313.65	.33	1336.54	.13
0.4	1313.98	.32	1336.67	.13
0.5	1314.30	.32	1336.80	.13
0.6	1314.62	.31	1336.93	.12
0.7	1314.93	.32	1337.05	.13
0.8	1315.25	.31	1337.18	.12
0.9	1315.56	.31	1337.30	.12
1.0	1315.87	.31	1337.42	.12
1.1	1316.18	.31	1337.54	.11
1.2	1316.49	.31	1337.65	.12
1.3	1316.80	.30	1337.77	.11
1.4	1317.10	.30	1337.88	.11
1.5	1317.40	.30	1337.99	.11
1.6	1317.70	.30	1338.10	.10
1.7	1318.00	.30	1338.20	.11
1.8	1318.30	.30	1338.31	.10
1.9	1318.60	.29	1338.41	.10
2.0	1318.89	.29	1338.51	.10
2.1	1319.18	.29	1338.61	.09
2.2	1319.47	.29	1338.70	.10
2.3	1319.76	.28	1338.80	.09
2.4	1320.04	.29	1338.89	.09
2.5	1320.33	.28	1338.98	.09
2.6	1320.61	.28	1339.07	.08
2.7	1320.89	.28	1339.15	.09
2.8	1321.17	.27	1339.24	.08
2.9	1321.44	.28	1339.32	.08
3.0	1321.72	.27	1339.40	.07
3.1	1321.99	.26	1339.47	.08
3.2	1322.25	.27	1339.55	.08
3.3	1322.52	.27	1339.63	.07
3.4	1322.79	.26	1339.70	.07
3.5	1323.05	.27	1339.77	.07
3.6	1323.32	.26	1339.84	.06
3.7	1323.58	.26	1339.90	.07
3.8	1323.84	.26	1339.97	.06
3.9	1324.10	.25	1340.03	.06
4.0	1324.35	.26	1340.09	.06
4.1	1324.61	.25	1340.15	.06
4.2	1324.86	.25	1340.21	.05
4.3	1325.11	.24	1340.26	.06
4.4	1325.35	.25	1340.32	.05
4.5	1325.60	.24	1340.37	.05
4.6	1325.84	.24	1340.42	.04
4.7	1326.08	.24	1340.46	.05
4.8	1326.32	.24	1340.51	.04
4.9	1326.56	.24	1340.55	.04
5.0	1326.80	+.23	1340.59	+.03

TABLE IX. ARGUMENT 4.

Equation = 670″.500 — 670″.3 sin. s — 7″.9 sin. 2s.

Period, 365.259687 days.

Days.	350		360	
Days.	Equation.	Difference.	Equation.	Difference.
d.	s.		s.	
5.0	1326.80	+.23	1340.59	+.03
5.1	1327.03	.23	1340.62	.04
5.2	1327.26	.24	1340.66	.04
5.3	1327.50	.22	1340.70	.03
5.4	1327.72	.23	1340.73	.03
5.5	1327.95	.22	1340.76	.03
5.6	1328.17	.23	1340.79	.03
5.7	1328.40	.22	1340.82	.02
5.8	1328.62	.22	1340.84	.03
5.9	1328.84	.21	1340.87	.02
6.0	1329.05	.22	1340.89	.02
6.1	1329.27	.21	1340.91	.01
6.2	1329.48	.21	1340.92	.02
6.3	1329.69	.21	1340.94	.01
6.4	1329.90	.21	1340.95	.01
6.5	1330.11	.20	1340.96	.01
6.6	1330.31	.21	1340.97	.01
6.7	1330.52	.20	1340.98	.00
6.8	1330.72	.20	1340.98	.00
6.9	1330.92	.20	1340.98	+.01
7.0	1331.12	.19	1340.99	.00
7.1	1331.31	.20	1340.99	−.01
7.2	1331.51	.19	1340.98	.00
7.3	1331.70	.19	1340.98	.01
7.4	1331.89	.18	1340.97	.00
7.5	1332.07	.19	1340.97	.01
7.6	1332.26	.18	1340.96	.02
7.7	1332.44	.19	1340.94	.01
7.8	1332.63	.18	1340.93	.02
7.9	1332.81	.17	1340.91	.02
8.0	1332.98	.18	1340.89	.02
8.1	1333.16	.17	1340.87	.03
8.2	1333.33	.18	1340.84	.02
8.3	1333.51	.17	1340.82	.03
8.4	1333.68	.17	1340.79	.03
8.5	1333.85	.16	1340.76	.03
8.6	1334.01	.17	1340.73	.03
8.7	1334.18	.16	1340.70	.04
8.8	1334.34	.16	1340.66	.04
8.9	1334.50	.16	1340.62	.04
9.0	1334.66	.16	1340.58	.04
9.1	1334.82	.15	1340.54	.04
9.2	1334.97	.15	1340.50	.04
9.3	1335.12	.15	1340.46	.05
9.4	1335.27	.15	1340.41	.05
9.5	1335.42	.15	1340.36	.05
9.6	1335.57	.15	1340.31	.05
9.7	1335.72	.14	1340.26	.05
9.8	1335.86	.14	1340.21	.06
9.9	1336.00	+.14	1340.15	−.06
10.0	1336.14		1340.09	

TABLE X. ARGUMENT 5.

Equation = 225″.300 + 18″.0 sin. $(x - t)$ — 212″.4 sin. 2 $(x - t)$.

Period, 411.78517 days.

Days.	0		10		20		30		40		50	
Days.	Equation.	Diff.	Equation.	Diff.	Equation.	Diff.	Equation.	Diff.	Equation.	Diff.	Equation.	Diff.
d.	″		″		″		″		″		″	
0.0	15.73	.24	0.73	.05	5.62	.15	29.82	.33	71.00	.48	125.26	.60
0.1	15.49	.25	0.68	.05	5.77	.15	30.15	.34	71.48	.49	125.86	.59
0.2	15.21	.21	0.63	.04	5.92	.15	30.49	.34	71.97	.49	126.45	.59
0.3	15.00	.24	0.59	.05	6.07	.16	30.83	.33	72.46	.49	127.04	.60
0.4	14.76	.24	0.54	.04	6.23	.15	31.16	.34	72.95	.49	127.61	.60
0.5	14.52	.23	0.50	.04	6.38	.16	31.50	.34	73.44	.49	128.24	.59
0.6	14.29	.24	0.46	.04	6.54	.16	31.84	.35	73.93	.49	128.83	.60
0.7	14.05	.23	0.42	.03	6.70	.17	32.19	.34	74.42	.50	129.43	.60
0.8	13.82	.23	0.39	.04	6.87	.16	32.53	.35	74.92	.49	130.03	.60
0.9	13.59	.23	0.35	.03	7.03	.17	32.88	.35	75.41	.50	130.63	.60
1.0	13.36	.23	0.32	.03	7.20	.16	33.23	.35	75.91	.50	131.23	.60
1.1	13.13	.22	0.29	.03	7.36	.17	33.58	.35	76.41	.50	131.83	.60
1.2	12.91	.23	0.26	.02	7.53	.18	33.93	.35	76.91	.50	132.43	.60
1.3	12.68	.22	0.24	.03	7.71	.17	34.28	.36	77.41	.50	133.03	.60
1.4	12.46	.22	0.21	.02	7.88	.18	34.64	.35	77.91	.50	133.63	.60
1.5	12.24	.21	0.19	.02	8.06	.17	34.99	.26	78.41	.51	134.23	.61
1.6	12.03	.22	0.17	.02	8.23	.18	35.35	.36	78.92	.51	134.84	.60
1.7	11.81	.22	0.15	.01	8.41	.19	35.71	.36	79.43	.50	135.44	.61
1.8	11.59	.21	0.14	.02	8.60	.18	36.07	.37	79.93	.51	136.05	.60
1.9	11.38	.21	0.12	.01	8.78	.19	36.44	.36	80.44	.51	136.65	.61
2.0	11.17	.20	0.11	.01	8.97	.18	36.80	.37	80.95	.51	137.26	.61
2.1	10.97	.21	0.10	.01	9.15	.19	37.17	.37	81.46	.51	137.87	.60
2.2	10.76	.20	0.09	.00	9.34	.19	37.54	.37	81.97	.52	138.47	.61
2.3	10.56	.20	0.09	.01	9.53	.20	37.91	.37	82.49	.51	139.08	.61
2.4	10.36	.20	0.08	.00	9.73	.19	38.28	.37	83.00	.52	139.69	.61
2.5	10.16	.20	0.08	.00	9.92	.20	38.65	.38	83.52	.51	140.30	.61
2.6	9.96	.20	0.08	.00	10.12	.20	39.03	.37	84.03	.52	140.91	.61
2.7	9.76	.19	0.08	.01	10.32	.20	39.40	.38	84.55	.52	141.52	.62
2.8	9.57	.20	0.09	.00	10.52	.20	39.78	.38	85.07	.52	142.14	.61
2.9	9.37	.19	0.09	.01	10.72	.20	40.16	.38	85.59	.52	142.75	.61
3.0	9.18	.19	0.10	.01	10.93	.20	40.54	.38	86.11	.53	143.36	.62
3.1	8.99	.18	0.11	.01	11.13	.21	40.92	.39	86.64	.52	143.98	.61
3.2	8.81	.19	0.12	.02	11.34	.21	41.31	.38	87.16	.52	144.59	.62
3.3	8.62	.18	0.14	.01	11.55	.21	41.69	.39	87.68	.53	145.21	.61
3.4	8.44	.18	0.15	.02	11.76	.22	42.08	.39	88.21	.53	145.82	.62
3.5	8.26	.18	0.17	.02	11.98	.21	42.47	.39	88.74	.53	146.44	.62
3.6	8.08	.18	0.19	.02	12.19	.22	42.86	.39	89.27	.53	147.06	.62
3.7	7.90	.17	0.21	.03	12.41	.22	43.25	.40	89.80	.53	147.68	.61
3.8	7.73	.18	0.24	.02	12.63	.22	43.65	.39	90.33	.53	148.29	.62
3.9	7.55	.17	0.26	.03	12.85	.23	44.04	.40	90.86	.53	148.91	.62
4.0	7.38	.16	0.29	.03	13.08	.22	44.44	.40	91.39	.54	149.53	.62
4.1	7.22	.17	0.32	.03	13.30	.23	44.84	.40	91.93	.53	150.15	.62
4.2	7.05	.17	0.35	.04	13.53	.23	45.24	.40	92.46	.54	150.77	.62
4.3	6.88	.16	0.39	.03	13.76	.23	45.64	.41	93.00	.54	151.39	.63
4.4	6.72	.16	0.42	.04	13.99	.23	46.05	.40	93.54	.54	152.02	.62
4.5	6.56	.16	0.46	.04	14.22	.24	46.45	.41	94.08	.54	152.64	.62
4.6	6.40	.16	0.50	.04	14.46	.23	46.86	.41	94.62	.54	153.26	.63
4.7	6.24	.15	0.54	.05	14.69	.24	47.27	.41	95.16	.54	153.89	.62
4.8	6.09	.16	0.59	.04	14.93	.24	47.68	.41	95.70	.54	154.51	.63
4.9	5.93	.15	0.63	.05	15.17	.24	48.09	.41	96.24	.55	155.14	.62
5.0	5.78	.15	0.68	.05	15.41	.25	48.50	.41	96.79	.54	155.76	.63

TABLE X. ARGUMENT 5.

Equation $= 225''.300 + 18''.0 \sin. (x - t) - 212''.4 \sin. 2 (x - t)$.

Period, 411.78517 days.

Days.	0		10		20		30		40		50	
Days.	Equation.	Diff.	Equation.	Diff.	Equation.	Diff.	Equation.	Diff.	Equation.	Diff.	Equation.	Diff.
d. 5.0	" 5.78	.15	" 0.68	.05	" 15.41	.25	" 48.50	.41	" 96.79	.54	" 155.76	.63
5.1	5.63	.15	0.73	.05	15.66	.24	48.91	.42	97.33	.55	156.39	.62
5.2	5.48	.14	0.78	.06	15.90	.25	49.33	.42	97.88	.55	157.01	.63
5.3	5.34	.15	0.84	.05	16.15	.25	49.75	.41	98.43	.55	157.64	.63
5.4	5.19	.14	0.89	.06	16.40	.25	50.16	.42	98.98	.55	158.27	.62
5.5	5.05	.14	0.95	.06	16.65	.25	50.58	.43	99.53	.55	158.89	.63
5.6	4.91	.14	1.01	.06	16.90	.26	51.01	.42	100.08	.55	159.52	.63
5.7	4.77	.13	1.07	.07	17.16	.26	51.43	.42	100.63	.55	160.15	.63
5.8	4.64	.14	1.14	.06	17.42	.25	51.85	.43	101.18	.56	160.78	.63
5.9	4.50	.13	1.20	.07	17.67	.26	52.28	.43	101.74	.55	161.41	.63
6.0	4.37	.13	1.27	.07	17.93	.27	52.71	.43	102.29	.56	162.04	.63
6.1	4.24	.12	1.34	.07	18.20	.26	53.14	.43	102.85	.55	162.67	.63
6.2	4.12	.13	1.41	.08	18.46	.27	53.57	.43	103.40	.56	163.30	.63
6.3	3.99	.12	1.49	.07	18.73	.27	54.00	.43	103.96	.56	163.93	.63
6.4	3.87	.13	1.56	.08	19.00	.27	54.43	.44	104.52	.56	164.56	.63
6.5	3.74	.12	1.64	.08	19.27	.27	54.87	.44	105.08	.56	165.19	.64
6.6	3.62	.11	1.72	.08	19.54	.27	55.31	.43	105.64	.56	165.83	.63
6.7	3.51	.12	1.80	.09	19.81	.27	55.74	.44	106.20	.57	166.46	.63
6.8	3.39	.11	1.89	.08	20.08	.28	56.18	.45	106.77	.56	167.09	.64
6.9	3.28	.12	1.97	.09	20.36	.28	56.63	.44	107.33	.56	167.73	.63
7.0	3.16	.11	2.06	.09	20.64	.28	57.07	.44	107.89	.57	168.36	.64
7.1	3.05	.10	2.15	.10	20.92	.28	57.51	.45	108.46	.57	169.00	.63
7.2	2.95	.11	2.25	.09	21.20	.29	57.96	.44	109.03	.56	169.63	.64
7.3	2.84	.10	2.34	.10	21.49	.28	58.40	.45	109.59	.57	170.27	.63
7.4	2.74	.10	2.44	.09	21.77	.29	58.85	.45	110.16	.57	170.90	.64
7.5	2.64	.10	2.53	.10	22.06	.29	59.30	.45	110.73	.57	171.54	.64
7.6	2.54	.10	2.63	.11	22.35	.29	59.75	.45	111.30	.57	172.18	.63
7.7	2.44	.10	2.74	.10	22.64	.29	60.20	.46	111.87	.58	172.81	.64
7.8	2.34	.09	2.84	.11	22.93	.30	60.65	.45	112.45	.57	173.45	.64
7.9	2.25	.10	2.95	.10	23.23	.29	61.11	.46	113.02	.57	174.09	.64
8.0	2.15	.09	3.05	.11	23.52	.30	61.57	.46	113.59	.58	174.73	.63
8.1	2.06	.08	3.16	.11	23.82	.30	62.03	.46	114.16	.57	175.36	.64
8.2	1.98	.09	3.27	.12	24.12	.30	62.49	.46	114.74	.58	176.00	.64
8.3	1.89	.08	3.39	.11	24.42	.31	62.95	.46	115.32	.58	176.64	.64
8.4	1.81	.09	3.50	.12	24.73	.30	63.41	.47	115.90	.58	177.28	.64
8.5	1.72	.08	3.62	.12	25.03	.31	63.88	.46	116.48	.58	177.92	.64
8.6	1.64	.08	3.74	.12	25.34	.31	64.31	.47	117.06	.58	178.56	.64
8.7	1.56	.07	3.86	.12	25.65	.31	64.81	.47	117.64	.58	179.20	.64
8.8	1.49	.08	3.98	.13	25.96	.31	65.28	.47	118.22	.58	179.84	.64
8.9	1.41	.07	4.11	.13	26.27	.32	65.75	.47	118.80	.58	180.48	.65
9.0	1.34	.07	4.24	.13	26.59	.31	66.22	.47	119.38	.59	181.13	.64
9.1	1.27	.07	4.37	.13	26.90	.32	66.69	.47	119.97	.58	181.77	.64
9.2	1.20	.06	4.50	.13	27.22	.32	67.16	.48	120.55	.59	182.41	.64
9.3	1.14	.07	4.63	.14	27.54	.32	67.64	.47	121.14	.59	183.05	.64
9.4	1.07	.06	4.77	.13	27.86	.32	68.11	.48	121.73	.58	183.69	.65
9.5	1.01	.06	4.90	.14	28.18	.33	68.59	.48	122.31	.59	184.34	.64
9.6	0.95	.06	5.04	.14	28.51	.32	69.07	.48	122.90	.59	184.98	.64
9.7	0.89	.05	5.18	.14	28.83	.33	69.55	.48	123.49	.59	185.62	.65
9.8	0.84	.06	5.32	.15	29.16	.33	70.03	.48	124.08	.59	186.27	.64
9.9	0.78	.05	5.47	.15	29.49	.33	70.51	.49	124.67	.59	186.91	.64
10.0	0.73	.05	5.62	.15	29.82	.33	71.00	.48	125.26	.60	187.55	.65

TABLE X. ARGUMENT 5.

Equation = 225″.300 + 18″.0 sin. $(x - t)$ — 212″.4 sin. 2 $(x - t)$.

Period, 411.78517 days.

Days.	60		70		80		90		100		110	
Days.	Equation.	Diff.	Equation.	Diff.	Equation.	Diff.	Equation.	Diff.	Equation.	Diff.	Equation.	Diff.
d.	″		″		″		″		″		″	
0.0	187.55	.65	252.09	.64	312.91	.57	364.43	.45	401.95	.29	422.10	.11
0.1	188.20	.64	252.73	.63	313.48	.57	364.88	.45	402.24	.29	422.21	.10
0.2	188.84	.65	253.36	.64	314.05	.57	365.33	.45	402.53	.29	422.31	.10
0.3	189.49	.64	254.00	.63	314.62	.56	365.78	.45	402.82	.29	422.41	.10
0.4	190.13	.65	254.63	.64	315.18	.57	366.23	.44	403.11	.28	422.51	.10
0.5	190.78	.64	255.27	.63	315.75	.57	366.67	.45	403.39	.29	422.61	.10
0.6	191.42	.65	255.90	.64	316.32	.56	367.12	.44	403.68	.28	422.71	.09
0.7	192.07	.65	256.54	.63	316.88	.56	367.56	.44	403.96	.28	422.80	.10
0.8	192.72	.64	257.17	.63	317.44	.56	368.00	.44	404.24	.28	422.90	.09
0.9	193.36	.65	257.80	.64	318.00	.57	368.44	.44	404.52	.27	422.99	.09
1.0	194.01	.64	258.44	.63	318.57	.56	368.88	.44	404.79	.28	423.08	.08
1.1	194.65	.65	259.07	.63	319.13	.55	369.32	.43	405.07	.27	423.16	.09
1.2	195.30	.65	259.70	.63	319.68	.56	369.75	.44	405.34	.27	423.25	.08
1.3	195.95	.64	260.33	.63	320.24	.56	370.19	.43	405.61	.27	423.33	.08
1.4	196.59	.65	260.96	.63	320.80	.56	370.62	.43	405.88	.26	423.41	.08
1.5	197.24	.65	261.59	.63	321.36	.55	371.05	.43	406.14	.27	423.49	.08
1.6	197.89	.64	262.22	.63	321.91	.55	371.48	.42	406.41	.26	423.57	.07
1.7	198.53	.65	262.85	.63	322.46	.56	371.90	.44	406.67	.26	423.64	.08
1.8	199.18	.65	263.48	.63	323.02	.55	372.34	.42	406.93	.26	423.72	.07
1.9	199.83	.65	264.11	.63	323.57	.55	372.76	.42	407.19	.26	423.79	.07
2.0	200.48	.64	264.74	.63	324.12	.55	373.18	.43	407.45	.26	423.86	.07
2.1	201.12	.65	265.37	.62	324.67	.55	373.61	.42	407.71	.25	423.93	.06
2.2	201.77	.65	265.99	.63	325.22	.55	374.03	.42	407.96	.26	423.99	.07
2.3	202.42	.65	266.62	.63	325.77	.55	374.45	.41	408.22	.25	424.06	.06
2.4	203.07	.64	267.25	.62	326.32	.54	374.86	.42	408.47	.25	424.12	.06
2.5	203.71	.65	267.87	.63	326.86	.55	375.28	.41	408.72	.24	424.18	.06
2.6	204.36	.65	268.50	.62	327.41	.54	375.69	.42	408.96	.25	424.24	.05
2.7	205.01	.65	269.12	.62	327.95	.55	376.11	.41	409.21	.24	424.29	.06
2.8	205.66	.65	269.74	.63	328.50	.54	376.52	.41	409.45	.25	424.35	.05
2.9	206.31	.64	270.37	.62	329.04	.54	376.93	.41	409.70	.23	424.40	.05
3.0	206.95	.65	270.99	.62	329.58	.54	377.34	.41	409.93	.24	424.45	.05
3.1	207.60	.65	271.61	.62	330.12	.54	377.75	.40	410.17	.24	424.50	.04
3.2	208.25	.65	272.23	.62	330.66	.53	378.15	.41	410.41	.23	424.54	.05
3.3	208.90	.64	272.85	.63	331.19	.54	378.56	.40	410.64	.24	424.59	.04
3.4	209.54	.65	273.48	.62	331.73	.53	378.96	.40	410.88	.23	424.63	.04
3.5	210.19	.65	274.10	.62	332.26	.54	379.36	.40	411.11	.23	424.67	.04
3.6	210.84	.65	274.72	.61	332.80	.53	379.76	.40	411.34	.22	424.71	.03
3.7	211.49	.65	275.33	.62	333.33	.53	380.16	.39	411.56	.23	424.74	.04
3.8	212.14	.65	275.95	.62	333.86	.53	380.55	.40	411.79	.22	424.78	.03
3.9	212.79	.65	276.57	.62	334.39	.53	380.95	.39	412.01	.22	424.81	.03
4.0	213.44	.64	277.19	.61	334.92	.53	381.34	.39	412.23	.22	424.84	.03
4.1	214.08	.65	277.80	.62	335.45	.53	381.73	.39	412.45	.22	424.87	.03
4.2	214.73	.65	278.42	.61	335.98	.52	382.12	.39	412.67	.22	424.90	.02
4.3	215.38	.65	279.03	.62	336.50	.53	382.51	.39	412.89	.21	424.92	.03
4.4	216.03	.65	279.65	.61	337.03	.52	382.90	.39	413.10	.21	424.95	.02
4.5	216.68	.64	280.26	.62	337.55	.52	383.29	.38	413.31	.22	424.97	.02
4.6	217.32	.65	280.88	.61	338.07	.53	383.67	.38	413.53	.20	424.99	.01
4.7	217.97	.65	281.49	.61	338.60	.52	384.05	.38	413.73	.21	425.00	.02
4.8	218.62	.65	282.10	.61	339.12	.52	384.43	.38	413.94	.21	425.02	.01
4.9	219.27	.65	282.71	.61	339.64	.51	384.81	.38	414.15	.20	425.03	.01
5.0	219.92	.64	283.32	.62	340.15	.52	385.19	.37	414.35	.20	425.04	.01

TABLE X. ARGUMENT 5.

Equation = 225″.300 + 18″.0 sin. $(x - t)$ — 212″.4 sin. 2 $(x - t)$.

Period, 411.78517 days.

Days,	60		70		80		90		100		110	
Days.	Equation	Diff.	Equation.	Diff.	Equation	Diff.	Equation	Diff.	Equation.	Diff.	Equation	Diff.
d.	″		″		″		″		″			
5.0	219.92	.64	283.32	.62	340.15	.52	385.19	.37	414.35	.20	425.04	.01
5.1	220.56	.65	283.94	.61	340.67	.52	385.56	.38	414.55	.20	425.05	.01
5.2	221.21	.65	284.55	.60	341.19	.51	385.94	.37	414.75	.20	425.06	.00
5.3	221.86	.65	285.15	.61	341.70	.51	386.31	.37	414.95	.19	425.06	.01
5.4	222.51	.65	285.76	.61	342.21	.52	386.68	.37	415.14	.20	425.07	.00
5.5	223.16	.64	286.37	.61	342.73	.51	387.05	.37	415.34	.19	425.07	.00
5.6	223.80	.65	286.98	.60	343.24	.50	387.42	.37	415.53	.19	425.07	.00
5.7	224.45	.65	287.58	.61	343.74	.51	387.79	.36	415.72	.19	425.07	.01
5.8	225.10	.64	288.19	.61	344.25	.51	388.15	.37	415.91	.18	425.06	.00
5.9	225.74	.65	288.80	.60	344.76	.50	388.52	.35	416.09	.19	425.06	.01
6.0	226.39	.65	289.40	.60	345.26	.51	388.87	.36	416.28	.18	425.05	.01
6.1	227.04	.64	290.00	.60	345.77	.50	389.23	.36	416.46	.18	425.04	.02
6.2	227.68	.65	290.60	.61	346.27	.50	389.59	.36	416.64	.18	425.02	.01
6.3	228.33	.65	291.21	.60	346.77	.51	389.95	.35	416.82	.18	425.01	.02
6.4	228.98	.64	291.81	.60	347.28	.49	390.30	.36	417.00	.17	424.99	.01
6.5	229.62	.65	292.41	.60	347.77	.50	390.66	.35	417.17	.17	424.98	.02
6.6	230.27	.64	293.01	.60	348.27	.50	391.01	.35	417.34	.18	424.96	.03
6.7	230.91	.65	293.61	.60	348.77	.49	391.36	.35	417.52	.16	424.93	.02
6.8	231.56	.65	294.21	.59	349.26	.50	391.71	.34	417.68	.17	424.91	.03
6.9	232.21	.64	294.80	.60	349.76	.49	392.05	.35	417.85	.17	424.88	.02
7.0	232.85	.64	295.40	.59	350.25	.49	392.40	.34	418.02	.16	424.86	.03
7.1	233.49	.65	295.99	.60	350.74	.49	392.74	.34	418.18	.16	424.83	.04
7.2	234.14	.64	296.59	.59	351.23	.49	393.08	.34	418.34	.16	424.79	.03
7.3	234.78	.65	297.18	.60	351.72	.49	393.42	.34	418.50	.16	424.76	.03
7.4	235.43	.64	297.78	.59	352.21	.49	393.76	.34	418.66	.16	424.73	.04
7.5	236.07	.65	298.37	.59	352.70	.48	394.10	.33	418.82	.15	424.69	.04
7.6	236.72	.64	298.96	.59	353.18	.49	394.43	.33	418.97	.15	424.65	.04
7.7	237.36	.65	299.55	.59	353.67	.48	394.76	.34	419.12	.15	424.61	.04
7.8	238.01	.64	300.14	.59	354.15	.48	395.10	.33	419.27	.15	424.57	.05
7.9	238.65	.64	300.73	.59	354.63	.48	395.43	.32	419.42	.15	424.52	.05
8.0	239.29	.64	301.32	.59	355.11	.48	395.75	.33	419.57	.14	424.47	.05
8.1	239.93	.65	301.91	.59	355.59	.48	396.08	.32	419.71	.15	424.42	.05
8.2	240.58	.64	302.50	.58	356.07	.47	396.40	.32	419.86	.14	424.37	.05
8.3	241.22	.64	303.08	.59	356.54	.48	396.72	.33	420.00	.14	424.32	.06
8.4	241.86	.64	303.67	.58	357.02	.47	397.05	.32	420.14	.14	424.26	.05
8.5	242.50	.64	304.25	.59	357.49	.48	397.37	.31	420.28	.13	424.21	.06
8.6	243.14	.64	304.84	.58	357.97	.47	397.68	.32	420.41	.13	424.15	.06
8.7	243.78	.65	305.42	.58	358.44	.47	398.00	.31	420.54	.14	424.09	.07
8.8	244.43	.64	306.00	.58	358.91	.46	398.31	.32	420.68	.12	424.02	.06
8.9	245.07	.64	306.58	.58	359.37	.47	398.63	.31	420.80	.13	423.96	.07
9.0	245.71	.64	307.16	.58	359.84	.46	398.94	.31	420.93	.13	423.89	.06
9.1	246.35	.64	307.74	.58	360.30	.47	399.25	.31	421.06	.12	423.83	.07
9.2	246.99	.63	308.32	.57	360.77	.46	399.56	.30	421.18	.12	423.76	.08
9.3	247.62	.64	308.89	.58	361.23	.46	399.86	.31	421.30	.12	423.68	.07
9.4	248.26	.64	309.47	.58	361.69	.46	400.17	.30	421.42	.12	423.61	.08
9.5	248.90	.64	310.05	.57	362.15	.46	400.47	.30	421.54	.12	423.53	.07
9.6	249.54	.64	310.62	.57	362.61	.46	400.77	.30	421.66	.11	423.46	.08
9.7	250.18	.64	311.19	.58	363.07	.45	401.07	.29	421.77	.11	423.38	.09
9.8	250.82	.63	311.77	.57	363.52	.46	401.36	.30	421.88	.11	423.29	.09
9.9	251.45	.64	312.34	.57	363.98	.45	401.66	.29	421.99	.11	423.21	.09
10.0	252.09	.64	312.91	.57	364.43	.45	401.95	.29	422.10	.11	423.12	.09

TABLE X. ARGUMENT 5.

Equation $= 225''.300 + 18''.0 \sin. (x - t) - 212''.4 \sin. 2 (x - t)$.

Period, 411.78517 days.

Days.	120		130		140		150		160		170	
Days.	Equation.	Diff.	Equation.	Diff.	Equation.	Diff.	Equation.	Diff.	Equation.	Diff.	Equation.	Diff.
d.	"		"		"		"		"		"	
0.0	423.12	.09	405.06	.27	369.74	.43	320.59	.55	262.33	.61	200.53	.61
0.1	423.03	.09	404.79	.28	369.31	.43	320.04	.55	261.72	.61	199.92	.62
0.2	422.94	.09	404.51	.27	368.88	.44	319.49	.55	261.11	.61	199.30	.61
0.3	422.85	.09	404.24	.28	368.44	.43	318.94	.54	260.50	.61	198.69	.62
0.4	422.76	.10	403.96	.28	368.01	.44	318.40	.55	259.89	.62	198.07	.61
0.5	422.66	.09	403.68	.28	367.57	.43	317.85	.56	259.27	.61	197.46	.62
0.6	422.57	.10	403.40	.28	367.14	.44	317.29	.55	258.66	.61	196.84	.61
0.7	422.47	.10	403.12	.29	366.70	.44	316.74	.55	258.05	.61	196.23	.62
0.8	422.37	.11	402.83	.28	366.26	.44	316.19	.55	257.44	.62	195.61	.61
0.9	422.26	.10	402.55	.29	365.82	.44	315.64	.56	256.82	.61	195.00	.61
1.0	422.16	.11	402.26	.29	365.38	.45	315.08	.55	256.21	.61	194.39	.62
1.1	422.05	.11	401.97	.29	364.93	.44	314.53	.56	255.60	.61	193.77	.61
1.2	421.94	.11	401.68	.29	364.49	.45	313.97	.56	254.99	.62	193.16	.61
1.3	421.83	.11	401.39	.30	364.04	.45	313.41	.55	254.37	.61	192.55	.61
1.4	421.72	.12	401.09	.29	363.59	.45	312.86	.56	253.76	.62	191.94	.61
1.5	421.60	.11	400.80	.30	363.14	.45	312.30	.56	253.14	.61	191.33	.62
1.6	421.49	.12	400.50	.30	362.69	.45	311.74	.56	252.53	.62	190.71	.61
1.7	421.37	.12	400.20	.30	362.24	.45	311.18	.56	251.91	.61	190.10	.61
1.8	421.25	.12	399.90	.31	361.79	.46	310.62	.56	251.30	.62	189.49	.61
1.9	421.13	.13	399.59	.30	361.33	.45	310.06	.57	250.68	.61	188.88	.61
2.0	421.00	.12	399.29	.31	360.88	.46	309.49	.56	250.07	.62	188.27	.61
2.1	420.88	.13	398.98	.31	360.42	.46	308.93	.56	249.45	.62	187.66	.61
2.2	420.75	.13	398.67	.31	359.96	.46	308.37	.57	248.83	.61	187.05	.61
2.3	420.62	.13	398.36	.31	359.50	.46	307.80	.56	248.22	.62	186.44	.61
2.4	420.49	.14	398.05	.31	359.04	.46	307.24	.57	247.60	.62	185.83	.60
2.5	420.35	.13	397.74	.32	358.58	.46	306.67	.57	246.98	.61	185.23	.61
2.6	420.22	.14	397.42	.31	358.12	.47	306.10	.57	246.37	.62	184.62	.61
2.7	420.08	.14	397.11	.32	357.65	.46	305.53	.57	245.75	.62	184.01	.61
2.8	419.94	.14	396.79	.32	357.19	.47	304.96	.57	245.13	.62	183.40	.60
2.9	419.80	.14	396.47	.32	356.72	.47	304.39	.57	244.51	.61	182.80	.61
3.0	419.66	.15	396.15	.32	356.25	.47	303.82	.57	243.90	.62	182.19	.61
3.1	419.51	.14	395.83	.33	355.78	.47	303.25	.57	243.28	.62	181.58	.60
3.2	419.37	.15	395.50	.33	355.31	.47	302.68	.57	242.66	.62	180.98	.61
3.3	419.22	.15	395.17	.32	354.84	.48	302.11	.58	242.04	.62	180.37	.60
3.4	419.07	.16	394.85	.33	354.36	.47	301.53	.57	241.42	.62	179.77	.61
3.5	418.91	.15	394.52	.33	353.89	.48	300.96	.58	240.80	.62	179.16	.60
3.6	418.76	.16	394.19	.34	353.41	.47	300.38	.57	240.18	.61	178.56	.60
3.7	418.60	.15	393.85	.33	352.94	.48	299.81	.58	239.57	.62	177.96	.61
3.8	418.45	.16	393.52	.34	352.46	.48	299.23	.58	238.95	.62	177.35	.60
3.9	418.29	.17	393.18	.34	351.98	.48	298.65	.57	238.33	.62	176.75	.60
4.0	418.12	.16	392.84	.34	351.50	.49	298.08	.58	237.71	.62	176.15	.60
4.1	417.96	.17	392.50	.34	351.01	.48	297.50	.58	237.09	.62	175.55	.60
4.2	417.79	.16	392.16	.34	350.53	.48	296.92	.58	236.47	.62	174.95	.60
4.3	417.63	.17	391.82	.34	350.05	.49	296.34	.58	235.85	.62	174.35	.60
4.4	417.46	.17	391.48	.35	349.56	.49	295.76	.58	235.23	.62	173.75	.60
4.5	417.29	.18	391.13	.35	349.07	.48	295.18	.58	234.61	.62	173.15	.60
4.6	417.11	.17	390.78	.35	348.59	.49	294.60	.59	233.99	.62	172.55	.60
4.7	416.94	.18	390.43	.35	348.10	.49	294.01	.58	233.37	.62	171.95	.60
4.8	416.76	.18	390.08	.35	347.61	.50	293.43	.58	232.75	.62	171.35	.60
4.9	416.58	.18	389.73	.35	347.11	.49	292.85	.59	232.13	.62	170.75	.59
5.0	416.40	.18	389.38	.36	346.62	.49	292.26	.58	231.51	.62	170.16	.60

TABLE X. ARGUMENT 5.

Equation $= 225''.300 + 18''.0 \sin. (x - t) - 212''.4 \sin. 2 (x - t)$.

Period, 411.78517 days.

Days.	120		130		140		150		160		170	
Days	Equation	Diff.	Equation	Diff.	Equation	Diff.	Equation	Diff.	Equation	Diff.	Equation	Diff.
d.	"				"		"		"			
5.0	416.40	.18	389.38	.36	346.62	.49	292.26	.58	231.51	.62	170.16	.60
5.1	416.22	.18	389.02	.35	346.13	.50	291.68	.59	230.89	.63	169.56	.59
5.2	416.04	.19	388.67	.36	345.63	.49	291.09	.59	230.26	.62	168.97	.60
5.3	415.85	.19	388.31	.36	345.14	.50	290.50	.58	229.64	.62	168.37	.60
5.4	415.66	.19	387.95	.37	344.64	.50	289.92	.59	229.02	.62	167.77	.59
5.5	415.47	.19	387.58	.36	344.14	.50	289.33	.59	228.40	.62	167.18	.59
5.6	415.28	.19	387.22	.36	343.64	.50	288.74	.59	227.78	.62	166.59	.60
5.7	415.09	.20	386.86	.37	343.14	.50	288.15	.59	227.16	.62	165.99	.59
5.8	414.89	.19	386.49	.37	342.64	.51	287.56	.59	226.54	.62	165.40	.59
5.9	414.70	.20	386.12	.37	342.13	.50	286.97	.59	225.92	.62	164.81	.59
6.0	414.50	.20	385.75	.37	341.63	.51	286.38	.59	225.30	.62	164.22	.59
6.1	414.30	.21	385.38	.37	341.12	.50	285.79	.59	224.68	.62	163.63	.59
6.2	414.09	.20	385.01	.38	340.62	.51	285.20	.60	224.06	.62	163.04	.59
6.3	413.89	.21	384.63	.37	340.11	.51	284.60	.59	223.44	.62	162.45	.59
6.4	413.68	.20	384.26	.38	339.60	.51	284.01	.59	222.82	.62	161.86	.59
6.5	413.48	.21	383.88	.38	339.09	.51	283.42	.60	222.20	.62	161.27	.59
6.6	413.27	.22	383.50	.38	338.58	.51	282.82	.59	221.58	.63	160.68	.58
6.7	413.05	.21	383.12	.38	338.07	.52	282.23	.60	220.95	.62	160.10	.59
6.8	412.84	.21	382.74	.38	337.55	.51	281.63	.59	220.33	.62	159.51	.59
6.9	412.63	.22	382.36	.39	337.04	.52	281.04	.60	219.71	.62	158.92	.58
7.0	412.41	.22	381.97	.39	336.52	.51	280.44	.60	219.09	.62	158.34	.59
7.1	412.19	.22	381.58	.38	336.01	.52	279.84	.59	218.47	.62	157.75	.58
7.2	411.97	.22	381.20	.39	335.49	.52	279.25	.60	217.85	.62	157.17	.58
7.3	411.75	.23	380.81	.39	334.97	.52	278.65	.60	217.23	.62	156.59	.59
7.4	411.52	.22	380.42	.40	334.45	.52	278.05	.60	216.61	.62	156.00	.58
7.5	411.30	.23	380.02	.39	333.93	.52	277.45	.60	215.99	.62	155.42	.58
7.6	411.07	.23	379.63	.39	333.41	.52	276.85	.60	215.37	.62	154.84	.58
7.7	410.84	.23	379.24	.40	332.89	.53	276.25	.60	214.75	.62	154.26	.58
7.8	410.61	.24	378.84	.40	332.36	.52	275.65	.60	214.13	.62	153.68	.58
7.9	410.37	.23	378.44	.40	331.84	.53	275.05	.60	213.51	.62	153.10	.58
8.0	410.14	.24	378.04	.40	331.31	.53	274.45	.60	212.89	.62	152.52	.58
8.1	409.90	.24	377.64	.40	330.78	.52	273.85	.61	212.27	.62	151.94	.57
8.2	409.66	.24	377.24	.41	330.26	.53	273.24	.60	211.65	.62	151.37	.58
8.3	409.42	.24	376.83	.40	329.73	.53	272.64	.60	211.03	.62	150.79	.58
8.4	409.18	.24	376.43	.41	329.20	.53	272.04	.61	210.41	.61	150.21	.57
8.5	408.94	.25	376.02	.41	328.67	.54	271.43	.60	209.80	.62	149.64	.57
8.6	408.69	.25	375.61	.41	328.13	.53	270.83	.60	209.18	.62	149.07	.58
8.7	408.44	.25	375.20	.41	327.60	.53	270.23	.61	208.56	.62	148.49	.57
8.8	408.19	.25	374.79	.41	327.07	.54	269.62	.61	207.94	.62	147.92	.57
8.9	407.94	.25	374.38	.42	326.53	.53	269.01	.60	207.32	.62	147.35	.57
9.0	407.69	.26	373.96	.42	326.00	.54	268.41	.61	206.70	.62	146.78	.58
9.1	407.43	.25	373.54	.41	325.46	.54	267.80	.60	206.08	.61	146.20	.57
9.2	407.18	.26	373.13	.42	324.92	.54	267.20	.61	205.47	.62	145.63	.56
9.3	406.92	.26	372.71	.42	324.38	.54	266.59	.61	204.85	.62	145.07	.57
9.4	406.66	.26	372.29	.42	323.84	.54	265.98	.61	204.23	.61	144.50	.57
9.5	406.40	.27	371.87	.43	323.30	.54	265.37	.60	203.62	.62	143.93	.57
9.6	406.13	.26	371.44	.42	322.76	.54	264.77	.61	203.00	.62	143.36	.56
9.7	405.87	.27	371.02	.43	322.22	.55	264.16	.61	202.38	.61	142.80	.57
9.8	405.60	.27	370.59	.42	321.67	.54	263.55	.61	201.77	.62	142.23	.56
9.9	405.33	.27	370.17	.43	321.13	.54	262.94	.61	201.15	.62	141.67	.56
10.0	405.06	.27	369.74	.43	320.59	.55	262.33	.61	200.53	.61	141.11	.57

TABLE X. ARGUMENT 5.

Equation $= 225''.300 + 18''.0 \sin. (x - t) - 212''.4 \sin. 2 (x - t)$.

Period, 411.78517 days.

Days.	180		190		200		210		220		230	
Days.	Equation.	Diff.	Equation.	Diff.	Equation.	Diff.	Equation.	Diff.	Equation.	Diff.	Equation.	Diff.
d.	"		"		"		"		"		"	
0.0	141.11	.57	89.72	.46	51.31	.31	29.60	.13	26.74	.07	43.15	.25
0.1	140.54	.56	89.26	.45	51.00	.30	29.47	.12	26.81	.07	43.40	.26
0.2	139.98	.56	88.81	.45	50.70	.30	29.35	.12	26.88	.08	43.66	.27
0.3	139.42	.56	88.36	.45	50.40	.30	29.23	.12	26.96	.07	43.93	.26
0.4	138.86	.56	87.91	.45	50.10	.30	29.11	.11	27.03	.08	44.19	.26
0.5	138.30	.56	87.46	.45	49.80	.29	29.00	.12	27.11	.08	44.45	.27
0.6	137.74	.56	87.01	.45	49.51	.30	28.88	.11	27.19	.08	44.72	.27
0.7	137.18	.55	86.56	.45	49.21	.29	28.77	.11	27.27	.08	44.99	.27
0.8	136.63	.56	86.11	.44	48.92	.29	28.66	.11	27.35	.09	45.26	.27
0.9	136.07	.55	85.67	.45	48.63	.29	28.55	.11	27.44	.08	45.53	.28
1.0	135.52	.56	85.22	.44	48.34	.29	28.44	.11	27.52	.09	45.81	.27
1.1	134.96	.55	84.78	.44	48.05	.29	28.33	.10	27.61	.09	46.08	.28
1.2	134.41	.55	84.34	.44	47.76	.28	28.23	.10	27.70	.10	46.36	.28
1.3	133.86	.56	83.90	.44	47.48	.28	28.13	.10	27.80	.09	46.64	.28
1.4	133.30	.55	83.46	.44	47.20	.28	28.03	.10	27.89	.10	46.92	.28
1.5	132.75	.55	83.02	.43	46.92	.28	27.93	.09	27.99	.10	47.20	.29
1.6	132.20	.55	82.59	.44	46.64	.28	27.84	.09	28.09	.10	47.49	.29
1.7	131.65	.54	82.15	.43	46.36	.28	27.75	.10	28.19	.10	47.78	.28
1.8	131.11	.55	81.72	.43	46.08	.27	27.65	.09	28.29	.10	48.06	.29
1.9	130.56	.55	81.29	.43	45.81	.27	27.56	.08	28.39	.11	48.35	.30
2.0	130.01	.54	80.86	.43	45.54	.27	27.48	.09	28.50	.11	48.65	.29
2.1	129.47	.54	80.43	.43	45.27	.27	27.39	.09	28.61	.11	48.94	29
2.2	128.93	.55	80.00	.42	45.00	.27	27.30	.08	28.72	.11	49.23	.30
2.3	128.38	.54	79.58	.43	44.73	.26	27.22	.08	28.83	.11	49.53	.30
2.4	127.84	.54	79.15	.42	44.47	.27	27.14	.08	28.94	.12	49.83	.30
2.5	127.30	.54	78.73	.42	44.20	.26	27.06	.07	29.06	.12	50.13	.30
2.6	126.76	.54	78.31	.42	43.94	.26	26.99	.08	29.18	.12	50.43	.30
2.7	126.22	.54	77.89	.42	43.68	.26	26.91	.07	29.30	.12	50.73	.31
2.8	125.68	.54	77.47	.42	43.42	.25	26.84	.07	29.42	.12	51.04	.31
2.9	125.14	.54	77.05	.41	43.17	.26	26.77	.06	29.54	.13	51.35	.31
3.0	124.60	.53	76.64	.42	42.91	.25	26.71	.07	29.67	.12	51.66	.31
3.1	124.07	.54	76.22	.41	42.66	.25	26.64	.07	29.79	.13	51.97	.31
3.2	123.53	.53	75.81	.41	42.41	.25	26.57	.06	29.92	.13	52.28	.32
3.3	123.00	.54	75.40	.41	42.16	.25	26.51	.06	30.05	.14	52.60	.32
3.4	122.46	.53	74.99	.41	41.91	.25	26.45	.06	30.19	.13	52.92	.34
3.5	121.93	.53	74.58	.41	41.66	.24	26.39	.06	30.32	.14	53.23	.32
3.6	121.40	.53	74.17	.40	41.42	.24	26.33	.05	30.46	.14	53.55	.32
3.7	120.87	.53	73.77	.41	41.18	.24	26.28	.05	30.60	.14	53.87	.33
3.8	120.34	.53	73.36	.40	40.94	.24	26.23	.05	30.74	.14	54.20	.32
3.9	119.81	.52	72.96	.40	40.70	.24	26.18	.05	30.88	.15	54.52	.33
4.0	119.29	.53	72.56	.40	40.46	.23	26.13	.05	31.03	.15	54.85	.32
4.1	118.76	.52	72.16	.40	40.23	.24	26.08	.05	31.18	.15	55.17	.33
4.2	118.24	.53	71.76	.40	39.99	.23	26.03	.04	31.33	.15	55.50	.33
4.3	117.71	.52	71.36	.39	39.76	.22	25.99	.04	31.48	.15	55.83	.34
4.4	117.19	.52	70.97	.40	39.54	.24	25.95	.04	31.63	.15	56.17	.33
4.5	116.67	.52	70.57	.39	39.30	.22	25.91	.04	31.78	.16	56.50	.34
4.6	116.15	.52	70.18	.39	39.08	.23	25.87	.03	31.94	.16	56.84	.34
4.7	115.63	.52	69.79	.39	38.85	.22	25.84	.04	32.10	.16	57.18	.34
4.8	115.11	.52	69.40	.39	38.63	.22	25.80	.03	32.26	.16	57.52	.34
4.9	114.59	.51	69.01	.38	38.41	.22	25.77	.03	32.42	.16	57.86	.34
5.0	114.08	.52	68.63	.39	38.19	.22	25.74	.02	32.58	.16	58.20	.35

TABLE X. ARGUMENT 5.

Equation = 225″.300 + 18″.0 sin. $(x - t)$ — 212″.4 sin. 2 $(x - t)$.

Period, 411.78517 days.

Days. 180			190		200		210		220		230	
Days.	Equation.	Diff.	Equation.	Diff.	Equation.	Diff.	Equation.	Diff.	Equation.	Diff.	Equation.	Diff.
d. 5.0	″ 114.08	.52	″ 68.63	.39	″ 38.19	.22	″ 25.74	.02	″ 32.58	.16	″ 58.20	.35
5.1	113.56	.51	68.24	.38	37.97	.21	25.72	.03	32.74	.17	58.55	.34
5.2	113.05	.52	67.86	.38	37.76	.21	25.69	.02	32.91	.17	58.89	.35
5.3	112.53	.51	67.48	.38	37.55	.22	25.67	.03	33.08	.17	59.24	.35
5.4	112.02	.51	67.10	.38	37.33	.21	25.64	.02	33.25	.18	59.59	.35
5.5	111.51	.51	66.72	.38	37.12	.20	25.62	.01	33.43	.17	59.94	.36
5.6	111.00	.51	66.34	.37	36.92	.21	25.61	.02	33.60	.18	60.30	.35
5.7	110.49	.51	65.97	.38	36.71	.20	25.59	.01	33.78	.18	60.65	.36
5.8	109.98	.50	65.59	.37	36.51	.21	25.58	.02	33.96	.18	61.01	.36
5.9	109.48	.51	65.22	.37	36.30	.20	25.56	.01	34.14	.18	61.37	.36
6.0	108.97	.50	64.85	.37	36.10	.20	25.55	.01	34.32	.19	61.73	.35
6.1	108.47	.51	64.48	.37	35.90	.19	25.54	.00	34.51	.18	62.08	.37
6.2	107.96	.50	64.11	.37	35.71	.20	25.54	.01	34.69	.19	62.45	.36
6.3	107.46	.50	63.74	.36	35.51	.19	25.53	.00	34.88	.19	62.81	.37
6.4	106.96	.50	63.38	.37	35.32	.19	25.53	.00	35.07	.19	63.18	.37
6.5	106.46	.50	63.01	.36	35.13	.19	25.53	.00	35.26	.20	63.55	.37
6.6	105.96	.50	62.65	.36	34.94	.19	25.53	.01	35.46	.19	63.92	.37
6.7	105.46	.49	62.29	.36	34.75	.19	25.54	.00	35.65	.20	64.29	.37
6.8	104.97	.50	61.93	.35	34.56	.18	25.54	.01	35.85	.20	64.66	.38
6.9	104.47	.49	61.58	.36	34.38	.18	25.55	.01	36.05	.20	65.04	.37
7.0	103.98	.49	61.22	.35	34.20	.18	25.56	.01	36.25	.20	65.41	.38
7.1	103.49	.50	60.87	.35	34.02	.18	25.57	.01	36.45	.21	65.79	.38
7.2	102.99	.49	60.52	.36	33.84	.18	25.58	.02	36.66	.20	66.17	.38
7.3	102.50	.49	60.16	.34	33.66	.17	25.60	.01	36.86	.21	66.55	.38
7.4	102.01	.48	59.82	.35	33.49	.18	25.61	.02	37.07	.21	66.93	.38
7.5	101.53	.49	59.47	.35	33.31	.17	25.63	.02	37.28	.22	67.31	.39
7.6	101.04	.49	59.12	.34	33.14	.17	25.65	.03	37.50	.21	67.70	.39
7.7	100.55	.48	58.78	.34	32.97	.16	25.68	.02	37.71	.22	68.09	.39
7.8	100.07	.49	58.44	.34	32.81	.17	25.70	.03	37.93	.22	68.48	.39
7.9	99.58	.48	58.10	.35	32.64	.16	25.73	.03	38.15	.22	68.87	.39
8.0	99.10	.48	57.75	.33	32.48	.17	25.76	.03	38.37	.22	69.26	.39
8.1	98.62	.48	57.42	.34	32.31	.16	25.79	.03	38.59	.22	69.65	.40
8.2	98.14	.48	57.08	.33	32.15	.15	25.82	.03	38.81	.23	70.05	.39
8.3	97.66	.47	56.75	.34	32.00	.16	25.85	.04	39.04	.22	70.44	.40
8.4	97.19	.48	56.41	.33	31.84	.15	25.89	.04	39.26	.23	70.84	.40
8.5	96.71	.47	56.08	.33	31.69	.16	25.93	.04	39.49	.23	71.24	.40
8.6	96.24	.48	55.75	.33	31.53	.15	25.97	.04	39.72	.23	71.64	.40
8.7	95.76	.47	55.42	.32	31.38	.15	26.01	.05	39.95	.24	72.04	.41
8.8	95.29	.47	55.10	.33	31.23	.14	26.06	.04	40.19	.24	72.45	.40
8.9	94.82	.47	54.77	.32	31.09	.15	26.10	.05	40.43	.24	72.85	.41
9.0	94.35	.47	54.45	.32	30.94	.14	26.15	.05	40.67	.23	73.26	.41
9.1	93.88	.47	54.13	.32	30.80	.14	26.20	.05	40.90	.25	73.67	.41
9.2	93.41	.46	53.81	.32	30.66	.14	26.25	.06	41.15	.24	74.08	.41
9.3	92.95	.47	53.49	.32	30.52	.14	26.31	.05	41.39	.25	74.49	.41
9.4	92.48	.46	53.17	.31	30.38	.13	26.36	.06	41.64	.24	74.90	.42
9.5	92.02	.46	52.86	.31	30.25	.14	26.42	.06	41.88	.25	75.32	.42
9.6	91.56	.46	52.55	.32	30.11	.13	26.48	.06	42.13	.25	75.74	.41
9.7	91.10	.46	52.23	.31	29.98	.13	26.54	.07	42.38	.26	76.15	.42
9.8	90.64	.46	51.92	.30	29.85	.13	26.61	.06	42.64	.25	76.57	.42
9.9	90.18	.46	51.62	.31	29.72	.12	26.67	.07	42.89	.26	76.99	.43
10.0	89.72	.46	51.31	.31	29.60	.13	26.74	.07	43.15	.25	77.42	.42

TABLE X. ARGUMENT 5.

Equation $= 225''.300 + 18''.0 \sin. (x - t) - 212''.4 \sin. 2 (x - t)$.

Period, 411.78517 days.

Days.	240		250		260		270		280		290	
Days	Equation.	Diff.	Equation.	Diff.	Equation.	Diff.	Equation.	Diff.	Equation.	Diff.	Equation.	Diff.
d.	"		"		"		"		"		"	
0.0	77.42	.42	126.48	.55	185.86	.63	250.12	.65	313.34	.61	369.65	.51
0.1	77.84	.42	127.03	.55	186.49	.63	250.77	.65	313.95	.60	370.16	.51
0.2	78.26	.43	127.58	.55	187.12	.63	251.42	.64	314.55	.61	370.67	.50
0.3	78.69	.43	128.13	.56	187.75	.63	252.06	.65	315.16	.60	371.17	.51
0.4	79.12	.43	128.69	.55	188.38	.63	252.71	.65	315.76	.61	371.68	.50
0.5	79.55	.43	129.24	.56	189.01	.63	253.36	.65	316.37	.60	372.18	.51
0.6	79.98	.43	129.80	.56	189.64	.63	254.01	.64	316.97	.60	372.69	.50
0.7	80.41	.44	130.36	.55	190.27	.63	254.65	.65	317.57	.60	373.19	.50
0.8	80.85	.43	130.91	.56	190.90	.63	255.30	.65	318.17	.60	373.69	.50
0.9	81.28	.44	131.47	.56	191.53	.63	255.95	.64	318.77	.60	374.19	.50
1.0	81.72	.44	132.03	.57	192.16	.63	256.59	.65	319.37	.60	374.69	.50
1.1	82.16	.44	132.60	.56	192.79	.64	257.24	.64	319.97	.60	375.19	.49
1.2	82.60	.44	133.16	.56	193.43	.63	257.88	.65	320.57	.60	375.68	.50
1.3	83.04	.44	133.72	.56	194.06	.64	258.53	.64	321.17	.60	376.18	.49
1.4	83.48	.45	134.28	.57	194.70	.63	259.17	.65	321.77	.59	376.67	.49
1.5	83.93	.44	134.85	.57	195.33	.64	259.82	.64	322.36	.60	377.16	.49
1.6	84.37	.45	135.42	.56	195.97	.63	260.46	.65	322.96	.59	377.65	.49
1.7	84.82	.45	135.98	.57	196.60	.64	261.11	.64	323.55	.60	378.14	.49
1.8	85.27	.45	136.55	.57	197.24	.63	261.75	.65	324.15	.59	378.63	.49
1.9	85.72	.45	137.12	.57	197.87	.64	262.40	.64	324.74	.59	379.12	.48
2.0	86.17	.45	137.69	.57	198.51	.64	263.04	.65	325.33	.60	379.60	.48
2.1	86.62	.46	138.26	.57	199.15	.63	263.69	.64	325.93	.59	380.08	.49
2.2	87.08	.45	138.83	.57	199.78	.64	264.33	.65	326.52	.59	380.57	.48
2.3	87.53	.46	139.40	.58	200.42	.64	264.98	.64	327.11	.59	381.05	.48
2.4	87.99	.46	139.98	.57	201.06	.64	265.62	.64	327.70	.58	381.53	.48
2.5	88.45	.46	140.55	.58	201.70	.64	266.26	.64	328.28	.59	382.01	.48
2.6	88.91	.46	141.13	.57	202.34	.63	266.90	.65	328.87	.59	382.49	.47
2.7	89.37	.46	141.70	.58	202.97	.64	267.55	.64	329.46	.59	382.96	.48
2.8	89.83	.46	142.28	.58	203.61	.64	268.19	.64	330.05	.58	383.44	.47
2.9	90.29	.47	142.86	.58	204.25	.64	268.83	.64	330.63	.58	383.91	.47
3.0	90.76	.47	143.44	.58	204.89	.64	269.47	.65	331.21	.59	384.38	.47
3.1	91.23	.46	144.02	.58	205.53	.64	270.12	.64	331.80	.58	384.85	.47
3.2	91.69	.47	144.60	.58	206.17	.64	270.76	.64	332.38	.58	385.32	.47
3.3	92.16	.47	145.18	.58	206.81	.65	271.40	.64	332.96	.58	385.79	.47
3.4	92.63	.48	145.76	.59	207.46	.64	272.04	.64	333.54	.58	386.26	.46
3.5	93.11	.47	146.35	.58	208.10	.64	272.68	.64	334.12	.58	386.72	.47
3.6	93.58	.47	146.93	.59	208.74	.64	273.32	.64	334.70	.58	387.19	.46
3.7	94.05	.48	147.52	.58	209.38	.64	273.96	.64	335.28	.58	387.65	.46
3.8	94.53	.48	148.10	.59	210.02	.65	274.60	.63	335.86	.57	388.11	.46
3.9	95.01	.48	148.69	.59	210.67	.64	275.23	.64	336.43	.57	388.57	.46
4.0	95.49	.48	149.28	.59	211.31	.64	275.87	.64	337.00	.58	389.03	.45
4.1	95.97	.48	149.87	.59	211.95	.64	276.51	.64	337.58	.57	389.48	.46
4.2	96.45	.48	150.46	.59	212.59	.65	277.15	.64	338.15	.58	389.94	.45
4.3	96.93	.48	151.05	.59	213.24	.64	277.79	.63	338.73	.57	390.39	.46
4.4	97.41	.49	151.64	.59	213.88	.65	278.42	.64	339.30	.57	390.85	.45
4.5	97.90	.49	152.23	.59	214.53	.64	279.06	.64	339.87	.57	391.30	.45
4.6	98.39	.49	152.82	.60	215.17	.64	279.70	.63	340.44	.57	391.75	.45
4.7	98.88	.48	153.42	.59	215.81	.65	280.33	.64	341.01	.56	392.20	.44
4.8	99.36	.50	154.01	.59	216.46	.64	280.97	.63	341.57	.57	392.64	.45
4.9	99.86	.49	154.60	.60	217.10	.65	281.60	.64	342.14	.56	393.09	.44
5.0	100.35	.49	155.20	.60	217.75	.64	282.24	.63	342.70	.57	393.53	.44

TABLE X. ARGUMENT 5.

Equation $= 225''.300 + 18''.0 \sin. (z - t) - 212''.4 \sin. 2 (z - t)$.

Period, 411.78517 days.

Days.	240		250		260		270		280		290	
Days.	Equation.	Diff.	Equation.	Diff.	Equation.	Diff.	Equation.	Diff.	Equation.	Diff.	Equation.	Diff.
d.	"		"		"		"		"		"	
5.0	100.35	.49	155.20	.60	217.75	.64	282.24	.63	342.70	.57	393.53	.44
5.1	100.84	.50	155.80	.59	218.39	.65	282.87	.64	343.27	.56	393.97	.44
5.2	101.34	.49	156.39	.60	219.04	.64	283.51	.63	343.83	.57	394.41	.44
5.3	101.83	.50	156.99	.60	219.68	.65	284.14	.63	344.40	.56	394.85	.44
5.4	102.33	.50	157.59	.60	220.33	.65	284.77	.64	344.96	.56	395.29	.44
5.5	102.83	.50	158.19	.60	220.98	.64	285.41	.63	345.52	.56	395.73	.43
5.6	103.33	.50	158.79	.60	221.62	.65	286.04	.63	346.08	.56	396.16	.44
5.7	103.83	.50	159.39	.60	222.27	.64	286.67	.63	346.64	.56	396.60	.43
5.8	104.33	.50	159.99	.61	222.91	.65	287.30	.63	347.20	.55	397.03	.43
5.9	104.83	.50	160.60	.60	223.56	.65	287.93	.63	347.75	.56	397.46	.43
6.0	105.33	.51	161.20	.60	224.21	.64	288.56	.63	348.31	.55	397.89	.43
6.1	105.84	.51	161.80	.61	224.85	.65	289.19	.63	348.86	.56	398.32	.42
6.2	106.35	.50	162.41	.60	225.50	.65	289.82	.63	349.42	.55	398.74	.43
6.3	106.85	.51	163.01	.61	226.15	.65	290.45	.63	349.97	.55	399.17	.42
6.4	107.36	.51	163.62	.61	226.80	.64	291.08	.63	350.52	.55	399.59	.42
6.5	107.87	.52	164.23	.61	227.44	.65	291.71	.62	351.07	.55	400.01	.42
6.6	108.39	.51	164.84	.60	228.09	.65	292.33	.63	351.62	.55	400.43	.42
6.7	108.90	.51	165.44	.61	228.74	.65	292.96	.63	352.17	.55	400.85	.42
6.8	109.41	.52	166.05	.61	229.39	.64	293.59	.62	352.72	.55	401.27	.42
6.9	109.93	.52	166.66	.61	230.03	.65	294.21	.63	353.27	.54	401.69	.41
7.0	110.45	.51	167.27	.61	230.68	.65	294.84	.62	353.81	.54	402.10	.41
7.1	110.96	.52	167.88	.62	231.33	.65	295.46	.63	354.35	.55	402.51	.41
7.2	111.48	.52	168.50	.61	231.98	.65	296.09	.62	354.90	.54	402.92	.41
7.3	112.00	.52	169.11	.61	232.63	.64	296.71	.62	355.44	.54	403.33	.40
7.4	112.52	.53	169.72	.62	233.27	.65	297.33	.63	355.98	.54	403.73	.42
7.5	113.05	.52	170.34	.61	233.92	.65	297.96	.62	356.52	.54	404.15	.40
7.6	113.57	.53	170.95	.61	234.57	.65	298.58	.62	357.06	.54	404.55	.41
7.7	114.10	.52	171.56	.62	235.22	.65	299.20	.63	357.60	.53	404.96	.40
7.8	114.62	.53	172.18	.62	235.87	.65	299.83	.62	358.13	.54	405.36	.40
7.9	115.15	.53	172.80	.61	236.52	.64	300.45	.62	358.67	.53	405.76	.40
8.0	115.68	.53	173.41	.62	237.16	.65	301.07	.62	359.20	.54	406.16	.39
8.1	116.21	.53	174.03	.62	237.81	.65	301.69	.61	359.74	.53	406.55	.40
8.2	116.74	.53	174.65	.61	238.46	.65	302.30	.62	360.27	.53	406.95	.39
8.3	117.27	.53	175.26	.62	239.11	.65	302.92	.62	360.80	.53	407.34	.40
8.4	117.80	.54	175.88	.62	239.76	.65	303.54	.62	361.33	.53	407.74	.39
8.5	118.34	.53	176.50	.62	240.41	.64	304.16	.62	361.86	.53	408.13	.39
8.6	118.87	.54	177.12	.62	241.05	.65	304.78	.61	362.39	.52	408.52	.38
8.7	119.41	.53	177.74	.62	241.70	.65	305.39	.62	362.91	.53	408.90	.39
8.8	119.94	.54	178.36	.63	242.35	.65	306.01	.61	363.44	.52	409.29	.38
8.9	120.48	.54	178.99	.62	243.00	.65	306.62	.62	363.96	.52	409.67	.39
9.0	121.02	.54	179.61	.62	243.65	.64	307.24	.61	364.48	.53	410.06	.38
9.1	121.56	.54	180.23	.62	244.29	.65	307.85	.61	365.01	.52	410.44	.38
9.2	122.10	.55	180.85	.63	244.94	.65	308.46	.61	365.53	.52	410.82	.38
9.3	122.65	.54	181.48	.62	245.59	.65	309.07	.62	366.05	.52	411.20	.37
9.4	123.19	.55	182.10	.63	246.24	.65	309.69	.61	366.57	.51	411.57	.38
9.5	123.74	.54	182.73	.62	246.89	.64	310.30	.61	367.08	.52	411.95	.37
9.6	124.28	.55	183.35	.63	247.53	.65	310.91	.61	367.60	.51	412.32	.37
9.7	124.83	.55	183.98	.63	248.18	.65	311.52	.60	368.11	.52	412.69	.37
9.8	125.38	.55	184.61	.62	248.83	.65	312.12	.61	368.63	.51	413.06	.37
9.9	125.93	.55	185.23	.63	249.48	.64	312.73	.61	369.14	.51	413.43	.36
10.0	126.48	.55	185.86	.63	250.12	.65	313.34	.61	369.65	.51	413.79	.37

TABLE X. ARGUMENT 5.

Equation = 225″.300 + 16″.0 sin. $(x - t)$ — 212″.4 sin. 2 $(x - t)$.

Period, 411.78517 days.

Days. 300			310		320		330		340		350	
Days.	Equation.	Diff.	Equation.	Diff.	Equation.	Diff.	Equation.	Diff.	Equation.	Diff.	Equation.	Diff.
d.	″		″		″		″		″		″	
0.0	413.79	.37	441.63	.19	450.49	.01	439.42	.20	409.31	.39	362.78	.53
0.1	414.16	.36	441.82	.18	450.48	.02	439.22	.22	408.92	.39	362.25	.54
0.2	414.52	.37	442.00	.19	450.46	.01	439.00	.21	408.53	.39	361.71	.54
0.3	414.89	.36	442.19	.18	450.45	.02	438.79	.22	408.14	.40	361.17	.54
0.4	415.25	.35	442.37	.17	450.43	.02	438.57	.21	407.74	.39	360.63	.54
0.5	415.60	.36	442.54	.18	450.41	.02	438.36	.22	407.35	.40	360.09	.54
0.6	415.96	.36	442.72	.17	450.39	.03	438.14	.23	406.95	.40	359.55	.54
0.7	416.32	.35	442.89	.17	450.36	.02	437.91	.22	406.55	.40	359.01	.55
0.8	416.67	.35	443.06	.16	450.34	.03	437.69	.22	406.15	.40	358.46	.54
0.9	417.02	.35	443.22	.16	450.31	.03	437.47	.23	405.75	.41	357.92	.55
1.0	417.37	.35	443.38	.17	450.28	.03	437.24	.23	405.34	.40	357.37	.55
1.1	417.72	.35	443.55	.18	450.25	.04	437.01	.23	404.94	.42	356.82	.55
1.2	418.07	.34	443.73	.17	450.21	.03	436.78	.23	404.52	.40	356.27	.55
1.3	418.41	.34	443.90	.16	450.18	.04	436.55	.24	404.12	.41	355.72	.55
1.4	418.75	.35	444.06	.16	450.14	.04	436.31	.23	403.71	.42	355.17	.55
1.5	419.10	.34	444.22	.15	450.10	.04	436.08	.24	403.29	.41	354.62	.55
1.6	419.44	.33	444.37	.16	450.06	.05	435.84	.24	402.88	.42	354.07	.56
1.7	419.77	.34	444.53	.15	450.01	.04	435.60	.24	402.46	.41	353.51	.55
1.8	420.11	.33	444.68	.15	449.97	.05	435.36	.25	402.05	.42	352.96	.56
1.9	420.44	.34	444.83	.15	449.92	.05	435.11	.24	401.63	.42	352.40	.56
2.0	420.78	.33	444.98	.15	449.87	.05	434.87	.25	401.21	.42	351.84	.56
2.1	421.11	.33	445.13	.14	449.82	.06	434.62	.25	400.79	.43	351.28	.56
2.2	421.44	.33	445.27	.15	449.76	.05	434.37	.25	400.36	.42	350.72	.56
2.3	421.77	.32	445.42	.14	449.71	.07	434.12	.26	399.94	.43	350.16	.56
2.4	422.09	.33	445.56	.14	449.64	.05	433.86	.25	399.51	.43	349.60	.56
2.5	422.42	.32	445.70	.13	449.59	.06	433.61	.26	399.08	.43	349.04	.57
2.6	422.74	.32	445.83	.14	449.53	.07	433.35	.26	398.65	.43	348.47	.57
2.7	423.06	.32	445.97	.13	449.46	.06	433.09	.26	398.22	.43	347.90	.56
2.8	423.38	.32	446.10	.13	449.40	.07	432.83	.26	397.79	.44	347.34	.57
2.9	423.70	.31	446.23	.13	449.33	.06	432.57	.27	397.35	.43	346.77	.57
3.0	424.01	.32	446.36	.13	449.26	.07	432.30	.26	396.92	.44	346.20	.56
3.1	424.33	.31	446.49	.13	449.19	.08	432.04	.27	396.48	.44	345.64	.58
3.2	424.64	.31	446.62	.12	449.11	.07	431.77	.27	396.04	.44	345.06	.57
3.3	424.95	.31	446.74	.12	449.04	.08	431.50	.27	395.60	.44	344.49	.58
3.4	425.26	.31	446.86	.12	448.96	.08	431.23	.28	395.16	.45	343.91	.57
3.5	425.57	.30	446.98	.12	448.88	.09	430.95	.27	394.71	.44	343.34	.58
3.6	425.87	.30	447.10	.11	448.79	.08	430.68	.28	394.27	.45	342.76	.57
3.7	426.17	.31	447.21	.12	448.71	.09	430.40	.28	393.82	.45	342.19	.58
3.8	426.48	.30	447.33	.11	448.62	.08	430.12	.28	393.37	.45	341.61	.58
3.9	426.78	.29	447.44	.11	448.54	.09	429.84	.28	392.92	.45	341.03	.58
4.0	427.07	.30	447.55	.10	448.45	.10	429.56	.29	392.47	.45	340.45	.58
4.1	427.37	.30	447.65	.11	448.35	.09	429.27	.28	392.02	.46	339.87	.58
4.2	427.67	.29	447.76	.10	448.26	.10	428.99	.29	391.56	.45	339.29	.59
4.3	427.96	.29	447.86	.11	448.16	.10	428.70	.29	391.11	.46	338.70	.58
4.4	428.25	.29	447.97	.10	448.06	.10	428.41	.30	390.65	.46	338.12	.59
4.5	428.54	.29	448.07	.09	447.96	.10	428.11	.29	390.19	.46	337.53	.58
4.6	428.83	.28	448.16	.10	447.86	.10	427.82	.30	389.73	.46	336.95	.59
4.7	429.11	.29	448.26	.09	447.76	.11	427.52	.29	389.27	.47	336.36	.59
4.8	429.40	.28	448.35	.10	447.65	.11	427.23	.30	388.80	.46	335.77	.59
4.9	429.68	.28	448.45	.09	447.54	.11	426.93	.31	388.34	.47	335.18	.59
5.0	429.96	.28	448.54	.09	447.43	.11	426.62	.30	387.87	.47	334.59	.58

TABLE X. ARGUMENT 5.

Equation = 225″.300 + 18″.0 sin. $(x - t)$ — 212″.4 sin. 2 $(x - t)$.

Period, 411.78517 days.

Days.	300		310		320		330		340		350	
Days.	Equation.	Diff.	Equation.	Diff.	Equation.	Diff.	Equation.	Diff.	Equation.	Diff.	Equation.	Diff.
d.	″		″		″		″		″		″	
5.0	429.96	.28	448.51	.09	447.43	.11	426.62	.30	387.87	.47	334.59	.58
5.1	430.24	.28	448.63	.08	447.32	.11	426.32	.30	387.40	.46	334.01	.60
5.2	430.52	.27	448.71	.09	447.21	.12	426.02	.31	386.94	.48	333.41	.59
5.3	430.79	.27	448.80	.08	447.09	.12	425.71	.31	386.46	.47	332.82	.59
5.4	431.06	.27	448.88	.08	446.97	.12	425.40	.31	385.99	.47	332.23	.60
5.5	431.33	.27	448.96	.07	446.85	.12	425.09	.31	385.52	.48	331.63	.59
5.6	431.60	.27	449.03	.08	446.73	.12	424.78	.32	385.04	.47	331.04	.60
5.7	431.87	.27	449.11	.07	446.61	.13	424.46	.31	384.57	.48	330.44	.60
5.8	432.14	.26	449.18	.08	446.48	.13	424.15	.32	384.09	.48	329.84	.60
5.9	432.40	.26	449.26	.07	446.35	.13	423.83	.32	383.61	.48	329.24	.60
6.0	432.66	.26	449.33	.07	446.22	.12	423.51	.32	383.13	.48	328.64	.60
6.1	432.92	.26	449.40	.06	446.10	.14	423.19	.32	382.65	.49	328.04	.60
6.2	433.18	.26	449.46	.07	445.96	.13	422.87	.33	382.16	.48	327.44	.60
6.3	433.44	.26	449.53	.06	445.83	.14	422.54	.32	381.68	.49	326.84	.61
6.4	433.70	.25	449.59	.06	445.69	.14	422.22	.33	381.19	.49	326.23	.60
6.5	433.95	.25	449.65	.06	445.55	.14	421.89	.33	380.70	.49	325.63	.60
6.6	434.20	.25	449.71	.05	445.41	.15	421.56	.33	380.21	.49	325.03	.61
6.7	434.45	.25	449.76	.06	445.26	.14	421.23	.34	379.72	.49	324.42	.61
6.8	434.70	.24	449.82	.05	445.12	.15	420.89	.33	379.23	.49	323.81	.61
6.9	434.94	.25	449.87	.05	444.97	.15	420.56	.34	378.74	.49	323.20	.60
7.0	435.19	.24	449.92	.05	444.82	.15	420.22	.34	378.25	.50	322.60	.61
7.1	435.43	.24	449.97	.04	444.67	.16	419.88	.34	377.75	.50	321.99	.61
7.2	435.67	.24	450.01	.05	444.51	.15	419.54	.34	377.25	.50	321.38	.62
7.3	435.91	.23	450.06	.04	444.36	.16	419.20	.34	376.75	.50	320.76	.61
7.4	436.14	.24	450.10	.04	444.20	.16	418.86	.35	376.25	.50	320.15	.61
7.5	436.38	.23	450.14	.04	444.04	.16	418.51	.35	375.75	.50	319.54	.61
7.6	436.61	.23	450.18	.03	443.88	.16	418.16	.35	375.25	.51	318.93	.62
7.7	436.84	.23	450.21	.04	443.72	.17	417.81	.35	374.74	.50	318.31	.62
7.8	437.07	.23	450.25	.03	443.55	.17	417.46	.35	374.24	.51	317.69	.61
7.9	437.30	.22	450.28	.03	443.38	.17	417.11	.35	373.73	.51	317.08	.62
8.0	437.52	.23	450.31	.03	443.21	.16	416.76	.36	373.22	.51	316.46	.62
8.1	437.75	.22	450.34	.02	443.05	.18	416.40	.36	372.71	.51	315.84	.62
8.2	437.97	.22	450.36	.03	442.87	.17	416.04	.36	372.20	.51	315.22	.61
8.3	438.19	.22	450.39	.02	442.70	.18	415.68	.36	371.69	.52	314.61	.63
8.4	438.41	.21	450.41	.02	442.52	.18	415.32	.36	371.17	.51	313.98	.62
8.5	438.62	.22	450.43	.02	442.34	.18	414.96	.37	370.66	.52	313.36	.62
8.6	438.84	.21	450.45	.01	442.16	.18	414.59	.36	370.14	.52	312.74	.62
8.7	439.05	.21	450.46	.02	441.98	.19	414.23	.37	369.62	.52	312.12	.63
8.8	439.26	.21	450.48	.01	441.79	.18	413.86	.37	369.10	.52	311.49	.62
8.9	439.47	.20	450.49	.01	441.61	.19	413.49	.37	368.58	.52	310.87	.63
9.0	439.67	.21	450.50	.01	441.42	.19	413.12	.38	368.06	.52	310.24	.62
9.1	439.88	.20	450.51	.00	441.23	.20	412.74	.37	367.54	.52	309.62	.63
9.2	440.08	.20	450.51	.01	441.03	.19	412.37	.38	367.02	.53	308.99	.62
9.3	440.28	.20	450.52	.00	440.84	.20	411.99	.37	366.49	.53	308.37	.63
9.4	440.48	.20	450.52	.00	440.64	.20	411.62	.38	365.96	.52	307.74	.63
9.5	440.68	.19	450.52	.00	440.44	.20	411.24	.39	365.44	.53	307.11	.63
9.6	440.87	.20	450.52	.01	440.24	.20	410.85	.38	364.91	.53	306.48	.63
9.7	441.07	.19	450.51	.00	440.04	.20	410.47	.38	364.38	.53	305.85	.63
9.8	441.26	.19	450.51	.01	439.84	.21	410.09	.39	363.85	.54	305.22	.64
9.9	441.45	.18	450.50	.01	439.63	.21	409.70	.39	363.31	.53	304.58	.63
10.0	441.63	.19	450.49	.01	439.42	.20	409.31	.39	362.78	.53	303.95	.63

TABLE X. ARGUMENT 5.

Equation = 225″.300 + 18″.0 sin. $(x - t)$ — 212″.4 sin. 2 $(x - t)$.

Period, 411.78517 days.

Days.	360		370		380		390		400		410		420	
Days	Equation.	Diff.	Equation.	Diff.	Equation.	Diff.	Equation.	Diff.	Equation.	Diff.	Equation.	Diff.	Equation.	Diff.
d.	″		″		″		″		″		″		″	
0.0	303.95	.63	238.08	.68	171.06	.66	108.90	.58	57.16	.45	20.44	.28	1.97	.09
0.1	303.32	.63	237.40	.67	170.40	.65	108.32	.57	56.71	.45	20.16	.28	1.88	.08
0.2	302.69	.64	236.73	.68	169.75	.66	107.75	.57	56.26	.44	19.88	.27	1.80	.09
0.3	302.05	.64	236.05	.67	169.09	.65	107.18	.57	55.82	.45	19.61	.28	1.71	.08
0.4	301.41	.63	235.38	.68	168.44	.65	106.61	.58	55.37	.44	19.33	.27	1.63	.08
0.5	300.78	.64	234.70	.67	167.79	.66	106.03	.58	54.93	.44	19.06	.27	1.55	.07
0.6	300.14	.64	234.03	.68	167.13	.65	105.45	.57	54.49	.44	18.79	.27	1.48	.08
0.7	299.50	.63	233.35	.67	166.48	.65	104.88	.56	54.05	.44	18.52	.26	1.40	.07
0.8	298.87	.64	232.68	.68	165.83	.65	104.32	.57	53.61	.43	18.26	.27	1.33	.07
0.9	298.23	.64	232.00	.67	165.18	.65	103.75	.57	53.18	.44	17.99	.26	1.26	.07
1.0	297.59	.64	231.33	.68	164.53	.65	103.18	.57	52.74	.43	17.73	.26	1.19	.06
1.1	296.95	.64	230.65	.67	163.88	.65	102.61	.56	52.31	.43	17.47	.26	1.13	.07
1.2	296.31	.64	229.98	.68	163.23	.65	102.05	.57	51.88	.43	17.21	.26	1.06	.06
1.3	295.67	.65	229.30	.67	162.58	.65	101.48	.56	51.45	.42	16.95	.25	1.00	.06
1.4	295.02	.64	228.63	.68	161.93	.65	100.92	.56	51.03	.43	16.70	.26	0.94	.06
1.5	294.38	.64	227.95	.67	161.28	.65	100.36	.57	50.60	.42	16.44	.25	0.88	.05
1.6	293.74	.65	227.28	.68	160.63	.64	99.79	.56	50.18	.43	16.19	.25	0.83	.06
1.7	293.09	.64	226.60	.68	159.99	.65	99.23	.55	49.75	.42	15.94	.25	0.77	.05
1.8	292.45	.65	225.92	.67	159.34	.64	98.68	.56	49.33	.42	15.69	.24	0.72	.05
1.9	291.80	.64	225.25	.68	158.70	.65	98.12	.56	48.91	.42	15.45	.25	0.67	.05
2.0	291.16	.65	224.57	.67	158.05	.64	97.56	.55	48.49	.42	15.20	.24	0.62	.04
2.1	290.51	.65	223.90	.68	157.41	.65	97.01	.56	48.07	.41	14.96	.24	0.58	.05
2.2	289.86	.64	223.22	.67	156.76	.64	96.45	.55	47.66	.41	14.72	.24	0.53	.04
2.3	289.22	.65	222.55	.68	156.12	.64	95.90	.55	47.25	.42	14.48	.23	0.49	.04
2.4	288.57	.65	221.87	.68	155.48	.65	95.35	.56	46.83	.41	14.25	.23	0.45	.04
2.5	287.92	.65	221.19	.67	154.83	.64	94.79	.55	46.42	.41	14.02	.23	0.41	.03
2.6	287.27	.65	220.52	.68	154.19	.64	94.24	.54	46.01	.41	13.79	.23	0.38	.04
2.7	286.62	.65	219.84	.67	153.55	.63	93.70	.55	45.60	.40	13.56	.23	0.34	.02
2.8	285.97	.65	219.17	.68	152.92	.64	93.15	.55	45.20	.41	13.33	.23	0.32	.03
2.9	285.32	.65	218.49	.67	152.28	.64	92.60	.54	44.79	.40	13.10	.22	0.29	.03
3.0	284.67	.65	217.82	.68	151.64	.63	92.06	.55	44.39	.40	12.88	.23	0.26	.02
3.1	284.02	.65	217.14	.67	151.01	.64	91.51	.54	43.99	.40	12.65	.22	0.24	.03
3.2	283.37	.66	216.47	.68	150.37	.64	90.97	.54	43.59	.40	12.43	.22	0.21	.02
3.3	282.71	.65	215.79	.67	149.73	.64	90.43	.54	43.19	.40	12.21	.21	0.19	.02
3.4	282.06	.66	215.12	.67	149.09	.63	89.89	.54	42.79	.39	12.00	.22	0.17	.02
3.5	281.40	.65	214.45	.68	148.46	.64	89.35	.54	42.40	.39	11.78	.22	0.15	.01
3.6	280.75	.66	213.77	.68	147.82	.63	88.81	.53	42.01	.39	11.56	.21	0.14	.02
3.7	280.09	.65	213.09	.67	147.19	.63	88.28	.54	41.62	.39	11.35	.21	0.12	.01
3.8	279.44	.66	212.42	.67	146.56	.64	87.74	.53	41.23	.39	11.14	.20	0.11	.01
3.9	278.78	.65	211.75	.68	145.92	.63	87.21	.54	40.84	.39	10.94	.21	0.10	.01
4.0	278.13	.66	211.07	.67	145.29	.63	86.67	.53	40.45	.38	10.73	.20	0.09	.00
4.1	277.47	.66	210.40	.68	144.66	.63	86.14	.53	40.07	.39	10.53	.20	0.09	.01
4.2	276.81	.66	209.72	.67	144.03	.63	85.61	.53	39.68	.38	10.33	.20	0.08	.00
4.3	276.15	.65	209.05	.67	143.40	.63	85.08	.53	39.30	.38	10.13	.20	0.08	.00
4.4	275.50	.66	208.38	.68	142.77	.63	84.55	.53	38.92	.38	9.93	.20	0.08	.00
4.5	274.84	.66	207.70	.67	142.14	.62	84.02	.52	38.54	.37	9.73	.19	0.08	.01
4.6	274.18	.66	207.03	.68	141.52	.63	83.50	.52	38.17	.38	9.54	.20	0.09	.00
4.7	273.52	.66	206.35	.67	140.89	.63	82.98	.52	37.79	.37	9.34	.19	0.09	.01
4.8	272.86	.66	205.68	.67	140.26	.62	82.46	.53	37.42	.37	9.15	.19	0.10	.01
4.9	272.20	.66	205.01	.67	139.64	.63	81.93	.52	37.05	.37	8.96	.18	0.11	.01
5.0	271.54	.66	204.34	.68	139.01	.62	81.41	.51	36.68	.36	8.78	.19	0.12	.02

TABLE X. ARGUMENT 5.

$$\text{Equation} = 225''.300 + 18''.0 \sin. (x - t) - 212''.4 \sin. 2 (x - t).$$

Period, 411.78517 days.

Days.	360		370		380		390		400		410		420	
Days.	Equation.	Diff.	Equation.	Diff.	Equation.	Diff.	Equation.	Diff.	Equation.	Diff.	Equation.	Diff.	Equation.	Diff.
d	″		″		″		″							
5.0	271.54	.66	204.34	.68	139.01	.62	81.41	.51	36.68	.36	8.78	.19	0.12	.02
5.1	270.88	.67	203.66	.67	138.39	.62	80.90	.52	36.32	.37	8.59	.18	0.14	.01
5.2	270.21	.66	202.99	.67	137.77	.62	80.38	.52	35.95	.36	8.41	.18	0.15	.02
5.3	269.55	.66	202.32	.67	137.15	.63	79.86	.51	35.59	.36	8.23	.18	0.17	.02
5.4	268.89	.66	201.65	.68	136.52	.62	79.35	.52	35.23	.36	8.05	.18	0.19	.02
5.5	268.23	.67	200.97	.67	135.90	.62	78.83	.51	34.87	.36	7.87	.17	0.21	.03
5.6	267.56	.66	200.30	.67	135.28	.62	78.32	.51	34.51	.36	7.70	.18	0.24	.02
5.7	266.90	.67	199.63	.67	134.66	.61	77.81	.51	34.15	.36	7.52	.17	0.26	.03
5.8	266.23	.66	198.96	.67	134.05	.62	77.30	.51	33.79	.35	7.35	.16	0.29	.03
5.9	265.57	.67	198.29	.67	133.43	.62	76.79	.51	33.44	.35	7.19	.16	0.32	.04
6.0	264.90	.66	197.62	.67	132.81	.61	76.28	.50	33.09	.35	7.03	.17	0.36	.04
6.1	264.24	.67	196.95	.67	132.20	.62	75.78	.51	32.74	.35	6.86	.16	0.40	.03
6.2	263.57	.67	196.28	.67	131.58	.61	75.27	.50	32.39	.35	6.70	.16	0.43	.04
6.3	262.90	.66	195.61	.67	130.97	.61	74.77	.50	32.04	.35	6.54	.16	0.47	.04
6.4	262.24	.67	194.94	.67	130.36	.62	74.27	.50	31.69	.34	6.38	.16	0.51	.04
6.5	261.57	.67	194.27	.67	129.74	.61	73.77	.50	31.35	.34	6.22	.15	0.55	.05
6.6	260.90	.66	193.60	.67	129.13	.61	73.27	.49	31.01	.34	6.07	.16	0.60	.04
6.7	260.24	.67	192.93	.67	128.52	.61	72.78	.50	30.67	.34	5.91	.15	0.64	.05
6.8	259.57	.67	192.26	.66	127.91	.61	72.28	.49	30.33	.34	5.76	.15	0.69	.05
6.9	258.90	.66	191.60	.67	127.30	.60	71.79	.49	29.99	.33	5.61	.15	0.74	.05
7.0	258.24	.67	190.93	.67	126.70	.61	71.30	.50	29.66	.34	5.46	.14	0.79	.06
7.1	257.57	.67	190.26	.66	126.09	.61	70.80	.49	29.32	.33	5.32	.15	0.85	.05
7.2	256.90	.67	189.60	.67	125.48	.60	70.31	.49	28.99	.33	5.17	.14	0.90	.06
7.3	256.23	.67	188.93	.67	124.88	.61	69.82	.48	28.66	.33	5.03	.14	0.96	.06
7.4	255.56	.67	188.26	.67	124.27	.60	69.34	.49	28.33	.32	4.89	.14	1.02	.06
7.5	254.89	.67	187.59	.66	123.67	.60	68.85	.48	28.01	.33	4.75	.13	1.08	.07
7.6	254.22	.67	186.93	.67	123.07	.60	68.37	.49	27.68	.32	4.62	.14	1.15	.06
7.7	253.55	.67	186.26	.66	122.47	.60	67.88	.48	27.36	.32	4.48	.13	1.21	.07
7.8	252.88	.67	185.60	.67	121.87	.60	67.40	.48	27.04	.32	4.35	.13	1.28	.07
7.9	252.21	.67	184.93	.67	121.27	.60	66.92	.48	26.72	.32	4.22	.12	1.35	.07
8.0	251.54	.67	184.26	.66	120.67	.60	66.44	.48	26.40	.31	4.10	.13	1.42	.08
8.1	250.87	.67	183.60	.66	120.07	.60	65.96	.47	26.09	.32	3.97	.12	1.50	.07
8.2	250.20	.67	182.94	.67	119.47	.60	65.49	.48	25.77	.31	3.85	.13	1.57	.08
8.3	249.53	.68	182.27	.66	118.87	.60	65.01	.47	25.46	.31	3.72	.12	1.65	.08
8.4	248.85	.67	181.61	.66	118.28	.59	64.54	.48	25.15	.31	3.60	.11	1.73	.08
8.5	248.18	.67	180.95	.66	117.69	.59	64.06	.47	24.84	.30	3.49	.12	1.81	.09
8.6	247.51	.67	180.29	.67	117.10	.60	63.59	.46	24.54	.30	3.37	.11	1.90	.08
8.7	246.84	.68	179.62	.66	116.50	.59	63.13	.47	24.24	.30	3.26	.12	1.98	.09
8.8	246.16	.67	178.96	.66	115.91	.59	62.66	.47	23.94	.31	3.14	.11	2.07	.09
8.9	245.49	.67	178.30	.66	115.32	.58	62.19	.46	23.63	.30	3.03	.10	2.16	.10
9.0	244.82	.67	177.64	.66	114.74	.59	61.73	.47	23.33	.29	2.93	.10	2.26	.09
9.1	244.15	.68	176.98	.66	114.15	.59	61.26	.46	23.04	.30	2.83	.10	2.35	.10
9.2	243.47	.67	176.32	.66	113.56	.59	60.80	.46	22.74	.29	2.73	.10	2.45	.09
9.3	242.80	.68	175.66	.66	112.97	.58	60.34	.46	22.45	.30	2.63	.10	2.54	.10
9.4	242.12	.67	175.00	.65	112.39	.58	59.88	.46	22.15	.29	2.53	.10	2.64	.11
9.5	241.45	.67	174.35	.66	111.81	.59	59.42	.45	21.86	.29	2.43	.10	2.75	.10
9.6	240.78	.68	173.69	.66	111.22	.58	58.97	.46	21.57	.28	2.33	.09	2.85	.11
9.7	240.10	.67	173.03	.66	110.64	.58	58.51	.45	21.29	.29	2.24	.10	2.96	.10
9.8	239.43	.68	172.37	.65	110.06	.58	58.06	.45	21.00	.28	2.14	.09	3.06	.11
9.9	238.75	.67	171.72	.66	109.48	.58	57.61	.45	20.72	.28	2.05	.08	3.17	.11
10.0	238.08	.68	171.06	.66	108.90	.58	57.16	.45	20.44	.28	1.97	.09	3.28	.12

TABLE XI. ARGUMENT 6.

Equation $= 206''.9\ [1 + \sin.\ (2t - z - x)]$.

Period, 34.846892 days.

Days.	0		1		2		3		4		5		Days.
Decimals of a Day.	Equation.	Diff.	Equation.	Diff.	Equation.	Diff.	Equation.	Diff.	Equation.	Diff.	Equation.	Diff.	Decimals of a Day.
.00	1.12″	.04	0.61″	.03	6.77″	.10	19.45″	.16	38.18″	.22	62.40″	.27	1.00
.01	1.08	.04	0.64	.03	6.87	.09	19.61	.16	38.40	.21	62.67	.26	.99
.02	1.04	.04	0.67	.03	6.96	.10	19.77	.16	38.61	.22	62.93	.27	.98
.03	1.00	.04	0.70	.03	7.06	.09	19.93	.16	38.83	.22	63.20	.27	.97
.04	0.96	.04	0.73	.03	7.15	.10	20.09	.16	39.05	.22	63.47	.27	.96
.05	0.92	.04	0.76	.03	7.25	.10	20.25	.16	39.27	.22	63.74	.27	.95
.06	0.88	.03	0.79	.04	7.35	.10	20.41	.16	39.49	.22	64.01	.27	.94
.07	0.85	.03	0.83	.03	7.45	.10	20.57	.16	39.71	.22	64.28	.27	.93
.08	0.82	.03	0.86	.03	7.55	.10	20.73	.16	39.93	.22	64.55	.27	.92
.09	0.79	.03	0.89	.04	7.65	.10	20.89	.16	40.15	.22	64.82	.27	.91
.10	0.76	.03	0.93	.03	7.75	.10	21.05	.16	40.37	.22	65.09	.27	.90
.11	0.73	.03	0.96	.04	7.85	.10	21.21	.16	40.59	.22	65.36	.27	.89
.12	0.70	.03	1.00	.03	7.95	.10	21.37	.17	40.81	.23	65.63	.27	.88
.13	0.67	.03	1.03	.04	8.05	.11	21.54	.16	41.04	.22	65.90	.28	.87
.14	0.64	.03	1.07	.04	8.16	.10	21.70	.16	41.26	.22	66.18	.27	.86
.15	0.61	.03	1.11	.04	8.26	.11	21.86	.17	41.48	.23	66.45	.27	.85
.16	0.58	.03	1.15	.04	8.37	.10	22.03	.17	41.71	.22	66.72	.28	.84
.17	0.55	.03	1.19	.04	8.47	.11	22.20	.17	41.93	.23	67.00	.27	.83
.18	0.52	.02	1.23	.04	8.58	.11	22.37	.17	42.16	.22	67.27	.28	.82
.19	0.50	.03	1.27	.04	8.69	.11	22.54	.17	42.38	.23	67.55	.28	.81
.20	0.47	.02	1.31	.04	8.80	.11	22.71	.17	42.61	.23	67.83	.27	.80
.21	0.45	.03	1.35	.04	8.91	.11	22.88	.17	42.84	.22	68.10	.28	.79
.22	0.42	.02	1.39	.04	9.02	.11	23.05	.17	43.06	.23	68.38	.28	.78
.23	0.40	.03	1.43	.05	9.13	.11	23.22	.17	43.29	.23	68.66	.27	.77
.24	0.37	.02	1.48	.04	9.24	.11	23.39	.17	43.52	.23	68.93	.28	.76
.25	0.35	.02	1.52	.05	9.35	.11	23.56	.18	43.75	.23	69.21	.28	.75
.26	0.33	.02	1.57	.04	9.46	.11	23.74	.17	43.98	.23	69.49	.28	.74
.27	0.31	.02	1.61	.05	9.57	.12	23.91	.18	44.21	.23	69.77	.28	.73
.28	0.29	.02	1.66	.05	9.69	.11	24.09	.17	44.44	.23	70.05	.28	.72
.29	0.27	.02	1.71	.05	9.80	.11	24.26	.18	44.67	.23	70.33	.28	.71
.30	0.25	.02	1.76	.05	9.91	.12	24.44	.18	44.90	.23	70.61	.28	.70
.31	0.23	.02	1.81	.05	10.03	.11	24.62	.17	45.13	.24	70.89	.28	.69
.32	0.21	.02	1.86	.05	10.14	.12	24.79	.18	45.37	.23	71.17	.28	.68
.33	0.19	.01	1.91	.05	10.26	.11	24.97	.18	45.60	.24	71.45	.28	.67
.34	0.18	.02	1.96	.05	10.37	.12	25.15	.18	45.84	.23	71.73	.29	.66
.35	0.16	.01	2.01	.06	10.49	.12	25.33	.18	46.07	.24	72.02	.28	.65
.36	0.15	.02	2.07	.05	10.61	.11	25.51	.18	46.31	.23	72.30	.28	.64
.37	0.13	.01	2.12	.05	10.72	.12	25.69	.18	46.54	.24	72.58	.29	.63
.38	0.12	.01	2.17	.06	10.84	.12	25.87	.18	46.78	.23	72.87	.28	.62
.39	0.11	.01	2.23	.05	10.96	.12	26.05	.18	47.01	.24	73.15	.29	.61
.40	0.10	.01	2.28	.06	11.08	.12	26.23	.18	47.25	.24	73.44	.29	.60
.41	0.09	.01	2.34	.05	11.20	.12	26.41	.18	47.49	.24	73.73	.28	.59
.42	0.08	.01	2.39	.06	11.32	.12	26.59	.19	47.73	.24	74.01	.29	.58
.43	0.07	.01	2.45	.06	11.44	.12	26.78	.18	47.97	.24	74.30	.29	.57
.44	0.06	.01	2.51	.05	11.56	.12	26.96	.19	48.21	.24	74.59	.28	.56
.45	0.05	.01	2.56	.06	11.68	.13	27.15	.18	48.45	.24	74.87	.29	.55
.46	0.04	.01	2.62	.06	11.81	.12	27.33	.19	48.69	.24	75.16	.29	.54
.47	0.03	.00	2.68	.06	11.93	.13	27.52	.19	48.93	.24	75.45	.29	.53
.48	0.03	.01	2.74	.06	12.06	.12	27.71	.19	49.17	.24	75.74	.29	.52
.49	0.02	.00	2.80	.06	12.18	.13	27.90	.19	49.41	.24	76.03	.29	.51
.50	0.02	.01	2.86	.06	12.31	.13	28.09	.19	49.65	.24	76.32	.29	.50
Days.	35		34		33		32		31		30		Days.

TABLE XI. ARGUMENT 6.

Equation = 206″.9 [1 + sin. (2t — z — x)].

Period, 34.846892 days.

Days.	0		1		2		3		4		5		Days.
Decimals of a Day	Equation	Diff.	Equation.	Diff.	Equation.	Diff.	Equation.	Diff.	Equation.	Diff.	Equation.	Diff.	Decimals of a Day
.50	0.02″	.01	2.86″	.06	12.31″	.13	28.09″	.19	49.65″	.24	76.32″	.29	.50
.51	0.01	.00	2.92	.06	12.44	.13	28.28	.19	49.89	.24	76.61	.29	.49
.52	0.01	.00	2.98	.06	12.57	.13	28.47	.19	50.13	.25	76.90	.29	.48
.53	0.01	.01	3.04	.07	12.70	.13	28.66	.19	50.38	.24	77.19	.29	.47
.54	0.00	.00	3.11	.06	12.83	.13	28.85	.19	50.62	.25	77.48	.29	.46
.55	0.00	.00	3.17	.07	12.96	.13	29.04	.19	50.87	.24	77.77	.30	.45
.56	0.00	.00	3.24	.06	13.09	.13	29.23	.19	51.11	.25	78.07	.29	.44
.57	0.00	.00	3.30	.07	13.22	.13	29.42	.19	51.36	.24	78.36	.29	.43
.58	0.00	.00	3.37	.07	13.35	.13	29.61	.20	51.60	.25	78.65	.30	.42
.59	0.00	.00	3.44	.07	13.48	.13	29.81	.19	51.85	.25	78.95	.29	.41
.60	0.00	.00	3.51	.07	13.61	.14	30.00	.19	52.10	.25	79.24	.29	.40
.61	0.00	.00	3.58	.07	13.75	.13	30.19	.19	52.35	.25	79.53	.30	.39
.62	0.00	.01	3.65	.07	13.88	.14	30.38	.20	52.60	.25	79.83	.29	.38
.63	0.01	.00	3.72	.07	14.02	.13	30.58	.19	52.85	.25	80.12	.29	.37
.64	0.01	.00	3.79	.07	14.15	.14	30.77	.20	53.10	.25	80.41	.30	.36
.65	0.01	.01	3.86	.07	14.29	.14	30.97	.19	53.35	.25	80.71	.29	.35
.66	0.02	.00	3.93	.08	14.43	.13	31.16	.20	53.60	.25	81.00	.30	.34
.67	0.02	.01	4.01	.07	14.56	.14	31.36	.20	53.85	.25	81.30	.29	.33
.68	0.03	.01	4.08	.07	14.70	.14	31.56	.20	54.10	.25	81.59	.30	.32
.69	0.04	.01	4.15	.08	14.84	.14	31.76	.20	54.35	.25	81.89	.30	.31
.70	0.05	.01	4.23	.08	14.98	.14	31.96	.20	54.60	.25	82.19	.30	.30
.71	0.06	.01	4.31	.07	15.12	.14	32.16	.20	54.85	.25	82.49	.29	.29
.72	0.07	.01	4.38	.08	15.26	.14	32.36	.20	55.10	.26	82.78	.30	.28
.73	0.08	.01	4.46	.07	15.40	.14	32.56	.20	55.36	.25	83.08	.30	.27
.74	0.09	.01	4.53	.08	15.54	.15	32.76	.20	55.61	.26	83.38	.30	.26
.75	0.10	.01	4.61	.08	15.69	.14	32.96	.20	55.87	.25	83.68	.30	.25
.76	0.11	.01	4.69	.08	15.83	.15	33.16	.21	56.12	.26	83.98	[illegible]	.24
.77	0.12	.02	4.77	.08	15.98	.14	33.37	.20	56.38	.25	84.28	.30	.23
.78	0.14	.01	4.85	.08	16.12	.15	33.57	.20	56.63	.26	84.58	.30	.22
.79	0.15	.02	4.93	.08	16.27	.15	33.77	.21	56.89	.26	84.88	.30	.21
.80	0.17	.01	5.01	.08	16.42	.15	33.98	.20	57.15	.26	85.18	.30	.20
.81	0.18	.02	5.09	.08	16.57	.14	34.18	.21	57.41	.26	85.48	.30	.19
.82	0.20	.01	5.17	.08	16.71	.15	34.39	.21	57.67	.26	85.78	.30	.18
.83	0.21	.02	5.25	.08	16.86	.15	34.60	.20	57.93	.26	86.08	.31	.17
.84	0.23	.02	5.33	.09	17.01	.15	34.80	.21	58.19	.26	86.39	.30	.16
.85	0.25	.02	5.42	.08	17.16	.15	35.01	.21	58.45	.26	86.69	.30	.15
.86	0.27	.02	5.50	.09	17.31	.15	35.22	.20	58.71	.26	86.99	.31	.14
.87	0.29	.02	5.59	.09	17.46	.15	35.42	.21	58.97	.26	87.30	.30	.13
.88	0.31	.02	5.68	.09	17.61	.15	35.63	.21	59.23	.26	87.60	.31	.12
.89	0.33	.02	5.77	.09	17.76	.15	35.84	.21	59.49	.26	87.91	.31	.11
.90	0.35	.02	5.86	.09	17.91	.15	36.05	.21	59.75	.26	88.22	.31	.10
.91	0.37	.02	5.95	.09	18.06	.15	36.26	.21	60.01	.26	88.53	.30	.09
.92	0.39	.02	6.04	.09	18.21	.15	36.47	.21	60.27	.27	88.83	.31	.08
.93	0.41	.03	6.13	.09	18.36	.15	36.68	.21	60.54	.26	89.14	.31	.07
.94	0.44	.02	6.22	.09	18.51	.16	36.89	.22	60.80	.27	89.45	.31	.06
.95	0.46	.03	6.31	.09	18.67	.15	37.11	.21	61.07	.26	89.76	.30	.05
.96	0.49	.03	6.40	.09	18.82	.16	37.32	.21	61.33	.27	90.06	.31	.04
.97	0.52	.03	6.49	.09	18.98	.15	37.53	.22	61.60	.26	90.37	.31	.03
.98	0.55	.03	6.58	.10	19.13	.16	37.75	.21	61.86	.27	90.68	.31	.02
.99	0.58	.03	6.68	.09	19.29	.16	37.96	.22	62.13	.27	90.99	.31	.01
1.00	0.61	.03	6.77	.10	19.45	.16	38.18	.22	62.40	.27	91.30	.31	.00
Days.	35		34		33		32		31		30		Days.

TABLE XI. ARGUMENT 6.

Equation $= 206''.9\ [1 + \sin.\ (2t - z - x)]$.

Period, 31.846892 days.

Decimals of a Day.	6 Equation.	Diff.	7 Equation.	Diff.	8 Equation.	Diff.	9 Equation.	Diff.	10 Equation.	Diff.	11 Equation.	Diff.	Decimals of a Day.
.00	91.30	.31	123.94	.34	159.28	.36	196.14	.37	233.37	.37	269.75	.36	1.00
.01	91.61	.31	124.28	.34	159.64	.36	196.51	.37	233.74	.37	270.11	.36	.99
.02	91.92	.31	124.62	.34	160.00	.37	196.88	.38	234.11	.37	270.47	.35	.98
.03	92.23	.31	124.96	.35	160.37	.36	197.26	.37	234.48	.37	270.82	.36	.97
.04	92.54	.31	125.31	.34	160.73	.36	197.63	.37	234.85	.37	271.18	.35	.96
.05	92.85	.31	125.65	.34	161.09	.37	198.00	.38	235.22	.37	271.53	.36	.95
.06	93.16	.31	125.99	.35	161.46	.36	198.38	.37	235.59	.37	271.89	.35	.94
.07	93.47	.32	126.34	.34	161.82	.36	198.75	.37	235.96	.37	272.24	.35	.93
.08	93.79	.31	126.68	.34	162.18	.37	199.12	.38	236.33	.37	272.59	.35	.92
.09	94.10	.31	127.02	.35	162.55	.36	199.50	.37	236.70	.37	272.94	.35	.91
.10	94.41	.31	127.37	.34	162.91	.36	199.87	.37	237.07	.37	273.29	.35	.90
.11	94.72	.32	127.71	.35	163.27	.37	200.24	.37	237.44	.37	273.64	.36	.89
.12	95.04	.31	128.06	.34	163.64	.36	200.61	.38	237.81	.37	274.00	.35	.88
.13	95.35	.31	128.40	.35	164.00	.36	200.99	.37	238.18	.37	274.35	.35	.87
.14	95.66	.32	128.75	.35	164.36	.37	201.36	.37	238.55	.37	274.70	.35	.86
.15	95.98	.31	129.10	.34	164.73	.36	201.73	.38	238.92	.37	275.05	.35	.85
.16	96.29	.32	129.44	.35	165.09	.37	202.11	.37	239.29	.36	275.40	.35	.84
.17	96.61	.31	129.79	.35	165.46	.36	202.48	.37	239.65	.37	275.75	.36	.83
.18	96.92	.32	130.14	.34	165.82	.37	202.85	.38	240.02	.37	276.11	.35	.82
.19	97.24	.32	130.48	.35	166.19	.37	203.23	.37	240.39	.37	276.46	.35	.81
.20	97.56	.32	130.83	.35	166.56	.36	203.60	.37	240.76	.37	276.81	.35	.80
.21	97.88	.31	131.18	.34	166.92	.37	203.97	.37	241.13	.36	277.16	.35	.79
.22	98.19	.32	131.52	.35	167.29	.37	204.34	.38	241.49	.37	277.51	.35	.78
.23	98.51	.32	131.87	.35	167.66	.36	204.72	.37	241.86	.37	277.86	.35	.77
.24	98.83	.32	132.22	.34	168.02	.37	205.09	.37	242.23	.37	278.21	.35	.76
.25	99.15	.31	132.56	.35	168.39	.37	205.46	.38	242.60	.37	278.56	.35	.75
.26	99.46	.32	132.91	.35	168.76	.36	205.84	.37	242.97	.36	278.91	.35	.74
.27	99.78	.32	133.26	.35	169.12	.37	206.21	.37	243.33	.37	279.26	.35	.73
.28	100.10	.32	133.61	.35	169.49	.37	206.58	.38	243.70	.37	279.61	.35	.72
.29	100.42	.32	133.96	.35	169.86	.36	206.96	.37	244.07	.36	279.96	.35	.71
.30	100.74	.32	134.31	.35	170.22	.37	207.33	.37	244.43	.37	280.31	.35	.70
.31	101.06	.32	134.66	.35	170.59	.37	207.70	.37	244.80	.37	280.66	.35	.69
.32	101.38	.32	135.01	.35	170.96	.36	208.07	.38	245.17	.36	281.01	.35	.68
.33	101.70	.32	135.36	.35	171.32	.37	208.45	.37	245.53	.37	281.36	.35	.67
.34	102.02	.32	135.71	.35	171.69	.37	208.82	.37	245.90	.36	281.71	.34	.66
.35	102.34	.33	136.06	.35	172.06	.37	209.19	.38	246.26	.37	282.05	.35	.65
.36	102.67	.32	136.41	.35	172.43	.36	209.57	.37	246.63	.36	282.40	.35	.64
.37	102.99	.32	136.76	.35	172.79	.37	209.94	.37	246.99	.37	282.75	.35	.63
.38	103.31	.33	137.11	.35	173.16	.37	210.31	.38	247.36	.36	283.10	.34	.62
.39	103.64	.32	137.46	.35	173.53	.37	210.69	.37	247.72	.37	283.44	.35	.61
.40	103.96	.32	137.81	.35	173.90	.37	211.06	.37	248.09	.36	283.79	.35	.60
.41	104.28	.33	138.16	.35	174.27	.37	211.43	.38	248.45	.37	284.14	.34	.59
.42	104.61	.32	138.51	.35	174.64	.36	211.81	.37	248.82	.36	284.48	.35	.58
.43	104.93	.33	138.86	.35	175.00	.37	212.18	.37	249.18	.37	284.83	.35	.57
.44	105.26	.32	139.21	.36	175.37	.37	212.55	.38	249.55	.36	285.18	.34	.56
.45	105.58	.33	139.57	.35	175.74	.37	212.93	.37	249.91	.37	285.52	.35	.55
.46	105.91	.33	139.92	.35	176.11	.37	213.30	.37	250.28	.36	285.87	.34	.54
.47	106.24	.32	140.27	.35	176.48	.37	213.67	.38	250.64	.37	286.21	.35	.53
.48	106.56	.33	140.62	.36	176.85	.37	214.05	.37	251.01	.36	286.56	.34	.52
.49	106.89	.33	140.98	.35	177.22	.37	214.42	.37	251.37	.37	286.90	.34	.51
.50	107.22	.32	141.33	.35	177.59	.37	214.79	.37	251.74	.36	287.24	.34	.50
Days.	29		28		27		26		25		24		Days.

TABLE XI. ARGUMENT 6.

Equation = 206″.9 [1 + sin. (2t — z — x)].

Period, 34.846892 days.

Days.	6		7		8		9		10		11		Days.
Decimals of a Day	Equation.	Diff.	Equation.	Diff.	Equation.	Diff.	Equation.	Diff.	Equation.	Diff.	Equation.	Diff.	Decimals of a Day
.50	107.22	.32	141.33	.35	177.59	.37	214.79	.37	251.74	.36	287.24	.34	.50
.51	107.54	.33	141.68	.36	177.96	.37	215.16	.38	252.10	.37	287.58	.35	.49
.52	107.87	.33	142.04	.35	178.33	.37	215.54	.37	252.47	.36	287.93	.34	.48
.53	108.20	.32	142.39	.35	178.70	.37	215.91	.37	252.83	.36	288.27	.34	.47
.54	108.52	.33	142.74	.36	179.07	.37	216.28	.38	253.19	.37	288.61	.34	.46
.55	108.85	.33	143.10	.35	179.44	.37	216.66	.37	253.56	.36	288.95	.34	.45
.56	109.18	.33	143.45	.36	179.81	.37	217.03	.37	253.92	.36	289.29	.35	.44
.57	109.51	.33	143.81	.35	180.18	.37	217.40	.38	254.28	.36	289.64	.34	.43
.58	109.84	.33	144.16	.36	180.55	.37	217.78	.37	254.64	.37	289.98	.34	.42
.59	110.17	.33	144.52	.36	180.92	.37	218.15	.37	255.01	.36	290.32	.34	.41
.60	110.50	.33	144.88	.35	181.29	.37	218.52	.37	255.37	.36	290.66	.34	.40
.61	110.83	.33	145.23	.36	181.66	.37	218.89	.38	255.73	.37	291.00	.34	.39
.62	111.16	.33	145.59	.36	182.03	.37	219.27	.37	256.10	.36	291.34	.34	.38
.63	111.49	.33	145.95	.35	182.40	.37	219.64	.37	256.46	.36	291.68	.34	.37
.64	111.82	.33	146.30	.36	182.77	.37	220.01	.37	256.82	.36	292.02	.34	.36
.65	112.15	.33	146.66	.36	183.14	.37	220.38	.37	257.18	.37	292.36	.34	.35
.66	112.48	.34	147.02	.35	183.51	.37	220.75	.38	257.55	.36	292.70	.34	.34
.67	112.82	.33	147.37	.36	183.88	.37	221.13	.37	257.91	.36	293.04	.34	.33
.68	113.15	.33	147.73	.36	184.25	.37	221.50	.37	258.27	.36	293.38	.34	.32
.69	113.48	.33	148.09	.36	184.62	.37	221.87	.37	258.63	.36	293.72	.34	.31
.70	113.81	.34	148.45	.36	184.99	.37	222.24	.37	258.99	.36	294.06	.34	.30
.71	114.15	.33	148.81	.35	185.36	.37	222.61	.38	259.35	.36	294.40	.34	.29
.72	114.48	.33	149.16	.36	185.73	.37	222.99	.37	259.71	.36	294.74	.33	.28
.73	114.81	.34	149.52	.36	186.10	.37	223.36	.37	260.07	.36	295.07	.34	.27
.74	115.15	.33	149.88	.36	186.47	.37	223.73	.37	260.43	.36	295.41	.34	.26
.75	115.48	.34	150.24	.36	186.84	.38	224.10	.37	260.79	.36	295.75	.34	.25
.76	115.82	.33	150.60	.36	187.22	.37	224.47	.38	261.15	.36	296.09	.33	.24
.77	116.15	.34	150.96	.36	187.59	.37	224.85	.37	261.51	.36	296.42	.34	.23
.78	116.49	.33	151.32	.36	187.96	.37	225.22	.37	261.87	.36	296.76	.34	.22
.79	116.82	.34	151.68	.36	188.33	.37	225.59	.37	262.23	.36	297.10	.33	.21
.80	117.16	.34	152.04	.36	188.70	.37	225.96	.37	262.59	.36	297.43	.34	.20
.81	117.50	.33	152.40	.36	189.07	.37	226.33	.37	262.95	.36	297.77	.34	.19
.82	117.83	.34	152.76	.36	189.44	.38	226.70	.37	263.31	.36	298.11	.33	.18
.83	118.17	.34	153.12	.36	189.82	.37	227.07	.38	263.67	.36	298.44	.34	.17
.84	118.51	.34	153.48	.36	190.19	.37	227.45	.37	264.03	.36	298.78	.33	.16
.85	118.85	.33	153.84	.36	190.56	.37	227.82	.37	264.39	.36	299.11	.34	.15
.86	119.18	.34	154.20	.37	190.93	.37	228.19	.37	264.75	.35	299.45	.33	.14
.87	119.52	.34	154.57	.36	191.30	.38	228.56	.37	265.10	.36	299.78	.33	.13
.88	119.86	.34	154.93	.36	191.68	.37	228.93	.37	265.46	.36	300.11	.33	.12
.89	120.20	.34	155.29	.36	192.05	.37	229.30	.37	265.82	.36	300.44	.33	.11
.90	120.54	.34	155.65	.36	192.42	.37	229.67	.37	266.18	.36	300.77	.33	.10
.91	120.88	.34	156.01	.37	192.79	.37	230.04	.37	266.54	.35	301.10	.33	.09
.92	121.22	.34	156.38	.36	193.16	.38	230.41	.37	266.89	.36	301.43	.34	.08
.93	121.56	.34	156.74	.36	193.54	.37	230.78	.37	267.25	.36	301.77	.33	.07
.94	121.90	.34	157.10	.36	193.91	.37	231.15	.37	267.61	.36	302.10	.33	.06
.95	122.24	.34	157.46	.37	194.28	.37	231.52	.37	267.97	.36	302.43	.33	.05
.96	122.58	.34	157.83	.36	194.65	.37	231.89	.37	268.33	.35	302.76	.33	.04
.97	122.92	.34	158.19	.36	195.02	.38	232.26	.37	268.68	.36	303.09	.33	.03
.98	123.26	.34	158.55	.37	195.40	.37	232.63	.37	269.04	.36	303.42	.33	.02
.99	123.60	.34	158.92	.36	195.77	.37	233.00	.37	269.40	.35	303.75	.33	.01
1.00	123.94	.34	159.28	.36	196.14	.37	233.37	.37	269.75	.36	304.08	.33	.00
Days.	29		28		27		26		25		24		Days.

TABLE XI. ARGUMENT 6.

Equation $= 206''.9\ [1 + \sin.(2t - z - x)]$.

Period, 34.846892 days.

Days.	12		13		14		15		16		17		Days.
Decimals of a Day	Equation.	Diff.	Equation.	Diff.	Equation.	Diff.	Equation.	Diff.	Equation.	Diff.	Equation.	Diff.	Decimals of a Day.
.00	304.08	.33	335.25	.29	362.27	.25	384.25	.19	400.19	.13	410.45	.07	1.00
.01	304.41	.33	335.54	.29	362.52	.25	384.44	.19	400.62	.13	410.52	.07	.99
.02	304.71	.33	335.83	.30	362.77	.24	384.63	.19	400.75	.13	410.59	.06	.98
.03	305.07	.33	336.13	.29	363.01	.25	384.82	.19	400.88	.13	410.65	.07	.97
.04	305.40	.32	336.42	.29	363.26	.24	385.01	.19	401.01	.13	410.72	.06	.96
.05	305.72	.33	336.71	.29	363.50	.25	385.20	.19	401.14	.13	410.78	.06	.95
.06	306.05	.33	337.00	.29	363.75	.24	385.39	.19	401.27	.13	410.84	.06	.94
.07	306.38	.33	337.29	.29	363.99	.24	385.58	.19	401.40	.13	410.90	.06	.93
.08	306.71	.32	337.58	.29	364.23	.24	385.77	.19	401.53	.12	410.96	.06	.92
.09	307.03	.33	337.87	.29	364.47	.24	385.96	.19	401.65	.13	411.02	.06	.91
.10	307.36	.33	338.16	.29	364.71	.24	386.15	.19	401.78	.13	411.08	.06	.90
.11	307.69	.32	338.45	.29	364.95	.24	386.34	.19	401.91	.12	411.14	.06	.89
.12	308.01	.33	338.74	.29	365.19	.24	386.53	.18	402.03	.12	411.20	.06	.88
.13	308.34	.33	339.03	.29	365.43	.24	386.71	.19	402.15	.13	411.26	.06	.87
.14	308.67	.32	339.32	.28	365.67	.24	386.90	.18	402.28	.12	411.32	.06	.86
.15	308.99	.32	339.60	.29	365.91	.24	387.08	.19	402.40	.12	411.38	.06	.85
.16	309.31	.33	339.89	.29	366.15	.23	387.27	.18	402.52	.12	411.44	.05	.84
.17	309.64	.32	340.18	.28	366.38	.24	387.45	.18	402.64	.12	411.49	.06	.83
.18	309.96	.32	340.46	.29	366.62	.24	387.63	.18	402.76	.12	411.55	.05	.82
.19	310.28	.32	340.75	.28	366.86	.24	387.81	.18	402.88	.12	411.60	.05	.81
.20	310.60	.32	341.03	.28	367.10	.24	387.99	.18	403.00	.12	411.65	.05	.80
.21	310.92	.33	341.31	.29	367.34	.23	388.17	.18	403.12	.12	411.70	.05	.79
.22	311.25	.32	341.60	.28	367.57	.24	388.35	.18	403.24	.12	411.75	.05	.78
.23	311.57	.32	341.88	.28	367.81	.24	388.53	.18	403.36	.12	411.80	.05	.77
.24	311.89	.32	342.16	.29	368.05	.23	388.71	.18	403.48	.12	411.85	.05	.76
.25	312.21	.32	342.45	.28	368.28	.24	388.89	.17	403.60	.11	411.90	.05	.75
.26	312.53	.32	342.73	.28	368.52	.23	389.06	.18	403.71	.12	411.95	.05	.74
.27	312.85	.33	343.01	.28	368.75	.23	389.24	.17	403.83	.11	412.00	.05	.73
.28	313.18	.32	343.29	.28	368.98	.23	389.41	.18	403.94	.11	412.05	.05	.72
.29	313.50	.32	343.57	.28	369.21	.23	389.59	.17	404.05	.11	412.10	.05	.71
.30	313.82	.32	343.85	.28	369.44	.23	389.76	.17	404.16	.11	412.15	.05	.70
.31	314.14	.32	344.13	.28	369.67	.23	389.93	.18	404.27	.11	412.20	.05	.69
.32	314.46	.32	344.41	.28	369.90	.23	390.11	.17	404.38	.11	412.25	.04	.68
.33	314.78	.32	344.69	.27	370.13	.23	390.28	.17	404.49	.11	412.29	.05	.67
.34	315.10	.31	344.96	.28	370.36	.23	390.45	.17	404.60	.11	412.34	.04	.66
.35	315.41	.32	345.24	.28	370.59	.22	390.62	.17	404.71	.11	412.38	.05	.65
.36	315.73	.32	345.52	.27	370.81	.23	390.79	.17	404.82	.11	412.43	.04	.64
.37	316.05	.32	345.79	.28	371.04	.23	390.96	.17	404.93	.11	412.47	.04	.63
.38	316.37	.31	346.07	.28	371.27	.22	391.13	.17	405.04	.11	412.51	.04	.62
.39	316.68	.32	346.35	.27	371.49	.23	391.30	.17	405.15	.11	412.55	.04	.61
.40	317.00	.32	346.62	.28	371.72	.23	391.47	.17	405.26	.11	412.59	.04	.60
.41	317.32	.32	346.90	.27	371.95	.22	391.64	.17	405.37	.10	412.63	.04	.59
.42	317.64	.31	347.17	.28	372.17	.22	391.81	.17	405.47	.11	412.67	.04	.58
.43	317.95	.32	347.45	.27	372.39	.23	391.98	.16	405.58	.10	412.71	.04	.57
.44	318.27	.31	347.72	.27	372.62	.22	392.14	.17	405.68	.11	412.75	.04	.56
.45	318.58	.32	347.99	.28	372.84	.22	392.31	.17	405.79	.10	412.79	.03	.55
.46	318.90	.31	348.27	.27	373.06	.23	392.48	.16	405.89	.10	412.82	.04	.54
.47	319.21	.31	348.54	.27	373.29	.22	392.64	.17	405.99	.10	412.86	.03	.53
.48	319.52	.31	348.81	.27	373.51	.22	392.81	.16	406.09	.10	412.89	.04	.52
.49	319.83	.31	349.08	.27	373.73	.22	392.97	.16	406.19	.10	412.93	.03	.51
.50	320.14	.31	349.35	.27	373.95	.22	393.13	.16	406.29	.10	412.96	.03	.50
Days.	23		22		21		20		19		18		Days.

TABLE XI. ARGUMENT 6t

Equation = $206''.9\ [1 + \sin.\ (2t - z - x)]$.

Period, 34.846892 days.

Days.	12		13		14		15		16		17		Days.
Decimals of a Day	Equation.	Diff.	Equation.	Diff.	Equation.	Diff.	Equation.	Diff.	Equation.	Diff.	Equation.	Diff.	Decimals of a Day
.50	320.14	.31	349.35	.27	373.95	.22	393.13	.16	406.29	.10	412.96	.03	.50
.51	320.45	.31	349.62	.27	374.17	.22	393.29	.16	406.39	.10	412.99	.03	.49
.52	320.76	.31	349.89	.27	374.39	.22	393.45	.16	406.49	.10	413.02	.03	.48
.53	321.07	.31	350.16	.27	374.61	.22	393.61	.16	406.59	.10	413.05	.03	.47
.54	321.38	.31	350.43	.27	374.83	.22	393.77	.16	406.69	.10	413.08	.03	.46
.55	321.69	.31	350.70	.27	375.05	.21	393.93	.16	406.79	.09	413.11	.03	.45
.56	322.00	.31	350.97	.26	375.26	.22	394.09	.16	406.88	.10	413.14	.03	.44
.57	322.31	.31	351.23	.27	375.48	.22	394.25	.16	406.98	.09	413.17	.03	.43
.58	322.62	.31	351.50	.27	375.70	.21	394.41	.16	407.07	.09	413.20	.03	.42
.59	322.93	.31	351.77	.27	375.91	.22	394.57	.16	407.16	.09	413.23	.03	.41
.60	323.24	.31	352.04	.27	376.13	.21	394.73	.16	407.25	.09	413.26	.03	.40
.61	323.55	.31	352.31	.26	376.34	.22	394.89	.15	407.34	.09	413.29	.03	.39
.62	323.86	.31	352.57	.27	376.56	.21	395.04	.16	407.43	.09	413.32	.02	.38
.63	324.17	.30	352.84	.27	376.77	.21	395.20	.15	407.52	.09	413.34	.03	.37
.64	324.47	.31	353.11	.26	376.98	.22	395.35	.16	407.61	.09	413.37	.02	.36
.65	324.78	.31	353.37	.27	377.20	.21	395.51	.15	407.70	.09	413.39	.02	.35
.66	325.09	.30	353.64	.26	377.41	.21	395.66	.15	407.79	.09	413.41	.02	.34
.67	325.39	.31	353.90	.26	377.62	.21	395.81	.15	407.88	.09	413.43	.02	.33
.68	325.70	.31	354.16	.26	377.83	.21	395.96	.15	407.97	.08	413.45	.02	.32
.69	326.01	.30	354.42	.26	378.04	.21	396.11	.15	408.05	.09	413.47	.02	.31
.70	326.31	.31	354.68	.26	378.25	.21	396.26	.15	408.14	.08	413.49	.02	.30
.71	326.62	.30	354.94	.26	378.46	.21	396.41	.15	408.22	.09	413.51	.02	.29
.72	326.92	.30	355.20	.26	378.67	.21	396.56	.15	408.31	.08	413.53	.02	.28
.73	327.22	.31	355.46	.26	378.88	.20	396.71	.15	408.39	.09	413.55	.02	.27
.74	327.53	.30	355.72	.25	379.08	.21	396.86	.15	408.48	.08	413.57	.02	.26
.75	327.83	.30	355.97	.26	379.29	.21	397.01	.14	408.56	.09	413.59	.01	.25
.76	328.13	.30	356.23	.26	379.50	.20	397.15	.15	408.65	.08	413.60	.02	.24
.77	328.43	.30	356.49	.26	379.70	.21	397.30	.14	408.73	.08	413.62	.01	.23
.78	328.73	.30	356.75	.25	379.91	.20	397.44	.15	408.81	.08	413.63	.01	.22
.79	329.03	.30	357.00	.26	380.11	.20	397.59	.14	408.89	.08	413.64	.01	.21
.80	329.33	.30	357.26	.26	380.31	.20	397.73	.14	408.97	.08	413.65	.01	.20
.81	329.63	.30	357.52	.25	380.51	.20	397.87	.15	409.05	.08	413.66	.01	.19
.82	329.93	.30	357.77	.25	380.71	.20	398.02	.14	409.13	.08	413.67	.01	.18
.83	330.23	.30	358.02	.26	380.91	.20	398.16	.14	409.21	.08	413.68	.01	.17
.84	330.53	.30	358.28	.25	381.11	.20	398.30	.14	409.29	.08	413.69	.01	.16
.85	330.83	.29	358.53	.25	381.31	.20	398.44	.14	409.37	.07	413.70	.01	.15
.86	331.12	.30	358.78	.25	381.51	.20	398.58	.14	409.44	.08	413.71	.01	.14
.87	331.42	.30	359.03	.26	381.71	.20	398.72	.14	409.52	.07	413.72	.01	.13
.88	331.72	.29	359.29	.25	381.91	.20	398.86	.14	409.59	.08	413.73	.01	.12
.89	332.01	.30	359.54	.25	382.11	.20	399.00	.14	409.67	.07	413.74	.01	.11
.90	332.31	.30	359.79	.25	382.31	.20	399.14	.14	409.74	.07	413.75	.01	.10
.91	332.61	.29	360.04	.25	382.51	.20	399.28	.14	409.81	.08	413.76	.01	.09
.92	332.90	.30	360.29	.25	382.71	.19	399.42	.14	409.89	.07	413.77	.01	.08
.93	333.20	.30	360.54	.25	382.90	.20	399.56	.13	409.96	.07	413.78	.01	.07
.94	333.50	.29	360.79	.24	383.10	.19	399.69	.14	410.03	.07	413.79	.00	.06
.95	333.79	.30	361.03	.25	383.29	.20	399.83	.13	410.10	.07	413.79	.01	.05
.96	334.09	.29	361.28	.25	383.49	.19	399.96	.14	410.17	.07	413.80	.00	.04
.97	334.38	.29	361.53	.25	383.68	.19	400.10	.13	410.24	.07	413.80	.00	.03
.98	334.67	.29	361.78	.24	383.87	.19	400.23	.13	410.31	.07	413.80	.00	.02
.99	334.96	.29	362.02	.25	384.06	.19	400.36	.13	410.38	.07	413.80	.00	.01
1.00	335.25	.29	362.27	.25	384.25	.19	400.49	.13	410.45	.07	413.80	.00	.00
Days.	23		22		21		20		19		18		Days.

TABLE XII. ARGUMENT 7.

Equation $= 192''.1\ [1 + \sin.\ (2t + x)]$.

Period, 9.613718 days.

Days.	0		1		2		3		4		Days.
Decimals of a Day	Equation.	Difference.	Equation.	Difference.	Equation.	Difference.	Equation.	Difference.	Equation.	Difference.	Decimals of a Day.
.00	1.53	.15	26.09	.64	119.09	1.16	242.17	1.21	344.61	.76	1.00
.01	1.38	.15	26.73	.65	120.25	1.16	243.38	1.21	345.37	.75	.99
.02	1.23	.14	27.38	.65	121.41	1.17	244.59	1.21	346.12	.75	.98
.03	1.09	.13	28.03	.66	122.58	1.17	245.80	1.20	346.87	.74	.97
.04	0.96	.12	28.69	.66	123.75	1.18	247.00	1.20	347.61	.73	.96
.05	0.84	.11	29.35	.67	124.93	1.18	248.20	1.20	348.34	.73	.95
.06	0.73	.11	30.02	.68	126.11	1.18	249.40	1.20	349.07	.72	.94
.07	0.62	.10	30.70	.68	127.29	1.18	250.60	1.20	349.79	.71	.93
.08	0.52	.09	31.38	.69	128.47	1.19	251.80	1.19	350.50	.71	.92
.09	0.43	.08	32.07	.70	129.66	1.19	252.99	1.19	351.21	.70	.91
.10	0.35	.07	32.77	.70	130.85	1.19	254.18	1.19	351.91	.69	.90
.11	0.28	.06	33.47	.71	132.04	1.19	255.37	1.18	352.60	.69	.89
.12	0.22	.06	34.18	.72	133.23	1.20	256.55	1.18	353.29	.68	.88
.13	0.16	.05	34.90	.73	134.43	1.20	257.73	1.18	353.97	.67	.87
.14	0.11	.04	35.63	.73	135.63	1.20	258.91	1.17	354.64	.67	.86
.15	0.07	.03	36.36	.74	136.83	1.20	260.08	1.17	355.31	.66	.85
.16	0.04	.02	37.10	.75	138.03	1.21	261.25	1.17	355.97	.65	.84
.17	0.02	.01	37.85	.75	139.24	1.21	262.42	1.17	356.62	.65	.83
.18	0.01	.01	38.60	.76	140.45	1.21	263.59	1.17	357.27	.64	.82
.19	0.00	.00	39.36	.76	141.66	1.21	264.76	1.16	357.91	.63	.81
.20	0.00	.01	40.12	.77	142.87	1.21	265.92	1.16	358.54	.62	.80
.21	0.01	.02	40.89	.78	144.08	1.22	267.08	1.15	359.16	.62	.79
.22	0.03	.03	41.67	.78	145.30	1.22	268.23	1.15	359.78	.61	.78
.23	0.06	.03	42.45	.79	146.52	1.22	269.38	1.15	360.39	.60	.77
.24	0.09	.04	43.24	.79	147.74	1.22	270.53	1.14	360.99	.60	.76
.25	0.13	.05	44.03	.80	148.96	1.22	271.67	1.14	361.59	.59	.75
.26	0.18	.06	44.83	.81	150.18	1.23	272.81	1.14	362.18	.58	.74
.27	0.24	.07	45.64	.81	151.41	1.23	273.95	1.13	362.76	.57	.73
.28	0.31	.08	46.45	.82	152.64	1.23	275.08	1.13	363.33	.56	.72
.29	0.39	.08	47.27	.83	153.87	1.23	276.21	1.13	363.89	.56	.71
.30	0.47	.09	48.10	.83	155.10	1.23	277.34	1.12	364.45	.55	.70
.31	0.56	.10	48.93	.84	156.33	1.23	278.46	1.12	365.00	.54	.69
.32	0.66	.11	49.77	.85	157.56	1.21	279.58	1.12	365.54	.54	.68
.33	0.77	.12	50.62	.85	158.80	1.24	280.70	1.11	366.08	.53	.67
.34	0.89	.12	51.47	.86	160.04	1.24	281.81	1.11	366.61	.52	.66
.35	1.01	.13	52.33	.87	161.28	1.24	282.92	1.10	367.13	.52	.65
.36	1.14	.14	53.20	.87	162.52	1.24	284.02	1.10	367.65	.51	.64
.37	1.28	.15	54.07	.88	163.76	1.24	285.12	1.10	368.16	.50	.63
.38	1.43	.16	54.95	.88	165.00	1.25	286.22	1.09	368.66	.49	.62
.39	1.59	.16	55.83	.89	166.25	1.24	287.31	1.09	369.15	.47	.61
.40	1.75	.17	56.72	.90	167.49	1.25	288.40	1.08	369.62	.48	.60
.41	1.92	.18	57.62	.90	168.74	1.24	289.48	1.08	370.10	.47	.59
.42	2.10	.19	58.52	.91	169.98	1.25	290.56	1.08	370.57	.46	.58
.43	2.29	.20	59.43	.91	171.23	1.25	291.64	1.07	371.03	.45	.57
.44	2.49	.21	60.34	.91	172.48	1.25	292.71	1.07	371.48	.44	.56
.45	2.70	.21	61.25	.92	173.73	1.25	293.78	1.06	371.92	.44	.55
.46	2.91	.22	62.17	.93	174.98	1.25	294.84	1.06	372.36	.43	.54
.47	3.13	.23	63.10	.93	176.23	1.25	295.90	1.05	372.79	.42	.53
.48	3.36	.24	64.03	.94	177.48	1.26	296.95	1.05	373.21	.41	.52
.49	3.60	.25	64.97	.94	178.74	1.25	298.00	1.05	373.62	.41	.51
.50	3.85	.25	65.91	.95	179.99	1.25	299.05	1.04	374.03	.40	.50
Days.	9		8		7		6		5		Days.

TABLE XII. ARGUMENT 7.

Equation = 192″.1 [1 + sin. (2t + x)].

Period, 9.613718 days.

Days.	0		1		2		3		4		Days.
Decimals of a Day.	Equation.	Difference	Equation.	Difference.	Equation.	Difference.	Equation.	Difference.	Equation	Difference	Decimals of a Day.
.50	3.85	.25	65.91	.95	179.99	1.25	299.05	1.04	374.03	.40	.50
.51	4.10	.26	66.86	.95	181.24	1.26	300.09	1.04	374.43	.39	.49
.52	4.36	.27	67.81	.96	182.50	1.25	301.13	1.03	374.82	.38	.48
.53	4.63	.28	68.77	.97	183.75	1.26	302.16	1.02	375.20	.38	.47
.54	4.91	.29	69.74	.97	185.01	1.25	303.18	1.02	375.58	.37	.46
.55	5.20	.29	70.71	.98	186.26	1.25	304.20	1.02	375.95	.36	.45
.56	5.49	.30	71.69	.98	187.51	1.26	305.22	1.01	376.31	.35	.44
.57	5.79	.31	72.67	.99	188.77	1.25	306.23	1.01	376.66	.35	.43
.58	6.10	.32	73.66	.99	190.02	1.26	307.24	1.00	377.01	.34	.42
.59	6.42	.33	74.65	.99	191.28	1.25	308.24	1.00	377.35	.33	.41
.60	6.75	.33	75.64	1.00	192.53	1.26	309.24	.99	377.68	.32	.40
.61	7.08	.34	76.64	1.01	193.79	1.25	310.23	.99	378.00	.31	.39
.62	7.42	.35	77.65	1.01	195.04	1.26	311.22	.98	378.31	.30	.38
.63	7.77	.36	78.66	1.02	196.30	1.25	312.20	.98	378.61	.30	.37
.64	8.13	.37	79.68	1.02	197.55	1.26	313.18	.97	378.91	.29	.36
.65	8.50	.37	80.70	1.03	198.81	1.25	314.15	.97	379.20	.28	.35
.66	8.87	.38	81.73	1.03	200.06	1.26	315.12	.96	379.48	.27	.34
.67	9.25	.39	82.76	1.03	201.32	1.25	316.08	.96	379.75	.26	.33
.68	9.64	.40	83.79	1.04	202.57	1.26	317.04	.95	380.01	.26	.32
.69	10.04	.41	84.83	1.04	203.83	1.25	317.99	.94	380.27	.25	.31
.70	10.45	.41	85.87	1.05	205.08	1.25	318.93	.94	380.52	.24	.30
.71	10.86	.42	86.92	1.05	206.33	1.25	319.87	.94	380.76	.23	.29
.72	11.28	.43	87.97	1.06	207.58	1.25	320.81	.93	380.99	.22	.28
.73	11.71	.44	89.03	1.06	208.83	1.25	321.74	.92	381.21	.22	.27
.74	12.15	.44	90.09	1.06	210.08	1.25	322.66	.92	381.43	.21	.26
.75	12.59	.45	91.15	1.07	211.33	1.25	323.58	.91	381.64	.20	.25
.76	13.04	.46	92.22	1.07	212.58	1.25	324.49	.91	381.84	.19	.24
.77	13.50	.47	93.29	1.08	213.83	1.25	325.40	.90	382.03	.18	.23
.78	13.97	.47	94.37	1.08	215.08	1.24	326.30	.90	382.21	.18	.22
.79	14.44	.48	95.45	1.09	216.32	1.25	327.20	.89	382.39	.17	.21
.80	14.92	.49	96.54	1.09	217.57	1.24	328.09	.89	382.56	.16	.20
.81	15.41	.50	97.63	1.10	218.81	1.25	328.98	.88	382.72	.15	.19
.82	15.91	.50	98.73	1.10	220.06	1.24	329.86	.87	382.87	.14	.18
.83	16.41	.51	99.83	1.10	221.30	1.24	330.73	.87	383.01	.14	.17
.84	16.92	.51	100.93	1.11	222.54	1.24	331.60	.86	383.15	.13	.16
.85	17.43	.52	102.04	1.11	223.78	1.24	332.46	.85	383.28	.12	.15
.86	17.95	.53	103.15	1.12	225.02	1.24	333.31	.85	383.40	.11	.14
.87	18.48	.54	104.27	1.12	226.26	1.23	334.16	.84	383.51	.10	.13
.88	19.02	.55	105.39	1.12	227.49	1.23	335.00	.84	383.61	.09	.12
.89	19.57	.56	106.51	1.13	228.72	1.23	335.84	.83	383.70	.09	.11
.90	20.13	.56	107.64	1.13	229.95	1.23	336.67	.82	383.79	.08	.10
.91	20.69	.57	108.77	1.13	231.18	1.23	337.49	.82	383.87	.07	.09
.92	21.26	.58	109.90	1.14	232.41	1.23	338.31	.81	383.94	.06	.08
.93	21.84	.59	111.04	1.14	233.64	1.22	339.12	.80	384.00	.05	.07
.94	22.43	.59	112.18	1.15	234.86	1.22	339.92	.80	384.05	.05	.06
.95	23.02	.60	113.33	1.15	236.08	1.22	340.72	.79	384.10	.04	.05
.96	23.62	.61	114.48	1.15	237.30	1.22	341.51	.78	384.14	.03	.04
.97	24.23	.61	115.63	1.15	238.52	1.22	342.29	.78	384.17	.02	.03
.98	24.84	.62	116.78	1.15	239.74	1.22	343.07	.77	384.19	.01	.02
.99	25.46	.63	117.93	1.16	240.96	1.21	343.84	.77	384.20	.00	.01
1.00	26.09	.64	119.09	1.16	242.17	1.21	344.61	.76	384.20	.00	.00
Days.	9		8		7		6		5		Days.

TABLE XIII. ARGUMENT 8.

Equation = 165″.9 [1 + sin. (2*t* — *z*)].

Period, 15.3873122 days.

Days.	0		1		2		3		4		5		Days.
Decimals of a Day.	Equation.	Diff.	Equation.	Diff.	Equation.	Diff.	Equation.	Diff.	Equation.	Diff.	Equation.	Diff.	Decimals of a Day.
.00	1″.29	.08	6″.62	.19	38″.12	.43	90″.64	.60	155″.53	.67	222″.13	.64	1.00
.01	1.21	.08	6.81	.19	38.55	.43	91.24	.60	156.20	.68	222.77	.63	.99
.02	1.13	.08	7.00	.19	38.98	.44	91.84	.61	156.88	.67	223.40	.64	.98
.03	1.05	.08	7.19	.20	39.42	.44	92.45	.60	157.55	.68	224.04	.63	.97
.04	0.97	.07	7.39	.20	39.86	.44	93.05	.61	158.23	.68	224.67	.63	.96
.05	0.90	.07	7.59	.20	40.30	.44	93.66	.61	158.91	.67	225.30	.63	.95
.06	0.83	.07	7.79	.21	40.74	.45	94.27	.61	159.58	.68	225.93	.63	.94
.07	0.76	.06	8.00	.21	41.19	.45	94.88	.61	160.26	.68	226.56	.63	.93
.08	0.70	.06	8.21	.21	41.64	.45	95.49	.62	160.94	.68	227.19	.63	.92
.09	0.64	.06	8.42	.21	42.09	.45	96.11	.62	161.62	.68	227.82	.63	.91
.10	0.58	.05	8.63	.22	42.54	.46	96.73	.62	162.30	.68	228.45	.63	.90
.11	0.53	.05	8.85	.22	43.00	.46	97.35	.62	162.98	.67	229.08	.62	.89
.12	0.48	.05	9.07	.22	43.46	.46	97.97	.62	163.65	.68	229.70	.63	.88
.13	0.43	.05	9.29	.22	43.92	.46	98.59	.62	164.33	.68	230.33	.62	.87
.14	0.38	.05	9.51	.23	44.38	.46	99.21	.62	165.01	.68	230.95	.62	.86
.15	0.33	.04	9.74	.23	44.84	.46	99.83	.62	165.69	.68	231.57	.62	.85
.16	0.29	.04	9.97	.23	45.30	.47	100.45	.62	166.37	.68	232.19	.62	.84
.17	0.25	.04	10.20	.24	45.77	.47	101.07	.62	167.05	.67	232.81	.62	.83
.18	0.21	.03	10.44	.24	46.24	.47	101.69	.63	167.72	.68	233.43	.62	.82
.19	0.18	.03	10.68	.24	46.71	.47	102.32	.63	168.40	.68	234.05	.62	.81
.20	0.15	.03	10.92	.24	47.18	.47	102.95	.63	169.08	.68	234.67	.62	.80
.21	0.12	.02	11.16	.25	47.65	.48	103.58	.63	169.76	.68	235.29	.62	.79
.22	0.10	.02	11.41	.25	48.13	.48	104.21	.63	170.44	.67	235.91	.61	.78
.23	0.08	.02	11.66	.25	48.61	.48	104.84	.63	171.11	.68	236.52	.62	.77
.24	0.06	.02	11.91	.25	49.09	.48	105.47	.63	171.79	.67	237.14	.61	.76
.25	0.04	.01	12.16	.26	49.57	.48	106.10	.63	172.46	.68	237.75	.61	.75
.26	0.03	.01	12.42	.26	50.05	.48	106.73	.63	173.14	.68	238.36	.61	.74
.27	0.02	.01	12.68	.26	50.53	.49	107.36	.64	173.82	.67	238.97	.61	.73
.28	0.01	.01	12.94	.26	51.02	.49	108.00	.63	174.49	.68	239.58	.60	.72
.29	0.00	.00	13.20	.26	51.51	.49	108.63	.64	175.17	.67	240.18	.61	.71
.30	0.00	.00	13.46	.27	52.00	.49	109.27	.63	175.84	.68	240.79	.60	.70
.31	0.00	.00	13.73	.27	52.49	.50	109.90	.64	176.52	.68	241.39	.60	.69
.32	0.00	.00	14.00	.27	52.99	.50	110.54	.64	177.20	.67	241.99	.60	.68
.33	0.00	.01	14.27	.28	53.49	.50	111.18	.64	177.87	.68	242.59	.60	.67
.34	0.01	.01	14.55	.28	53.99	.50	111.82	.64	178.55	.67	243.19	.60	.66
.35	0.02	.01	14.83	.28	54.49	.50	112.46	.64	179.22	.68	243.79	.60	.65
.36	0.03	.02	15.11	.29	54.99	.51	113.10	.64	179.90	.68	244.39	.59	.64
.37	0.05	.02	15.40	.29	55.50	.51	113.74	.64	180.58	.67	244.98	.60	.63
.38	0.07	.02	15.69	.29	56.01	.51	114.38	.65	181.25	.68	245.58	.59	.62
.39	0.09	.02	15.98	.29	56.52	.51	115.03	.64	181.93	.67	246.17	.59	.61
.40	0.11	.03	16.27	.30	57.03	.51	115.67	.65	182.60	.67	246.76	.59	.60
.41	0.14	.03	16.57	.30	57.54	.51	116.32	.64	183.27	.68	247.35	.59	.59
.42	0.17	.03	16.87	.30	58.05	.51	116.96	.65	183.95	.67	247.94	.59	.58
.43	0.20	.04	17.17	.30	58.56	.52	117.61	.65	184.62	.67	248.53	.59	.57
.44	0.24	.04	17.47	.30	59.08	.52	118.26	.65	185.29	.67	249.12	.59	.56
.45	0.28	.04	17.77	.31	59.60	.52	118.91	.65	185.96	.67	249.71	.58	.55
.46	0.32	.05	18.08	.31	60.12	.52	119.56	.65	186.63	.68	250.29	.59	.54
.47	0.37	.05	18.39	.31	60.64	.52	120.21	.66	187.31	.67	250.88	.58	.53
.48	0.42	.05	18.70	.31	61.16	.53	120.87	.65	187.98	.67	251.46	.58	.52
.49	0.47	.05	19.01	.32	61.69	.53	121.52	.66	188.65	.67	252.04	.58	.51
.50	0.52	.06	19.33	.32	62.22	.53	122.18	.65	189.32	.67	252.62	.57	.50
Days.	15		14		13		12		11		10		Days.

TABLE XIII. ARGUMENT 8.

Equation = 165″.9 [1 + sin. (2t — z)].

Period, 15.3873122 days.

Days.	0		1		2		3		4		5		Days.
Decimals of a Day.	Equation.	Diff.	Equation.	Diff.	Equation.	Diff.	Equation.	Diff.	Equation.	Diff.	Equation.	Diff.	Decimals of a Day.
.50	0.52″	.06	19.33″	.32	62.22″	.53	122.18″	.65	189.32″	.67	252.62″	.57	.50
.51	0.58	.06	19.65	.32	62.75	.53	122.83	.65	189.99	.67	253.19	.58	.49
.52	0.64	.06	19.97	.32	63.28	.53	123.48	.65	190.66	.67	253.77	.57	.48
.53	0.70	.06	20.29	.32	63.81	.54	124.13	.66	191.33	.67	254.34	.57	.47
.54	0.76	.06	20.61	.33	64.35	.54	124.79	.66	192.00	.67	254.91	.57	.46
.55	0.82	.07	20.94	.33	64.89	.54	125.45	.65	192.67	.67	255.48	.57	.45
.56	0.89	.07	21.27	.33	65.43	.54	126.10	.66	193.34	.66	256.05	.57	.44
.57	0.96	.07	21.60	.34	65.97	.54	126.76	.66	194.00	.67	256.62	.56	.43
.58	1.03	.08	21.94	.34	66.51	.54	127.42	.66	194.67	.66	257.18	.57	.42
.59	1.11	.08	22.28	.34	67.05	.55	128.08	.66	195.33	.67	257.75	.56	.41
.60	1.19	.08	22.62	.34	67.60	.55	128.74	.66	196.00	.67	258.31	.56	.40
.61	1.27	.09	22.96	.34	68.15	.55	129.40	.66	196.67	.66	258.87	.56	.39
.62	1.36	.09	23.30	.35	68.70	.55	130.06	.67	197.33	.67	259.43	.56	.38
.63	1.45	.09	23.65	.35	69.25	.55	130.73	.66	198.00	.66	259.99	.56	.37
.64	1.54	.09	24.00	.35	69.80	.55	131.39	.67	198.66	.67	260.55	.56	.36
.65	1.63	.10	24.35	.35	70.35	.55	132.06	.66	199.33	.66	261.11	.55	.35
.66	1.73	.10	24.70	.36	70.90	.56	132.72	.67	199.99	.67	261.66	.55	.34
.67	1.83	.10	25.06	.36	71.46	.56	133.39	.66	200.66	.66	262.21	.55	.33
.68	1.93	.10	25.42	.36	72.02	.56	134.05	.67	201.32	.66	262.76	.55	.32
.69	2.03	.11	25.78	.36	72.58	.56	134.72	.67	201.98	.66	263.31	.55	.31
.70	2.14	.11	26.14	.37	73.14	.56	135.39	.67	202.64	.66	263.86	.54	.30
.71	2.25	.11	26.51	.37	73.70	.56	136.06	.66	203.30	.66	264.40	.55	.29
.72	2.36	.12	26.88	.37	74.26	.57	136.72	.67	203.96	.66	264.95	.54	.28
.73	2.48	.12	27.25	.37	74.83	.57	137.39	.67	204.62	.65	265.49	.54	.27
.74	2.60	.12	27.62	.38	75.40	.57	138.06	.67	205.27	.66	266.03	.54	.26
.75	2.72	.13	28.00	.38	75.97	.57	138.73	.67	205.93	.66	266.57	.54	.25
.76	2.85	.13	28.38	.38	76.54	.57	139.40	.66	[illegible]	.66	[illegible]	[illegible]	.24
.77	2.98	.13	28.76	.38	77.11	.57	140.06	.67	207.25	.65	267.64	.54	.23
.78	3.11	.13	29.14	.38	77.68	.57	140.73	.67	207.90	.66	268.18	.53	.22
.79	3.24	.13	29.52	.39	78.25	.58	141.40	.67	208.56	.65	268.71	.53	.21
.80	3.37	.14	29.91	.39	78.83	.58	142.07	.67	209.21	.65	269.24	.53	.20
.81	3.51	.14	30.30	.39	79.41	.58	142.74	.67	209.86	.66	269.77	.53	.19
.82	3.65	.14	30.69	.39	79.99	.58	143.41	.67	210.52	.65	270.30	.52	.18
.83	3.79	.14	31.08	.40	80.57	.58	144.08	.67	211.17	.65	270.82	.53	.17
.84	3.93	.15	31.48	.40	81.15	.58	144.75	.67	211.82	.65	271.35	.52	.16
.85	4.08	.15	31.88	.40	81.73	.58	145.42	.67	212.47	.65	271.87	.52	.15
.86	4.23	.15	32.28	.40	82.31	.59	146.09	.67	213.12	.65	272.39	.52	.14
.87	4.38	.16	32.68	.41	82.90	.59	146.76	.67	213.77	.65	272.91	.51	.13
.88	4.54	.16	33.09	.41	83.49	.59	147.43	.68	214.42	.64	273.42	.52	.12
.89	4.70	.16	33.50	.41	84.08	.59	148.11	.67	215.06	.65	273.94	.51	.11
.90	4.86	.17	33.91	.41	84.67	.59	148.78	.67	215.71	.65	274.45	.51	.10
.91	5.03	.17	34.32	.41	85.26	.59	149.45	.67	216.36	.64	274.96	.51	.09
.92	5.20	.17	34.73	.42	85.85	.59	150.12	.68	217.00	.65	275.47	.51	.08
.93	5.37	.17	35.15	.42	86.44	.60	150.80	.67	217.65	.64	275.98	.50	.07
.94	5.54	.17	35.57	.42	87.04	.60	151.47	.68	218.29	.64	276.48	.51	.06
.95	5.71	.18	35.99	.42	87.64	.60	152.15	.67	218.93	.64	276.99	.50	.05
.96	5.89	.18	36.41	.42	88.24	.60	152.82	.68	219.57	.64	277.49	.50	.04
.97	6.07	.18	36.83	.43	88.84	.60	153.50	.67	220.21	.64	277.99	.50	.03
.98	6.25	.18	37.26	.43	89.44	.60	154.17	.68	220.85	.64	278.49	.50	.02
.99	6.43	.19	37.69	.43	90.04	.60	154.85	.68	221.49	.64	278.99	.50	.01
1.00	6.62	.19	38.12	.43	90.64	.60	155.53	.67	222.13	.64	279.49	.49	.00
Days.	15		14		13		12		11		10		Days.

TABLE XIII.

(Continued.)

Period, 15.3873122 days.

TABLE XIV. ARGUMENT 9.

Equation = 148″.1 [1 + sin. (x — z)].

Period, 29.802826 days.

Days.	6		7		0		1		2		Days.
Decimal of a Day	Equation.	Diff.	Equation.	Diff.	Equation.	Diff.	Equation.	Diff.	Equation.	Diff.	Decimals of a Day.
.00	279.49	.49	318.17	.27	0.01	.01	2.67	.06	11.71	.12	1.00
.01	279.98	.49	318.44	.26	0.02	.01	2.73	.06	11.86	.12	.99
.02	280.47	.49	318.70	.26	0.02	.00	2.79	.06	11.98	.13	.98
.03	280.96	.49	318.96	.26	0.02	.01	2.85	.06	12.11	.12	.97
.04	281.45	.48	319.22	.26	0.01	.00	2.91	.06	12.23	.12	.96
.05	281.93	.48	319.48	.25	0.01	.00	2.97	.07	12.35	.13	.95
.06	282.41	.48	319.73	.25	0.01	.01	3.04	.06	12.48	.13	.94
.07	282.89	.48	319.98	.25	0.00	.00	3.10	.06	12.61	.12	.93
.08	283.37	.48	320.23	.25	0.00	.00	3.16	.07	12.73	.13	.92
.09	283.85	.47	320.48	.24	0.00	.00	3.23	.06	12.86	.13	.91
.10	284.32	.47	320.72	.24	0.00	.00	3.29	.07	12.99	.13	.90
.11	284.79	.47	320.96	.24	0.00	.00	3.36	.06	13.12	.13	.89
.12	285.26	.47	321.20	.24	0.00	.00	3.42	.07	13.25	.13	.88
.13	285.73	.47	321.44	.24	0.00	.00	3.49	.07	13.38	.13	.87
.14	286.20	.47	321.68	.23	0.00	.01	3.56	.06	13.51	.13	.86
.15	286.67	.46	321.91	.23	0.01	.00	3.62	.07	13.64	.13	.85
.16	287.13	.46	322.14	.23	0.01	.00	3.69	.07	13.77	.14	.84
.17	287.59	.46	322.37	.22	0.01	.01	3.76	.07	13.91	.13	.83
.18	288.05	.46	322.59	.22	0.02	.00	3.83	.07	14.04	.13	.82
.19	288.51	.46	322.81	.22	0.02	.01	3.90	.07	14.17	.13	.81
.20	288.97	.45	323.03	.22	0.03	.01	3.97	.07	14.30	.13	.80
.21	289.42	.45	323.25	.21	0.04	.01	4.04	.07	14.43	.14	.79
.22	289.87	.45	323.46	.21	0.05	.01	4.11	.07	14.57	.13	.78
.23	290.32	.45	323.67	.21	0.06	.01	4.18	.08	14.70	.13	.77
.24	290.77	.45	323.88	.20	0.07	.01	4.26	.07	14.83	.14	.76
.25	291.22	.44	324.08	.20	0.08	.01	4.33	.08	14.97	.13	.75
.26	291.66	.44	324.28	.20	0.09	.01	4.41	.07	15.10	.14	.74
.27	292.10	.44	324.48	.20	0.10	.01	4.48	.08	15.24	.14	.73
.28	292.54	.44	324.68	.19	0.11	.01	4.56	.08	15.38	.14	.72
.29	292.98	.43	324.87	.19	0.12	.01	4.64	.08	15.52	.14	.71
.30	293.41	.43	325.06	.19	0.13	.01	4.72	.08	15.66	.14	.70
.31	293.84	.43	325.25	.19	0.14	.02	4.80	.08	15.80	.14	.69
.32	294.27	.43	325.44	.18	0.16	.01	4.88	.08	15.94	.14	.68
.33	294.70	.43	325.62	.18	0.17	.02	4.96	.08	16.08	.15	.67
.34	295.13	.42	325.80	.18	0.19	.02	5.04	.08	16.23	.14	.66
.35	295.55	.42	325.98	.18	0.21	.01	5.12	.08	16.37	.14	.65
.36	295.97	.42	326.16	.17	0.22	.02	5.20	.08	16.51	.15	.64
.37	296.39	.42	326.33	.17	0.24	.02	5.28	.09	16.66	.14	.63
.38	296.81	.41	326.50	.17	0.26	.01	5.37	.08	16.80	.14	.62
.39	297.22	.41	326.67	.16	0.27	.02	5.45	.08	16.94	.15	.61
.40	297.63	.41	326.83	.16	0.29	.02	5.53	.09	17.09	.15	.60
.41	298.04	.41	326.99	.16	0.31	.02	5.62	.08	17.24	.15	.59
.42	298.45	.41	327.15	.16	0.33	.02	5.70	.09	17.39	.14	.58
.43	298.86	.40	327.31	.16	0.35	.02	5.79	.09	17.53	.15	.57
.44	299.26	.40	327.47	.15	0.37	.03	5.88	.08	17.68	.15	.56
.45	299.66	.40	327.62	.15	0.40	.02	5.96	.09	17.83	.15	.55
.46	300.06	.40	327.77	.15	0.42	.02	6.05	.09	17.98	.15	.54
.47	300.46	.40	327.92	.14	0.44	.03	6.14	.09	18.13	.15	.53
.48	300.86	.39	328.06	.14	0.47	.02	6.23	.09	18.28	.15	.52
.49	301.25	.39	328.20	.14	0.49	.03	6.32	.09	18.43	.15	.51
.50	301.64	.39	328.34	.14	0.52	.03	6.41	.09	18.58	.15	.50
Days.	9		8		29		28		27		Days.

TABLE XIII.

(*Continued.*)

Period, 15.3873122 days.

TABLE XIV. ARGUMENT 9.

Equation = 148″.1 [1 + sin. $(x - z)$].

Period, 29.802826 days.

Days.	6		7		0		1		2		Days.
Decimal of a Day.	Equation.	Diff.	Equation	Diff.	Equation.	Diff.	Equation.	Diff.	Equation.	Diff.	Decimal of a Day.
.50	301.64	.39	328.34	.14	0.52	.03	6.41	.09	18.58	.15	.50
.51	302.03	.39	328.48	.13	0.55	.03	6.50	.09	18.73	.15	.49
.52	302.42	.38	328.61	.13	0.58	.03	6.59	.09	18.88	.15	.48
.53	302.80	.38	328.74	.13	0.61	.03	6.68	.10	19.03	.16	.47
.54	303.18	.38	328.87	.13	0.64	.03	6.78	.09	19.19	.15	.46
.55	303.56	.38	329.00	.12	0.67	.03	6.87	.09	19.34	.15	.45
.56	303.94	.37	329.12	.12	0.70	.03	6.96	.10	19.49	.16	.44
.57	304.31	.37	329.24	.12	0.73	.03	7.06	.09	19.65	.15	.43
.58	304.68	.37	329.36	.12	0.76	.03	7.15	.10	19.80	.16	.42
.59	305.05	.37	329.48	.11	0.79	.03	7.25	.10	19.96	.16	.41
.60	305.42	.37	329.59	.11	0.82	.03	7.35	.10	20.12	.16	.40
.61	305.79	.36	329.70	.11	0.85	.04	7.45	.10	20.28	.16	.39
.62	306.15	.36	329.81	.10	0.89	.03	7.55	.10	20.44	.16	.38
.63	306.51	.36	329.91	.10	0.92	.04	7.65	.10	20.60	.16	.37
.64	306.87	.36	330.01	.10	0.96	.04	7.75	.10	20.76	.16	.36
.65	307.23	.35	330.11	.10	1.00	.03	7.85	.10	20.92	.16	.35
.66	307.58	.35	330.21	.09	1.03	.04	7.95	.10	21.08	.16	.34
.67	307.93	.35	330.30	.09	1.07	.04	8.05	.10	21.24	.16	.33
.68	308.28	.35	330.39	.09	1.11	.04	8.15	.10	21.40	.16	.32
.69	308.63	.34	330.48	.08	1.15	.04	8.25	.10	21.56	.16	.31
.70	308.97	.34	330.56	.08	1.19	.04	8.35	.10	21.72	.16	.30
.71	309.31	.34	330.64	.08	1.23	.04	8.45	.11	21.88	.16	.29
.72	309.65	.34	330.72	.08	1.27	.04	8.56	.10	22.04	.17	.28
.73	309.99	.33	330.80	.07	1.31	.04	8.66	.11	22.21	.16	.27
.74	310.32	.33	330.87	.07	1.35	.05	8.77	.11	22.37	.16	.26
.75	310.65	.33	330.94	.07	1.40	.04	8.88	.10	22.53	.17	.25
.76	310.98	.33	331.01	.06	1.44	.04	8.98	.11	22.70	.16	.24
.77	311.31	.33	331.07	.06	1.48	.05	9.09	.11	22.86	.17	.23
.78	311.64	.32	331.13	.06	1.53	.04	9.20	.11	23.03	.17	.22
.79	311.96	.32	331.19	.06	1.57	.05	9.31	.11	23.20	.17	.21
.80	312.28	.32	331.25	.05	1.62	.05	9.42	.11	23.37	.17	.20
.81	312.60	.31	331.30	.05	1.67	.04	9.53	.11	23.54	.17	.19
.82	312.91	.31	331.35	.05	1.71	.05	9.64	.11	23.71	.17	.18
.83	313.22	.31	331.40	.05	1.76	.05	9.75	.11	23.88	.17	.17
.84	313.53	.31	331.45	.04	1.81	.05	9.86	.12	24.05	.17	.16
.85	313.84	.30	331.49	.04	1.86	.05	9.98	.11	24.22	.17	.15
.86	314.14	.30	331.53	.04	1.91	.05	10.09	.11	24.39	.18	.14
.87	314.44	.30	331.57	.04	1.96	.05	10.20	.12	24.57	.17	.13
.88	314.74	.30	331.61	.03	2.01	.05	10.32	.11	24.74	.17	.12
.89	315.04	.30	331.64	.03	2.06	.05	10.43	.12	24.91	.17	.11
.90	315.34	.29	331.67	.03	2.11	.05	10.55	.12	25.08	.18	.10
.91	315.63	.29	331.70	.02	2.16	.05	10.67	.11	25.26	.17	.09
.92	315.92	.29	331.72	.02	2.21	.06	10.78	.12	25.43	.18	.08
.93	316.21	.29	331.74	.02	2.27	.05	10.90	.12	25.61	.18	.07
.94	316.50	.29	331.76	.01	2.32	.06	11.02	.12	25.79	.17	.06
.95	316.79	.28	331.77	.01	2.38	.05	11.14	.12	25.96	.18	.05
.96	317.07	.28	331.78	.01	2.43	.06	11.26	.12	26.14	.18	.04
.97	317.35	.28	331.79	.01	2.49	.06	11.38	.12	26.32	.18	.03
.98	317.63	.27	331.80	.00	2.55	.06	11.50	.12	26.50	.17	.02
.99	317.90	.27	331.80	.00	2.61	.06	11.62	.12	26.67	.18	.01
1.00	318.17	.27	331.80	.00	2.67	.06	11.74	.12	26.85	.18	.00
Days.	9		8		29		28		27		Days.

TABLE XIV. ARGUMENT 9.

Equation $= 148''.1\ [1 + \sin.\,(x - z)]$.

Period, 29.802826 days.

Days.	3		4		5		6		7		8		Days.
Decimals of a Day.	Equation.	Diff.	Equation.	Diff.	Equation.	Diff.	Equation.	Diff.	Equation.	Diff.	Equation.	Diff.	Decimals of a Day.
.00	26.85	.18	47.33	.23	72.28	.27	100.58	.29	130.99	.31	162.15	.31	1.00
.01	27.03	.18	47.56	.23	72.55	.27	100.87	.30	131.30	.31	162.46	.31	.99
.02	27.21	.18	47.79	.23	72.82	.27	101.17	.30	131.61	.31	162.77	.31	.98
.03	27.39	.18	48.02	.23	73.09	.27	101.47	.29	131.92	.31	163.08	.31	.97
.04	27.57	.18	48.25	.23	73.36	.27	101.76	.30	132.23	.31	163.39	.31	.96
.05	27.75	.19	48.48	.23	73.63	.27	102.06	.30	132.54	.31	163.70	.31	.95
.06	27.94	.18	48.71	.24	73.90	.27	102.36	.29	132.85	.31	164.01	.31	.94
.07	28.12	.18	48.95	.23	74.17	.27	102.65	.30	133.16	.31	164.32	.31	.93
.08	28.30	.19	49.18	.23	74.44	.27	102.95	.30	133.47	.31	164.63	.31	.92
.09	28.49	.18	49.41	.23	74.71	.27	103.25	.30	133.78	.31	164.94	.31	.91
.10	28.67	.18	49.64	.23	74.98	.27	103.55	.30	134.09	.31	165.25	.31	.90
.11	28.85	.19	49.87	.23	75.25	.27	103.85	.29	134.40	.31	165.56	.31	.89
.12	29.04	.18	50.10	.23	75.52	.27	104.14	.30	134.71	.31	165.87	.31	.88
.13	29.22	.19	50.33	.24	75.79	.28	104.44	.30	135.02	.31	166.18	.31	.87
.14	29.41	.19	50.57	.23	76.07	.27	104.74	.30	135.33	.31	166.49	.31	.86
.15	29.60	.18	50.80	.24	76.34	.27	105.04	.30	135.64	.32	166.80	.31	.85
.16	29.78	.19	51.04	.24	76.61	.28	105.34	.30	135.96	.31	167.11	.31	.84
.17	29.97	.19	51.28	.23	76.89	.27	105.64	.30	136.27	.31	167.42	.31	.83
.18	30.16	.19	51.51	.24	77.16	.28	105.94	.30	136.58	.31	167.73	.31	.82
.19	30.35	.19	51.75	.24	77.44	.27	106.24	.30	136.89	.31	168.04	.31	.81
.20	30.54	.19	51.99	.24	77.71	.28	106.54	.30	137.20	.31	168.35	.31	.80
.21	30.73	.19	52.23	.24	77.99	.27	106.84	.30	137.51	.31	168.66	.31	.79
.22	30.92	.19	52.47	.24	78.26	.28	107.14	.30	137.82	.31	168.97	.31	.78
.23	31.11	.20	52.71	.24	78.54	.27	107.44	.30	138.13	.32	169.28	.31	.77
.24	31.31	.19	52.95	.24	78.81	.28	107.74	.30	138.45	.31	169.59	.31	.76
.25	31.50	.19	53.19	.24	79.09	.27	108.04	.30	138.76	.31	169.90	.31	.75
.26	31.69	.20	53.43	.24	79.36	.28	108.34	.30	139.07	.31	170.21	.30	.74
.27	31.89	.19	53.67	.24	79.64	.27	108.64	.31	139.38	.31	170.51	.31	.73
.28	32.08	.19	53.91	.24	79.91	.28	108.95	.30	139.69	.32	170.82	.31	.72
.29	32.27	.20	54.15	.24	80.19	.28	109.25	.30	140.01	.31	171.13	.31	.71
.30	32.47	.20	54.39	.24	80.47	.28	109.55	.30	140.32	.31	171.44	.31	.70
.31	32.67	.19	54.63	.24	80.75	.27	109.85	.30	140.63	.31	171.75	.31	.69
.32	32.86	.20	54.87	.25	81.02	.28	110.15	.30	140.94	.31	172.06	.30	.68
.33	33.06	.20	55.12	.24	81.30	.28	110.45	.31	141.25	.32	172.36	.31	.67
.34	33.26	.20	55.36	.25	81.58	.28	110.76	.30	141.57	.31	172.67	.31	.66
.35	33.46	.19	55.61	.24	81.86	.28	111.06	.30	141.88	.31	172.98	.31	.65
.36	33.65	.20	55.85	.25	82.14	.28	111.36	.30	142.19	.31	173.29	.31	.64
.37	33.85	.20	56.10	.24	82.42	.28	111.66	.30	142.50	.31	173.60	.30	.63
.38	34.05	.20	56.34	.25	82.70	.28	111.96	.31	142.81	.32	173.90	.31	.62
.39	34.25	.20	56.59	.25	82.98	.28	112.27	.30	143.13	.31	174.21	.31	.61
.40	34.45	.20	56.84	.25	83.26	.28	112.57	.30	143.44	.31	174.52	.31	.60
.41	34.65	.20	57.09	.24	83.54	.28	112.87	.31	143.75	.31	174.83	.31	.59
.42	34.85	.21	57.33	.25	83.82	.28	113.18	.30	144.06	.31	175.14	.30	.58
.43	35.06	.20	57.58	.25	84.10	.28	113.48	.30	144.37	.32	175.44	.31	.57
.44	35.26	.20	57.83	.25	84.38	.28	113.78	.31	144.69	.31	175.75	.31	.56
.45	35.46	.20	58.08	.24	84.66	.29	114.09	.30	145.00	.31	176.06	.30	.55
.46	35.66	.21	58.32	.25	84.95	.28	114.39	.31	145.31	.31	176.36	.31	.54
.47	35.87	.20	58.57	.25	85.23	.28	114.70	.30	145.62	.31	176.67	.31	.53
.48	36.07	.20	58.82	.25	85.51	.29	115.00	.31	145.93	.32	176.98	.30	.52
.49	36,27	.21	59.07	.25	85.80	.28	115.31	.30	146.25	.31	177.28	.31	.51
.50	36.48	.21	59.32	.25	86.08	.28	115.61	.31	146.56	.31	177.59	.31	.50
Days.	26		25		24		23		22		21		Days.

TABLE XIV. ARGUMENT 9.

Equation = 148″.1 [1 + sin. (x — z)].

Period, 29.802826 days.

Days.	3		4		5		6		7		8		Days.
Decimals of a Day	Equation.	Diff.	Equation.	Diff.	Equation.	Diff.	Equation.	Diff.	Equation.	Diff.	Equation.	Diff.	Decimals of a Day.
.50	36.48	.21	59.32	.25	86.08	.28	115.61	.31	146.56	.31	177.59	.31	.50
.51	36.69	.20	59.57	.25	86.36	.29	115.92	.30	146.87	.31	177.90	.30	.49
.52	36.89	.21	59.82	.25	86.65	.28	116.22	.31	147.18	.32	178.20	.31	.48
.53	37.10	.21	60.07	.25	86.93	.29	116.53	.30	147.50	.31	178.51	.30	.47
.54	37.31	.21	60.32	.25	87.22	.28	116.83	.31	147.81	.31	178.81	.31	.46
.55	37.52	.20	60.57	.26	87.50	.29	117.14	.30	148.12	.32	179.12	.30	.45
.56	37.72	.21	60.83	.25	87.79	.28	117.44	.31	148.44	.31	179.42	.31	.44
.57	37.93	.21	61.08	.25	88.07	.29	117.75	.30	148.75	.31	179.73	.30	.43
.58	38.14	.21	61.33	.26	88.36	.28	118.05	.31	149.06	.32	180.03	.31	.42
.59	38.35	.21	61.59	.25	88.64	.29	118.36	.30	149.38	.31	180.34	.30	.41
.60	38.56	.21	61.84	.25	88.93	.29	118.66	.31	149.69	.31	180.64	.31	.40
.61	38.77	.21	62.09	.26	89.22	.28	118.97	.30	150.00	.31	180.95	.30	.39
.62	38.98	.21	62.35	.25	89.50	.29	119.27	.31	150.31	.31	181.25	.31	.38
.63	39.19	.21	62.60	.25	89.79	.29	119.58	.30	150.62	.32	181.56	.30	.37
.64	39.40	.22	62.85	.26	90.08	.29	119.88	.31	150.94	.31	181.86	.31	.36
.65	39.62	.21	63.11	.25	90.37	.28	120.19	.30	151.25	.31	182.17	.30	.35
.66	39.83	.21	63.36	.26	90.65	.29	120.49	.31	151.56	.31	182.47	.30	.34
.67	40.04	.21	63.62	.26	90.94	.29	120.80	.30	151.87	.31	182.77	.31	.33
.68	40.25	.22	63.88	.25	91.23	.29	121.10	.31	152.18	.32	183.08	.30	.32
.69	40.47	.21	64.13	.26	91.52	.29	121.41	.31	152.50	.31	183.38	.30	.31
.70	40.68	.21	64.39	.26	91.81	.29	121.72	.30	152.81	.31	183.68	.30	.30
.71	40.89	.22	64.65	.26	92.10	.29	122.03	.30	153.12	.31	183.98	.30	.29
.72	41.11	.21	64.91	.25	92.39	.29	122.33	.31	153.43	.31	184.28	.31	.28
.73	41.32	.22	65.16	.26	92.68	.29	122.64	.31	153.74	.32	184.59	.30	.27
.74	41.54	.22	65.42	.26	92.97	.29	122.95	.31	154.06	.31	184.89	.30	.26
.75	41.76	.21	65.68	.26	93.26	.29	123.26	.31	154.37	.31	185.19	.30	.25
.76	41.97	.22	65.94	.26	93.55	.29	123.57	.30	154.68	.31	185.49	.30	.24
.77	42.19	.22	66.20	.26	93.84	.29	123.87	.31	154.99	.31	185.79	.31	.23
.78	42.41	.22	66.46	.26	94.13	.29	124.18	.31	155.30	.32	186.10	.30	.22
.79	42.63	.22	66.72	.26	94.42	.29	124.49	.31	155.62	.31	186.40	.30	.21
.80	42.85	.22	66.98	.26	94.71	.29	124.80	.31	155.93	.31	186.70	.30	.20
.81	43.07	.22	67.24	.26	95.00	.29	125.11	.31	156.24	.31	187.00	.30	.19
.82	43.29	.22	67.50	.26	95.29	.29	125.42	.31	156.55	.31	187.30	.30	.18
.83	43.51	.22	67.76	.27	95.58	.29	125.73	.30	156.86	.31	187.60	.30	.17
.84	43.73	.23	68.03	.26	95.87	.29	126.03	.31	157.17	.32	187.90	.30	.16
.85	43.96	.22	68.29	.26	96.16	.30	126.34	.31	157.49	.31	188.20	.30	.15
.86	44.18	.22	68.55	.27	96.46	.29	126.65	.31	157.80	.31	188.50	.30	.14
.87	44.40	.22	68.82	.26	96.75	.29	126.96	.31	158.11	.31	188.80	.30	.13
.88	44.62	.23	69.08	.26	97.04	.30	127.27	.31	158.42	.31	189.10	.30	.12
.89	44.85	.22	69.34	.27	97.34	.29	127.58	.31	158.73	.31	189.40	.30	.11
.90	45.07	.22	69.61	.27	97.63	.29	127.89	.31	159.04	.31	189.70	.30	.10
.91	45.29	.22	69.88	.26	97.92	.30	128.20	.31	159.35	.31	190.00	.30	.09
.92	45.51	.23	70.14	.27	98.22	.29	128.51	.31	159.66	.32	190.30	.30	.08
.93	45.74	.22	70.41	.27	98.51	.30	128.82	.31	159.98	.31	190.60	.30	.07
.94	45.96	.23	70.68	.26	98.81	.29	129.13	.31	160.29	.31	190.90	.30	.06
.95	46.19	.22	70.94	.27	99.10	.30	129.44	.31	160.60	.31	191.20	.29	.05
.96	46.41	.23	71.21	.27	99.40	.29	129.75	.31	160.91	.31	191.49	.30	.04
.97	46.64	.23	71.48	.26	99.69	.30	130.06	.31	161.22	.31	191.79	.30	.03
.98	46.87	.23	71.74	.27	99.99	.29	130.37	.31	161.53	.31	192.09	.30	.02
.99	47.10	.23	72.01	.27	100.28	.30	130.68	.31	161.84	.31	192.39	.30	.01
1.00	47.33	.23	72.28	.27	100.58	.29	130.99	.31	162.15	.31	192.69	.30	.00
Days.	26		25		24		23		22		21		Days.

TABLE XIV ARGUMENT 9.

Equation = 148″.1 [1 + sin. $(x - z)$]

Period, 29.802826 days.

Days.	9		10		11		12		13		14		Days.
Decimals of a Day	Equation.	Diff.	Equation.	Diff.	Equation.	Diff.	Equation.	Diff.	Equation.	Diff.	Equation.	Diff.	Decimals of a Day.
.00	192.69	.30	221.26	.27	246.59	.23	267.56	.18	283.23	.13	292.92	.06	1.00
.01	192.99	.29	221.53	.27	246.82	.24	267.74	.19	283.36	.13	292.98	.07	.99
.02	193.28	.30	221.80	.27	247.06	.23	267.93	.18	283.49	.13	293.05	.06	.98
.03	193.58	.30	222.07	.27	247.29	.23	268.11	.18	283.62	.12	293.11	.06	.97
.04	193.88	.29	222.34	.27	247.52	.23	268.29	.18	283.74	.13	293.17	.07	.96
.05	194.17	.30	222.61	.27	247.75	.23	268.47	.19	283.87	.12	293.24	.06	.95
.06	194.47	.30	222.88	.27	247.98	.23	268.66	.18	283.99	.13	293.30	.06	.94
.07	194.77	.29	223.15	.27	248.21	.23	268.84	.18	284.12	.12	293.36	.06	.93
.08	195.06	.30	223.42	.27	248.44	.23	269.02	.18	284.24	.12	293.42	.06	.92
.09	195.36	.30	223.69	.27	248.67	.23	269.20	.18	284.36	.12	293.48	.06	.91
.10	195.66	.30	223.96	.27	248.90	.23	269.38	.18	284.48	.12	293.54	.06	.90
.11	195.96	.29	224.23	.27	249.13	.23	269.56	.18	284.60	.12	293.60	.06	.89
.12	196.25	.30	224.50	.26	249.36	.22	269.74	.18	284.72	.12	293.66	.06	.88
.13	196.55	.29	224.76	.27	249.58	.23	269.92	.17	284.84	.12	293.72	.05	.87
.14	196.84	.30	225.03	.27	249.81	.23	270.09	.18	284.96	.12	293.77	.06	.86
.15	197.14	.29	225.30	.26	250.04	.22	270.27	.18	285.08	.12	293.83	.06	.85
.16	197.43	.30	225.56	.27	250.26	.23	270.45	.17	285.20	.12	293.89	.05	.84
.17	197.73	.29	225.83	.27	250.49	.22	270.62	.18	285.32	.11	293.94	.06	.83
.18	198.02	.30	226.10	.26	250.71	.23	270.80	.17	285.43	.12	294.00	.05	.82
.19	198.32	.29	226.36	.27	250.94	.22	270.97	.17	285.55	.12	294.05	.05	.81
.20	198.61	.29	226.63	.26	251.16	.22	271.14	.17	285.67	.11	294.10	.05	.80
.21	198.90	.30	226.89	.27	251.38	.23	271.31	.17	285.78	.12	294.15	.05	.79
.22	199.20	.29	227.16	.26	251.61	.22	271.48	.18	285.90	.11	294.20	.05	.78
.23	199.49	.29	227.42	.26	251.83	.22	271.66	.17	286.01	.12	294.25	.05	.77
.24	199.78	.30	227.68	.27	252.05	.22	271.83	.17	286.13	.11	294.30	.05	.76
.25	200.08	.29	227.95	.26	252.27	.23	272.00	.17	286.24	.11	294.35	.05	.75
.26	200.37	.29	228.21	.26	252.50	.22	272.17	.17	286.35	.12	294.40	.05	.74
.27	200.66	.29	228.47	.27	252.72	.22	272.34	.17	286.47	.11	294.45	.05	.73
.28	200.95	.29	228.74	.26	252.94	.22	272.51	.17	286.58	.11	294.50	.04	.72
.29	201.24	.29	229.00	.26	253.16	.22	272.68	.17	286.69	.11	294.54	.05	.71
.30	201.53	.29	229.26	.26	253.38	.22	272.85	.17	286.80	.11	294.59	.05	.70
.31	201.82	.29	229.52	.26	253.60	.22	273.02	.17	286.91	.11	294.64	.04	.69
.32	202.11	.29	229.78	.26	253.82	.22	273.19	.17	287.02	.11	294.68	.05	.68
.33	202.40	.29	230.04	.26	254.04	.21	273.36	.16	287.13	.10	294.73	.04	.67
.34	202.69	.29	230.30	.26	254.25	.22	273.52	.17	287.23	.11	294.77	.04	.66
.35	202.98	.29	230.56	.26	254.47	.22	273.69	.16	287.34	.11	294.81	.05	.65
.36	203.27	.29	230.82	.26	254.69	.21	273.85	.17	287.45	.10	294.86	.04	.64
.37	203.56	.29	231.08	.25	254.90	.22	274.02	.16	287.55	.11	294.90	.04	.63
.38	203.85	.29	231.33	.26	255.12	.21	274.18	.16	287.66	.10	294.94	.04	.62
.39	204.14	.29	231.59	.26	255.33	.22	274.34	.16	287.76	.10	294.98	.04	.61
.40	204.43	.29	231.85	.26	255.55	.21	274.50	.16	287.86	.10	295.02	.04	.60
.41	204.72	.29	232.11	.25	255.76	.21	274.66	.16	287.96	.10	295.06	.04	.59
.42	205.01	.29	232.36	.26	255.97	.22	274.82	.16	288.06	.10	295.10	.04	.58
.43	205.30	.28	232.62	.25	256.19	.21	274.98	.16	288.16	.10	295.14	.03	.57
.44	205.58	.29	232.87	.26	256.40	.21	275.14	.16	288.26	.10	295.17	.04	.56
.45	205.87	.29	233.13	.25	256.61	.21	275.30	.16	288.36	.10	295.21	.04	.55
.46	206.16	.29	233.38	.26	256.82	.22	275.46	.16	288.46	.10	295.25	.03	.54
.47	206.45	.28	233.64	.25	257.04	.21	275.62	.16	288.56	.10	295.28	.04	.53
.48	206.73	.29	233.89	.26	257.25	.21	275.78	.16	288.66	.10	295.32	.03	.52
.49	207.02	.29	234.15	.25	257.46	.21	275.94	.16	288.76	.10	295.35	.03	.51
.50	207.31	.29	234.40	.25	257.67	.21	276.10	.16	288.86	.10	295.38	.03	.50
Days.	20		19		18		17		16		15		Days.

TABLE XIV. ARGUMENT 9.

Equation $= 148''.1\ [1 + \sin.\ (z - z)]$.

Period, 29.802826 days.

Days.	9		10		11		12		13		14		Days.
Decimals of a Day.	Equation.	Diff.	Equation.	Diff.	Equation.	Diff.	Equation.	Diff.	Equation.	Diff.	Equation.	Diff.	Decimals of a Day.
.50	207.31	.29	234.40	.25	257.67	.21	276.10	.16	288.86	.10	295.38	.03	.50
.51	207.60	.28	234.65	.26	257.88	.21	276.26	.16	288.96	.10	295.41	.03	.49
.52	207.88	.29	234.91	.25	258.09	.21	276.42	.15	289.06	.09	295.44	.03	.48
.53	208.17	.28	235.16	.25	258.30	.21	276.57	.16	289.15	.09	295.47	.03	.47
.54	208.45	.29	235.41	.25	258.51	.20	276.73	.15	289.24	.10	295.50	.03	.46
.55	208.74	.28	235.66	.26	258.71	.21	276.88	.15	289.34	.09	295.53	.03	.45
.56	209.02	.29	235.92	.25	258.92	.21	277.04	.15	289.43	.10	295.56	.03	.44
.57	209.31	.28	236.17	.25	259.13	.21	277.19	.15	289.53	.09	295.59	.03	.43
.58	209.59	.29	236.42	.25	259.34	.20	277.34	.15	289.62	.09	295.62	.03	.42
.59	209.88	.28	236.67	.25	259.54	.21	277.49	.15	289.71	.09	295.65	.03	.41
.60	210.16	.28	236.92	.25	259.75	.21	277.64	.15	289.80	.09	295.68	.03	.40
.61	210.44	.29	237.17	.25	259.96	.20	277.79	.15	289.89	.09	295.71	.02	.39
.62	210.73	.28	237.42	.25	260.16	.21	277.94	.15	289.98	.09	295.73	.03	.38
.63	211.01	.28	237.67	.25	260.37	.20	278.09	.15	290.07	.09	295.76	.02	.37
.64	211.29	.28	237.92	.24	260.57	.21	278.24	.15	290.16	.09	295.78	.02	.36
.65	211.57	.29	238.16	.25	260.78	.20	278.39	.15	290.25	.08	295.80	.03	.35
.66	211.86	.28	238.41	.25	260.98	.20	278.54	.15	290.33	.09	295.83	.02	.34
.67	212.14	.28	238.66	.25	261.18	.20	278.69	.14	290.42	.09	295.85	.02	.33
.68	212.42	.28	238.91	.24	261.38	.20	278.83	.15	290.51	.08	295.87	.02	.32
.69	212.70	.28	239.15	.25	261.58	.20	278.98	.15	290.59	.09	295.89	.02	.31
.70	212.98	.28	239.40	.25	261.78	.20	279.13	.15	290.68	.08	295.91	.02	.30
.71	213.26	.28	239.65	.24	261.98	.20	279.28	.14	290.76	.08	295.93	.02	.29
.72	213.54	.28	239.89	.25	262.18	.20	279.42	.15	290.84	.09	295.95	.02	.28
.73	213.82	.28	240.14	.25	262.38	.20	279.57	.15	290.93	.08	295.97	.02	.27
.74	214.10	.28	240.39	.24	262.58	.20	279.72	.14	291.01	.08	295.99	.01	.26
.75	214.38	.28	240.63	.25	262.78	.19	279.86	.14	291.09	.08	296.00	.02	.25
.76	214.66	.27	240.88	.24	262.97	.20	280.00	.14	291.17	.08	296.02	.01	.24
.77	214.93	.28	241.12	.24	263.17	.20	280.14	.14	291.25	.08	296.03	.02	.23
.78	215.21	.28	241.36	.24	263.37	.19	280.28	.14	291.33	.08	296.05	.01	.22
.79	215.49	.28	241.60	.24	263.56	.20	280.42	.14	291.41	.08	296.06	.01	.21
.80	215.77	.28	241.84	.24	263.76	.20	280.56	.14	291.49	.08	296.07	.01	.20
.81	216.05	.27	242.08	.24	263.96	.19	280.70	.14	291.57	.08	296.08	.01	.19
.82	216.32	.28	242.32	.24	264.15	.20	280.84	.14	291.65	.07	296.09	.01	.18
.83	216.60	.28	242.56	.24	264.35	.19	280.98	.14	291.72	.08	296.10	.01	.17
.84	216.88	.27	242.80	.24	264.54	.19	281.12	.13	291.80	.08	296.11	.01	.16
.85	217.15	.28	243.04	.24	264.73	.20	281.25	.14	291.88	.07	296.12	.01	.15
.86	217.43	.27	243.28	.24	264.93	.19	281.39	.13	291.95	.08	296.13	.01	.14
.87	217.70	.28	243.52	.24	265.12	.19	281.52	.14	292.03	.07	296.14	.00	.13
.88	217.98	.27	243.76	.24	265.31	.19	281.66	.13	292.10	.07	296.14	.01	.12
.89	218.25	.28	244.00	.24	265.50	.19	281.79	.13	292.17	.07	296.15	.01	.11
.90	218.53	.27	244.24	.24	265.69	.19	281.92	.13	292.24	.07	296.16	.00	.10
.91	218.80	.28	244.48	.24	265.88	.19	282.05	.13	292.31	.07	296.16	.01	.09
.92	219.08	.27	244.72	.23	266.07	.19	282.18	.14	292.38	.07	296.17	.01	.08
.93	219.35	.27	244.95	.24	266.26	.19	282.32	.13	292.45	.07	296.18	.00	.07
.94	219.62	.28	245.19	.23	266.45	.19	282.45	.13	292.52	.07	296.18	.01	.06
.95	219.90	.27	245.42	.24	266.64	.19	282.58	.13	292.59	.07	296.19	.00	.05
.96	220.17	.27	245.66	.23	266.83	.18	282.71	.13	292.66	.06	296.19	.01	.04
.97	220.44	.28	245.89	.24	267.01	.18	282.84	.13	292.72	.07	296.20	.00	.03
.98	220.72	.27	246.13	.23	267.19	.19	282.97	.13	292.79	.07	296.20	.00	.02
.99	220.99	.27	246.36	.23	267.38	.18	283.10	.13	292.86	.06	296.20	.00	.01
1.00	221.26	.27	246.59	.23	267.56	.18	283.23	.13	292.92	.06	296.20	.00	.00
Days.	20		19		18		17		16		15		Days.

TABLE XV. ARGUMENT 10.

Equation = 110″.0 [1 — sin. $(\alpha + z)$].

Period, 25.621696 days.

Days. Decimals of a Day	0 Equation.	0 Diff.	1 Equation.	1 Diff.	2 Equation.	2 Diff.	3 Equation.	3 Diff.	4 Equation.	4 Diff.	5 Equation.	5 Diff.	Days. Decimals of a Day.
.00	219.88	.01	217.84	.05	209.32	.12	194.88	.17	175.36	.22	151.92	.25	1.00
.01	219.89	.01	217.79	.06	209.20	.11	194.71	.17	175.14	.22	151.67	.25	.99
.02	219.90	.01	217.73	.05	209.09	.12	194.54	.17	174.92	.22	151.42	.25	.98
.03	219.91	.01	217.68	.06	208.97	.12	194.37	.17	174.70	.21	151.17	.25	.97
.04	219.92	.01	217.62	.06	208.85	.12	194.20	.18	174.49	.22	150.92	.25	.96
.05	219.93	.01	217.56	.05	208.73	.12	194.02	.17	174.27	.22	150.67	.25	.95
.06	219.94	.01	217.51	.06	208.61	.12	193.85	.18	174.05	.22	150.42	.26	.94
.07	219.95	.01	217.45	.06	208.49	.12	193.67	.18	173.83	.22	150.16	.25	.93
.08	219.96	.01	217.39	.06	208.37	.12	193.49	.18	173.61	.22	149.91	.25	.92
.09	219.97	.01	217.33	.06	208.25	.12	193.31	.18	173.39	.22	149.66	.25	.91
.10	219.98	.00	217.27	.06	208.13	.12	193.13	.17	173.17	.22	149.41	.25	.90
.11	219.98	.01	217.21	.06	208.01	.12	192.96	.18	172.95	.22	149.16	.26	.89
.12	219.99	.00	217.15	.06	207.89	.12	192.78	.18	172.73	.23	148.90	.25	.88
.13	219.99	.00	217.09	.06	207.77	.12	192.60	.18	172.50	.22	148.65	.25	.87
.14	219.99	.01	217.03	.07	207.65	.13	192.42	.18	172.28	.22	148.40	.26	.86
.15	220.00	.00	216.96	.06	207.52	.12	192.24	.18	172.06	.23	148.14	.25	.85
.16	220.00	.00	216.90	.06	207.40	.13	192.06	.18	171.83	.22	147.89	.25	.84
.17	220.00	.00	216.84	.07	207.27	.12	191.88	.18	171.61	.22	147.64	.26	.83
.18	220.00	.00	216.77	.06	207.15	.13	191.70	.18	171.39	.23	147.38	.25	.82
.19	220.00	.00	216.71	.07	207.02	.13	191.52	.18	171.16	.22	147.13	.25	.81
.20	220.00	.00	216.64	.07	206.89	.13	191.34	.18	170.94	.22	146.88	.26	.80
.21	220.00	.00	216.57	.07	206.76	.13	191.16	.18	170.72	.23	146.62	.25	.79
.22	220.00	.00	216.50	.07	206.63	.13	190.98	.18	170.49	.22	146.37	.25	.78
.23	220.00	.00	216.43	.07	206.50	.13	190.80	.18	170.27	.23	146.12	.26	.77
.24	220.00	.01	216.36	.07	206.37	.13	190.62	.19	170.04	.22	145.86	.25	.76
.25	219.99	.00	216.29	.07	206.24	.13	190.43	.18	169.82	.23	145.61	.25	.75
.26	219.99	.01	216.22	.07	206.11	.13	190.25	.18	169.59	.23	145.36	.26	.74
.27	219.98	.00	216.15	.07	205.98	.13	190.07	.19	169.36	.22	145.10	.25	.73
.28	219.98	.01	216.08	.07	205.85	.13	189.88	.18	169.14	.23	144.85	.26	.72
.29	219.97	.01	216.01	.07	205.72	.13	189.70	.19	168.91	.23	144.59	.25	.71
.30	219.96	.01	215.94	.07	205.59	.13	189.51	.18	168.68	.23	144.34	.26	.70
.31	219.95	.01	215.87	.07	205.46	.13	189.33	.19	168.45	.23	144.08	.25	.69
.32	219.94	.01	215.80	.08	205.33	.14	189.14	.19	168.22	.23	143.83	.26	.68
.33	219.93	.01	215.72	.07	205.19	.13	188.95	.18	167.99	.23	143.57	.26	.67
.34	219.92	.01	215.65	.08	205.06	.14	188.77	.19	167.76	.23	143.31	.25	.66
.35	219.91	.01	215.57	.07	204.92	.13	188.58	.19	167.53	.23	143.06	.26	.65
.36	219.90	.01	215.50	.08	204.79	.14	188.39	.19	167.30	.23	142.80	.26	.64
.37	219.89	.01	215.42	.08	204.65	.14	188.20	.19	167.07	.23	142.54	.25	.63
.38	219.88	.02	215.34	.08	204.51	.14	188.01	.19	166.84	.23	142.29	.26	.62
.39	219.86	.01	215.26	.08	204.37	.14	187.82	.19	166.61	.23	142.03	.26	.61
.40	219.85	.02	215.18	.08	204.23	.14	187.63	.19	166.38	.23	141.77	.26	.60
.41	219.83	.01	215.10	.08	204.09	.14	187.44	.19	166.15	.23	141.51	.26	.59
.42	219.82	.02	215.02	.08	203.95	.14	187.25	.19	165.92	.24	141.25	.25	.58
.43	219.80	.01	214.94	.08	203.81	.14	187.06	.20	165.68	.23	141.00	.26	.57
.44	219.79	.02	214.86	.08	203.67	.14	186.86	.19	165.45	.23	140.74	.26	.56
.45	219.77	.01	214.78	.08	203.53	.14	186.67	.19	165.22	.24	140.48	.26	.55
.46	219.76	.02	214.70	.09	203.39	.15	186.48	.20	164.98	.23	140.22	.26	.54
.47	219.74	.02	214.61	.08	203.24	.14	186.28	.19	164.75	.23	139.96	.26	.53
.48	219.72	.02	214.53	.09	203.10	.15	186.09	.20	164.52	.24	139.70	.26	.52
.49	219.70	.02	214.44	.08	202.95	.14	185.89	.19	164.28	.23	139.44	.26	.51
.50	219.68	.02	214.36	.09	202.81	.15	185.70	.20	164.05	.24	139.18	.26	.50
Days.	25		24		23		22		21		20		Days.

TABLE XV. ARGUMENT 10.

Equation = 110″.0 [1 — sin. $(x+z)$].

Period, 25.621696 days.

Days.	0		1		2		3		4		5		Days.
Decimals of a Day	Equation.	Diff.	Equation.	Diff.	Equation.	Diff.	Equation.	Diff.	Equation.	Diff.	Equation.	Diff.	Decimals of a Day.
.50	219.68″	.02	214.36″	.09	202.81″	.15	185.70″	.20	164.05″	.24	139.18″	.26	.50
.51	219.66	.02	214.27	.08	202.66	.14	185.50	.19	163.81	.23	138.92	.26	.49
.52	219.64	.02	214.19	.09	202.52	.15	185.31	.20	163.58	.24	138.66	.26	.48
.53	219.62	.02	214.10	.09	202.37	.14	185.11	.20	163.34	.24	138.40	.26	.47
.54	219.60	.03	214.01	.08	202.23	.15	184.91	.19	163.10	.23	138.14	.26	.46
.55	219.57	.02	213.93	.09	202.08	.15	184.72	.20	162.87	.24	137.88	.26	.45
.56	219.55	.03	213.84	.09	201.93	.15	184.52	.20	162.63	.24	137.62	.26	.44
.57	219.52	.02	213.75	.09	201.78	.15	184.32	.20	162.39	.23	137.36	.27	.43
.58	219.50	.03	213.66	.09	201.63	.15	184.12	.20	162.16	.24	137.09	.26	.42
.59	219.47	.03	213.57	.09	201.48	.15	183.92	.20	161.92	.24	136.83	.26	.41
.60	219.44	.03	213.48	.09	201.33	.15	183.72	.20	161.68	.24	136.57	.26	.40
.61	219.41	.03	213.39	.09	201.18	.15	183.52	.20	161.44	.24	136.31	.27	.39
.62	219.38	.03	213.30	.09	201.03	.15	183.32	.20	161.20	.24	136.04	.26	.38
.63	219.35	.03	213.21	.09	200.88	.15	183.12	.21	160.96	.24	135.78	.26	.37
.64	219.32	.03	213.12	.10	200.73	.15	182.91	.20	160.72	.24	135.52	.27	.36
.65	219.29	.03	213.02	.09	200.58	.16	182.71	.20	160.48	.24	135.25	.26	.35
.66	219.26	.03	212.93	.10	200.42	.15	182.51	.21	160.24	.24	134.99	.26	.34
.67	219.23	.03	212.83	.09	200.27	.16	182.30	.20	160.00	.24	134.73	.27	.33
.68	219.20	.03	212.74	.10	200.11	.15	182.10	.21	159.76	.24	134.46	.26	.32
.69	219.17	.03	212.64	.10	199.96	.16	181.89	.20	159.52	.24	134.20	.26	.31
.70	219.14	.04	212.54	.10	199.80	.16	181.69	.21	159.28	.24	133.94	.27	.30
.71	219.10	.03	212.44	.10	199.64	.15	181.48	.20	159.04	.24	133.67	.26	.29
.72	219.07	.04	212.34	.10	199.49	.16	181.28	.21	158.80	.24	133.41	.27	.28
.73	219.03	.03	212.24	.10	199.33	.16	181.07	.21	158.56	.24	133.14	.26	.27
.74	219.00	.04	212.14	.10	199.17	.15	180.86	.20	158.32	.25	132.88	.27	.26
.75	218.96	.03	212.04	.10	199.02	.16	180.66	.21	158.07	.24	132.61	.26	.25
.76	218.93	.04	211.94	.10	198.86	.16	180.45	.21	157.83	.24	132.35	.27	.24
.77	218.89	.04	211.84	.11	198.70	.16	180.24	.20	157.59	.24	132.08	.26	.23
.78	218.85	.04	211.73	.10	198.54	.16	180.04	.21	157.35	.25	131.82	.27	.22
.79	218.81	.04	211.63	.10	198.38	.16	179.83	.21	157.10	.24	131.55	.26	.21
.80	218.77	.04	211.53	.11	198.22	.16	179.62	.21	156.86	.24	131.29	.27	.20
.81	218.73	.04	211.42	.10	198.06	.16	179.41	.21	156.62	.25	131.02	.26	.19
.82	218.69	.04	211.32	.11	197.90	.16	179.20	.21	156.37	.24	130.76	.27	.18
.83	218.65	.04	211.21	.10	197.74	.16	178.99	.21	156.13	.25	130.49	.26	.17
.84	218.61	.04	211.11	.11	197.58	.17	178.78	.21	155.88	.24	130.23	.27	.16
.85	218.57	.05	211.00	.11	197.41	.16	178.57	.21	155.64	.25	129.96	.26	.15
.86	218.52	.04	210.89	.10	197.25	.17	178.36	.21	155.39	.24	129.70	.27	.14
.87	218.48	.05	210.79	.11	197.08	.16	178.15	.22	155.15	.25	129.43	.26	.13
.88	218.43	.04	210.68	.11	196.92	.17	177.93	.21	154.90	.24	129.17	.27	.12
.89	218.39	.05	210.57	.11	196.75	.17	177.72	.21	154.66	.25	128.90	.26	.11
.90	218.34	.05	210.46	.11	196.58	.17	177.51	.22	154.41	.25	128.64	.27	.10
.91	218.29	.05	210.35	.11	196.41	.17	177.29	.21	154.16	.25	128.37	.26	.09
.92	218.24	.05	210.24	.11	196.24	.17	177.08	.22	153.91	.24	128.11	.27	.08
.93	218.19	.05	210.13	.11	196.07	.17	176.86	.21	153.67	.25	127.84	.26	.07
.94	218.14	.05	210.02	.12	195.90	.17	176.65	.22	153.42	.25	127.58	.27	.06
.95	218.09	.05	209.90	.11	195.73	.17	176.43	.21	153.17	.25	127.31	.27	.05
.96	218.04	.05	209.79	.12	195.56	.17	176.22	.22	152.92	.25	127.04	.26	.04
.97	217.99	.05	209.67	.11	195.39	.17	176.00	.21	152.67	.25	126.78	.27	.03
.98	217.94	.05	209.56	.12	195.22	.17	175.79	.22	152.42	.25	126.51	.27	.02
.99	217.89	.05	209.44	.12	195.05	.17	175.57	.21	152.17	.25	126.24	.26	.01
1.00	217.84	.05	209.32	.12	194.88	.17	175.36	.22	151.92	.25	125.98	.27	.00
Days.	25		24		23		22		21		20		Days.

TABLE XV. ARGUMENT 10.

Equation = 110″.0 [1 − sin. (x + z)].

Period, 25.621696 days.

Days.	6		7		8		9		10		11		12		Days.
Decimals of a Day.	Equation.	Diff.	Equation.	Diff.	Equation.	Diff.	Equation.	Diff.	Equation.	Diff.	Equation.	Diff.	Equation.	Diff.	Decimals of a Day.
.00	125.98	.27	99.08	.27	72.81	.25	48.81	.22	28.45	.18	12.97	.13	3.29	.06	1.00
.01	125.71	.26	98.81	.27	72.59	.26	48.59	.23	28.27	.18	12.84	.12	3.23	.07	.99
.02	125.45	.27	98.54	.27	72.33	.25	48.36	.22	28.09	.18	12.72	.12	3.16	.06	.98
.03	125.18	.27	98.27	.27	72.08	.25	48.14	.22	27.91	.18	12.60	.13	3.10	.07	.97
.04	124.91	.26	98.00	.26	71.83	.26	47.92	.23	27.73	.17	12.47	.12	3.03	.06	.96
.05	124.65	.27	97.74	.27	71.57	.25	47.69	.22	27.56	.18	12.35	.12	2.97	.06	.95
.06	124.38	.27	97.47	.27	71.32	.25	47.47	.22	27.38	.18	12.23	.13	2.91	.06	.94
.07	124.11	.26	97.20	.27	71.07	.26	47.25	.23	27.20	.18	12.10	.12	2.85	.06	.93
.08	123.85	.27	96.93	.26	70.81	.25	47.02	.22	27.02	.17	11.98	.12	2.79	.06	.92
.09	123.58	.27	96.67	.27	70.56	.25	46.80	.22	26.85	.18	11.86	.12	2.73	.06	.91
.10	123.31	.27	96.40	.27	70.31	.25	46.58	.22	26.67	.18	11.74	.12	2.67	.06	.90
.11	123.04	.26	96.13	.27	70.06	.25	46.36	.22	26.49	.17	11.62	.12	2.61	.06	.89
.12	122.78	.27	95.86	.26	69.81	.25	46.14	.22	26.32	.18	11.50	.12	2.55	.06	.88
.13	122.51	.27	95.60	.27	69.56	.26	45.92	.22	26.14	.18	11.38	.12	2.49	.06	.87
.14	122.24	.27	95.33	.27	69.30	.25	45.70	.22	25.96	.17	11.26	.12	2.43	.05	.86
.15	121.97	.27	95.06	.26	69.05	.25	45.48	.22	25.79	.17	11.14	.12	2.38	.06	.85
.16	121.70	.26	94.80	.27	68.80	.25	45.26	.21	25.62	.18	11.02	.12	2.32	.05	.84
.17	121.44	.27	94.53	.27	68.55	.25	45.05	.22	25.44	.17	10.90	.11	2.27	.06	.83
.18	121.17	.27	94.26	.26	68.30	.25	44.83	.22	25.27	.17	10.79	.12	2.21	.05	.82
.19	120.90	.27	94.00	.27	68.05	.25	44.61	.22	25.10	.17	10.67	.12	2.16	.05	.81
.20	120.63	.27	93.73	.27	67.80	.25	44.39	.22	24.93	.17	10.55	.12	2.11	.05	.80
.21	120.36	.27	93.46	.26	67.55	.25	44.17	.21	24.76	.17	10.43	.11	2.06	.05	.79
.22	120.09	.26	93.20	.27	67.30	.25	43.96	.22	24.59	.17	10.32	.12	2.01	.05	.78
.23	119.83	.27	92.93	.27	67.05	.24	43.74	.21	24.42	.17	10.20	.11	1.96	.05	.77
.24	119.56	.27	92.66	.26	66.81	.25	43.53	.22	24.25	.17	10.09	.12	1.91	.05	.76
.25	119.29	.27	92.40	.27	66.56	.25	43.31	.21	24.08	.17	9.97	.11	1.86	.05	.75
.26	119.02	.27	92.13	.26	66.31	.25	43.10	.21	23.91	.17	9.86	.11	1.81	.05	.74
.27	118.75	.27	91.87	.27	66.06	.25	42.89	.22	23.74	.16	9.75	.11	1.76	.05	.73
.28	118.48	.27	91.60	.26	65.81	.24	42.67	.21	23.58	.17	9.64	.11	1.71	.04	.72
.29	118.21	.27	91.34	.27	65.57	.25	42.46	.21	23.41	.17	9.53	.11	1.67	.05	.71
.30	117.94	.27	91.07	.27	65.32	.25	42.25	.21	23.24	.16	9.42	.11	1.62	.04	.70
.31	117.67	.27	90.80	.26	65.07	.24	42.04	.21	23.08	.17	9.31	.11	1.58	.05	.69
.32	117.40	.26	90.54	.27	64.83	.25	41.83	.21	22.91	.16	9.20	.11	1.53	.04	.68
.33	117.14	.27	90.27	.26	64.58	.24	41.62	.21	22.75	.16	9.09	.11	1.49	.05	.67
.34	116.87	.27	90.01	.27	64.34	.25	41.41	.21	22.59	.17	8.98	.10	1.44	.04	.66
.35	116.60	.27	89.74	.27	64.09	.24	41.20	.21	22.42	.16	8.88	.11	1.40	.05	.65
.36	116.33	.27	89.47	.26	63.85	.25	40.99	.21	22.26	.16	8.77	.10	1.35	.04	.64
.37	116.06	.27	89.21	.27	63.60	.24	40.78	.21	22.10	.17	8.67	.11	1.31	.04	.63
.38	115.79	.27	88.94	.26	63.36	.25	40.57	.21	21.93	.16	8.56	.10	1.27	.04	.62
.39	115.52	.27	88.68	.27	63.11	.24	40.36	.21	21.77	.16	8.46	.10	1.23	.04	.61
.40	115.25	.27	88.41	.26	62.87	.24	40.15	.21	21.61	.16	8.36	.10	1.19	.04	.60
.41	114.98	.27	88.15	.27	62.63	.25	39.94	.21	21.45	.16	8.26	.10	1.15	.04	.59
.42	114.71	.27	87.88	.26	62.38	.24	39.73	.21	21.29	.16	8.16	.10	1.11	.04	.58
.43	114.44	.27	87.62	.27	62.14	.24	39.52	.20	21.13	.16	8.06	.10	1.07	.04	.57
.44	114.17	.27	87.35	.26	61.90	.24	39.32	.21	20.97	.16	7.96	.10	1.03	.03	.56
.45	113.90	.27	87.09	.26	61.66	.25	39.11	.21	20.81	.15	7.86	.10	1.00	.04	.55
.46	113.63	.27	86.83	.27	61.41	.24	38.90	.20	20.66	.16	7.76	.10	0.96	.03	.54
.47	113.36	.27	86.56	.26	61.17	.24	38.70	.21	20.50	.16	7.66	.10	0.93	.04	.53
.48	113.09	.27	86.30	.26	60.93	.24	38.49	.20	20.34	.15	7.56	.10	0.89	.03	.52
.49	112.82	.27	86.04	.27	60.69	.24	38.29	.20	20.19	.16	7.46	.10	0.86	.03	.51
.50	112.55	.27	85.77	.26	60.45	.24	38.09	.20	20.03	.15	7.36	.10	0.83	.03	.50
Days.	19		18		17		16		15		14		13		Days.

TABLE XV. ARGUMENT 10.

Equation = 110″.0 [1 — sin. $(c + z)$].

Period, 25.621696 days.

Days.	6		7		8		9		10		11		12		Days.
Decimals of a Day.	Equation.	Diff.	Equation.	Diff.	Equation.	Diff.	Equation.	Diff.	Equation.	Diff.	Equation.	Diff.	Equation.	Diff.	Decimals of a Day.
.50	112.55″	.27	85.77″	.26	60.45″	.24	38.09″	.20	20.03″	.15	7.36″	.10	0.83″	.03	.50
.51	112.28	.27	85.51	.26	60.21	.24	37.89	.21	19.88	.15	7.26	.10	0.80	.03	.49
.52	112.01	.27	85.25	.27	59.97	.24	37.68	.20	19.73	.16	7.16	.09	0.77	.03	.48
.53	111.74	.27	84.98	.26	59.73	.24	37.48	.20	19.57	.15	7.07	.10	0.74	.03	.47
.54	111.47	.27	84.72	.26	59.49	.24	37.28	.20	19.42	.15	6.97	.09	0.71	.03	.46
.55	111.20	.27	84.46	.26	59.25	.23	37.08	.20	19.27	.16	6.88	.10	0.68	.03	.45
.56	110.93	.27	84.20	.27	59.02	.24	36.88	.21	19.11	.15	6.78	.09	0.65	.03	.44
.57	110.66	.27	83.93	.26	58.78	.24	36.67	.20	18.96	.15	6.69	.09	0.62	.03	.43
.58	110.39	.27	83.67	.26	58.54	.24	36.47	.20	18.81	.15	6.60	.09	0.59	.03	.42
.59	110.12	.27	83.41	.26	58.30	.24	36.27	.20	18.66	.15	6.51	.09	0.56	.03	.41
.60	109.85	.27	83.15	.26	58.06	.24	36.07	.20	18.51	.15	6.42	.09	0.53	.03	.40
.61	109.58	.27	82.89	.26	57.82	.23	35.87	.20	18.36	.15	6.33	.09	0.50	.02	.39
.62	109.31	.27	82.63	.26	57.59	.24	35.67	.20	18.21	.15	6.24	.09	0.48	.03	.38
.63	109.04	.27	82.37	.26	57.35	.24	35.47	.20	18.06	.14	6.15	.09	0.45	.02	.37
.64	108.77	.27	82.11	.26	57.11	.23	35.27	.20	17.92	.15	6.06	.09	0.43	.03	.36
.65	108.50	.27	81.85	.26	56.88	.24	35.07	.19	17.77	.15	5.97	.09	0.40	.02	.35
.66	108.23	.27	81.59	.26	56.64	.24	34.88	.20	17.62	.14	5.88	.08	0.38	.02	.34
.67	107.96	.27	81.33	.26	56.40	.23	34.68	.20	17.48	.15	5.80	.09	0.36	.02	.33
.68	107.69	.27	81.07	.26	56.17	.24	34.48	.20	17.33	.14	5.71	.08	0.34	.02	.32
.69	107.42	.27	80.81	.26	55.93	.23	34.28	.19	17.19	.15	5.63	.09	0.32	.02	.31
.70	107.15	.27	80.55	.26	55.70	.23	34.09	.19	17.04	.14	5.54	.08	0.30	.02	.30
.71	106.88	.27	80.29	.26	55.47	.24	33.90	.20	16.90	.14	5.46	.08	0.28	.02	.29
.72	106.61	.27	80.03	.26	55.23	.23	33.70	.19	16.76	.15	5.38	.09	0.26	.02	.28
.73	106.34	.27	79.77	.26	55.00	.23	33.51	.19	16.61	.14	5.29	.08	0.24	.02	.27
.74	106.07	.27	79.51	.26	54.77	.24	33.32	.20	16.47	.14	5.21	.08	0.22	.02	.26
.75	105.80	.26	79.25	.26	54.53	.23	33.12	.19	16.33	.14	5.13	.08	0.20	.01	.25
.76	105.54	.27	78.99	.25	54.30	.23	32.93	.19	16.19	.15	5.05	.08	0.19	.02	.24
.77	105.27	.27	78.74	.26	54.07	.24	32.74	.20	16.04	.14	4.97	.08	0.17	.01	.23
.78	105.00	.27	78.48	.26	53.83	.23	32.54	.19	15.90	.14	4.89	.08	0.16	.02	.22
.79	104.73	.27	78.22	.26	53.60	.23	32.35	.19	15.76	.14	4.81	.08	0.14	.01	.21
.80	104.46	.27	77.96	.26	53.37	.23	32.16	.19	15.62	.14	4.73	.08	0.13	.01	.20
.81	104.19	.27	77.70	.26	53.14	.23	31.97	.19	15.48	.14	4.65	.08	0.12	.01	.19
.82	103.92	.27	77.44	.25	52.91	.23	31.78	.19	15.34	.13	4.57	.08	0.11	.01	.18
.83	103.65	.27	77.19	.26	52.68	.23	31.59	.19	15.21	.14	4.49	.07	0.10	.01	.17
.84	103.38	.27	76.93	.26	52.45	.23	31.40	.18	15.07	.14	4.42	.08	0.09	.01	.16
.85	103.11	.27	76.67	.25	52.22	.23	31.22	.19	14.93	.13	4.34	.07	0.08	.01	.15
.86	102.84	.26	76.42	.26	51.99	.23	31.03	.19	14.80	.14	4.27	.08	0.07	.01	.14
.87	102.58	.27	76.16	.26	51.76	.23	30.84	.19	14.66	.13	4.19	.07	0.06	.01	.13
.88	102.31	.27	75.90	.25	51.53	.23	30.65	.18	14.53	.14	4.12	.07	0.05	.01	.12
.89	102.04	.27	75.65	.26	51.30	.23	30.47	.19	14.39	.13	4.05	.07	0.04	.01	.11
.90	101.77	.27	75.39	.25	51.07	.23	30.28	.18	14.26	.13	3.98	.07	0.03	.01	.10
.91	101.50	.27	75.14	.26	50.84	.23	30.10	.19	14.13	.13	3.91	.07	0.02	.00	.09
.92	101.23	.27	74.88	.25	50.61	.22	29.91	.18	14.00	.13	3.84	.07	0.02	.01	.08
.93	100.96	.27	74.63	.26	50.39	.23	29.73	.18	13.87	.13	3.77	.07	0.01	.00	.07
.94	100.69	.27	74.37	.25	50.16	.23	29.55	.19	13.74	.13	3.70	.07	0.01	.01	.06
.95	100.42	.27	74.12	.26	49.93	.22	29.36	.18	13.61	.13	3.63	.07	0.00	.00	.05
.96	100.15	.26	73.86	.25	49.71	.23	29.18	.18	13.48	.13	3.56	.07	0.00	.00	.04
.97	99.89	.27	73.61	.26	49.48	.22	29.00	.19	13.35	.13	3.49	.07	0.00	.00	.03
.98	99.62	.27	73.35	.25	49.26	.23	28.81	.18	13.22	.12	3.42	.06	0.00	.00	.02
.99	99.35	.27	73.10	.26	49.03	.22	28.63	.18	13.10	.13	3.36	.07	0.00	.00	.01
1.00	99.08	.27	72.84	.25	48.81	.22	28.45	.18	12.97	.13	3.29	.06	0.00	.00	.00
Days.	19		18		17		16		15		14		13		Days.

TABLE XVI. ARGUMENT 11.

Equation = 85″.0 [1 — sin. (2y — x)].

Period, 26.878290 days.

Days.	0	1	2	3	4	5	6	7	8	9	Days.
Days. .00	169″.30	169″.56	165″.24	156″.55	143″.97	128″.18	110″.07	90″.57	70″.76	51″.74	Days. 1.00
.01	169.32	169.54	165.17	156.44	143.83	128.01	109.88	90.37	70.56	51.56	.99
.02	169.35	169.52	165.10	156.33	143.68	127.84	109.69	90.17	70.37	51.38	.98
.03	169.37	169.50	165.03	156.22	143.54	127.67	109.49	89.98	70.17	51.20	.97
.04	169.40	169.48	164.96	156.11	143.39	127.50	109.30	89.78	69.98	51.02	.96
.05	169.42	169.46	164.89	156.00	143.25	127.33	109.11	89.58	69.78	50.84	.95
.06	169.44	169.44	164.82	155.89	143.10	127.16	108.92	89.38	69.59	50.66	.94
.07	169.46	169.41	164.76	155.78	142.96	126.99	108.73	89.18	69.39	50.48	.93
.08	169.48	169.39	164.69	155.67	142.81	126.81	108.53	88.99	69.20	50.29	.92
.09	169.50	169.36	164.62	155.56	142.67	126.64	108.34	88.79	69.00	50.11	.91
.10	169.52	169.33	164.55	155.45	142.52	126.47	108.15	88.59	68.81	49.93	.90
.11	169.54	169.31	164.48	155.34	142.37	126.30	107.96	88.39	68.61	49.75	.89
.12	169.56	169.28	164.41	155.23	142.23	126.12	107.77	88.19	68.42	49.57	.88
.13	169.58	169.26	164.34	155.11	142.08	125.95	107.58	88.00	68.22	49.38	.87
.14	169.60	169.23	164.27	155.00	141.93	125.77	107.39	87.80	68.03	49.20	.86
.15	169.62	169.21	164.20	154.89	141.78	125.60	107.20	87.60	67.83	49.02	.85
.16	169.64	169.18	164.13	154.78	141.63	125.43	107.01	87.40	67.64	48.84	.84
.17	169.66	169.15	164.05	154.66	141.48	125.25	106.82	87.20	67.44	48.66	.83
.18	169.67	169.12	163.98	154.55	141.33	125.08	106.62	87.00	67.25	48.48	.82
.19	169.69	169.09	163.90	154.43	141.18	124.90	106.43	86.80	67.05	48.30	.81
.20	169.70	169.06	163.83	154.32	141.03	124.73	106.24	86.60	66.86	48.12	.80
.21	169.72	169.03	163.76	154.20	140.88	124.55	106.05	86.40	66.67	47.94	.79
.22	169.73	169.00	163.68	154.09	140.73	124.38	105.86	86.20	66.48	47.76	.78
.23	169.75	168.97	163.61	153.97	140.58	124.20	105.66	86.00	66.28	47.59	.77
.24	169.76	168.94	163.53	153.86	140.43	124.03	105.47	85.80	66.09	47.41	.76
.25	169.78	168.91	163.46	153.74	140.28	123.85	105.28	85.60	65.90	47.23	.75
.26	169.79	168.87	163.38	153.62	140.13	123.67	105.09	85.40	65.71	47.05	.74
.27	169.81	168.84	163.31	153.50	139.98	123.49	104.90	85.20	65.51	46.87	.73
.28	169.82	168.80	163.23	153.39	139.82	123.32	104.70	85.01	65.32	46.70	.72
.29	169.83	168.77	163.15	153.27	139.67	123.14	104.51	84.81	65.12	46.52	.71
.30	169.84	168.73	163.07	153.15	139.52	122.96	104.32	84.61	64.93	46.34	.70
.31	169.85	168.70	162.99	153.03	139.37	122.78	104.13	84.41	64.74	46.16	.69
.32	169.86	168.66	162.91	152.91	139.22	122.60	103.93	84.21	64.55	45.99	.68
.33	169.87	168.63	162.82	152.79	139.06	122.43	103.74	84.01	64.35	45.81	.67
.34	169.88	168.59	162.74	152.67	138.91	122.25	103.54	83.82	64.16	45.64	.66
.35	169.89	168.56	162.66	152.55	138.76	122.07	103.35	83.62	63.97	45.46	.65
.36	169.90	168.52	162.58	152.43	138.60	121.89	103.15	83.42	63.78	45.28	.64
.37	169.91	168.48	162.50	152.31	138.45	121.71	102.96	83.22	63.59	45.11	.63
.38	169.92	168.44	162.42	152.18	138.29	121.53	102.76	83.02	63.39	44.93	.62
.39	169.93	168.40	162.34	152.06	138.14	121.35	102.57	82.82	63.20	44.76	.61
.40	169.94	168.36	162.26	151.94	137.99	121.17	102.37	82.62	63.01	44.58	.60
.41	169.95	168.32	162.18	151.82	137.83	120.99	102.18	82.42	62.82	44.41	.59
.42	169.96	168.28	162.09	151.70	137.68	120.81	101.98	82.23	62.63	44.24	.58
.43	169.97	168.24	162.01	151.57	137.52	120.63	101.79	82.03	62.44	44.07	.57
.44	169.97	168.20	161.92	151.45	137.37	120.45	101.59	81.83	62.25	43.89	.56
.45	169.98	168.16	161.84	151.33	137.21	120.27	101.40	81.63	62.06	43.72	.55
.46	169.98	168.12	161.75	151.20	137.05	120.09	101.20	81.43	61.87	43.55	.54
.47	169.98	168.08	161.67	151.08	136.90	119.91	101.01	81.23	61.68	43.37	.53
.48	169.99	168.04	161.58	150.95	136.74	119.73	100.81	81.04	61.48	43.20	.52
.49	169.99	167.99	161.50	150.83	136.59	119.54	100.62	80.84	61.29	43.02	.51
.50	169.99	167.95	161.41	150.70	136.43	119.36	100.42	80.64	61.10	42.85	.50
Days.	27	26	25	24	23	22	21	20	19	18	Days.

TABLE XVI. ARGUMENT 11.

Equation = 85″.0 [1 — sin. (2y — x)].

Period, 26.878290 days.

Days.	0	1	2	3	4	5	6	7	8	9	Days.
.50	169.99	167.95	161.41	150.70	136.43	119.36	100.42	80.61	61.10	42.85	.50
.51	169.99	167.90	161.32	150.57	136.27	119.18	100.22	80.41	60.91	42.68	.49
.52	170.00	167.86	161.25	150.41	136.11	119.00	100.02	80.24	60.72	42.51	.48
.53	170.00	167.81	161.16	150.32	135.96	118.81	99.83	80.05	60.53	42.33	.47
.54	170.00	167.77	161.07	150.19	135.80	118.63	99.63	79.85	60.34	42.16	.46
.55	170.00	167.72	160.98	150.06	135.64	118.45	99.43	79.65	60.15	41.99	.45
.56	170.00	167.68	160.89	149.93	135.48	118.27	99.24	79.45	59.96	41.82	.44
.57	170.00	167.63	160.80	149.80	135.32	118.09	99.01	79.25	59.77	41.65	.43
.58	170.00	167.59	160.71	149.68	135.15	117.90	98.84	79.06	59.58	41.48	.42
.59	170.00	167.54	160.61	149.55	134.99	117.72	98.65	78.86	59.39	41.31	.41
.60	169.99	167.49	160.52	149.42	134.83	117.54	98.45	78.66	59.20	41.14	.40
.61	169.99	167.44	160.43	149.29	134.67	117.36	98.25	78.46	59.01	40.97	.39
.62	169.99	167.39	160.34	149.16	134.51	117.17	98.06	78.26	58.83	40.80	.38
.63	169.99	167.34	160.25	149.03	134.34	116.99	97.86	78.06	58.63	40.63	.37
.64	169.98	167.29	160.16	148.90	134.18	116.80	97.67	77.87	58.44	40.46	.36
.65	169.98	167.24	160.07	148.77	134.02	116.62	97.47	77.67	58.25	40.29	.35
.66	169.97	167.19	159.97	148.64	133.86	116.44	97.27	77.47	58.06	40.12	.34
.67	169.97	167.14	159.88	148.51	133.69	116.25	97.08	77.27	57.87	39.95	.33
.68	169.96	167.09	159.78	148.37	133.53	116.07	96.88	77.07	57.69	39.79	.32
.69	169.96	167.04	159.69	148.24	133.36	115.88	96.69	76.87	57.50	39.62	.31
.70	169.95	166.99	159.59	148.11	133.20	115.70	96.49	76.67	57.31	39.45	.30
.71	169.94	166.94	159.49	147.98	133.04	115.51	96.29	76.47	57.12	39.28	.29
.72	169.94	166.89	159.39	147.84	132.87	115.33	96.10	76.28	56.93	39.12	.28
.73	169.93	166.84	159.30	147.71	132.71	115.14	95.90	76.08	56.75	38.95	.27
.74	169.92	166.79	159.20	147.57	132.54	114.96	95.71	75.88	56.56	38.79	.26
.75	169.91	166.74	159.10	147.44	132.38	114.77	95.50	75.68	56.37	38.62	.25
.76	169.90	166.68	159.00	147.31	132.21	114.58	95.31	75.49	56.18	38.45	.24
.77	169.89	166.63	158.90	147.17	132.05	114.40	95.12	75.29	55.99	38.29	.23
.78	169.88	166.57	158.81	147.04	131.88	114.21	94.92	75.09	55.81	38.12	.22
.79	169.87	166.51	158.71	146.90	131.72	114.03	94.73	74.89	55.62	37.96	.21
.80	169.86	166.46	158.61	146.77	131.55	113.84	94.53	74.70	55.43	37.79	.20
.81	169.85	166.40	158.51	146.63	131.38	113.65	94.33	74.50	55.24	37.62	.19
.82	169.84	166.34	158.41	146.49	131.21	113.46	94.13	74.30	55.06	37.46	.18
.83	169.83	166.28	158.31	146.36	131.05	113.28	93.94	74.11	54.87	37.29	.17
.84	169.82	166.23	158.21	146.22	130.88	113.09	93.74	73.91	54.69	37.13	.16
.85	169.81	166.17	158.11	146.08	130.71	112.90	93.54	73.71	54.50	36.96	.15
.86	169.80	166.11	158.01	145.94	130.54	112.71	93.34	73.51	54.32	36.80	.14
.87	169.78	166.05	157.90	145.80	130.37	112.52	93.14	73.31	54.13	36.63	.13
.88	169.76	165.99	157.80	145.66	130.21	112.34	92.95	73.12	53.95	36.47	.12
.89	169.75	165.93	157.69	145.52	130.04	112.15	92.75	72.92	53.76	36.30	.11
.90	169.73	165.87	157.59	145.38	129.87	111.96	92.55	72.72	53.58	36.14	.10
.91	169.72	165.81	157.48	145.24	129.70	111.77	92.35	72.52	53.40	35.98	.09
.92	169.70	165.75	157.38	145.10	129.53	111.58	92.15	72.33	53.21	35.82	.08
.93	169.69	165.69	157.27	144.96	129.37	111.40	91.96	72.13	53.03	35.65	.07
.94	169.67	165.63	157.17	144.82	129.20	111.21	91.76	71.93	52.84	35.49	.06
.95	169.66	165.57	157.06	144.68	129.03	111.02	91.56	71.74	52.66	35.33	.05
.96	169.64	165.51	156.96	144.54	128.86	110.83	91.36	71.54	52.48	35.17	.04
.97	169.62	165.41	156.86	144.40	128.69	110.64	91.16	71.35	52.29	35.01	.03
.98	169.60	165.37	156.75	144.25	128.52	110.45	90.97	71.15	52.11	34.84	.02
.99	169.58	165.31	156.65	144.11	128.35	110.26	90.77	70.95	51.92	34.68	.01
1.00	169.56	165.24	156.55	143.97	128.18	110.07	90.57	70.76	51.74	34.52	.00
Days.	27	26	25	24	23	22	21	20	19	18	Days.

TABLE XVI.

(Continued).

Period, 26.878290 days.

Days.	10	11	12	13	Days.
Days	″	″	″	″	Days.
.00	34.52	20.06	9.11	2.33	1.00
.01	34.36	19.93	9.02	2.28	.99
.02	34.20	19.81	8.93	2.24	.98
.03	34.05	19.68	8.85	2.19	.97
.04	33.89	19.56	8.76	2.15	.96
.05	33.73	19.43	8.67	2.10	.95
.06	33.57	19.30	8.59	2.06	.94
.07	33.41	19.18	8.50	2.01	.93
.08	33.26	19.05	8.42	1.97	.92
.09	33.10	18.93	8.33	1.93	.91
.10	32.94	18.80	8.25	1.89	.90
.11	32.78	18.68	8.17	1.85	.89
.12	32.63	18.56	8.08	1.81	.88
.13	32.47	18.43	8.00	1.77	.87
.14	32.32	18.31	7.91	1.73	.86
.15	32.16	18.19	7.83	1.69	.85
.16	32.01	18.07	7.75	1.65	.84
.17	31.85	17.95	7.67	1.60	.83
.18	31.70	17.82	7.58	1.56	.82
.19	31.54	17.70	7.50	1.52	.81
.20	31.39	17.58	7.42	1.48	.80
.21	31.24	17.46	7.34	1.44	.79
.22	31.09	17.34	7.26	1.40	.78
.23	30.93	17.22	7.18	1.37	.77
.24	30.78	17.10	7.10	1.33	.76
.25	30.63	16.98	7.02	1.30	.75
.26	30.48	16.86	6.94	1.26	.74
.27	30.33	16.74	6.86	1.23	.73
.28	30.17	16.63	6.79	1.19	.72
.29	30.02	16.51	6.71	1.16	.71
.30	29.87	16.39	6.63	1.13	.70
.31	29.72	16.27	6.55	1.10	.69
.32	29.57	16.15	6.48	1.06	.68
.33	29.42	16.04	6.40	1.03	.67
.34	29.27	15.92	6.33	1.00	.66
.35	29.12	15.80	6.25	0.97	.65
.36	28.97	15.69	6.17	0.94	.64
.37	28.82	15.57	6.10	0.91	.63
.38	28.67	15.46	6.02	0.88	.62
.39	28.52	15.34	5.95	0.85	.61
.40	28.37	15.23	5.87	0.82	.60
.41	28.22	15.12	5.80	0.79	.59
.42	28.07	15.00	5.73	0.76	.58
.43	27.93	14.89	5.65	0.73	.57
.44	27.78	14.77	5.58	0.71	.56
.45	27.63	14.66	5.51	0.68	.55
.46	27.48	14.55	5.44	0.65	.54
.47	27.34	14.44	5.37	0.63	.53
.48	27.19	14.33	5.30	0.60	.52
.49	27.05	14.22	5.23	0.58	.51
.50	26.90	14.11	5.16	0.56	.50
Days.	17	16	15	14	Days.

TABLE XVII. ARG. 12′.

Equation = $58''.7\ [1 - \sin.\ (2y - 2t)]$.

Period, 173.31006 days.

Days.	0	10	20	30	40	Days.
Days.	″	″	″	″	″	Days.
0.0	0.41	1.71	10.38	25.33	44.63	10.0
0.1	0.39	1.76	10.50	25.51	44.84	9.9
0.2	0.37	1.81	10.62	25.69	45.05	9.8
0.3	0.35	1.87	10.74	25.86	45.25	9.7
0.4	0.33	1.92	10.87	26.04	45.46	9.6
0.5	0.31	1.97	10.99	26.22	45.67	9.5
0.6	0.29	2.03	11.12	26.40	45.88	9.4
0.7	0.27	2.08	11.24	26.58	46.09	9.3
0.8	0.25	2.14	11.37	26.76	46.30	9.2
0.9	0.23	2.19	11.49	26.94	46.51	9.1
1.0	0.21	2.25	11.62	27.12	46.72	9.0
1.1	0.19	2.31	11.75	27.30	46.93	8.9
1.2	0.17	2.37	11.87	27.48	47.14	8.8
1.3	0.16	2.43	12.00	27.66	47.34	8.7
1.4	0.14	2.49	12.13	27.84	47.55	8.6
1.5	0.13	2.55	12.26	28.02	47.76	8.5
1.6	0.12	2.61	12.39	28.20	47.97	8.4
1.7	0.10	2.68	12.52	28.38	48.18	8.3
1.8	0.09	2.74	12.65	28.56	48.38	8.2
1.9	0.08	2.81	12.79	28.75	48.59	8.1
2.0	0.07	2.87	12.92	28.93	48.80	8.0
2.1	0.06	2.94	13.06	29.11	49.01	7.9
2.2	0.05	3.00	13.19	29.30	49.22	7.8
2.3	0.04	3.07	13.33	29.48	49.43	7.7
2.4	0.03	3.14	13.46	29.67	49.64	7.6
2.5	0.02	3.21	13.60	29.85	49.85	7.5
2.6	0.02	3.28	13.74	30.04	50.06	7.4
2.7	0.01	3.35	13.87	30.22	50.27	7.3
2.8	0.01	3.42	14.01	30.41	50.48	7.2
2.9	0.00	3.49	14.15	30.59	50.70	7.1
3.0	0.00	3.56	14.29	30.78	50.91	7.0
3.1	0.00	3.63	14.43	30.97	51.12	6.9
3.2	0.00	3.71	14.57	31.15	51.33	6.8
3.3	0.00	3.78	14.71	31.34	51.54	6.7
3.4	0.00	3.86	14.85	31.53	51.75	6.6
3.5	0.01	3.93	14.99	31.72	51.96	6.5
3.6	0.01	4.01	15.13	31.91	52.17	6.4
3.7	0.01	4.08	15.28	32.10	52.38	6.3
3.8	0.02	4.16	15.42	32.29	52.59	6.2
3.9	0.02	4.24	15.57	32.48	52.81	6.1
4.0	0.02	4.32	15.71	32.67	53.02	6.0
4.1	0.03	4.40	15.86	32.86	53.23	5.9
4.2	0.04	4.48	16.00	33.05	53.44	5.8
4.3	0.05	4.56	16.15	33.24	53.65	5.7
4.4	0.06	4.65	16.29	33.44	53.87	5.6
4.5	0.07	4.73	16.44	33.63	54.08	5.5
4.6	0.08	4.81	16.59	33.82	54.29	5.4
4.7	0.09	4.90	16.73	34.02	54.50	5.3
4.8	0.10	4.98	16.88	34.21	54.71	5.2
4.9	0.11	5.07	17.03	34.41	54.93	5.1
5.0	0.12	5.16	17.18	34.60	55.14	5.0
Days.	170	160	150	140	130	Days.

NOTE. — Arg. 12′ = Arg. 12.

TABLE XVI.

(*Continued.*)

Period, 26.878290 days.

Days.	10	11	12	13	Days.
Days	"	"	"	"	Days
.50	26.90	14.11	5.16	0.56	.50
.51	26.76	14.00	5.09	0.54	.49
.52	26.61	13.89	5.02	0.52	.48
.53	26.47	13.78	4.96	0.50	.47
.54	26.32	13.67	4.89	0.48	.46
.55	26.18	13.56	4.82	0.46	.45
.56	26.04	13.45	4.76	0.44	.44
.57	25.89	13.35	4.69	0.42	.43
.58	25.75	13.24	4.63	0.40	.42
.59	25.60	13.14	4.56	0.38	.41
.60	25.46	13.03	4.50	0.36	.40
.61	25.32	12.93	4.44	0.34	.39
.62	25.18	12.82	4.38	0.32	.38
.63	25.03	12.72	4.31	0.31	.37
.64	24.89	12.61	4.25	0.29	.36
.65	24.75	12.51	4.19	0.28	.35
.66	24.61	12.41	4.13	0.26	.34
.67	24.47	12.30	4.07	0.25	.33
.68	24.34	12.20	4.01	0.23	.32
.69	24.20	12.09	3.95	0.22	.31
.70	24.06	11.99	3.89	0.21	.30
.71	23.92	11.89	3.83	0.20	.29
.72	23.78	11.79	3.77	0.18	.28
.73	23.65	11.69	3.72	0.17	.27
.74	23.51	11.59	3.66	0.16	.26
.75	23.37	11.49	3.60	0.15	.25
.76	23.23	11.39	3.54	0.14	.24
.77	23.10	11.29	3.49	0.13	.23
.78	22.96	11.20	3.43	0.12	.22
.79	22.83	11.10	3.38	0.11	.21
.80	22.69	11.00	3.32	0.10	.20
.81	22.56	10.90	3.27	0.09	.19
.82	22.42	10.80	3.21	0.08	.18
.83	22.29	10.71	3.16	0.07	.17
.84	22.15	10.61	3.10	0.06	.16
.85	22.02	10.51	3.05	0.05	.15
.86	21.89	10.41	3.00	0.04	.14
.87	21.76	10.32	2.95	0.04	.13
.88	21.62	10.22	2.90	0.03	.12
.89	21.49	10.13	2.85	0.03	.11
.90	21.36	10.03	2.80	0.03	.10
.91	21.23	9.94	2.75	0.02	.09
.92	21.10	9.85	2.70	0.02	.08
.93	20.97	9.75	2.66	0.01	.07
.94	20.84	9.66	2.61	0.01	.06
.95	20.71	9.57	2.56	0.01	.05
.96	20.58	9.48	2.51	0.00	.04
.97	20.45	9.39	2.47	0.00	.03
.98	20.32	9.29	2.42	0.00	.02
.99	20.19	9.20	2.38	0.00	.01
1.00	20.06	9.11	2.33	0.00	.00
Days.	17	16	15	14	Days.

TABLE XVII. ARG. 12′.

Equation = 58″.7 [1 — sin. (2y — 2t)].

Period, 173.31006 days.

Days.	0	10	20	30	40	Days.
Days.	"	"	"	"	"	Days.
5.0	0.12	5.16	17.18	34.60	55.14	5.0
5.1	0.13	5.25	17.33	34.80	55.35	4.9
5.2	0.14	5.33	17.48	34.99	55.56	4.8
5.3	0.16	5.42	17.63	35.19	55.77	4.7
5.4	0.17	5.51	17.79	35.38	55.99	4.6
5.5	0.19	5.60	17.94	35.58	56.20	4.5
5.6	0.21	5.69	18.09	35.78	56.41	4.4
5.7	0.23	5.79	18.25	35.97	56.63	4.3
5.8	0.25	5.88	18.40	36.17	56.84	4.2
5.9	0.27	5.98	18.56	36.36	57.06	4.1
6.0	0.29	6.07	18.71	36.56	57.27	4.0
6.1	0.31	6.17	18.86	36.76	57.48	3.9
6.2	0.33	6.26	19.02	36.95	57.69	3.8
6.3	0.35	6.36	19.17	37.15	57.91	3.7
6.4	0.38	6.45	19.33	37.34	58.12	3.6
6.5	0.40	6.55	19.49	37.54	58.33	3.5
6.6	0.42	6.65	19.65	37.74	58.54	3.4
6.7	0.44	6.75	19.81	37.93	58.76	3.3
6.8	0.47	6.85	19.97	38.13	58.97	3.2
6.9	0.49	6.95	20.13	38.33	59.19	3.1
7.0	0.52	7.05	20.29	38.53	59.40	3.0
7.1	0.55	7.15	20.45	38.73	59.61	2.9
7.2	0.58	7.26	20.61	38.93	59.82	2.8
7.3	0.61	7.36	20.77	39.13	60.04	2.7
7.4	0.64	7.47	20.94	39.34	60.25	2.6
7.5	0.67	7.57	21.10	39.54	60.46	2.5
7.6	0.70	7.68	21.26	39.74	60.67	2.4
7.7	0.73	7.78	21.43	39.94	60.88	2.3
7.8	0.77	7.89	21.59	40.14	61.10	2.2
7.9	0.80	7.99	21.76	40.35	61.31	2.1
8.0	0.84	8.10	21.93	40.55	61.52	2.0
8.1	0.88	8.21	22.10	40.75	61.73	1.9
8.2	0.92	8.31	22.26	40.95	61.94	1.8
8.3	0.96	8.42	22.43	41.15	62.16	1.7
8.4	1.00	8.53	22.60	41.36	62.37	1.6
8.5	1.04	8.64	22.77	41.56	62.58	1.5
8.6	1.08	8.75	22.94	41.76	62.79	1.4
8.7	1.12	8.86	23.10	41.97	63.00	1.3
8.8	1.17	8.97	23.27	42.17	63.22	1.2
8.9	1.21	9.09	23.44	42.38	63.43	1.1
9.0	1.25	9.20	23.61	42.58	63.64	1.0
9.1	1.29	9.32	23.78	42.78	63.85	0.9
9.2	1.33	9.43	23.95	42.99	64.06	0.8
9.3	1.38	9.55	24.12	43.19	64.28	0.7
9.4	1.42	9.66	24.29	43.40	64.49	0.6
9.5	1.46	9.78	24.46	43.60	64.70	0.5
9.6	1.51	9.90	24.63	43.81	64.91	0.4
9.7	1.56	10.02	24.81	44.01	65.12	0.3
9.8	1.61	10.14	24.98	44.22	65.34	0.2
9.9	1.66	10.26	25.16	44.42	65.55	0.1
10.0	1.71	10.38	25.33	44.63	65.76	0.0
Days.	170	160	150	140	130	Days.

NOTE. — Arg. 12′ = Arg. 12.

TABLE XVII.

(*Continued*).

Period, 173.31006 days.

Days.	50	60	70	80	Days.
0.0	65.76	85.98	102.63	113.58	10.0
0.1	65.97	86.17	102.77	113.65	9.9
0.2	66.18	86.36	102.91	113.73	9.8
0.3	66.39	86.55	103.05	113.80	9.7
0.4	66.61	86.73	103.18	113.88	9.6
0.5	66.82	86.92	103.32	113.95	9.5
0.6	67.03	87.11	103.46	114.02	9.4
0.7	67.24	87.29	103.59	114.09	9.3
0.8	67.45	87.48	103.73	114.16	9.2
0.9	67.66	87.66	103.86	114.23	9.1
1.0	67.87	87.85	104.00	114.30	9.0
1.1	68.08	88.03	104.13	114.37	8.9
1.2	68.29	88.22	104.27	114.43	8.8
1.3	68.50	88.40	104.40	114.50	8.7
1.4	68.71	88.59	104.54	114.56	8.6
1.5	68.92	88.77	104.67	114.63	8.5
1.6	69.13	88.95	104.80	114.69	8.4
1.7	69.34	89.14	104.94	114.76	8.3
1.8	69.55	89.32	105.07	114.82	8.2
1.9	69.76	89.50	105.20	114.88	8.1
2.0	69.97	89.68	105.33	114.94	8.0
2.1	70.18	89.86	105.46	115.00	7.9
2.2	70.39	90.04	105.59	115.06	7.8
2.3	70.60	90.22	105.72	115.12	7.7
2.4	70.80	90.39	105.84	115.17	7.6
2.5	71.01	90.57	105.97	115.23	7.5
2.6	71.22	90.75	106.10	115.29	7.4
2.7	71.43	90.92	106.22	115.34	7.3
2.8	71.64	91.10	106.35	115.40	7.2
2.9	71.84	91.27	106.47	115.45	7.1
3.0	72.05	91.45	106.60	115.51	7.0
3.1	72.26	91.62	106.72	115.56	6.9
3.2	72.46	91.80	106.84	115.62	6.8
3.3	72.67	91.97	106.96	115.67	6.7
3.4	72.87	92.15	107.08	115.72	6.6
3.5	73.08	92.32	107.20	115.77	6.5
3.6	73.29	92.49	107.32	115.82	6.4
3.7	73.49	92.67	107.44	115.87	6.3
3.8	73.70	92.84	107.56	115.92	6.2
3.9	73.90	93.01	107.67	115.96	6.1
4.0	74.11	93.18	107.79	116.01	6.0
4.1	74.31	93.35	107.91	116.05	5.9
4.2	74.52	93.53	108.02	116.10	5.8
4.3	74.72	93.70	108.14	116.14	5.7
4.4	74.93	93.87	108.25	116.19	5.6
4.5	75.13	94.04	108.37	116.23	5.5
4.6	75.33	94.21	108.48	116.27	5.4
4.7	75.54	94.38	108.60	116.31	5.3
4.8	75.74	94.55	108.71	116.35	5.2
4.9	75.95	94.72	108.82	116.39	5.1
5.0	76.15	94.89	108.93	116.43	5.0
Days.	120	110	100	90	Days.

TABLE XVIII. ARG. 13.

Equation = 38″.0 [1 + sin. (4*t* — *x*)].

Period, 10.084597 days.

Days.	0	1	2	3	4	5	Days.
.00	6.57	0.02	7.74	26.82	50.13	68.86	1.00
.01	6.44	0.03	7.89	27.05	50.35	69.00	.99
.02	6.31	0.04	8.03	27.27	50.58	69.13	.98
.03	6.18	0.05	8.18	27.50	50.80	69.27	.97
.04	6.05	0.06	8.32	27.73	51.02	69.40	.96
.05	5.92	0.07	8.47	27.96	51.24	69.54	.95
.06	5.79	0.09	8.62	28.19	51.46	69.67	.94
.07	5.67	0.10	8.77	28.42	51.68	69.80	.93
.08	5.54	0.12	8.92	28.65	51.90	69.93	.92
.09	5.42	0.13	9.08	28.88	52.12	70.05	.91
.10	5.30	0.15	9.23	29.11	52.34	70.18	.90
.11	5.18	0.17	9.39	29.34	52.56	70.30	.89
.12	5.06	0.20	9.54	29.57	52.78	70.43	.88
.13	4.94	0.22	9.70	29.80	53.00	70.55	.87
.14	4.83	0.25	9.86	30.04	53.21	70.67	.86
.15	4.71	0.27	10.02	30.27	53.43	70.79	.85
.16	4.60	0.30	10.18	30.50	53.64	70.91	.84
.17	4.49	0.33	10.34	30.73	53.86	71.02	.83
.18	4.38	0.36	10.50	30.96	54.07	71.14	.82
.19	4.27	0.40	10.67	31.20	54.29	71.25	.81
.20	4.16	0.43	10.83	31.43	54.50	71.37	.80
.21	4.06	0.47	11.00	31.66	54.71	71.48	.79
.22	3.95	0.51	11.17	31.90	54.93	71.59	.78
.23	3.85	0.55	11.33	32.13	55.14	71.70	.77
.24	3.74	0.59	11.50	32.37	55.35	71.81	.76
.25	3.64	0.63	11.67	32.60	55.56	71.92	.75
.26	3.54	0.68	11.84	32.83	55.77	72.02	.74
.27	3.44	0.72	12.01	33.07	55.98	72.13	.73
.28	3.34	0.77	12.18	33.30	56.19	72.23	.72
.29	3.25	0.81	12.36	33.54	56.40	72.31	.71
.30	3.15	0.86	12.53	33.77	56.61	72.41	.70
.31	3.06	0.91	12.71	34.01	56.81	72.51	.69
.32	2.96	0.97	12.88	34.24	57.02	72.64	.68
.33	2.87	1.02	13.06	34.48	57.22	72.74	.67
.34	2.78	1.08	13.24	34.71	57.43	72.83	.66
.35	2.69	1.13	13.42	34.95	57.63	72.93	.65
.36	2.61	1.19	13.60	35.19	57.83	73.02	.64
.37	2.52	1.25	13.78	35.42	58.03	73.11	.63
.38	2.44	1.31	13.96	35.66	58.23	73.20	.62
.39	2.35	1.37	14.15	35.89	58.43	73.29	.61
.40	2.27	1.43	14.33	36.13	58.63	73.38	.60
.41	2.19	1.50	14.52	36.37	58.83	73.46	.59
.42	2.11	1.56	14.70	36.61	59.03	73.55	.58
.43	2.03	1.63	14.89	36.84	59.23	73.63	.57
.44	1.96	1.70	15.08	37.08	59.42	73.71	.56
.45	1.88	1.77	15.27	37.32	59.62	73.79	.55
.46	1.81	1.84	15.46	37.56	59.81	73.87	.54
.47	1.74	1.92	15.65	37.79	60.00	73.94	.53
.48	1.67	1.99	15.84	38.03	60.20	74.02	.52
.49	1.60	2.07	16.04	38.26	60.39	74.09	.51
.50	1.53	2.15	16.23	38.50	60.58	74.17	.50
Days.	11	10	9	8	7	6	Days.

TABLE XVII.

(Continued.)

Period, 173.31006 days.

Days.	50	60	70	80	Days.
5.0	76.15	94.89	108.93	116.43	5.0
5.1	76.35	95.06	109.01	116.47	4.9
5.2	76.56	95.22	109.15	116.50	4.8
5.3	76.76	95.39	109.26	116.54	4.7
5.4	76.96	95.55	109.36	116.57	4.6
5.5	77.16	95.72	109.47	116.61	4.5
5.6	77.36	95.88	109.58	116.64	4.4
5.7	77.57	96.05	109.68	116.68	4.3
5.8	77.77	96.21	109.79	116.71	4.2
5.9	77.97	96.38	109.89	116.75	4.1
6.0	78.17	96.54	110.00	116.78	4.0
6.1	78.37	96.70	110.10	116.81	3.9
6.2	78.57	96.87	110.21	116.84	3.8
6.3	78.77	97.03	110.31	116.87	3.7
6.4	78.97	97.19	110.41	116.90	3.6
6.5	79.17	97.35	110.51	116.93	3.5
6.6	79.37	97.51	110.61	116.95	3.4
6.7	79.57	97.67	110.70	116.98	3.3
6.8	79.77	97.83	110.80	117.00	3.2
6.9	79.96	97.99	110.90	117.02	3.1
7.0	80.16	98.15	111.00	117.04	3.0
7.1	80.36	98.31	111.09	117.06	2.9
7.2	80.56	98.46	111.19	117.08	2.8
7.3	80.76	98.62	111.28	117.10	2.7
7.4	80.95	98.77	111.38	117.12	2.6
7.5	81.15	98.93	111.47	117.14	2.5
7.6	81.35	99.08	111.56	117.16	2.4
7.7	81.54	99.24	111.66	117.18	2.3
7.8	81.74	99.39	111.75	117.20	2.2
7.9	81.93	99.55	111.84	117.21	2.1
8.0	82.13	99.70	111.93	117.23	2.0
8.1	82.32	99.85	112.02	117.25	1.9
8.2	82.52	100.00	112.11	117.26	1.8
8.3	82.71	100.15	112.20	117.28	1.7
8.4	82.91	100.30	112.28	117.29	1.6
8.5	83.10	100.45	112.37	117.31	1.5
8.6	83.29	100.60	112.45	117.32	1.4
8.7	83.48	100.75	112.54	117.33	1.3
8.8	83.68	100.90	112.62	117.34	1.2
8.9	83.87	101.04	112.71	117.35	1.1
9.0	84.06	101.19	112.79	117.36	1.0
9.1	84.25	101.34	112.87	117.37	0.9
9.2	84.45	101.48	112.95	117.38	0.8
9.3	84.64	101.63	113.03	117.38	0.7
9.4	84.83	101.77	113.11	117.39	0.6
9.5	85.02	101.92	113.19	117.39	0.5
9.6	85.21	102.06	113.27	117.40	0.4
9.7	85.41	102.21	113.35	117.40	0.3
9.8	85.60	102.35	113.43	117.40	0.2
9.9	85.79	102.49	113.50	117.40	0.1
10.0	85.98	102.63	113.58	117.40	0.0
Days.	120	110	100	90	Days.

TABLE XVIII. ARG. 13.

Equation = 38″.0 [1 + sin. (4t — x)].

Period, 10.084597 days.

Days.	0	1	2	3	4	5	Days.
.50	1.53	2.15	16.23	38.50	60.58	74.17	.50
.51	1.47	2.23	16.43	38.74	60.77	74.24	.49
.52	1.40	2.31	16.62	38.97	60.96	74.31	.48
.53	1.34	2.39	16.82	39.21	61.15	74.38	.47
.54	1.28	2.48	17.02	39.44	61.33	74.45	.46
.55	1.22	2.56	17.22	39.68	61.52	74.52	.45
.56	1.16	2.65	17.42	39.92	61.70	74.58	.44
.57	1.11	2.73	17.62	40.15	61.89	74.65	.43
.58	1.05	2.82	17.82	40.39	62.07	74.71	.42
.59	1.00	2.91	18.02	40.62	62.26	74.77	.41
.60	0.94	3.00	18.22	40.86	62.44	74.83	.40
.61	0.89	3.10	18.42	41.10	62.62	74.89	.39
.62	0.84	3.19	18.63	41.33	62.80	74.94	.38
.63	0.79	3.29	18.83	41.57	62.98	75.00	.37
.64	0.75	3.38	19.04	41.80	63.15	75.05	.36
.65	0.70	3.48	19.24	42.04	63.33	75.10	.35
.66	0.66	3.58	19.45	42.28	63.50	75.15	.34
.67	0.61	3.68	19.65	42.51	63.68	75.20	.33
.68	0.57	3.78	19.86	42.75	63.85	75.25	.32
.69	0.53	3.89	20.07	42.98	64.03	75.29	.31
.70	0.49	3.99	20.28	43.22	64.20	75.34	.30
.71	0.45	4.10	20.49	43.45	64.37	75.38	.29
.72	0.42	4.20	20.70	43.69	64.54	75.42	.28
.73	0.38	4.31	20.91	43.92	64.71	75.46	.27
.74	0.35	4.42	21.12	44.16	64.88	75.50	.26
.75	0.32	4.53	21.33	44.39	65.05	75.54	.25
.76	0.29	4.65	21.54	44.62	65.21	75.57	.24
.77	0.27	4.76	21.76	44.85	65.38	75.61	.23
.78	0.24	4.88	21.97	45.09	65.54	75.64	.22
.79	0.22	4.99	22.19	45.32	65.71	75.67	.21
.80	0.19	5.11	22.40	45.55	65.87	75.70	.20
.81	0.17	5.23	22.62	45.78	66.03	75.73	.19
.82	0.15	5.36	22.83	46.01	66.18	75.75	.18
.83	0.13	5.48	23.05	46.25	66.34	75.78	.17
.84	0.11	5.61	23.26	46.48	66.49	75.80	.16
.85	0.09	5.73	23.48	46.71	66.65	75.83	.15
.86	0.08	5.86	23.70	46.94	66.80	75.85	.14
.87	0.06	5.98	23.92	47.17	66.96	75.87	.13
.88	0.05	6.11	24.14	47.40	67.11	75.89	.12
.89	0.04	6.24	24.36	47.63	67.26	75.90	.11
.90	0.03	6.37	24.58	47.86	67.41	75.92	.10
.91	0.02	6.50	24.80	48.09	67.56	75.93	.09
.92	0.02	6.64	25.02	48.32	67.71	75.95	.08
.93	0.01	6.77	25.24	48.55	67.86	75.96	.07
.94	0.01	6.91	25.47	48.77	68.00	75.97	.06
.95	0.00	7.04	25.69	49.00	68.15	75.98	.05
.96	0.00	7.18	25.92	49.23	68.29	75.99	.04
.97	0.00	7.32	26.14	49.45	68.44	75.99	.03
.98	0.01	7.46	26.37	49.68	68.58	76.00	.02
.99	0.01	7.60	26.59	49.90	68.72	76.00	.01
1.00	0.02	7.74	26.82	50.13	68.86	76.00	.00
Days.	11	10	9	8	7	6	Days.

TABLE XIX. ARGUMENT 14.

Equation $= 28''.8\ [1 - \sin.\ (2t + z - x)]$.

Period, 29.263279 days.

Days.	0	1	2	3	4	5	6	7	8	9	Days.
.00	57.50	57.33	55.86	53.13	49.27	44.50	38.99	33.02	26.85	20.77	1.00
.01	57.51	57.32	55.84	53.10	49.23	44.45	38.93	32.96	26.79	20.71	.99
.02	57.52	57.31	55.82	53.07	49.19	44.40	38.87	32.90	26.73	20.65	.98
.03	57.52	57.30	55.79	53.03	49.14	44.34	38.82	32.84	26.67	20.59	.97
.04	57.53	57.30	55.77	53.00	49.10	44.29	38.76	32.78	26.60	20.53	.96
.05	57.53	57.29	55.75	52.97	49.05	44.24	38.70	32.72	26.54	20.47	.95
.06	57.54	57.28	55.73	52.93	49.01	44.18	38.64	32.65	26.48	20.42	.94
.07	57.54	57.27	55.71	52.90	48.96	44.13	38.59	32.59	26.41	20.36	.93
.08	57.54	57.26	55.68	52.87	48.92	44.08	38.53	32.53	26.35	20.30	.92
.09	57.55	57.25	55.66	52.83	48.87	44.02	38.47	32.47	26.29	20.24	.91
.10	57.55	57.24	55.64	52.80	48.83	43.97	38.41	32.41	26.23	20.18	.90
.11	57.55	57.23	55.62	52.77	48.78	43.92	38.35	32.35	26.17	20.12	.89
.12	57.56	57.22	55.60	52.73	48.74	43.86	38.29	32.29	26.11	20.06	.88
.13	57.56	57.21	55.57	52.70	48.69	43.81	38.23	32.23	26.05	20.00	.87
.14	57.56	57.20	55.55	52.66	48.65	43.76	38.17	32.16	25.98	19.94	.86
.15	57.57	57.19	55.53	52.63	48.60	43.71	38.11	32.10	25.92	19.88	.85
.16	57.57	57.18	55.50	52.59	48.56	43.65	38.06	32.04	25.86	19.83	.84
.17	57.57	57.17	55.48	52.56	48.51	43.60	38.00	31.97	25.79	19.77	.83
.18	57.58	57.16	55.46	52.52	48.47	43.55	37.94	31.91	25.75	19.71	.82
.19	57.58	57.15	55.43	52.49	48.42	43.49	37.88	31.85	25.67	19.65	.81
.20	57.58	57.14	55.41	52.45	48.38	43.44	37.82	31.79	25.61	19.59	.80
.21	57.58	57.13	55.39	52.41	48.33	43.39	37.76	31.73	25.55	19.53	.79
.22	57.58	57.12	55.36	52.38	48.29	43.34	37.70	31.67	25.49	19.47	.78
.23	57.59	57.11	55.34	52.34	48.24	43.28	37.64	31.61	25.43	19.41	.77
.24	57.59	57.10	55.31	52.30	48.20	43.23	37.58	31.54	25.36	19.36	.76
.25	57.59	57.09	55.29	52.27	48.15	43.18	37.52	31.48	25.30	19.30	.75
.26	57.59	57.08	55.26	52.23	48.11	43.12	37.47	31.42	25.24	19.24	.74
.27	57.59	57.06	55.24	52.19	48.06	43.07	37.41	31.35	25.17	19.19	.73
.28	57.59	57.05	55.21	52.16	48.02	43.02	37.35	31.29	25.11	19.13	.72
.29	57.60	57.04	55.19	52.12	47.97	42.96	37.29	31.23	25.05	19.07	.71
.30	57.60	57.03	55.16	52.08	47.93	42.91	37.23	31.17	24.99	19.01	.70
.31	57.60	57.02	55.14	52.04	47.88	42.86	37.17	31.11	24.93	18.95	.69
.32	57.60	57.01	55.11	52.00	47.84	42.80	37.11	31.05	24.87	18.89	.68
.33	57.60	56.99	55.09	51.97	47.79	42.75	37.05	30.99	24.81	18.83	.67
.34	57.60	56.98	55.06	51.93	47.75	42.70	36.99	30.93	24.74	18.78	.66
.35	57.60	56.97	55.03	51.89	47.70	42.64	36.93	30.87	24.68	18.72	.65
.36	57.60	56.95	55.01	51.85	47.66	42.59	36.88	30.80	24.62	18.66	.64
.37	57.60	56.94	54.98	51.81	47.61	42.53	36.82	30.74	24.55	18.61	.63
.38	57.60	56.93	54.96	51.78	47.56	42.48	36.76	30.68	24.49	18.55	.62
.39	57.60	56.91	54.93	51.74	47.52	42.42	36.70	30.62	24.43	18.49	.61
.40	57.60	56.90	54.90	51.70	47.47	42.37	36.64	30.56	24.37	18.43	.60
.41	57.60	56.89	54.87	51.66	47.43	42.31	36.58	30.50	24.31	18.37	.59
.42	57.59	56.87	54.85	51.62	47.38	42.26	36.52	30.44	24.25	18.31	.58
.43	57.59	56.86	54.82	51.59	47.33	42.20	36.46	30.38	24.19	18.26	.57
.44	57.59	56.84	54.79	51.55	47.29	42.15	36.40	30.31	24.13	18.20	.56
.45	57.59	56.83	54.77	51.51	47.24	42.09	36.34	30.25	24.07	18.14	.55
.46	57.59	56.81	54.74	51.47	47.19	42.04	36.28	30.19	24.00	18.09	.54
.47	57.59	56.80	54.71	51.43	47.14	41.98	36.22	30.12	23.94	18.03	.53
.48	57.58	56.78	54.69	51.40	47.09	41.93	36.16	30.06	23.88	17.98	.52
.49	57.58	56.77	54.66	51.36	47.04	41.87	36.10	30.00	23.82	17.92	.51
.50	57.58	56.75	54.63	51.32	46.99	41.82	36.04	29.94	23.76	17.86	.50
Days.	29	28	27	26	25	24	23	22	21	20	Days.

TABLE XIX. ARGUMENT 14.

Equation $= 28''.8 \, [1 - \sin. \, (2t + z - x)]$.

Period, 29.263279 days.

Days.	0	1	2	3	4	5	6	7	8	9	Days.
Days.											Days.
.50	57.58	56.75	54.63	51.32	46.99	41.82	36.01	29.94	23.76	17.86	.50
.51	57.58	56.74	54.60	51.28	46.94	41.76	35.98	29.88	23.70	17.80	.49
.52	57.58	56.72	54.57	51.24	46.89	41.71	35.92	29.82	23.64	17.74	.48
.53	57.57	56.71	54.55	51.20	46.85	41.65	35.86	29.76	23.58	17.69	.47
.54	57.57	56.69	54.52	51.17	46.80	41.60	35.80	29.69	23.52	17.63	.46
.55	57.57	56.68	54.49	51.13	46.75	41.54	35.74	29.63	23.46	17.57	.45
.56	57.57	56.66	54.46	51.09	46.70	41.49	35.68	29.57	23.40	17.52	.44
.57	57.57	56.65	54.44	51.05	46.65	41.43	35.62	29.50	23.34	17.46	.43
.58	57.56	56.63	54.41	51.01	46.60	41.38	35.56	29.44	23.28	17.41	.42
.59	57.56	56.62	54.38	50.97	46.55	41.32	35.50	29.38	23.22	17.35	.41
.60	57.56	56.60	54.35	50.93	46.50	41.26	35.44	29.32	23.16	17.29	.40
.61	57.56	56.58	54.32	50.89	46.45	41.21	35.38	29.26	23.10	17.23	.39
.62	57.55	56.57	54.29	50.85	46.40	41.15	35.32	29.20	23.04	17.17	.38
.63	57.55	56.55	54.27	50.81	46.36	41.10	35.26	29.14	22.98	17.12	.37
.64	57.55	56.53	54.24	50.77	46.31	41.01	35.20	29.07	22.92	17.06	.36
.65	57.54	56.52	54.21	50.73	46.26	40.98	35.14	29.01	22.86	17.00	.35
.66	57.54	56.50	54.18	50.69	46.21	40.93	35.08	28.95	22.80	16.95	.34
.67	57.54	56.48	54.15	50.65	46.16	40.87	35.02	28.88	22.74	16.89	.33
.68	57.53	56.47	54.13	50.61	46.11	40.81	34.96	28.82	22.68	16.84	.32
.69	57.53	56.45	54.10	50.57	46.06	40.76	34.90	28.76	22.62	16.78	.31
.70	57.53	56.43	54.07	50.53	46.01	40.70	34.84	28.70	22.56	16.72	.30
.71	57.52	56.41	54.04	50.49	45.96	40.64	34.78	28.64	22.50	16.66	.29
.72	57.52	56.39	54.01	50.45	45.91	40.59	34.72	28.58	22.44	16.61	.28
.73	57.52	56.38	53.98	50.41	45.86	40.53	34.66	28.52	22.38	16.55	.27
.74	57.51	56.36	53.95	50.37	45.82	40.47	34.60	28.46	22.32	16.50	.26
.75	57.51	56.34	53.92	50.33	45.77	40.42	34.54	28.40	22.26	16.44	.25
.76	57.50	56.32	53.89	50.28	45.72	40.36	34.47	28.33	22.20	16.38	.24
.77	57.50	56.30	53.86	50.24	45.67	40.30	34.41	28.27	22.14	16.33	.23
.78	57.49	56.29	53.83	50.20	45.62	40.25	34.35	28.21	22.08	16.27	.22
.79	57.49	56.27	53.80	50.16	45.57	40.19	34.29	28.15	22.02	16.22	.21
.80	57.48	56.25	53.77	50.12	45.52	40.13	34.23	28.09	21.96	16.16	.20
.81	57.47	56.23	53.74	50.08	45.47	40.07	34.17	28.03	21.90	16.10	.19
.82	57.47	56.21	53.71	50.04	45.42	40.02	34.11	27.97	21.84	16.05	.18
.83	57.46	56.19	53.68	50.00	45.37	39.96	34.05	27.91	21.78	15.99	.17
.84	57.46	56.18	53.65	49.95	45.32	39.90	33.99	27.84	21.72	15.94	.16
.85	57.45	56.16	53.62	49.91	45.27	39.85	33.93	27.78	21.66	15.88	.15
.86	57.45	56.14	53.59	49.87	45.22	39.79	33.86	27.72	21.60	15.83	.14
.87	57.44	56.12	53.55	49.83	45.17	39.73	33.80	27.65	21.54	15.77	.13
.88	57.43	56.10	53.52	49.78	45.12	39.68	33.74	27.59	21.48	15.72	.12
.89	57.42	56.08	53.49	49.74	45.07	39.62	33.68	27.53	21.42	15.66	.11
.90	57.41	56.06	53.46	49.70	45.02	39.56	33.62	27.47	21.36	15.61	.10
.91	57.40	56.04	53.43	49.66	44.97	39.50	33.56	27.41	21.30	15.55	.09
.92	57.40	56.02	53.40	49.62	44.92	39.45	33.50	27.35	21.24	15.50	.08
.93	57.39	56.00	53.36	49.57	44.86	39.39	33.44	27.29	21.18	15.44	.07
.94	57.39	55.98	53.33	49.53	44.81	39.33	33.38	27.22	21.12	15.39	.06
.95	57.38	55.96	53.30	49.49	44.76	39.28	33.32	27.16	21.07	15.33	.05
.96	57.37	55.94	53.26	49.44	44.71	39.22	33.26	27.10	21.01	15.28	.04
.97	57.36	55.92	53.23	49.40	44.65	39.16	33.20	27.03	20.95	15.22	.03
.98	57.35	55.90	53.20	49.36	44.60	39.11	33.14	26.97	20.89	15.17	.02
.99	57.34	55.88	53.16	49.31	44.55	39.05	33.08	26.91	20.83	15.11	.01
1.00	57.33	55.86	53.13	49.27	44.50	38.99	33.02	26.85	20.77	15.06	.00
Days.	29	28	27	26	25	24	23	22	21	20	Days.

TABLE XIX.

(*Continued.*)

Period, 29.263279 days.

TABLE XX. ARG. 15.

Equation = 25″.0 [1 — sin. (2t + z)].

Period, 14.1916109 days.

Days.	10	11	12	13	14	0	1	2	3	Days.
.00	15.06″	9.98″	5.77″	2.62″	0.66″	48.06″	49.98″	47.12″	39.98″	1.00
.01	15.01	9.93	5.73	2.59	0.65	48.10	49.98	47.07	39.89	.99
.02	14.95	9.89	5.69	2.57	0.64	48.15	49.97	47.01	39.80	.98
.03	14.90	9.84	5.66	2.54	0.62	48.19	49.96	46.96	39.71	.97
.04	14.84	9.79	5.62	2.52	0.61	48.23	49.96	46.91	39.62	.96
.05	14.79	9.75	5.58	2.49	0.60	48.27	49.95	46.86	39.53	.95
.06	14.74	9.70	5.55	2.46	0.59	48.31	49.95	46.80	39.44	.94
.07	14.68	9.66	5.51	2.44	0.57	48.35	49.94	46.75	39.35	.93
.08	14.63	9.61	5.48	2.41	0.56	48.39	49.93	46.69	39.26	.92
.09	14.57	9.57	5.44	2.39	0.55	48.43	49.92	46.64	39.17	.91
.10	14.52	9.52	5.40	2.36	0.54	48.47	49.91	46.58	39.08	.90
.11	14.47	9.47	5.36	2.34	0.53	48.51	49.90	46.52	38.99	.89
.12	14.41	9.43	5.33	2.31	0.52	48.54	49.89	46.46	38.90	.88
.13	14.36	9.38	5.29	2.29	0.51	48.58	49.88	46.41	38.81	.87
.14	14.31	9.34	5.25	2.26	0.50	48.61	49.87	46.35	38.71	.86
.15	14.25	9.29	5.22	2.24	0.48	48.65	49.86	46.29	38.62	.85
.16	14.20	9.25	5.18	2.22	0.47	48.68	49.84	46.24	38.53	.84
.17	14.15	9.21	5.15	2.19	0.46	48.71	49.83	46.18	38.44	.83
.18	14.09	9.16	5.11	2.17	0.45	48.75	49.82	46.12	38.34	.82
.19	14.04	9.12	5.08	2.14	0.44	48.78	49.80	46.06	38.25	.81
.20	13.99	9.07	5.04	2.12	0.43	48.82	49.79	46.00	38.16	.80
.21	13.94	9.02	5.01	2.10	0.42	48.85	49.78	45.94	38.07	.79
.22	13.88	8.98	4.97	2.08	0.41	48.88	49.76	45.88	37.97	.78
.23	13.83	8.93	4.94	2.05	0.40	48.92	49.74	45.82	37.88	.77
.24	13.78	8.89	4.90	2.03	0.39	48.95	49.73	45.76	37.78	.76
.25	13.72	8.84	4.87	2.00	0.38	48.98	49.71	45.69	37.69	.75
.26	13.67	8.80	4.83	1.98	0.37	49.01	49.69	45.63	37.59	.74
.27	13.62	8.76	4.80	1.95	0.36	49.04	49.68	45.57	37.50	.73
.28	13.56	8.71	4.77	1.93	0.35	49.07	49.66	45.50	37.40	.72
.29	13.51	8.67	4.73	1.91	0.34	49.10	49.64	45.44	37.31	.71
.30	13.46	8.62	4.70	1.89	0.33	49.13	49.62	45.38	37.21	.70
.31	13.41	8.58	4.67	1.87	0.32	49.16	49.60	45.31	37.11	.69
.32	13.36	8.53	4.64	1.85	0.31	49.19	49.58	45.25	37.01	.68
.33	13.31	8.49	4.60	1.83	0.30	49.22	49.56	45.18	36.92	.67
.34	13.25	8.44	4.57	1.81	0.29	49.25	49.54	45.11	36.82	.66
.35	13.20	8.40	4.53	1.78	0.28	49.27	49.51	45.05	36.72	.65
.36	13.15	8.36	4.50	1.76	0.28	49.30	49.49	44.98	36.62	.64
.37	13.09	8.31	4.47	1.74	0.27	49.33	49.47	44.91	36.53	.63
.38	13.04	8.27	4.43	1.72	0.26	49.35	49.45	44.85	36.43	.62
.39	12.99	8.22	4.40	1.70	0.25	49.38	49.42	44.78	36.33	.61
.40	12.94	8.18	4.37	1.68	0.24	49.40	49.40	44.71	36.23	.60
.41	12.89	8.14	4.34	1.66	0.23	49.42	49.37	44.64	36.13	.59
.42	12.84	8.10	4.31	1.64	0.22	49.45	49.35	44.57	36.03	.58
.43	12.79	8.06	4.28	1.62	0.22	49.47	49.32	44.50	35.93	.57
.44	12.73	8.01	4.24	1.60	0.21	49.49	49.29	44.43	35.83	.56
.45	12.68	7.97	4.21	1.58	0.20	49.52	49.27	44.36	35.73	.55
.46	12.63	7.93	4.18	1.56	0.20	49.54	49.24	44.29	35.63	.54
.47	12.57	7.88	4.14	1.54	0.19	49.56	49.21	44.22	35.53	.53
.48	12.52	7.84	4.11	1.52	0.19	49.58	49.19	44.15	35.43	.52
.49	12.47	7.80	4.08	1.50	0.18	49.60	49.16	44.08	35.33	.51
.50	12.42	7.76	4.05	1.48	0.17	49.62	49.13	44.01	35.23	.50
Days.	19	18	17	16	15	15	14	13	12	Days.

TABLE XIX.

(Continued.)

Period, 29.263279 days.

TABLE XX. ARG. 15.

Equation = 25″.0 [1 — sin. $(2t + z)$].

Period, 14.1916109 days.

Days.	10	11	12	13	14	0	1	2	3	Days.
.50	12.42	7.76	4.05	1.48	0.17	49.62	49.13	44.01	35.23	.50
.51	12.37	7.72	4.02	1.46	0.16	49.64	49.10	43.94	35.13	.49
.52	12.32	7.68	3.99	1.44	0.15	49.66	49.07	43.86	35.03	.48
.53	12.27	7.64	3.96	1.42	0.15	49.68	49.04	43.79	34.93	.47
.54	12.22	7.60	3.93	1.40	0.14	49.70	49.01	43.72	34.82	.46
.55	12.17	7.56	3.89	1.38	0.14	49.71	48.98	43.64	34.72	.45
.56	12.12	7.51	3.86	1.37	0.13	49.73	48.94	43.57	34.62	.44
.57	12.07	7.47	3.83	1.35	0.13	49.74	48.91	43.50	34.52	.43
.58	12.02	7.43	3.80	1.33	0.12	49.75	48.88	43.42	34.41	.42
.59	11.97	7.39	3.77	1.31	0.12	49.77	48.85	43.35	34.31	.41
.60	11.92	7.35	3.74	1.29	0.11	49.78	48.82	43.28	34.21	.40
.61	11.87	7.31	3.71	1.27	0.11	49.79	48.79	43.20	34.11	.39
.62	11.82	7.27	3.68	1.25	0.10	49.81	48.75	43.13	34.00	.38
.63	11.77	7.23	3.65	1.24	0.10	49.82	48.72	43.05	33.90	.37
.64	11.72	7.19	3.62	1.22	0.09	49.83	48.69	42.97	33.79	.36
.65	11.67	7.15	3.59	1.20	0.08	49.85	48.65	42.90	33.69	.35
.66	11.63	7.10	3.56	1.19	0.08	49.86	48.61	42.82	33.59	.34
.67	11.58	7.06	3.53	1.17	0.07	49.87	48.58	42.75	33.48	.33
.68	11.53	7.02	3.50	1.15	0.07	49.88	48.54	42.67	33.38	.32
.69	11.48	6.98	3.47	1.13	0.06	49.89	48.50	42.59	33.27	.31
.70	11.43	6.94	3.44	1.11	0.06	49.90	48.46	42.51	33.17	.30
.71	11.38	6.90	3.41	1.09	0.06	49.91	48.42	42.43	33.07	.29
.72	11.33	6.86	3.38	1.08	0.05	49.92	48.38	42.35	32.96	.28
.73	11.28	6.82	3.35	1.06	0.05	49.93	48.34	42.27	32.86	.27
.74	11.23	6.78	3.33	1.04	0.05	49.94	48.30	42.19	32.75	.26
.75	11.18	6.74	3.30	1.03	0.04	49.94	48.26	42.11	32.65	.25
.76	11.14	6.70	3.27	1.01	0.04	49.95	48.22	42.02	32.54	.24
.77	11.09	6.66	3.25	1.00	0.04	49.95	48.18	41.94	32.43	.23
.78	11.04	6.62	3.22	0.98	0.03	49.96	48.14	41.86	32.33	.22
.79	10.99	6.58	3.19	0.97	0.03	49.97	48.10	41.78	32.22	.21
.80	10.94	6.54	3.16	0.95	0.03	49.97	48.06	41.70	32.12	.20
.81	10.89	6.50	3.13	0.93	0.03	49.97	48.02	41.62	32.01	.19
.82	10.84	6.46	3.10	0.92	0.03	49.98	47.97	41.53	31.90	.18
.83	10.79	6.42	3.08	0.90	0.03	49.98	47.93	41.45	31.80	.17
.84	10.74	6.38	3.05	0.89	0.02	49.98	47.88	41.36	31.69	.16
.85	10.69	6.34	3.02	0.87	0.02	49.99	47.84	41.28	31.59	.15
.86	10.65	6.31	3.00	0.86	0.02	49.99	47.79	41.19	31.49	.14
.87	10.60	6.27	2.97	0.85	0.02	49.99	47.75	41.10	31.38	.13
.88	10.55	6.23	2.95	0.83	0.01	50.00	47.70	41.02	31.28	.12
.89	10.50	6.19	2.92	0.82	0.01	50.00	47.66	40.93	31.17	.11
.90	10.45	6.15	2.89	0.80	0.01	50.00	47.61	40.85	31.06	.10
.91	10.40	6.11	2.86	0.79	0.01	50.00	47.56	40.76	30.95	.09
.92	10.35	6.07	2.83	0.77	0.01	50.00	47.51	40.68	30.84	.08
.93	10.31	6.03	2.81	0.76	0.00	50.00	47.46	40.59	30.73	.07
.94	10.26	6.00	2.78	0.74	0.00	49.99	47.41	40.51	30.63	.06
.95	10.22	5.96	2.75	0.73	0.00	49.99	47.37	40.42	30.52	.05
.96	10.17	5.92	2.73	0.72	0.00	49.99	47.32	40.33	30.41	.04
.97	10.12	5.89	2.70	0.70	0.00	49.99	47.27	40.25	30.31	.03
.98	10.08	5.85	2.68	0.69	0.00	49.98	47.22	40.16	30.20	.02
.99	10.03	5.81	2.65	0.67	0.00	49.98	47.17	40.07	30.09	.01
1.00	9.98	5.77	2.62	0.66	0.00	49.98	47.12	39.98	29.98	.00
Days.	19	18	17	16	15	15	14	13	12	Days.

TABLE XX.

(*Continued.*)

Period, 14.1916109 days.

Days.	4	5	6	7	Days.
.00	29.98	18.99	9.17	2.41	1.00
.01	29.87	18.88	9.09	2.36	.99
.02	29.76	18.78	9.00	2.32	.98
.03	29.65	18.67	8.92	2.27	.97
.04	29.54	18.56	8.83	2.23	.96
.05	29.43	18.46	8.75	2.18	.95
.06	29.32	18.35	8.67	2.14	.94
.07	29.21	18.24	8.58	2.10	.93
.08	29.10	18.14	8.50	2.05	.92
.09	28.99	18.03	8.41	2.01	.91
.10	28.88	17.92	8.33	1.96	.90
.11	28.77	17.82	8.25	1.92	.89
.12	28.66	17.71	8.17	1.88	.88
.13	28.55	17.61	8.09	1.84	.87
.14	28.44	17.50	8.01	1.80	.86
.15	28.33	17.39	7.92	1.75	.85
.16	28.22	17.29	7.84	1.71	.84
.17	28.11	17.18	7.76	1.67	.83
.18	28.00	17.08	7.68	1.63	.82
.19	27.89	16.97	7.60	1.59	.81
.20	27.78	16.87	7.52	1.55	.80
.21	27.67	16.77	7.44	1.51	.79
.22	27.56	16.66	7.36	1.48	.78
.23	27.45	16.56	7.29	1.44	.77
.24	27.34	16.45	7.21	1.40	.76
.25	27.23	16.35	7.13	1.37	.75
.26	27.12	16.24	7.06	1.33	.74
.27	27.01	16.14	6.98	1.30	.73
.28	26.90	16.04	6.91	1.26	.72
.29	26.79	15.93	6.83	1.23	.71
.30	26.68	15.83	6.75	1.19	.70
.31	26.57	15.73	6.68	1.16	.69
.32	26.46	15.63	6.60	1.13	.68
.33	26.35	15.52	6.53	1.10	.67
.34	26.24	15.42	6.45	1.07	.66
.35	26.13	15.32	6.38	1.03	.65
.36	26.02	15.22	6.30	1.00	.64
.37	25.91	15.11	6.23	0.97	.63
.38	25.80	15.01	6.16	0.94	.62
.39	25.69	14.91	6.08	0.91	.61
.40	25.58	14.81	6.01	0.88	.60
.41	25.47	14.71	5.94	0.85	.59
.42	25.36	14.61	5.87	0.82	.58
.43	25.25	14.51	5.80	0.80	.57
.44	25.14	14.41	5.73	0.77	.56
.45	25.03	14.31	5.66	0.74	.55
.46	24.92	14.21	5.59	0.72	.54
.47	24.81	14.11	5.52	0.69	.53
.48	24.70	14.01	5.45	0.67	.52
.49	24.59	13.91	5.38	0.64	.51
.50	24.48	13.81	5.31	0.61	.50
Days.	11	10	9	8	Days.

TABLE XXI. ARG. 16.

Equation = 21″. — 8″.2 sin. $(t+x)$ + 14″.1 sin. 2 $(t+x)$.

Period, 14.2541847 days.

Days.	0	1	2	3	4
.00	36.71	41.06	34.34	22.23	13.49
.01	36.82	41.04	34.23	22.11	13.45
.02	36.92	41.02	34.12	21.99	13.40
.03	37.02	41.00	34.01	21.88	13.36
.04	37.11	40.98	33.90	21.76	13.31
.05	37.21	40.96	33.79	21.64	13.27
.06	37.30	40.94	33.68	21.53	13.23
.07	37.40	40.91	33.57	21.41	13.19
.08	37.49	40.89	33.46	21.29	13.15
.09	37.58	40.86	33.34	21.18	13.11
.10	37.67	40.84	33.23	21.07	13.07
.11	37.76	40.81	33.12	20.95	13.04
.12	37.85	40.78	33.01	20.84	13.00
.13	37.94	40.75	32.89	20.73	12.97
.14	38.02	40.72	32.78	20.61	12.93
.15	38.11	40.69	32.66	20.50	12.90
.16	38.19	40.65	32.55	20.39	12.87
.17	38.28	40.62	32.43	20.28	12.84
.18	38.36	40.58	32.31	20.17	12.81
.19	38.44	40.55	32.19	20.06	12.78
.20	38.52	40.51	32.07	19.95	12.75
.21	38.60	40.47	31.95	19.84	12.73
.22	38.68	40.43	31.83	19.73	12.70
.23	38.75	40.39	31.71	19.62	12.68
.24	38.83	40.35	31.59	19.52	12.65
.25	38.90	40.30	31.47	19.41	12.63
.26	38.98	40.26	31.35	19.30	12.61
.27	39.05	40.21	31.23	19.19	12.59
.28	39.12	40.17	31.11	19.08	12.57
.29	39.19	40.12	30.99	18.98	12.55
.30	39.26	40.07	30.87	18.88	12.53
.31	39.32	40.02	30.75	18.78	12.51
.32	39.39	39.97	30.63	18.68	12.50
.33	39.45	39.92	30.51	18.58	12.48
.34	39.51	39.86	30.39	18.48	12.47
.35	39.57	39.81	30.27	18.38	12.45
.36	39.63	39.75	30.15	18.28	12.44
.37	39.69	39.70	30.02	18.18	12.43
.38	39.75	39.64	29.90	18.08	12.42
.39	39.81	39.58	29.78	17.98	12.41
.40	39.87	39.52	29.65	17.88	12.40
.41	39.92	39.46	29.53	17.78	12.39
.42	39.97	39.40	29.41	17.69	12.39
.43	40.02	39.34	29.29	17.59	12.38
.44	40.07	39.27	29.17	17.50	12.38
.45	40.12	39.21	29.04	17.40	12.38
.46	40.17	39.14	28.92	17.31	12.37
.47	40.22	39.08	28.80	17.22	12.37
.48	40.27	39.01	28.67	17.12	12.37
.49	40.31	38.94	28.54	17.03	12.37
.50	40.36	38.87	28.42	16.94	12.37

TABLE XX.

(*Continued.*)

Period, 14.1916109 days.

Days.	4	5	6	7	Days.
Days.	"	"	"	"	Days.
.50	24.48	13.81	5.31	0.61	.50
.51	24.37	13.71	5.24	0.59	.49
.52	24.26	13.61	5.18	0.57	.48
.53	24.15	13.51	5.11	0.55	.47
.54	24.04	13.42	5.05	0.52	.46
.55	23.93	13.32	4.98	0.50	.45
.56	23.82	13.22	4.92	0.48	.44
.57	23.71	13.12	4.85	0.45	.43
.58	23.60	13.03	4.78	0.43	.42
.59	23.49	12.93	4.72	0.41	.41
.60	23.38	12.83	4.65	0.39	.40
.61	23.27	12.73	4.59	0.37	.39
.62	23.16	12.63	4.53	0.35	.38
.63	23.05	12.54	4.47	0.34	.37
.64	22.94	12.44	4.40	0.32	.36
.65	22.83	12.35	4.34	0.30	.35
.66	22.72	12.25	4.27	0.29	.34
.67	22.61	12.16	4.21	0.27	.33
.68	22.50	12.06	4.14	0.26	.32
.69	22.39	11.97	4.08	0.24	.31
.70	22.28	11.87	4.02	0.22	.30
.71	22.17	11.78	3.96	0.21	.29
.72	22.06	11.69	3.90	0.20	.28
.73	21.95	11.59	3.85	0.19	.27
.74	21.84	11.50	3.79	0.17	.26
.75	21.73	11.41	3.73	0.16	.25
.76	21.62	11.31	3.67	0.15	.24
.77	21.51	11.22	3.62	0.13	.23
.78	21.40	11.12	3.56	0.12	.22
.79	21.29	11.03	3.50	0.11	.21
.80	21.18	10.94	3.44	0.10	.20
.81	21.07	10.85	3.39	0.09	.19
.82	20.96	10.76	3.33	0.08	.18
.83	20.85	10.67	3.28	0.08	.17
.84	20.74	10.58	3.22	0.07	.16
.85	20.63	10.49	3.17	0.06	.15
.86	20.52	10.40	3.11	0.06	.14
.87	20.41	10.31	3.06	0.05	.13
.88	20.30	10.22	3.01	0.05	.12
.89	20.19	10.13	2.95	0.04	.11
.90	20.08	10.04	2.90	0.03	.10
.91	19.97	9.95	2.85	0.03	.09
.92	19.86	9.86	2.80	0.03	.08
.93	19.75	9.78	2.75	0.02	.07
.94	19.64	9.69	2.70	0.02	.06
.95	19.53	9.60	2.65	0.01	.05
.96	19.42	9.52	2.61	0.01	.04
.97	19.32	9.43	2.56	0.01	.03
.98	19.21	9.35	2.51	0.00	.02
.99	19.10	9.26	2.46	0.00	.01
1.00	18.99	9.17	2.41	0.00	.00
Days.	11	10	9	8	Days.

TABLE XXI. ARG. 16.

Equation $= 21''. - 8''.2 \sin. (t + x) + 14''.1 \sin. 2 (t + x)$.

Period, 14.2541847 days.

Days.	0	1	2	3	4
Days.	"	"	"	"	"
.50	40.36	38.87	28.42	16.94	12.37
.51	40.40	38.80	28.29	16.85	12.37
.52	40.44	38.73	28.17	16.76	12.38
.53	40.48	38.66	28.05	16.67	12.38
.54	40.52	38.59	27.92	16.59	12.38
.55	40.56	38.51	27.80	16.50	12.39
.56	40.60	38.44	27.66	16.41	12.40
.57	40.63	38.36	27.54	16.33	12.41
.58	40.67	38.29	27.42	16.24	12.42
.59	40.70	38.21	27.29	16.16	12.43
.60	40.74	38.13	27.16	16.08	12.44
.61	40.77	38.05	27.04	16.00	12.45
.62	40.80	37.97	26.91	15.92	12.46
.63	40.84	37.89	26.79	15.84	12.48
.64	40.86	37.81	26.66	15.76	12.49
.65	40.88	37.73	26.54	15.68	12.50
.66	40.91	37.64	26.41	15.61	12.52
.67	40.93	37.56	26.29	15.53	12.54
.68	40.96	37.47	26.16	15.46	12.56
.69	40.98	37.39	26.04	15.38	12.58
.70	41.00	37.30	25.91	15.31	12.60
.71	41.02	37.21	25.78	15.24	12.62
.72	41.04	37.12	25.66	15.17	12.64
.73	41.05	37.03	25.53	15.10	12.67
.74	41.07	36.94	25.41	15.03	12.69
.75	41.08	36.85	25.28	14.96	12.72
.76	41.10	36.76	25.16	14.89	12.74
.77	41.11	36.67	25.03	14.82	12.77
.78	41.12	36.57	24.91	14.76	12.79
.79	41.13	36.48	24.78	14.69	12.82
.80	41.14	36.38	24.66	14.62	12.85
.81	41.14	36.28	24.53	14.56	12.88
.82	41.15	36.19	24.41	14.49	12.92
.83	41.15	36.09	24.29	14.43	12.95
.84	41.16	35.99	24.16	14.37	12.99
.85	41.16	35.89	24.04	14.31	13.02
.86	41.16	35.79	23.92	14.25	13.06
.87	41.16	35.69	23.79	14.19	13.09
.88	41.16	35.59	23.67	14.13	13.12
.89	41.16	35.49	23.55	14.07	13.16
.90	41.16	35.39	23.43	14.01	13.19
.91	41.15	35.29	23.31	13.96	13.23
.92	41.14	35.19	23.19	13.90	13.27
.93	41.13	35.09	23.07	13.85	13.31
.94	41.12	34.98	22.95	13.79	13.35
.95	41.11	34.88	22.83	13.74	13.40
.96	41.10	34.77	22.71	13.69	13.44
.97	41.09	34.67	22.59	13.64	13.48
.98	41.08	34.56	22.47	13.59	13.53
.99	41.07	34.45	22.35	13.54	13.57
1.00	41.06	34.34	22.23	13.49	13.62

TABLE XXI. ARGUMENT 16.

Equation = $21''. - 8''.2 \sin. (t + x) + 14''.1 \sin. 2 (t + x)$.

Period, 14.2541847 days.

Days.	5	6	7	8	9	10	11	12	13	14
Days.	"	"	"	"	"	"	"	"	"	"
.00	13.62	21.00	28.38	28.51	19.77	7.66	0.94	5.28	18.97	33.88
.01	13.67	21.09	28.43	28.46	19.65	7.55	0.93	5.38	19.13	34.00
.02	13.72	21.17	28.47	28.41	19.53	7.44	0.92	5.48	19.29	34.12
.03	13.77	21.26	28.52	28.36	19.41	7.34	0.91	5.58	19.45	34.24
.04	13.82	21.35	28.56	28.31	19.29	7.23	0.90	5.69	19.60	34.36
.05	13.87	21.44	28.60	28.26	19.17	7.13	0.89	5.79	19.76	34.48
.06	13.92	21.53	28.65	28.20	19.05	7.02	0.88	5.90	19.92	34.60
.07	13.97	21.61	28.69	28.15	18.93	6.92	0.87	6.00	20.08	34.71
.08	14.02	21.70	28.73	28.10	18.81	6.81	0.86	6.11	20.24	34.83
.09	14.08	21.79	28.77	28.04	18.69	6.71	0.85	6.22	20.40	34.95
.10	14.13	21.88	28.81	27.99	18.57	6.61	0.84	6.33	20.56	35.06
.11	14.18	21.97	28.85	27.93	18.45	6.51	0.84	6.44	20.72	35.17
.12	14.23	22.05	28.89	27.87	18.33	6.41	0.83	6.55	20.88	35.28
.13	14.29	22.14	28.92	27.81	18.21	6.31	0.83	6.66	21.04	35.39
.14	14.34	22.23	28.96	27.75	18.08	6.21	0.83	6.77	21.20	35.50
.15	14.40	22.32	28.99	27.69	17.96	6.11	0.84	6.89	21.36	35.61
.16	14.46	22.40	29.02	27.63	17.84	6.01	0.84	7.00	21.52	35.72
.17	14.51	22.49	29.06	27.57	17.71	5.91	0.84	7.12	21.68	35.83
.18	14.57	22.58	29.09	27.50	17.59	5.82	0.85	7.23	21.84	35.94
.19	14.63	22.66	29.12	27.44	17.47	5.72	0.85	7.35	22.00	36.05
.20	14.69	22.75	29.15	27.38	17.34	5.62	0.86	7.47	22.16	36.16
.21	14.75	22.84	29.18	27.32	17.22	5.53	0.87	7.59	22.32	36.26
.22	14.81	22.92	29.21	27.25	17.09	5.43	0.88	7.71	22.48	36.36
.23	14.87	23.01	29.24	27.18	16.97	5.34	0.89	7.83	22.64	36.46
.24	14.93	23.09	29.26	27.11	16.84	5.25	0.90	7.95	22.80	36.56
.25	14.99	23.18	29.29	27.04	16.72	5.15	0.92	8.07	22.96	36.66
.26	15.06	23.27	29.32	26.97	16.59	5.06	0.93	8.19	23.12	36.76
.27	15.12	23.35	29.34	26.90	16.47	4.97	0.94	8.32	23.28	36.86
.28	15.18	23.44	29.36	26.83	16.34	4.88	0.96	8.44	23.44	36.96
.29	15.25	23.53	29.38	26.76	16.22	4.79	0.98	8.56	23.60	37.06
.30	15.31	23.61	29.40	26.69	16.09	4.70	1.00	8.69	23.76	37.16
.31	15.38	23.70	29.42	26.62	15.97	4.61	1.02	8.82	23.92	37.25
.32	15.44	23.78	29.44	26.54	15.84	4.53	1.04	8.94	24.08	37.34
.33	15.51	23.86	29.46	26.47	15.72	4.44	1.06	9.07	24.24	37.43
.34	15.58	23.95	29.47	26.39	15.59	4.36	1.09	9.20	24.39	37.52
.35	15.65	24.03	29.49	26.31	15.47	4.28	1.11	9.33	24.55	37.61
.36	15.72	24.11	29.51	26.24	15.34	4.19	1.14	9.46	24.71	37.70
.37	15.79	24.20	29.52	26.16	15.22	4.11	1.17	9.59	24.87	37.79
.38	15.86	24.28	29.54	26.08	15.09	4.03	1.20	9.72	25.02	37.88
.39	15.93	24.36	29.55	26.00	14.97	3.95	1.23	9.85	25.18	37.97
.40	16.00	24.44	29.56	25.92	14.84	3.87	1.26	9.98	25.34	38.06
.41	16.07	24.52	29.57	25.84	14.72	3.79	1.30	10.11	25.50	38.14
.42	16.14	24.60	29.58	25.76	14.59	3.71	1.33	10.25	25.65	38.22
.43	16.22	24.68	29.59	25.67	14.47	3.64	1.37	10.38	25.81	38.30
.44	16.30	24.76	29.59	25.59	14.34	3.56	1.41	10.52	25.97	38.38
.45	16.37	24.84	29.60	25.50	14.22	3.49	1.44	10.66	26.12	38.46
.46	16.45	24.92	29.61	25.41	14.09	3.41	1.48	10.79	26.28	38.54
.47	16.53	25.00	29.61	25.33	13.97	3.34	1.52	10.93	26.43	38.62
.48	16.60	25.08	29.62	25.24	13.84	3.27	1.56	11.07	26.59	38.69
.49	16.68	25.16	29.63	25.15	13.72	3.20	1.60	11.21	26.74	38.77
.50	16.76	25.24	29.63	25.06	13.59	3.13	1.64	11.35	26.89	38.85

TABLE XXI. ARGUMENT 16.

Equation = $21''.-8''.2 \sin.(t+x)+14''.1 \sin. 2 (t+x)$.

Period, 14.2541847 days.

Days.	5	6	7	8	9	10	11	12	13	14
Days.	″	″	″	″	″	″	″	″	″	″
.50	16.76	25.24	29.63	25.06	13.59	3.13	1.64	11.35	26.89	38.85
.51	16.84	25.32	29.63	24.97	13.47	3.06	1.69	11.49	27.04	38.92
.52	16.92	25.40	29.63	24.88	13.34	2.99	1.74	11.63	27.20	38.99
.53	17.00	25.47	29.62	24.78	13.22	2.92	1.78	11.77	27.35	39.06
.54	17.08	25.55	29.62	24.69	13.09	2.86	1.83	11.91	27.50	39.13
.55	17.16	25.63	29.62	24.60	12.97	2.79	1.88	12.06	27.65	39.20
.56	17.24	25.70	29.61	24.50	12.84	2.73	1.93	12.20	27.80	39.26
.57	17.32	25.78	29.61	24.41	12.72	2.66	1.98	12.35	27.95	39.33
.58	17.40	25.85	29.61	24.31	12.60	2.60	2.03	12.49	28.10	39.40
.59	17.48	25.93	29.60	24.22	12.47	2.54	2.08	12.64	28.25	39.46
.60	17.56	26.00	29.60	24.12	12.35	2.48	2.13	12.79	28.40	39.53
.61	17.64	26.07	29.59	24.02	12.23	2.42	2.19	12.94	28.55	39.59
.62	17.72	26.14	29.58	23.92	12.10	2.36	2.25	13.08	28.70	39.65
.63	17.80	26.21	29.57	23.82	11.98	2.30	2.31	13.23	28.85	39.71
.64	17.88	26.28	29.56	23.72	11.86	2.25	2.37	13.38	29.00	39.77
.65	17.97	26.35	29.54	23.62	11.74	2.19	2.43	13.53	29.14	39.83
.66	18.05	26.42	29.53	23.52	11.61	2.14	2.49	13.68	29.29	39.88
.67	18.13	26.49	29.51	23.42	11.49	2.08	2.55	13.83	29.44	39.94
.68	18.22	26.55	29.50	23.32	11.37	2.05	2.61	13.98	29.58	39.99
.69	18.30	26.62	29.48	23.22	11.25	1.98	2.68	14.13	29.73	40.05
.70	18.39	26.69	29.47	23.12	11.13	1.93	2.74	14.28	29.88	40.10
.71	18.47	26.75	29.45	23.01	11.01	1.88	2.81	14.43	30.02	40.15
.72	18.55	26.82	29.43	22.91	10.89	1.83	2.88	14.58	30.16	40.20
.73	18.64	26.88	29.41	22.80	10.77	1.79	2.95	14.73	30.30	40.25
.74	18.72	26.94	29.39	22.70	10.65	1.74	3.02	14.88	30.44	40.29
.75	18.81	27.01	29.37	22.59	10.53	1.70	3.10	15.04	30.58	40.34
.76	18.90	27.07	29.35	22.48	10.41	1.66	3.17	15.19	30.72	40.38
.77	18.98	27.13	29.32	22.38	10.29	1.61	3.25	15.34	30.86	40.43
.78	19.07	27.19	29.30	22.27	10.17	1.57	3.32	15.50	31.00	40.47
.79	19.16	27.25	29.28	22.16	10.05	1.53	3.40	15.65	31.14	40.51
.80	19.25	27.31	29.25	22.05	9.93	1.49	3.48	15.81	31.28	40.55
.81	19.34	27.37	29.22	21.94	9.81	1.45	3.56	15.96	31.42	40.59
.82	19.43	27.43	29.19	21.83	9.69	1.42	3.64	16.12	31.56	40.62
.83	19.51	27.49	29.16	21.72	9.58	1.38	3.72	16.27	31.69	40.66
.84	19.59	27.54	29.13	21.61	9.46	1.35	3.81	16.43	31.83	40.69
.85	19.68	27.60	29.10	21.50	9.35	1.32	3.89	16.59	31.96	40.72
.86	19.77	27.65	29.06	21.38	9.23	1.28	3.97	16.74	32.09	40.76
.87	19.85	27.71	29.03	21.27	9.12	1.25	4.06	16.90	32.23	40.79
.88	19.94	27.77	29.00	21.16	9.00	1.22	4.15	17.06	32.36	40.82
.89	20.03	27.82	28.96	21.04	8.89	1.19	4.24	17.22	32.49	40.85
.90	20.12	27.87	28.93	20.93	8.77	1.16	4.33	17.38	32.62	40.88
.91	20.21	27.92	28.89	20.82	8.66	1.14	4.42	17.54	32.75	40.90
.92	20.30	27.97	28.85	20.70	8.54	1.11	4.51	17.70	32.88	40.92
.93	20.38	28.02	28.81	20.59	8.43	1.09	4.60	17.86	33.01	40.95
.94	20.47	28.07	28.77	20.47	8.32	1.07	4.70	18.01	33.14	40.97
.95	20.56	28.12	28.73	20.35	8.21	1.04	4.79	18.17	33.26	40.99
.96	20.65	28.17	28.68	20.24	8.10	1.02	4.89	18.33	33.39	41.01
.97	20.74	28.22	28.64	20.12	7.99	1.00	4.98	18.49	33.51	41.03
.98	20.82	28.28	28.60	20.00	7.88	0.98	5.08	18.65	33.64	41.05
.99	20.91	28.33	28.55	19.89	7.77	0.96	5.18	18.81	33.76	41.07
1.00	21.00	28.38	28.51	19.77	7.66	0.94	5.28	18.97	33.88	41.08

TABLE XXII. ARGUMENT 17.

Equation = 17″.2 [1 + sin. $(t + z)$].

Period, 27.3216794 days.

Days.	0	1	2	3	4	5	6	7	8	9	Days.
Days. .00	″ 0.05	″ 0.20	″ 1.23	″ 3.12	″ 5.74	″ 8.97	″ 12.64	″ 16.52	″ 20.47	″ 24.24	Days. 1.00
.01	0.05	0.21	1.24	3.14	5.77	9.00	12.68	16.56	20.51	24.28	.99
.02	0.04	0.21	1.26	3.16	5.80	9.04	12.72	16.60	20.55	24.31	.98
.03	0.04	0.22	1.27	3.19	5.83	9.07	12.76	16.64	20.59	24.35	.97
.04	0.04	0.23	1.29	3.21	5.86	9.11	12.79	16.68	20.63	24.38	.96
.05	0.03	0.23	1.30	3.23	5.89	9.15	12.83	16.71	20.67	24.42	.95
.06	0.03	0.24	1.32	3.26	5.92	9.18	12.87	16.75	20.70	24.46	.94
.07	0.03	0.25	1.33	3.28	5.95	9.22	12.90	16.79	20.74	24.49	.93
.08	0.03	0.25	1.35	3.30	5.98	9.25	12.94	16.83	20.78	24.53	.92
.09	0.02	0.26	1.36	3.33	6.01	9.29	12.98	16.87	20.82	24.56	.91
.10	0.02	0.27	1.38	3.35	6.04	9.32	13.02	16.91	20.86	24.60	.90
.11	0.02	0.28	1.40	3.37	6.07	9.35	13.06	16.95	20.90	24.64	.89
.12	0.02	0.28	1.41	3.39	6.10	9.39	13.10	16.99	20.94	24.67	.88
.13	0.01	0.29	1.43	3.42	6.13	9.42	13.14	17.03	20.98	24.71	.87
.14	0.01	0.30	1.44	3.44	6.16	9.46	13.17	17.07	21.01	24.74	.86
.15	0.01	0.30	1.46	3.47	6.19	9.50	13.21	17.10	21.05	24.78	.85
.16	0.01	0.31	1.47	3.49	6.23	9.53	13.25	17.14	21.09	24.81	.84
.17	0.01	0.32	1.49	3.52	6.26	9.57	13.28	17.18	21.12	24.84	.83
.18	0.00	0.32	1.51	3.54	6.29	9.60	13.32	17.22	21.16	24.88	.82
.19	0.00	0.33	1.52	3.57	6.32	9.64	13.36	17.26	21.20	24.91	.81
.20	0.00	0.34	1.54	3.59	6.35	9.67	13.40	17.30	21.24	24.95	.80
.21	0.00	0.35	1.56	3.61	6.38	9.71	13.44	17.34	21.28	24.99	.79
.22	0.00	0.36	1.58	3.64	6.41	9.74	13.48	17.38	21.32	25.02	.78
.23	0.00	0.36	1.59	3.66	6.44	9.78	13.52	17.42	21.36	25.06	.77
.24	0.00	0.37	1.61	3.69	6.47	9.81	13.55	17.46	21.39	25.09	.76
.25	0.00	0.38	1.63	3.71	6.50	9.85	13.59	17.50	21.43	25.13	.75
.26	0.00	0.39	1.65	3.74	6.54	9.89	13.63	17.54	21.47	25.16	.74
.27	0.00	0.40	1.67	3.76	6.57	9.92	13.66	17.58	21.50	25.19	.73
.28	0.00	0.40	1.68	3.79	6.60	9.96	13.70	17.62	21.54	25.23	.72
.29	0.00	0.41	1.70	3.81	6.63	9.99	13.74	17.66	21.58	25.26	.71
.30	0.00	0.42	1.72	3.84	6.66	10.03	13.78	17.70	21.62	25.30	.70
.31	0.00	0.43	1.74	3.86	6.69	10.07	13.82	17.74	21.66	25.33	.69
.32	0.00	0.44	1.76	3.89	6.72	10.10	13.86	17.78	21.70	25.37	.68
.33	0.00	0.45	1.77	3.91	6.75	10.14	13.90	17.82	21.73	25.40	.67
.34	0.00	0.46	1.79	3.94	6.78	10.17	13.94	17.86	21.77	25.44	.66
.35	0.00	0.46	1.81	3.96	6.81	10.21	13.97	17.90	21.81	25.47	.65
.36	0.00	0.47	1.83	3.99	6.85	10.25	14.01	17.94	21.85	25.50	.64
.37	0.00	0.48	1.85	4.01	6.88	10.28	14.05	17.98	21.88	25.54	.63
.38	0.00	0.49	1.86	4.04	6.91	10.32	14.09	18.02	21.92	25.57	.62
.39	0.00	0.50	1.88	4.06	6.94	10.35	14.13	18.06	21.96	25.61	.61
.40	0.00	0.51	1.90	4.09	6.97	10.39	14.17	18.10	22.00	25.64	.60
.41	0.00	0.52	1.92	4.12	7.00	10.43	14.21	18.14	22.04	25.67	.59
.42	0.00	0.53	1.94	4.14	7.03	10.47	14.25	18.18	22.08	25.71	.58
.43	0.00	0.54	1.95	4.17	7.07	10.50	14.29	18.22	22.12	25.74	.57
.44	0.00	0.55	1.97	4.19	7.10	10.54	14.33	18.26	22.15	25.78	.56
.45	0.00	0.56	1.99	4.22	7.13	10.57	14.36	18.30	22.19	25.81	.55
.46	0.01	0.57	2.01	4.24	7.16	10.61	14.40	18.34	22.23	25.84	.54
.47	0.01	0.58	2.03	4.27	7.20	10.65	14.44	18.38	22.26	25.88	.53
.48	0.01	0.59	2.04	4.29	7.23	10.68	14.48	18.42	22.30	25.91	.52
.49	0.01	0.60	2.06	4.32	7.26	10.72	14.52	18.46	22.34	25.95	.51
.50	0.01	0.61	2.08	4.35	7.29	10.76	14.56	18.50	22.38	25.98	.50
Days.	27	26	25	24	23	22	21	20	19	18	Days.

TABLE XXII. ARGUMENT 17.

Equation = $17''.2\ [1 + \sin.\ (t + z)]$.

Period, 27.3216794 days.

Days.	0	1	2	3	4	5	6	7	8	9	Days.
Days.	"	"	"	"	"	"	"	"	"	"	Days.
.50	0.01	0.61	2.08	4.35	7.29	10.76	14.56	18.50	22.38	25.98	.50
.51	0.01	0.62	2.10	4.39	7.32	10.80	14.60	18.51	22.42	26.01	.49
.52	0.01	0.63	2.12	4.40	7.35	10.84	14.64	18.58	22.46	26.05	.48
.53	0.02	0.64	2.14	4.42	7.39	10.87	14.68	18.62	22.50	26.08	.47
.54	0.02	0.65	2.16	4.45	7.42	10.91	14.72	18.66	22.53	26.12	.46
.55	0.02	0.66	2.17	4.48	7.45	10.94	14.75	18.70	22.57	26.15	.45
.56	0.02	0.68	2.19	4.50	7.48	10.98	14.79	18.74	22.61	26.18	.44
.57	0.03	0.69	2.21	4.53	7.52	11.02	14.83	18.78	22.64	26.22	.43
.58	0.03	0.70	2.23	4.55	7.55	11.05	14.87	18.82	22.68	26.25	.42
.59	0.03	0.71	2.25	4.58	7.58	11.09	14.91	18.86	22.72	26.29	.41
.60	0.03	0.72	2.27	4.61	7.61	11.13	14.95	18.90	22.76	26.32	.40
.61	0.03	0.73	2.29	4.64	7.64	11.17	14.99	18.94	22.80	26.35	.39
.62	0.04	0.74	2.31	4.66	7.67	11.21	15.03	18.98	22.84	26.39	.38
.63	0.04	0.75	2.33	4.69	7.71	11.24	15.07	19.02	22.87	26.42	.37
.64	0.04	0.77	2.35	4.72	7.74	11.28	15.11	19.06	22.91	26.46	.36
.65	0.04	0.78	2.37	4.74	7.77	11.31	15.14	19.10	22.95	26.49	.35
.66	0.05	0.79	2.39	4.77	7.81	11.35	15.18	19.14	22.98	26.52	.34
.67	0.05	0.80	2.41	4.80	7.84	11.39	15.22	19.18	23.02	26.56	.33
.68	0.05	0.82	2.43	4.82	7.87	11.42	15.26	19.22	23.05	26.59	.32
.69	0.06	0.83	2.45	4.85	7.91	11.46	15.30	19.26	23.09	26.63	.31
.70	0.06	0.84	2.47	4.88	7.94	11.50	15.34	19.30	23.13	26.66	.30
.71	0.06	0.85	2.49	4.91	7.97	11.54	15.38	19.34	23.17	26.69	.29
.72	0.07	0.87	2.51	4.94	8.00	11.58	15.42	19.38	23.21	26.72	.28
.73	0.07	0.88	2.53	4.96	8.04	11.61	15.46	19.42	23.24	26.76	.27
.74	0.08	0.89	2.55	4.99	8.07	11.65	15.50	19.46	23.28	26.79	.26
.75	0.08	0.90	2.57	5.02	8.11	11.69	15.54	19.50	23.32	26.82	.25
.76	0.08	0.91	2.59	5.05	8.14	11.73	15.58	19.53	23.35	26.86	.24
.77	0.09	0.93	2.62	5.08	8.18	11.77	15.62	19.57	23.39	26.89	.23
.78	0.09	0.94	2.64	5.10	8.21	11.80	15.66	19.61	23.42	26.93	.22
.79	0.10	0.95	2.66	5.13	8.25	11.84	15.70	19.65	23.46	26.96	.21
.80	0.10	0.96	2.68	5.16	8.28	11.88	15.74	19.69	23.50	26.99	.20
.81	0.11	0.97	2.70	5.19	8.31	11.92	15.78	19.73	23.54	27.02	.19
.82	0.11	0.99	2.72	5.22	8.35	11.96	15.82	19.77	23.58	27.05	.18
.83	0.11	1.00	2.75	5.24	8.38	12.00	15.86	19.81	23.61	27.09	.17
.84	0.12	1.01	2.77	5.27	8.42	12.03	15.90	19.85	23.65	27.12	.16
.85	0.12	1.03	2.79	5.30	8.45	12.07	15.93	19.89	23.69	27.15	.15
.86	0.13	1.04	2.81	5.34	8.48	12.11	15.97	19.92	23.72	27.18	.14
.87	0.13	1.05	2.83	5.36	8.52	12.14	16.01	19.96	23.76	27.22	.13
.88	0.13	1.07	2.86	5.38	8.55	12.18	16.05	20.00	23.79	27.25	.12
.89	0.14	1.08	2.88	5.41	8.58	12.22	16.09	20.04	23.83	27.28	.11
.90	0.14	1.09	2.90	5.44	8.62	12.26	16.13	20.08	23.87	27.31	.10
.91	0.15	1.10	2.92	5.47	8.65	12.30	16.17	20.12	23.91	27.34	.09
.92	0.15	1.12	2.94	5.50	8.69	12.34	16.21	20.16	23.95	27.37	.08
.93	0.16	1.13	2.97	5.53	8.72	12.38	16.25	20.20	23.98	27.40	.07
.94	0.16	1.15	2.99	5.56	8.76	12.41	16.29	20.24	24.02	27.43	.06
.95	0.17	1.16	3.01	5.59	8.79	12.45	16.32	20.28	24.06	27.47	.05
.96	0.17	1.17	3.03	5.62	8.83	12.49	16.36	20.31	24.09	27.50	.04
.97	0.18	1.19	3.05	5.65	8.86	12.52	16.40	20.35	24.13	27.53	.03
.98	0.19	1.20	3.08	5.68	8.89	12.56	16.44	20.39	24.16	27.56	.02
.99	0.19	1.21	3.10	5.71	8.93	12.60	16.48	20.43	24.20	27.59	.01
1.00	0.20	1.23	3.12	5.74	8.97	12.64	16.52	20.47	24.24	27.62	.00
Days.	27	26	25	24	23	22	21	20	19	18	Days.

TABLE XXII.

(Continued.)

Period, 27.3216794 days.

TABLE XXIII. ARG. 18.

Equation = 14″.0 [1 + sin. (2*t* + *x* — *z*)].

Period, 9.873593 days.

Days.	10	11	12	13	0	1	2	3	4	Days.
Days .00	27.62	30.47	32.61	33.94	0.01	2.42	9.35	18.11	25.25	Days 1.00
.01	27.65	30.49	32.63	33.95	0.01	2.47	9.43	18.19	25.30	.99
.02	27.68	30.52	32.65	33.96	0.01	2.52	9.52	18.28	25.35	.98
.03	27.71	30.54	32.66	33.97	0.01	2.58	9.60	18.36	25.41	.97
.04	27.74	30.57	32.68	33.98	0.00	2.63	9.69	18.45	25.46	.96
.05	27.78	30.59	32.70	33.99	0.00	2.68	9.77	18.53	25.51	.95
.06	27.81	30.61	32.72	33.99	0.00	2.73	9.86	18.61	25.56	.94
.07	27.84	30.64	32.74	34.00	0.00	2.78	9.94	18.70	25.61	.93
.08	27.87	30.66	32.75	34.01	0.00	2.84	10.03	18.78	25.66	.92
.09	27.90	30.69	32.77	34.02	0.00	2.89	10.11	18.87	25.71	.91
.10	27.93	30.71	32.79	34.03	0.00	2.94	10.20	18.95	25.76	.90
.11	27.96	30.73	32.81	34.04	0.00	3.00	10.29	19.03	25.81	.89
.12	27.99	30.76	32.83	34.05	0.01	3.05	10.37	19.12	25.86	.88
.13	28.02	30.78	32.84	34.05	0.01	3.11	10.46	19.20	25.90	.87
.14	28.05	30.81	32.86	34.06	0.02	3.16	10.54	19.29	25.95	.86
.15	28.09	30.83	32.88	34.07	0.02	3.22	10.63	19.37	26.00	.85
.16	28.12	30.85	32.89	34.08	0.03	3.28	10.72	19.45	26.04	.84
.17	28.15	30.88	32.91	34.09	0.03	3.34	10.80	19.53	26.09	.83
.18	28.18	30.90	32.92	34.09	0.04	3.39	10.89	19.62	26.13	.82
.19	28.21	30.93	32.94	34.10	0.04	3.45	10.97	19.70	26.18	.81
.20	28.24	30.95	32.96	34.11	0.05	3.51	11.06	19.78	26.22	.80
.21	28.27	30.97	32.98	34.12	0.06	3.57	11.15	19.86	26.26	.79
.22	28.30	31.00	32.99	34.13	0.07	3.63	11.24	19.94	26.30	.78
.23	28.33	31.02	33.01	34.13	0.08	3.69	11.32	20.02	26.35	.77
.24	28.36	31.05	33.02	34.14	0.09	3.75	11.41	20.10	26.39	.76
.25	28.39	31.07	33.04	34.15	0.10	3.81	11.50	20.18	26.43	.75
.26	28.42	31.09	33.05	34.15	0.11	3.87	11.59	20.26	26.47	.74
.27	28.45	31.12	33.06	34.16	0.12	3.93	11.68	20.34	26.51	.73
.28	28.48	31.14	33.08	34.16	0.14	4.00	11.76	20.42	26.55	.72
.29	28.51	31.17	33.09	34.17	0.15	4.06	11.85	20.50	26.59	.71
.30	28.54	31.19	33.11	34.18	0.16	4.12	11.94	20.58	26.63	.70
.31	28.57	31.21	33.12	34.19	0.17	4.18	12.03	20.66	26.67	.69
.32	28.60	31.23	33.14	34.19	0.19	4.25	12.12	20.74	26.71	.68
.33	28.63	31.26	33.15	34.20	0.20	4.31	12.20	20.81	26.74	.67
.34	28.66	31.28	33.17	34.20	0.22	4.38	12.29	20.89	26.78	.66
.35	28.69	31.30	33.18	34.21	0.23	4.44	12.38	20.97	26.82	.65
.36	28.72	31.33	33.19	34.22	0.25	4.51	12.47	21.05	26.85	.64
.37	28.75	31.35	33.21	34.22	0.27	4.57	12.56	21.12	26.89	.63
.38	28.78	31.38	33.22	34.23	0.28	4.64	12.65	21.20	26.92	.62
.39	28.81	31.40	33.24	34.23	0.30	4.70	12.74	21.27	26.96	.61
.40	28.84	31.42	33.25	34.24	0.32	4.77	12.83	21.35	26.99	.60
.41	28.87	31.44	33.26	34.24	0.34	4.84	12.92	21.42	27.02	.59
.42	28.90	31.46	33.28	34.25	0.36	4.91	13.01	21.50	27.05	.58
.43	28.93	31.49	33.29	34.25	0.38	4.97	13.10	21.57	27.09	.57
.44	28.96	31.51	33.31	34.26	0.40	5.04	13.19	21.65	27.12	.56
.45	28.99	31.53	33.32	34.27	0.42	5.11	13.27	21.72	27.15	.55
.46	29.01	31.55	33.33	34.27	0.44	5.18	13.36	21.79	27.18	.54
.47	29.04	31.58	33.35	34.28	0.47	5.25	13.45	21.87	27.21	.53
.48	29.07	31.60	33.36	34.28	0.49	5.32	13.54	21.94	27.23	.52
.49	29.10	31.62	33.38	34.29	0.52	5.39	13.63	22.02	27.26	.51
.50	29.13	31.64	33.39	34.29	0.54	5.46	13.72	22.09	27.29	.50
Days.	17	16	15	14	9	8	7	6	5	Days.

TABLE XXII.

(*Continued.*)

Period, 27.3216794 days.

TABLE XXIII. ARG. 18.

Equation = 14″.0 [1 + sin. (2*t* + *z* — *z*)].

Period, 9.873593 days.

Days.	10	11	12	13	0	1	2	3	4	Days.
.50	29.13	31.64	33.39	34.29	0.54	5.46	13.72	22.09	27.29	.50
.51	29.16	31.66	33.40	34.29	0.57	5.53	13.81	22.16	27.32	.49
.52	29.19	31.68	33.41	34.30	0.59	5.60	13.90	22.23	27.34	.48
.53	29.21	31.70	33.43	34.30	0.62	5.68	13.99	22.31	27.37	.47
.54	29.24	31.72	33.44	34.31	0.64	5.75	14.08	22.38	27.39	.46
.55	29.27	31.75	33.46	34.31	0.67	5.82	14.16	22.45	27.42	.45
.56	29.30	31.77	33.47	34.31	0.70	5.89	14.25	22.52	27.44	.44
.57	29.32	31.79	33.48	34.32	0.73	5.96	14.34	22.59	27.47	.43
.58	29.35	31.81	33.50	34.32	0.75	6.04	14.43	22.66	27.49	.42
.59	29.38	31.83	33.51	34.33	0.78	6.11	14.52	22.73	27.52	.41
.60	29.41	31.85	33.52	34.33	0.81	6.18	14.61	22.80	27.54	.40
.61	29.44	31.87	33.53	34.33	0.84	6.25	14.70	22.87	27.56	.39
.62	29.47	31.89	33.54	34.33	0.87	6.33	14.79	22.94	27.58	.38
.63	29.49	31.91	33.56	34.34	0.90	6.40	14.88	23.01	27.61	.37
.64	29.52	31.93	33.57	34.34	0.93	6.48	14.97	23.08	27.63	.36
.65	29.55	31.95	33.58	34.35	0.96	6.55	15.06	23.15	27.65	.35
.66	29.57	31.97	33.59	34.35	0.99	6.63	15.14	23.22	27.67	.34
.67	29.60	31.99	33.60	34.35	1.03	6.70	15.23	23.28	27.69	.33
.68	29.62	32.01	33.62	34.36	1.06	6.78	15.32	23.35	27.70	.32
.69	29.65	32.03	33.63	34.36	1.10	6.85	15.41	23.41	27.72	.31
.70	29.68	32.05	33.64	34.36	1.13	6.93	15.50	23.48	27.74	.30
.71	29.71	32.07	33.65	34.36	1.17	7.01	15.59	23.54	27.76	.29
.72	29.74	32.09	33.66	34.36	1.20	7.09	15.68	23.61	27.77	.28
.73	29.76	32.11	33.67	34.37	1.24	7.16	15.77	23.67	27.79	.27
.74	29.79	32.13	33.68	34.37	1.27	7.24	15.86	23.74	27.80	.26
.75	29.82	32.15	33.70	34.38	1.31	7.32	15.95	23.80	27.82	.25
.76	29.84	32.16	33.71	34.38	1.35	7.40	16.04	23.86	27.83	.24
.77	29.87	32.18	33.72	34.38	1.39	7.48	16.13	23.92	27.85	.23
.78	29.89	32.20	33.73	34.39	1.43	7.55	16.21	23.99	27.86	.22
.79	29.92	32.22	33.74	34.39	1.47	7.63	16.30	24.05	27.88	.21
.80	29.95	32.24	33.75	34.39	1.51	7.71	16.39	24.11	27.89	.20
.81	29.98	32.26	33.76	34.39	1.55	7.79	16.48	24.17	27.90	.19
.82	30.00	32.28	33.77	34.39	1.59	7.87	16.57	24.23	27.91	.18
.83	30.03	32.30	33.78	34.39	1.64	7.95	16.65	24.29	27.92	.17
.84	30.05	32.32	33.79	34.39	1.68	8.03	16.74	24.35	27.93	.16
.85	30.08	32.34	33.80	34.39	1.72	8.11	16.83	24.41	27.94	.15
.86	30.11	32.35	33.81	34.40	1.76	8.19	16.92	24.47	27.95	.14
.87	30.13	32.37	33.82	34.40	1.81	8.27	17.00	24.53	27.96	.13
.88	30.16	32.39	33.83	34.40	1.85	8.36	17.09	24.58	27.96	.12
.89	30.18	32.41	33.84	34.40	1.90	8.44	17.17	24.64	27.97	.11
.90	30.21	32.43	33.85	34.40	1.94	8.52	17.26	24.70	27.98	.10
.91	30.24	32.45	33.86	34.40	1.99	8.60	17.35	24.76	27.98	.09
.92	30.26	32.47	33.87	34.40	2.03	8.68	17.43	24.81	27.99	.08
.93	30.29	32.48	33.88	34.40	2.08	8.77	17.52	24.87	27.99	.07
.94	30.31	32.50	33.89	34.40	2.12	8.85	17.60	24.92	28.00	.06
.95	30.34	32.52	33.90	34.40	2.17	8.93	17.69	24.98	28.00	.05
.96	30.37	32.54	33.90	34.40	2.22	9.01	17.77	25.03	28.00	.04
.97	30.39	32.55	33.91	34.40	2.27	9.10	17.86	25.09	28.00	.03
.98	30.42	32.57	33.92	34.40	2.32	9.18	17.94	25.14	28.00	.02
.99	30.44	32.59	33.93	34.40	2.37	9.27	18.03	25.20	28.00	.01
1.00	30.47	32.61	33.94	34.40	2.42	9.35	18.11	25.25	28.00	.00
Days.	17	16	15	14	9	8	7	6	5	Days.

TABLE XXIV. ARGUMENT 19.

Equation = 12″.8 [1 — sin. (3x — 2t)].

Period, 24.302196 days.

Days.	0	1	2	3	4	5	6	7	8	Days.
Days	″	″	″	″	″	″	″	″	″	Days.
.00	25.29	25.59	25.03	23.67	21.59	18.91	15.83	12.55	9.29	1.00
.01	25.30	25.59	25.02	23.65	21.56	18.88	15.80	12.52	9.26	.99
.02	25.31	25.59	25.01	23.63	21.53	18.85	15.77	12.49	9.23	.98
.03	25.31	25.58	25.00	23.62	21.51	18.82	15.73	12.45	9.20	.97
.04	25.32	25.58	24.99	23.60	21.48	18.79	15.70	12.42	9.16	.96
.05	25.33	25.58	24.98	23.58	21.46	18.76	15.67	12.39	9.13	.95
.06	25.33	25.58	24.97	23.56	21.44	18.74	15.63	12.35	9.10	.94
.07	25.34	25.58	24.96	23.54	21.41	18.71	15.60	12.32	9.06	.93
.08	25.35	25.57	24.95	23.53	21.39	18.68	15.56	12.28	9.03	.92
.09	25.35	25.57	24.94	23.51	21.36	18.65	15.53	12.25	9.00	.91
.10	25.36	25.57	24.93	23.49	21.34	18.62	15.50	12.22	8.97	.90
.11	25.37	25.57	24.92	23.47	21.32	18.59	15.47	12.19	8.94	.89
.12	25.38	25.57	24.91	23.45	21.29	18.56	15.44	12.16	8.91	.88
.13	25.38	25.56	24.90	23.44	21.27	18.53	15.41	12.12	8.88	.87
.14	25.39	25.56	24.89	23.42	21.24	18.50	15.37	12.09	8.84	.86
.15	25.39	25.56	24.88	23.40	21.22	18.47	15.34	12.06	8.81	.85
.16	25.40	25.55	24.86	23.38	21.20	18.44	15.31	12.02	8.78	.84
.17	25.40	25.55	24.85	23.36	21.17	18.41	15.27	11.99	8.74	.83
.18	25.41	25.55	24.84	23.35	21.15	18.38	15.24	11.95	8.71	.82
.19	25.41	25.54	24.83	23.33	21.12	18.35	15.21	11.92	8.68	.81
.20	25.42	25.54	24.82	23.31	21.10	18.32	15.18	11.89	8.65	.80
.21	25.42	25.54	24.81	23.29	21.08	18.29	15.15	11.86	8.62	.79
.22	25.43	25.54	24.80	23.27	21.05	18.26	15.12	11.82	8.59	.78
.23	25.43	25.53	24.79	23.25	21.03	18.23	15.09	11.79	8.56	.77
.24	25.44	25.53	24.78	23.23	21.00	18.20	15.05	11.75	8.53	.76
.25	25.44	25.53	24.77	23.21	20.98	18.17	15.02	11.72	8.50	.75
.26	25.45	25.52	24.75	23.20	20.95	18.13	14.99	11.69	8.48	.74
.27	25.45	25.52	24.74	23.18	20.92	18.10	14.95	11.65	8.45	.73
.28	25.46	25.52	24.73	23.16	20.90	18.07	14.92	11.62	8.40	.72
.29	25.46	25.51	24.72	23.14	20.87	18.04	14.89	11.58	8.37	.71
.30	25.47	25.51	24.71	23.12	20.85	18.01	14.86	11.55	8.34	.70
.31	25.47	25.51	24.70	23.10	20.82	17.98	14.83	11.52	8.31	.69
.32	25.48	25.50	24.69	23.08	20.80	17.95	14.80	11.49	8.28	.68
.33	25.48	25.50	24.68	23.06	20.77	17.92	14.76	11.45	8.25	.67
.34	25.49	25.49	24.66	23.04	20.75	17.89	14.73	11.42	8.22	.66
.35	25.49	25.49	24.65	23.02	20.72	17.86	14.70	11.39	8.19	.65
.36	25.50	25.49	24.64	23.00	20.69	17.82	14.66	11.35	8.15	.64
.37	25.50	25.48	24.63	22.98	20.67	17.79	14.63	11.32	8.12	.63
.38	25.50	25.48	24.62	22.96	20.64	17.76	14.59	11.28	8.09	.62
.39	25.51	25.47	24.60	22.94	20.62	17.73	14.56	11.25	8.06	.61
.40	25.51	25.47	24.59	22.92	20.59	17.70	14.53	11.22	8.03	.60
.41	25.51	25.46	24.58	22.90	20.56	17.67	14.50	11.19	8.00	.59
.42	25.52	25.46	24.57	22.88	20.53	17.61	14.47	11.16	7.97	.58
.43	25.52	25.45	24.55	22.86	20.51	17.61	14.43	11.12	7.94	.57
.44	25.52	25.45	24.54	22.84	20.48	17.58	14.40	11.09	7.91	.56
.45	25.53	25.44	24.52	22.82	20.46	17.55	14.37	11.06	7.88	.55
.46	25.53	25.44	24.51	22.79	20.43	17.52	14.33	11.02	7.85	.54
.47	25.53	25.43	24.49	22.77	20.40	17.49	14.30	10.99	7.82	.53
.48	25.54	25.43	24.48	22.75	20.38	17.46	14.26	10.95	7.79	.52
.49	25.54	25.42	24.46	22.73	20.35	17.43	14.23	10.92	7.76	.51
.50	25.54	25.42	24.45	22.71	20.32	17.40	14.20	10.89	7.73	.50
Days.	25	24	23	22	21	20	19	18	17	Days.

TABLE XXIV. ARGUMENT 19.

Equation = 12″.8 [1 — sin. (3x — 2t)].

Period, 24.302196 days.

Days.	0	1	2	3	4	5	6	7	8	Days.
Days	″	″	″	″	″	″	″	″	″	Days.
.50	25.54	25.42	24.45	22.71	20.32	17.40	14.20	10.89	7.73	.50
.51	25.54	25.41	24.44	22.69	20.29	17.37	14.17	10.86	7.70	.49
.52	25.55	25.41	24.42	22.67	20.26	17.34	14.14	10.83	7.67	.48
.53	25.55	25.40	24.41	22.65	20.24	17.31	14.10	10.80	7.64	.47
.54	25.55	25.40	24.39	22.63	20.21	17.28	14.07	10.76	7.61	.46
.55	25.55	25.39	24.38	22.61	20.18	17.25	14.04	10.73	7.58	.45
.56	25.56	25.39	24.37	22.58	20.15	17.21	14.00	10.70	7.55	.44
.57	25.56	25.38	24.35	22.56	20.12	17.18	13.97	10.66	7.52	.43
.58	25.56	25.37	24.34	22.54	20.10	17.15	13.93	10.63	7.49	.42
.59	25.56	25.37	24.32	22.52	20.07	17.12	13.90	10.60	7.46	.41
.60	25.57	25.36	24.31	22.50	20.04	17.09	13.87	10.57	7.43	.40
.61	25.57	25.35	24.29	22.48	20.01	17.06	13.84	10.54	7.40	.39
.62	25.57	25.35	24.28	22.46	19.98	17.03	13.81	10.51	7.37	.38
.63	25.57	25.34	24.26	22.43	19.95	17.00	13.77	10.48	7.34	.37
.64	25.58	25.33	24.25	22.41	19.93	16.97	13.74	10.44	7.31	.36
.65	25.58	25.33	24.23	22.39	19.90	16.94	13.71	10.41	7.28	.35
.66	25.58	25.32	24.22	22.37	19.87	16.90	13.67	10.38	7.25	.34
.67	25.58	25.31	24.20	22.35	19.85	16.87	13.64	10.34	7.22	.33
.68	25.59	25.31	24.19	22.32	19.82	16.84	13.60	10.31	7.19	.32
.69	25.59	25.30	24.17	22.30	19.79	16.81	13.57	10.28	7.16	.31
.70	25.59	25.29	24.16	22.28	19.76	16.78	13.54	10.25	7.13	.30
.71	25.59	25.28	24.14	22.26	19.73	16.75	13.51	10.22	7.10	.29
.72	25.59	25.27	24.13	22.24	19.70	16.72	13.48	10.19	7.07	.28
.73	25.59	25.27	24.11	22.21	19.67	16.69	13.44	10.16	7.04	.27
.74	25.60	25.26	24.10	22.19	19.65	16.66	13.41	10.12	7.01	.26
.75	25.60	25.25	24.08	22.17	19.62	16.63	13.38	10.09	6.98	.25
.76	25.60	25.24	24.07	22.15	19.59	16.59	13.34	10.06	6.95	.24
.77	25.60	25.23	24.05	22.13	19.57	16.56	13.31	10.02	6.92	.23
.78	25.60	25.23	24.04	22.10	19.54	16.53	13.27	9.99	6.89	.22
.79	25.60	25.22	24.02	22.08	19.51	16.50	13.24	9.96	6.86	.21
.80	25.60	25.21	24.01	22.06	19.48	16.47	13.21	9.93	6.83	.20
.81	25.60	25.20	23.99	22.04	19.45	16.44	13.18	9.90	6.80	.19
.82	25.60	25.19	23.98	22.02	19.42	16.41	13.15	9.87	6.77	.18
.83	25.60	25.18	23.96	21.99	19.39	16.38	13.11	9.84	6.74	.17
.84	25.60	25.18	23.95	21.97	19.37	16.34	13.08	9.80	6.71	.16
.85	25.60	25.17	23.93	21.95	19.34	16.31	13.05	9.77	6.68	.15
.86	25.60	25.16	23.92	21.92	19.31	16.28	13.01	9.74	6.66	.14
.87	25.60	25.15	23.90	21.90	19.29	16.24	12.98	9.70	6.63	.13
.88	25.60	25.14	23.88	21.88	19.26	16.21	12.94	9.67	6.60	.12
.89	25.60	25.13	23.87	21.85	19.23	16.18	12.91	9.64	6.57	.11
.90	25.60	25.12	23.85	21.83	19.20	16.15	12.88	9.61	6.54	.10
.91	25.60	25.11	23.83	21.81	19.17	16.12	12.85	9.58	6.51	.09
.92	25.60	25.10	23.82	21.79	19.14	16.09	12.82	9.55	6.48	.08
.93	25.60	25.10	23.80	21.76	19.11	16.06	12.78	9.52	6.45	.07
.94	25.60	25.09	23.78	21.74	19.08	16.02	12.75	9.48	6.42	.06
.95	25.60	25.08	23.76	21.72	19.05	15.99	12.72	9.45	6.39	.05
.96	25.60	25.07	23.74	21.69	19.03	15.96	12.68	9.42	6.37	.04
.97	25.59	25.06	23.73	21.67	19.00	15.92	12.65	9.38	6.34	.03
.98	25.59	25.05	23.71	21.65	18.97	15.89	12.61	9.35	6.31	.02
.99	25.59	25.04	23.69	21.62	18.94	15.86	12.58	9.32	6.28	.01
1.00	25.59	25.03	23.67	21.59	18.91	15.83	12.55	9.29	6.25	.00
Days.	25	24	23	22	21	20	19	18	17	Days.

TABLE XXIV.

(*Continued.*)

Period, 24.302196 days.

Days.	9	10	11	12
.00	6.25	3.66	1.67	0.43
.01	6.22	3.64	1.65	0.42
.02	6.19	3.62	1.64	0.41
.03	6.16	3.59	1.62	0.40
.04	6.14	3.57	1.61	0.40
.05	6.11	3.54	1.59	0.39
.06	6.08	3.52	1.57	0.38
.07	6.06	3.50	1.56	0.38
.08	6.03	3.47	1.54	0.37
.09	6.00	3.45	1.53	0.36
.10	5.97	3.43	1.51	0.35
.11	5.94	3.41	1.49	0.34
.12	5.91	3.39	1.48	0.33
.13	5.88	3.37	1.46	0.32
.14	5.86	3.34	1.45	0.32
.15	5.83	3.32	1.43	0.31
.16	5.80	3.30	1.42	0.30
.17	5.78	3.27	1.41	0.30
.18	5.75	3.25	1.39	0.29
.19	5.72	3.23	1.38	0.28
.20	5.69	3.21	1.36	0.27
.21	5.66	3.19	1.35	0.26
.22	5.63	3.17	1.33	0.26
.23	5.61	3.15	1.32	0.25
.24	5.58	3.12	1.30	0.25
.25	5.55	3.10	1.29	0.24
.26	5.53	3.08	1.28	0.23
.27	5.50	3.05	1.26	0.23
.28	5.48	3.03	1.25	0.22
.29	5.45	3.01	1.23	0.22
.30	5.42	2.99	1.22	0.21
.31	5.39	2.97	1.21	0.20
.32	5.36	2.95	1.19	0.20
.33	5.34	2.93	1.18	0.19
.34	5.31	2.91	1.16	0.19
.35	5.28	2.88	1.15	0.18
.36	5.26	2.86	1.14	0.18
.37	5.23	2.84	1.12	0.18
.38	5.21	2.82	1.11	0.17
.39	5.18	2.80	1.09	0.17
.40	5.15	2.78	1.08	0.16
.41	5.12	2.76	1.07	0.15
.42	5.10	2.74	1.05	0.15
.43	5.07	2.72	1.04	0.14
.44	5.05	2.70	1.03	0.14
.45	5.02	2.68	1.01	0.13
.46	4.99	2.66	1.00	0.13
.47	4.97	2.64	0.99	0.13
.48	4.94	2.62	0.97	0.12
.49	4.92	2.60	0.96	0.12
.50	4.89	2.58	0.95	0.11
Days.	16	15	14	13

TABLE XXV. ARG. 20.

Equation = 9″.6 [1 + sin. (2x — z)].

Period, 14.317313 days.

0	1	2	3	4	Days.
0.65	0.02	1.21	4.00	7.84	1.00
0.64	0.02	1.23	4.03	7.88	.99
0.62	0.03	1.25	4.07	7.92	.98
0.61	0.03	1.27	4.10	7.96	.97
0.59	0.04	1.30	4.14	8.01	.96
0.58	0.04	1.32	4.17	8.05	.95
0.57	0.04	1.34	4.21	8.09	.94
0.55	0.05	1.37	4.25	8.14	.93
0.54	0.05	1.39	4.28	8.18	.92
0.52	0.06	1.41	4.32	8.22	.91
0.51	0.06	1.43	4.35	8.26	.90
0.50	0.07	1.45	4.39	8.30	.89
0.48	0.07	1.47	4.42	8.34	.88
0.47	0.08	1.50	4.46	8.38	.87
0.45	0.08	1.52	4.49	8.43	.86
0.44	0.08	1.54	4.53	8.47	.85
0.43	0.09	1.57	4.57	8.51	.84
0.42	0.09	1.59	4.60	8.56	.83
0.40	0.10	1.62	4.64	8.60	.82
0.39	0.11	1.64	4.67	8.64	.81
0.38	0.12	1.66	4.71	8.68	.80
0.37	0.13	1.68	4.75	8.72	.79
0.36	0.14	1.71	4.78	8.76	.78
0.35	0.14	1.73	4.82	8.80	.77
0.34	0.15	1.76	4.85	8.85	.76
0.32	0.16	1.78	4.89	8.89	.75
0.31	0.17	1.81	4.93	8.93	.74
0.30	0.17	1.84	4.96	8.98	.73
0.29	0.18	1.86	5.00	9.02	.72
0.28	0.19	1.89	5.03	9.06	.71
0.27	0.20	1.91	5.07	9.10	.70
0.26	0.21	1.94	5.11	9.14	.69
0.25	0.22	1.96	5.15	9.18	.68
0.24	0.23	1.99	5.18	9.22	.67
0.23	0.24	2.01	5.22	9.27	.66
0.22	0.24	2.04	5.25	9.31	.65
0.22	0.25	2.07	5.29	9.35	.64
0.21	0.26	2.09	5.33	9.40	.63
0.20	0.27	2.12	5.36	9.44	.62
0.19	0.28	2.14	5.40	9.48	.61
0.18	0.29	2.17	5.44	9.52	.60
0.17	0.30	2.20	5.48	9.56	.59
0.16	0.31	2.22	5.52	9.60	.58
0.16	0.32	2.25	5.55	9.64	.57
0.15	0.33	2.27	5.59	9.69	.56
0.14	0.34	2.29	5.63	9.73	.55
0.13	0.36	2.32	5.67	9.77	.54
0.13	0.37	2.34	5.70	9.81	.53
0.12	0.38	2.37	5.74	9.86	.52
0.12	0.39	2.41	5.78	9.90	.51
0.11	0.40	2.44	5.82	9.94	.50
15	14	13	12	11	Days.

TABLE XXIV.

(*Continued.*)

Period, 24.302196 days.

TABLE XXV. ARG. 20.

Equation = 9″.6 [1 + sin. (2x — z)].

Period, 14.317313 days.

Days.	9	10	11	12	0	1	2	3	4	Days.
.50	4.89	2.58	0.95	0.11	0.11	0.40	2.41	5.82	9.91	.50
.51	4.86	2.56	0.94	0.11	0.11	0.41	2.47	5.86	9.98	.49
.52	4.84	2.54	0.93	0.10	0.10	0.42	2.50	5.90	10.02	.48
.53	4.81	2.52	0.92	0.10	0.10	0.44	2.53	5.94	10.06	.47
.54	4.79	2.50	0.90	0.09	0.09	0.45	2.55	5.98	10.11	.46
.55	4.76	2.48	0.89	0.09	0.08	0.46	2.58	6.01	10.15	.45
.56	4.73	2.47	0.88	0.09	0.08	0.48	2.61	6.05	10.19	.44
.57	4.71	2.45	0.86	0.08	0.07	0.49	2.63	6.09	10.24	.43
.58	4.68	2.43	0.85	0.08	0.07	0.51	2.66	6.13	10.28	.42
.59	4.66	2.41	0.84	0.07	0.06	0.52	2.69	6.17	10.32	.41
.60	4.63	2.39	0.83	0.07	0.06	0.53	2.72	6.21	10.36	.40
.61	4.60	2.37	0.82	0.07	0.06	0.54	2.75	6.25	10.40	.39
.62	4.58	2.35	0.81	0.07	0.05	0.56	2.78	6.29	10.44	.38
.63	4.55	2.33	0.80	0.06	0.05	0.57	2.81	6.33	10.48	.37
.64	4.53	2.31	0.79	0.06	0.04	0.59	2.84	6.37	10.52	.36
.65	4.50	2.29	0.77	0.06	0.04	0.60	2.87	6.41	10.57	.35
.66	4.48	2.28	0.76	0.05	0.04	0.62	2.90	6.45	10.61	.34
.67	4.46	2.26	0.75	0.05	0.03	0.64	2.93	6.49	10.65	.33
.68	4.43	2.24	0.74	0.04	0.03	0.65	2.96	6.53	10.69	.32
.69	4.41	2.22	0.73	0.04	0.02	0.67	2.99	6.57	10.73	.31
.70	4.38	2.20	0.72	0.04	0.02	0.68	3.02	6.61	10.77	.30
.71	4.35	2.18	0.71	0.04	0.02	0.70	3.05	6.65	10.81	.29
.72	4.33	2.16	0.70	0.04	0.02	0.71	3.08	6.69	10.85	.28
.73	4.30	2.14	0.69	0.03	0.01	0.73	3.11	6.73	10.89	.27
.74	4.28	2.13	0.68	0.03	0.01	0.74	3.14	6.77	10.94	.26
.75	4.25	2.11	0.67	0.03	0.01	0.76	3.17	6.81	10.98	.25
.76	4.23	2.09	0.66	0.02	0.01	0.78	3.21	6.85	11.02	.24
.77	4.21	2.08	0.65	0.02	0.00	0.79	3.24	6.89	11.07	.23
.78	4.18	2.06	0.64	0.01	0.00	0.81	3.27	6.93	11.11	.22
.79	4.16	2.04	0.63	0.01	0.00	0.82	3.30	6.97	11.15	.21
.80	4.13	2.02	0.62	0.01	0.00	0.84	3.33	7.01	11.19	.20
.81	4.11	2.00	0.61	0.01	0.00	0.86	3.36	7.05	11.23	.19
.82	4.08	1.98	0.60	0.01	0.00	0.88	3.39	7.09	11.27	.18
.83	4.06	1.96	0.59	0.01	0.00	0.89	3.43	7.13	11.31	.17
.84	4.03	1.95	0.58	0.01	0.00	0.91	3.46	7.17	11.36	.16
.85	4.01	1.93	0.57	0.01	0.00	0.92	3.49	7.21	11.40	.15
.86	3.99	1.91	0.56	0.00	0.00	0.94	3.53	7.26	11.44	.14
.87	3.96	1.90	0.55	0.00	0.00	0.96	3.56	7.30	11.49	.13
.88	3.94	1.88	0.54	0.00	0.00	0.97	3.60	7.34	11.53	.12
.89	3.91	1.86	0.53	0.00	0.00	0.99	3.63	7.38	11.57	.11
.90	3.89	1.84	0.52	0.00	0.00	1.01	3.66	7.42	11.61	.10
.91	3.87	1.82	0.51	0.00	0.00	1.03	3.69	7.46	11.65	.09
.92	3.85	1.80	0.50	0.00	0.00	1.05	3.73	7.50	11.69	.08
.93	3.82	1.79	0.49	0.00	0.00	1.07	3.76	7.54	11.73	.07
.94	3.80	1.77	0.48	0.00	0.01	1.09	3.80	7.59	11.78	.06
.95	3.77	1.75	0.47	0.00	0.01	1.11	3.83	7.63	11.82	.05
.96	3.75	1.74	0.47	0.00	0.01	1.13	3.86	7.67	11.86	.04
.97	3.73	1.72	0.46	0.00	0.01	1.15	3.90	7.71	11.91	.03
.98	3.70	1.71	0.45	0.00	0.02	1.17	3.93	7.76	11.95	.02
.99	3.68	1.69	0.44	0.00	0.02	1.19	3.97	7.80	11.99	.01
1.00	3.66	1.67	0.43	0.00	0.02	1.21	4.00	7.84	12.03	.00
Days.	16	15	14	13	15	14	13	12	11	Days.

TABLE XXV.

(Continued.)

Period, 14.317313 days

Days.	5	6	7	Days.
Days.	"	"	"	Days.
.00	12.03	15.73	18.29	1.00
.01	12.07	15.76	18.31	.99
.02	12.11	15.79	18.33	.98
.03	12.15	15.83	18.34	.97
.04	12.19	15.86	18.36	.96
.05	12.24	15.89	18.38	.95
.06	12.28	15.92	18.39	.94
.07	12.32	15.96	18.41	.93
.08	12.36	15.99	18.42	.92
.09	12.40	16.02	18.44	.91
.10	12.44	16.05	18.46	.90
.11	12.48	16.08	18.48	.89
.12	12.52	16.11	18.49	.88
.13	12.56	16.14	18.51	.87
.14	12.60	16.17	18.52	.86
.15	12.64	16.21	18.54	.85
.16	12.68	16.24	18.55	.84
.17	12.72	16.27	18.56	.83
.18	12.76	16.30	18.58	.82
.19	12.80	16.33	18.59	.81
.20	12.84	16.36	18.61	.80
.21	12.88	16.39	18.62	.79
.22	12.92	16.42	18.64	.78
.23	12.96	16.45	18.65	.77
.24	13.00	16.48	18.67	.76
.25	13.04	16.51	18.68	.75
.26	13.07	16.53	18.69	.74
.27	13.11	16.56	18.71	.73
.28	13.15	16.59	18.72	.72
.29	13.19	16.62	18.74	.71
.30	13.23	16.65	18.75	.70
.31	13.27	16.68	18.76	.69
.32	13.31	16.71	18.77	.68
.33	13.35	16.74	18.79	.67
.34	13.38	16.76	18.80	.66
.35	13.42	16.79	18.81	.65
.36	13.46	16.82	18.82	.64
.37	13.49	16.84	18.84	.63
.38	13.53	16.87	18.85	.62
.39	13.57	16.90	18.86	.61
.40	13.61	16.93	18.87	.60
.41	13.65	16.96	18.88	.59
.42	13.69	16.98	18.89	.58
.43	13.72	17.01	18.90	.57
.44	13.76	17.03	18.91	.56
.45	13.80	17.06	18.92	.55
.46	13.83	17.09	18.93	.54
.47	13.87	17.11	18.94	.53
.48	13.90	17.14	18.95	.52
.49	13.94	17.16	18.96	.51
.50	13.98	17.19	18.97	.50
Days.	10	9	8	Days.

TABLE XXVI. ARG. 21.

Equation $= 9''.2\ [1 - \sin.\ (2x + z - 2t)]$.

Period, 131.671123 days.

Days.	0	10	20	30	40	50	60	Days.
Days	"	"	"	"	"	"	"	Days
0.0	18.22	18.01	15.89	12.25	7.92	3.88	1.02	10.0
0.1	18.23	18.03	15.86	12.21	7.88	3.85	1.00	9.9
0.2	18.24	18.02	15.83	12.17	7.83	3.81	0.98	9.8
0.3	18.25	18.00	15.80	12.12	7.79	3.78	0.96	9.7
0.4	18.25	17.99	15.77	12.08	7.74	3.74	0.94	9.6
0.5	18.26	17.98	15.74	12.04	7.70	3.71	0.92	9.5
0.6	18.27	17.96	15.71	12.00	7.66	3.67	0.91	9.4
0.7	18.28	17.95	15.68	11.96	7.61	3.64	0.89	9.3
0.8	18.28	17.94	15.65	11.91	7.57	3.60	0.87	9.2
0.9	18.29	17.92	15.62	11.87	7.52	3.57	0.85	9.1
1.0	18.30	17.91	15.59	11.83	7.48	3.53	0.83	9.0
1.1	18.30	17.89	15.56	11.79	7.44	3.50	0.81	8.9
1.2	18.31	17.88	15.53	11.75	7.39	3.46	0.79	8.8
1.3	18.32	17.86	15.50	11.70	7.35	3.43	0.78	8.7
1.4	18.32	17.85	15.46	11.66	7.31	3.39	0.76	8.6
1.5	18.33	17.83	15.43	11.62	7.26	3.36	0.74	8.5
1.6	18.33	17.82	15.40	11.58	7.22	3.33	0.73	8.4
1.7	18.34	17.80	15.37	11.54	7.18	3.29	0.71	8.3
1.8	18.35	17.78	15.33	11.49	7.13	3.26	0.69	8.2
1.9	18.36	17.76	15.30	11.45	7.09	3.22	0.68	8.1
2.0	18.36	17.77	15.27	11.41	7.05	3.19	0.66	8.0
2.1	18.36	17.75	15.24	11.37	7.01	3.16	0.65	7.9
2.2	18.36	17.74	15.20	11.32	6.97	3.12	0.63	7.8
2.3	18.37	17.72	15.17	11.28	6.92	3.09	0.62	7.7
2.4	18.37	17.70	15.13	11.24	6.88	3.06	0.60	7.6
2.5	18.37	17.69	15.10	11.19	6.84	3.02	0.59	7.5
2.6	18.38	17.67	15.07	11.15	6.80	2.99	0.57	7.4
2.7	18.38	17.65	15.03	11.11	6.75	2.96	0.56	7.3
2.8	18.38	17.64	15.00	11.06	6.71	2.92	0.54	7.2
2.9	18.39	17.62	14.96	11.02	6.67	2.89	0.53	7.1
3.0	18.39	17.60	14.93	10.98	6.63	2.86	0.51	7.0
3.1	18.39	17.58	14.89	10.94	6.59	2.83	0.50	6.9
3.2	18.39	17.56	14.86	10.89	6.55	2.80	0.48	6.8
3.3	18.39	17.54	14.82	10.85	6.50	2.76	0.47	6.7
3.4	18.40	17.53	14.79	10.81	6.46	2.73	0.46	6.6
3.5	18.40	17.51	14.75	10.76	6.42	2.70	0.44	6.5
3.6	18.40	17.49	14.72	10.72	6.38	2.67	0.43	6.4
3.7	18.40	17.47	14.68	10.67	6.33	2.64	0.42	6.3
3.8	18.40	17.45	14.65	10.63	6.29	2.61	0.40	6.2
3.9	18.40	17.43	14.61	10.58	6.25	2.58	0.39	6.1
4.0	18.40	17.41	14.58	10.54	6.21	2.55	0.38	6.0
4.1	18.40	17.39	14.54	10.49	6.17	2.52	0.37	5.9
4.2	18.40	17.37	14.51	10.45	6.13	2.49	0.36	5.8
4.3	18.40	17.35	14.47	10.41	6.09	2.46	0.34	5.7
4.4	18.40	17.33	14.44	10.36	6.04	2.43	0.33	5.6
4.5	18.40	17.31	14.40	10.32	6.00	2.40	0.32	5.5
4.6	18.40	17.29	14.37	10.28	5.96	2.37	0.31	5.4
4.7	18.40	17.26	14.33	10.23	5.92	2.35	0.30	5.3
4.8	18.39	17.24	14.30	10.19	5.88	2.32	0.29	5.2
4.9	18.39	17.22	14.26	10.15	5.84	2.29	0.28	5.1
5.0	18.39	17.20	14.22	10.11	5.80	2.26	0.27	5.0
Days.	130	120	110	100	90	80	70	Days.

TABLE XXV.

(Continued.)

Period, 14.317313 days

Days.	5	6	7	Days.
.50	13.98	17.19	18.97	.50
.51	14.02	17.22	18.98	.49
.52	14.06	17.24	18.99	.48
.53	14.09	17.27	19.00	.47
.54	14.13	17.29	19.00	.46
.55	14.17	17.32	19.01	.45
.56	14.20	17.34	19.02	.44
.57	14.24	17.36	19.03	.43
.58	14.27	17.39	19.03	.42
.59	14.31	17.41	19.04	.41
.60	14.35	17.44	19.05	.40
.61	14.39	17.46	19.06	.39
.62	14.42	17.49	19.06	.38
.63	14.46	17.51	19.07	.37
.64	14.49	17.54	19.07	.36
.65	14.53	17.56	19.08	.35
.66	14.57	17.58	19.09	.34
.67	14.60	17.61	19.09	.33
.68	14.64	17.63	19.10	.32
.69	14.67	17.66	19.10	.31
.70	14.71	17.68	19.11	.30
.71	14.75	17.70	19.12	.29
.72	14.78	17.72	19.12	.28
.73	14.82	17.75	19.13	.27
.74	14.85	17.77	19.13	.26
.75	14.89	17.79	19.14	.25
.76	14.92	17.81	19.14	.24
.77	14.95	17.84	19.15	.23
.78	14.99	17.86	19.15	.22
.79	15.02	17.88	19.16	.21
.80	15.06	17.90	19.16	.20
.81	15.09	17.92	19.16	.19
.82	15.13	17.94	19.17	.18
.83	15.16	17.96	19.17	.17
.84	15.20	17.98	19.17	.16
.85	15.23	18.00	19.18	.15
.86	15.26	18.02	19.18	.14
.87	15.30	18.04	19.18	.13
.88	15.33	18.06	19.19	.12
.89	15.37	18.08	19.19	.11
.90	15.40	18.10	19.19	.10
.91	15.43	18.12	19.19	.09
.92	15.46	18.14	19.19	.08
.93	15.50	18.16	19.19	.07
.94	15.53	18.18	19.20	.06
.95	15.57	18.20	19.20	.05
.96	15.60	18.21	19.20	.04
.97	15.63	18.23	19.20	.03
.98	15.67	18.25	19.20	.02
.99	15.70	18.27	19.20	.01
1.00	15.73	18.29	19.20	.00
Days.	10	9	8	Days.

TABLE XXVI. ARG. 21.

Equation = 9″.2 [1 — sin. (2x + z — 2t)].

Period, 131.671123 days.

Days.	0	10	20	30	40	50	60	Days.
5.0	18.39	17.20	14.22	10.11	5.80	2.26	0.27	5.0
5.1	18.39	17.18	14.18	10.06	5.76	2.23	0.26	4.9
5.2	18.39	17.15	14.15	10.02	5.72	2.20	0.25	4.8
5.3	18.38	17.13	14.11	9.97	5.68	2.18	0.24	4.7
5.4	18.38	17.11	14.07	9.93	5.64	2.15	0.23	4.6
5.5	18.38	17.08	14.04	9.88	5.60	2.12	0.22	4.5
5.6	18.37	17.06	14.00	9.84	5.56	2.09	0.21	4.4
5.7	18.37	17.04	13.96	9.79	5.52	2.07	0.20	4.3
5.8	18.37	17.01	13.93	9.75	5.48	2.04	0.19	4.2
5.9	18.36	16.99	13.89	9.71	5.44	2.01	0.18	4.1
6.0	18.36	16.97	13.85	9.67	5.40	1.98	0.17	4.0
6.1	18.36	16.95	13.81	9.62	5.36	1.95	0.16	3.9
6.2	18.35	16.92	13.77	9.58	5.32	1.93	0.15	3.8
6.3	18.35	16.90	13.74	9.53	5.28	1.90	0.14	3.7
6.4	18.35	16.87	13.70	9.49	5.24	1.87	0.14	3.6
6.5	18.34	16.85	13.66	9.44	5.20	1.85	0.13	3.5
6.6	18.34	16.82	13.62	9.40	5.17	1.82	0.12	3.4
6.7	18.33	16.80	13.58	9.36	5.13	1.79	0.11	3.3
6.8	18.33	16.77	13.54	9.32	5.09	1.77	0.11	3.2
6.9	18.32	16.75	13.50	9.27	5.05	1.74	0.10	3.1
7.0	18.32	16.72	13.46	9.23	5.01	1.71	0.09	3.0
7.1	18.31	16.70	13.42	9.19	4.97	1.69	0.09	2.9
7.2	18.30	16.67	13.38	9.14	4.93	1.66	0.08	2.8
7.3	18.30	16.65	13.34	9.10	4.89	1.64	0.08	2.7
7.4	18.29	16.62	13.30	9.06	4.85	1.61	0.07	2.6
7.5	18.28	16.60	13.26	9.01	4.82	1.59	0.07	2.5
7.6	18.28	16.57	13.22	8.97	4.78	1.56	0.06	2.4
7.7	18.27	16.54	13.18	8.93	4.74	1.54	0.06	2.3
7.8	18.26	16.52	13.14	8.88	4.70	1.51	0.05	2.2
7.9	18.26	16.49	13.10	8.84	4.66	1.49	0.05	2.1
8.0	18.25	16.46	13.06	8.80	4.62	1.46	0.04	2.0
8.1	18.24	16.43	13.02	8.76	4.58	1.44	0.04	1.9
8.2	18.23	16.41	12.98	8.71	4.54	1.41	0.03	1.8
8.3	18.22	16.38	12.94	8.67	4.51	1.39	0.03	1.7
8.4	18.21	16.35	12.90	8.62	4.47	1.37	0.03	1.6
8.5	18.20	16.32	12.86	8.58	4.43	1.34	0.02	1.5
8.6	18.19	16.30	12.82	8.54	4.39	1.32	0.02	1.4
8.7	18.18	16.27	12.78	8.49	4.35	1.30	0.02	1.3
8.8	18.17	16.24	12.74	8.45	4.32	1.27	0.02	1.2
8.9	18.16	16.21	12.70	8.40	4.28	1.25	0.01	1.1
9.0	18.15	16.18	12.66	8.36	4.24	1.23	0.01	1.0
9.1	18.14	16.15	12.62	8.31	4.20	1.21	0.01	0.9
9.2	18.13	16.12	12.58	8.27	4.17	1.19	0.01	0.8
9.3	18.12	16.10	12.54	8.23	4.13	1.16	0.01	0.7
9.4	18.11	16.07	12.50	8.18	4.09	1.14	0.00	0.6
9.5	18.10	16.04	12.46	8.14	4.06	1.12	0.00	0.5
9.6	18.09	16.01	12.41	8.10	4.02	1.10	0.00	0.4
9.7	18.07	15.98	12.37	8.05	3.99	1.08	0.00	0.3
9.8	18.06	15.95	12.33	8.01	3.95	1.06	0.00	0.2
9.9	18.05	15.92	12.29	7.96	3.92	1.04	0.00	0.1
10.0	18.04	15.89	12.25	7.92	3.88	1.02	0.00	0.0
Days.	130	120	110	100	90	80	70	Days.

TABLE XXVII. ARGUMENT 22.

Equation = $8''.1 - 0''.4 \sin. (t - z) + 7''.8 \sin. 2 (t - z)$.

Period, 32.128086 days.

Days.	0	1	2	3	4	5	6	7	8	9
Days.	"	"	"	"	"	"	"	"	"	"
.00	0.02	0.78	2.59	5.23	8.30	11.33	13.89	15.59	16.18	15.59
.01	0.02	0.79	2.62	5.26	8.33	11.36	13.92	15.60	16.18	15.58
.02	0.03	0.80	2.64	5.29	8.36	11.39	13.94	15.62	16.18	15.56
.03	0.03	0.82	2.66	5.32	8.39	11.42	13.96	15.63	16.18	15.55
.04	0.03	0.83	2.68	5.35	8.42	11.45	13.98	15.61	16.18	15.54
.05	0.03	0.84	2.70	5.38	8.45	11.48	14.00	15.65	16.18	15.53
.06	0.03	0.86	2.73	5.41	8.48	11.51	14.02	15.66	16.18	15.52
.07	0.04	0.87	2.75	5.44	8.51	11.53	14.04	15.67	16.18	15.51
.08	0.04	0.89	2.78	5.47	8.51	11.56	14.06	15.69	16.18	15.49
.09	0.04	0.90	2.80	5.50	8.58	11.59	14.08	15.70	16.18	15.[illegible]8
.10	0.05	0.91	2.82	5.53	8.61	11.62	14.11	15.71	16.18	15.47
.11	0.05	0.93	2.85	5.56	8.64	11.65	14.13	15.72	16.17	15.45
.12	0.05	0.94	2.87	5.59	8.67	11.67	14.15	15.73	16.17	15.44
.13	0.06	0.96	2.90	5.62	8.70	11.70	14.17	15.74	16.17	15.43
.14	0.06	0.97	2.92	5.65	8.73	11.73	14.19	15.75	16.17	15.41
.15	0.06	0.99	2.95	5.68	8.76	11.76	14.21	15.76	16.17	15.40
.16	0.07	1.00	2.97	5.71	8.79	11.79	14.23	15.77	16.17	15.39
.17	0.07	1.02	3.00	5.74	8.82	11.81	14.25	15.78	16.17	15.37
.18	0.08	1.03	3.02	5.77	8.85	11.84	14.27	15.79	16.16	15.36
.19	0.08	1.05	3.05	5.80	8.89	11.87	14.29	15.80	16.16	15.35
.20	0.08	1.06	3.07	5.83	8.92	11.90	14.31	15.81	16.16	15.33
.21	0.09	1.08	3.10	5.86	8.95	11.93	14.33	15.81	16.16	15.32
.22	0.09	1.09	3.12	5.89	8.98	11.95	14.35	15.82	16.15	15.31
.23	0.10	1.11	3.15	5.92	9.01	11.98	14.37	15.83	16.15	15.29
.24	0.10	1.12	3.17	5.95	9.01	12.01	14.39	15.84	16.15	15.28
.25	0.11	1.11	3.20	5.98	9.07	12.01	14.41	15.85	16.15	15.26
.26	0.11	1.15	3.22	6.01	9.10	12.07	14.43	15.86	16.15	15.25
.27	0.12	1.17	3.25	6.01	9.13	12.09	14.45	15.87	16.14	15.23
.28	0.12	1.18	3.27	6.07	9.16	12.12	14.47	15.87	16.14	15.22
.29	0.13	1.20	3.30	6.10	9.20	12.15	14.48	15.88	16.14	15.20
.30	0.13	1.22	3.32	6.13	9.23	12.17	14.50	15.89	16.13	15.19
.31	0.14	1.23	3.35	6.16	9.26	12.20	14.52	15.90	16.13	15.17
.32	0.14	1.25	3.37	6.19	9.29	12.23	14.51	15.90	16.13	15.16
.33	0.15	1.26	3.10	6.22	9.32	12.25	14.56	15.91	16.12	15.14
.34	0.15	1.28	3.42	6.25	9.35	12.28	14.58	15.92	16.12	15.13
.35	0.16	1.30	3.15	6.28	9.38	12.31	14.60	15.93	16.11	15.11
.36	0.16	1.31	3.47	6.31	9.41	12.33	14.62	15.94	16.11	15.10
.37	0.17	1.33	3.50	6.31	9.44	12.36	14.64	15.94	16.10	15.08
.38	0.17	1.35	3.52	6.37	9.47	12.38	14.66	15.95	16.10	15.07
.39	0.18	1.36	3.55	6.41	9.51	12.41	14.68	15.96	16.09	15.05
.40	0.19	1.38	3.58	6.44	9.54	12.44	14.69	15.96	16.09	15.03
.41	0.20	1.40	3.60	6.47	9.57	12.46	14.71	15.97	16.08	15.02
.42	0.20	1.41	3.63	6.50	9.60	12.49	14.73	15.98	16.08	15.00
.43	0.21	1.43	3.65	6.53	9.63	12.51	14.75	15.98	16.07	14.99
.44	0.21	1.45	3.68	6.56	9.66	12.54	14.77	15.99	16.07	14.97
.45	0.22	1.47	3.71	6.59	9.69	12.57	14.79	16.00	16.06	14.95
.46	0.22	1.49	3.73	6.62	9.72	12.59	14.81	16.00	16.06	14.94
.47	0.23	1.51	3.76	6.65	9.75	12.62	14.82	16.01	16.05	14.92
.48	0.24	1.53	3.78	6.68	9.78	12.64	14.84	16.01	16.05	14.90
.49	0.24	1.54	3.81	6.72	9.82	12.67	14.86	16.02	16.04	14.89
.50	0.25	1.56	3.84	6.75	9.85	12.70	14.87	16.03	16.03	14.87

TABLE XXVII. ARGUMENT 22.

Equation $= 8''.1 - 0''.4 \sin. (t - z) + 7''.8 \sin. 2 (t - z)$.

Period, 32.128086 days.

Days.	0	1	2	3	4	5	6	7	8	9
Days.										
.50	0.25″	1.56″	3.84″	6.75″	9.85″	12.70″	14.87″	16.03″	16.03″	14.87″
.51	0.26	1.58	3.86	6.78	9.88	12.72	14.89	16.03	16.03	14.85
.52	0.26	1.60	3.89	6.81	9.91	12.75	14.91	16.01	16.02	14.84
.53	0.27	1.62	3.91	6.84	9.94	12.77	14.92	16.04	16.02	14.82
.54	0.28	1.64	3.94	6.87	9.97	12.80	14.94	16.05	16.01	14.80
.55	0.29	1.66	3.97	6.90	10.00	12.82	14.96	16.05	16.00	14.78
.56	0.30	1.68	3.99	6.93	10.03	12.85	14.97	16.06	16.00	14.76
.57	0.31	1.70	4.02	6.96	10.06	12.87	14.99	16.06	15.99	14.75
.58	0.32	1.71	4.05	6.99	10.09	12.90	15.00	16.07	15.98	14.73
.59	0.32	1.73	4.07	7.03	10.12	12.92	15.02	16.07	15.98	14.71
.60	0.33	1.75	4.10	7.06	10.15	12.95	15.04	16.08	15.97	14.69
.61	0.34	1.77	4.13	7.09	10.18	12.97	15.05	16.08	15.96	14.67
.62	0.35	1.79	4.15	7.12	10.21	13.00	15.07	16.09	15.96	14.66
.63	0.36	1.81	4.18	7.15	10.24	13.02	15.08	16.09	15.95	14.64
.64	0.37	1.83	4.21	7.18	10.27	13.05	15.10	16.10	15.94	14.62
.65	0.38	1.85	4.24	7.21	10.30	13.07	15.11	16.10	15.93	14.60
.66	0.39	1.87	4.26	7.24	10.33	13.10	15.13	16.11	15.92	14.58
.67	0.40	1.89	4.29	7.27	10.36	13.12	15.14	16.11	15.91	14.56
.68	0.41	1.91	4.32	7.30	10.39	13.15	15.16	16.12	15.91	14.54
.69	0.42	1.93	4.35	7.34	10.42	13.17	15.17	16.12	15.90	14.53
.70	0.43	1.95	4.38	7.37	10.45	13.20	15.19	16.12	15.89	14.51
.71	0.44	1.97	4.40	7.40	10.48	13.22	15.20	16.13	15.88	14.49
.72	0.45	1.99	4.43	7.43	10.51	13.25	15.22	16.13	15.88	14.47
.73	0.46	2.01	4.46	7.46	10.54	13.27	15.23	16.13	15.87	14.45
.74	0.47	2.03	4.49	7.49	10.57	13.30	15.25	16.14	15.86	14.43
.75	0.48	2.05	4.52	7.52	10.60	13.32	15.26	16.14	15.85	14.41
.76	0.49	2.07	4.54	7.55	10.63	13.35	15.28	16.14	15.84	14.39
.77	0.50	2.09	4.57	7.58	10.66	13.37	15.29	16.15	15.83	14.37
.78	0.51	2.12	4.60	7.61	10.69	13.40	15.31	16.15	15.82	14.36
.79	0.53	2.14	4.63	7.65	10.72	13.42	15.32	16.15	15.81	14.34
.80	0.54	2.16	4.66	7.68	10.75	13.44	15.33	16.16	15.80	14.32
.81	0.55	2.18	4.68	7.71	10.78	13.47	15.34	16.16	15.79	14.30
.82	0.56	2.20	4.71	7.74	10.81	13.49	15.36	16.16	15.78	14.28
.83	0.57	2.22	4.74	7.77	10.84	13.52	15.38	16.17	15.77	14.26
.84	0.58	2.24	4.77	7.80	10.87	13.54	15.39	16.17	15.76	14.24
.85	0.59	2.26	4.80	7.83	10.90	13.56	15.40	16.17	15.75	14.22
.86	0.61	2.29	4.83	7.86	10.93	13.59	15.42	16.17	15.74	14.20
.87	0.62	2.31	4.86	7.89	10.96	13.61	15.43	16.18	15.73	14.18
.88	0.63	2.33	4.88	7.93	10.98	13.63	15.44	16.18	15.72	14.16
.89	0.64	2.35	4.91	7.96	11.01	13.65	15.46	16.18	15.71	14.14
.90	0.65	2.37	4.94	7.99	11.04	13.67	15.47	16.18	15.70	14.11
.91	0.67	2.40	4.97	8.02	11.07	13.70	15.48	16.18	15.69	14.09
.92	0.68	2.42	5.00	8.05	11.10	13.72	15.50	16.18	15.68	14.07
.93	0.69	2.44	5.03	8.08	11.13	13.74	15.51	16.18	15.67	14.05
.94	0.70	2.46	5.06	8.11	11.16	13.76	15.52	16.18	15.66	14.03
.95	0.71	2.48	5.09	8.14	11.19	13.78	15.53	16.18	15.65	14.01
.96	0.72	2.51	5.12	8.17	11.22	13.81	15.54	16.18	15.64	13.99
.97	0.74	2.53	5.15	8.20	11.25	13.83	15.56	16.18	15.63	13.97
.98	0.75	2.55	5.17	8.23	11.27	13.85	15.57	16.18	15.61	13.94
.99	0.76	2.57	5.20	8.27	11.30	13.87	15.58	16.18	15.60	13.92
1.00	0.78	2.59	5.23	8.30	11.33	13.89	15.59	16.18	15.59	13.90

TABLE XXVII. ARGUMENT 22.

Equation $= 8''.1 - 0''.4 \sin. (t - z) + 7''.8 \sin. 2 (t - z)$.

Period, 32.128086 days.

Days.	10	11	12	13	14	15	16	17	18	19
Days. .00	″ 13.90	″ 11.38	″ 8.40	″ 5.42	″ 2.90	″ 1.20	″ 0.58	″ 1.13	″ 2.76	″ 5.21
.01	13.88	11.36	8.37	5.39	2.88	1.19	0.58	1.14	2.78	5.23
.02	13.85	11.33	8.34	5.36	2.86	1.18	0.58	1.15	2.80	5.26
.03	13.83	11.30	8.31	5.34	2.84	1.16	0.58	1.17	2.82	5.29
.04	13.81	11.27	8.27	5.31	2.81	1.15	0.58	1.18	2.84	5.31
.05	13.79	11.24	8.24	5.28	2.79	1.14	0.58	1.19	2.86	5.34
.06	13.77	11.21	8.21	5.26	2.77	1.13	0.58	1.20	2.89	5.37
.07	13.75	11.18	8.18	5.23	2.75	1.12	0.59	1.21	2.91	5.39
.08	13.72	11.15	8.15	5.20	2.73	1.10	0.59	1.23	2.93	5.43
.09	13.70	11.13	8.12	5.18	2.71	1.09	0.59	1.24	2.95	5.45
.10	13.68	11.10	8.09	5.15	2.69	1.08	0.59	1.25	2.97	5.48
.11	13.66	11.07	8.06	5.12	2.67	1.07	0.59	1.26	2.99	5.51
.12	13.63	11.04	8.03	5.10	2.65	1.06	0.59	1.27	3.01	5.54
.13	13.61	11.01	8.00	5.07	2.63	1.05	0.59	1.29	3.04	5.56
.14	13.59	10.98	7.97	5.04	2.61	1.04	0.59	1.30	3.06	5.59
.15	13.57	10.94	7.94	5.02	2.58	1.03	0.59	1.31	3.08	5.62
.16	13.55	10.91	7.91	4.99	2.56	1.02	0.60	1.33	3.10	5.65
.17	13.52	10.89	7.88	4.96	2.54	1.01	0.60	1.34	3.12	5.68
.18	13.50	10.86	7.85	4.94	2.52	1.00	0.60	1.36	3.15	5.70
.19	13.48	10.84	7.82	4.91	2.50	0.99	0.60	1.37	3.17	5.73
.20	13.45	10.81	7.79	4.88	2.48	0.98	0.60	1.38	3.19	5.76
.21	13.43	10.78	7.76	4.86	2.46	0.97	0.60	1.39	3.21	5.79
.22	13.41	10.75	7.73	4.83	2.44	0.96	0.60	1.41	3.23	5.82
.23	13.38	10.72	7.70	4.80	2.42	0.95	0.61	1.42	3.26	5.85
.24	13.36	10.69	7.67	4.78	2.40	0.94	0.61	1.43	3.28	5.87
.25	13.33	10.66	7.64	4.75	2.38	0.93	0.61	1.45	3.30	5.90
.26	13.31	10.63	7.61	4.71	2.37	0.93	0.62	1.46	3.33	5.93
.27	13.28	10.60	7.58	4.69	2.35	0.92	0.62	1.48	3.35	5.96
.28	13.26	10.57	7.55	4.66	2.33	0.91	0.62	1.49	3.37	5.99
.29	13.23	10.55	7.52	4.64	2.31	0.90	0.63	1.51	3.40	6.02
.30	13.21	10.52	7.49	4.61	2.29	0.89	0.63	1.52	3.42	6.05
.31	13.18	10.49	7.46	4.58	2.27	0.88	0.63	1.53	3.44	6.08
.32	13.16	10.46	7.43	4.56	2.25	0.87	0.64	1.55	3.47	6.11
.33	13.13	10.43	7.40	4.53	2.24	0.86	0.64	1.57	3.49	6.14
.34	13.11	10.40	7.37	4.50	2.22	0.86	0.65	1.58	3.51	6.17
.35	13.08	10.37	7.34	4.48	2.20	0.85	0.65	1.60	3.54	6.19
.36	13.06	10.34	7.31	4.45	2.18	0.84	0.65	1.61	3.56	6.22
.37	13.03	10.31	7.28	4.43	2.17	0.83	0.66	1.63	3.59	6.25
.38	13.01	10.28	7.25	4.40	2.15	0.83	0.66	1.64	3.61	6.28
.39	12.98	10.25	7.22	4.38	2.13	0.82	0.67	1.66	3.64	6.31
.40	12.96	10.22	7.19	4.35	2.11	0.81	0.67	1.67	3.66	6.34
.41	12.93	10.19	7.16	4.33	2.09	0.80	0.67	1.69	3.68	6.37
.42	12.91	10.16	7.13	4.30	2.07	0.79	0.68	1.70	3.71	6.40
.43	12.88	10.13	7.10	4.28	2.06	0.79	0.68	1.72	3.73	6.43
.44	12.86	10.10	7.07	4.25	2.04	0.78	0.69	1.73	3.75	6.46
.45	12.83	10.07	7.04	4.23	2.02	0.78	0.69	1.75	3.78	6.48
.46	12.81	10.04	7.01	4.21	2.00	0.77	0.69	1.76	3.80	6.51
.47	12.78	10.01	6.98	4.18	1.98	0.76	0.70	1.78	3.82	6.54
.48	12.76	9.98	6.95	4.16	1.97	0.76	0.70	1.79	3.85	6.57
.49	12.73	9.95	6.92	4.12	1.95	0.75	0.71	1.81	3.87	6.60
.50	12.71	9.92	6.89	4.09	1.93	0.75	0.71	1.82	3.90	6.63

TABLE XXVII. ARGUMENT 22.

Equation = 8″.1 — 0″.4 sin. $(t-z)$ + 7″.8 sin. 2 $(t-z)$.

Period, 32.128086 days.

Days.	10	11	12	13	14	15	16	17	18	19
Days	″	″	″	″	″	″	″	″	″	″
.50	12.71	9.92	6.89	4.09	1.93	0.75	0.71	1.82	3.90	6.63
.51	12.68	9.89	6.86	4.06	1.91	0.74	0.72	1.84	3.92	6.66
.52	12.66	9.86	6.83	4.04	1.89	0.74	0.72	1.85	3.95	6.69
.53	12.63	9.83	6.80	4.01	1.88	0.73	0.73	1.87	3.97	6.72
.54	12.61	9.80	6.77	3.99	1.86	0.73	0.74	1.88	4.00	6.75
.55	12.58	9.77	6.74	3.96	1.84	0.72	0.74	1.90	4.02	6.77
.56	12.56	9.74	6.71	3.94	1.83	0.72	0.75	1.92	4.05	6.80
.57	12.53	9.71	6.68	3.91	1.81	0.71	0.75	1.93	4.07	6.83
.58	12.51	9.68	6.65	3.89	1.79	0.71	0.76	1.95	4.10	6.86
.59	12.48	9.65	6.62	3.86	1.78	0.70	0.77	1.96	4.12	6.89
.60	12.45	9.62	6.59	3.84	1.76	0.70	0.77	1.98	4.15	6.92
.61	12.43	9.59	6.56	3.81	1.74	0.69	0.78	2.00	4.17	6.95
.62	12.40	9.56	6.53	3.79	1.73	0.69	0.79	2.02	4.20	6.98
.63	12.38	9.53	6.50	3.76	1.71	0.68	0.79	2.03	4.22	7.01
.64	12.35	9.50	6.47	3.74	1.70	0.68	0.80	2.05	4.25	7.04
.65	12.32	9.47	6.44	3.71	1.68	0.67	0.81	2.07	4.28	7.06
.66	12.30	9.44	6.41	3.69	1.67	0.67	0.81	2.09	4.30	7.09
.67	12.27	9.41	6.38	3.66	1.65	0.66	0.82	2.11	4.33	7.12
.68	12.25	9.38	6.35	3.64	1.64	0.66	0.83	2.12	4.35	7.15
.69	12.22	9.35	6.32	3.61	1.62	0.65	0.83	2.14	4.38	7.18
.70	12.19	9.32	6.29	3.59	1.61	0.65	0.84	2.16	4.41	7.21
.71	12.17	9.29	6.26	3.56	1.59	0.64	0.85	2.18	4.43	7.24
.72	12.14	9.26	6.23	3.54	1.58	0.64	0.86	2.20	4.46	7.27
.73	12.12	9.23	6.20	3.51	1.56	0.63	0.86	2.22	4.48	7.30
.74	12.09	9.20	6.17	3.49	1.55	0.63	0.87	2.24	4.51	7.33
.75	12.06	9.17	6.14	3.47	1.54	0.63	0.88	2.25	4.54	7.36
.76	12.04	9.14	6.11	3.44	1.52	0.62	0.89	2.27	4.56	7.39
.77	12.01	9.11	6.08	3.42	1.51	0.62	0.90	2.29	4.59	7.42
.78	11.98	9.08	6.05	3.39	1.49	0.61	0.90	2.31	4.61	7.45
.79	11.96	9.05	6.02	3.37	1.48	0.61	0.91	2.33	4.64	7.48
.80	11.93	9.02	5.99	3.35	1.47	0.61	0.92	2.35	4.67	7.51
.81	11.90	8.99	5.96	3.32	1.45	0.61	0.93	2.37	4.70	7.54
.82	11.88	8.96	5.93	3.30	1.44	0.60	0.94	2.39	4.72	7.57
.83	11.85	8.93	5.90	3.28	1.42	0.60	0.95	2.41	4.75	7.60
.84	11.82	8.90	5.87	3.25	1.41	0.60	0.96	2.43	4.78	7.63
.85	11.79	8.87	5.84	3.23	1.40	0.60	0.97	2.45	4.80	7.66
.86	11.77	8.84	5.82	3.21	1.38	0.60	0.98	2.47	4.83	7.69
.87	11.74	8.81	5.79	3.19	1.37	0.59	0.99	2.49	4.86	7.72
.88	11.71	8.78	5.76	3.16	1.35	0.59	1.00	2.51	4.88	7.75
.89	11.69	8.75	5.73	3.14	1.34	0.59	1.01	2.53	4.91	7.78
.90	11.66	8.71	5.70	3.12	1.33	0.59	1.02	2.55	4.94	7.81
.91	11.63	8.68	5.67	3.10	1.32	0.59	1.03	2.57	4.96	7.84
.92	11.61	8.65	5.64	3.07	1.30	0.59	1.04	2.59	4.99	7.87
.93	11.58	8.62	5.61	3.05	1.29	0.59	1.05	2.61	5.02	7.90
.94	11.55	8.59	5.59	3.03	1.27	0.58	1.06	2.63	5.04	7.93
.95	11.52	8.56	5.56	3.01	1.26	0.58	1.07	2.65	5.07	7.96
.96	11.49	8.53	5.53	2.99	1.25	0.58	1.08	2.68	5.10	7.98
.97	11.46	8.50	5.50	2.96	1.24	0.58	1.10	2.70	5.12	8.01
.98	11.44	8.46	5.48	2.94	1.22	0.58	1.11	2.72	5.15	8.04
.99	11.41	8.43	5.45	2.92	1.21	0.58	1.12	2.74	5.18	8.07
1.00	11.38	8.40	5.42	2.90	1.20	0.58	1.13	2.76	5.21	8.10

TABLE XXVII. ARGUMENT 22.

Equation $= 8''.1 - 0''.4 \sin.\ (t - z) + 7''.8 \sin.\ 2\ (t - z)$.

Period, 32.128086 days.

Days.	20	21	22	23	24	25	26	27	28	29
Days.	"	"	"	"	"	"	"	"	"	"
.00	8.10	10.99	13.44	15.07	15.62	15.00	13.30	10.78	7.80	4.82
.01	8.13	11.02	13.46	15.08	15.62	14.99	13.28	10.75	7.77	4.79
.02	8.16	11.04	13.48	15.09	15.62	14.98	13.26	10.72	7.74	4.76
.03	8.19	11.07	13.50	15.10	15.62	14.96	13.23	10.70	7.70	4.74
.04	8.22	11.10	13.52	15.11	15.62	14.95	13.21	10.67	7.67	4.71
.05	8.25	11.12	13.55	15.13	15.62	14.94	13.19	10.64	7.64	4.68
.06	8.27	11.15	13.57	15.14	15.62	14.92	13.17	10.61	7.61	4.65
.07	8.30	11.18	13.59	15.15	15.61	14.91	13.15	10.58	7.58	4.62
.08	8.33	11.20	13.61	15.16	15.61	14.90	13.12	10.56	7.55	4.60
.09	8.36	11.23	13.63	15.17	15.61	14.88	13.10	10.53	7.52	4.57
.10	8.39	11.26	13.65	15.18	15.61	14.87	13.08	10.50	7.49	4.54
.11	8.42	11.28	13.67	15.19	15.61	14.86	13.06	10.47	7.45	4.52
.12	8.45	11.31	13.69	15.20	15.61	14.84	13.04	10.44	7.42	4.49
.13	8.48	11.34	13.71	15.21	15.60	14.83	13.01	10.41	7.39	4.46
.14	8.51	11.36	13.73	15.22	15.60	14.81	12.99	10.38	7.36	4.44
.15	8.54	11.39	13.75	15.23	15.60	14.80	12.97	10.36	7.33	4.41
.16	8.57	11.42	13.77	15.24	15.60	14.79	12.94	10.33	7.30	4.38
.17	8.60	11.44	13.79	15.25	15.60	14.77	12.92	10.30	7.27	4.35
.18	8.63	11.47	13.81	15.26	15.59	14.76	12.90	10.27	7.24	4.33
.19	8.66	11.50	13.83	15.27	15.59	14.74	12.87	10.24	7.21	4.30
.20	8.69	11.53	13.85	15.28	15.59	14.73	12.85	10.21	7.18	4.27
.21	8.72	11.55	13.87	15.29	15.59	14.72	12.83	10.18	7.15	4.25
.22	8.75	11.58	13.89	15.30	15.58	14.70	12.80	10.15	7.12	4.22
.23	8.78	11.61	13.91	15.30	15.58	14.69	12.78	10.12	7.09	4.19
.24	8.81	11.63	13.93	15.31	15.57	14.67	12.75	10.09	7.06	4.17
.25	8.84	11.66	13.95	15.32	15.57	14.66	12.73	10.06	7.03	4.14
.26	8.87	11.68	13.96	15.33	15.57	14.65	12.71	10.03	7.00	4.11
.27	8.90	11.71	13.98	15.34	15.56	14.63	12.68	10.00	6.97	4.08
.28	8.93	11.74	14.00	15.34	15.56	14.62	12.66	9.97	6.94	4.06
.29	8.96	11.76	14.02	15.35	15.55	14.60	12.63	9.94	6.91	4.03
.30	8.99	11.79	14.04	15.36	15.55	14.59	12.61	9.91	6.88	4.01
.31	9.03	11.82	14.06	15.37	15.55	14.58	12.59	9.88	6.85	3.98
.32	9.05	11.84	14.08	15.38	15.54	14.56	12.56	9.85	6.82	3.95
.33	9.08	11.87	14.09	15.38	15.54	14.55	12.54	9.82	6.79	3.93
.34	9.11	11.90	14.11	15.39	15.53	14.53	12.51	9.79	6.76	3.90
.35	9.14	11.92	14.13	15.40	15.53	14.52	12.49	9.76	6.73	3.88
.36	9.16	11.95	14.15	15.40	15.52	14.50	12.46	9.73	6.70	3.85
.37	9.19	11.98	14.17	15.41	15.52	14.49	12.44	9.70	6.67	3.82
.38	9.22	12.00	14.18	15.42	15.51	14.47	12.41	9.67	6.64	3.80
.39	9.25	12.03	14.20	15.42	15.51	14.46	12.39	9.64	6.61	3.77
.40	9.28	12.05	14.22	15.43	15.50	14.44	12.36	9.61	6.58	3.75
.41	9.31	12.08	14.24	15.44	15.50	14.42	12.34	9.58	6.55	3.72
.42	9.34	12.10	14.25	15.44	15.49	14.41	12.31	9.55	6.52	3.69
.43	9.37	12.13	14.27	15.45	15.49	14.39	12.29	9.52	6.49	3.67
.44	9.40	12.15	14.28	15.45	15.48	14.37	12.26	9 49	6.46	3.64
.45	9.43	12.18	14.30	15.46	15.48	14.36	12.24	9.46	6.43	3.62
.46	9.45	12.20	14.32	15.47	15.47	14.34	12.21	9.43	6.40	3.59
.47	9.48	12.23	14.33	15.47	15.47	14.32	12.19	9.40	6.37	3.57
.48	9.51	12.25	14.35	15.48	15.46	14.31	12.16	9.37	6.34	3.54
.49	9.54	12.28	14.36	15.48	15.46	14.29	12.14	9.34	6.31	3.52
.50	9.57	12.30	14.38	15.49	15.45	14.27	12.11	9.31	6.28	3.49

TABLE XXVII. ARGUMENT 22.

Equation = 8″.1 — 0″.4 sin. $(t-z)$ + 7″.8 sin. 2 $(t-z)$.

Period, 32.128086 days.

Days.	20	21	22	23	24	25	26	27	28	29
Days .50	″ 9.57	″ 12.30	″ 14.38	″ 15.49	″ 15.45	″ 14.27	″ 12.11	9.31	″ 6.28	″ 3.49
.51	9.60	12.33	14.40	15.49	15.44	14.25	12.08	9.28	6.25	3.47
.52	9.63	12.35	14.41	15.50	15.44	14.23	12.06	9.25	6.22	3.44
.53	9.66	12.38	14.43	15.50	15.43	14.22	12.03	9.22	6.19	3.42
.54	9.69	12.40	14.44	15.51	15.43	14.20	12.01	9.19	6.16	3.39
.55	9.72	12.42	14.46	15.51	15.42	14.18	11.98	9.16	6.13	3.37
.56	9.76	12.45	14.47	15.51	15.41	14.16	11.95	9.13	6.10	3.34
.57	9.79	12.47	14.49	15.52	15.41	14.14	11.93	9.10	6.07	3.32
.58	9.82	12.50	14.50	15.52	15.40	14.13	11.90	9.07	6.04	3.29
.59	9.85	12.52	14.52	15.53	15.40	14.11	11.88	9.04	6.01	3.27
.60	9.86	12.54	14.53	15.53	15.39	14.09	11.85	9.01	5.98	3.24
.61	9.89	12.57	14.55	15.53	15.38	14.07	11.82	8.98	5.95	3.22
.62	9.92	12.59	14.56	15.54	15.37	14.05	11.80	8.95	5.92	3.19
.63	9.95	12.62	14.58	15.54	15.37	14.04	11.77	8.92	5.89	3.17
.64	9.98	12.64	14.59	15.55	15.36	14.02	11.75	8.89	5.86	3.14
.65	10.01	12.66	14.61	15.55	15.35	14.00	11.72	8.86	5.83	3.12
.66	10.03	12.69	14.62	15.55	15.34	13.98	11.69	8.83	5.80	3.09
.67	10.06	12.71	14.64	15.56	15.33	13.96	11.67	8.80	5.77	3.07
.68	10.09	12.74	14.65	15.56	15.33	13.95	11.64	8.77	5.74	3.04
.69	10.12	12.76	14.67	15.57	15.32	13.93	11.62	8.74	5.71	3.02
.70	10.15	12.78	14.68	15.57	15.31	13.91	11.59	8.71	5.68	2.99
71	10.18	12.80	14.69	15.57	15.30	13.89	11.56	8.68	5.65	2.97
.72	10.21	12.83	14.71	15.58	15.29	13.87	11.54	8.65	5.62	2.94
.73	10.24	12.85	14.72	15.58	15.28	13.85	11.51	8.62	5.60	2.92
.74	10.27	12.87	14.74	15.58	15.27	13.83	11.48	8.59	5.57	2.89
.75	10.30	12.90	14.75	15.59	15.27	13.82	11.46	8.56	5.54	2.87
.76	10.32	12.92	14.76	15.59	15.26	13.80	11.43	8.53	5.51	2.84
.77	10.35	12.94	14.78	15.59	15.25	13.78	11.40	8.50	5.48	2.82
.78	10.38	12.97	14.79	15.60	15.24	13.76	11.38	8.47	5.45	2.79
.79	10.41	12.99	14.81	15.60	15.23	13.74	11.35	8.44	5.42	2.77
.80	10.44	13.01	14.82	15.60	15.22	13.72	11.32	8.41	5.39	2.75
.81	10.47	13.03	14.83	15.60	15.21	13.70	11.29	8.38	5.36	2.72
.82	10.50	13.05	14.85	15.60	15.20	13.68	11.27	8.35	5.33	2.70
.83	10.52	13.08	14.86	15.60	15.19	13.66	11.24	8.32	5.31	2.68
.84	10.55	13.10	14.87	15.60	15.18	13.64	11.21	8.29	5.28	2.65
.85	10.58	13.12	14.89	15.61	15.17	13.62	11.19	8.26	5.25	2.63
.86	10.61	13.14	14.90	15.61	15.16	13.59	11.16	8.23	5.22	2.61
.87	10.64	13.16	14.91	15.61	15.15	13.57	11.13	8.20	5.19	2.59
.88	10.66	13.19	14.93	15.61	15.14	13.55	11.11	8.17	5.16	2.57
.89	10.69	13.21	14.94	15.61	15.13	13.52	11.08	8.14	5.13	2.54
.90	10.72	13.23	14.95	15.61	15.12	13.51	11.05	8.11	5.10	2.52
.91	10.75	13.25	14.96	15.61	15.11	13.49	11.02	8.08	5.07	2.50
.92	10.78	13.27	14.97	15.61	15.10	13.47	11.00	8.05	5.04	2.48
.93	10.80	13.29	14.99	15.61	15.08	13.45	10.97	8.02	5.02	2.46
.94	10.83	13.31	15.00	15.61	15.07	13.43	10.94	7.99	4.99	2.43
.95	10.86	13.34	15.01	15.62	15.06	13.41	10.92	7.96	4.96	2.41
.96	10.88	13.36	15.02	15.62	15.05	13.38	10.89	7.93	4.93	2.39
.97	10.91	13.38	15.03	15.62	15.04	13.36	10.86	7.90	4.90	2.37
.98	10.94	13.40	15.05	15.62	15.02	13.34	10.84	7.86	4.87	2.35
.99	10.96	13.42	15.06	15.62	15.01	13.32	10.81	7.83	4.85	2.32
1.00	10.99	13.44	15.07	15.62	15.00	13.30	10.78	7.80	4.82	2.30

TABLE XXVII.

(Continued.)

Period, 32.128086 days.

Days.	30	31	32
.00	2.30	0.61	0.02
.01	2.28	0.60	0.02
.02	2.26	0.58	0.02
.03	2.24	0.57	0.02
.04	2.21	0.56	0.02
.05	2.19	0.54	0.02
.06	2.17	0.53	0.02
.07	2.15	0.52	0.02
.08	2.13	0.51	0.02
.09	2.11	0.50	0.02
.10	2.09	0.49	0.02
.11	2.07	0.48	0.02
.12	2.05	0.47	0.02
.13	2.02	0.46	0.03
.14	2.00	0.45	0.03
.15	1.98	0.44	0.03
.16	1.96	0.43	0.03
.17	1.94	0.42	0.03
.18	1.92	0.41	0.04
.19	1.90	0.40	0.04
.20	1.88	0.39	0.04
.21	1.86	0.38	0.05
.22	1.84	0.38	0.05
.23	1.83	0.37	0.05
.24	1.81	0.36	0.06
.25	1.79	0.35	0.06
.26	1.77	0.34	0.06
.27	1.75	0.33	0.07
.28	1.73	0.32	0.07
.29	1.71	0.32	0.07
.30	1.69	0.31	0.08
.31	1.67	0.30	0.08
.32	1.65	0.29	0.08
.33	1.64	0.28	0.09
.34	1.62	0.28	0.09
.35	1.60	0.27	0.10
.36	1.58	0.26	0.10
.37	1.56	0.25	0.11
.38	1.54	0.25	0.11
.39	1.53	0.24	0.12
.40	1.51	0.23	0.12
.41	1.49	0.23	0.13
.42	1.47	0.22	0.13
.43	1.45	0.21	0.14
.44	1.44	0.21	0.14
.45	1.42	0.20	0.15
.46	1.40	0.19	0.15
.47	1.38	0.19	0.16
.48	1.37	0.18	0.16
.49	1.35	0.17	0.17
.50	1.33	0.17	0.17

TABLE XXVIII. ARG. 23.

Equation = 9″.8 [1 — sin. $(x + 2y - 2t)$].

Period, 23.774626 days.

Days.	0	1	2	3	4	5	Days.
.00	19.59	19.33	18.40	16.88	14.87	12.49	1.00
.01	19.59	19.32	18.39	16.86	14.85	12.46	.99
.02	19.59	19.32	18.38	16.84	14.82	12.44	.98
.03	19.60	19.31	18.36	16.83	14.80	12.41	.97
.04	19.60	19.31	18.35	16.81	14.77	12.39	.96
.05	19.60	19.30	18.34	16.79	14.75	12.36	.95
.06	19.60	19.29	18.33	16.77	14.73	12.34	.94
.07	19.60	19.29	18.32	16.75	14.71	12.31	.93
.08	19.60	19.28	18.30	16.74	14.68	12.29	.92
.09	19.60	19.28	18.29	16.72	14.66	12.26	.91
.10	19.60	19.27	18.28	16.70	14.64	12.24	.90
.11	19.60	19.26	18.27	16.68	14.62	12.21	.89
.12	19.60	19.25	18.25	16.66	14.59	12.19	.88
.13	19.60	19.25	18.24	16.65	14.57	12.16	.87
.14	19.60	19.24	18.22	16.63	14.54	12.14	.86
.15	19.60	19.23	18.21	16.61	14.52	12.11	.85
.16	19.60	19.22	18.20	16.59	14.50	12.09	.84
.17	19.60	19.21	18.18	16.57	14.48	12.06	.83
.18	19.60	19.21	18.17	16.55	14.45	12.04	.82
.19	19.60	19.20	18.15	16.53	14.43	12.01	.81
.20	19.60	19.19	18.14	16.51	14.41	11.99	.80
.21	19.60	19.18	18.13	16.49	14.39	11.96	.79
.22	19.60	19.17	18.11	16.47	14.36	11.94	.78
.23	19.59	19.17	18.10	16.46	14.31	11.91	.77
.24	19.59	19.16	18.08	16.44	14.31	11.89	.76
.25	19.59	19.15	18.07	16.42	14.29	11.86	.75
.26	19.59	19.14	18.05	16.40	14.27	11.84	.74
.27	19.59	19.14	18.04	16.38	14.25	11.81	.73
.28	19.59	19.13	18.02	16.37	14.22	11.79	.72
.29	19.59	19.13	18.01	16.35	14.20	11.76	.71
.30	19.59	19.12	17.99	16.33	14.18	11.74	.70
.31	19.59	19.11	17.98	16.31	14.16	11.71	.69
.32	19.59	19.10	17.96	16.29	14.13	11.69	.68
.33	19.58	19.10	17.95	16.27	14.11	11.66	.67
.34	19.58	19.09	17.93	16.25	14.08	11.64	.66
.35	19.58	19.08	17.92	16.23	14.06	11.61	.65
.36	19.58	19.07	17.91	16.21	14.04	11.59	.64
.37	19.58	19.06	17.89	16.19	14.02	11.56	.63
.38	19.57	19.06	17.88	16.17	13.99	11.54	.62
.39	19.57	19.05	17.86	16.15	13.97	11.51	.61
.40	19.57	19.04	17.85	16.13	13.95	11.49	.60
.41	19.57	19.03	17.84	16.11	13.93	11.46	.59
.42	19.57	19.02	17.82	16.09	13.90	11.44	.58
.43	19.56	19.02	17.81	16.07	13.88	11.41	.57
.44	19.56	19.01	17.79	16.05	13.85	11.39	.56
.45	19.56	19.00	17.78	16.03	13.83	11.36	.55
.46	19.56	18.99	17.76	16.01	13.81	11.33	.54
.47	19.56	18.98	17.75	15.99	13.78	11.31	.53
.48	19.55	18.97	17.73	15.97	13.76	11.28	.52
.49	19.55	18.96	17.72	15.95	13.73	11.26	.51
.50	19.55	18.95	17.70	15.93	13.71	11.23	.50
Days.	23	22	21	20	19	18	Days.

TABLE XXVII.

(Continued.)

Period, 32.128086 days.

Days.	30	31	32
Days.	"	"	"
.50	1.33	0.17	0.17
.51	1.32	0.16	0.18
.52	1.30	0.15	0.19
.53	1.28	0.15	0.19
.54	1.27	0.14	0.20
.55	1.25	0.13	0.20
.56	1.23	0.13	0.21
.57	1.21	0.12	0.22
.58	1.20	0.12	0.22
.59	1.18	0.11	0.23
.60	1.17	0.11	0.24
.61	1.15	0.10	0.24
.62	1.13	0.10	0.25
.63	1.12	0.09	0.26
.64	1.10	0.09	0.26
.65	1.09	0.08	0.27
.66	1.07	0.08	0.28
.67	1.06	0.08	0.29
.68	1.04	0.07	0.30
.69	1.03	0.07	0.30
.70	1.01	0.07	0.31
.71	1.00	0.06	0.32
.72	0.98	0.06	0.33
.73	0.97	0.06	0.34
.74	0.95	0.05	0.34
.75	0.94	0.05	0.35
.76	0.92	0.05	0.36
.77	0.91	0.05	0.37
.78	0.89	0.04	0.38
.79	0.88	0.04	0.39
.80	0.87	0.04	0.40
.81	0.85	0.04	0.41
.82	0.84	0.04	0.42
.83	0.83	0.03	0.43
.84	0.81	0.03	0.44
.85	0.80	0.03	0.45
.86	0.79	0.03	0.46
.87	0.77	0.03	0.47
.88	0.76	0.03	0.48
.89	0.75	0.03	0.49
.90	0.73	0.03	0.50
.91	0.72	0.02	0.51
.92	0.71	0.02	0.53
.93	0.70	0.02	0.54
.94	0.68	0.02	0.55
.95	0.67	0.02	0.56
.96	0.66	0.02	0.57
.97	0.65	0.02	0.58
.98	0.63	0.02	0.59
.99	0.62	0.02	0.61
1.00	0.61	0.02	0.62

TABLE XXVIII. ARG. 23.

Equation $= 9''.8\ [1 - \sin.\ (x + 2y - 2t)]$.

Period, 23.774626 days.

Days.	0	1	2	3	4	5	Days.
Days.	"	"	"	"	"	"	Days.
.50	19.55	18.95	17.70	15.93	13.71	11.23	.50
.51	19.55	18.94	17.68	15.91	13.69	11.20	.49
.52	19.55	18.93	17.67	15.89	13.66	11.18	.48
.53	19.54	18.92	17.65	15.87	13.64	11.15	.47
.54	19.54	18.91	17.64	15.85	13.61	11.13	.46
.55	19.54	18.90	17.62	15.83	13.59	11.10	.45
.56	19.54	18.89	17.60	15.81	13.57	11.08	.44
.57	19.53	18.88	17.59	15.79	13.54	11.05	.43
.58	19.53	18.88	17.57	15.77	13.52	11.03	.42
.59	19.52	18.87	17.56	15.75	13.49	11.00	.41
.60	19.52	18.86	17.54	15.73	13.47	10.98	.40
.61	19.52	18.85	17.52	15.71	13.45	10.95	.39
.62	19.51	18.84	17.51	15.69	13.42	10.93	.38
.63	19.51	18.83	17.49	15.67	13.40	10.90	.37
.64	19.50	18.82	17.48	15.65	13.37	10.88	.36
.65	19.50	18.81	17.46	15.63	13.35	10.85	.35
.66	19.50	18.80	17.45	15.61	13.33	10.82	.34
.67	19.49	18.79	17.43	15.59	13.30	10.80	.33
.68	19.49	18.78	17.42	15.56	13.28	10.77	.32
.69	19.48	18.77	17.40	15.54	13.25	10.75	.31
.70	19.48	18.76	17.39	15.52	13.23	10.72	.30
.71	19.47	18.75	17.37	15.50	13.21	10.69	.29
.72	19.47	18.74	17.36	15.18	13.18	10.67	.28
.73	19.46	18.72	17.34	15.46	13.16	10.64	.27
.74	19.46	18.71	17.33	15.44	13.13	10.62	.26
.75	19.45	18.70	17.31	15.42	13.11	10.59	.25
.76	19.45	18.69	17.29	15.40	13.09	10.57	.24
.77	19.41	18.68	17.27	15.38	13.06	10.54	.23
.78	19.44	18.66	17.26	15.35	13.04	10.52	.22
.79	19.43	18.65	17.24	15.33	13.01	10.49	.21
.80	19.43	18.64	17.22	15.31	12.99	10.47	.20
.81	19.42	18.63	17.20	15.29	12.97	10.44	.19
.82	19.42	18.62	17.19	15.27	12.94	10.42	.18
.83	19.41	18.61	17.17	15.24	12.92	10.39	.17
.84	19.41	18.60	17.16	15.22	12.89	10.37	.16
.85	19.40	18.59	17.14	15.20	12.87	10.34	.15
.86	19.40	18.58	17.12	15.18	12.84	10.31	.14
.87	19.39	18.57	17.11	15.16	12.82	10.29	.13
.88	19.39	18.55	17.09	15.13	12.79	10.26	.12
.89	19.38	18.54	17.08	15.11	12.77	10.24	.11
.90	19.38	18.53	17.06	15.09	12.74	10.21	.10
.91	19.37	18.52	17.04	15.07	12.71	10.18	.09
.92	19.37	18.50	17.02	15.05	12.69	10.16	.08
.93	19.36	18.49	17.01	15.02	12.66	10.13	.07
.94	19.36	18.47	16.99	15.00	12.64	10.11	.06
.95	19.35	18.46	16.97	14.98	12.61	10.08	.05
.96	19.35	18.45	16.95	14.96	12.59	10.05	.04
.97	19.34	18.44	16.93	14.94	12.56	10.03	.03
.98	19.34	18.42	16.92	14.91	12.54	10.00	.02
.99	19.33	18.41	16.90	14.89	12.51	9.98	.01
1.00	19.33	18.40	16.88	14.87	12.49	9.95	.00
Days.	23	22	21	20	19	18	Days.

TABLE XXVIII.

(Continued.)

Period, 23.774626 days.

TABLE XXIX. ARG. 24.

Equation = 7″.3 [1 — sin. (2x + z)].

Period, 13.276498 days.

Days.	6	7	8	9	10	11	0	1	2	Days.
.00	9.95	7.39	4.98	2.92	1.33	0.35	14.50	14.27	12.51	1.00
.01	9.92	7.36	4.96	2.90	1.32	0.34	14.50	14.26	12.49	.99
.02	9.90	7.34	4.94	2.88	1.31	0.33	14.51	14.25	12.46	.98
.03	9.87	7.31	4.91	2.87	1.29	0.33	14.52	14.24	12.44	.97
.04	9.85	7.29	4.89	2.85	1.28	0.32	14.52	14.23	12.41	.96
.05	9.82	7.26	4.87	2.83	1.27	0.31	14.53	14.22	12.39	.95
.06	9.79	7.23	4.85	2.81	1.26	0.30	14.53	14.20	12.36	.94
.07	9.77	7.21	4.82	2.79	1.25	0.30	14.53	14.19	12.33	.93
.08	9.74	7.18	4.80	2.78	1.23	0.29	14.54	14.18	12.31	.92
.09	9.72	7.16	4.77	2.76	1.22	0.29	14.54	14.17	12.28	.91
.10	9.69	7.13	4.75	2.74	1.21	0.28	14.55	14.16	12.26	.90
.11	9.66	7.11	4.73	2.72	1.20	0.27	14.55	14.15	12.23	.89
.12	9.64	7.08	4.71	2.70	1.19	0.27	14.56	14.14	12.21	.88
.13	9.61	7.06	4.68	2.69	1.17	0.26	14.56	14.12	12.18	.87
.14	9.59	7.03	4.66	2.67	1.16	0.26	14.56	14.11	12.16	.86
.15	9.56	7.01	4.64	2.65	1.15	0.25	14.57	14.10	12.13	.85
.16	9.53	6.98	4.62	2.63	1.14	0.24	14.57	14.08	12.10	.84
.17	9.51	6.96	4.60	2.61	1.12	0.24	14.57	14.07	12.08	.83
.18	9.48	6.93	4.57	2.60	1.11	0.23	14.58	14.05	12.05	.82
.19	9.46	6.91	4.55	2.58	1.09	0.23	14.58	14.04	12.03	.81
.20	9.43	6.88	4.53	2.56	1.08	0.22	14.58	14.03	12.00	.80
.21	9.40	6.86	4.51	2.54	1.07	0.21	14.58	14.02	11.97	.79
.22	9.38	6.83	4.49	2.53	1.06	0.21	14.58	14.00	11.94	.78
.23	9.35	6.81	4.46	2.51	1.04	0.20	14.59	13.99	11.92	.77
.24	9.33	6.78	4.44	2.50	1.03	0.20	14.59	13.97	11.89	.76
.25	9.30	6.76	4.42	2.48	1.02	0.19	14.59	13.96	11.87	.75
.26	9.27	6.73	4.40	2.46	1.01	0.19	14.59	13.95	11.84	.74
.27	9.25	6.71	4.38	2.45	1.00	0.18	14.60	13.93	11.81	.73
.28	9.22	6.68	4.36	2.43	0.99	0.18	14.60	13.92	11.79	.72
.29	9.20	6.66	4.34	2.42	0.98	0.17	14.60	13.90	11.76	.71
.30	9.17	6.63	4.32	2.40	0.97	0.17	14.60	13.89	11.73	.70
.31	9.15	6.61	4.30	2.38	0.96	0.17	14.60	13.88	11.70	.69
.32	9.12	6.58	4.28	2.36	0.95	0.16	14.60	13.86	11.67	.68
.33	9.10	6.56	4.25	2.35	0.93	0.16	14.60	13.85	11.65	.67
.34	9.07	6.53	4.23	2.33	0.92	0.15	14.60	13.83	11.62	.66
.35	9.05	6.51	4.21	2.31	0.91	0.15	14.60	13.82	11.60	.65
.36	9.02	6.49	4.19	2.29	0.90	0.15	14.60	13.80	11.57	.64
.37	9.00	6.46	4.17	2.28	0.89	0.14	14.60	13.78	11.54	.63
.38	8.97	6.44	4.15	2.26	0.87	0.14	14.60	13.77	11.52	.62
.39	8.95	6.41	4.13	2.25	0.86	0.13	14.60	13.75	11.49	.61
.40	8.92	6.39	4.11	2.23	0.85	0.13	14.60	13.74	11.46	.60
.41	8.89	6.37	4.09	2.21	0.84	0.12	14.60	13.72	11.43	.59
.42	8.87	6.34	4.07	2.20	0.83	0.12	14.60	13.70	11.40	.58
.43	8.84	6.32	4.04	2.18	0.82	0.11	14.60	13.69	11.37	.57
.44	8.82	6.29	4.02	2.17	0.81	0.11	14.60	13.67	11.34	.56
.45	8.79	6.27	4.00	2.15	0.80	0.10	14.60	13.66	11.32	.55
.46	8.76	6.25	3.98	2.13	0.79	0.10	14.59	13.64	11.29	.54
.47	8.74	6.22	3.96	2.11	0.78	0.09	14.59	13.62	11.26	.53
.48	8.71	6.20	3.94	2.10	0.77	0.09	14.59	13.61	11.23	.52
.49	8.69	6.17	3.92	2.08	0.76	0.08	14.59	13.59	11.20	.51
.50	8.66	6.15	3.90	2.06	0.75	0.08	14.59	13.57	11.17	.50
Days.	17	16	15	14	13	12	13	12	11	Days.

TABLE XXVIII.

(Continued.)

Period, 23.774626 days.

TABLE XXIX. ARG. 24.

Equation = 7″.3 [1 — sin. (2x + z)].

Period, 13.276498 days.

Days.	6	7	8	9	10	11	0	1	2	Days.
.50	8.66	6.15	3.90	2.06	0.75	0.08	14.59	13.57	11.17	.50
.51	8.63	6.13	3.88	2.04	0.74	0.08	14.59	13.55	11.14	.49
.52	8.61	6.10	3.86	2.03	0.73	0.07	14.59	13.53	11.11	.48
.53	8.58	6.07	3.83	2.01	0.72	0.07	14.58	13.51	11.08	.47
.54	8.56	6.05	3.81	2.00	0.71	0.06	14.58	13.49	11.05	.46
.55	8.53	6.03	3.79	1.98	0.70	0.06	14.58	13.48	11.02	.45
.56	8.50	6.01	3.77	1.97	0.69	0.06	14.57	13.46	10.99	.44
.57	8.48	5.98	3.75	1.95	0.68	0.06	14.57	13.44	10.96	.43
.58	8.45	5.96	3.73	1.94	0.68	0.05	14.57	13.42	10.93	.42
.59	8.43	5.93	3.71	1.92	0.67	0.05	14.56	13.40	10.90	.41
.60	8.40	5.91	3.69	1.91	0.66	0.05	14.56	13.38	10.87	.40
.61	8.37	5.89	3.67	1.89	0.65	0.05	14.56	13.36	10.84	.39
.62	8.35	5.86	3.65	1.88	0.64	0.05	14.55	13.34	10.81	.38
.63	8.32	5.84	3.63	1.86	0.64	0.04	14.55	13.32	10.78	.37
.64	8.30	5.81	3.61	1.85	0.63	0.04	14.54	13.30	10.75	.36
.65	8.27	5.79	3.59	1.83	0.62	0.04	14.54	13.28	10.72	.35
.66	8.24	5.77	3.57	1.82	0.61	0.04	14.53	13.26	10.69	.34
.67	8.22	5.74	3.55	1.80	0.60	0.04	14.53	13.24	10.66	.33
.68	8.19	5.72	3.53	1.79	0.59	0.03	14.52	13.22	10.63	.32
.69	8.17	5.69	3.51	1.77	0.58	0.03	14.52	13.20	10.60	.31
.70	8.14	5.67	3.49	1.76	0.57	0.03	14.51	13.18	10.57	.30
.71	8.12	5.65	3.47	1.75	0.56	0.03	14.50	13.16	10.54	.29
.72	8.09	5.63	3.45	1.73	0.55	0.03	14.50	13.14	10.51	.28
.73	8.07	5.60	3.43	1.72	0.55	0.02	14.49	13.12	10.48	.27
.74	8.04	5.58	3.41	1.70	0.54	0.02	14.48	13.10	10.45	.26
.75	8.02	5.56	3.39	1.69	0.53	0.02	14.48	13.08	10.42	.25
.76	7.99	5.54	3.37	1.68	0.52	0.02	14.47	13.05	10.38	.24
.77	7.97	5.51	3.35	1.66	0.51	0.02	14.46	13.03	10.35	.23
.78	7.94	5.49	3.33	1.65	0.51	0.01	14.46	13.01	10.32	.22
.79	7.92	5.46	3.31	1.63	0.50	0.01	14.45	12.99	10.29	.21
.80	7.89	5.44	3.29	1.62	0.49	0.01	14.44	12.97	10.26	.20
.81	7.87	5.42	3.27	1.60	0.48	0.01	14.43	12.95	10.23	.19
.82	7.84	5.40	3.25	1.59	0.47	0.01	14.42	12.93	10.20	.18
.83	7.82	5.37	3.24	1.57	0.47	0.01	14.41	12.91	10.17	.17
.84	7.79	5.35	3.22	1.56	0.46	0.00	14.41	12.88	10.13	.16
.85	7.77	5.33	3.20	1.54	0.45	0.00	14.40	12.86	10.10	.15
.86	7.74	5.31	3.18	1.53	0.44	0.00	14.39	12.84	10.07	.14
.87	7.72	5.28	3.16	1.51	0.44	0.00	14.39	12.81	10.03	.13
.88	7.69	5.26	3.15	1.50	0.43	0.00	14.38	12.79	10.00	.12
.89	7.67	5.23	3.13	1.48	0.43	0.00	14.37	12.77	9.97	.11
.90	7.64	5.21	3.11	1.47	0.42	0.00	14.36	12.75	9.94	.10
.91	7.62	5.19	3.09	1.46	0.41	0.00	14.35	12.73	9.91	.09
.92	7.59	5.17	3.07	1.44	0.40	0.00	14.34	12.70	9.88	.08
.93	7.57	5.14	3.05	1.43	0.40	0.00	14.33	12.68	9.84	.07
.94	7.54	5.12	3.03	1.41	0.39	0.00	14.32	12.65	9.81	.06
.95	7.52	5.10	3.01	1.40	0.38	0.00	14.32	12.63	9.78	.05
.96	7.49	5.08	2.99	1.39	0.37	0.00	14.31	12.61	9.74	.04
.97	7.47	5.05	2.97	1.37	0.37	0.00	14.30	12.58	9.71	.03
.98	7.44	5.03	2.96	1.36	0.36	0.00	14.29	12.56	9.67	.02
.99	7.42	5.00	2.94	1.34	0.36	0.00	14.28	12.53	9.64	.01
1.00	7.39	4.98	2.92	1.33	0.35	0.00	14.27	12.51	9.61	.00
Days.	17	16	15	14	13	12	13	12	11	Days.

TABLE XXIX.

(*Continued.*)

Period, 13.276498 days.

TABLE XXX. ARG. 25.

Equation = 2″.9 [1 — sin. (2*t* + *z* + *x*)].

Period, 9.36717 days.

Days.	3	4	5	6	0	1	2	3	4	Days.
Days	″	″	″	″	″	″	″	″	″	Days.
.00	9.61	6.20	3.03	0.80	5.73	5.50	4.14	2.24	0.63	1.00
.01	9.58	6.17	3.00	0.78	5.74	5.49	4.12	2.22	0.62	.99
.02	9.55	6.13	2.97	0.77	5.74	5.48	4.11	2.20	0.61	.98
.03	9.51	6.10	2.95	0.75	5.75	5.48	4.09	2.18	0.59	.97
.04	9.48	6.06	2.92	0.74	5.75	5.47	4.07	2.16	0.58	.96
.05	9.45	6.03	2.89	0.72	5.76	5.46	4.05	2.14	0.57	.95
.06	9.41	6.00	2.87	0.71	5.76	5.45	4.03	2.13	0.56	.94
.07	9.38	5.96	2.84	0.69	5.76	5.44	4.02	2.11	0.55	.93
.08	9.34	5.93	2.82	0.68	5.77	5.43	4.00	2.09	0.53	.92
.09	9.31	5.89	2.79	0.66	5.77	5.42	3.98	2.07	0.52	.91
.10	9.28	5.86	2.76	0.65	5.77	5.41	3.96	2.05	0.51	.90
.11	9.25	5.83	2.73	0.64	5.77	5.40	3.94	2.03	0.50	.89
.12	9.22	5.79	2.70	0.62	5.78	5.39	3.93	2.01	0.49	.88
.13	9.18	5.76	2.68	0.61	5.78	5.38	3.91	1.99	0.48	.87
.14	9.15	5.72	2.65	0.59	5.78	5.37	3.89	1.98	0.47	.86
.15	9.12	5.69	2.62	0.57	5.78	5.36	3.87	1.96	0.46	.85
.16	9.08	5.66	2.60	0.56	5.79	5.35	3.85	1.94	0.44	.84
.17	9.05	5.62	2.57	0.54	5.79	5.33	3.84	1.92	0.43	.83
.18	9.01	5.59	2.55	0.53	5.79	5.32	3.82	1.90	0.42	.82
.19	8.98	5.55	2.52	0.52	5.79	5.31	3.80	1.89	0.41	.81
.20	8.95	5.52	2.49	0.51	5.79	5.30	3.78	1.87	0.40	.80
.21	8.92	5.49	2.46	0.50	5.79	5.29	3.76	1.85	0.39	.79
.22	8.88	5.46	2.44	0.49	5.79	5.28	3.74	1.83	0.38	.78
.23	8.85	5.42	2.41	0.47	5.80	5.27	3.72	1.82	0.37	.77
.24	8.81	5.39	2.39	0.46	5.80	5.26	3.70	1.80	0.36	.76
.25	8.78	5.35	2.36	0.45	5.80	5.25	3.68	1.78	0.35	.75
.26	8.75	5.32	2.33	0.44	5.80	5.23	3.67	1.76	0.35	.74
.27	8.71	5.29	2.31	0.42	5.80	5.22	3.65	1.74	0.34	.73
.28	8.68	5.25	2.28	0.41	5.80	5.21	3.63	1.73	0.33	.72
.29	8.64	5.22	2.26	0.40	5.80	5.20	3.61	1.71	0.32	.71
.30	8.61	5.19	2.23	0.39	5.80	5.19	3.59	1.69	0.31	.70
.31	8.58	5.16	2.21	0.38	5.80	5.18	3.57	1.67	0.30	.69
.32	8.54	5.13	2.18	0.37	5.80	5.17	3.55	1.65	0.29	.68
.33	8.51	5.09	2.16	0.36	5.80	5.15	3.53	1.64	0.28	.67
.34	8.47	5.06	2.13	0.35	5.80	5.14	3.51	1.62	0.28	.66
.35	8.44	5.02	2.11	0.34	5.80	5.13	3.49	1.60	0.27	.65
.36	8.41	4.99	2.09	0.33	5.80	5.12	3.48	1.58	0.26	.64
.37	8.37	4.96	2.06	0.32	5.80	5.11	3.46	1.56	0.25	.63
.38	8.34	4.92	2.04	0.31	5.79	5.09	3.44	1.55	0.24	.62
.39	8.30	4.89	2.01	0.30	5.79	5.08	3.42	1.53	0.24	.61
.40	8.27	4.86	1.99	0.29	5.79	5.07	3.40	1.51	0.23	.60
.41	8.24	4.83	1.97	0.28	5.79	5.06	3.38	1.49	0.22	.59
.42	8.20	4.80	1.95	0.27	5.79	5.04	3.36	1.47	0.21	.58
.43	8.17	4.76	1.92	0.26	5.78	5.03	3.34	1.46	0.21	.57
.44	8.13	4.73	1.90	0.25	5.78	5.01	3.32	1.44	0.20	.56
.45	8.10	4.69	1.87	0.24	5.78	5.00	3.30	1.42	0.19	.55
.46	8.07	4.66	1.85	0.24	5.78	4.99	3.28	1.41	0.19	.54
.47	8.03	4.63	1.82	0.23	5.78	4.97	3.26	1.39	0.18	.53
.48	8.00	4.59	1.80	0.22	5.77	4.96	3.24	1.36	0.17	.52
.49	7.96	4.56	1.78	0.21	5.77	4.94	3.22	1.36	0.17	.51
.50	7.93	4.53	1.76	0.20	5.77	4.93	3.20	1.34	0.16	.50
Days.	10	9	8	7	9	8	7	6	5	Days.

TABLE XXIX.

(*Continued.*)

Period, 13.276498 days.

TABLE XXX. ARG. 25.

Equation = 2″.9 [1 — sin. (2t + z + x)].

Period, 9.36717 days.

Days.	3	4	5	6	0	1	2	3	4	Days.
Days	″	″	″	″	″	″	″	″	″	Days
.50	7.93	4.53	1.76	0.20	5.77	4.93	3.20	1.34	0.16	.50
.51	7.90	4.50	1.74	0.19	5.77	4.92	3.18	1.32	0.15	.49
.52	7.86	4.47	1.72	0.18	5.76	4.90	3.16	1.31	0.15	.48
.53	7.83	4.44	1.70	0.18	5.76	4.89	3.14	1.29	0.14	.47
.54	7.79	4.40	1.67	0.17	5.76	4.87	3.12	1.28	0.14	.46
.55	7.76	4.37	1.65	0.16	5.75	4.86	3.10	1.26	0.13	.45
.56	7.73	4.34	1.63	0.16	5.75	4.85	3.09	1.25	0.12	.44
.57	7.69	4.30	1.60	0.15	5.75	4.83	3.07	1.23	0.12	.43
.58	7.66	4.27	1.58	0.14	5.74	4.82	3.05	1.22	0.11	.42
.59	7.62	4.24	1.56	0.14	5.74	4.80	3.03	1.20	0.11	.41
.60	7.59	4.21	1.54	0.13	5.74	4.79	3.01	1.19	0.10	.40
.61	7.56	4.18	1.52	0.12	5.73	4.77	2.99	1.17	0.10	.39
.62	7.52	4.15	1.50	0.12	5.73	4.76	2.97	1.16	0.09	.38
.63	7.49	4.12	1.48	0.11	5.73	4.74	2.95	1.14	0.09	.37
.64	7.45	4.09	1.46	0.11	5.73	4.73	2.93	1.13	0.08	.36
.65	7.42	4.06	1.44	0.10	5.72	4.71	2.91	1.11	0.08	.35
.66	7.38	4.03	1.42	0.09	5.72	4.70	2.90	1.10	0.08	.34
.67	7.34	4.00	1.40	0.09	5.71	4.68	2.88	1.08	0.07	.33
.68	7.31	3.97	1.38	0.08	5.71	4.67	2.86	1.07	0.07	.32
.69	7.27	3.94	1.36	0.08	5.70	4.65	2.84	1.05	0.06	.31
.70	7.24	3.91	1.34	0.07	5.70	4.64	2.82	1.04	0.06	.30
.71	7.21	3.88	1.32	0.07	5.69	4.62	2.80	1.02	0.06	.29
.72	7.17	3.85	1.30	0.06	5.69	4.61	2.78	1.01	0.05	.28
.73	7.14	3.82	1.28	0.06	5.68	4.59	2.76	0.99	0.05	.27
.74	7.10	3.79	1.26	0.05	5.68	4.58	2.74	0.98	0.05	.26
.75	7.07	3.76	1.24	0.05	5.67	4.56	2.72	0.96	0.04	.25
.76	7.03	3.73	1.23	0.05	5.67	4.54	2.71	0.95	0.04	.24
.77	6.99	3.70	1.21	0.04	5.66	4.53	2.69	0.93	0.04	.23
.78	6.96	3.67	1.19	0.04	5.66	4.51	2.67	0.92	0.04	.22
.79	6.92	3.64	1.17	0.03	5.65	4.50	2.65	0.90	0.03	.21
.80	6.89	3.61	1.15	0.03	5.65	4.48	2.63	0.89	0.03	.20
.81	6.86	3.58	1.13	0.03	5.64	4.46	2.61	0.88	0.03	.19
.82	6.82	3.55	1.11	0.02	5.63	4.45	2.59	0.86	0.03	.18
.83	6.79	3.52	1.09	0.02	5.63	4.43	2.57	0.85	0.02	.17
.84	6.75	3.49	1.08	0.02	5.62	4.41	2.55	0.83	0.02	.16
.85	6.72	3.46	1.06	0.02	5.61	4.40	2.53	0.82	0.02	.15
.86	6.68	3.43	1.04	0.01	5.61	4.38	2.51	0.81	0.02	.14
.87	6.64	3.40	1.03	0.01	5.60	4.36	2.49	0.79	0.02	.13
.88	6.61	3.37	1.01	0.01	5.59	4.35	2.47	0.78	0.01	.12
.89	6.57	3.34	0.99	0.01	5.59	4.33	2.45	0.76	0.01	.11
.90	6.54	3.31	0.97	0.01	5.58	4.31	2.43	0.75	0.01	.10
.91	6.51	3.28	0.95	0.01	5.57	4.29	2.41	0.74	0.01	.09
.92	6.47	3.25	0.93	0.01	5.56	4.28	2.39	0.73	0.01	.08
.93	6.44	3.22	0.92	0.00	5.56	4.26	2.37	0.71	0.01	.07
.94	6.40	3.20	0.90	0.00	5.55	4.24	2.35	0.70	0.00	.06
.95	6.37	3.17	0.88	0.00	5.54	4.23	2.33	0.69	0.00	.05
.96	6.34	3.14	0.87	0.00	5.53	4.21	2.32	0.68	0.00	.04
.97	6.30	3.12	0.85	0.00	5.52	4.19	2.30	0.67	0.00	.03
.98	6.27	3.09	0.83	0.00	5.52	4.18	2.28	0.65	0.00	.02
.99	6.23	3.06	0.82	0.00	5.51	4.16	2.26	0.64	0.00	.01
1.00	6.20	3.03	0.80	0.00	5.50	4.14	2.24	0.63	0.00	.00
Days.	10	9	8	7	9	8	7	6	5	Days.

TABLES XXXI-LXXXI.

ARGUMENTS, PERIODS, AND EQUATIONS.

TABLE.	ARGUMENT.	PERIOD.	EQUATION.
		Days.	
XXXI.	26	5.822606	$1''.9 + 1''.9 \sin. (4t + x)$.
XXXII.	27	25.82638	$1.6 + 0.8 \sin. (2x - t) - 0''.9 \sin. (4x - 2t)$.
XXXIII.	28	38.52204	$7.5 + 7.5 \sin. (2t - 2z - x)$.
XXXIV.	29	32.76364	$6.3 - 6.3 \sin. (2t + x - 2y)$.
XXXV.	30	10.37093	$3.8 + 3.8 \sin. (4t - x - z)$.
XXXVI.	31	15.31442	$3.0 - 3.0 \sin. (3t - x)$.
XXXVII.	32	16.63016	$3.0 + 3.0 \sin. (4t - 2x - z)$.
XXXVIII.	33	32.45058	$2.1 + 2.1 \sin. (x - 2z)$.
XXXIX.	34	27.09271	$2.0 - 2.0 \sin. (2t + 2z - x)$.
XL.	35	23.94223	$1.2 - 1.2 \sin. (x + 2z)$.
XLI.	36	13.71881	$1.0 + 1.0 \sin. (t + z + x)$.
XLII.	37	37.62533	$0.5 + 0.5 \sin. (4t - 3x)$.
XLIII.	38	18.78878	$0.4 - 0.4 \sin. (t + 2x)$.
XLIV.	39	31.47528	$0.6 - 0.6 \sin. (3t - 2x)$.
XLV.	40	18.84354	$0.4 + 0.4 \sin. (3x - z)$.
XLVI.	41	17.91910	$0.4 - 0.4 \sin. (3x + z)$.
XLVII.	42	10.14791	$0.3 + 0.3 \sin. (2t + x - 2z)$.
XLVIII.	43	11.53786	$0.6 + 0.6 \sin. (2t + 2x - z)$.
XLIX.	44	22.78614	$0.3 - 0.3 \sin. (3x - 2t + z)$.
L.	45	16.90004	$0.2 - 0.2 \sin. (2t + x + 2y)$.
LI.	46	29.01328	$0.4 - 0.4 \sin. (x + z - 2y)$.
LII.	47	11.13256	$1.3 + 1.3 \sin. (2y - z)$.
LIII.	48′	11.96709	$0.5 - 0.5 \sin. (2t + 2x - 2y)$.
LIV.	49	12.76271	$0.2 - 0.2 \sin. (2x - 2t + 2y)$.
LV.	50	25.03597	$0.4 - 0.4 \sin. (2y + z - x)$.
LVI.	51	18.21692	$0.2 - 0.2 \sin. (x + 2y)$.
LVII.	52	25.23137	$1.0 + 1.0 \sin. (2y - t)$.
LVIII.	53	117.53942	$2.6 - 2.6 \sin. (2y - 2t + z)$.
LIX.	54	15.06989	$1.2 + 1.2 \sin. (4t - z)$.
LX.	55	19.62730	$0.9 - 0.9 \sin. (4t - x + z)$.
LXI.	56	13.11748	$0.7 - 0.7 \sin. (2y + z)$.
LXII.	57	16.98741	$0.8 + 0.8 \sin. (2t + 3x)$.
LXIII.	58	15.24221	$0.5 - 0.5 \sin. (4t - 2x + z)$.
LXIV.	59	38.96397	$0.6 - 0.6 \sin. (x - 4t + 2y)$.
LXV.	60	22.32171	$0.4 - 0.4 \sin. (x - 2t + 2y + z)$.
LXVI.	61	13.60611	$0.4 + 0.4 \sin. (4y)$.
LXVII.	62	14.44200	$0.3 + 0.3 \sin. (2t + 2y - z)$.
LXVIII.	63	13.63340	$0.3 + 0.3 \sin. (u + y)$.
LXIX.	64	35.99212	$0.3 - 0.3 \sin. (2t + x - 2y - z)$.
LXX.	65	14.47277	$0.2 - 0.2 \sin. (4t + z)$.
LXXI.	66	35.59582	$0.2 - 0.2 \sin. (3t - 2y)$.
LXXII.	67	27.44332	$1.0 - 1.0 \sin. (x + y - u)$.
LXXIII.	68	27.66669	$0.7 + 0.7 \sin. (x - y + u)$.
LXXIV.	69	471.89326	$2.2 - 2.2 \sin. (2x - 2t - z)$.
LXXV.	70	329.79056	$1.5 + 1.5 \sin. (2y - z - 2t)$.
LXXVI.	71	583.921	1.3 − 1.1 sin. (♀ − ⊕) + 0″.4 sin. 2 (♀ − ⊕).
LXXVII.	72	398.884	0.8 + 0.7 sin. (⊕ − ♃) − 0″.2 sin. 2 (⊕ − ♃).
LXXVIII.	73′	1095.1653	$2.0 + 2.0 \sin. (2y - 2x)$.
LXXIX.	74	3232.8202	$0.5 + 0.5 \sin. (t - x + z)$.
LXXX.	75	84753.24	$53.2 + 23.2 \sin. (8g'' - 13z + 315^\circ 30')$.
LXXXI.	76	95489.94	$65.64 + 27.4 \sin. (18g'' - 16z - x + 35^\circ 20'.2)$.

TABLE XXXI. ARGUMENT 26.

Equation = $1''.9\ [1 + \sin.\ (4t + x)]$.

Period, 5.822606 days.

Days.	0	1	2	Days.	Days.	0	1	2	Days.
Days	″	″	″	Days.	Days.	″	″	″	Days
.00	0.01	0.85	2.80	1.00	.50	0.18	1.81	3.53	.50
.01	0.01	0.87	2.82	.99	.51	0.19	1.83	3.54	.49
.02	0.01	0.89	2.84	.98	.52	0.20	1.85	3.55	.48
.03	0.01	0.91	2.85	.97	.53	0.21	1.87	3.56	.47
.04	0.01	0.92	2.87	.96	.54	0.22	1.89	3.57	.46
.05	0.01	0.94	2.89	.95	.55	0.23	1.91	3.58	.45
.06	0.00	0.96	2.90	.94	.56	0.24	1.93	3.58	.44
.07	0.00	0.97	2.92	.93	.57	0.25	1.95	3.59	.43
.08	0.00	0.99	2.93	.92	.58	0.26	1.97	3.60	.42
.09	0.00	1.01	2.95	.91	.59	0.27	1.99	3.61	.41
.10	0.00	1.03	2.97	.90	.60	0.28	2.01	3.62	.40
.11	0.00	1.05	2.99	.89	.61	0.29	2.03	3.63	.39
.12	0.00	1.07	3.00	.88	.62	0.30	2.05	3.64	.38
.13	0.00	1.09	3.02	.87	.63	0.31	2.07	3.65	.37
.14	0.00	1.11	3.03	.86	.64	0.33	2.09	3.65	.36
.15	0.00	1.12	3.05	.85	.65	0.34	2.11	3.66	.35
.16	0.01	1.14	3.07	.84	.66	0.35	2.14	3.67	.34
.17	0.01	1.16	3.08	.83	.67	0.37	2.16	3.68	.33
.18	0.01	1.18	3.10	.82	.68	0.38	2.18	3.68	.32
.19	0.01	1.20	3.11	.81	.69	0.39	2.20	3.69	.31
.20	0.01	1.22	3.13	.80	.70	0.40	2.22	3.70	.30
.21	0.01	1.24	3.15	.79	.71	0.41	2.24	3.71	.29
.22	0.01	1.26	3.16	.78	.72	0.42	2.26	3.71	.28
.23	0.02	1.28	3.18	.77	.73	0.44	2.28	3.72	.27
.24	0.02	1.30	3.19	.76	.74	0.45	2.30	3.72	.26
.25	0.02	1.31	3.21	.75	.75	0.46	2.32	3.73	.25
.26	0.03	1.33	3.22	.74	.76	0.48	2.34	3.74	.24
.27	0.03	1.35	3.23	.73	.77	0.49	2.36	3.74	.23
.28	0.04	1.37	3.25	.72	.78	0.51	2.38	3.75	.22
.29	0.04	1.39	3.26	.71	.79	0.52	2.40	3.75	.21
.30	0.04	1.41	3.28	.70	.80	0.53	2.42	3.76	.20
.31	0.05	1.43	3.29	.69	.81	0.54	2.44	3.76	.19
.32	0.05	1.45	3.31	.68	.82	0.56	2.46	3.77	.18
.33	0.06	1.47	3.32	.67	.83	0.57	2.48	3.77	.17
.34	0.06	1.49	3.34	.66	.84	0.59	2.50	3.77	.16
.35	0.07	1.51	3.35	.65	.85	0.60	2.52	3.78	.15
.36	0.08	1.53	3.36	.64	.86	0.62	2.54	3.78	.14
.37	0.08	1.55	3.38	.63	.87	0.64	2.56	3.78	.13
.38	0.09	1.57	3.39	.62	.88	0.65	2.58	3.79	.12
.39	0.09	1.59	3.41	.61	.89	0.67	2.60	3.79	.11
.40	0.10	1.61	3.42	.60	.90	0.68	2.62	3.79	.10
.41	0.11	1.63	3.43	.59	.91	0.70	2.64	3.79	.09
.42	0.12	1.65	3.44	.58	.92	0.72	2.66	3.79	.08
.43	0.13	1.67	3.45	.57	.93	0.73	2.68	3.79	.07
.44	0.13	1.69	3.46	.56	.94	0.75	2.69	3.79	.06
.45	0.14	1.71	3.48	.55	.95	0.76	2.71	3.80	.05
.46	0.15	1.73	3.49	.54	.96	0.78	2.73	3.80	.04
.47	0.15	1.75	3.50	.53	.97	0.80	2.74	3.80	.03
.48	0.16	1.77	3.51	.52	.98	0.81	2.76	3.80	.02
.49	0.17	1.79	3.52	.51	.99	0.83	2.78	3.80	.01
.50	0.18	1.81	3.53	.50	1.00	0.85	2.80	3.80	.00
Days.	5	4	3	Days.	Days.	5	4	3	Days.

TABLES XXXII - XXXVI.

Table	XXXII.		
Argument	27.		
Days.	0	10	20
Days	"	"	"
0.0	1.65	1.96	2.62
0.1	1.59	1.97	2.66
0.2	1.52	1.98	2.70
0.3	1.46	1.98	2.73
0.4	1.39	1.99	2.77
0.5	1.33	1.99	2.80
0.6	1.27	1.99	2.83
0.7	1.21	1.99	2.86
0.8	1.15	1.99	2.89
0.9	1.09	1.98	2.92
1.0	1.03	1.98	2.95
1.1	0.97	1.97	2.97
1.2	0.92	1.96	3.00
1.3	0.86	1.95	3.02
1.4	0.81	1.93	3.04
1.5	0.76	1.92	3.06
1.6	0.71	1.90	3.07
1.7	0.66	1.89	3.08
1.8	0.61	1.87	3.09
1.9	0.57	1.85	3.10
2.0	0.52	1.83	3.11
2.1	0.48	1.81	3.11
2.2	0.44	1.79	3.11
2.3	0.40	1.77	3.10
2.4	0.37	1.74	3.10
2.5	0.33	1.72	3.09
2.6	0.30	1.70	3.08
2.7	0.27	1.67	3.06
2.8	0.24	1.65	3.05
2.9	0.22	1.62	3.03
3.0	0.19	1.60	3.01
3.1	0.17	1.58	2.98
3.2	0.15	1.55	2.96
3.3	0.14	1.53	2.93
3.4	0.12	1.50	2.90
3.5	0.11	1.48	2.87
3.6	0.10	1.46	2.83
3.7	0.10	1.43	2.80
3.8	0.09	1.41	2.76
3.9	0.09	1.39	2.72
4.0	0.09	1.37	2.68
4.1	0.10	1.35	2.63
4.2	0.11	1.33	2.59
4.3	0.12	1.31	2.54
4.4	0.13	1.30	2.49
4.5	0.14	1.28	2.44
4.6	0.16	1.27	2.39
4.7	0.18	1.25	2.34
4.8	0.20	1.24	2.28
4.9	0.23	1.23	2.23
5.0	0.25	1.22	2.17

Table	XXXIII.		XXXIV.		XXXV.	XXXVI.	
Argument	28.		29.		30.	31.	
Days.	0	10	0	10	0	0	Days.
Days.	"	"	"	"	"	"	Days.
0.0	0.05	7.05	1.45	4.15	7.50	4.72	10.0
0.1	0.04	7.17	1.38	4.26	7.44	4.81	9.9
0.2	0.03	7.29	1.30	4.38	7.37	4.91	9.8
0.3	0.02	7.41	1.23	4.49	7.30	5.00	9.7
0.4	0.01	7.53	1.16	4.61	7.20	5.09	9.6
0.5	0.01	7.65	1.09	4.72	7.08	5.18	9.5
0.6	0.00	7.77	1.02	4.84	6.96	5.26	9.4
0.7	0.00	7.89	0.96	4.96	6.83	5.31	9.3
0.8	0.00	8.02	0.89	5.07	6.68	5.42	9.2
0.9	0.01	8.14	0.83	5.19	6.53	5.49	9.1
1.0	0.01	8.27	0.77	5.31	6.36	5.56	9.0
1.1	0.02	8.39	0.71	5.43	6.18	5.62	8.9
1.2	0.03	8.51	0.66	5.55	6.00	5.68	8.8
1.3	0.04	8.63	0.60	5.68	5.81	5.73	8.7
1.4	0.05	8.76	0.55	5.80	5.61	5.78	8.6
1.5	0.06	8.88	0.50	5.92	5.41	5.82	8.5
1.6	0.07	9.00	0.46	6.04	5.20	5.86	8.4
1.7	0.09	9.12	0.41	6.16	4.98	5.89	8.3
1.8	0.11	9.24	0.37	6.29	4.76	5.92	8.2
1.9	0.14	9.36	0.33	6.41	4.54	5.95	8.1
2.0	0.16	9.47	0.29	6.53	4.31	5.97	8.0
2.1	0.19	9.59	0.26	6.65	4.08	5.99	7.9
2.2	0.22	9.70	0.23	6.77	3.85	6.00	7.8
2.3	0.25	9.82	0.20	6.89	3.62	6.00	7.7
2.4	0.28	9.94	0.17	7.01	3.39	6.00	7.6
2.5	0.31	10.06	0.14	7.13	3.16	5.99	7.5
2.6	0.34	10.17	0.12	7.25	2.93	5.98	7.4
2.7	0.38	10.29	0.10	7.37	2.71	5.96	7.3
2.8	0.42	10.40	0.08	7.49	2.49	5.94	7.2
2.9	0.47	10.52	0.06	7.61	2.28	5.92	7.1
3.0	0.50	10.63	0.04	7.73	2.07	5.89	7.0
3.1	0.55	10.74	0.03	7.85	1.87	5.86	6.9
3.2	0.60	10.85	0.02	7.97	1.68	5.82	6.8
3.3	0.65	10.96	0.01	8.08	1.49	5.77	6.7
3.4	0.70	11.06	0.01	8.20	1.31	5.72	6.6
3.5	0.75	11.17	0.00	8.31	1.14	5.67	6.5
3.6	0.80	11.27	0.00	8.42	0.99	5.61	6.4
3.7	0.86	11.38	0.00	8.54	0.84	5.55	6.3
3.8	0.92	11.48	0.01	8.65	0.70	5.48	6.2
3.9	0.98	11.59	0.01	8.76	0.57	5.41	6.1
4.0	1.03	11.69	0.02	8.87	0.45	5.33	6.0
4.1	1.09	11.79	0.03	8.98	0.35	5.25	5.9
4.2	1.16	11.89	0.04	9.09	0.26	5.17	5.8
4.3	1.23	11.99	0.06	9.20	0.18	5.09	5.7
4.4	1.30	12.08	0.07	9.30	0.12	5.00	5.6
4.5	1.37	12.18	0.09	9.41	0.07	4.91	5.5
4.6	1.44	12.27	0.11	9.51	0.03	4.81	5.4
4.7	1.51	12.37	0.13	9.61	0.01	4.71	5.3
4.8	1.59	12.46	0.16	9.71	0.00	4.60	5.2
4.9	1.66	12.55	0.18	9.81	0.01	4.49	5.1
5.0	1.74	12.64	0.21	9.91	0.03	4.38	5.0
Days.	30	20	30	20	10	10	Days.

TABLES XXXII - XXXVI.

Table	XXXII.			XXXIII.			XXXIV.		XXXV.	XXXVI.	
Argument	27.			28.			29.		30.	31.	
Days.	0	10	20	Days.	0	10	0	10	0	0	Days.
5.0	0.25	1.22	2.17	5.0	1.74	12.64	0.21	9.91	0.03	4.38	5.0
5.1	0.28	1.22	2.11	5.1	1.82	12.73	0.24	10.01	0.06	4.27	4.9
5.2	0.31	1.21	2.05	5.2	1.90	12.82	0.27	10.10	0.11	4.16	4.8
5.3	0.34	1.21	1.99	5.3	1.98	12.91	0.31	10.20	0.17	4.05	4.7
5.4	0.37	1.21	1.93	5.4	2.06	12.99	0.35	10.29	0.24	3.93	4.6
5.5	0.40	1.21	1.87	5.5	2.14	13.07	0.39	10.38	0.32	3.81	4.5
5.6	0.43	1.21	1.81	5.6	2.23	13.15	0.44	10.47	0.42	3.69	4.4
5.7	0.47	1.22	1.74	5.7	2.32	13.23	0.48	10.56	0.54	3.57	4.3
5.8	0.50	1.22	1.67	5.8	2.41	13.31	0.53	10.65	0.66	3.45	4.2
5.9	0.54	1.23	1.61	5.9	2.50	13.39	0.58	10.73	0.79	3.33	4.1
6.0	0.58	1.24	1.55	6.0	2.59	13.46	0.63	10.82	0.94	3.21	4.0
6.1	0.62	1.25	1.49	6.1	2.68	13.54	0.67	10.90	1.10	3.09	3.9
6.2	0.66	1.26	1.42	6.2	2.78	13.61	0.75	10.98	1.27	2.97	3.8
6.3	0.70	1.28	1.36	6.3	2.88	13.68	0.81	11.06	1.45	2.85	3.7
6.4	0.75	1.29	1.30	6.4	2.98	13.75	0.87	11.14	1.63	2.72	3.6
6.5	0.79	1.31	1.24	6.5	3.08	13.82	0.93	11.22	1.82	2.60	3.5
6.6	0.83	1.33	1.18	6.6	3.18	13.88	1.00	11.29	2.02	2.47	3.4
6.7	0.88	1.35	1.12	6.7	3.28	13.94	1.07	11.36	2.23	2.35	3.3
6.8	0.92	1.37	1.06	6.8	3.38	14.00	1.14	11.43	2.44	2.23	3.2
6.9	0.97	1.40	1.00	6.9	3.48	14.06	1.21	11.50	2.65	2.11	3.1
7.0	1.01	1.42	0.94	7.0	3.59	14.12	1.28	11.57	2.87	2.00	3.0
7.1	1.05	1.45	0.88	7.1	3.69	14.18	1.36	11.63	3.09	1.88	2.9
7.2	1.10	1.48	0.83	7.2	3.80	14.23	1.43	11.69	3.32	1.77	2.8
7.3	1.14	1.51	0.77	7.3	3.90	14.28	1.51	11.75	3.55	1.66	2.7
7.4	1.19	1.54	0.72	7.4	4.01	14.33	1.59	11.81	3.78	1.55	2.6
7.5	1.23	1.57	0.67	7.5	4.12	14.38	1.67	11.87	4.01	1.44	2.5
7.6	1.27	1.61	0.62	7.6	4.23	14.43	1.75	11.92	4.24	1.34	2.4
7.7	1.32	1.64	0.57	7.7	4.34	14.48	1.83	11.98	4.47	1.24	2.3
7.8	1.36	1.68	0.53	7.8	4.45	14.52	1.92	12.03	4.70	1.14	2.2
7.9	1.40	1.72	0.48	7.9	4.56	14.56	2.00	12.08	4.92	1.04	2.1
8.0	1.44	1.76	0.44	8.0	4.67	14.60	2.09	12.13	5.14	0.95	2.0
8.1	1.48	1.80	0.40	8.1	4.78	14.64	2.18	12.17	5.35	0.86	1.9
8.2	1.52	1.84	0.37	8.2	4.90	14.67	2.27	12.21	5.56	0.78	1.8
8.3	1.56	1.88	0.33	8.3	5.02	14.71	[illegible]	12.25	5.76	0.70	1.7
8.4	1.59	1.93	0.30	8.4	5.14	14.74	2.46	12.29	5.95	0.62	1.6
8.5	1.63	1.97	0.27	8.5	5.26	14.77	2.55	12.33	6.13	0.55	1.5
8.6	1.66	2.01	0.25	8.6	5.38	14.80	2.65	12.36	6.31	0.48	1.4
8.7	1.69	2.06	0.22	8.7	5.50	14.83	2.75	12.39	6.48	0.42	1.3
8.8	1.72	2.10	0.20	8.8	5.62	14.85	2.85	12.42	6.63	0.36	1.2
8.9	1.75	2.15	0.18	8.9	5.74	14.88	2.95	12.45	6.78	0.30	1.1
9.0	1.78	2.19	0.16	9.0	5.85	14.90	3.05	12.48	6.92	0.25	1.0
9.1	1.80	2.23	0.14	9.1	5.97	14.92	3.16	12.50	7.05	0.20	0.9
9.2	1.83	2.28	0.13	9.2	6.08	14.94	3.26	12.52	7.17	0.16	0.8
9.3	1.85	2.32	0.12	9.3	6.20	14.95	3.37	12.54	7.26	0.12	0.7
9.4	1.87	2.37	0.11	9.4	6.32	14.96	3.48	12.55	7.35	0.09	0.6
9.5	1.89	2.41	0.11	9.5	6.44	14.97	3.59	12.57	7.43	0.06	0.5
9.6	1.91	2.45	0.10	9.6	6.56	14.98	3.70	12.58	7.49	0.04	0.4
9.7	1.92	2.50	0.10	9.7	6.68	14.99	3.81	12.59	7.54	0.02	0.3
9.8	1.94	2.54	0.10	9.8	6.81	14.99	3.92	12.59	7.57	0.01	0.2
9.9	1.95	2.58	0.10	9.9	6.93	15.00	4.04	12.60	7.59	0.00	0.1
10.0	1.96	2.62	0.11	10.0	7.05	15.00	4.15	12.60	7.60	0.00	0.0
				Days.	30	20	30	20	10	10	Days.

TABLES XXXVII.-XLII.

Tables	XXXVII.	XXXVIII.		XXXIX.		XL.		XLI.	XLII.		
Arguments	32.	33.		34.		35.		36.	37.		
Days.	0	0	10	0	10	0	10	0	0	10	Days.
Days. 0.0	″ 0.58	″ 0.54	″ 1.35	″ 2.15	″ 3.36	″ 0.58	″ 2.24	″ 0.86	″ 0.01	″ 0.45	Days. 10.0
0.1	0.52	0.52	1.39	2.20	3.33	0.61	2.22	0.82	0.01	0.46	9.9
0.2	0.46	0.48	1.43	2.24	3.29	0.64	2.21	0.77	0.01	0.47	9.8
0.3	0.40	0.46	1.47	2.29	3.25	0.67	2.19	0.73	0.01	0.47	9.7
0.4	0.35	0.43	1.51	2.33	3.22	0.70	2.17	0.69	0.01	0.48	9.6
0.5	0.30	0.41	1.55	2.38	3.18	0.73	2.15	0.65	0.00	0.49	9.5
0.6	0.25	0.39	1.59	2.42	3.15	0.76	2.13	0.61	0.00	0.50	9.4
0.7	0.21	0.36	1.63	2.47	3.11	0.79	2.11	0.57	0.00	0.51	9.3
0.8	0.17	0.34	1.67	2.51	3.07	0.82	2.09	0.53	0.00	0.51	9.2
0.9	0.13	0.32	1.71	2.56	3.03	0.85	2.07	0.49	0.00	0.52	9.1
1.0	0.10	0.30	1.75	2.60	2.99	0.88	2.05	0.45	0.00	0.53	9.0
1.1	0.07	0.28	1.79	2.65	2.95	0.91	2.03	0.41	0.00	0.54	8.9
1.2	0.05	0.26	1.83	2.69	2.91	0.94	2.01	0.37	0.00	0.55	8.8
1.3	0.03	0.24	1.87	2.74	2.87	0.97	1.99	0.33	0.00	0.56	8.7
1.4	0.02	0.22	1.91	2.78	2.82	1.00	1.96	0.30	0.00	0.57	8.6
1.5	0.01	0.20	1.95	2.82	2.78	1.03	1.94	0.27	0.00	0.58	8.5
1.6	0.00	0.18	1.99	2.86	2.74	1.06	1.91	0.24	0.00	0.58	8.4
1.7	0.00	0.16	2.03	2.90	2.69	1.09	1.89	0.20	0.00	0.59	8.3
1.8	0.00	0.15	2.07	2.94	2.65	1.12	1.86	0.18	0.00	0.60	8.2
1.9	0.01	0.13	2.11	2.98	2.60	1.16	1.84	0.16	0.00	0.61	8.1
2.0	0.02	0.12	2.15	3.02	2.56	1.19	1.81	0.13	0.00	0.62	8.0
2.1	0.04	0.11	2.19	3.06	2.51	1.22	1.78	0.11	0.00	0.63	7.9
2.2	0.06	0.09	2.23	3.10	2.47	1.26	1.75	0.09	0.00	0.64	7.8
2.3	0.08	0.08	2.27	3.14	2.42	1.29	1.72	0.07	0.00	0.64	7.7
2.4	0.11	0.07	2.31	3.17	2.38	1.32	1.69	0.06	0.01	0.65	7.6
2.5	0.14	0.06	2.35	3.21	2.33	1.35	1.66	0.04	0.01	0.66	7.5
2.6	0.18	0.05	2.39	3.25	2.28	1.38	1.63	0.03	0.01	0.67	7.4
2.7	0.22	0.04	2.43	3.29	2.24	1.41	1.60	0.02	0.01	0.68	7.3
2.8	0.27	0.03	2.47	3.32	2.19	1.44	1.57	0.01	0.02	0.68	7.2
2.9	0.32	0.03	2.51	3.36	2.15	1.47	1.54	0.01	0.02	0.69	7.1
3.0	0.37	0.02	2.55	3.39	2.10	1.50	1.51	0.00	0.02	0.70	7.0
3.1	0.42	0.02	2.59	3.42	2.05	1.53	1.48	0.00	0.02	0.71	6.9
3.2	0.48	0.01	2.63	3.46	2.01	1.56	1.45	0.00	0.03	0.72	6.8
3.3	0.54	0.01	2.67	3.49	1.96	1.59	1.42	0.00	0.03	0.72	6.7
3.4	0.61	0.00	2.71	3.52	1.92	1.62	1.39	0.01	0.03	0.73	6.6
3.5	0.68	0.00	2.75	3.55	1.87	1.65	1.36	0.02	0.03	0.74	6.5
3.6	0.75	0.00	2.79	3.58	1.82	1.68	1.33	0.03	0.04	0.74	6.4
3.7	0.83	0.00	2.83	3.61	1.78	1.71	1.30	0.04	0.04	0.75	6.3
3.8	0.91	0.00	2.87	3.64	1.73	1.74	1.27	0.05	0.04	0.76	6.2
3.9	0.98	0.00	2.90	3.66	1.69	1.76	1.23	0.06	0.05	0.76	6.1
4.0	1.08	0.00	2.94	3.69	1.64	1.79	1.20	0.08	0.05	0.77	6.0
4.1	1.17	0.00	2.98	3.71	1.60	1.82	1.17	0.10	0.05	0.78	5.9
4.2	1.26	0.00	3.01	3.74	1.55	1.84	1.13	0.12	0.06	0.78	5.8
4.3	1.35	0.01	3.05	3.76	1.51	1.87	1.10	0.14	0.06	0.79	5.7
4.4	1.45	0.01	3.08	3.78	1.46	1.89	1.07	0.17	0.07	0.79	5.6
4.5	1.55	0.02	3.12	3.80	1.42	1.92	1.04	0.19	0.07	0.80	5.5
4.6	1.65	0.02	3.16	3.82	1.38	1.94	1.01	0.22	0.08	0.81	5.4
4.7	1.75	0.03	3.19	3.84	1.33	1.97	0.98	0.25	0.08	0.81	5.3
4.8	1.85	0.03	3.23	3.86	1.29	1.99	0.95	0.28	0.09	0.82	5.2
4.9	1.95	0.04	3.26	3.87	1.24	2.02	0.92	0.31	0.09	0.82	5.1
5.0	2.06	0.05	3.29	3.89	1.20	2.04	0.89	0.34	0.10	0.83	5.0
Days.	10	30	20	30	20	30	20	10	30	20	Days.

TABLES XXXVII.-XLII.

Tables	XXXVII.	XXXVIII.		XXXIX.		XL.		XLI.	XLII.		
Arguments	32.	33.		34.		35.		36.	37.		
Days.	0	0	10	0	10	0	10	0	0	10	Days.
Days.											Days.
5.0	2.06	0.05	3.29	3.89	1.20	2.04	0.89	0.34	0.10	0.83	5.0
5.1	2.17	0.06	3.32	3.90	1.16	2.06	0.86	0.37	0.11	0.84	4.9
5.2	2.28	0.07	3.36	3.92	1.11	2.08	0.83	0.41	0.11	0.84	4.8
5.3	2.39	0.08	3.39	3.93	1.07	2.10	0.80	0.45	0.12	0.85	4.7
5.4	2.50	0.10	3.42	3.94	1.03	2.12	0.78	0.49	0.12	0.85	4.6
5.5	2.61	0.11	3.45	3.95	0.99	2.14	0.75	0.53	0.13	0.86	4.5
5.6	2.73	0.12	3.48	3.96	0.95	2.16	0.72	0.57	0.14	0.87	4.4
5.7	2.84	0.14	3.51	3.97	0.91	2.18	0.69	0.61	0.14	0.87	4.3
5.8	2.96	0.15	3.54	3.98	0.87	2.20	0.67	0.65	0.15	0.88	4.2
5.9	3.07	0.16	3.57	3.98	0.84	2.21	0.64	0.69	0.15	0.88	4.1
6.0	3.19	0.18	3.60	3.99	0.80	2.23	0.61	0.74	0.16	0.89	4.0
6.1	3.30	0.20	3.63	3.99	0.76	2.24	0.58	0.78	0.17	0.89	3.9
6.2	3.41	0.22	3.66	4.00	0.72	2.26	0.56	0.83	0.17	0.90	3.8
6.3	3.52	0.24	3.68	4.00	0.69	2.27	0.53	0.87	0.18	0.90	3.7
6.4	3.63	0.26	3.71	4.00	0.65	2.29	0.51	0.92	0.18	0.91	3.6
6.5	3.74	0.28	3.73	4.00	0.62	2.30	0.48	0.96	0.19	0.91	3.5
6.6	3.85	0.30	3.76	4.00	0.59	2.31	0.46	1.01	0.20	0.92	3.4
6.7	3.96	0.32	3.78	4.00	0.55	2.33	0.43	1.06	0.20	0.92	3.3
6.8	4.06	0.35	3.81	3.99	0.52	2.34	0.41	1.11	0.21	0.93	3.2
6.9	4.17	0.37	3.83	3.99	0.49	2.35	0.38	1.16	0.21	0.93	3.1
7.0	4.27	0.40	3.85	3.98	0.46	2.36	0.36	1.20	0.22	0.94	3.0
7.1	4.37	0.42	3.87	3.97	0.43	2.37	0.34	1.25	0.23	0.94	2.9
7.2	4.47	0.45	3.89	3.97	0.40	2.37	0.31	1.29	0.23	0.95	2.8
7.3	4.57	0.47	3.91	3.96	0.37	2.38	0.29	1.33	0.24	0.95	2.7
7.4	4.67	0.50	3.93	3.95	0.35	2.38	0.27	1.37	0.25	0.95	2.6
7.5	4.76	0.53	3.95	3.94	0.32	2.39	0.25	1.41	0.25	0.96	2.5
7.6	4.85	0.56	3.97	3.93	0.30	2.39	0.23	1.45	0.26	0.96	2.4
7.7	4.94	0.58	3.99	3.92	0.27	2.40	0.21	1.49	0.27	0.96	2.3
7.8	5.02	0.61	4.01	3.90	0.25	2.40	0.19	1.53	0.27	0.97	2.2
7.9	5.10	0.64	4.02	3.89	0.23	2.40	0.18	1.57	0.28	0.97	2.1
8.0	5.18	0.67	4.04	3.87	0.21	2.40	0.16	1.61	0.29	0.97	2.0
8.1	5.26	0.70	4.06	3.85	0.19	2.40	0.15	1.65	0.30	0.97	1.9
8.2	5.33	0.73	4.07	3.84	0.17	2.40	0.13	1.68	0.31	0.98	1.8
8.3	5.40	0.76	4.09	3.82	0.15	2.40	0.12	1.71	0.31	0.98	1.7
8.4	5.47	0.79	4.10	3.80	0.14	2.39	0.10	1.74	0.32	0.98	1.6
8.5	5.53	0.82	4.11	3.78	0.12	2.39	0.09	1.77	0.33	0.98	1.5
8.6	5.59	0.85	4.12	3.76	0.10	2.38	0.08	1.80	0.34	0.98	1.4
8.7	5.65	0.88	4.13	3.73	0.09	2.38	0.07	1.83	0.35	0.99	1.3
8.8	5.70	0.91	4.14	3.71	0.07	2.37	0.06	1.85	0.35	0.99	1.2
8.9	5.75	0.95	4.15	3.69	0.06	2.37	0.05	1.87	0.36	0.99	1.1
9.0	5.79	0.98	4.16	3.66	0.05	2.36	0.04	1.89	0.37	0.99	1.0
9.1	5.83	1.01	4.17	3.64	0.04	2.35	0.03	1.91	0.38	0.99	0.9
9.2	5.86	1.05	4.18	3.61	0.03	2.34	0.03	1.93	0.39	0.99	0.8
9.3	5.89	1.08	4.18	3.58	0.02	2.33	0.02	1.95	0.39	0.99	0.7
9.4	5.92	1.12	4.19	3.55	0.02	2.32	0.02	1.96	0.40	1.00	0.6
9.5	5.94	1.16	4.19	3.52	0.01	2.31	0.01	1.97	0.41	1.00	0.5
9.6	5.96	1.20	4.19	3.49	0.01	2.30	0.01	1.98	0.42	1.00	0.4
9.7	5.98	1.23	4.20	3.46	0.00	2.28	0.01	1.99	0.43	1.00	0.3
9.8	5.99	1.27	4.20	3.43	0.00	2.27	0.00	1.99	0.43	1.00	0.2
9.9	6.00	1.31	4.20	3.39	0.00	2.25	0.00	2.00	0.44	1.00	0.1
10.0	6.00	1.35	4.20	3.36	0.00	2.24	0.00	2.00	0.45	1.00	0.0
Days.	10	30	20	30	20	30	20	10	30	20	Days.

TABLES XLIII.-L.

Tables	XLIII.	XLIV.		XLV.	XLVI.	XLVII.	XLVIII.	XLIX.		L.	
Arguments	38.	39.		40.	41.	42.	43.	44.		45.	
Days.	0	0	10	0	0	0	0	0	10	0	Days.
Days.	"	"	"	"	"	"	"	"	"	"	Days.
0.0	0.03	1.12	0.75	0.77	0.10	0.60	0.17	0.09	0.58	0.17	10.0
0.1	0.02	1.13	0.74	0.78	0.08	0.60	0.21	0.10	0.58	0.19	9.9
0.2	0.01	1.13	0.73	0.79	0.07	0.59	0.25	0.10	0.57	0.21	9.8
0.3	0.01	1.14	0.71	0.80	0.05	0.59	0.30	0.11	0.57	0.24	9.7
0.4	0.00	1.15	0.70	0.80	0.04	0.58	0.35	0.11	0.56	0.26	9.6
0.5	0.00	1.15	0.69	0.80	0.03	0.58	0.40	0.12	0.56	0.28	9.5
0.6	0.00	1.15	0.68	0.80	0.02	0.57	0.45	0.13	0.56	0.30	9.4
0.7	0.00	1.16	0.67	0.80	0.01	0.56	0.50	0.13	0.55	0.32	9.3
0.8	0.00	1.16	0.66	0.79	0.01	0.55	0.55	0.14	0.55	0.34	9.2
0.9	0.01	1.17	0.65	0.79	0.00	0.54	0.60	0.14	0.54	0.35	9.1
1.0	0.01	1.17	0.64	0.78	0.00	0.53	0.65	0.15	0.54	0.36	9.0
1.1	0.02	1.17	0.63	0.77	0.00	0.52	0.70	0.16	0.53	0.37	8.9
1.2	0.03	1.17	0.62	0.76	0.00	0.50	0.75	0.17	0.53	0.38	8.8
1.3	0.04	1.18	0.61	0.75	0.00	0.48	0.80	0.17	0.52	0.39	8.7
1.4	0.05	1.18	0.60	0.74	0.01	0.47	0.85	0.18	0.52	0.40	8.6
1.5	0.06	1.18	0.59	0.73	0.02	0.45	0.89	0.19	0.51	0.40	8.5
1.6	0.08	1.18	0.58	0.71	0.03	0.44	0.93	0.20	0.50	0.40	8.4
1.7	0.10	1.18	0.57	0.69	0.04	0.42	0.97	0.21	0.50	0.40	8.3
1.8	0.12	1.19	0.55	0.67	0.06	0.41	1.01	0.21	0.49	0.39	8.2
1.9	0.14	1.19	0.54	0.65	0.07	0.39	1.05	0.22	0.49	0.38	8.1
2.0	0.16	1.19	0.53	0.63	0.09	0.37	1.08	0.23	0.48	0.37	8.0
2.1	0.18	1.19	0.52	0.61	0.11	0.35	1.11	0.24	0.47	0.36	7.9
2.2	0.21	1.19	0.51	0.59	0.13	0.33	1.14	0.25	0.46	0.35	7.8
2.3	0.23	1.20	0.50	0.57	0.15	0.31	1.16	0.25	0.46	0.33	7.7
2.4	0.26	1.20	0.49	0.54	0.17	0.30	1.17	0.26	0.45	0.31	7.6
2.5	0.28	1.20	0.48	0.52	0.19	0.28	1.18	0.27	0.44	0.29	7.5
2.6	0.31	1.20	0.47	0.49	0.22	0.27	1.19	0.28	0.43	0.27	7.4
2.7	0.33	1.20	0.46	0.46	0.24	0.25	1.20	0.29	0.42	0.25	7.3
2.8	0.36	1.20	0.45	0.43	0.27	0.23	1.20	0.29	0.42	0.23	7.2
2.9	0.38	1.20	0.44	0.40	0.29	0.21	1.20	0.30	0.41	0.21	7.1
3.0	0.41	1.20	0.43	0.38	0.32	0.19	1.19	0.31	0.40	0.19	7.0
3.1	0.44	1.20	0.42	0.35	0.35	0.17	1.17	0.32	0.39	0.17	6.9
3.2	0.47	1.20	0.41	0.33	0.38	0.15	1.15	0.33	0.38	0.14	6.8
3.3	0.50	1.19	0.40	0.30	0.41	0.13	1.13	0.33	0.38	0.12	6.7
3.4	0.52	1.19	0.39	0.28	0.43	0.12	1.10	0.34	0.37	0.10	6.6
3.5	0.55	1.19	0.38	0.25	0.46	0.10	1.07	0.35	0.36	0.08	6.5
3.6	0.57	1.19	0.37	0.23	0.49	0.09	1.04	0.36	0.35	0.06	6.4
3.7	0.59	1.19	0.36	0.20	0.52	0.08	1.00	0.37	0.34	0.05	6.3
3.8	0.61	1.18	0.35	0.18	0.54	0.07	0.96	0.37	0.34	0.04	6.2
3.9	0.63	1.18	0.34	0.16	0.57	0.06	0.92	0.38	0.33	0.08	6.1
4.0	0.65	1.18	0.33	0.14	0.59	0.05	0.87	0.39	0.32	0.02	6.0
4.1	0.67	1.18	0.32	0.12	0.62	0.04	0.82	0.40	0.31	0.01	5.9
4.2	0.69	1.18	0.31	0.10	0.64	0.03	0.77	0.41	0.30	0.01	5.8
4.3	0.71	1.17	0.30	0.08	0.66	0.02	0.72	0.41	0.30	0.00	5.7
4.4	0.73	1.17	0.29	0.07	0.68	0.02	0.67	0.42	0.29	0.00	5.6
4.5	0.75	1.17	0.28	0.05	0.70	0.01	0.62	0.43	0.28	0.01	5.5
4.6	0.76	1.17	0.27	0.04	0.72	0.01	0.57	0.44	0.27	0.01	5.4
4.7	0.77	1.16	0.26	0.03	0.74	0.00	0.52	0.45	0.26	0.02	5.3
4.8	0.78	1.16	0.25	0.02	0.75	0.00	0.47	0.45	0.26	0.03	5.2
4.9	0.79	1.15	0.24	0.01	0.76	0.00	0.42	0.46	0.25	0.04	5.1
5.0	0.79	1.15	0.23	0.01	0.77	0.00	0.37	0.47	0.24	0.05	5.0
Days.	10	30	20	10	10	10	10	30	20	10	Days.

TABLES XLIII. - L.

Tables	XLIII.	XLIV.		XLV.	XLVI.	XLVII.	XLVIII.	XLIX.		L.	
Arguments	38.	39.		40.	41.	42.	43.	44.		45.	
Days.	**0**	**0**	**10**	**0**	**0**	**0**	**0**	**0**	**10**	**0**	**Days.**
5.0	0.79	1.15	0.23	0.01	0.77	0.00	0.37	0.47	0.21	0.05	5.0
5.1	0.80	1.15	0.22	0.00	0.78	0.00	0.32	0.48	0.23	0.06	4.9
5.2	0.80	1.14	0.21	0.00	0.79	0.00	0.28	0.48	0.22	0.08	4.8
5.3	0.80	1.14	0.21	0.00	0.80	0.00	0.24	0.49	0.22	0.10	4.7
5.4	0.80	1.13	0.20	0.00	0.80	0.01	0.20	0.49	0.21	0.12	4.6
5.5	0.80	1.13	0.19	0.00	0.80	0.01	0.16	0.50	0.20	0.14	4.5
5.6	0.79	1.12	0.18	0.01	0.80	0.02	0.13	0.51	0.19	0.16	4.4
5.7	0.79	1.12	0.17	0.02	0.80	0.03	0.10	0.51	0.18	0.18	4.3
5.8	0.78	1.11	0.17	0.03	0.79	0.04	0.07	0.52	0.18	0.20	4.2
5.9	0.77	1.10	0.16	0.04	0.78	0.05	0.05	0.52	0.17	0.23	4.1
6.0	0.76	1.10	0.15	0.05	0.77	0.06	0.03	0.53	0.16	0.25	4.0
6.1	0.75	1.09	0.14	0.06	0.76	0.07	0.02	0.53	0.15	0.27	3.9
6.2	0.73	1.09	0.11	0.08	0.75	0.09	0.01	0.54	0.15	0.29	3.8
6.3	0.72	1.08	0.13	0.09	0.74	0.10	0.00	0.54	0.14	0.31	3.7
6.4	0.70	1.07	0.13	0.11	0.72	0.12	0.00	0.55	0.14	0.33	3.6
6.5	0.68	1.07	0.12	0.13	0.71	0.13	0.00	0.55	0.13	0.35	3.5
6.6	0.66	1.06	0.12	0.15	0.69	0.15	0.01	0.55	0.12	0.36	3.4
6.7	0.64	1.05	0.11	0.17	0.67	0.16	0.02	0.56	0.12	0.37	3.3
6.8	0.62	1.05	0.10	0.19	0.65	0.18	0.04	0.56	0.11	0.38	3.2
6.9	0.60	1.04	0.09	0.21	0.63	0.19	0.06	0.57	0.11	0.39	3.1
7.0	0.57	1.03	0.09	0.24	0.60	0.21	0.09	0.57	0.10	0.40	3.0
7.1	0.55	1.02	0.08	0.26	0.58	0.23	0.12	0.57	0.09	0.40	2.9
7.2	0.52	1.01	0.07	0.29	0.55	0.25	0.15	0.58	0.09	0.40	2.8
7.3	0.50	1.01	0.07	0.31	0.53	0.27	0.19	0.58	0.08	0.40	2.7
7.4	0.47	1.00	0.06	0.34	0.50	0.29	0.23	0.59	0.08	0.39	2.6
7.5	0.44	0.99	0.06	0.36	0.47	0.31	0.27	0.59	0.07	0.38	2.5
7.6	0.41	0.98	0.06	0.39	0.44	0.33	0.31	0.59	0.06	0.37	2.4
7.7	0.38	0.97	0.05	0.41	0.41	0.35	0.36	0.59	0.06	0.36	2.3
7.8	0.36	0.97	0.05	0.44	0.39	0.37	0.41	0.60	0.05	0.35	2.2
7.9	0.33	0.96	0.04	0.46	0.36	0.39	0.46	0.60	0.05	0.34	2.1
8.0	0.31	0.95	0.04	0.49	0.34	0.40	0.51	0.60	0.04	0.32	2.0
8.1	0.28	0.94	0.04	0.52	0.31	0.42	0.56	0.60	0.04	0.30	1.9
8.2	0.26	0.93	0.03	0.55	0.28	0.43	0.61	0.60	0.03	0.28	1.8
8.3	0.23	0.92	0.03	0.58	0.25	0.45	0.66	0.60	0.03	0.26	1.7
8.4	0.21	0.91	0.02	0.60	0.23	0.46	0.71	0.60	0.02	0.24	1.6
8.5	0.18	0.90	0.02	0.62	0.20	0.48	0.76	0.60	0.02	0.22	1.5
8.6	0.16	0.89	0.02	0.64	0.18	0.49	0.81	0.60	0.02	0.20	1.4
8.7	0.14	0.88	0.02	0.66	0.15	0.51	0.86	0.60	0.02	0.18	1.3
8.8	0.12	0.87	0.01	0.68	0.13	0.52	0.91	0.60	0.01	0.15	1.2
8.9	0.10	0.86	0.01	0.70	0.11	0.53	0.95	0.60	0.01	0.13	1.1
9.0	0.09	0.85	0.01	0.71	0.09	0.54	0.99	0.60	0.01	0.11	1.0
9.1	0.07	0.84	0.01	0.73	0.07	0.55	1.03	0.60	0.01	0.09	0.9
9.2	0.06	0.83	0.01	0.74	0.06	0.56	1.06	0.60	0.01	0.07	0.8
9.3	0.04	0.82	0.00	0.75	0.05	0.57	1.09	0.59	0.00	0.05	0.7
9.4	0.03	0.81	0.00	0.76	0.04	0.58	1.12	0.59	0.00	0.04	0.6
9.5	0.02	0.80	0.00	0.77	0.03	0.59	1.14	0.59	0.00	0.03	0.5
9.6	0.01	0.79	0.00	0.78	0.02	0.59	1.16	0.59	0.00	0.02	0.4
9.7	0.00	0.78	0.00	0.79	0.01	0.60	1.18	0.59	0.00	0.01	0.3
9.8	0.00	0.77	0.00	0.79	0.01	0.60	1.19	0.58	0.00	0.01	0.2
9.9	0.00	0.76	0.00	0.80	0.00	0.60	1.20	0.58	0.00	0.00	0.1
10.0	0.00	0.75	0.00	0.80	0.00	0.60	1.20	0.58	0.00	0.00	0.0
Days.	**10**	**30**	**20**	**10**	**10**	**10**	**10**	**30**	**20**	**10**	**Days.**

TABLES LI.-LVII.

Tables	LI.		LII.	LIII.	LIV.	LV.		LVI.	LVII.		
Arguments	46.		47.	48′.	49.	50.		51.	52.		
Days.	0	10	0	0	0	0	10	0	0	10	Days.
0.0	0.54″	0.61″	0.96″	0.75″	0.16″	0.28″	0.73″	0.03″	1.26″	0.21″	10.0
0.1	0.55	0.60	0.90	0.76	0.17	0.29	0.72	0.03	1.23	0.22	9.9
0.2	0.56	0.60	0.85	0.78	0.18	0.30	0.72	0.02	1.21	0.24	9.8
0.3	0.56	0.59	0.79	0.79	0.19	0.31	0.71	0.02	1.18	0.25	9.7
0.4	0.57	0.59	0.74	0.81	0.20	0.32	0.70	0.01	1.16	0.27	9.6
0.5	0.58	0.58	0.69	0.82	0.21	0.33	0.70	0.01	1.13	0.29	9.5
0.6	0.59	0.57	0.64	0.83	0.22	0.34	0.69	0.01	1.11	0.31	9.4
0.7	0.59	0.56	0.59	0.85	0.23	0.35	0.68	0.00	1.08	0.32	9.3
0.8	0.60	0.56	0.54	0.86	0.24	0.36	0.68	0.00	1.06	0.34	9.2
0.9	0.60	0.55	0.50	0.88	0.25	0.37	0.67	0.00	1.03	0.36	9.1
1.0	0.61	0.54	0.45	0.89	0.26	0.38	0.66	0.00	1.01	0.38	9.0
1.1	0.62	0.53	0.41	0.90	0.27	0.39	0.65	0.00	0.98	0.40	8.9
1.2	0.63	0.52	0.37	0.91	0.28	0.40	0.64	0.01	0.96	0.42	8.8
1.3	0.63	0.52	0.33	0.93	0.28	0.41	0.63	0.01	0.93	0.44	8.7
1.4	0.64	0.51	0.30	0.94	0.29	0.42	0.63	0.01	0.91	0.46	8.6
1.5	0.65	0.50	0.26	0.95	0.30	0.43	0.62	0.02	0.88	0.48	8.5
1.6	0.66	0.49	0.23	0.96	0.31	0.44	0.61	0.02	0.86	0.50	8.4
1.7	0.66	0.48	0.20	0.97	0.32	0.45	0.60	0.03	0.83	0.52	8.3
1.8	0.67	0.48	0.17	0.97	0.32	0.46	0.59	0.03	0.81	0.54	8.2
1.9	0.67	0.47	0.15	0.98	0.33	0.47	0.58	0.04	0.78	0.57	8.1
2.0	0.68	0.46	0.12	0.99	0.34	0.48	0.57	0.05	0.76	0.59	8.0
2.1	0.69	0.45	0.10	0.99	0.35	0.49	0.56	0.06	0.73	0.61	7.9
2.2	0.69	0.44	0.08	0.99	0.35	0.50	0.55	0.07	0.71	0.64	7.8
2.3	0.70	0.41	0.06	1.00	0.36	0.51	0.54	0.08	0.68	0.66	7.7
2.4	0.70	0.43	0.05	1.00	0.36	0.52	0.53	0.09	0.66	0.69	7.6
2.5	0.71	0.42	0.03	1.00	0.37	0.53	0.52	0.10	0.64	0.71	7.5
2.6	0.72	0.41	0.02	1.00	0.37	0.53	0.51	0.11	0.62	0.73	7.4
2.7	0.72	0.40	0.01	1.00	0.38	0.54	0.50	0.13	0.59	0.76	7.3
2.8	0.73	0.40	0.01	0.99	0.38	0.55	0.49	0.14	0.57	0.78	7.2
2.9	0.73	0.39	0.00	0.99	0.39	0.56	0.48	0.15	0.55	0.81	7.1
3.0	0.74	0.38	0.00	0.99	0.39	0.57	0.47	0.16	0.53	0.83	7.0
3.1	0.74	0.37	0.01	0.98	0.39	0.58	0.46	0.18	0.50	0.85	6.9
3.2	0.75	0.36	0.01	0.98	0.39	0.59	0.45	0.19	0.49	0.88	6.8
3.3	0.75	0.36	0.02	0.97	0.40	0.59	0.44	0.21	0.47	0.90	6.7
3.4	0.76	0.35	0.03	0.97	0.40	0.60	0.43	0.22	0.45	0.93	6.6
3.5	0.76	0.34	0.04	0.96	0.40	0.61	0.42	0.24	0.43	0.95	6.5
3.6	0.76	0.33	0.06	0.95	0.40	0.62	0.41	0.26	0.41	0.98	6.4
3.7	0.77	0.32	0.08	0.94	0.40	0.63	0.40	0.27	0.39	1.00	6.3
3.8	0.77	0.32	0.10	0.93	0.40	0.63	0.39	0.28	0.37	1.03	6.2
3.9	0.78	0.31	0.12	0.92	0.40	0.64	0.38	0.29	0.35	1.05	6.1
4.0	0.78	0.30	0.14	0.91	0.40	0.65	0.37	0.30	0.33	1.08	6.0
4.1	0.78	0.29	0.17	0.90	0.40	0.66	0.36	0.31	0.31	1.10	5.9
4.2	0.78	0.28	0.20	0.88	0.39	0.66	0.35	0.32	0.29	1.13	5.8
4.3	0.79	0.28	0.23	0.87	0.39	0.67	0.34	0.33	0.28	1.15	5.7
4.4	0.79	0.27	0.27	0.85	0.38	0.68	0.33	0.34	0.26	1.18	5.6
4.5	0.79	0.26	0.30	0.84	0.38	0.68	0.32	0.35	0.25	1.20	5.5
4.6	0.79	0.25	0.34	0.82	0.37	0.69	0.32	0.36	0.23	1.22	5.4
4.7	0.79	0.24	0.38	0.80	0.37	0.70	0.31	0.37	0.21	1.25	5.3
4.8	0.80	0.24	0.42	0.79	0.36	0.70	0.30	0.38	0.20	1.27	5.2
4.9	0.80	0.23	0.47	0.77	0.36	0.71	0.29	0.38	0.18	1.30	5.1
5.0	0.80	0.22	0.51	0.75	0.35	0.72	0.28	0.39	0.17	1.32	5.0
Days.	30	20	10	10	10	30	20	10	30	20	Days.

Note. — Arg. 48′ = Arg. 48 — 5.26.

TABLES LI. - LVII.

Tables	LI.		LII.	LIII.	LIV.	LV.		LVI.	LVII.		
Arguments	46.		47.	48'.	49.	50.		51.	52.		
Days.	0	10	0	0	0	0	10	0	0	10	Days.
5.0	0.80	0.22	0.51	0.75	0.35	0.72	0.28	0.39	0.17	1.32	5.0
5.1	0.80	0.21	0.56	0.73	0.34	0.72	0.27	0.39	0.16	1.34	4.9
5.2	0.80	0.20	0.61	0.71	0.34	0.73	0.26	0.39	0.14	1.36	4.8
5.3	0.80	0.20	0.66	0.69	0.33	0.73	0.25	0.40	0.13	1.39	4.7
5.4	0.80	0.19	0.71	0.67	0.33	0.74	0.24	0.40	0.12	1.41	4.6
5.5	0.80	0.18	0.76	0.65	0.32	0.74	0.23	0.40	0.11	1.43	4.5
5.6	0.80	0.17	0.81	0.63	0.31	0.75	0.23	0.40	0.10	1.45	4.4
5.7	0.80	0.16	0.86	0.61	0.30	0.75	0.22	0.39	0.09	1.47	4.3
5.8	0.80	0.16	0.92	0.59	0.30	0.76	0.21	0.39	0.08	1.50	4.2
5.9	0.80	0.15	0.97	0.57	0.29	0.76	0.20	0.39	0.07	1.52	4.1
6.0	0.80	0.14	1.03	0.55	0.28	0.77	0.19	0.38	0.06	1.54	4.0
6.1	0.80	0.13	1.09	0.53	0.27	0.77	0.18	0.38	0.05	1.56	3.9
6.2	0.80	0.13	1.14	0.51	0.26	0.78	0.17	0.37	0.04	1.58	3.8
6.3	0.79	0.12	1.20	0.49	0.25	0.78	0.17	0.36	0.03	1.60	3.7
6.4	0.79	0.12	1.26	0.47	0.24	0.78	0.16	0.35	0.03	1.62	3.6
6.5	0.79	0.11	1.32	0.45	0.23	0.79	0.15	0.34	0.02	1.64	3.5
6.6	0.79	0.10	1.38	0.43	0.22	0.79	0.14	0.33	0.02	1.66	3.4
6.7	0.79	0.10	1.44	0.41	0.21	0.79	0.14	0.32	0.01	1.68	3.3
6.8	0.78	0.09	1.50	0.39	0.20	0.80	0.13	0.31	0.01	1.69	3.2
6.9	0.78	0.09	1.55	0.37	0.19	0.80	0.12	0.30	0.01	1.71	3.1
7.0	0.78	0.08	1.61	0.35	0.18	0.80	0.11	0.29	0.00	1.73	3.0
7.1	0.78	0.08	1.67	0.33	0.17	0.80	0.10	0.28	0.00	1.75	2.9
7.2	0.77	0.07	1.72	0.31	0.16	0.80	0.10	0.27	0.00	1.76	2.8
7.3	0.77	0.07	1.78	0.29	0.15	0.80	0.09	0.26	0.00	1.78	2.7
7.4	0.76	0.06	1.83	0.27	0.14	0.80	0.09	0.24	0.00	1.79	2.6
7.5	0.76	0.06	1.88	0.25	0.13	0.80	0.08	0.22	0.00	1.81	2.5
7.6	0.76	0.06	1.93	0.23	0.12	0.80	0.07	0.21	0.00	1.82	2.4
7.7	0.75	0.05	1.98	0.21	0.11	0.80	0.07	0.19	0.01	[illegible]	2.3
7.8	0.75	0.05	2.03	0.20	0.11	0.80	0.06	0.18	0.01	1.85	2.2
7.9	0.74	0.04	2.07	0.18	0.10	0.80	0.06	0.16	0.01	1.87	2.1
8.0	0.74	0.04	2.12	0.16	0.09	0.80	0.05	0.15	0.01	1.88	2.0
8.1	0.73	0.04	2.16	0.15	0.08	0.80	0.05	0.14	0.02	1.89	1.9
8.2	0.73	0.03	2.20	0.13	0.07	0.80	0.04	0.13	0.02	1.90	1.8
8.3	0.72	0.03	2.24	0.12	0.07	0.80	0.04	0.11	0.03	1.91	1.7
8.4	0.72	0.02	2.28	0.10	0.06	0.79	0.03	0.10	0.03	1.92	1.6
8.5	0.71	0.02	2.32	0.09	0.05	0.79	0.03	0.09	0.04	1.93	1.5
8.6	0.70	0.02	2.35	0.08	0.04	0.79	0.02	0.08	0.05	1.94	1.4
8.7	0.70	0.02	2.38	0.07	0.04	0.79	0.02	0.07	0.05	1.95	1.3
8.8	0.69	0.01	2.41	0.06	0.03	0.78	0.02	0.06	0.06	1.96	1.2
8.9	0.69	0.01	2.44	0.05	0.03	0.78	0.01	0.05	0.07	1.96	1.1
9.0	0.68	0.01	2.47	0.04	0.02	0.78	0.01	0.04	0.08	1.97	1.0
9.1	0.67	0.01	2.49	0.03	0.02	0.77	0.01	0.03	0.09	1.97	0.9
9.2	0.67	0.01	2.51	0.03	0.02	0.77	0.01	0.03	0.10	1.98	0.8
9.3	0.66	0.00	2.53	0.02	0.01	0.76	0.01	0.02	0.11	1.98	0.7
9.4	0.66	0.00	2.55	0.02	0.01	0.76	0.00	0.02	0.13	1.99	0.6
9.5	0.65	0.00	2.57	0.01	0.01	0.75	0.00	0.01	0.14	1.99	0.5
9.6	0.64	0.00	2.58	0.01	0.01	0.75	0.00	0.01	0.15	1.99	0.4
9.7	0.63	0.00	2.59	0.01	0.00	0.74	0.00	0.01	0.16	2.00	0.3
9.8	0.63	0.00	2.59	0.00	0.00	0.74	0.00	0.00	0.18	2.00	0.2
9.9	0.62	0.00	2.60	0.00	0.00	0.73	0.00	0.00	0.19	2.00	0.1
10.0	0.61	0.00	2.60	0.00	0.00	0.73	0.00	0.00	0.21	2.00	0.0
Days.	30	20	10	10	10	30	20	10	30	20	Days.

NOTE. — Arg. 46' = Arg. 46 — 6.26.

TABLES LVIII.-LXII.

TABLES	LVIII.						LIX.	LX.	LXI.	LXII.	
ARGUMENTS	53.						54.	55.	56.	57.	
Days.	**0**	**10**	**20**	**30**	**40**	**50**	**0**	**0**	**0**	**0**	**Days.**
0.0	5.19	4.92	4.00	2.69	1.35	0.37	0.64	0.01	0.65	0.88	10.0
0.1	5.19	4.91	3.99	2.67	1.34	0.36	0.73	0.00	0.68	0.80	9.9
0.2	5.19	4.91	3.98	2.66	1.33	0.36	0.83	0.00	0.71	0.71	9.8
0.3	5.19	4.90	3.97	2.65	1.31	0.35	0.92	0.00	0.75	0.62	9.7
0.4	5.19	4.90	3.95	2.63	1.30	0.34	1.02	0.01	0.78	0.53	9.6
0.5	5.19	4.89	3.94	2.62	1.29	0.34	1.12	0.02	0.81	0.45	9.5
0.6	5.20	4.88	3.93	2.61	1.28	0.33	1.22	0.04	0.84	0.37	9.4
0.7	5.20	4.87	3.92	2.59	1.27	0.32	1.32	0.05	0.87	0.30	9.3
0.8	5.20	4.87	3.91	2.58	1.25	0.32	1.42	0.07	0.91	0.23	9.2
0.9	5.20	4.86	3.89	2.57	1.24	0.31	1.52	0.10	0.94	0.17	9.1
1.0	5.20	4.85	3.88	2.55	1.23	0.30	1.61	0.12	0.97	0.12	9.0
1.1	5.20	4.84	3.87	2.54	1.22	0.30	1.70	0.15	1.00	0.08	8.9
1.2	5.20	4.84	3.86	2.52	1.21	0.29	1.79	0.19	1.03	0.04	8.8
1.3	5.20	4.83	3.85	2.51	1.19	0.28	1.88	0.22	1.06	0.02	8.7
1.4	5.20	4.83	3.83	2.50	1.18	0.28	1.96	0.26	1.09	0.01	8.6
1.5	5.20	4.82	3.82	2.48	1.17	0.27	2.03	0.30	1.12	0.00	8.5
1.6	5.20	4.81	3.81	2.47	1.16	0.26	2.10	0.35	1.15	0.00	8.4
1.7	5.20	4.80	3.80	2.45	1.15	0.26	2.16	0.39	1.17	0.01	8.3
1.8	5.20	4.80	3.79	2.44	1.13	0.25	2.22	0.44	1.20	0.04	8.2
1.9	5.20	4.79	3.77	2.43	1.12	0.24	2.27	0.49	1.22	0.08	8.1
2.0	5.20	4.78	3.76	2.41	1.11	0.24	2.31	0.54	1.24	0.12	8.0
2.1	5.20	4.77	3.75	2.40	1.10	0.23	2.34	0.59	1.26	0.17	7.9
2.2	5.20	4.76	3.74	2.38	1.09	0.22	2.37	0.65	1.28	0.23	7.8
2.3	5.20	4.76	3.72	2.37	1.08	0.22	2.39	0.70	1.30	0.29	7.7
2.4	5.20	4.75	3.71	2.36	1.07	0.21	2.40	0.76	1.31	0.36	7.6
2.5	5.20	4.74	3.70	2.34	1.06	0.20	2.40	0.82	1.33	0.44	7.5
2.6	5.19	4.73	3.68	2.33	1.04	0.20	2.39	0.88	1.34	0.52	7.4
2.7	5.19	4.73	3.67	2.31	1.03	0.19	2.37	0.94	1.35	0.60	7.3
2.8	5.19	4.72	3.66	2.30	1.02	0.19	2.35	1.00	1.36	0.69	7.2
2.9	5.19	4.71	3.64	2.29	1.01	0.18	2.32	1.06	1.37	0.78	7.1
3.0	5.19	4.70	3.63	2.27	1.00	0.18	2.28	1.12	1.38	0.87	7.0
3.1	5.19	4.69	3.62	2.26	0.99	0.17	2.23	1.17	1.38	0.96	6.9
3.2	5.19	4.68	3.60	2.24	0.98	0.17	2.18	1.23	1.39	1.04	6.8
3.3	5.18	4.67	3.59	2.23	0.97	0.16	2.12	1.28	1.40	1.12	6.7
3.4	5.18	4.66	3.58	2.22	0.96	0.16	2.05	1.33	1.40	1.20	6.6
3.5	5.18	4.65	3.56	2.20	0.95	0.15	1.98	1.38	1.39	1.28	6.5
3.6	5.18	4.65	3.55	2.19	0.93	0.15	1.90	1.42	1.39	1.35	6.4
3.7	5.18	4.64	3.54	2.17	0.92	0.14	1.82	1.47	1.38	1.41	6.3
3.8	5.17	4.63	3.52	2.16	0.91	0.14	1.73	1.51	1.38	1.46	6.2
3.9	5.17	4.62	3.51	2.15	0.90	0.13	1.64	1.55	1.37	1.50	6.1
4.0	5.17	4.61	3.50	2.13	0.89	0.13	1.54	1.59	1.37	1.54	6.0
4.1	5.17	4.60	3.48	2.12	0.88	0.12	1.45	1.63	1.36	1.57	5.9
4.2	5.17	4.59	3.47	2.10	0.87	0.12	1.35	1.66	1.35	1.59	5.8
4.3	5.16	4.58	3.46	2.09	0.86	0.11	1.25	1.69	1.34	1.60	5.7
4.4	5.16	4.57	3.44	2.08	0.85	0.11	1.15	1.72	1.32	1.60	5.6
4.5	5.16	4.57	3.43	2.06	0.84	0.10	1.05	1.74	1.31	1.59	5.5
4.6	5.16	4.56	3.42	2.05	0.83	0.10	0.95	1.76	1.29	1.57	5.4
4.7	5.16	4.55	3.40	2.03	0.82	0.10	0.85	1.77	1.27	1.54	5.3
4.8	5.15	4.54	3.39	2.02	0.81	0.09	0.76	1.78	1.25	1.50	5.2
4.9	5.15	4.53	3.38	2.01	0.80	0.09	0.67	1.79	1.23	1.45	5.1
5.0	5.15	4.52	3.37	1.99	0.79	0.09	0.58	1.80	1.21	1.39	5.0
Days.	**110**	**100**	**90**	**80**	**70**	**60**	**10**	**10**	**10**	**10**	**Days.**

TABLES LVIII.-LXII.

Tables	LVIII.						LIX.	LX.	LXI.	LXII.	
Arguments	52.						54.	55.	56.	57.	
Days.	0	10	20	30	40	50	0	0	0	0	Days.
5.0	5.15	4.52	3.37	1.99	0.79	0.09	0.58	1.80	1.21	1.39	5.0
5.1	5.15	4.51	3.35	1.98	0.78	0.08	0.50	1.80	1.19	1.33	4.9
5.2	5.14	4.50	3.34	1.97	0.77	0.08	0.42	1.80	1.16	1.26	4.8
5.3	5.14	4.49	3.33	1.95	0.76	0.08	0.35	1.79	1.14	1.18	4.7
5.4	5.14	4.48	3.31	1.94	0.75	0.07	0.28	1.78	1.11	1.10	4.6
5.5	5.14	4.48	3.30	1.93	0.74	0.07	0.22	1.77	1.08	1.02	4.5
5.6	5.13	4.47	3.29	1.91	0.73	0.07	0.17	1.75	1.05	0.94	4.4
5.7	5.13	4.46	3.27	1.90	0.72	0.06	0.12	1.72	1.02	0.85	4.3
5.8	5.13	4.45	3.26	1.89	0.71	0.06	0.08	1.70	0.99	0.76	4.2
5.9	5.12	4.44	3.24	1.87	0.70	0.06	0.05	1.67	0.96	0.67	4.1
6.0	5.12	4.43	3.23	1.86	0.69	0.06	0.02	1.64	0.93	0.58	4.0
6.1	5.12	4.42	3.22	1.85	0.68	0.06	0.01	1.61	0.90	0.50	3.9
6.2	5.11	4.41	3.20	1.83	0.67	0.05	0.00	1.57	0.87	0.42	3.8
6.3	5.11	4.40	3.19	1.82	0.66	0.05	0.00	1.54	0.83	0.34	3.7
6.4	5.10	4.39	3.18	1.81	0.65	0.05	0.01	1.50	0.80	0.27	3.6
6.5	5.10	4.38	3.16	1.79	0.64	0.05	0.03	1.46	0.77	0.21	3.5
6.6	5.10	4.37	3.15	1.78	0.64	0.05	0.05	1.41	0.74	0.15	3.4
6.7	5.09	4.36	3.14	1.77	0.63	0.04	0.09	1.36	0.70	0.11	3.3
6.8	5.09	4.35	3.12	1.75	0.62	0.04	0.13	1.31	0.67	0.07	3.2
6.9	5.08	4.34	3.11	1.74	0.61	0.04	0.18	1.26	0.63	0.04	3.1
7.0	5.08	4.33	3.10	1.73	0.60	0.04	0.24	1.21	0.60	0.01	3.0
7.1	5.08	4.32	3.08	1.72	0.59	0.04	0.30	1.15	0.57	0.00	2.9
7.2	5.07	4.31	3.07	1.70	0.58	0.03	0.37	1.09	0.54	0.00	2.8
7.3	5.07	4.30	3.06	1.69	0.57	0.03	0.44	1.03	0.50	0.01	2.7
7.4	5.06	4.29	3.04	1.68	0.57	0.03	0.52	0.97	0.47	0.03	2.6
7.5	5.06	4.28	3.03	1.66	0.56	0.03	0.60	0.91	0.44	0.06	2.5
7.6	5.05	4.26	3.02	1.65	0.55	0.03	0.69	0.85	0.41	0.09	2.4
7.7	5.05	4.25	3.00	1.64	0.54	0.03	0.78	0.80	0.38	0.14	2.3
7.8	5.04	4.24	2.99	1.62	0.53	0.02	0.88	0.74	0.35	0.19	2.2
7.9	5.04	4.23	2.98	1.61	0.53	0.02	0.98	0.69	0.33	0.25	2.1
8.0	5.03	4.22	2.96	1.60	0.52	0.02	1.08	0.64	0.30	0.32	2.0
8.1	5.03	4.21	2.95	1.58	0.51	0.02	1.18	0.59	0.27	0.39	1.9
8.2	5.02	4.20	2.93	1.57	0.50	0.02	1.28	0.53	0.25	0.47	1.8
8.3	5.02	4.19	2.92	1.56	0.50	0.02	1.38	0.48	0.22	0.55	1.7
8.4	5.01	4.18	2.91	1.54	0.49	0.02	1.48	0.43	0.20	0.64	1.6
8.5	5.01	4.17	2.89	1.53	0.48	0.01	1.58	0.38	0.18	0.72	1.5
8.6	5.00	4.15	2.88	1.52	0.47	0.01	1.67	0.34	0.16	0.81	1.4
8.7	5.00	4.14	2.86	1.50	0.46	0.01	1.76	0.30	0.14	0.90	1.3
8.8	4.99	4.13	2.85	1.49	0.46	0.01	1.85	0.26	0.12	0.99	1.2
8.9	4.99	4.12	2.84	1.48	0.45	0.01	1.93	0.22	0.10	1.08	1.1
9.0	4.98	4.11	2.82	1.47	0.44	0.01	2.01	0.18	0.08	1.16	1.0
9.1	4.97	4.10	2.81	1.46	0.43	0.01	2.08	0.15	0.07	1.23	0.9
9.2	4.97	4.09	2.80	1.45	0.42	0.01	2.14	0.12	0.05	1.30	0.8
9.3	4.96	4.08	2.78	1.43	0.42	0.01	2.20	0.10	0.04	1.37	0.7
9.4	4.96	4.07	2.77	1.42	0.41	0.00	2.25	0.07	0.03	1.43	0.6
9.5	4.95	4.06	2.76	1.41	0.40	0.00	2.29	0.05	0.02	1.48	0.5
9.6	4.94	4.04	2.74	1.40	0.40	0.00	2.33	0.03	0.01	1.52	0.4
9.7	4.94	4.03	2.73	1.39	0.39	0.00	2.36	0.02	0.01	1.55	0.3
9.8	4.93	4.02	2.72	1.37	0.38	0.00	2.38	0.01	0.00	1.58	0.2
9.9	4.93	4.01	2.70	1.36	0.38	0.00	2.39	0.00	0.00	1.59	0.1
10.0	4.92	4.00	2.69	1.35	0.37	0.00	2.40	0.00	0.00	1.60	0.0
Days.	110	100	90	80	70	60	10	10	10	10	Days.

TABLES LXIII. - LXIX.

Tables	LXIII.	LXIV.		LXV.		LXVI.	LXVII.	LXVIII.	LXIX.		
Arguments	**58.**	**59.**		**60.**		**61.**	**62.**	**63.**	**64.**		
Days.	**0**	**0**	**10**	**0**	**10**	**0**	**0**	**0**	**0**	**10**	Days.
Days.											Days.
0.0	0.78	1.20	0.62	0.08	0.78	0.01	0.08	0.27	0.58	0.35	10.0
0.1	0.79	1.20	0.61	0.09	0.77	0.02	0.10	0.26	0.58	0.35	9.9
0.2	0.81	1.20	0.60	0.09	0.77	0.03	0.12	0.24	0.58	0.34	9.8
0.3	0.82	1.20	0.59	0.10	0.76	0.05	0.14	0.23	0.58	0.34	9.7
0.4	0.84	1.20	0.58	0.11	0.76	0.07	0.16	0.21	0.58	0.33	9.6
0.5	0.85	1.20	0.57	0.12	0.75	0.09	0.18	0.20	0.59	0.33	9.5
0.6	0.86	1.20	0.57	0.13	0.75	0.11	0.20	0.19	0.59	0.32	9.4
0.7	0.88	1.20	0.56	0.13	0.74	0.13	0.23	0.17	0.59	0.32	9.3
0.8	0.89	1.20	0.55	0.14	0.74	0.16	0.25	0.16	0.59	0.31	9.2
0.9	0.91	1.20	0.54	0.15	0.73	0.19	0.28	0.15	0.59	0.31	9.1
1.0	0.92	1.20	0.53	0.16	0.73	0.22	0.31	0.14	0.59	0.30	9.0
1.1	0.93	1.20	0.52	0.17	0.72	0.25	0.34	0.13	0.59	0.30	8.9
1.2	0.91	1.20	0.51	0.18	0.71	0.28	0.37	0.12	0.59	0.29	8.8
1.3	0.95	1.19	0.50	0.19	0.71	0.32	0.39	0.11	0.59	0.29	8.7
1.4	0.96	1.19	0.49	0.20	0.70	0.36	0.42	0.10	0.59	0.28	8.6
1.5	0.97	1.19	0.48	0.21	0.69	0.10	0.44	0.09	0.60	0.28	8.5
1.6	0.97	1.19	0.47	0.22	0.68	0.44	0.46	0.08	0.60	0.27	8.4
1.7	0.98	1.19	0.46	0.23	0.68	0.17	0.48	0.07	0.60	0.27	8.3
1.8	0.98	1.18	0.15	0.24	0.67	0.51	0.50	0.06	0.60	0.26	8.2
1.9	0.99	1.18	0.44	0.25	0.66	0.55	0.51	0.05	0.60	0.26	8.1
2.0	0.99	1.18	0.43	0.26	0.65	0.59	0.53	0.04	0.60	0.25	8.0
2.1	0.99	1.18	0.42	0.27	0.64	0.62	0.54	0.03	0.60	0.25	7.9
2.2	1.00	1.17	0.41	0.28	0.63	0.65	0.55	0.03	0.60	0.24	7.8
2.3	1.00	1.17	0.40	0.29	0.62	0.68	0.56	0.02	0.60	0.24	7.7
2.4	1.00	1.17	0.39	0.30	0.61	0.70	0.57	0.02	0.60	0.23	7.6
2.5	0.99	1.16	0.38	0.31	0.60	0.72	0.58	0.01	0.60	0.23	7.5
2.6	0.99	1.16	0.38	0.33	0.59	0.74	0.59	0.01	0.59	0.22	7.4
2.7	0.99	1.16	0.37	0.34	0.58	0.75	0.60	0.01	0.59	0.22	7.3
2.8	0.98	1.15	0.36	0.35	0.57	0.77	0.59	0.00	0.59	0.21	7.2
2.9	0.98	1.15	0.35	0.36	0.56	0.78	0.58	0.00	0.59	0.21	7.1
3.0	0.98	1.15	0.34	0.37	0.55	0.79	0.57	0.00	0.59	0.20	7.0
3.1	0.97	1.14	0.33	0.38	0.54	0.80	0.56	0.00	0.59	0.20	6.9
3.2	0.97	1.14	0.32	0.39	0.53	0.80	0.55	0.00	0.59	0.19	6.8
3.3	0.96	1.13	0.32	0.40	0.52	0.80	0.54	0.00	0.59	0.19	6.7
3.4	0.96	1.13	0.31	0.41	0.51	0.79	0.53	0.00	0.59	0.18	6.6
3.5	0.95	1.13	0.30	0.42	0.50	0.78	0.52	0.00	0.59	0.18	6.5
3.6	0.94	1.12	0.29	0.44	0.49	0.76	0.50	0.00	0.58	0.17	6.4
3.7	0.93	1.12	0.28	0.45	0.48	0.75	0.49	0.01	0.58	0.17	6.3
3.8	0.92	1.12	0.28	0.46	0.47	0.73	0.47	0.01	0.58	0.16	6.2
3.9	0.90	1.11	0.27	0.47	0.46	0.71	0.45	0.01	0.58	0.16	6.1
4.0	0.89	1.11	0.26	0.48	0.45	0.69	0.43	0.02	0.58	0.15	6.0
4.1	0.88	1.10	0.25	0.49	0.44	0.66	0.41	0.02	0.58	0.15	5.9
4.2	0.86	1.10	0.24	0.50	0.43	0.63	0.39	0.03	0.58	0.14	5.8
4.3	0.85	1.09	0.24	0.51	0.42	0.60	0.36	0.03	0.57	0.14	5.7
4.4	0.84	1.09	0.23	0.52	0.41	0.57	0.34	0.04	0.57	0.13	5.6
4.5	0.82	1.08	0.22	0.53	0.40	0.54	0.31	0.05	0.57	0.13	5.5
4.6	0.80	1.07	0.21	0.55	0.38	0.50	0.28	0.06	0.57	0.12	5.4
4.7	0.79	1.07	0.20	0.56	0.37	0.47	0.25	0.07	0.57	0.12	5.3
4.8	0.77	1.06	0.20	0.57	0.36	0.43	0.23	0.08	0.56	0.11	5.2
4.9	0.75	1.06	0.19	0.58	0.35	0.39	0.21	0.09	0.56	0.11	5.1
5.0	0.73	1.05	0.18	0.59	0.34	0.35	0.19	0.10	0.56	0.10	5.0
Days.	**10**	**30**	**20**	**30**	**20**	**10**	**10**	**10**	**30**	**20**	Days.

TABLES LXIII. - LXIX.

TABLES	LXIII.	LXIV.		LXV.		LXVI.	LXVII.	LXVIII.	LXIX.		
ARGUMENTS	58.	59.		60.		61.	62.	63.	64.		
Days.	0	0	10	0	10	0	0	0	0	10	Days.
Days.											Days.
5.0	0.73	1.05	0.18	0.59	0.34	0.35	0.19	0.10	0.56	0.10	5.0
5.1	0.71	1.04	0.17	0.60	0.33	0.31	0.17	0.11	0.56	0.10	4.9
5.2	0.69	1.04	0.17	0.61	0.32	0.28	0.15	0.12	0.56	0.09	4.8
5.3	0.68	1.03	0.16	0.62	0.31	0.24	0.13	0.13	0.55	0.09	4.7
5.4	0.66	1.02	0.15	0.63	0.30	0.21	0.11	0.14	0.55	0.09	4.6
5.5	0.64	1.02	0.15	0.64	0.29	0.18	0.09	0.15	0.55	0.08	4.5
5.6	0.62	1.01	0.14	0.64	0.27	0.15	0.07	0.16	0.55	0.08	4.4
5.7	0.60	1.00	0.14	0.65	0.26	0.13	0.06	0.18	0.55	0.08	4.3
5.8	0.58	1.00	0.13	0.66	0.25	0.10	0.04	0.19	0.54	0.07	4.2
5.9	0.56	0.99	0.12	0.67	0.24	0.08	0.03	0.21	0.54	0.07	4.1
6.0	0.54	0.98	0.12	0.68	0.23	0.06	0.02	0.22	0.54	0.07	4.0
6.1	0.52	0.97	0.11	0.69	0.22	0.04	0.01	0.23	0.54	0.07	3.9
6.2	0.50	0.96	0.11	0.70	0.21	0.03	0.01	0.25	0.53	0.06	3.8
6.3	0.48	0.96	0.10	0.70	0.20	0.02	0.00	0.26	0.53	0.06	3.7
6.4	0.46	0.95	0.10	0.71	0.19	0.01	0.00	0.28	0.53	0.06	3.6
6.5	0.44	0.94	0.09	0.72	0.18	0.00	0.00	0.29	0.52	0.05	3.5
6.6	0.42	0.93	0.09	0.72	0.18	0.00	0.01	0.30	0.52	0.05	3.4
6.7	0.40	0.92	0.08	0.73	0.17	0.01	0.01	0.32	0.52	0.05	3.3
6.8	0.38	0.92	0.08	0.74	0.16	0.01	0.02	0.33	0.51	0.05	3.2
6.9	0.36	0.91	0.07	0.74	0.15	0.02	0.03	0.35	0.51	0.04	3.1
7.0	0.34	0.90	0.07	0.75	0.14	0.03	0.04	0.36	0.50	0.04	3.0
7.1	0.32	0.89	0.06	0.75	0.13	0.05	0.06	0.37	0.50	0.04	2.9
7.2	0.30	0.88	0.06	0.76	0.12	0.07	0.07	0.39	0.49	0.04	2.8
7.3	0.28	0.87	0.05	0.76	0.12	0.09	0.09	0.40	0.49	0.03	2.7
7.4	0.26	0.86	0.05	0.77	0.11	0.11	0.11	0.41	0.48	0.03	2.6
7.5	0.24	0.85	0.04	0.77	0.10	0.13	0.13	0.42	0.48	0.03	2.5
7.6	0.22	0.85	0.04	0.78	0.09	0.16	0.15	0.43	0.47	0.03	2.4
7.7	0.21	0.84	0.04	0.78	0.08	0.19	0.17	0.45	0.47	0.03	2.3
7.8	0.19	0.83	0.03	0.78	0.08	0.22	0.20	0.46	0.46	0.02	2.2
7.9	0.18	0.82	0.03	0.79	0.07	0.25	0.22	0.47	0.46	0.02	2.1
8.0	0.16	0.81	0.03	0.79	0.06	0.29	0.25	0.48	0.45	0.02	2.0
8.1	0.15	0.80	0.03	0.79	0.05	0.32	0.27	0.49	0.45	0.02	1.9
8.2	0.13	0.79	0.02	0.79	0.05	0.36	0.30	0.50	0.44	0.02	1.8
8.3	0.12	0.78	0.02	0.79	0.04	0.39	0.32	0.51	0.44	0.02	1.7
8.4	0.10	0.77	0.02	0.80	0.04	0.43	0.35	0.52	0.43	0.01	1.6
8.5	0.09	0.76	0.02	0.80	0.04	0.47	0.38	0.53	0.43	0.01	1.5
8.6	0.08	0.76	0.02	0.80	0.03	0.51	0.40	0.54	0.42	0.01	1.4
8.7	0.07	0.75	0.01	0.80	0.03	0.54	0.43	0.55	0.42	0.01	1.3
8.8	0.06	0.74	0.01	0.80	0.03	0.58	0.45	0.56	0.41	0.01	1.2
8.9	0.05	0.73	0.01	0.80	0.02	0.61	0.47	0.56	0.41	0.01	1.1
9.0	0.04	0.72	0.01	0.80	0.02	0.64	0.49	0.57	0.40	0.01	1.0
9.1	0.03	0.71	0.01	0.80	0.02	0.67	0.51	0.57	0.40	0.01	0.9
9.2	0.03	0.70	0.01	0.79	0.01	0.69	0.53	0.58	0.39	0.01	0.8
9.3	0.02	0.69	0.01	0.79	0.01	0.72	0.54	0.58	0.39	0.00	0.7
9.4	0.02	0.68	0.00	0.79	0.01	0.74	0.56	0.59	0.38	0.00	0.6
9.5	0.01	0.67	0.00	0.79	0.01	0.76	0.57	0.59	0.38	0.00	0.5
9.6	0.01	0.66	0.00	0.79	0.01	0.77	0.58	0.59	0.37	0.00	0.4
9.7	0.01	0.65	0.00	0.78	0.00	0.78	0.59	0.59	0.37	0.00	0.3
9.8	0.00	0.64	0.00	0.78	0.00	0.79	0.59	0.60	0.36	0.00	0.2
9.9	0.00	0.63	0.00	0.78	0.00	0.80	0.60	0.60	0.36	0.00	0.1
10.0	0.00	0.62	0.00	0.78	0.00	0.80	0.60	0.60	0.35	0.00	0.0
Days.	10	30	20	30	20	10	10	10	30	20	Days.

TABLES LXX.-LXXIV.

Tables	LXX.	LXXI.		LXXII.		LXXIII.		
Arguments	**65.**	**66.**		**67.**		**68.**		
Days.	**0**	**0**	**10**	**0**	**10**	**0**	**10**	Days.
0.0	0.35	0.38	0.24	1.13	1.66	0.58	0.25	10.0
0.1	0.34	0.38	0.23	1.15	1.64	0.56	0.26	9.9
0.2	0.32	0.38	0.23	1.17	1.62	0.55	0.27	9.8
0.3	0.31	0.38	0.23	1.19	1.61	0.53	0.29	9.7
0.4	0.29	0.38	0.22	1.22	1.59	0.52	0.30	9.6
0.5	0.28	0.38	0.22	1.24	1.57	0.50	0.31	9.5
0.6	0.26	0.39	0.22	1.26	1.55	0.49	0.33	9.4
0.7	0.24	0.39	0.21	1.28	1.53	0.47	0.34	9.3
0.8	0.23	0.39	0.21	1.31	1.51	0.46	0.35	9.2
0.9	0.21	0.39	0.20	1.33	1.49	0.44	0.37	9.1
1.0	0.19	0.39	0.20	1.35	1.47	0.43	0.38	9.0
1.1	0.17	0.39	0.20	1.37	1.45	0.41	0.39	8.9
1.2	0.16	0.39	0.19	1.39	1.43	0.40	0.41	8.8
1.3	0.14	0.39	0.19	1.41	1.41	0.38	0.42	8.7
1.4	0.13	0.39	0.19	1.43	1.39	0.37	0.44	8.6
1.5	0.11	0.40	0.18	1.46	1.37	0.36	0.45	8.5
1.6	0.10	0.40	0.18	1.48	1.34	0.34	0.47	8.4
1.7	0.08	0.40	0.18	1.50	1.32	0.33	0.48	8.3
1.8	0.07	0.40	0.17	1.52	1.30	0.31	0.50	8.2
1.9	0.05	0.40	0.17	1.54	1.28	0.30	0.51	8.1
2.0	0.04	0.40	0.17	1.56	1.26	0.29	0.53	8.0
2.1	0.03	0.40	0.16	1.58	1.24	0.28	0.55	7.9
2.2	0.03	0.40	0.16	1.59	1.22	0.26	0.56	7.8
2.3	0.02	0.40	0.16	1.61	1.19	0.25	0.58	7.7
2.4	0.02	0.40	0.15	1.63	1.17	0.24	0.59	7.6
2.5	0.01	0.40	0.15	1.64	1.15	0.23	0.61	7.5
2.6	0.01	0.40	0.15	1.66	1.12	0.21	0.63	7.4
2.7	0.01	0.40	0.14	1.68	1.10	0.20	0.64	7.3
2.8	0.00	0.40	0.14	1.69	1.08	0.19	0.66	7.2
2.9	0.00	0.40	0.14	1.71	1.05	0.18	0.67	7.1
3.0	0.00	0.40	0.13	1.73	1.03	0.17	0.69	7.0
3.1	0.01	0.40	0.13	1.75	1.01	0.16	0.70	6.9
3.2	0.01	0.40	0.13	1.76	0.98	0.15	0.72	6.8
3.3	0.02	0.40	0.12	1.78	0.96	0.14	0.73	6.7
3.4	0.03	0.40	0.12	1.79	0.94	0.13	0.75	6.6
3.5	0.04	0.40	0.12	1.81	0.91	0.12	0.76	6.5
3.6	0.05	0.39	0.11	1.82	0.89	0.11	0.78	6.4
3.7	0.06	0.39	0.11	1.83	0.87	0.11	0.79	6.3
3.8	0.08	0.39	0.11	1.85	0.84	0.10	0.81	6.2
3.9	0.09	0.39	0.10	1.86	0.82	0.09	0.82	6.1
4.0	0.10	0.39	0.10	1.87	0.80	0.08	0.84	6.0
4.1	0.12	0.39	0.10	1.88	0.78	0.07	0.85	5.9
4.2	0.13	0.39	0.09	1.89	0.76	0.07	0.87	5.8
4.3	0.15	0.39	0.09	1.90	0.74	0.06	0.88	5.7
4.4	0.17	0.39	0.09	1.91	0.71	0.05	0.90	5.6
4.5	0.19	0.39	0.08	1.92	0.69	0.05	0.91	5.5
4.6	0.21	0.38	0.08	1.93	0.67	0.04	0.93	5.4
4.7	0.22	0.38	0.08	1.94	0.65	0.04	0.94	5.3
4.8	0.24	0.38	0.07	1.94	0.63	0.03	0.96	5.2
4.9	0.25	0.38	0.07	1.95	0.61	0.03	0.97	5.1
5.0	0.27	0.38	0.07	1.96	0.59	0.02	0.99	5.0
Days.	**10**	**30**	**20**	**30**	**20**	**30**	**20**	Days.

Tables	LXXIV.			
Arguments	**69.**			
Days.	**0**	**100**	**200**	Days.
0	3.65	4.15	1.68	100
1	3.67	4.14	1.65	99
2	3.69	4.13	1.62	98
3	3.71	4.11	1.60	97
4	3.73	4.10	1.57	96
5	3.75	4.08	1.54	95
6	3.77	4.07	1.51	94
7	3.79	4.05	1.48	93
8	3.81	4.04	1.46	92
9	3.83	4.02	1.43	91
10	3.85	4.00	1.40	90
11	3.87	3.98	1.37	89
12	3.89	3.97	1.34	88
13	3.90	3.95	1.32	87
14	3.92	3.93	1.29	86
15	3.94	3.91	1.26	85
16	3.96	3.89	1.24	84
17	3.98	3.88	1.21	83
18	3.99	3.86	1.18	82
19	4.01	3.84	1.16	81
20	4.03	3.82	1.13	80
21	4.05	3.80	1.11	79
22	4.07	3.78	1.08	78
23	4.08	3.76	1.06	77
24	4.10	3.74	1.03	76
25	4.11	3.72	1.00	75
26	4.13	3.69	0.98	74
27	4.14	3.67	0.95	73
28	4.16	3.65	0.93	72
29	4.17	3.63	0.91	71
30	4.18	3.61	0.89	70
31	4.19	3.59	0.87	69
32	4.20	3.56	0.84	68
33	4.22	3.54	0.82	67
34	4.23	3.52	0.80	66
35	4.24	3.49	0.78	65
36	4.25	3.47	0.76	64
37	4.26	3.44	0.73	63
38	4.27	3.42	0.71	62
39	4.28	3.39	0.69	61
40	4.29	3.37	0.67	60
41	4.30	3.35	0.65	59
42	4.31	3.32	0.63	58
43	4.31	3.30	0.61	57
44	4.32	3.27	0.59	56
45	4.33	3.25	0.57	55
46	4.33	3.22	0.55	54
47	4.34	3.19	0.53	53
48	4.35	3.17	0.51	52
49	4.35	3.14	0.49	51
50	4.36	3.11	0.47	50
Days.	**500**	**400**	**300**	Days.

TABLES LXX. - LXXIV.

Tables	LXX.	LXXI.		LXXII.		LXXIII.				LXXIV.			
Arguments	65.	66.		67.		68.				69.			
Days.	0	0	10	0	10	0	10	Days.	Days.	0	100	200	Days.
5.0	0.27	0.38	0.07	1.96	0.59	0.02	0.99	5.0	50	4.36	3.11	0.47	50
5.1	0.28	0.38	0.07	1.96	0.57	0.02	1.00	4.9	51	4.36	3.08	0.45	49
5.2	0.30	0.38	0.06	1.97	0.55	0.02	1.02	4.8	52	4.37	3.06	0.44	48
5.3	0.31	0.37	0.06	1.97	0.53	0.01	1.03	4.7	53	4.37	3.03	0.42	47
5.4	0.33	0.37	0.06	1.98	0.51	0.01	1.05	4.6	54	4.38	3.00	0.41	46
5.5	0.34	0.37	0.06	1.98	0.49	0.01	1.06	4.5	55	4.38	2.98	0.39	45
5.6	0.35	0.37	0.06	1.98	0.47	0.01	1.07	4.4	56	4.39	2.95	0.38	44
5.7	0.36	0.37	0.05	1.99	0.45	0.00	1.09	4.3	57	4.39	2.92	0.36	43
5.8	0.37	0.36	0.05	1.99	0.43	0.00	1.10	4.2	58	4.39	2.90	0.34	42
5.9	0.38	0.36	0.05	2.00	0.41	0.00	1.12	4.1	59	4.40	2.87	0.33	41
6.0	0.39	0.36	0.05	2.00	0.39	0.00	1.13	4.0	60	4.40	2.84	0.31	40
6.1	0.39	0.36	0.05	2.00	0.37	0.00	1.14	3.9	61	4.40	2.81	0.29	39
6.2	0.39	0.35	0.05	2.00	0.35	0.00	1.15	3.8	62	4.40	2.78	0.28	38
6.3	0.40	0.35	0.04	2.00	0.34	0.00	1.17	3.7	63	4.40	2.75	0.26	37
6.4	0.40	0.35	0.04	2.00	0.32	0.00	1.18	3.6	64	4.40	2.73	0.25	36
6.5	0.40	0.34	0.04	2.00	0.30	0.00	1.19	3.5	65	4.40	2.70	0.23	35
6.6	0.39	0.34	0.04	2.00	0.29	0.00	1.20	3.4	66	4.40	2.67	0.22	34
6.7	0.39	0.34	0.04	1.99	0.27	0.01	1.21	3.3	67	4.39	2.64	0.20	33
6.8	0.38	0.33	0.04	1.99	0.26	0.01	1.23	3.2	68	4.39	2.61	0.19	32
6.9	0.38	0.33	0.03	1.99	0.24	0.01	1.24	3.1	69	4.39	2.58	0.18	31
7.0	0.37	0.33	0.03	1.99	0.23	0.01	1.25	3.0	70	4.39	2.55	0.17	30
7.1	0.36	0.33	0.03	1.98	0.21	0.01	1.26	2.9	71	4.38	2.52	0.16	29
7.2	0.35	0.32	0.03	1.98	0.20	0.02	1.27	2.8	72	4.38	2.49	0.15	28
7.3	0.34	0.32	0.03	1.97	0.18	0.02	1.28	2.7	73	4.38	2.46	0.14	27
7.4	0.32	0.32	0.02	1.97	0.17	0.02	1.28	2.6	74	4.37	2.43	0.13	26
7.5	0.31	0.31	0.02	1.96	0.16	0.03	1.29	2.5	75	4.37	2.41	0.12	25
7.6	0.29	0.31	0.02	1.95	0.14	0.03	1.30	2.4	76	4.36	2.38	0.11	24
7.7	0.28	0.31	0.02	1.95	0.13	0.04	1.31	2.3	77	4.36	2.35	0.11	23
7.8	0.26	0.30	0.02	1.94	0.12	0.04	1.32	2.2	78	4.36	2.32	0.10	22
7.9	0.25	0.30	0.01	1.93	0.11	0.05	1.32	2.1	79	4.35	2.29	0.09	21
8.0	0.23	0.30	0.01	1.92	0.10	0.06	1.33	2.0	80	4.35	2.26	0.08	20
8.1	0.21	0.30	0.01	1.91	0.09	0.07	1.34	1.9	81	4.34	2.23	0.07	19
8.2	0.19	0.29	0.01	1.90	0.08	0.07	1.34	1.8	82	4.33	2.20	0.07	18
8.3	0.18	0.29	0.01	1.89	0.07	0.08	1.35	1.7	83	4.33	2.17	0.06	17
8.4	0.16	0.29	0.01	1.88	0.07	0.09	1.35	1.6	84	4.32	2.14	0.06	16
8.5	0.14	0.28	0.01	1.87	0.06	0.10	1.36	1.5	85	4.31	2.11	0.05	15
8.6	0.13	0.28	0.01	1.86	0.05	0.10	1.36	1.4	86	4.30	2.09	0.04	14
8.7	0.11	0.28	0.00	1.84	0.05	0.11	1.37	1.3	87	4.30	2.06	0.04	13
8.8	0.10	0.27	0.00	1.83	0.04	0.12	1.37	1.2	88	4.29	2.03	0.03	12
8.9	0.08	0.27	0.00	1.82	0.03	0.13	1.38	1.1	89	4.28	2.00	0.03	11
9.0	0.07	0.27	0.00	1.81	0.03	0.14	1.38	1.0	90	4.27	1.97	0.02	10
9.1	0.06	0.27	0.00	1.80	0.03	0.15	1.38	0.9	91	4.26	1.94	0.02	09
9.2	0.05	0.26	0.00	1.79	0.02	0.16	1.39	0.8	92	4.25	1.91	0.02	08
9.3	0.04	0.26	0.00	1.77	0.02	0.17	1.39	0.7	93	4.23	1.88	0.01	07
9.4	0.03	0.26	0.00	1.76	0.02	0.18	1.39	0.6	94	4.22	1.85	0.01	06
9.5	0.02	0.25	0.00	1.74	0.01	0.19	1.39	0.5	95	4.21	1.82	0.01	05
9.6	0.02	0.25	0.00	1.73	0.01	0.21	1.39	0.4	96	4.20	1.80	0.01	04
9.7	0.01	0.25	0.00	1.71	0.01	0.22	1.40	0.3	97	4.19	1.77	0.00	03
9.8	0.01	0.24	0.00	1.70	0.00	0.23	1.40	0.2	98	4.17	1.74	0.00	02
9.9	0.00	0.24	0.00	1.68	0.00	0.24	1.40	0.1	99	4.16	1.71	0.00	01
10.0	0.00	0.24	0.00	1.66	0.00	0.25	1.40	0.0	100	4.15	1.68	0.00	00
Days.	10	30	20	30	20	30	20	Days.	Days.	500	400	300	Days.

TABLES LXXV. LXXVI.

Tables	LXXV.	LXXVI.
Arguments	70.	71.

Days.	0	100	Days.
0	0.32	1.01	100
1	0.30	1.04	99
2	0.28	1.07	98
3	0.27	1.09	97
4	0.25	1.12	96
5	0.24	1.15	95
6	0.22	1.18	94
7	0.21	1.21	93
8	0.19	1.23	92
9	0.18	1.26	91
10	0.17	1.29	90
11	0.16	1.32	89
12	0.14	1.35	88
13	0.13	1.37	87
14	0.12	1.40	86
15	0.11	1.43	85
16	0.10	1.46	84
17	0.09	1.49	83
18	0.08	1.51	82
19	0.07	1.54	81
20	0.06	1.57	80
21	0.05	1.60	79
22	0.05	1.63	78
23	0.04	1.65	77
24	0.04	1.68	76
25	0.03	1.71	75
26	0.03	1.74	74
27	0.02	1.77	73
28	0.02	1.79	72
29	0.01	1.82	71
30	0.01	1.85	70
31	0.01	1.88	69
32	0.00	1.91	68
33	0.00	1.93	67
34	0.00	1.96	66
35	0.00	1.99	65
36	0.00	2.01	64
37	0.00	2.04	63
38	0.01	2.07	62
39	0.01	2.09	61
40	0.01	2.12	60
41	0.02	2.15	59
42	0.02	2.17	58
43	0.03	2.20	57
44	0.03	2.22	56
45	0.04	2.25	55
46	0.04	2.27	54
47	0.05	2.30	53
48	0.05	2.32	52
49	0.06	2.35	51
50	0.06	2.37	50
Days.	300	200	Days.

Days.	0	100	200	300	400	500
0	1.14	2.57	1.93	1.30	0.67	0.03
1	1.16	2.57	1.92	1.30	0.66	0.04
2	1.18	2.58	1.91	1.29	0.65	0.04
3	1.20	2.58	1.90	1.29	0.64	0.04
4	1.22	2.58	1.89	1.29	0.63	0.05
5	1.24	2.58	1.88	1.28	0.62	0.05
6	1.26	2.58	1.87	1.28	0.60	0.06
7	1.28	2.59	1.86	1.28	0.59	0.06
8	1.30	2.59	1.85	1.27	0.58	0.06
9	1.32	2.59	1.84	1.27	0.57	0.07
10	1.34	2.59	1.83	1.27	0.56	0.07
11	1.36	2.59	1.82	1.27	0.55	0.08
12	1.38	2.59	1.81	1.26	0.54	0.08
13	1.40	2.59	1.80	1.26	0.53	0.09
14	1.42	2.59	1.79	1.26	0.52	0.10
15	1.44	2.59	1.78	1.25	0.51	0.10
16	1.46	2.58	1.78	1.25	0.49	0.11
17	1.48	2.58	1.77	1.25	0.48	0.12
18	1.50	2.58	1.76	1.24	0.47	0.12
19	1.52	2.58	1.75	1.24	0.46	0.13
20	1.54	2.58	1.74	1.24	0.45	0.14
21	1.56	2.58	1.73	1.23	0.44	0.15
22	1.58	2.57	1.72	1.23	0.43	0.16
23	1.60	2.57	1.71	1.23	0.42	0.17
24	1.62	2.57	1.71	1.22	0.41	0.18
25	1.64	2.56	1.70	1.22	0.40	0.19
26	1.65	2.56	1.69	1.21	0.39	0.20
27	1.67	2.56	1.68	1.21	0.38	0.21
28	1.69	2.55	1.68	1.21	0.37	0.22
29	1.71	2.55	1.67	1.20	0.36	0.23
30	1.73	2.55	1.66	1.20	0.35	0.24
31	1.75	2.54	1.65	1.19	0.34	0.25
32	1.77	2.54	1.64	1.19	0.33	0.26
33	1.79	2.53	1.63	1.18	0.32	0.28
34	1.80	2.53	1.63	1.18	0.31	0.29
35	1.82	2.52	1.62	1.17	0.30	0.30
36	1.84	2.52	1.61	1.17	0.30	0.32
37	1.86	2.51	1.60	1.16	0.29	0.33
38	1.88	2.51	1.59	1.16	0.28	0.34
39	1.89	2.50	1.59	1.15	0.27	0.36
40	1.91	2.49	1.58	1.15	0.26	0.37
41	1.93	2.48	1.57	1.14	0.25	0.38
42	1.95	2.48	1.56	1.14	0.24	0.40
43	1.96	2.47	1.56	1.13	0.24	0.41
44	1.98	2.46	1.55	1.13	0.23	0.43
45	2.00	2.46	1.54	1.12	0.22	0.44
46	2.01	2.45	1.54	1.12	0.21	0.46
47	2.03	2.44	1.53	1.11	0.20	0.47
48	2.05	2.44	1.52	1.10	0.20	0.49
49	2.06	2.43	1.52	1.10	0.19	0.50
50	2.08	2.42	1.51	1.09	0.18	0.52

TABLES LXXV. LXXVI.

TABLES	LXXV.			LXXVI.
ARGUMENTS	70.			71.

Table LXXV. Argument 70.

Days.	0	100	Days.
Days.	″	″	Days.
50	0.06	2.37	50
51	0.07	2.39	49
52	0.08	2.41	48
53	0.09	2.44	47
54	0.10	2.46	46
55	0.11	2.48	45
56	0.12	2.50	44
57	0.13	2.52	43
58	0.15	2.54	42
59	0.16	2.56	41
60	0.17	2.58	40
61	0.18	2.60	39
62	0.19	2.62	38
63	0.21	2.64	37
64	0.22	2.65	36
65	0.24	2.67	35
66	0.25	2.69	34
67	0.27	2.71	33
68	0.28	2.72	32
69	0.30	2.74	31
70	0.32	2.76	30
71	0.34	2.77	29
72	0.36	2.79	28
73	0.38	2.80	27
74	0.40	2.81	26
75	0.42	2.83	25
76	0.44	2.84	24
77	0.46	2.85	23
78	0.48	2.87	22
79	0.50	2.88	21
80	0.52	2.89	20
81	0.54	2.90	19
82	0.56	2.91	18
83	0.59	2.92	17
84	0.61	2.93	16
85	0.63	2.93	15
86	0.66	2.94	14
87	0.68	2.95	13
88	0.70	2.96	12
89	0.73	2.96	11
90	0.75	2.97	10
91	0.77	2.97	09
92	0.80	2.98	08
93	0.82	2.98	07
94	0.85	2.98	06
95	0.88	2.99	05
96	0.90	2.99	04
97	0.93	2.99	03
98	0.95	3.00	02
99	0.98	3.00	01
100	1.01	3.00	00
Days.	300	200	Days.

Table LXXVI. Argument 71.

Days.	0	100	200	300	400	500
Days.	″	″	″	″	″	
50	2.08	2.42	1.51	1.09	0.18	0.52
51	2.10	2.41	1.50	1.08	0.17	0.54
52	2.11	2.40	1.50	1.08	0.16	0.55
53	2.13	2.40	1.49	1.07	0.16	0.57
54	2.15	2.39	1.49	1.06	0.15	0.59
55	2.16	2.38	1.48	1.06	0.14	0.60
56	2.18	2.37	1.47	1.05	0.14	0.62
57	2.19	2.37	1.47	1.04	0.13	0.64
58	2.20	2.36	1.46	1.04	0.12	0.65
59	2.22	2.35	1.46	1.03	0.12	0.67
60	2.23	2.34	1.45	1.02	0.11	0.69
61	2.24	2.33	1.45	1.01	0.10	0.71
62	2.26	2.32	1.44	1.00	0.10	0.73
63	2.27	2.31	1.44	1.00	0.09	0.74
64	2.28	2.30	1.43	0.99	0.09	0.76
65	2.30	2.30	1.43	0.98	0.08	0.78
66	2.31	2.29	1.42	0.97	0.07	0.80
67	2.32	2.28	1.42	0.97	0.07	0.82
68	2.34	2.27	1.41	0.96	0.06	0.83
69	2.35	2.26	1.41	0.95	0.06	0.85
70	2.36	2.25	1.40	0.94	0.05	0.87
71	2.37	2.24	1.40	0.93	0.05	0.89
72	2.38	2.23	1.39	0.92	0.04	0.91
73	2.39	2.22	1.39	0.92	0.04	0.93
74	2.40	2.21	1.38	0.91	0.04	0.95
75	2.41	2.20	1.38	0.90	0.03	0.96
76	2.42	2.19	1.37	0.89	0.03	0.98
77	2.43	2.18	1.37	0.88	0.03	1.00
78	2.44	2.17	1.37	0.88	0.02	1.02
79	2.45	2.16	1.36	0.87	0.02	1.04
80	2.46	2.15	1.36	0.86	0.02	1.06
81	2.47	2.14	1.36	0.85	0.02	1.08
82	2.48	2.13	1.35	0.84	0.02	1.10
83	2.48	2.12	1.35	0.83	0.02	1.12
84	2.49	2.11	1.35	0.82	0.01	1.14
85	2.50	2.10	1.34	0.81	0.01	1.16
86	2.50	2.08	1.34	0.81	0.01	1.18
87	2.51	2.07	1.34	0.80	0.01	1.20
88	2.52	2.06	1.33	0.79	0.01	1.22
89	2.52	2.05	1.33	0.78	0.01	1.24
90	2.53	2.04	1.33	0.77	0.01	1.26
91	2.53	2.03	1.33	0.76	0.01	1.28
92	2.54	2.02	1.32	0.75	0.01	1.30
93	2.54	2.01	1.32	0.74	0.02	1.32
94	2.55	2.00	1.32	0.73	0.02	1.34
95	2.55	1.99	1.31	0.72	0.02	1.36
96	2.56	1.97	1.31	0.71	0.02	1.38
97	2.56	1.96	1.31	0.70	0.02	1.40
98	2.56	1.95	1.30	0.69	0.03	1.42
99	2.57	1.94	1.30	0.68	0.03	1.44
100	2.57	1.93	1.30	0.67	0.03	1.46

TABLES LXXVII.-LXXIX.

Tables	LXXVII.				LXXVIII.		LXXIX.		
Arguments	72.				73'.		74.		

Days.	0	100	200	300
0	0.80″	1.50″	0.80″	0.10″
1	0.80	1.50	0.78	0.11
2	0.81	1.51	0.77	0.12
3	0.81	1.51	0.75	0.12
4	0.82	1.52	0.73	0.13
5	0.82	1.52	0.72	0.14
6	0.83	1.53	0.70	0.15
7	0.83	1.53	0.68	0.16
8	0.81	1.54	0.67	0.16
9	0.84	1.51	0.65	0.17
10	0.85	1.55	0.63	0.18
11	0.85	1.55	0.61	0.19
12	0.86	1.56	0.60	0.19
13	0.86	1.56	0.58	0.20
14	0.87	1.56	0.56	0.21
15	0.88	1.57	0.55	0.21
16	0.88	1.57	0.53	0.22
17	0.89	1.58	0.51	0.23
18	0.89	1.58	0.50	0.23
19	0.90	1.59	0.48	0.24
20	0.91	1.59	0.46	0.25
21	0.91	1.59	0.11	0.26
22	0.92	1.59	0.43	0.27
23	0.92	1.59	0.41	0.28
24	0.93	1.59	0.40	0.29
25	0.93	1.59	0.38	0.29
26	0.94	1.59	0.36	0.30
27	0.91	1.58	0.35	0.31
28	0.95	1.58	0.34	0.32
29	0.95	1.58	0.33	0.33
30	0.96	1.58	0.32	0.34
31	0.96	1.58	0.30	0.35
32	0.97	1.57	0.29	0.36
33	0.98	1.57	0.28	0.36
34	0.98	1.57	0.26	0.37
35	0.99	1.57	0.25	0.38
36	0.99	1.57	0.24	0.39
37	1.00	1.56	0.22	0.40
38	1.00	1.56	0.21	0.40
39	1.01	1.56	0.20	0.41
40	1.02	1.56	0.19	0.42
41	1.03	1.55	0.18	0.43
42	1.03	1.55	0.17	0.44
43	1.04	1.54	0.16	0.45
44	1.05	1.54	0.15	0.46
45	1.05	1.53	0.14	0.46
46	1.06	1.53	0.13	0.47
47	1.07	1.52	0.13	0.48
48	1.07	1.51	0.12	0.49
49	1.08	1.51	0.11	0.50
50	1.09	1.50	0.10	0.51

Days.	0	0	1000	Days.
0	3.71″	0.13″	0.32″	1000
10	3.64	0.12	0.33	990
20	3.57	0.12	0.34	980
30	3.50	0.11	0.35	970
40	3.42	0.10	0.36	960
50	3.34	0.10	0.36	950
60	3.25	0.09	0.37	940
70	3.16	0.09	0.38	930
80	3.07	0.08	0.39	920
90	2.97	0.08	0.40	910
100	2.87	0.07	0.41	900
110	2.77	0.07	0.42	890
120	2.66	0.06	0.43	880
130	2.55	0.06	0.44	870
140	2.14	0.05	0.45	860
150	2.33	0.05	0.46	850
160	2.22	0.04	0.47	840
170	2.11	0.01	0.48	830
180	2.00	0.01	0.49	820
190	1.88	0.03	0.50	810
200	1.77	0.03	0.51	800
210	1.66	0.03	0.52	790
220	1.55	0.02	0.53	780
230	1.11	0.02	0.54	770
240	1.33	0.02	0.55	760
250	1.22	0.02	0.56	750
260	1.12	0.02	0.56	740
270	1.03	0.01	0.57	730
280	0.93	0.01	0.58	720
290	0.83	0.01	0.59	710
300	0.73	0.01	0.60	700
310	0.61	0.01	0.61	690
320	0.56	0.01	0.62	680
330	0.48	0.01	0.63	670
310	0.41	0.01	0.64	660
350	0.34	0.00	0.65	650
360	0.28	0.00	0.65	640
370	0.22	0.00	0.66	630
380	0.17	0.00	0.67	620
390	0.13	0.00	0.68	610
400	0.09	0.00	0.69	600
410	0.06	0.00	0.70	590
420	0.03	0.00	0.71	580
430	0.01	0.00	0.72	570
440	0.00	0.00	0.73	560
450	0.00	0.00	0.74	550
460	0.01	0.01	0.74	540
470	0.02	0.01	0.75	530
480	0.03	0.01	0.76	520
490	0.05	0.01	0.77	510
500	0.08	0.01	0.78	500
Days.	1000	3000	2000	Days.

Note.—Arg. 73′ = Arg. 73 + 173.792.

TABLES LXXVII.-LXXIX.

TABLES	LXXVII.	LXXVIII.	LXXIX.
ARGUMENTS	72.	73′.	74.

Table LXXVII.

Days.	0	100	200	300
50	1.09	1.50	0.10	0.51
51	1.10	1.49	0.09	0.52
52	1.11	1.48	0.09	0.52
53	1.12	1.47	0.08	0.53
54	1.12	1.47	0.07	0.54
55	1.13	1.46	0.07	0.54
56	1.14	1.45	0.06	0.55
57	1.15	1.44	0.06	0.56
58	1.16	1.43	0.05	0.56
59	1.17	1.42	0.05	0.57
60	1.18	1.41	0.04	0.58
61	1.19	1.40	0.04	0.59
62	1.20	1.39	0.04	0.60
63	1.20	1.37	0.03	0.60
64	1.21	1.36	0.03	0.61
65	1.22	1.35	0.03	0.62
66	1.23	1.33	0.03	0.62
67	1.23	1.32	0.02	0.63
68	1.24	1.31	0.02	0.63
69	1.25	1.29	0.02	0.64
70	1.26	1.28	0.02	0.64
71	1.27	1.27	0.02	0.65
72	1.28	1.25	0.02	0.65
73	1.29	1.24	0.01	0.66
74	1.30	1.22	0.01	0.66
75	1.30	1.21	0.01	0.67
76	1.31	1.20	0.01	0.67
77	1.32	1.18	0.01	0.68
78	1.33	1.17	0.01	0.68
79	1.34	1.15	0.01	0.69
80	1.35	1.14	0.01	0.69
81	1.36	1.12	0.01	0.70
82	1.37	1.11	0.02	0.70
83	1.37	1.09	0.02	0.71
84	1.38	1.07	0.02	0.71
85	1.39	1.06	0.03	0.72
86	1.39	1.04	0.03	0.73
87	1.40	1.02	0.04	0.73
88	1.41	1.01	0.04	0.74
89	1.41	0.99	0.05	0.74
90	1.42	0.97	0.05	0.75
91	1.43	0.95	0.06	0.75
92	1.44	0.94	0.06	0.76
93	1.44	0.92	0.07	0.76
94	1.45	0.90	0.07	0.77
95	1.46	0.89	0.08	0.77
96	1.47	0.87	0.08	0.78
97	1.47	0.85	0.09	0.78
98	1.48	0.84	0.09	0.79
99	1.49	0.82	0.10	0.79
100	1.50	0.80	0.10	0.80

Tables LXXVIII. and LXXIX.

Days.	0	0	1000	Days.
500	0.08	0.01	0.78	500
510	0.12	0.01	0.79	490
520	0.16	0.02	0.80	480
530	0.21	0.02	0.80	470
540	0.26	0.02	0.81	460
550	0.31	0.02	0.82	450
560	0.37	0.03	0.83	440
570	0.44	0.03	0.84	430
580	0.51	0.03	0.84	420
590	0.59	0.04	0.85	410
600	0.68	0.04	0.86	400
610	0.77	0.05	0.87	390
620	0.86	0.05	0.87	380
630	0.96	0.06	0.88	370
640	1.06	0.06	0.88	360
650	1.16	0.07	0.89	350
660	1.26	0.07	0.90	340
670	1.37	0.08	0.90	330
680	1.48	0.08	0.91	320
690	1.59	0.09	0.91	310
700	1.70	0.09	0.92	300
710	1.81	0.10	0.92	290
720	1.92	0.10	0.93	280
730	2.01	0.11	0.93	270
740	2.15	0.12	0.94	260
750	2.27	0.12	0.94	250
760	2.38	0.13	0.95	240
770	2.49	0.14	0.95	230
780	2.60	0.14	0.95	220
790	2.71	0.15	0.96	210
800	2.82	0.16	0.96	200
810	2.92	0.17	0.96	190
820	3.02	0.18	0.97	180
830	3.11	0.18	0.97	170
840	3.21	0.19	0.97	160
850	3.30	0.20	0.98	150
860	3.38	0.21	0.98	140
870	3.46	0.21	0.98	130
880	3.54	0.22	0.98	120
890	3.61	0.23	0.99	110
900	3.68	0.24	0.99	100
910	3.74	0.25	0.99	90
920	3.79	0.26	0.99	80
930	3.84	0.26	0.99	70
940	3.88	0.27	0.99	60
950	3.92	0.28	1.00	50
960	3.95	0.29	1.00	40
970	3.97	0.30	1.00	30
980	3.98	0.30	1.00	20
990	3.99	0.31	1.00	10
1000	4.00	0.32	1.00	0
Days.	1000	3000	2000	Days.

NOTE. — Arg. 73′ = Arg. 73 + 173.792.

TABLES LXXX. LXXXI.

Tables	LXXX.					LXXXI.					
Arguments	75.					76.					
Days.	0	10000	20000	30000	40000	0	10000	20000	30000	40000	Days.
0	33.61	30.35	39.10	55.24	70.31	38.56	41.74	54.89	72.56	87.33	10000
100	33.52	30.38	39.24	55.41	70.43	38.53	41.83	55.06	72.74	87.44	9900
200	33.43	30.41	39.37	55.58	70.54	38.51	41.92	55.22	72.91	87.55	9800
300	33.34	30.45	39.51	55.76	70.66	38.48	42.01	55.39	73.08	87.66	9700
400	33.25	30.48	39.65	55.93	70.77	38.46	42.10	55.56	73.26	87.76	9600
500	33.16	30.52	39.79	56.10	70.88	38.44	42.19	55.73	73.43	87.87	9500
600	33.07	30.56	39.93	56.27	70.99	38.42	42.28	55.90	73.60	87.97	9400
700	32.99	30.60	40.07	56.44	71.10	38.40	42.38	56.07	73.77	88.08	9300
800	32.90	30.64	40.21	56.61	71.21	38.38	42.47	56.24	73.94	88.18	9200
900	32.82	30.68	40.36	56.78	71.32	38.37	42.57	56.42	74.11	88.28	9100
1000	32.74	30.72	40.50	56.95	71.43	38.35	42.67	56.59	74.28	88.38	9000
1100	32.66	30.76	40.64	57.12	71.54	38.34	42.77	56.76	74.45	88.48	8900
1200	32.58	30.81	40.79	57.29	71.64	38.32	42.87	56.93	74.62	88.58	8800
1300	32.50	30.85	40.93	57.45	71.75	38.31	42.97	57.10	74.79	88.68	8700
1400	32.43	30.90	41.08	57.62	71.85	38.30	43.07	57.27	74.96	88.77	8600
1500	32.35	30.95	41.23	57.79	71.95	38.29	43.17	57.44	75.13	88.87	8500
1600	32.28	31.00	41.38	57.96	72.05	38.28	43.27	57.61	75.30	88.96	8400
1700	32.20	31.05	41.52	58.13	72.15	38.28	43.37	57.78	75.47	89.06	8300
1800	32.13	31.10	41.67	58.29	72.25	38.27	43.48	57.96	75.64	89.15	8200
1900	32.06	31.16	41.82	58.46	72.34	38.27	43.58	58.13	75.81	89.24	8100
2000	31.99	31.21	41.97	58.63	72.44	38.26	43.69	58.30	75.98	89.33	8000
2100	31.92	31.27	42.12	58.80	72.54	38.26	43.80	58.47	76.15	89.42	7900
2200	31.85	31.32	42.27	58.97	72.63	38.25	43.91	58.65	76.31	89.51	7800
2300	31.78	31.38	42.42	59.13	72.72	38.25	44.02	58.82	76.48	89.60	7700
2400	31.72	31.44	42.58	59.30	72.82	38.26	44.13	59.00	76.64	89.68	7600
2500	31.65	31.50	42.73	59.46	72.91	38.26	44.24	59.17	76.81	89.77	7500
2600	31.59	31.56	42.88	59.63	73.00	38.27	44.35	59.35	76.97	89.86	7400
2700	31.53	31.62	43.04	59.79	73.09	38.27	44.47	59.52	77.14	89.94	7300
2800	31.47	31.68	43.19	59.96	73.18	38.28	44.58	59.70	77.30	90.03	7200
2900	31.41	31.75	43.35	60.12	73.26	38.28	44.70	59.87	77.46	90.11	7100
3000	31.35	31.81	43.51	60.29	73.35	38.29	44.82	60.05	77.62	90.19	7000
3100	31.29	31.88	43.67	60.46	73.43	38.30	44.94	60.23	77.78	90.27	6900
3200	31.24	31.95	43.83	60.62	73.52	38.31	45.06	60.41	77.94	90.35	6800
3300	31.18	32.02	43.98	60.79	73.60	38.32	45.18	60.58	78.10	90.43	6700
3400	31.13	32.09	44.14	60.95	73.68	38.34	45.30	60.76	78.26	90.50	6600
3500	31.08	32.16	44.30	61.11	73.76	38.35	45.42	60.94	78.42	90.58	6500
3600	31.03	32.23	44.46	61.27	73.84	38.36	45.54	61.12	78.58	90.65	6400
3700	30.98	32.31	44.62	61.43	73.92	38.38	45.66	61.30	78.74	90.73	6300
3800	30.93	32.38	44.78	61.59	73.99	38.39	45.79	61.47	78.90	90.80	6200
3900	30.89	32.46	44.94	61.75	74.07	38.41	45.91	61.65	79.05	90.87	6100
4000	30.84	32.54	45.10	61.91	74.14	38.43	46.04	61.83	79.21	90.94	6000
4100	30.80	32.62	45.26	62.07	74.21	38.45	46.17	62.01	79.37	91.01	5900
4200	30.75	32.70	45.42	62.23	74.29	38.47	46.29	62.19	79.52	91.08	5800
4300	30.71	32.78	45.59	62.39	74.36	38.50	46.42	62.36	79.68	91.15	5700
4400	30.67	32.87	45.75	62.54	74.43	38.52	46.55	62.54	79.83	91.21	5600
4500	30.63	32.95	45.91	62.70	74.50	38.55	46.68	62.72	79.99	91.28	5500
4600	30.59	33.03	46.07	62.86	74.57	38.58	46.81	62.90	80.14	91.34	5400
4700	30.55	33.12	46.24	63.01	74.64	38.61	46.94	63.08	80.30	91.40	5300
4800	30.51	33.20	46.40	63.17	74.70	38.64	47.07	63.26	80.45	91.46	5200
4900	30.48	33.29	46.57	63.32	74.77	38.67	47.21	63.44	80.60	91.52	5100
5000	30.44	33.38	46.73	63.48	74.83	38.70	47.34	63.62	80.75	91.58	5000
Days.	90000	80000	70000	60000	50000	90000	80000	70000	60000	50000	Days.

TABLES LXXX. LXXXI.

TABLES	LXXX.	LXXXI.
ARGUMENTS	75.	76.

Table LXXX. (Argument 75.)

Days.	0	10000	20000	30000	40000
5000	30.14	33.38	46.73	63.48	74.83
5100	30.41	33.47	46.90	63.63	74.89
5200	30.37	33.56	47.06	63.79	74.95
5300	30.34	33.65	47.23	63.94	75.01
5400	30.31	33.75	47.39	64.09	75.06
5500	30.28	33.84	47.56	64.24	75.12
5600	30.25	33.94	47.73	64.39	75.18
5700	30.23	34.03	47.90	64.54	75.23
5800	30.20	34.13	48.06	64.69	75.29
5900	30.18	34.23	48.23	64.84	75.34
6000	30.16	34.33	48.40	64.99	75.39
6100	30.14	34.43	48.57	65.14	75.44
6200	30.12	34.53	48.74	65.29	75.49
6300	30.10	34.63	48.90	65.43	75.54
6400	30.09	34.74	49.07	65.58	75.58
6500	30.07	34.84	49.24	65.72	75.63
6600	30.06	34.95	49.41	65.86	75.67
6700	30.05	35.05	49.58	66.01	75.71
6800	30.04	35.16	49.75	66.15	75.75
6900	30.03	35.27	49.92	66.29	75.79
7000	30.02	35.38	50.09	66.43	75.83
7100	30.01	35.49	50.26	66.57	75.87
7200	30.01	35.60	50.43	66.71	75.90
7300	30.00	35.71	50.60	66.85	75.94
7400	30.00	35.83	50.77	66.99	75.97
7500	30.00	35.94	50.94	67.13	76.00
7600	30.00	36.06	51.11	67.27	76.03
7700	30.00	36.17	51.28	67.41	76.06
7800	30.00	36.29	51.46	67.54	76.09
7900	30.01	36.41	51.63	67.68	76.12
8000	30.01	36.53	51.80	67.81	76.15
8100	30.02	36.65	51.97	67.94	76.17
8200	30.02	36.77	52.14	68.08	76.20
8300	30.03	36.89	52.32	68.21	76.22
8400	30.04	37.02	52.49	68.34	76.24
8500	30.05	37.14	52.66	68.47	76.26
8600	30.06	37.26	52.83	68.60	76.28
8700	30.07	37.39	53.00	68.73	76.30
8800	30.09	37.51	53.18	68.86	76.31
8900	30.10	37.64	53.35	68.98	76.33
9000	30.12	37.77	53.52	69.11	76.34
9100	30.14	37.90	53.69	69.23	76.35
9200	30.16	38.03	53.86	69.36	76.36
9300	30.18	38.16	54.04	69.48	76.37
9400	30.20	38.29	54.21	69.60	76.37
9500	30.22	38.42	54.38	69.72	76.38
9600	30.24	38.55	54.55	69.84	76.38
9700	30.27	38.69	54.72	69.96	76.39
9800	30.29	38.82	54.90	70.08	76.39
9900	30.32	38.96	55.07	70.19	76.40
10000	30.35	39.10	55.21	70.31	76.40
Days.	90000	80000	70000	60000	50000

Table LXXXI. (Argument 76.)

0	10000	20000	30000	40000	Days.
38.70	47.31	63.62	80.75	91.58	5000
38.73	47.47	63.80	80.90	91.64	4900
38.77	47.61	63.98	81.05	91.69	4800
38.80	47.74	64.16	81.20	91.75	4700
38.84	47.88	64.34	81.35	91.80	4600
38.88	48.02	64.52	81.50	91.85	4500
38.92	48.15	64.70	81.65	91.90	4400
38.96	48.30	64.88	81.80	91.95	4300
39.00	48.44	65.06	81.94	92.00	4200
39.04	48.58	65.24	82.09	92.05	4100
39.08	48.72	65.42	82.23	92.10	4000
39.13	48.86	65.60	82.37	92.15	3900
39.18	49.00	65.78	82.51	92.19	3800
39.22	49.14	65.96	82.65	92.24	3700
39.27	49.29	66.14	82.79	92.28	3600
39.32	49.43	66.32	82.93	92.32	3500
39.37	49.57	66.50	83.07	92.36	3400
39.42	49.72	66.68	83.21	92.40	3300
39.47	49.86	66.86	83.35	92.44	3200
39.53	50.01	67.04	83.48	92.47	3100
39.58	50.16	67.22	83.62	92.51	3000
39.64	50.31	67.40	83.76	92.55	2900
39.69	50.46	67.58	83.89	92.58	2800
39.75	50.61	67.76	84.03	92.62	2700
39.81	50.77	67.94	84.16	92.65	2600
39.87	50.92	68.12	84.29	92.68	2500
39.93	51.07	68.30	84.42	92.71	2400
39.99	51.23	68.48	84.55	92.74	2300
40.06	51.38	68.66	84.68	92.77	2200
40.12	51.53	68.84	84.81	92.79	2100
40.19	51.69	69.02	84.94	92.82	2000
40.26	51.85	69.20	85.07	92.84	1900
40.32	52.00	69.38	85.20	92.86	1800
40.39	52.16	69.55	85.32	92.88	1700
40.46	52.31	69.73	85.45	92.90	1600
40.53	52.47	69.91	85.57	92.92	1500
40.60	52.63	70.09	85.69	92.94	1400
40.67	52.78	70.27	85.81	92.95	1300
40.75	52.94	70.44	85.93	92.97	1200
40.82	53.10	70.62	86.05	92.98	1100
40.90	53.26	70.80	86.17	92.99	1000
40.98	53.42	70.98	86.29	93.00	900
41.06	53.58	71.15	86.41	93.01	800
41.14	53.74	71.33	86.53	93.02	700
41.23	53.91	71.50	86.64	93.02	600
41.31	54.07	71.68	86.76	93.03	500
41.39	54.23	71.86	86.88	93.03	400
41.48	54.40	72.03	86.99	93.03	300
41.56	54.56	72.21	87.11	93.04	200
41.65	54.73	72.38	87.22	93.04	100
41.74	54.89	72.56	87.33	93.04	0
90000	80000	70000	60000	50000	Days.

TABLE LXXXI*A*.

Argument $y-u$.

Equation $= -6''.38 \sin(y-u) - 0''.97 \cos(y-u)$

$y-u$		0°+ 180°−	10°+ 190°−	20°+ 200°−	30°+ 210°−	40°+ 220°−	50°+ 230°−	60°+ 240°−	70°+ 250°−	80°+ 260°−
0°	0″	−0″.97	−2″.07	−3″.09	−4″.03	−4″.84	−5″.51	−6″.01	−6″.32	−6″.46
0	1000	1.00	2.10	3.12	4.05	4.86	5.53	6.02	6.33	6.46
0	2000	1.03	2.13	3.14	4.08	4.88	5.54	6.03	6.33	6.46
0	3000	1.06	2.16	3.17	4.10	4.90	5.56	6.04	6.34	6.46
1	400	1.09	2.19	3.20	4.12	4.92	5.57	6.05	6.34	6.46
1	1400	1.12	2.22	3.22	4.15	4.94	5.59	6.06	6.35	6.47
1	2400	1.15	2.24	3.25	4.17	4.96	5.60	6.07	6.35	6.47
1	3400	1.19	2.27	3.28	4.19	4.98	5.62	6.08	6.36	6.47
2	800	1.22	2.30	3.30	4.22	5.00	5.63	6.09	6.36	6.47
2	1800	1.25	2.33	3.33	4.24	5.02	5.65	6.10	6.37	6.47
2	2800	1.28	2.36	3.36	4.26	5.04	5.66	6.11	6.37	6.47
3	200	1.31	2.39	3.38	4.29	5.06	5.68	6.12	6.38	6.47
3	1200	1.34	2.42	3.41	4.31	5.08	5.69	6.13	6.38	6.47
3	2200	1.38	2.44	3.44	4.33	5.10	5.71	6.14	6.39	6.46
3	3200	1.41	2.47	3.46	4.36	5.12	5.72	6.15	6.39	6.46
4	600	1.44	2.50	3.49	4.38	5.14	5.74	6.16	6.40	6.46
4	1600	1.47	2.53	3.52	4.40	5.16	5.75	6.17	6.40	6.46
4	2600	1.50	2.56	3.54	4.43	5.18	5.77	6.18	6.41	6.46
5	0	1.53	2.59	3.57	4.45	5.20	5.78	6.19	6.41	6.46
5	1000	1.56	2.62	3.60	4.47	5.22	5.79	6.20	6.41	6.46
5	2000	1.59	2.65	3.62	4.49	5.24	5.81	6.20	6.42	6.45
5	3000	1.62	2.67	3.65	4.52	5.25	5.82	6.21	6.42	6.45
6	400	1.65	2.70	3.67	4.54	5.27	5.84	6.22	6.42	6.45
6	1400	1.68	2.73	3.70	4.56	5.29	5.85	6.22	6.43	6.44
6	2400	1.71	2.76	3.72	4.58	5.30	5.86	6.23	6.43	6.44
6	3400	1.74	2.78	3.75	4.60	5.32	5.88	6.24	6.43	6.43
7	800	1.77	2.81	3.78	4.63	5.34	5.89	6.24	6.43	6.43
7	1800	1.80	2.84	3.80	4.65	5.36	5.90	6.25	6.44	6.42
7	2800	1.83	2.87	3.83	4.67	5.37	5.92	6.26	6.44	6.42
8	200	1.86	2.89	3.85	4.69	5.39	5.93	6.27	6.44	6.41
8	1200	1.89	2.92	3.88	4.71	5.41	5.95	6.27	6.44	6.41
8	2200	1.92	2.95	3.90	4.74	5.42	5.96	6.28	6.44	6.40
8	3200	1.95	2.98	3.93	4.76	5.44	5.97	6.29	6.45	6.40
9	600	1.98	3.00	3.95	4.78	5.46	5.98	6.30	6.45	6.39
9	1600	2.01	3.03	3.98	4.80	5.47	5.99	6.30	6.45	6.39
9	2600	2.04	3.06	4.00	4.82	5.49	6.00	6.31	6.46	6.38
10	0	−2.07	−3.09	−4.03	−4.84	−5.51	−6.01	−6.32	−6.46	−6.38

NOTE. — When $y-u$ exceeds 180° the Sign of the Equation is to be reversed.

TABLE LXXXIA.

Argument $y-u$.

Equation $= -6''.38 \sin(y-u) - 0''.97 \cos(y-u)$.

$y-u$	90°+ 270°−	100°+ 280°−	110°+ 290°−	120°+ 300°−	130°+ 310°−	140°+ 320°−	150°+ 330°−	160°+ 340°−	170°+ 350°−
° ″	″	″	″	″	″	″	″	″	″
0 0	−6.38	−6.12	−5.66	−5.04	−4.26	−3.36	−2.35	−1.27	−0.15
0 1000	6.38	6.11	5.64	5.02	4.24	3.33	2.32	1.24	0.12
0 2000	6.37	6.10	5.63	5.00	4.21	3.30	2.29	1.21	0.09
0 3000	6.37	6.08	5.61	4.98	4.19	3.28	2.26	1.18	0.06
1 400	6.36	6.07	5.59	4.96	4.16	3.25	2.23	1.15	−0.02
1 1400	6.36	6.06	5.58	4.94	4.14	3.22	2.20	1.11	+0.01
1 2400	6.35	6.05	5.56	4.92	4.11	3.20	2.17	1.08	0.04
1 3400	6.35	6.04	5.55	4.90	4.09	3.17	2.14	1.05	0.07
2 800	6.34	6.02	5.53	4.88	4.07	3.14	2.11	1.02	0.10
2 1800	6.34	6.01	5.51	4.85	4.04	3.11	2.08	0.99	0.13
2 2800	6.33	6.00	5.50	4.83	4.02	3.09	2.06	0.96	0.16
3 200	6.33	5.99	5.48	4.81	3.99	3.06	2.03	0.93	0.20
3 1200	6.32	5.98	5.47	4.79	3.97	3.03	2.00	0.90	0.23
3 2200	6.31	5.97	5.45	4.77	3.94	3.01	1.97	0.86	0.26
3 3200	6.31	5.95	5.43	4.75	3.92	2.98	1.94	0.83	0.29
4 600	6.30	5.94	5.42	4.73	3.89	2.95	1.91	0.80	0.32
4 1600	6.29	5.93	5.40	4.71	3.87	2.93	1.88	0.77	0.35
4 2600	6.29	5.92	5.39	4.69	3.84	2.90	1.85	0.74	0.38
5 0	6.28	5.91	5.37	4.67	3.82	2.87	1.82	0.71	0.41
5 1000	6.27	5.90	5.35	4.65	3.79	2.84	1.79	0.68	0.44
5 2000	6.26	5.88	5.33	4.63	3.77	2.81	1.76	0.65	0.47
5 3000	6.26	5.87	5.32	4.60	3.74	2.78	1.73	0.62	0.50
6 400	6.25	5.86	5.30	4.58	3.71	2.76	1.70	0.59	0.53
6 1400	6.24	5.84	5.28	4.56	3.69	2.73	1.67	0.55	0.56
6 2400	6.23	5.83	5.26	4.54	3.66	2.70	1.64	0.52	0.60
6 3400	6.22	5.81	5.24	4.51	3.64	2.67	1.61	0.49	0.63
7 800	6.21	5.80	5.23	4.49	3.61	2.64	1.58	0.46	0.66
7 1800	6.20	5.78	5.21	4.47	3.58	2.61	1.54	0.43	0.69
7 2800	6.20	5.77	5.19	4.45	3.56	2.58	1.51	0.40	0.72
8 200	6.19	5.75	5.17	4.42	3.53	2.56	1.48	0.37	0.75
8 1200	6.18	5.74	5.15	4.40	3.50	2.53	1.45	0.33	0.79
8 2200	6.17	5.73	5.14	4.38	3.48	2.50	1.42	0.30	0.82
8 3200	6.16	5.71	5.12	4.35	3.45	2.47	1.39	0.27	0.85
9 600	6.15	5.70	5.10	4.33	3.43	2.44	1.36	0.24	0.88
9 1600	6.14	5.68	5.08	4.31	3.40	2.41	1.33	0.21	0.91
9 2600	6.13	5.67	5.06	4.28	3.38	2.38	1.30	0.18	0.94
10 0	−6.12	−5.66	−5.04	−4.26	−3.36	−2.35	−1.27	−0.15	+0.97

Note. — When $y-u$ exceeds 180° the Sign of the Equation is to be reversed.

TABLE LXXXII. ARGUMENT 77.

Equation = — 416″.9 sin. 2g.

g	270° +	271° +	272° +	273° +	274° +	275° +	276° +	277° +	278° +	g
	180 —	181 —	182 —	183 —	184 —	185 —	186 —	187 —	188 —	
	90 +	91 +	92 +	93 +	94 +	95 +	96 +	97 +	98 +	
	0 —	1 —	2 —	3 —	4 —	5 —	6 —	7 —	8 —	
0″	0″.00	14″.55	29″.08	43″.58	58″.02	72″.39	86″.68	100″.86	114″.91	3600″
100	0.40	14.95	29.48	43.98	58.42	72.79	87.07	101.25	115.30	3500
200	0.81	15.36	29.89	44.38	58.82	73.19	87.47	101.64	115.69	3400
300	1.21	15.76	30.29	44.78	59.22	73.59	87.86	102.03	116.08	3300
400	1.62	16.16	30.69	45.18	59.62	73.99	88.26	102.42	116.47	3200
500	2.02	16.57	31.10	45.59	60.02	74.38	88.65	102.82	116.85	3100
600	2.43	16.97	31.50	45.99	60.42	74.78	89.05	103.21	117.24	3000
700	2.83	17.37	31.90	46.39	60.82	75.18	89.44	103.60	117.63	2900
800	3.24	17.78	32.31	46.79	61.22	75.57	89.84	103.99	118.01	2800
900	3.64	18.18	32.71	47.19	61.62	75.97	90.23	104.38	118.40	2700
1000	4.04	18.59	33.11	47.59	62.02	76.37	90.63	104.77	118.79	2600
1100	4.45	18.99	33.52	48.00	62.42	76.77	91.02	105.17	119.18	2500
1200	4.85	19.40	33.92	48.40	62.82	77.17	91.42	105.56	119.57	2400
1300	5.25	19.80	34.32	48.80	63.22	77.57	91.81	105.95	119.96	2300
1400	5.66	20.21	34.73	49.20	63.62	77.96	92.21	106.34	120.34	2200
1500	6.06	20.61	35.13	49.60	64.02	78.36	92.60	106.73	120.73	2100
1600	6.46	21.01	35.53	50.00	64.42	78.76	92.99	107.12	121.12	2000
1700	6.87	21.42	35.94	50.41	64.82	79.15	93.39	107.51	121.50	1900
1800	7.28	21.82	36.34	50.81	65.22	79.55	93.78	107.90	121.89	1800
1900	7.68	22.22	36.74	51.21	65.62	79.95	94.17	108.29	122.28	1700
2000	8.09	22.63	37.15	51.61	66.01	80.34	94.57	108.68	122.66	1600
2100	8.49	23.03	37.55	52.01	66.41	80.74	94.96	109.07	123.05	1500
2200	8.89	23.43	37.95	52.41	66.81	81.14	95.35	109.46	123.43	1400
2300	9.30	23.84	38.35	52.81	67.21	81.53	95.75	109.85	123.82	1300
2400	9.70	24.24	38.75	53.21	67.61	81.93	96.14	110.24	124.20	1200
2500	10.10	24.64	39.15	53.61	68.01	82.33	96.53	110.63	124.59	1100
2600	10.51	25.05	39.56	54.02	68.41	82.72	96.93	111.02	124.97	1000
2700	10.91	25.45	39.96	54.42	68.81	83.12	97.32	111.41	125.36	900
2800	11.32	25.85	40.36	54.82	69.21	83.51	97.71	111.80	125.75	800
2900	11.72	26.26	40.77	55.22	69.60	83.91	98.11	112.19	126.13	700
3000	12.13	26.66	41.17	55.62	70.00	84.30	98.50	112.58	126.52	600
3100	12.53	27.06	41.57	56.02	70.40	84.70	98.89	112.97	126.90	500
3200	12.94	27.47	41.98	56.42	70.80	85.09	99.29	113.36	127.29	400
3300	13.34	27.87	42.38	56.82	71.20	85.49	99.68	113.75	127.67	300
3400	13.74	28.27	42.78	57.22	71.60	85.89	100.07	114.14	128.06	200
3500	14.15	28.68	43.18	57.62	71.99	86.28	100.47	114.52	128.44	100
3600	14.55	29.08	43.58	58.02	72.39	86.68	100.86	114.91	128.83	0
g	89° —	88° —	87° —	86° —	85° —	84° —	83° —	82° —	81° —	g
	179 +	178 +	177 +	176 +	175 +	174 +	173 +	172 +	171 +	
	269 —	268 —	267 —	266 —	265 —	264 —	263 —	262 —	261 —	
	359 +	358 +	357 +	356 +	355 +	354 +	353 +	352 +	351 +	

TABLE LXXXII. ARGUMENT 77.

Equation = — 416″.9 sin. 2*g*.

g	279° +	280° +	281° +	282° +	283° +	284° +	285° +	286° +	287° +	*g*
	189 —	190 —	191 —	192 —	193 —	194 —	195 —	196 —	197 —	
	99 +	100 +	101 +	102 +	103 +	104 +	105 +	106 +	107 +	
	9 —	10 —	11 —	12 —	13 —	14 —	15 —	16 —	17 —	
0″	128″.83	142″.59	156″.17	169″.57	182″.76	195″.72	208″.45	220″.92	233″.13	3600″
100	129.21	142.97	156.55	169.94	183.12	196.08	208.80	221.26	233.46	3500
200	129.60	143.35	156.92	170.31	183.48	196.43	209.15	221.61	233.80	3400
300	129.98	143.73	157.30	170.68	183.84	196.79	209.50	221.95	234.13	3300
400	130.36	144.11	157.67	171.05	184.20	197.15	209.85	222.29	234.46	3200
500	130.75	144.48	158.05	171.42	184.57	197.50	210.20	222.64	234.80	3100
600	131.13	144.86	158.42	171.79	184.93	197.86	210.55	222.98	235.13	3000
700	131.51	145.24	158.79	172.16	185.29	198.22	210.90	223.32	235.46	2900
800	131.90	145.62	159.17	172.52	185.66	198.57	211.24	223.66	235.80	2800
900	132.28	146.00	159.54	172.89	186.02	198.93	211.59	224.00	236.13	2700
1000	132.66	146.38	159.91	173.26	186.38	199.28	211.94	224.34	236.46	2600
1100	133.05	146.76	160.29	173.62	186.75	199.64	212.29	224.68	236.80	2500
1200	133.43	147.14	160.66	173.99	187.11	199.99	212.64	225.02	237.13	2400
1300	133.81	147.52	161.03	174.36	187.47	200.34	212.99	225.36	237.46	2300
1400	134.20	147.89	161.41	174.72	187.83	200.70	213.33	225.70	237.80	2200
1500	134.58	148.27	161.78	175.09	188.19	201.05	213.68	226.04	238.13	2100
1600	134.96	148.65	162.15	175.46	188.55	201.41	214.03	226.38	238.46	2000
1700	135.35	149.02	162.53	175.82	188.91	201.76	214.37	226.72	238.79	1900
1800	135.73	149.40	162.90	176.19	189.27	202.12	214.72	227.06	239.12	1800
1900	136.11	149.78	163.27	176.56	189.63	202.47	215.06	227.40	239.45	1700
2000	136.50	150.15	163.64	176.92	189.99	202.83	215.41	227.73	239.79	1600
2100	136.88	150.53	164.01	177.29	190.35	203.18	215.75	228.07	240.12	1500
2200	137.26	150.91	164.38	177.66	190.71	203.53	216.10	228.41	240.45	1400
2300	137.64	151.28	164.75	178.02	191.07	203.89	216.44	228.74	240.78	1300
2400	138.02	151.66	165.12	178.39	191.43	204.24	216.79	229.08	241.11	1200
2500	138.40	152.04	165.49	178.75	191.79	204.59	217.14	229.42	241.44	1100
2600	138.78	152.41	165.87	179.12	192.15	204.94	217.48	229.76	241.77	1000
2700	139.16	152.79	166.24	179.48	192.51	205.29	217.83	230.10	242.10	900
2800	139.54	153.17	166.61	179.84	192.87	205.64	218.17	230.44	242.43	800
2900	139.93	153.54	166.98	180.21	193.22	206.00	218.52	230.77	242.75	700
3000	140.31	153.92	167.35	180.57	193.58	206.35	218.86	231.11	243.08	600
3100	140.69	154.30	167.72	180.94	193.94	206.70	219.20	231.45	243.41	500
3200	141.07	154.67	168.09	181.30	194.29	207.05	219.55	231.78	243.73	400
3300	141.45	155.05	168.46	181.67	194.65	207.40	219.89	232.12	244.06	300
3400	141.83	155.42	168.83	182.03	195.01	207.75	220.23	232.46	244.39	200
3500	142.21	155.80	169.20	182.40	195.36	208.10	220.58	232.79	244.71	100
3600	142.59	156.17	169.57	182.76	195.72	208.45	220.92	233.13	245.04	0
g	80° —	79° —	78° —	77° —	76° —	75° —	74° —	73° —	72° —	*g*
	170 +	169 +	168 +	167 +	166 +	165 +	164 +	163 +	162 +	
	260 —	259 —	258 —	257 —	256 —	255 —	254 —	253 —	252 —	
	350 +	349 +	348 +	347 +	346 +	345 +	344 +	343 +	342 +	

TABLE LXXXII. ARGUMENT 77.

Equation = — 416″.9 sin. 2ŷ.

ŷ	288° + / 198 − / 108 + / 18 −	289° + / 199 − / 109 + / 19 −	290° + / 200 − / 110 + / 20 −	291° + / 201 − / 111 + / 21 −	292° + / 202 − / 112 + / 22 −	293° + / 203 − / 113 + / 23 −	294° + / 204 − / 114 + / 24 −	295° + / 205 − / 115 + / 25 −	296° + / 206 − / 116 + / 26 −	ŷ
0″	245.04	256.67	267.98	278.96	289.60	299.88	309.82	319.36	328.52	3600″
100	245.37	256.99	268.29	279.26	289.89	300.16	310.09	319.62	328.77	3500
200	245.69	257.30	268.59	279.56	290.18	300.45	310.36	319.88	329.02	3400
300	246.02	257.62	268.90	279.86	290.47	300.73	310.63	320.14	329.27	3300
400	246.35	257.91	269.21	280.16	290.76	301.01	310.90	320.40	329.52	3200
500	246.67	258.25	269.52	280.46	291.05	301.29	311.17	320.65	329.76	3100
600	247.00	258.57	269.83	280.76	291.34	301.57	311.44	320.91	330.01	3000
700	247.33	258.88	270.14	281.06	291.63	301.85	311.71	321.17	330.26	2900
800	247.65	259.20	270.44	281.35	291.91	302.13	311.97	321.42	330.50	2800
900	247.98	259.52	270.75	281.65	292.20	302.41	312.24	321.68	330.75	2700
1000	248.31	259.81	271.06	281.95	292.49	302.69	312.51	321.94	330.99	2600
1100	248.63	260.15	271.37	282.24	292.77	302.96	312.77	322.19	331.24	2500
1200	248.96	260.47	271.68	282.54	293.06	303.24	313.04	322.45	331.48	2400
1300	249.28	260.79	271.99	282.84	293.35	303.52	313.31	322.71	331.73	2300
1400	249.61	261.10	272.29	283.13	293.64	303.79	313.57	322.96	331.97	2200
1500	249.93	261.42	272.60	283.43	293.93	304.07	313.84	323.22	332.22	2100
1600	250.25	261.73	272.90	283.73	294.22	304.35	314.11	323.47	332.46	2000
1700	250.58	262.05	273.21	284.02	294.50	304.62	314.37	323.73	332.71	1900
1800	250.90	262.36	273.51	284.32	294.79	304.90	314.64	323.98	332.95	1800
1900	251.22	262.67	273.81	284.61	295.08	305.17	314.90	324.24	333.19	1700
2000	251.54	262.99	274.12	284.91	295.36	305.45	315.17	324.49	333.44	1600
2100	251.86	263.30	274.42	285.20	295.65	305.72	315.43	324.75	333.68	1500
2200	252.18	263.61	274.72	285.50	295.93	306.00	315.69	325.00	333.92	1400
2300	252.51	263.93	275.03	285.79	296.22	306.27	315.96	325.26	334.16	1300
2400	252.83	264.24	275.33	286.09	296.50	306.55	316.22	325.51	334.40	1200
2500	253.15	264.55	275.63	286.38	296.78	306.82	316.48	325.76	334.64	1100
2600	253.47	264.87	275.94	286.68	297.07	307.10	316.75	326.01	334.89	1000
2700	253.79	265.18	276.24	286.97	297.35	307.37	317.01	326.26	335.13	900
2800	254.11	265.49	276.54	287.26	297.63	307.64	317.27	326.51	335.37	800
2900	254.43	265.81	276.85	287.56	297.92	307.92	317.54	326.77	335.61	700
3000	254.75	266.12	277.15	287.85	298.20	308.19	317.80	327.02	335.85	600
3100	255.07	266.43	277.45	288.14	298.48	308.46	318.06	327.27	336.09	500
3200	255.39	266.74	277.76	288.44	298.76	308.73	318.32	327.52	336.33	400
3300	255.71	267.05	278.06	288.73	299.04	309.00	318.58	327.77	336.57	300
3400	256.03	267.36	278.36	289.02	299.32	309.27	318.84	328.02	336.81	200
3500	256.35	267.67	278.66	289.31	299.60	309.55	319.10	328.27	337.04	100
3600	256.67	267.98	278.96	289.60	299.88	309.82	319.36	328.52	337.28	0
ŷ	71° − / 161 + / 251 − / 341 +	70° − / 160 + / 250 − / 340 +	69° − / 159 + / 249 − / 339 +	68° − / 158 + / 248 − / 338 +	67° − / 157 + / 247 − / 337 +	66° − / 156 + / 246 − / 336 +	65° − / 155 + / 245 − / 335 +	64° − / 154 + / 244 − / 334 +	63° − / 153 + / 243 − / 333 +	ŷ

TABLE LXXXII. ARGUMENT 77.

Equation = — 416″.9 sin. 2ϑ.

ϑ	297° +	298° +	299° +	300° +	301° +	302° +	303° +	304° +	305° +	ϑ
	207 −	208 −	209 −	210 −	211 −	212 −	213 −	214 −	215 −	
	117 +	118 +	119 +	120 +	121 +	122 +	123 +	124 +	125 +	
	27 −	28 −	29 −	30 −	31 −	32 −	33 −	34 −	35 −	
0″	337.28″	345.62″	353.55″	361.05″	368.09″	374.71″	380.85″	386.54″	391.76″	3600″
100	337.51	345.85	353.76	361.25	368.28	374.89	381.02	386.69	391.90	3500
200	337.75	346.07	353.98	361.45	368.47	375.06	381.18	386.84	392.03	3400
300	337.98	346.30	354.19	361.65	368.66	375.24	381.35	386.99	392.17	3300
400	338.22	346.53	354.40	361.85	368.85	375.42	381.51	387.14	392.31	3200
500	338.45	346.75	354.62	362.05	369.04	375.59	381.68	387.29	392.44	3100
600	338.69	346.98	354.83	362.25	369.23	375.77	381.84	387.44	392.58	3000
700	338.93	347.20	355.04	362.45	369.42	375.94	382.00	387.59	392.72	2900
800	339.16	347.43	355.26	362.65	369.60	376.12	382.16	387.74	392.85	2800
900	339.40	347.65	355.47	362.85	369.79	376.29	382.32	387.89	392.99	2700
1000	339.63	347.87	355.68	363.05	369.98	376.46	382.48	388.04	393.12	2600
1100	339.87	348.09	355.89	363.25	370.16	376.64	382.64	388.18	393.26	2500
1200	340.10	348.31	356.10	363.45	370.35	376.81	382.80	388.33	393.39	2400
1300	340.33	348.53	356.31	363.65	370.53	376.99	382.96	388.48	393.52	2300
1400	340.57	348.76	356.51	363.84	370.72	377.16	383.12	388.62	393.65	2200
1500	340.80	348.98	356.72	364.04	370.90	377.33	383.28	388.77	393.78	2100
1600	341.03	349.20	356.93	364.24	371.09	377.50	383.44	388.92	393.91	2000
1700	341.27	349.42	357.14	364.43	371.27	377.67	383.60	389.06	394.05	1900
1800	341.50	349.64	357.35	364.63	371.46	377.84	383.76	389.21	394.18	1800
1900	341.73	349.86	357.56	364.83	371.64	378.01	383.92	389.35	394.31	1700
2000	341.97	350.08	357.76	365.02	371.83	378.18	384.07	389.50	394.44	1600
2100	342.20	350.30	357.97	365.22	372.01	378.35	384.23	389.64	394.57	1500
2200	342.43	350.52	358.18	365.41	372.19	378.52	384.39	389.78	394.70	1400
2300	342.66	350.73	358.38	365.61	372.37	378.68	384.54	389.93	394.83	1300
2400	342.89	350.95	358.59	365.80	372.55	378.85	384.70	390.07	394.96	1200
2500	343.12	351.17	358.80	365.99	372.73	379.02	384.85	390.21	395.09	1100
2600	343.35	351.39	359.00	366.19	372.91	379.18	385.01	390.36	395.22	1000
2700	343.58	351.61	359.21	366.38	373.09	379.35	385.16	390.50	395.35	900
2800	343.81	351.83	359.42	366.57	373.27	379.52	385.32	390.64	395.48	800
2900	344.03	352.04	359.62	366.76	373.45	379.69	385.47	390.78	395.61	700
3000	344.26	352.26	359.83	366.95	373.63	379.86	385.63	390.92	395.74	600
3100	344.49	352.48	360.03	367.14	373.81	380.03	385.78	391.06	395.87	500
3200	344.71	352.69	360.24	367.33	373.99	380.19	385.94	391.20	395.99	400
3300	344.94	352.91	360.44	367.52	374.17	380.36	386.09	391.34	396.12	300
3400	345.17	353.12	360.64	367.71	374.35	380.52	386.24	391.48	396.25	200
3500	345.39	353.34	360.85	367.90	374.53	380.69	386.39	391.62	396.37	100
3600	345.62	353.55	361.05	368.09	374.71	380.85	386.54	391.76	396.50	0
ϑ	62° −	61° −	60° −	59° −	58° −	57° −	56° −	55° −	54° −	ϑ
	152 +	151 +	150 +	149 +	148 +	147 +	146 +	145 +	144 +	
	242 −	241 −	240 −	239 −	238 −	237 −	236 −	235 −	234 −	
	332 +	331 +	330 +	329 +	328 +	327 +	326 +	325 +	324 +	

TABLE LXXXII. ARGUMENT 77.

Equation = — 416″.9 sin. 2g.

g	306° + 216 — 126 + 36 —	307° + 217 — 127 + 37 —	308° + 218 — 128 + 38 —	309° + 219 — 129 + 39 —	310° + 220 — 130 + 40 —	311° + 221 — 131 + 41 —	312° + 222 — 132 + 42 —	313° + 223 — 133 + 43 —	314° + 224 — 134 + 44 —	g
0″	396.″50	400.″75	404.″51	407.″78	410.″56	412.″84	414.″61	415.″89	416.″65	3600″
100	396.62	400.86	401.61	407.86	410.63	412.90	414.65	415.92	416.66	3500
200	396.75	400.97	404.71	407.95	410.70	412.95	414.70	415.94	416.68	3400
300	396.87	401.08	404.81	408.03	410.77	413.01	414.74	415.97	416.69	3300
400	396.99	401.19	404.91	408.11	410.81	413.06	414.78	416.00	416.70	3200
500	397.12	401.30	405.00	408.20	410.91	413.12	414.82	416.02	416.72	3100
600	397.24	401.41	405.10	408.28	410.98	413.17	414.86	416.05	416.73	3000
700	397.36	401.52	405.19	408.36	411.05	413.22	414.90	416.07	416.74	2900
800	397.48	401.63	405.29	408.15	411.11	413.28	414.94	416.10	416.75	2800
900	397.60	401.74	405.38	408.53	411.18	413.33	414.98	416.12	416.76	2700
1000	397.72	401.85	405.47	408.61	411.25	413.38	415.02	416.14	416.77	2600
1100	397.85	401.95	405.57	408.69	411.31	413.43	415.05	416.17	416.78	2500
1200	397.97	402.06	405.66	408.77	411.38	413.48	415.09	416.19	416.79	2400
1300	398.09	402.16	405.75	408.85	411.41	413 54	415.13	416.21	416.80	2300
1400	398.21	402.27	405.85	408.93	411.51	413.59	415.16	416.21	416.80	2200
1500	398.33	402.37	405.94	409.01	411.57	413.61	415.20	416.26	416.81	2100
1600	398.45	402.48	406.03	409.09	411.63	413.69	415.21	416.28	416.82	2000
1700	398.56	402.58	406.12	409.16	411.70	413.71	415.27	416.30	416.82	1900
1800	398.68	402.69	406.21	409.24	411.76	413.79	415.31	416.32	416.83	1800
1900	398.80	402.79	406.30	409.32	411.82	413.84	415.34	416.34	416.84	1700
2000	398.91	402.90	406.39	409.39	411.89	413.88	415.38	416.37	416.84	1600
2100	399.03	403.00	406.48	409.47	411.95	413.93	415.41	416.39	416.85	1500
2200	399.15	403.10	406.57	409.51	412.01	413.98	415.44	416.41	416.86	1400
2300	399.26	403.21	406.66	409.62	412.08	411.02	415.48	416.43	416.86	1300
2400	399.38	403.31	406.75	409.69	412.14	414.07	415.51	416.45	416.87	1200
2500	399.50	403.41	406.84	409.77	412.20	414.12	415.54	416.47	416.87	1100
2600	399.61	403.52	406.93	409.84	412.26	414.16	415.58	416.48	416.88	1000
2700	399.73	403.62	407.02	409.92	412.32	414.21	415.61	416.50	416.88	900
2800	399.84	403.72	407.11	409.99	412.38	414.26	415.64	416.52	416.88	800
2900	399.96	403.82	407.19	410.07	412.44	414.30	415.68	416.53	416.89	700
3000	400.07	403.92	407.28	410.14	412.50	414.35	415.71	416.55	416.89	600
3100	400.18	401.02	407.36	410.21	412.56	414.39	415.74	416.57	416.89	500
3200	400.30	404.12	407.45	410.28	412.61	414.44	415.77	416.58	416.90	400
3300	400.41	404 22	407.53	410.35	412.67	414.48	415.80	416.60	416.90	300
3400	400.52	404.32	407.61	410.42	412.73	414.52	415.83	416.62	416.90	200
3500	400.64	404.41	407.70	410.49	412.78	414.57	415.86	416.63	416.90	100
3600	400.75	404.51	407.78	410.56	412.84	414.61	415.89	416.65	416.90	0
g	53° — 143 + 233 — 323 +	52° — 142 + 232 — 322 +	51° — 141 + 231 — 321 +	50° — 140 + 230 — 320 +	49° — 139 + 229 — 319 +	48° — 138 + 228 — 318 +	47° — 137 + 227 — 317 +	46° — 136 + 226 — 316 +	45° — 135 + 225 — 315 +	g

TABLES VI''. - XXI.IV.

ARGUMENTS AND EQUATIONS.

TABLE.	ARGUMENT.	EQUATION.
VI''.	1	$76''.0 - 73''.563 \sin. x - 9''.993 \sin. 2x - 1''.071 \sin. 3x - 0''.104 \sin. 4x - 0''.008 \sin. 5x.$
VII''.	2	$11.2 - 11.18 \sin. (2t - x) - 0''.304 \sin. (4t - 2x).$
VIII''.	3	$31.16 + 0.345 \sin. t - 20''.808 \sin. 2t - 0''.023 \sin. 3t - 0''.65 \sin. 4t.$
XI''.	6	$0.42 - 0.420 \sin. (2t - z - x).$
XII''.	7	$5.12 - 5.117 \sin. (2t + x).$
XIII''.	8	$1.73 - 1.727 \sin. (2t - z).$
XIV''.	9	$0.41 - 0.411 \sin. (x - z).$
XV''.	10	$0.41 + 0.413 \sin. (x + z).$
XVI''.	11	$0.29 + 0.288 \sin. (2y - x).$
XVIII''.	13	$0.92 - 0.920 \sin. (4t - x).$
XIX''.	14	$0.08 + 0.083 \sin. (2t + z - x).$
XX''.	15	$0.31 + 0.306 \sin. (2t + z).$
XXI''.	16	$0.80 - 0.100 \sin. (2t + 2x).$
XXII''.	17	$0.06 - 0.057 \sin. (t + z).$
XXIII''.	18	$0.35 - 0.354 \sin. (2t + x - z).$
XXIV''.	19	$0.05 + 0.054 \sin. (3x - 2t).$
XXV''.	20	$0.11 - 0.115 \sin. (2x - z).$
XXVII''.	22	$0.07 - 0.074 \sin. (2t - 2z).$
XXIX''.	24	$0.10 + 0.102 \sin. (2x + z).$
XXX''.	25	$0.08 + 0.081 \sin. (2t + z + x).$
XXXI''.	26	$0.14 - 0.137 \sin. (4t + x).$
XXXV''.	30	$0.19 - 0.087 \sin. (4t - x - z).$
VI.IV.	1	$0.45 + 0.239 \sin. x + 0''.130 \sin. 2x + 0''.031 \sin. 3x + 0''.005 \sin. 4x.$
VII.IV.	2	$0.03 + 0.027 \sin. (2t - x) + 0''.003 \sin. (4t - 2x).$
VIII.IV.	3	$0.31 + 0.303 \sin. 2t + 0''.006 \sin. 3t + 0''.029 \sin. 4t.$
XII.IV.	7	$0.14 + 0.136 \sin. (2t + x).$
XIII.IV.	8	$0.02 + 0.018 \sin. (2t - x).$
XVIII.IV.	13	$0.02 + 0.022 \sin. (4t - x).$
XXI.IV.	16	$0.03 + 0.033 \sin. (2t + 2x).$

TABLE VI'. ARGUMENT 1.

Equation $= 76''. - 73''.563 \sin. x - 9''.993 \sin. 2x - 1''.071 \sin. 3x - 0''.104 \sin. 4x - 0''.008 \sin. 5x.$

Period, 27.55455246 days.

Days.	0	1	2	3	4	5	6	7	8	9
Days	"	"	"	"	"	"	"	"	"	"
.00	148.41	151.14	149.71	143.45	132.05	115.85	95.92	74.04	52.34	32.97
.01	148.45	151.15	149.67	143.36	131.91	115.67	95.71	73.82	52.13	32.80
.02	148.50	151.16	149.63	143.27	131.77	115.48	95.50	73.60	51.92	32.62
.03	148.54	151.16	149.59	143.18	131.63	115.30	95.28	73.38	51.72	32.45
.04	148.58	151.17	149.55	143.09	131.49	115.11	95.07	73.15	51.51	32.27
.05	148.63	151.17	149.51	143.00	131.34	114.93	94.86	72.93	51.30	32.10
.06	148.67	151.18	149.47	142.91	131.20	114.74	94.64	72.71	51.10	31.93
.07	148.72	151.18	149.43	142.82	131.06	114.56	94.43	72.48	50.89	31.75
.08	148.76	151.18	149.39	142.72	130.91	114.37	94.22	72.26	50.69	31.58
.09	148.80	151.19	149.35	142.63	130.77	114.19	94.01	72.04	50.48	31.40
.10	148.84	151.19	149.31	142.54	130.63	114.00	93.79	71.82	50.27	31.23
.11	148.88	151.19	149.27	142.44	130.48	113.81	93.58	71.60	50.06	31.06
.12	148.92	151.19	149.23	142.35	130.34	113.62	93.36	71.38	49.86	30.89
.13	148.97	151.20	149.18	142.26	130.19	113.44	93.15	71.16	49.65	30.72
.14	149.01	151.20	149.14	142.17	130.05	113.25	92.93	70.94	49.45	30.55
.15	149.05	151.20	149.10	142.07	129.90	113.06	92.72	70.71	49.24	30.38
.16	149.09	151.20	149.05	141.97	129.76	112.87	92.50	70.49	49.04	30.21
.17	149.13	151.21	149.01	141.87	129.61	112.68	92.29	70.27	48.83	30.04
.18	149.17	151.21	148.96	141.78	129.47	112.49	92.07	70.05	48.63	29.87
.19	149.21	151.21	148.92	141.68	129.32	112.30	91.86	69.83	48.42	29.70
.20	149.25	151.21	148.87	141.58	129.17	112.11	91.64	69.61	48.22	29.53
.21	149.29	151.21	148.82	141.48	129.02	111.92	91.42	69.39	48.02	29.36
.22	149.33	151.21	148.78	141.38	128.87	111.73	91.21	69.17	47.82	29.20
.23	149.37	151.20	148.73	141.28	128.72	111.53	90.99	68.95	47.62	29.03
.24	149.40	151.20	148.68	141.18	128.57	111.34	90.78	68.73	47.41	28.87
.25	149.44	151.20	148.63	141.08	128.42	111.15	90.56	68.50	47.21	28.70
.26	149.47	151.19	148.58	140.98	128.27	110.96	90.34	68.28	47.01	28.53
.27	149.51	151.19	148.53	140.88	128.11	110.77	90.12	68.06	46.80	28.37
.28	149.54	151.19	148.47	140.77	127.96	110.57	89.91	67.84	46.60	28.20
.29	149.58	151.18	148.42	140.67	127.81	110.38	89.69	67.62	46.40	28.04
.30	149.62	151.18	148.37	140.57	127.66	110.19	89.47	67.40	46.20	27.87
.31	149.65	151.17	148.31	140.47	127.51	109.99	89.25	67.18	46.00	27.71
.32	149.68	151.17	148.26	140.36	127.35	109.80	89.03	66.96	45.80	27.55
.33	149.72	151.16	148.20	140.25	127.20	109.61	88.82	66.74	45.60	27.38
.34	149.75	151.16	148.15	140.14	127.04	109.42	88.60	66.52	45.40	27.22
.35	149.79	151.15	148.09	140.04	126.89	109.22	88.38	66.30	45.20	27.06
.36	149.82	151.14	148.04	139.93	126.73	109.02	88.16	66.09	45.01	26.90
.37	149.86	151.14	147.98	139.82	126.58	108.83	87.94	65.87	44.81	26.74
.38	149.89	151.13	147.93	139.72	126.42	108.63	87.73	65.65	44.61	26.58
.39	149.92	151.12	147.87	139.61	126.26	108.44	87.51	65.43	44.41	26.42
.40	149.95	151.11	147.82	139.50	126.10	108.24	87.29	65.21	44.21	26.26
.41	149.98	151.10	147.76	139.39	125.94	108.04	87.07	64.99	44.01	26.10
.42	150.01	151.09	147.71	139.28	125.78	107.84	86.85	64.77	43.82	25.94
.43	150.04	151.08	147.65	139.17	125.62	107.65	86.63	64.55	43.62	25.79
.44	150.07	151.07	147.59	139.06	125.46	107.45	86.41	64.34	43.43	25.63
.45	150.10	151.06	147.53	138.95	125.30	107.25	86.19	64.12	43.23	25.47
.46	150.13	151.05	147.47	138.84	125.14	107.05	85.97	63.90	43.03	25.31
.47	150.16	151.03	147.41	138.73	124.98	106.85	85.75	63.69	42.84	25.16
.48	150.19	151.02	147.34	138.61	124.82	106.65	85.53	63.47	42.64	25.00
.49	150.22	151.01	147.28	138.50	124.66	106.45	85.31	63.25	42.45	24.85
.50	150.25	151.00	147.22	138.39	124.50	106.25	85.09	63.03	42.25	24.69

TABLE VI'. ARGUMENT 1.

Equation = 76″. — 73″.563 sin. x — 9″.993 sin. $2x$ — 1″.071 sin. $3x$ — 0″.104 sin. $4x$ — 0″.008 sin. $5x$.

Period, 27.55455246 days.

Days.	0	1	2	3	4	5	6	7	8	9
Days										
.50	150.25	151.00	147.22	138.39	124.50	106.25	85.09	63.03	42.25	24.69
.51	150.28	150.98	147.16	138.27	124.34	106.05	84.87	62.81	42.06	24.54
.52	150.30	150.97	147.09	138.16	124.17	105.85	84.65	62.59	41.86	24.38
.53	150.33	150.95	147.03	138.04	124.01	105.65	84.43	62.38	41.67	24.23
.54	150.35	150.94	146.96	137.93	123.84	105.45	84.21	62.16	41.47	24.07
.55	150.38	150.92	146.90	137.81	123.68	105.25	83.99	61.94	41.28	23.92
.56	150.40	150.91	146.83	137.69	123.51	105.05	83.77	61.73	41.09	23.77
.57	150.43	150.89	146.77	137.57	123.35	104.85	83.55	61.51	40.90	23.62
.58	150.45	150.87	146.70	137.46	123.18	104.64	83.33	61.30	40.70	23.47
.59	150.48	150.85	146.64	137.34	123.02	104.44	83.11	61.08	40.51	23.32
.60	150.50	150.83	146.57	137.22	122.85	104.24	82.89	60.86	40.32	23.17
.61	150.52	150.81	146.50	137.10	122.68	104.04	82.67	60.64	40.13	23.02
.62	150.54	150.79	146.43	136.98	122.51	103.83	82.45	60.43	39.94	22.87
.63	150.57	150.77	146.37	136.86	122.35	103.63	82.23	60.21	39.75	22.73
.64	150.59	150.75	146.30	136.74	122.18	103.42	82.01	60.00	39.56	22.58
.65	150.62	150.73	146.23	136.62	122.01	103.22	81.79	59.78	39.37	22.43
.66	150.64	150.71	146.16	136.50	121.84	103.02	81.57	59.56	39.18	22.28
.67	150.66	150.68	146.09	136.37	121.67	102.81	81.35	59.35	39.00	22.13
.68	150.68	150.66	146.01	136.25	121.50	102.61	81.12	59.13	38.81	21.99
.69	150.70	150.64	145.94	136.12	121.33	102.40	80.90	58.92	38.62	21.84
.70	150.72	150.62	145.87	136.00	121.16	102.20	80.68	58.70	38.43	21.69
.71	150.74	150.60	145.79	135.88	120.99	101.99	80.46	58.49	38.24	21.55
.72	150.76	150.58	145.72	135.75	120.82	101.78	80.24	58.27	38.06	21.40
.73	150.78	150.55	145.64	135.63	120.64	101.58	80.02	58.06	37.87	21.26
.74	150.80	150.53	145.57	135.50	120.47	101.38	79.80	57.84	37.69	21.11
.75	150.81	150.50	145.49	135.38	120.30	101.17	79.58	57.63	37.50	20.97
.76	150.83	150.47	145.41	135.25	120.13	100.96	79.36	57.42	37.31	20.83
.77	150.85	150.45	145.34	135.12	119.95	100.75	79.14	57.20	37.13	20.69
.78	150.86	150.42	145.26	135.00	119.78	100.55	78.91	56.99	36.94	20.55
.79	150.88	150.40	145.19	134.87	119.60	100.34	78.69	56.77	36.76	20.41
.80	150.90	150.37	145.11	134.74	119.43	100.13	78.47	56.56	36.57	20.27
.81	150.91	150.34	145.03	134.61	119.25	99.92	78.25	56.35	36.39	20.13
.82	150.93	150.31	144.95	134.48	119.08	99.71	78.03	56.14	36.21	19.99
.83	150.94	150.28	144.88	134.35	118.90	99.51	77.80	55.92	36.02	19.86
.84	150.95	150.25	144.80	134.22	118.73	99.30	77.58	55.71	35.84	19.72
.85	150.97	150.22	144.72	134.09	118.55	99.09	77.36	55.50	35.66	19.58
.86	150.98	150.19	144.64	133.96	118.37	98.88	77.14	55.29	35.48	19.44
.87	151.00	150.15	144.56	133.82	118.19	98.67	76.92	55.08	35.30	19.30
.88	151.01	150.12	144.47	133.69	118.02	98.46	76.69	54.86	35.11	19.17
.89	151.02	150.09	144.39	133.55	117.84	98.25	76.47	54.65	34.93	19.03
.90	151.04	150.06	144.31	133.42	117.66	98.04	76.25	54.44	34.75	18.89
.91	151.05	150.03	144.23	133.28	117.48	97.83	76.03	54.23	34.57	18.76
.92	151.06	150.00	144.14	133.15	117.30	97.62	75.81	54.02	34.39	18.62
.93	151.07	149.96	144.06	133.01	117.12	97.40	75.59	53.81	34.21	18.49
.94	151.08	149.93	143.97	132.88	116.94	97.19	75.37	53.60	34.01	18.35
.95	151.09	149.89	143.89	132.74	116.76	96.98	75.15	53.39	33.86	18.22
.96	151.10	149.86	143.80	132.60	116.58	96.77	74.93	53.18	33.68	18.09
.97	151.11	149.82	143.71	132.46	116.40	96.56	74.71	52.97	33.51	17.96
.98	151.12	149.79	143.63	132.33	116.21	96.34	74.48	52.76	33.33	17.82
.99	151.13	149.75	143.54	132.19	116.03	96.13	74.26	52.55	33.15	17.69
1.00	151.14	149.71	143.45	132.05	115.85	95.92	74.04	52.34	32.97	17.56

TABLE VI″. ARGUMENT 1.

Equation = 76″ — 73″.563 sin. x — 9″.993 sin. $2x$ — 1″.071 sin. $3x$ — 0″.104 sin. $4x$ — 0″.008 sin. $5x$.

Period, 27.55455246 days.

Days.	10	11	12	13	14	15	16	17	18
Days. .00	17″.56	7″.07	1″.70	1″.07	4″.46	11″.01	19″.91	30″.46	42″.11
.01	17.43	6.99	1.67	1.09	4.51	11.09	20.01	30.57	42.23
.02	17.30	6.91	1.64	1.10	4.56	11.17	20.11	30.68	42.35
.03	17.17	6.84	1.62	1.12	4.62	11.25	20.21	30.80	42.47
.04	17.04	6.76	1.59	1.13	4.67	11.33	20.31	30.91	42.59
.05	16.91	6.68	1.56	1.15	4.72	11.41	20.41	31.02	42.71
.06	16.78	6.61	1.53	1.17	4.77	11.49	20.50	31.14	42.84
.07	16.66	6.53	1.51	1.19	4.83	11.57	20.60	31.25	42.96
.08	16.53	6.46	1.48	1.20	4.88	11.65	20.70	31.37	43.08
.09	16.41	6.38	1.46	1.22	4.94	11.73	20.80	31.48	43.20
.10	16.28	6.31	1.43	1.24	4.99	11.81	20.90	31.59	43.32
.11	16.16	6.24	1.41	1.26	5.04	11.89	21.00	31.70	43.44
.12	16.03	6.16	1.38	1.28	5.10	11.97	21.10	31.81	43.56
.13	15.91	6.09	1.36	1.30	5.15	12.06	21.20	31.93	43.68
.14	15.78	6.01	1.33	1.32	5.21	12.14	21.30	32.04	43.80
.15	15.66	5.94	1.31	1.34	5.26	12.22	21.40	32.15	43.92
.16	15.54	5.87	1.29	1.36	5.32	12.30	21.51	32.27	44.05
.17	15.42	5.80	1.27	1.38	5.37	12.38	21.61	32.38	44.17
.18	15.29	5.73	1.25	1.41	5.43	12.47	21.71	32.50	44.29
.19	15.17	5.66	1.23	1.43	5.48	12.55	21.81	32.61	44.41
.20	15.05	5.59	1.21	1.45	5.54	12.63	21.91	32.72	44.53
.21	14.93	5.52	1.19	1.47	5.60	12.71	22.01	32.83	44.65
.22	14.81	5.45	1.17	1.50	5.66	12.80	22.11	32.94	44.77
.23	14.69	5.39	1.16	1.52	5.71	12.88	22.21	33.06	44.89
.24	14.57	5.32	1.14	1.55	5.77	12.97	22.32	33.17	45.02
.25	14.45	5.25	1.12	1.57	5.83	13.05	22.42	33.29	45.14
.26	14.33	5.19	1.10	1.60	5.89	13.13	22.52	33.40	45.26
.27	14.22	5.12	1.09	1.62	5.95	13.22	22.63	33.52	45.39
.28	14.10	5.06	1.07	1.65	6.01	13.30	22.73	33.63	45.51
.29	13.99	4.99	1.06	1.67	6.07	13.39	22.83	33.75	45.63
.30	13.87	4.93	1.04	1.70	6.13	13.47	22.93	33.86	45.75
.31	13.76	4.87	1.03	1.73	6.19	13.56	23.03	33.98	45.87
.32	13.64	4.81	1.01	1.76	6.25	13.64	23.14	34.09	45.99
.33	13.53	4.74	1.00	1.78	6.32	13.73	23.24	34.21	46.12
.34	13.41	4.68	0.98	1.81	6.38	13.81	23.34	34.32	46.24
.35	13.30	4.62	0.97	1.84	6.44	13.90	23.45	34.44	46.36
.36	13.19	4.56	0.96	1.87	6.50	13.99	23.55	34.56	46.49
.37	13.08	4.50	0.95	1.90	6.56	14.07	23.66	34.67	46.61
.38	12.97	4.44	0.93	1.93	6.63	14.16	23.76	34.79	46.74
.39	12.86	4.38	0.92	1.96	6.69	14.24	23.87	34.90	46.86
.40	12.75	4.32	0.91	1.99	6.75	14.33	23.97	35.02	46.98
.41	12.64	4.26	0.90	2.02	6.81	14.42	24.07	35.14	47.10
.42	12.53	4.20	0.89	2.05	6.88	14.51	24.18	35.25	47.22
.43	12.42	4.15	0.88	2.09	6.94	14.59	24.28	35.37	47.35
.44	12.31	4.09	0.87	2.12	7.01	14.68	24.39	35.48	47.47
.45	12.20	4.03	0.86	2.15	7.07	14.77	24.49	35.60	47.59
.46	12.09	3.97	0.85	2.18	7.13	14.86	24.60	35.72	47.72
.47	11.99	3.92	0.85	2.22	7.20	14.95	24.70	35.83	47.84
.48	11.88	3.86	0.84	2.25	7.26	15.03	24.81	35.95	47.97
.49	11.78	3.81	0.84	2.29	7.33	15.12	24.91	36.06	48.09
.50	11.67	3.75	0.83	2.32	7.39	15.21	25.02	36.18	48.21

TABLE VI″. ARGUMENT 1.

Equation = 76″. — 73″.563 sin. x — 9″.993 sin. $2x$ — 1″.071 sin. $3x$ — 0″.104 sin. $4x$ — 0″.008 sin. $5x$.

Period, 27.55455246 days.

Days.	10	11	12	13	14	15	16	17	18
Days. .50	11.67″	3.75″	0.83″	2.32″	7.39″	15.21″	25.02″	36.18″	48.21″
.51	11.57	3.70	0.83	2.35	7.46	15.30	25.13	36.30	48.33
.52	11.46	3.65	0.82	2.39	7.52	15.39	25.23	36.42	48.46
.53	11.36	3.59	0.82	2.42	7.59	15.48	25.34	36.53	48.58
.54	11.25	3.54	0.81	2.46	7.65	15.57	25.44	36.65	48.71
.55	11.15	3.49	0.81	2.49	7.72	15.66	25.55	36.76	48.83
.56	11.05	3.44	0.81	2.53	7.79	15.75	25.66	36.88	48.95
.57	10.95	3.39	0.81	2.56	7.86	15.84	25.76	37.00	49.08
.58	10.85	3.35	0.80	2.60	7.92	15.93	25.87	37.11	49.20
.59	10.75	3.30	0.80	2.63	7.99	16.02	25.97	37.23	49.33
.60	10.65	3.25	0.80	2.67	8.06	16.11	26.08	37.35	49.45
.61	10.55	3.20	0.80	2.71	8.13	16.20	26.19	37.47	49.57
.62	10.45	3.15	0.80	2.75	8.20	16.29	26.30	37.59	49.70
.63	10.36	3.11	0.80	2.78	8.27	16.39	26.40	37.71	49.82
.64	10.26	3.06	0.79	2.82	8.34	16.48	26.51	37.82	49.95
.65	10.16	3.01	0.79	2.86	8.41	16.57	26.62	37.94	50.07
.66	10.06	2.97	0.79	2.90	8.48	16.66	26.73	38.06	50.19
.67	9.97	2.92	0.80	2.94	8.55	16.76	26.84	38.17	50.32
.68	9.87	2.88	0.80	2.99	8.62	16.85	26.94	38.29	50.44
.69	9.78	2.83	0.80	3.03	8.69	16.95	27.05	38.41	50.57
.70	9.68	2.79	0.80	3.07	8.76	17.04	27.16	38.53	50.69
.71	9.59	2.75	0.80	3.11	8.83	17.13	27.27	38.65	50.81
.72	9.49	2.71	0.81	3.15	8.90	17.23	27.38	38.77	50.94
.73	9.40	2.66	0.81	3.20	8.98	17.32	27.48	38.89	51.06
.74	9.30	2.62	0.82	3.24	9.05	17.42	27.59	39.00	51.19
.75	9.21	2.58	0.82	3.28	9.12	17.51	27.70	39.12	51.31
.76	9.12	2.54	0.83	3.32	9.19	17.60	27.81	39.24	51.44
.77	9.03	2.50	0.83	3.37	9.27	17.70	27.92	39.35	51.56
.78	8.94	2.46	0.84	3.41	9.34	17.79	28.03	39.47	51.69
.79	8.85	2.42	0.84	3.46	9.42	17.89	28.14	39.59	51.81
.80	8.76	2.38	0.85	3.50	9.49	17.98	28.25	39.71	51.94
.81	8.67	2.34	0.86	3.55	9.56	18.08	28.36	39.83	52.06
.82	8.58	2.30	0.87	3.59	9.64	18.17	28.47	39.95	52.19
.83	8.49	2.27	0.87	3.64	9.71	18.27	28.58	40.07	52.31
.84	8.40	2.23	0.88	3.68	9.79	18.36	28.69	40.19	52.44
.85	8.31	2.19	0.89	3.73	9.86	18.46	28.80	40.31	52.56
.86	8.22	2.15	0.90	3.78	9.94	18.56	28.91	40.43	52.69
.87	8.14	2.12	0.91	3.82	10.01	18.65	29.02	40.55	52.82
.88	8.05	2.08	0.92	3.87	10.09	18.75	29.13	40.67	52.94
.89	7.97	2.05	0.93	3.91	10.16	18.84	29.24	40.79	53.07
.90	7.88	2.01	0.94	3.96	10.24	18.94	29.35	40.91	53.19
.91	7.80	1.98	0.95	4.01	10.32	19.04	29.46	41.03	53.31
.92	7.72	1.95	0.96	4.06	10.39	19.14	29.57	41.15	53.44
.93	7.63	1.91	0.98	4.11	10.47	19.23	29.68	41.27	53.56
.94	7.55	1.88	0.99	4.16	10.54	19.33	29.79	41.39	53.69
.95	7.47	1.85	1.00	4.21	10.62	19.42	29.90	41.51	53.81
.96	7.39	1.82	1.01	4.26	10.70	19.52	30.01	41.63	53.94
.97	7.31	1.79	1.03	4.31	10.78	19.62	30.13	41.75	54.07
.98	7.23	1.76	1.04	4.36	10.85	19.71	30.24	41.87	54.19
.99	7.15	1.73	1.06	4.41	10.93	19.81	30.35	41.99	54.32
1.00	7.07	1.70	1.07	4.46	11.01	19.91	30.46	42.11	54.44

TABLE VI″. ARGUMENT 1.

Equation = 76″. — 73″.563 sin. x — 9″.993 sin. $2x$ — 1″.071 sin. $3x$ — 0″.104 sin. $4x$ — 0″.008 sin. $5x$.

Period, 27.55455246 days.

Days.	19	20	21	22	23	24	25	26	27
.00	54.44	67.15	80.01	92.79	105.30	117.25	128.27	137.87	145.39
.01	54.57	67.28	80.14	92.92	105.42	117.37	128.37	137.96	145.45
.02	54.69	67.41	80.27	93.05	105.54	117.48	128.47	138.04	145.51
.03	54.82	67.54	80.40	93.17	105.67	117.60	128.58	138.13	145.57
.04	54.94	67.67	80.52	93.30	105.79	117.71	128.68	138.21	145.63
.05	55.07	67.80	80.65	93.42	105.92	117.83	128.78	138.30	145.70
.06	55.20	67.92	80.78	93.55	106.04	117.95	128.89	138.39	145.76
.07	55.32	68.05	80.90	93.68	106.16	118.05	128.99	138.47	145.82
.08	55.45	68.18	81.03	93.80	106.29	118.17	129.10	138.56	145.88
.09	55.57	68.31	81.16	93.93	106.41	118.28	129.20	138.64	145.94
.10	55.70	68.44	81.29	94.06	106.53	118.40	129.30	138.73	146.00
.11	55.83	68.57	81.42	94.19	106.65	118.51	129.40	138.81	146.06
.12	55.95	68.70	81.55	94.32	106.77	118.63	129.50	138.89	146.12
.13	56.08	68.83	81.68	94.44	106.89	118.74	129.60	138.98	146.18
.14	56.20	68.95	81.80	94.57	107.02	118.86	129.71	139.06	146.23
.15	56.33	69.08	81.93	94.69	107.14	118.97	129.81	139.15	146.29
.16	56.46	69.21	82.06	94.82	107.26	119.08	129.91	139.23	146.35
.17	56.58	69.33	82.18	94.95	107.39	119.20	130.02	139.31	146.40
.18	56.71	69.46	82.31	95.07	107.51	119.31	130.12	139.40	146.46
.19	56.83	69.59	82.44	95.20	107.63	119.43	130.22	139.48	146.52
.20	56.96	69.72	82.57	95.33	107.75	119.54	130.32	139.56	146.58
.21	57.09	69.85	82.70	95.46	107.87	119.65	130.42	139.64	146.64
.22	57.22	69.98	82.83	95.58	107.99	119.76	130.52	139.72	146.69
.23	57.34	70.11	82.96	95.71	108.11	119.88	130.62	139.80	146.75
.24	57.47	70.23	83.09	95.83	108.23	119.99	130.72	139.89	146.80
.25	57.59	70.36	83.22	95.96	108.35	120.11	130.82	139.97	146.86
.26	57.72	70.49	83.34	96.09	108.48	120.22	130.92	140.05	146.91
.27	57.85	70.61	83.47	96.21	108.60	120.34	131.02	140.14	146.97
.28	57.97	70.74	83.60	96.34	108.72	120.45	131.12	140.22	147.02
.29	58.10	70.87	83.73	96.46	108.84	120.56	131.22	140.30	147.08
.30	58.23	71.00	83.86	96.59	108.96	120.67	131.32	140.38	147.13
.31	58.36	71.13	83.99	96.72	109.08	120.78	131.42	140.46	147.18
.32	58.48	71.26	84.12	96.85	109.20	120.89	131.52	140.54	147.23
.33	58.61	71.39	84.25	96.97	109.32	121.00	131.62	140.62	147.28
.34	58.73	71.52	84.37	97.10	109.44	121.12	131.72	140.70	147.34
.35	58.86	71.65	84.50	97.22	109.56	121.23	131.82	140.78	147.39
.36	58.99	71.77	84.63	97.34	109.68	121.34	131.91	140.85	147.45
.37	59.11	71.90	84.75	97.46	109.80	121.46	132.01	140.93	147.50
.38	59.24	72.03	84.88	97.59	109.92	121.57	132.11	141.01	147.55
.39	59.36	72.16	85.01	97.71	110.04	121.68	132.21	141.09	147.60
.40	59.49	72.29	85.14	97.84	110.16	121.79	132.31	141.17	147.65
.41	59.62	72.42	85.27	97.97	110.28	121.90	132.41	141.25	147.70
.42	59.75	72.55	85.40	98.09	110.40	122.01	132.51	141.32	147.75
.43	59.87	72.68	85.53	98.22	110.52	122.12	132.60	141.40	147.80
.44	60.00	72.80	85.65	98.34	110.64	122.23	132.70	141.47	147.85
.45	60.12	72.93	85.78	98.47	110.76	122.35	132.80	141.55	147.90
.46	60.25	73.06	85.91	98.60	110.88	122.46	132.89	141.63	147.94
.47	60.38	73.18	86.03	98.72	111.00	122.57	132.99	147.70	147.99
.48	60.50	73.31	86.16	98.85	111.12	122.68	133.09	141.78	148.04
.49	60.63	73.44	86.29	98.97	111.24	122.79	133.18	141.85	148.09
.50	60.76	73.57	86.42	99.10	111.36	122.90	133.28	141.93	148.14

TABLE VI″. ARGUMENT 1.

Equation = 76″. — 73″.563 sin. x — 9″.993 sin. $2x$ — 1″.071 sin. $3x$ — 0″.104 sin. $4x$ — 0″.008 sin. $5x$.

Period, 27.55455246 days.

Days.	19	20	21	22	23	24	25	26	27
Days. .50	60.76″	73.57″	86.42″	99.10″	111.36″	122.90″	133.28″	141.93″	148.14″
.51	60.89	73.70	86.55	99.23	111.48	123.01	133.38	142.00	148.19
.52	61.02	73.83	86.68	99.35	111.60	123.12	133.47	142.08	148.23
.53	61.15	73.96	86.81	99.48	111.72	123.23	133.57	142.15	148.28
.54	61.27	74.09	86.93	99.60	111.84	123.34	133.66	142.23	148.32
.55	61.40	74.22	87.06	99.73	111.96	123.45	133.76	142.30	148.37
.56	61.53	74.34	87.19	99.85	112.07	123.56	133.86	142.37	148.41
.57	61.65	74.47	87.31	99.98	112.19	123.67	133.95	142.45	148.46
.58	61.78	74.60	87.44	100.10	112.31	123.78	134.05	142.52	148.50
.59	61.91	74.73	87.57	100.23	112.43	123.89	134.14	142.60	148.55
.60	62.04	74.86	87.70	100.35	112.55	124.00	134.24	142.67	148.59
.61	62.17	74.99	87.83	100.47	112.67	124.11	134.33	142.74	148.63
.62	62.30	75.12	87.96	100.60	112.79	124.22	134.43	142.81	148.67
.63	62.42	75.25	88.09	100.72	112.91	124.33	134.52	142.88	148.72
.64	62.55	75.38	88.21	100.85	113.03	124.44	134.62	142.96	148.76
.65	62.67	75.51	88.34	100.97	113.15	124.55	134.71	143.03	148.80
.66	62.80	75.63	88.47	101.09	113.26	124.65	134.80	143.10	148.85
.67	62.93	75.76	88.59	101.22	113.38	124.76	134.89	143.18	148.89
.68	63.05	75.89	88.72	101.34	113.50	124.87	134.99	143.25	148.93
.69	63.18	76.02	88.85	101.47	113.62	124.98	135.07	143.32	148.97
.70	63.31	76.15	88.98	101.59	113.74	125.09	135.17	143.39	149.01
.71	63.44	76.28	89.11	101.71	113.86	125.20	135.26	143.46	149.05
.72	63.57	76.41	89.24	101.84	113.98	125.31	135.35	143.53	149.09
.73	63.70	76.54	89.36	101.96	114.10	125.41	135.44	143.60	149.13
.74	63.82	76.66	89.49	102.09	114.21	125.52	135.54	143.67	149.17
.75	63.95	76.79	89.61	102.21	114.33	125.63	135.63	143.74	149.21
.76	64.08	76.92	89.74	102.33	114.45	125.73	135.72	143.80	149.25
.77	64.20	77.04	89.87	102.46	114.56	125.84	135.82	143.87	149.29
.78	64.33	77.17	89.99	102.58	114.68	125.94	135.91	143.94	149.33
.79	64.46	77.30	90.12	102.71	114.80	126.05	136.00	144.01	149.37
.80	64.59	77.43	90.25	102.83	114.92	126.16	136.09	144.08	149.41
.81	64.72	77.56	90.38	102.95	115.04	126.27	136.18	144.15	149.45
.82	64.85	77.69	90.51	103.08	115.16	126.37	136.27	144.22	149.48
.83	64.98	77.82	90.63	103.20	115.27	126.48	136.36	144.28	149.52
.84	65.10	77.95	90.76	103.33	115.39	126.58	136.45	144.35	149.55
.85	65.23	78.08	90.88	103.45	115.51	126.69	136.54	144.42	149.59
.86	65.36	78.20	91.01	103.57	115.62	126.80	136.63	144.48	149.63
.87	65.48	78.33	91.14	103.70	115.74	126.90	136.72	144.55	149.66
.88	65.61	78.46	91.26	103.82	115.85	127.01	136.81	144.61	149.69
.89	65.74	78.59	91.40	103.95	115.97	127.11	136.90	144.68	149.73
.90	65.87	78.72	91.52	104.07	116.09	127.22	136.99	144.75	149.76
.91	66.00	78.85	91.65	104.19	116.21	127.33	137.08	144.81	149.80
.92	66.13	78.98	91.78	104.31	116.32	127.43	137.17	144.88	149.83
.93	66.26	79.11	91.90	104.44	116.44	127.54	137.26	144.94	149.87
.94	66.38	79.24	92.03	104.56	116.55	127.64	137.34	145.01	149.90
.95	66.51	79.37	92.15	104.69	116.67	127.75	137.43	145.07	149.93
.96	66.64	79.49	92.28	104.81	116.79	127.85	137.52	145.13	149.96
.97	66.76	79.62	92.40	104.93	116.90	127.96	137.60	145.20	149.99
.98	66.89	79.75	92.53	105.06	117.02	128.06	137.69	145.26	150.02
.99	67.02	79.88	92.66	105.18	117.13	128.17	137.78	145.33	150.05
1.00	67.15	80.01	92.79	105.30	117.25	128.27	137.87	145.39	150.08

TABLE VII′. ARGUMENT 2.

Equation $= 11''.2 - 11''.18 \sin. (2t - x) - 0''.304 \sin. (4t - 2x)$.

Period, 31.81193574 days.

Days.	0	1	2	3	4	5	6	7	8	9
Days. .00	22″.37	22″.31	21″.81	20″.87	19″.52	17″.81	15″.82	13″.62	11″.31	9″.00
.01	22.37	22.30	21.80	20.86	19.51	17.80	15.80	13.60	11.29	8.98
.02	22.37	22.30	21.79	20.85	19.49	17.78	15.78	13.58	11.27	8.96
.03	22.37	22.30	21.78	20.84	19.48	17.76	15.76	13.55	11.24	8.93
.04	22.37	22.29	21.77	20.83	19.46	17.74	15.74	13.53	11.22	8.91
.05	22.37	22.29	21.77	20.81	19.44	17.72	15.72	13.51	11.20	8.89
.06	22.38	22.29	21.76	20.80	19.43	17.70	15.70	13.48	11.17	8.86
.07	22.38	22.29	21.75	20.79	19.41	17.68	15.68	13.46	11.15	8.84
.08	22.38	22.28	21.74	20.78	19.40	17.66	15.66	13.44	11.13	8.82
.09	22.38	22.28	21.74	20.77	19.38	17.64	15.64	13.42	11.11	8.80
.10	22.38	22.28	21.73	20.75	19.36	17.62	15.60	13.39	11.08	8.77
.11	22.38	22.27	21.72	20.74	19.35	17.61	15.59	13.37	11.06	8.75
.12	22.38	22.27	21.71	20.73	19.33	17.59	15.57	13.35	11.04	8.73
.13	22.38	22.27	21.70	20.72	19.32	17.57	15.55	13.32	11.01	8.70
.14	22.38	22.26	21.69	20.71	19.30	17.55	15.53	13.30	10.99	8.68
.15	22.38	22.26	21.69	20.69	19.28	17.53	15.50	13.28	10.96	8.66
.16	22.39	22.26	21.68	20.68	19.27	17.51	15.48	13.25	10.91	8.63
.17	22.39	22.25	21.67	20.67	19.25	17.19	15.16	13.23	10.91	8.61
.18	22.39	22.25	21.66	20.66	19.21	17.47	15.11	13.21	10.89	8.59
.19	22.39	22.25	21.66	20.65	19.22	17.15	15.42	13.19	10.87	8.57
.20	22.39	22.24	21.65	20.63	19.20	17.43	15.39	13.16	10.84	8.54
.21	22.39	22.24	21.64	20.62	19.19	17.42	15.37	13.14	10.82	8.52
.22	22.39	22.24	21.63	20.61	19.17	17.40	15.35	13.12	10.80	8.50
.23	22.39	22.23	21.62	20.59	19.16	17.38	15.33	13.09	10.77	8.48
.24	22.39	22.23	21.61	20.58	19.14	17.36	15.31	13.07	10.75	8.46
.25	22.39	22.23	21.61	20.57	19.12	17.34	15.28	13.05	10.73	8.44
.26	22.39	22.22	21.60	20.55	19.11	17.32	15.26	13.02	10.70	8.41
.27	22.39	22.22	21.59	20.54	19.09	17.30	15.24	13.00	10.68	8.39
.28	22.40	22.21	21.58	20.53	19.08	17.28	15.22	12.98	10.66	8.37
.29	22.40	22.21	21.57	20.52	19.06	17.26	15.20	12.96	10.64	8.35
.30	22.10	22.20	21.57	20.50	19.04	17.24	15.17	12.93	10.61	8.32
.31	22.40	22.20	21.56	20.49	19.03	17.23	15.15	12.91	10.59	8.30
.32	22.10	22.20	21.55	20.48	19.01	17.21	15.13	12.89	10.57	8.28
.33	22.40	22.19	21.54	20.47	19.00	17.19	15.11	12.86	10.54	8.25
.34	22.40	22.19	21.53	20.46	18.98	17.17	15.09	12.84	10.52	8.23
.35	22.40	22.18	21.52	20.44	18.96	17.15	15.07	12.82	10.50	8.21
.36	22.10	22.18	21.51	20.43	18.95	17.13	15.05	12.79	10.47	8.18
.37	22.40	22.17	21.50	20.42	18.93	17.11	15.03	12.77	10.45	8.16
.38	22.40	22.17	21.49	20.41	18.92	17.09	15.01	12.75	10.43	8.14
.39	22.40	22.16	21.49	20.40	18.90	17.07	14.99	12.73	10.41	8.12
.40	22.40	22.16	21.48	20.38	18.88	17.05	14.96	12.70	10.38	8.09
.41	22.40	22.15	21.47	20.37	18.87	17.03	14.94	12.68	10.36	8.07
.42	22.40	22.15	21.46	20.35	18.85	17.01	14.92	12.66	10.34	8.05
.43	22.40	22.15	21.45	20.34	18.83	16.99	14.90	12.63	10.31	8.03
.44	22.39	22.14	21.44	20.33	18.82	16.97	14.88	12.61	10.29	8.01
.45	22.39	22.14	21.43	20.31	18.80	16.95	14.85	12.59	10.27	7.98
.46	22.39	22.13	21.42	20.30	18.78	16.93	14.83	12.56	10.24	7.96
.47	22.39	22.13	21.41	20.28	18.77	16.91	14.81	12.54	10.22	7.94
.48	22.39	22.12	21.40	20.27	18.75	16.89	14.79	12.52	10.20	7.92
.49	22.39	22.12	21.40	20.26	18.73	16.87	14.77	12.50	10.18	7.90
.50	22.39	22.11	21.39	20.24	18.71	16.85	14.74	12.47	10.15	7.87

TABLE VII'. ARGUMENT 2.

Equation $= 11''.2 - 11''.18 \sin. (2t - x) - 0''.304 \sin. (4t - 2x)$.

Period, 31.81193574 days.

Days.	0	1	2	3	4	5	6	7	8	9
Days.										
.50	22.39″	22.11″	21.39″	20.24″	18.71″	16.85″	14.74	12.47″	10.15″	7.87″
.51	22.39	22.11	21.38	20.23	18.70	16.83	14.72	12.45	10.13	7.85
.52	22.39	22.10	21.37	20.22	18.68	16.81	14.70	12.43	10.11	7.83
.53	22.39	22.10	21.36	20.20	18.66	16.79	14.68	12.40	10.08	7.81
.54	22.39	22.09	21.35	20.19	18.65	16.77	14.66	12.38	10.06	7.79
.55	22.39	22.09	21.34	20.17	18.63	16.75	14.63	12.36	10.03	7.76
.56	22.39	22.08	21.33	20.16	18.61	16.73	14.61	12.33	10.01	7.74
.57	22.39	22.08	21.32	20.14	18.60	16.71	14.59	12.31	9.98	7.72
.58	22.39	22.07	21.31	20.13	18.58	16.69	14.57	12.29	9.96	7.70
.59	22.39	22.07	21.30	20.12	18.56	16.67	14.55	12.27	9.94	7.68
.60	22.38	22.06	21.29	20.10	18.54	16.65	14.52	12.24	9.91	7.65
.61	22.38	22.06	21.28	20.09	18.53	16.63	14.50	12.22	9.89	7.63
.62	22.38	22.05	21.27	20.08	18.51	16.61	14.48	12.20	9.87	7.61
.63	22.38	22.05	21.26	20.06	18.49	16.59	14.45	12.17	9.84	7.58
.64	22.38	22.04	21.25	20.05	18.47	16.57	14.43	12.15	9.82	7.56
.65	22.38	22.04	21.24	20.03	18.45	16.55	14.41	12.13	9.80	7.54
.66	22.38	22.03	21.23	20.02	18.44	16.53	14.38	12.10	9.77	7.52
.67	22.38	22.02	21.22	20.00	18.42	16.51	14.36	12.08	9.75	7.49
.68	22.37	22.02	21.21	19.99	18.40	16.49	14.34	12.06	9.73	7.47
.69	22.37	22.01	21.20	19.98	18.38	16.47	14.32	12.04	9.71	7.45
.70	22.37	22.00	21.19	19.96	18.36	16.44	14.29	12.01	9.68	7.43
.71	22.37	22.00	21.18	19.95	18.35	16.42	14.27	11.99	9.66	7.40
.72	22.37	21.99	21.17	19.94	18.33	16.40	14.25	11.97	9.64	7.38
.73	22.37	21.99	21.16	19.92	18.31	16.38	14.23	11.94	9.61	7.36
.74	22.37	21.98	21.15	19.91	18.29	16.36	14.21	11.92	9.59	7.34
.75	22.36	21.98	21.14	19.89	18.27	16.34	14.18	11.89	9.57	7.32
.76	22.36	21.97	21.13	19.88	18.26	16.32	14.16	11.87	9.54	7.29
.77	22.36	21.96	21.12	19.86	18.24	16.30	14.14	11.85	9.52	7.27
.78	22.36	21.96	21.11	19.85	18.22	16.28	14.12	11.81	9.50	7.25
.79	22.36	21.95	21.10	19.84	18.20	16.26	14.10	11.79	9.48	7.23
.80	22.35	21.94	21.09	19.82	18.18	16.24	14.07	11.77	9.45	7.21
.81	22.35	21.94	21.08	19.81	18.17	16.22	14.05	11.75	9.43	7.18
.82	22.35	21.93	21.07	19.79	18.15	16.20	14.03	11.73	9.41	7.16
.83	22.35	21.93	21.06	19.78	18.13	16.18	14.00	11.70	9.38	7.14
.84	22.35	21.92	21.05	19.76	18.11	16.16	13.98	11.68	9.36	7.12
.85	22.34	21.91	21.04	19.75	18.09	16.14	13.96	11.66	9.34	7.10
.86	22.34	21.91	21.03	19.73	18.08	16.12	13.93	11.63	9.31	7.08
.87	22.34	21.90	21.02	19.72	18.06	16.10	13.91	11.61	9.29	7.06
.88	22.34	21.90	21.01	19.70	18.04	16.08	13.89	11.59	9.27	7.04
.89	22.34	21.89	21.00	19.69	18.02	16.06	13.87	11.57	9.25	7.02
.90	22.33	21.88	20.98	19.67	18.00	16.03	13.84	11.54	9.22	7.00
.91	22.33	21.87	20.97	19.66	17.99	16.01	13.82	11.52	9.20	6.97
.92	22.33	21.87	20.96	19.64	17.97	15.99	13.80	11.50	9.18	6.95
.93	22.33	21.86	20.95	19.63	17.95	15.97	13.78	11.47	9.16	6.93
.94	22.32	21.86	20.94	19.61	17.93	15.95	13.76	11.45	9.14	6.91
.95	22.32	21.85	20.93	19.60	17.91	15.93	13.73	11.43	9.11	6.89
.96	22.32	21.84	20.92	19.58	17.89	15.91	13.71	11.40	9.09	6.86
.97	22.32	21.83	20.91	19.57	17.87	15.89	13.69	11.38	9.07	6.84
.98	22.31	21.82	20.90	19.55	17.85	15.87	13.67	11.36	9.05	6.82
.99	22.31	21.81	20.89	19.54	17.83	15.85	13.65	11.34	9.03	6.80
1.00	22.31	21.81	20.87	19.52	17.81	15.82	13.62	11.31	9.00	6.78

TABLE VII″. ARGUMENT 2.

Equation = 11″.2 — 11″.18 sin. $(2t - x)$ — 0″.304 sin. $(4t - 2x)$.

Period, 31.81193574 days.

Days.	10	11	12	13	14	15	16	17	18	19	20
Days	″	″	″	″	″	″	″	″	″	″	″
.00	6.78	4.76	3.03	1.64	0.66	0.12	0.02	0.35	1.10	2.21	3.63
.01	6.75	4.74	3.01	1.62	0.66	0.11	0.02	0.35	1.10	2.22	3.64
.02	6.73	4.72	2.99	1.61	0.65	0.11	0.02	0.36	1.11	2.23	3.66
.03	6.71	4.70	2.98	1.60	0.64	0.11	0.02	0.36	1.12	2.25	3.67
.04	6.69	4.68	2.96	1.59	0.64	0.10	0.02	0.37	1.13	2.26	3.69
.05	6.67	4.66	2.95	1.58	0.63	0.10	0.03	0.37	1.14	2.27	3.71
.06	6.65	4.64	2.93	1.56	0.62	0.10	0.03	0.38	1.15	2.28	3.72
.07	6.63	4.62	2.91	1.55	0.62	0.09	0.03	0.39	1.16	2.30	3.73
.08	6.61	4.60	2.90	1.54	0.61	0.09	0.03	0.39	1.17	2.31	3.75
.09	6.59	4.58	2.88	1.53	0.60	0.09	0.03	0.40	1.18	2.32	3.77
.10	6.57	4.57	2.87	1.52	0.59	0.09	0.03	0.41	1.19	2.34	3.79
.11	6.54	4.55	2.85	1.50	0.59	0.08	0.04	0.41	1.20	2.35	3.80
.12	6.52	4.53	2.84	1.49	0.58	0.08	0.04	0.42	1.21	2.36	3.82
.13	6.50	4.51	2.82	1.48	0.57	0.08	0.04	0.42	1.22	2.37	3.83
.14	6.48	4.49	2.81	1.47	0.57	0.07	0.04	0.43	1.23	2.39	3.85
.15	6.46	4.48	2.79	1.46	0.56	0.07	0.04	0.43	1.24	2.40	3.87
.16	6.44	4.46	2.78	1.45	0.55	0.07	0.05	0.44	1.25	2.41	3.88
.17	6.42	4.44	2.76	1.44	0.55	0.06	0.05	0.45	1.26	2.43	3.90
.18	6.40	4.42	2.75	1.43	0.54	0.06	0.05	0.45	1.27	2.44	3.91
.19	6.38	4.40	2.73	1.42	0.53	0.06	0.05	0.46	1.28	2.45	3.93
.20	6.36	4.39	2.72	1.41	0.52	0.06	0.05	0.47	1.29	2.47	3.95
.21	6.33	4.37	2.70	1.40	0.52	0.05	0.06	0.47	1.30	2.48	3.96
.22	6.31	4.35	2.69	1.39	0.51	0.05	0.06	0.48	1.31	2.49	3.98
.23	6.29	4.33	2.67	1.38	0.50	0.05	0.06	0.48	1.32	2.50	4.00
.24	6.27	4.31	2.66	1.37	0.50	0.05	0.06	0.49	1.33	2.52	4.01
.25	6.25	4.30	2.64	1.36	0.49	0.05	0.06	0.50	1.34	2.54	4.03
.26	6.23	4.28	2.63	1.35	0.48	0.04	0.07	0.50	1.35	2.55	4.05
.27	6.21	4.26	2.61	1.34	0.48	0.04	0.07	0.51	1.36	2.56	4.06
.28	6.19	4.24	2.59	1.33	0.47	0.04	0.07	0.51	1.37	2.57	4.08
.29	6.17	4.22	2.58	1.32	0.46	0.04	0.07	0.52	1.38	2.58	4.10
.30	6.15	4.21	2.57	1.31	0.45	0.04	0.07	0.53	1.39	2.60	4.12
.31	6.13	4.19	2.55	1.29	0.44	0.04	0.08	0.53	1.40	2.61	4.13
.32	6.11	4.17	2.54	1.28	0.43	0.04	0.08	0.54	1.41	2.62	4.15
.33	6.09	4.15	2.52	1.27	0.43	0.04	0.08	0.55	1.42	2.64	4.16
.34	6.07	4.13	2.51	1.26	0.42	0.04	0.08	0.55	1.43	2.65	4.18
.35	6.05	4.12	2.49	1.25	0.41	0.03	0.09	0.56	1.44	2.66	4.20
.36	6.03	4.10	2.48	1.24	0.41	0.03	0.09	0.57	1.45	2.68	4.21
.37	6.01	4.08	2.47	1.23	0.40	0.03	0.09	0.57	1.46	2.69	4.23
.38	5.99	4.06	2.45	1.22	0.40	0.03	0.10	0.58	1.47	2.71	4.24
.39	5.97	4.04	2.44	1.21	0.39	0.03	0.10	0.59	1.48	2.72	4.26
.40	5.95	4.03	2.43	1.20	0.39	0.03	0.10	0.60	1.50	2.74	4.28
.41	5.92	4.01	2.41	1.19	0.38	0.03	0.11	0.60	1.51	2.75	4.29
.42	5.90	3.99	2.40	1.18	0.38	0.03	0.11	0.61	1.52	2.77	4.31
.43	5.88	3.97	2.38	1.17	0.37	0.03	0.11	0.62	1.53	2.78	4.32
.44	5.86	3.95	2.37	1.16	0.37	0.02	0.11	0.63	1.54	2.79	4.34
.45	5.84	3.94	2.36	1.15	0.36	0.02	0.12	0.64	1.55	2.81	4.36
.46	5.82	3.92	2.34	1.14	0.36	0.02	0.12	0.64	1.56	2.82	4.37
.47	5.80	3.90	2.33	1.13	0.35	0.02	0.12	0.65	1.57	2.84	4.39
.48	5.78	3.88	2.31	1.12	0.35	0.02	0.13	0.66	1.58	2.85	4.41
.49	5.76	3.86	2.30	1.11	0.34	0.02	0.13	0.67	1.59	2.86	4.43
.50	5.74	3.85	2.29	1.10	0.34	0.01	0.13	0.68	1.61	2.88	4.45

TABLE VII″. ARGUMENT 2.

Equation = 11″.2 — 11″.18 sin. (2t — x) — 0″.304 sin. (4t — 2x).

Period, 31.81193574 days.

Days.	10	11	12	13	14	15	16	17	18	19	20
Days.	″	″	″	″	″	″	″	″	″	″	″
.50	5.74	3.85	2.29	1.10	0.34	0.01	0.13	0.68	1.61	2.88	4.45
.51	5.72	3.83	2.27	1.09	0.33	0.01	0.13	0.68	1.62	2.89	4.46
.52	5.70	3.81	2.26	1.08	0.33	0.01	0.13	0.69	1.63	2.91	4.48
.53	5.68	3.79	2.24	1.07	0.32	0.01	0.14	0.70	1.64	2.92	4.50
.54	5.66	3.78	2.23	1.06	0.32	0.01	0.14	0.70	1.65	2.94	4.51
.55	5.64	3.76	2.22	1.05	0.31	0.01	0.15	0.71	1.66	2.95	4.53
.56	5.62	3.74	2.20	1.04	0.31	0.01	0.15	0.72	1.67	2.97	4.55
.57	5.60	3.73	2.19	1.03	0.30	0.01	0.15	0.72	1.68	2.98	4.56
.58	5.58	3.71	2.17	1.02	0.30	0.01	0.16	0.73	1.69	3.00	4.58
.59	5.56	3.69	2.16	1.01	0.29	0.01	0.16	0.74	1.70	3.01	4.60
.60	5.54	3.68	2.15	1.01	0.29	0.01	0.17	0.75	1.72	3.03	4.62
.61	5.52	3.66	2.13	1.00	0.28	0.01	0.17	0.75	1.73	3.04	4.63
.62	5.50	3.64	2.12	0.99	0.28	0.01	0.17	0.76	1.74	3.06	4.65
.63	5.48	3.62	2.11	0.98	0.27	0.01	0.18	0.77	1.75	3.07	4.67
.64	5.46	3.61	2.09	0.97	0.27	0.01	0.18	0.78	1.76	3.09	4.68
.65	5.44	3.59	2.08	0.96	0.26	0.01	0.19	0.79	1.78	3.10	4.70
.66	5.42	3.57	2.06	0.95	0.26	0.00	0.19	0.79	1.79	3.12	4.71
.67	5.40	3.56	2.05	0.94	0.25	0.00	0.19	0.80	1.80	3.13	4.73
.68	5.38	3.54	2.04	0.93	0.25	0.00	0.20	0.81	1.81	3.15	4.75
.69	5.36	3.52	2.03	0.92	0.24	0.00	0.20	0.82	1.82	3.16	4.77
.70	5.34	3.51	2.02	0.91	0.24	0.00	0.21	0.83	1.84	3.18	4.79
.71	5.32	3.49	2.00	0.90	0.23	0.00	0.21	0.83	1.85	3.19	4.80
.72	5.30	3.47	1.99	0.89	0.23	0.00	0.21	0.84	1.86	3.21	4.82
.73	5.28	3.46	1.98	0.88	0.22	0.00	0.22	0.85	1.87	3.22	4.84
.74	5.26	3.44	1.96	0.87	0.22	0.00	0.22	0.86	1.88	3.24	4.86
.75	5.24	3.43	1.95	0.87	0.21	0.00	0.23	0.87	1.90	3.25	4.88
.76	5.22	3.41	1.93	0.86	0.21	0.00	0.23	0.88	1.91	3.27	4.89
.77	5.20	3.39	1.92	0.85	0.20	0.00	0.23	0.89	1.92	3.28	4.91
.78	5.18	3.38	1.91	0.84	0.20	0.00	0.24	0.90	1.93	3.30	4.93
.79	5.16	3.36	1.90	0.83	0.19	0.00	0.24	0.91	1.94	3.31	4.95
.80	5.15	3.35	1.89	0.83	0.19	0.00	0.25	0.92	1.96	3.33	4.97
.81	5.13	3.33	1.87	0.82	0.18	0.00	0.25	0.92	1.97	3.34	4.98
.82	5.11	3.31	1.86	0.81	0.18	0.00	0.26	0.93	1.98	3.36	5.00
.83	5.09	3.29	1.85	0.80	0.17	0.00	0.26	0.94	1.99	3.37	5.02
.84	5.07	3.28	1.83	0.79	0.17	0.01	0.27	0.95	2.00	3.39	5.03
.85	5.05	3.26	1.82	0.78	0.17	0.01	0.27	0.96	2.02	3.40	5.05
.86	5.03	3.24	1.81	0.77	0.16	0.01	0.28	0.96	2.03	3.42	5.07
.87	5.01	3.23	1.79	0.76	0.16	0.01	0.28	0.97	2.04	3.43	5.08
.88	4.99	3.21	1.78	0.75	0.15	0.01	0.29	0.98	2.05	3.45	5.10
.89	4.97	3.19	1.77	0.74	0.15	0.01	0.29	0.99	2.06	3.46	5.12
.90	4.95	3.18	1.76	0.74	0.15	0.01	0.30	1.00	2.08	3.48	5.14
.91	4.93	3.16	1.74	0.73	0.14	0.01	0.30	1.01	2.09	3.49	5.15
.92	4.91	3.15	1.73	0.72	0.14	0.01	0.31	1.02	2.10	3.51	5.17
.93	4.89	3.13	1.72	0.71	0.14	0.01	0.31	1.03	2.12	3.52	5.19
.94	4.87	3.12	1.71	0.70	0.13	0.01	0.32	1.04	2.13	3.54	5.21
.95	4.85	3.10	1.70	0.70	0.13	0.01	0.32	1.05	2.14	3.55	5.23
.96	4.83	3.09	1.68	0.69	0.13	0.02	0.33	1.06	2.16	3.57	5.24
.97	4.81	3.07	1.67	0.68	0.12	0.02	0.33	1.07	2.17	3.58	5.26
.98	4.79	3.06	1.66	0.67	0.12	0.02	0.34	1.08	2.18	3.60	5.28
.99	4.77	3.04	1.65	0.66	0.12	0.02	0.34	1.09	2.19	3.61	5.30
1.00	4.76	3.03	1.64	0.66	0.12	0.02	0.35	1.10	2.21	3.63	5.32

TABLE VII″. ARGUMENT 2.

Equation $= 11''.2 - 11''.18 \sin. (2t - x) - 0''.304 \sin. (4t - 2x)$.

Period, 31.81193574 days.

Days.	21	22	23	24	25	26	27	28	29	30	31
Days. .00	″ 5.32	″ 7.21	″ 9.22	″ 11.30	″ 13.37	″ 15.38	″ 17.25	″ 18.91	″ 20.31	″ 21.39	″ 22.10
.01	5.33	7.22	9.24	11.32	13.39	15.40	17.27	18.93	20.33	21.40	22.11
.02	5.35	7.24	9.26	11.34	13.41	15.42	17.29	18.94	20.34	21.41	22.11
.03	5.37	7.26	9.28	11.36	13.43	15.44	17.30	18.96	20.35	21.42	22.12
.04	5.39	7.28	9.30	11.38	13.45	15.46	17.32	18.97	20.36	21.43	22.12
.05	5.41	7.30	9.32	11.40	13.47	15.48	17.34	18.98	20.37	21.44	22.13
.06	5.42	7.32	9.34	11.42	13.49	15.50	17.36	19.00	20.39	21.45	22.13
.07	5.44	7.34	9.36	11.44	13.51	15.52	17.37	19.01	20.40	21.46	22.13
.08	5.46	7.36	9.38	11.46	13.53	15.54	17.39	19.03	20.41	21.47	22.14
.09	5.48	7.38	9.40	11.48	13.55	15.56	17.41	19.04	20.42	21.48	22.14
.10	5.50	7.40	9.43	11.51	13.58	15.57	17.42	19.05	20.43	21.48	22.14
.11	5.51	7.42	9.45	11.53	13.60	15.59	17.44	19.07	20.45	21.49	22.15
.12	5.53	7.44	9.47	11.55	13.62	15.61	17.46	19.09	20.46	21.50	22.15
.13	5.55	7.46	9.49	11.57	13.64	15.63	17.48	19.10	20.47	21.51	22.16
.14	5.57	7.48	9.51	11.59	13.66	15.65	17.50	19.12	20.48	21.52	22.16
.15	5.59	7.50	9.53	11.61	13.68	15.67	17.51	19.13	20.49	21.52	22.17
.16	5.60	7.52	9.55	11.63	13.70	15.69	17.53	19.15	20.51	21.53	22.17
.17	5.62	7.54	9.57	11.65	13.72	15.71	17.55	19.17	20.52	21.54	22.17
.18	5.64	7.56	9.59	11.67	13.74	15.73	17.57	19.18	20.53	21.55	22.18
.19	5.66	7.58	9.61	11.69	13.76	15.75	17.59	19.20	20.54	21.56	22.18
.20	5.68	7.60	9.63	11.71	13.78	15.77	17.60	19.21	20.55	21.56	22.19
.21	5.69	7.62	9.65	11.73	13.80	15.79	17.62	19.23	20.57	21.57	22.19
.22	5.71	7.64	9.67	11.75	13.82	15.81	17.64	19.24	20.58	21.58	22.20
.23	5.73	7.66	9.69	11.77	13.84	15.83	17.66	19.26	20.59	21.59	22.20
.24	5.75	7.68	9.71	11.79	13.86	15.85	17.67	19.27	20.60	21.60	22.20
.25	5.77	7.70	9.73	11.81	13.88	15.87	17.69	19.29	20.61	21.60	22.21
.26	5.79	7.72	9.75	11.83	13.90	15.89	17.71	19.30	20.63	21.61	22.21
.27	5.81	7.74	9.77	11.85	13.92	15.91	17.72	19.32	20.64	21.62	22.22
.28	5.83	7.76	9.79	11.87	13.94	15.93	17.74	19.33	20.65	21.63	22.22
.29	5.85	7.78	9.81	11.89	13.96	15.95	17.76	19.35	20.66	21.64	22.22
.30	5.87	7.80	9.84	11.92	13.99	15.96	17.77	19.36	20.67	21.64	22.23
.31	5.88	7.82	9.86	11.94	14.01	15.98	17.79	19.38	20.69	21.65	22.23
.32	5.90	7.84	9.88	11.96	14.03	16.00	17.81	19.39	20.70	21.66	22.23
.33	5.92	7.86	9.90	11.98	14.05	16.02	17.83	19.41	20.71	21.67	22.24
.34	5.94	7.88	9.92	12.00	14.07	16.04	17.84	19.42	20.72	21.68	22.24
.35	5.96	7.90	9.94	12.02	14.09	16.06	17.86	19.43	20.73	21.68	22.24
.36	5.97	7.92	9.96	12.04	14.11	16.08	17.88	19.45	20.75	21.69	22.25
.37	5.99	7.94	9.98	12.06	14.13	16.10	17.89	19.47	20.76	21.70	22.25
.38	6.01	7.96	10.00	12.08	14.15	16.12	17.91	19.48	20.77	21.71	22.25
.39	6.03	7.98	10.02	12.10	14.17	16.14	17.93	19.49	20.78	21.72	22.25
.40	6.05	8.00	10.05	12.13	14.19	16.15	17.94	19.50	20.79	21.72	22.26
.41	6.06	8.02	10.07	12.15	14.21	16.17	17.96	19.52	20.81	21.73	22.26
.42	6.08	8.04	10.09	12.17	14.23	16.19	17.98	19.53	20.82	21.74	22.26
.43	6.10	8.06	10.11	12.19	14.25	16.21	18.00	19.55	20.83	21.75	22.27
.44	6.12	8.08	10.13	12.21	14.27	16.23	18.01	19.56	20.84	21.75	22.27
.45	6.14	8.10	10.15	12.23	14.29	16.24	18.03	19.58	20.85	21.76	22.28
.46	6.16	8.12	10.17	12.25	14.31	16.26	18.05	19.59	20.86	21.77	22.28
.47	6.18	8.14	10.19	12.27	14.33	16.28	18.06	19.61	20.87	21.77	22.28
.48	6.20	8.16	10.21	12.29	14.35	16.30	18.08	19.62	20.88	21.78	22.29
.49	6.22	8.18	10.23	12.31	14.37	16.32	18.10	19.64	20.89	21.79	22.29
.50	6.24	8.20	10.25	12.34	14.39	16.33	18.11	19.65	20.90	21.79	22.30

TABLE VII″. ARGUMENT 2.

Equation = 11″.2 — 11″.18 sin. $(2t - x)$ — 0″.304 sin. $(4t - 2x)$.

Period, 31.81193574 days.

Days.	21	22	23	24	25	26	27	28	29	30	31
Days.	″	″	″	″	″	″	″	″	″	″	″
.50	6.24	8.20	10.25	12.34	14.39	16.33	18.11	19.65	20.90	21.79	22.30
.51	6.25	8.22	10.27	12.36	14.41	16.35	18.13	19.67	20.91	21.80	22.30
.52	6.27	8.24	10.29	12.38	14.43	16.37	18.15	19.68	20.92	21.81	22.30
.53	6.29	8.26	10.31	12.40	14.45	16.39	18.16	19.70	20.93	21.81	22.30
.54	6.31	8.28	10.33	12.42	14.47	16.41	18.18	19.71	20.94	21.82	22.30
.55	6.33	8.30	10.35	12.44	14.49	16.43	18.19	19.72	20.95	21.83	22.31
.56	6.35	8.32	10.37	12.46	14.51	16.45	18.21	19.74	20.96	21.83	22.31
.57	6.37	8.34	10.39	12.48	14.53	16.47	18.23	19.75	20.97	21.84	22.31
.58	6.39	8.36	10.41	12.50	14.55	16.49	18.24	19.77	20.98	21.85	22.31
.59	6.41	8.38	10.43	12.52	14.57	16.51	18.26	19.78	20.99	21.86	22.31
.60	6.43	8.40	10.46	12.55	14.59	16.52	18.27	19.79	21.00	21.86	22.32
.61	6.44	8.42	10.48	12.57	14.61	16.54	18.29	19.81	21.01	21.87	22.32
.62	6.46	8.44	10.50	12.59	14.63	16.56	18.31	19.82	21.02	21.88	22.32
.63	6.48	8.46	10.52	12.61	14.65	16.58	18.33	19.83	21.03	21.88	22.33
.64	6.50	8.48	10.54	12.63	14.67	16.60	18.34	19.85	21.04	21.89	22.33
.65	6.52	8.50	10.56	12.65	14.69	16.61	18.36	19.86	21.05	21.90	22.33
.66	6.54	8.52	10.58	12.67	14.71	16.63	18.37	19.87	21.06	21.90	22.34
.67	6.56	8.54	10.60	12.69	14.73	16.65	18.39	19.89	21.07	21.91	22.34
.68	6.58	8.56	10.62	12.71	14.75	16.67	18.40	19.90	21.08	21.92	22.34
.69	6.60	8.58	10.64	12.73	14.77	16.69	18.42	19.91	21.09	21.93	22.34
.70	6.62	8.61	10.67	12.76	14.79	16.70	18.44	19.92	21.10	21.93	22.35
.71	6.64	8.63	10.69	12.78	14.81	16.72	18.46	19.94	21.11	21.94	22.35
.72	6.66	8.65	10.71	12.80	14.83	16.74	18.48	19.95	21.12	21.95	22.35
.73	6.68	8.67	10.73	12.82	14.85	16.76	18.49	19.96	21.13	21.95	22.35
.74	6.70	8.69	10.75	12.84	14.87	16.78	18.51	19.98	21.14	21.96	22.35
.75	6.72	8.71	10.77	12.86	14.89	16.80	18.52	19.99	21.15	21.96	22.36
.76	6.74	8.73	10.79	12.88	14.91	16.82	18.54	20.01	21.16	21.97	22.36
.77	6.76	8.75	10.81	12.90	14.93	16.84	18.56	20.02	21.17	21.98	22.36
.78	6.78	8.77	10.83	12.92	14.95	16.86	18.57	20.03	21.18	21.98	22.36
.79	6.80	8.79	10.85	12.94	14.97	16.88	18.59	20.04	21.19	21.99	22.36
.80	6.82	8.81	10.88	12.96	14.99	16.89	18.60	20.05	21.20	21.99	22.36
.81	6.83	8.83	10.90	12.98	15.01	16.91	18.62	20.07	21.21	22.00	22.37
.82	6.85	8.85	10.92	13.00	15.03	16.93	18.61	20.08	21.22	22.00	22.37
.83	6.87	8.87	10.94	13.02	15.05	16.95	18.65	20.09	21.23	22.01	22.37
.84	6.89	8.89	10.96	13.04	15.07	16.97	18.67	20.11	21.24	22.01	22.37
.85	6.91	8.91	10.98	13.06	15.09	16.98	18.68	20.12	21.25	22.02	22.38
.86	6.93	8.93	11.00	13.08	15.11	17.00	18.70	20.13	21.26	22.02	22.38
.87	6.95	8.95	11.02	13.10	15.13	17.02	18.72	20.15	21.27	22.03	22.38
.88	6.97	8.97	11.04	13.12	15.15	17.04	18.73	20.16	21.28	22.03	22.38
.89	6.99	8.99	11.06	13.14	15.17	17.06	18.75	20.17	21.29	22.04	22.38
.90	7.01	9.01	11.09	13.17	15.18	17.07	18.76	20.18	21.30	22.04	22.38
.91	7.03	9.03	11.11	13.19	15.20	17.09	18.78	20.20	21.31	22.05	22.38
.92	7.05	9.05	11.13	13.21	15.22	17.11	18.79	20.21	21.32	22.05	22.38
.93	7.07	9.07	11.15	13.23	15.24	17.13	18.81	20.22	21.33	22.06	22.38
.94	7.09	9.09	11.17	13.25	15.26	17.15	18.82	20.24	21.34	22.06	22.38
.95	7.11	9.11	11.19	13.27	15.28	17.16	18.84	20.25	21.35	22.07	22.39
.96	7.13	9.13	11.21	13.29	15.30	17.18	18.85	20.26	21.36	22.08	22.39
.97	7.15	9.15	11.23	13.31	15.32	17.20	18.87	20.28	21.37	22.08	22.39
.98	7.17	9.17	11.25	13.33	15.34	17.22	18.88	20.29	21.38	22.09	22.39
.99	7.19	9.19	11.27	13.35	15.36	17.24	18.90	20.30	21.39	22.10	22.39
1.00	7.21	9.22	11.30	13.37	15.38	17.25	18.91	20.31	21.39	22.10	22.39

TABLE VIII″. ARGUMENT 3.

Equation $= 31''.16 + 0''.345 \sin. t - 20''.808 \sin. 2t - 0''.023 \sin. 3t - 0''.65 \sin. 4t$.

Period, 29.530588 days.

Days.	0	1	2	3	4	5	6	7	8	9
Days.	″	″	″	″	″	″	″	″	″	″
.00	33.97	22.14	12.11	5.78	4.18	7.33	14.43	24.16	34.93	45.10
.01	33.85	22.03	12.03	5.74	4.19	7.38	14.52	24.27	35.04	45.19
.02	33.73	21.92	11.94	5.70	4.20	7.44	14.61	24.37	35.15	45.28
.03	33.62	21.80	11.86	5.66	4.20	7.49	14.69	24.48	35.25	45.38
.04	33.50	21.69	11.77	5.62	4.21	7.55	14.78	24.58	35.36	45.47
.05	33.38	21.58	11.69	5.58	4.22	7.60	14.87	24.69	35.47	45.56
.06	33.26	21.47	11.61	5.54	4.23	7.66	14.96	24.80	35.58	45.65
.07	33.14	21.36	11.53	5.51	4.24	7.71	15.05	24.90	35.68	45.74
.08	33.02	21.24	11.45	5.47	4.26	7.77	15.13	25.01	35.79	45.83
.09	32.90	21.13	11.37	5.44	4.27	7.82	15.22	25.11	35.89	45.92
.10	32.78	21.02	11.29	5.40	4.28	7.88	15.31	25.22	36.00	46.01
.11	32.66	20.91	11.21	5.36	4.30	7.94	15.40	25.33	36.11	46.10
.12	32.54	20.80	11.13	5.33	4.31	8.00	15.49	25.43	36.21	46.19
.13	32.42	20.69	11.05	5.29	4.33	8.05	15.58	25.54	36.32	46.28
.14	32.30	20.58	10.97	5.26	4.34	8.11	15.67	25.64	36.42	46.37
.15	32.18	20.47	10.89	5.22	4.36	8.17	15.76	25.75	36.53	46.46
.16	32.06	20.36	10.81	5.19	4.38	8.23	15.85	25.86	36.64	46.55
.17	31.94	20.25	10.73	5.16	4.39	8.29	15.94	25.96	36.74	46.64
.18	31.82	20.14	10.66	5.12	4.41	8.35	16.04	26.07	36.85	46.73
.19	31.70	20.03	10.58	5.09	4.42	8.41	16.13	26.17	36.95	46.82
.20	31.58	19.92	10.50	5.06	4.44	8.47	16.22	26.28	37.06	46.91
.21	31.46	19.81	10.42	5.03	4.46	8.53	16.31	26.39	37.17	47.00
.22	31.34	19.70	10.35	5.00	4.48	8.59	16.40	26.50	37.27	47.09
.23	31.22	19.60	10.27	4.98	4.50	8.66	16.50	26.60	37.38	47.17
.24	31.10	19.49	10.20	4.95	4.52	8.72	16.59	26.71	37.48	47.26
.25	30.98	19.38	10.12	4.92	4.54	8.78	16.68	26.82	37.59	47.35
.26	30.86	19.27	10.05	4.89	4.56	8.84	16.77	26.93	37.70	47.44
.27	30.74	19.16	9.98	4.86	4.58	8.90	16.87	27.04	37.80	47.52
.28	30.62	19.06	9.90	4.84	4.60	8.97	16.96	27.14	37.91	47.61
.29	30.50	18.95	9.83	4.81	4.62	9.03	17.06	27.25	38.01	47.69
.30	30.38	18.84	9.76	4.78	4.64	9.09	17.15	27.36	38.12	47.78
.31	30.26	18.73	9.69	4.76	4.66	9.16	17.24	27.47	38.22	47.86
.32	30.14	18.63	9.62	4.73	4.69	9.22	17.34	27.58	38.33	47.95
.33	30.01	18.52	9.54	4.71	4.71	9.29	17.43	27.68	38.43	48.03
.34	29.89	18.42	9.47	4.68	4.74	9.35	17.53	27.79	38.54	48.12
.35	29.77	18.31	9.40	4.66	4.76	9.42	17.62	27.90	38.64	48.20
.36	29.65	18.21	9.33	4.64	4.79	9.49	17.72	28.01	38.74	48.28
.37	29.53	18.10	9.26	4.62	4.82	9.56	17.81	28.12	38.85	48.37
.38	29.41	18.00	9.19	4.59	4.84	9.62	17.91	28.22	38.95	48.45
.39	29.29	17.89	9.12	4.57	4.87	9.69	18.00	28.33	39.06	48.54
.40	29.17	17.79	9.05	4.55	4.90	9.75	18.10	28.44	39.16	48.62
.41	29.05	17.69	8.98	4.53	4.93	9.83	18.20	28.55	39.26	48.70
.42	28.93	17.59	8.92	4.51	4.96	9.90	18.29	28.66	39.37	48.78
.43	28.81	17.48	8.85	4.49	4.98	9.96	18.39	28.76	39.47	48.87
.44	28.69	17.38	8.79	4.47	5.01	10.03	18.48	28.87	39.58	48.95
.45	28.57	17.28	8.72	4.45	5.04	10.10	18.58	28.98	39.68	49.03
.46	28.45	17.18	8.65	4.43	5.07	10.17	18.68	29.09	39.78	49.11
.47	28.33	17.08	8.59	4.41	5.10	10.24	18.77	29.20	39.88	49.19
.48	28.21	16.97	8.52	4.40	5.13	10.31	18.87	29.30	39.99	49.28
.49	28.09	16.87	8.46	4.38	5.16	10.38	18.96	29.41	40.09	49.36
.50	27.97	16.77	8.39	4.36	5.19	10.45	19.06	29.52	40.19	49.44

TABLE VIII″. ARGUMENT 3.

Equation = 31″.16 + 0″.345 sin. t — 20″.808 sin. $2t$ — 0″.023 sin. $3t$ — 0″.65 sin. $4t$.

Period, 29.530588 days.

Days.	0	1	2	3	4	5	6	7	8	9
Days. .50	27.97″	16.77″	8.39″	4.36″	5.19″	10.45″	19.06″	29.52″	40.19″	49.44″
.51	27.85	16.67	8.33	4.35	5.22	10.52	19.16	29.63	40.29	49.52
.52	27.73	16.57	8.27	4.33	5.26	10.59	19.26	29.74	40.39	49.60
.53	27.62	16.46	8.20	4.32	5.29	10.67	19.36	29.84	40.50	49.67
.54	27.50	16.36	8.14	4.30	5.33	10.74	19.46	29.95	40.60	49.75
.55	27.38	16.26	8.08	4.29	5.36	10.81	19.56	30.06	40.70	49.83
.56	27.26	16.16	8.02	4.28	5.39	10.89	19.66	30.17	40.80	49.91
.57	27.14	16.06	7.96	4.27	5.43	10.96	19.76	30.28	40.90	49.99
.58	27.03	15.97	7.90	4.25	5.46	11.01	19.85	30.38	41.01	50.06
.59	26.91	15.87	7.84	4.24	5.50	11.11	19.95	30.49	41.11	50.14
.60	26.79	15.77	7.78	4.23	5.53	11.19	20.05	30.60	41.21	50.22
.61	26.67	15.67	7.72	4.22	5.57	11.27	20.15	30.71	41.31	50.30
.62	26.55	15.57	7.66	4.21	5.61	11.34	20.25	30.82	41.41	50.37
.63	26.44	15.48	7.61	4.20	5.65	11.42	20.35	30.93	41.51	50.45
.64	26.32	15.38	7.55	4.19	5.69	11.49	20.45	31.04	41.61	50.52
.65	26.20	15.28	7.49	4.18	5.73	11.57	20.55	31.15	41.71	50.60
.66	26.08	15.18	7.43	4.17	5.77	11.65	20.65	31.26	41.81	50.67
.67	25.96	15.09	7.38	4.16	5.81	11.73	20.75	31.37	41.91	50.75
.68	25.85	14.99	7.32	4.16	5.85	11.80	20.86	31.47	42.01	50.82
.69	25.73	14.90	7.27	4.15	5.89	11.88	20.96	31.58	42.11	50.90
.70	25.61	14.80	7.21	4.14	5.93	11.96	21.07	31.69	42.21	50.97
.71	25.49	14.71	7.16	4.13	5.97	12.04	21.16	31.80	42.31	51.04
.72	25.37	14.61	7.10	4.13	6.01	12.12	21.26	31.91	42.41	51.12
.73	25.26	14.52	7.05	4.12	6.05	12.19	21.37	32.02	42.50	51.19
.74	25.14	14.42	6.99	4.12	6.09	12.27	21.47	32.13	42.60	51.27
.75	25.02	14.33	6.94	4.11	6.13	12.35	21.57	32.23	42.70	51.34
.76	24.90	14.24	6.89	4.11	6.17	12.43	21.67	32.34	42.80	51.41
.77	24.79	14.15	6.84	4.11	6.22	12.51	21.77	32.45	42.90	51.48
.78	24.67	14.05	6.79	4.10	6.26	12.59	21.88	32.55	42.99	51.56
.79	24.56	13.96	6.74	4.10	6.31	12.67	21.98	32.66	43.09	51.63
.80	24.44	13.87	6.69	4.10	6.35	12.75	22.08	32.77	43.19	51.70
.81	24.32	13.78	6.64	4.10	6.40	12.83	22.18	32.88	43.29	51.77
.82	24.21	13.69	6.59	4.10	6.44	12.91	22.28	32.99	43.38	51.84
.83	24.09	13.60	6.54	4.10	6.49	13.00	22.39	33.09	43.48	51.90
.84	23.98	13.51	6.49	4.10	6.53	13.08	22.49	33.20	43.57	51.97
.85	23.86	13.42	6.44	4.10	6.58	13.16	22.59	33.31	43.67	52.04
.86	23.74	13.33	6.39	4.10	6.63	13.24	22.69	33.42	43.77	52.11
.87	23.63	13.24	6.35	4.11	6.68	13.32	22.80	33.53	43.86	52.18
.88	23.51	13.15	6.30	4.11	6.72	13.41	22.90	33.63	43.96	52.24
.89	23.40	13.06	6.26	4.12	6.77	13.49	23.01	33.74	44.05	52.31
.90	23.28	12.97	6.21	4.12	6.82	13.57	23.11	33.85	44.15	52.38
.91	23.17	12.88	6.17	4.12	6.87	13.66	23.22	33.96	44.25	52.45
.92	23.05	12.80	6.12	4.13	6.92	13.74	23.32	34.07	44.34	52.51
.93	22.94	12.71	6.08	4.13	6.97	13.83	23.43	34.17	44.44	52.58
.94	22.82	12.63	6.03	4.14	7.02	13.91	23.53	34.28	44.53	52.64
.95	22.71	12.54	5.99	4.14	7.07	14.00	23.64	34.39	44.63	52.71
.96	22.60	12.45	5.95	4.15	7.12	14.09	23.74	34.50	44.72	52.77
.97	22.48	12.37	5.91	4.16	7.17	14.17	23.85	34.61	44.82	52.84
.98	22.37	12.28	5.86	4.16	7.23	14.26	23.95	34.71	44.91	52.90
.99	22.25	12.20	5.82	4.17	7.28	14.34	24.06	34.82	45.01	52.97
1.00	22.14	12.11	5.78	4.18	7.33	14.43	24.16	34.93	45.10	53.03

TABLE VIII″. ARGUMENT 3.

Equation $= 31''.16 + 0''.345 \sin. t - 20''.808 \sin. 2t - 0''.023 \sin. 3t - 0''.65 \sin. 4t.$

Period, 29.530588 days.

Days.	10	11	12	13	14	15	16	17	18	19
Days.										
.00	53.03	57.32	56.99	51.84	42.65	31.16	19.67	10.48	5.33	5.00
.01	53.09	57.34	56.96	51.77	42.51	31.04	19.56	10.41	5.30	5.02
.02	53.15	57.36	56.93	51.69	42.44	30.92	19.45	10.34	5.28	5.04
.03	53.22	57.38	56.91	51.62	42.33	30.80	19.35	10.26	5.25	5.07
.04	53.28	57.40	56.88	51.54	42.23	30.68	19.24	10.19	5.23	5.09
.05	53.34	57.42	56.85	51.47	42.12	30.56	19.13	10.12	5.20	5.11
.06	53.40	57.44	56.82	51.39	42.01	30.44	19.03	10.05	5.18	5.13
.07	53.46	57.46	56.79	51.31	41.90	30.32	18.92	9.98	5.15	5.16
.08	53.53	57.47	56.75	51.24	41.79	30.21	18.82	9.90	5.13	5.18
.09	53.59	57.49	56.72	51.16	41.68	30.09	18.71	9.83	5.10	5.21
.10	53.65	57.51	56.69	51.08	41.57	29.97	18.61	9.76	5.08	5.23
.11	53.71	57.52	56.66	51.00	41.46	29.85	18.51	9.69	5.06	5.26
.12	53.77	57.54	56.62	50.92	41.35	29.73	18.40	9.62	5.04	5.28
.13	53.82	57.55	56.59	50.85	41.24	29.62	18.30	9.56	5.01	5.31
.14	53.88	57.57	56.55	50.77	41.13	29.50	18.19	9.49	4.99	5.33
.15	53.94	57.58	56.52	50.69	41.02	29.38	18.09	9.42	4.97	5.36
.16	54.00	57.59	56.48	50.61	40.91	29.26	17.99	9.35	4.95	5.39
.17	54.05	57.61	56.45	50.53	40.80	29.14	17.89	9.29	4.93	5.42
.18	54.11	57.62	56.41	50.44	40.69	29.02	17.78	9.22	4.91	5.45
.19	54.16	57.64	56.38	50.36	40.58	28.90	17.68	9.16	4.89	5.48
.20	54.22	57.65	56.31	50.28	40.47	28.78	17.58	9.09	4.87	5.51
.21	54.28	57.66	56.30	50.20	40.36	28.66	17.48	9.02	4.85	5.54
.22	54.33	57.67	56.26	50.12	40.25	28.51	17.38	8.96	4.83	5.57
.23	54.39	57.68	56.23	50.03	40.13	28.43	17.27	8.89	4.82	5.60
.24	54.44	57.69	56.19	49.95	40.02	28.31	17.17	8.83	4.80	5.63
.25	54.50	57.70	56.15	49.87	39.91	28.19	17.07	8.76	4.78	5.66
.26	54.55	57.71	56.11	49.79	39.80	28.07	16.97	8.70	4.77	5.69
.27	54.60	57.71	56.07	49.70	39.69	27.95	16.87	8.64	4.75	5.73
.28	54.66	57.72	56.02	49.62	39.57	27.84	16.77	8.57	4.74	5.76
.29	54.71	57.72	55.98	49.53	39.46	27.72	16.67	8.51	4.72	5.80
.30	54.76	57.73	55.94	49.45	39.35	27.60	16.57	8.45	4.71	5.83
.31	54.81	57.74	55.90	49.36	39.24	27.48	16.47	8.39	4.70	5.87
.32	54.86	57.74	55.85	49.28	39.12	27.37	16.37	8.33	4.69	5.90
.33	54.91	57.75	55.81	49.19	39.01	27.25	16.28	8.28	4.68	5.94
.34	54.96	57.75	55.76	49.11	38.89	27.14	16.18	8.22	4.67	5.97
.35	55.01	57.76	55.72	49.02	38.78	27.02	16.08	8.16	4.66	6.01
.36	55.06	57.76	55.67	48.93	38.67	26.90	15.98	8.10	4.65	6.05
.37	55.11	57.77	55.63	48.84	38.55	26.78	15.89	8.04	4.64	6.09
.38	55.16	57.77	55.58	48.75	38.44	26.67	15.79	7.99	4.63	6.12
.39	55.21	57.78	55.54	48.66	38.32	26.55	15.70	7.93	4.62	6.16
.40	55.26	57.78	55.49	48.57	38.21	26.43	15.60	7.87	4.61	6.20
.41	55.31	57.78	55.44	48.48	38.10	26.31	15.50	7.81	4.60	6.24
.42	55.35	57.78	55.39	48.39	37.98	26.19	15.41	7.76	4.60	6.28
.43	55.40	57.78	55.35	48.30	37.87	26.08	15.31	7.70	4.59	6.32
.44	55.44	57.78	55.30	48.21	37.75	25.96	15.22	7.65	4.59	6.36
.45	55.49	57.78	55.25	48.12	37.64	25.84	15.12	7.59	4.58	6.40
.46	55.53	57.78	55.20	48.03	37.52	25.72	15.03	7.54	4.57	6.44
.47	55.58	57.78	55.15	47.94	37.41	25.61	14.94	7.49	4.57	6.48
.48	55.62	57.77	55.09	47.85	37.29	25.49	14.84	7.43	4.56	6.53
.49	55.67	57.77	55.04	47.76	37.18	25.38	14.75	7.38	4.56	6.57
.50	55.71	57.77	54.99	47.67	37.06	25.26	14.66	7.33	4.55	6.61

TABLE VIII″. ARGUMENT 3.

Equation = 31″.16 + 0″.345 sin. t — 20″.808 sin. $2t$ — 0″.023 sin. $3t$ — 0″.65 sin. $4t$.

Period, 29.530588 days.

Days.	10	11	12	13	14	15	16	17	18	19
Days.										
.50	55.71	57.77	54.99	47.67	37.06	25.26	14.66	7.33	4.55	6.61
.51	55.75	57.77	54.94	47.58	36.94	25.14	14.57	7.28	4.55	6.65
.52	55.79	57.76	54.89	47.48	36.82	25.03	14.48	7.23	4.55	6.70
.53	55.84	57.76	54.83	47.39	36.71	24.91	14.38	7.17	4.54	6.74
.54	55.88	57.75	54.78	47.29	36.60	24.80	14.29	7.12	4.54	6.79
.55	55.92	57.75	54.73	47.20	36.48	24.68	14.20	7.07	4.54	6.83
.56	55.96	57.74	54.67	47.10	36.36	24.57	14.11	7.02	4.54	6.88
.57	56.00	57.73	54.62	47.01	36.24	24.45	14.02	6.97	4.54	6.92
.58	56.04	57.73	54.56	46.91	36.13	24.34	13.93	6.93	4.54	6.97
.59	56.08	57.72	54.51	46.82	36.01	24.22	13.84	6.88	4.54	7.01
.60	56.12	57.71	54.45	46.72	35.89	24.11	13.75	6.83	4.54	7.06
.61	56.16	57.70	54.39	46.62	35.77	24.00	13.66	6.78	4.54	7.11
.62	56.20	57.69	54.33	46.53	35.65	23.88	13.57	6.74	4.55	7.16
.63	56.23	57.68	54.28	46.43	35.54	23.77	13.48	6.69	4.55	7.21
.64	56.27	57.67	54.22	46.34	35.42	23.65	13.39	6.65	4.56	7.26
.65	56.31	57.66	54.16	46.24	35.30	23.54	13.30	6.60	4.56	7.31
.66	56.35	57.65	54.10	46.14	35.18	23.43	13.21	6.56	4.57	7.36
.67	56.38	57.64	54.04	46.04	35.06	23.31	13.13	6.51	4.57	7.41
.68	56.42	57.63	53.99	45.95	34.95	23.20	13.04	6.47	4.58	7.46
.69	56.45	57.62	53.93	45.85	34.83	23.08	12.96	6.42	4.58	7.51
.70	56.49	57.61	53.87	45.75	34.71	22.97	12.87	6.38	4.59	7.56
.71	56.52	57.60	53.81	45.65	34.59	22.86	12.79	6.34	4.60	7.61
.72	56.56	57.58	53.75	45.55	34.48	22.75	12.70	6.30	4.60	7.66
.73	56.59	57.57	53.69	45.45	34.36	22.63	12.62	6.26	4.61	7.72
.74	56.63	57.55	53.63	45.35	34.25	22.52	12.53	6.22	4.61	7.77
.75	56.66	57.54	53.57	45.25	34.13	22.41	12.45	6.18	4.62	7.82
.76	56.69	57.52	53.50	45.15	34.01	22.30	12.37	6.14	4.63	7.87
.77	56.72	57.50	53.44	45.05	33.89	22.19	12.29	6.10	4.64	7.93
.78	56.75	57.49	53.37	44.94	33.78	22.07	12.20	6.06	4.65	7.98
.79	56.78	57.47	53.31	44.84	33.66	21.96	12.12	6.02	4.66	8.04
.80	56.81	57.45	53.24	44.74	33.54	21.85	12.04	5.98	4.67	8.10
.81	56.84	57.43	53.17	44.64	33.42	21.74	11.96	5.94	4.68	8.16
.82	56.87	57.41	53.10	44.54	33.30	21.63	11.88	5.91	4.70	8.21
.83	56.90	57.39	53.04	44.43	33.18	21.52	11.79	5.87	4.71	8.27
.84	56.93	57.37	52.97	44.33	33.06	21.41	11.71	5.84	4.73	8.32
.85	56.96	57.35	52.90	44.23	32.94	21.30	11.63	5.80	4.74	8.38
.86	56.99	57.33	52.83	44.13	32.82	21.19	11.55	5.77	4.75	8.44
.87	57.01	57.31	52.76	44.02	32.70	21.08	11.47	5.73	4.77	8.50
.88	57.04	57.29	52.70	43.92	32.59	20.97	11.39	5.70	4.78	8.55
.89	57.06	57.27	52.63	43.81	32.47	20.86	11.31	5.66	4.80	8.61
.90	57.09	57.25	52.56	43.71	32.35	20.75	11.23	5.63	4.81	8.67
.91	57.11	57.23	52.49	43.61	32.23	20.64	11.15	5.60	4.83	8.73
.92	57.14	57.20	52.42	43.50	32.11	20.53	11.08	5.57	4.85	8.79
.93	57.16	57.18	52.35	43.40	32.00	20.43	11.00	5.53	4.86	8.86
.94	57.19	57.15	52.28	43.29	31.88	20.32	10.93	5.50	4.88	8.92
.95	57.21	57.13	52.21	43.19	31.76	20.21	10.85	5.47	4.90	8.98
.96	57.23	57.10	52.14	43.08	31.64	20.10	10.78	5.44	4.92	9.04
.97	57.25	57.07	52.06	42.97	31.52	19.99	10.70	5.41	4.94	9.10
.98	57.28	57.05	51.99	42.87	31.40	19.89	10.63	5.39	4.96	9.17
.99	57.30	57.02	51.91	42.76	31.28	19.78	10.55	5.36	4.98	9.23
1.00	57.32	56.99	51.84	42.65	31.16	19.67	10.48	5.33	5.00	9.29

TABLE VIII″. ARGUMENT 3.

Equation = 31″.16 + 0″.345 sin. t — 20″.808 sin. $2t$ — 0″.023 sin. $3t$ — 0″.65 sin. $4t$.

Period, 29.530588 days.

Days.	20	21	22	23	24	25	26	27	28	29
Days.	″	″	″	″	″	″	″	″	″	″
.00	9.29	17.23	27.39	38.16	47.89	51.99	58.14	56.54	50.21	40.18
.01	9.35	17.32	27.50	38.26	47.98	55.04	58.15	56.50	50.12	40.07
.02	9.42	17.42	27.61	38.37	48.06	55.09	58.16	56.46	50.04	39.95
.03	9.48	17.51	27.71	38.47	48.15	55.15	58.16	56.41	49.95	39.84
.04	9.55	17.61	27.82	38.58	48.23	55.20	58.17	56.37	49.87	39.72
.05	9.61	17.70	27.93	38.68	48.32	55.25	58.18	56.33	49.78	39.61
.06	9.68	17.79	28.01	38.79	48.10	55.30	58.18	56.29	49.69	39.50
.07	9.74	17.89	28.14	38.89	48.19	55.35	58.19	56.24	49.60	39.38
.08	9.81	17.98	28.25	39.00	48.57	55.40	58.19	56.20	49.52	39.27
.09	9.87	18.08	28.35	39.10	48.66	55.45	58.20	56.15	49.43	39.15
.10	9.94	18.17	28.46	39.21	48.74	55.50	58.20	56.11	49.34	39.04
.11	10.01	18.27	28.57	39.31	48.82	55.55	58.20	56.06	49.25	38.92
.12	10.08	18.36	28.68	39.42	48.91	55.60	58.21	56.02	49.16	38.81
.13	10.14	18.46	28.78	39.52	48.99	55.64	58.21	55.97	49.08	38.69
.14	10.21	18 55	28.89	39.63	49.08	55.69	58.22	55.93	48.99	38.58
.15	10.28	18.65	29.00	39.73	49.16	55.74	58.22	55.88	48.90	38.46
.16	10.35	18.75	29.11	39.83	49.24	55.79	58.22	55.83	48.81	38.34
.17	10.42	18.84	29.22	39.93	49.32	55.83	58.22	55.78	48.72	38.23
.18	10.49	18.94	29.33	40.04	49.41	55.88	58.22	55.73	48.63	38.11
.19	10.56	19.03	29.44	40.14	49.49	55.92	58.22	55.68	48.54	38.00
.20	10.63	19.13	29.55	40.24	49.57	55.97	58.22	55.63	48.45	37.88
.21	10.70	19.23	29.66	40.34	49.65	56.01	58.22	55.58	48.36	37.76
.22	10.77	19.33	29.77	40.44	49.73	56.06	58.22	55.53	48.27	37.65
.23	10.84	19.42	29.87	40.55	49.81	56.10	58.21	55.48	48.17	37.53
.24	10.91	19.52	29.98	40.65	49.89	56.15	58.21	55.43	48.08	37.42
.25	10.98	19.62	30.09	40.75	49.97	56.19	58.21	55.38	47.99	37.30
.26	11.05	19.72	30.20	40.85	50.05	56.23	58.20	55.33	47.89	37.18
.27	11.13	19.82	30.31	40.95	50.13	56.27	58.20	55.27	47.80	37.06
.28	11.20	19.91	30.11	41.06	50.20	56.32	58.19	55.22	47.70	36.95
.29	11.28	20.01	30.52	41.16	50.28	56.36	58.19	55.16	47.61	36.83
.30	11.35	20.11	30.63	41.26	50.36	56.40	58.18	55.11	47.52	36.71
.31	11.42	20.21	30.74	41.36	50.44	56.44	58.17	55.05	47.42	36.59
.32	11.50	20.31	30.85	41.46	50.52	56.48	58.16	55.00	47.33	36.47
.33	11.57	20.11	30.95	41.57	50.59	56.52	58.16	54.94	47.23	36.36
.34	11.65	20.51	31.06	41.67	50.67	56.56	58.15	54.89	47.14	36.24
.35	11.72	20.61	31.17	41.77	50.75	56.60	58.14	54.83	47.04	36.12
.36	11.80	20.71	31.28	41.87	50.83	56.64	58.13	54.77	46.94	36.00
.37	11.87	20.81	31.39	41.97	50.90	56.68	58.12	54.71	46.84	35.88
.38	11.95	20.91	31.50	42.07	50.98	56.71	58.11	54.66	46.75	35.77
.39	12.02	21.01	31.61	42.17	51.05	56.75	58.10	54.60	46.65	35.65
.40	12.10	21.11	31.72	42.27	51.13	56.79	58.09	54.54	46.55	35.53
.41	12.18	21.21	31.83	42.37	51.20	56.82	58.08	54.48	46.45	35.41
.42	12.26	21.31	31.93	42.47	51.28	56.86	58.07	54.42	46.35	35.29
.43	12.33	21.42	32.04	42.56	51.35	56.89	58.05	54.36	46.26	35.18
.44	12.41	21.52	32.15	42.66	51.43	56.93	58.04	54.30	46.16	35.06
.45	12.49	21.62	32.26	42.76	51.50	56.96	58.03	54.24	46.06	34.94
.46	12.57	21.72	32.37	42.86	51.57	56.99	58.02	54.18	45.96	34.82
.47	12.65	21.82	32.48	42.96	51.64	57.02	58.00	54.12	45.86	34.70
.48	12.72	21.93	32.58	43.05	51.72	57.06	57.99	54.05	45.76	34.58
.49	12.80	22.03	32.69	43.15	51.79	57.09	57.97	53.99	45.66	34.46
.50	12.88	22.13	32.80	43.25	51.86	57.13	57.96	53.93	45.55	34.34

TABLE VIII″. ARGUMENT 3.

Equation $= 31''.16 + 0''.345 \sin. t - 20''.808 \sin. 2t - 0''.023 \sin. 3t - 0''.65 \sin. 4t$.

Period, 29.530588 days.

Days.	20	21	22	23	24	25	26	27	28	29
Days .50	12.88	22.13	32.80	43.25	51.86	57.13	57.96	53.93	45.55	34.34
.51	12.96	22.23	32.91	43.35	51.93	57.15	57.94	53.86	45.45	34.22
.52	13.04	22.33	33.02	43.45	52.00	57.18	57.92	53.80	45.35	34.10
.53	13.13	22.44	33.12	43.54	52.08	57.21	57.91	53.73	45.24	33.98
.54	13.21	22.54	33.23	43.64	52.15	57.24	57.89	53.67	45.14	33.86
.55	13.29	22.64	33.34	43.74	52.22	57.27	57.87	53.60	45.04	33.74
.56	13.37	22.74	33.45	43.84	52.29	57.30	57.85	53.53	44.94	33.62
.57	13.45	22.85	33.56	43.93	52.36	57.33	57.83	53.47	44.84	33.50
.58	13.54	22.95	33.66	44.03	52.43	57.36	57.80	53.40	44.73	33.39
.59	13.62	23.06	33.77	44.12	52.50	57.39	57.78	53.34	44.63	33.27
.60	13.70	23.16	33.88	44.22	52.57	57.42	57.76	53.27	44.53	33.15
.61	13.78	23.26	33.99	44.32	52.64	57.45	57.74	53.20	44.43	33.03
.62	13.87	23.37	34.10	44.41	52.70	57.48	57.72	53.13	44.32	32.91
.63	13.95	23.47	34.20	44.51	52.77	57.50	57.69	53.06	44.22	32.79
.64	14.04	23.58	34.31	44.60	52.83	57.53	57.67	52.99	44.11	32.67
.65	14.12	23.68	34.42	44.70	52.90	57.56	57.65	52.92	44.01	32.55
.66	14.20	23.78	34.53	44.79	52.96	57.58	57.63	52.85	43.90	32.43
.67	14.29	23.89	34.64	44.89	53.03	57.61	57.60	52.78	43.80	32.31
.68	14.37	23.99	34.74	44.98	53.09	57.63	57.58	52.70	43.69	32.18
.69	14.46	24.10	34.85	45.08	53.16	57.66	57.55	52.63	43.59	32.06
.70	14.54	24.20	34.96	45.17	53.22	57.68	57.53	52.56	43.48	31.94
.71	14.63	24.31	35.07	45.26	53.28	57.70	57.50	52.49	43.37	31.82
.72	14.71	24.41	35.18	45.36	53.35	57.72	57.48	52.42	43.26	31.70
.73	14.80	24.52	35.28	45.45	53.41	57.74	57.45	52.34	43.16	31.58
.74	14.88	24.62	35.39	45.55	53.48	57.76	57.43	52.27	43.05	31.46
.75	14.97	24.73	35.50	45.64	53.54	57.78	57.40	52.20	42.94	31.34
.76	15.06	24.84	35.61	45.73	53.60	57.80	57.37	52.12	42.83	31.22
.77	15.15	24.94	35.72	45.82	53.66	57.82	57.34	52.05	42.72	31.10
.78	15.23	25.05	35.82	45.92	53.73	57.83	57.32	51.97	42.62	30.98
.79	15.32	25.15	35.93	46.01	53.79	57.85	57.29	51.90	42.51	30.86
.80	15.41	25.26	36.04	46.10	53.85	57.87	57.26	51.82	42.40	30.74
.81	15.50	25.37	36.15	46.19	53.91	57.89	57.23	51.74	42.29	30.62
.82	15.59	25.47	36.25	46.28	53.97	57.91	57.20	51.66	42.18	30.50
.83	15.68	25.58	36.36	46.38	54.03	57.92	57.16	51.59	42.07	30.38
.84	15.77	25.68	36.46	46.47	54.09	57.94	57.13	51.51	41.96	30.26
.85	15.86	25.79	36.57	46.56	54.15	57.96	57.10	51.43	41.85	30.14
.86	15.95	25.90	36.68	46.65	54.21	57.97	57.06	51.35	41.74	30.02
.87	16.04	26.00	36.78	46.74	54.27	57.99	57.03	51.27	41.63	29.90
.88	16.13	26.11	36.89	46.83	54.32	58.00	56.99	51.19	41.52	29.78
.89	16.22	26.21	36.99	46.92	54.38	58.02	56.96	51.11	41.41	29.66
.90	16.31	26.32	37.10	47.01	54.44	58.03	56.92	51.03	41.30	29.54
.91	16.40	26.43	37.21	47.10	54.50	58.04	56.88	50.95	41.19	29.42
.92	16.49	26.53	37.31	47.19	54.55	58.05	56.84	50.87	41.08	29.30
.93	16.59	26.64	37.42	47.27	54.61	58.07	56.81	50.79	40.96	29.18
.94	16.68	26.74	37.52	47.36	54.66	58.08	56.77	50.71	40.85	29.06
.95	16.77	26.85	37.63	47.45	54.72	58.09	56.73	50.63	40.74	28.94
.96	16.86	26.96	37.74	47.54	54.77	58.10	56.69	50.55	40.63	28.82
.97	16.95	27.07	37.84	47.63	54.83	58.11	56.65	50.46	40.52	28.70
.98	17.05	27.17	37.95	47.71	54.88	58.12	56.62	50.38	40.40	28.58
.99	17.14	27.28	38.05	47.80	54.94	58.13	56.58	50.29	40.29	28.46
1.00	17.23	27.39	38.16	47.89	54.99	58.14	56.54	50.21	40.18	28.34

TABLES XI″. XII″.

Tables	XI″.				XII″.						
Arguments	6.				7.						

Days.	0	10	20	30
0.0	0.84″	0.35″	0.02″	0.65″
0.1	0.84	0.34	0.02	0.65
0.2	0.84	0.33	0.03	0.66
0.3	0.84	0.33	0.03	0.66
0.4	0.84	0.32	0.03	0.67
0.5	0.84	0.31	0.04	0.68
0.6	0.84	0.30	0.04	0.68
0.7	0.84	0.30	0.04	0.69
0.8	0.84	0.29	0.05	0.69
0.9	0.84	0.28	0.05	0.70
1.0	0.84	0.27	0.05	0.70
1.1	0.84	0.26	0.06	0.71
1.2	0.84	0.25	0.06	0.71
1.3	0.83	0.25	0.07	0.72
1.4	0.83	0.24	0.07	0.72
1.5	0.83	0.23	0.07	0.73
1.6	0.83	0.23	0.08	0.73
1.7	0.83	0.22	0.08	0.74
1.8	0.82	0.21	0.08	0.74
1.9	0.82	0.21	0.09	0.75
2.0	0.82	0.20	0.09	0.75
2.1	0.82	0.19	0.10	0.76
2.2	0.81	0.19	0.10	0.76
2.3	0.81	0.18	0.11	0.77
2.4	0.81	0.18	0.11	0.77
2.5	0.80	0.17	0.12	0.78
2.6	0.80	0.16	0.12	0.78
2.7	0.80	0.16	0.13	0.78
2.8	0.79	0.15	0.13	0.79
2.9	0.79	0.15	0.14	0.79
3.0	0.79	0.14	0.14	0.79
3.1	0.78	0.14	0.15	0.80
3.2	0.78	0.13	0.15	0.80
3.3	0.77	0.13	0.16	0.80
3.4	0.77	0.12	0.16	0.81
3.5	0.77	0.12	0.17	0.81
3.6	0.76	0.11	0.18	0.81
3.7	0.76	0.11	0.18	0.82
3.8	0.75	0.10	0.19	0.82
3.9	0.75	0.10	0.20	0.82
4.0	0.75	0.09	0.20	0.82
4.1	0.74	0.09	0.21	0.83
4.2	0.74	0.08	0.21	0.83
4.3	0.73	0.08	0.22	0.83
4.4	0.73	0.07	0.23	0.83
4.5	0.72	0.07	0.23	0.83
4.6	0.72	0.06	0.24	0.84
4.7	0.71	0.06	0.25	0.84
4.8	0.71	0.06	0.26	0.84
4.9	0.70	0.05	0.27	0.84
5.0	0.70	0.05	0.27	0.84

Days.	0	1	2	3	4	Days.
.00	10.19″	9.55″	7.06″	3.79″	1.06″	1.00
.01	10.19	9.53	7.03	3.76	1.04	.99
.02	10.20	9.51	7.00	3.72	1.02	.98
.03	10.20	9.50	6.97	3.69	1.00	.97
.04	10.21	9.48	6.94	3.65	0.98	.96
.05	10.21	9.46	6.91	3.62	0.96	.95
.06	10.21	9.44	6.88	3.59	0.94	.94
.07	10.21	9.42	6.85	3.56	0.92	.93
.08	10.22	9.41	6.81	3.52	0.90	.92
.09	10.22	9.39	6.78	3.49	0.88	.91
.10	10.22	9.37	6.75	3.46	0.86	.90
.11	10.22	9.35	6.72	3.43	0.84	.89
.12	10.22	9.33	6.69	3.40	0.82	.88
.13	10.23	9.31	6.65	3.36	0.81	.87
.14	10.23	9.29	6.62	3.33	0.79	.86
.15	10.23	9.27	6.59	3.30	0.77	.85
.16	10.23	9.25	6.56	3.27	0.75	.84
.17	10.23	9.23	6.53	3.24	0.73	.83
.18	10.24	9.21	6.49	3.21	0.72	.82
.19	10.24	9.19	6.46	3.18	0.70	.81
.20	10.24	9.17	6.43	3.15	0.68	.80
.21	10.24	9.15	6.40	3.12	0.66	.79
.22	10.24	9.13	6.37	3.09	0.65	.78
.23	10.23	9.11	6.33	3.06	0.63	.77
.24	10.23	9.09	6.30	3.03	0.62	.76
.25	10.23	9.07	6.27	3.00	0.60	.75
.26	10.23	9.05	6.24	2.97	0.59	.74
.27	10.23	9.03	6.21	2.94	0.57	.73
.28	10.22	9.00	6.17	2.91	0.56	.72
.29	10.22	8.98	6.14	2.88	0.54	.71
.30	10.22	8.96	6.11	2.85	0.53	.70
.31	10.22	8.94	6.08	2.82	0.52	.69
.32	10.21	8.92	6.04	2.79	0.50	.68
.33	10.21	8.89	6.01	2.76	0.49	.67
.34	10.21	8.87	5.97	2.73	0.47	.66
.35	10.20	8.85	5.94	2.70	0.46	.65
.36	10.20	8.83	5.91	2.67	0.45	.64
.37	10.20	8.80	5.88	2.64	0.43	.63
.38	10.19	8.78	5.84	2.62	0.42	.62
.39	10.19	8.75	5.81	2.59	0.40	.61
.40	10.19	8.73	5.78	2.56	0.39	.60
.41	10.18	8.71	5.75	2.53	0.38	.59
.42	10.18	8.68	5.71	2.50	0.37	.58
.43	10.17	8.66	5.68	2.47	0.35	.57
.44	10.17	8.63	5.64	2.44	0.34	.56
.45	10.16	8.61	5.61	2.41	0.33	.55
.46	10.15	8.58	5.58	2.38	0.32	.54
.47	10.15	8.56	5.54	2.35	0.31	.53
.48	10.14	8.53	5.51	2.33	0.29	.52
.49	10.14	8.51	5.47	2.30	0.28	.51
.50	10.13	8.48	5.44	2.27	0.27	.50
Days.	9	8	7	6	5	Days.

TABLES XI″. XII″.

Tables	XI″.			
Arguments	6.			
Days.	0	10	20	30
5.0	0.70	0.05	0.27	0.84
5.1	0.69	0.05	0.28	0.84
5.2	0.69	0.04	0.29	0.84
5.3	0.68	0.04	0.29	0.84
5.4	0.68	0.04	0.30	0.84
5.5	0.67	0.03	0.31	0.84
5.6	0.66	0.03	0.32	0.84
5.7	0.66	0.03	0.32	0.84
5.8	0.65	0.03	0.33	0.84
5.9	0.65	0.02	0.35	0.84
6.0	0.64	0.02	0.36	0.84
6.1	0.63	0.02	0.37	0.83
6.2	0.62	0.02	0.37	0.83
6.3	0.62	0.01	0.38	0.83
6.4	0.61	0.01	0.39	0.83
6.5	0.60	0.01	0.39	0.83
6.6	0.59	0.01	0.40	0.82
6.7	0.58	0.01	0.40	0.82
6.8	0.58	0.00	0.41	0.82
6.9	0.57	0.00	0.42	0.82
7.0	0.56	0.00	0.42	0.81
7.1	0.55	0.00	0.43	0.81
7.2	0.54	0.00	0.43	0.81
7.3	0.54	0.00	0.44	0.80
7.4	0.53	0.00	0.45	0.80
7.5	0.52	0.00	0.45	0.80
7.6	0.52	0.00	0.46	0.79
7.7	0.51	0.00	0.47	0.79
7.8	0.50	0.00	0.48	0.79
7.9	0.50	0.00	0.49	0.78
8.0	0.49	0.00	0.50	0.78
8.1	0.48	0.00	0.50	0.77
8.2	0.48	0.00	0.51	0.77
8.3	0.47	0.00	0.52	0.77
8.4	0.46	0.00	0.52	0.76
8.5	0.46	0.00	0.53	0.76
8.6	0.45	0.00	0.54	0.75
8.7	0.44	0.00	0.54	0.75
8.8	0.44	0.00	0.55	0.75
8.9	0.43	0.00	0.56	0.74
9.0	0.42	0.00	0.57	0.74
9.1	0.41	0.00	0.58	0.73
9.2	0.40	0.00	0.58	0.73
9.3	0.40	0.01	0.59	0.72
9.4	0.39	0.01	0.60	0.72
9.5	0.38	0.01	0.61	0.71
9.6	0.38	0.01	0.61	0.71
9.7	0.37	0.01	0.62	0.70
9.8	0.36	0.02	0.63	0.70
9.9	0.36	0.02	0.64	0.69
10.0	0.35	0.02	0.65	0.69

XII″.						
7.						
Days.	0	1	2	3	4	Days.
.50	10.13	8.48	5.44	2.27	0.27	.50
.51	10.12	8.45	5.41	2.24	0.26	.49
.52	10.11	8.43	5.37	2.21	0.25	.48
.53	10.11	8.40	5.34	2.19	0.24	.47
.54	10.10	8.38	5.30	2.16	0.23	.46
.55	10.09	8.35	5.27	2.13	0.22	.45
.56	10.08	8.32	5.24	2.10	0.21	.44
.57	10.07	8.30	5.20	2.08	0.20	.43
.58	10.07	8.27	5.17	2.05	0.20	.42
.59	10.06	8.25	5.13	2.03	0.19	.41
.60	10.05	8.22	5.10	2.00	0.18	.40
.61	10.04	8.19	5.07	1.97	0.17	.39
.62	10.03	8.17	5.04	1.95	0.16	.38
.63	10.03	8.14	5.00	1.92	0.16	.37
.64	10.02	8.12	4.97	1.90	0.15	.36
.65	10.01	8.09	4.94	1.87	0.14	.35
.66	10.00	8.06	4.91	1.84	0.13	.34
.67	9.99	8.03	4.87	1.82	0.12	.33
.68	9.98	8.01	4.84	1.79	0.12	.32
.69	9.97	7.98	4.80	1.77	0.11	.31
.70	9.96	7.95	4.77	1.74	0.10	.30
.71	9.95	7.92	4.74	1.72	0.09	.29
.72	9.94	7.89	4.71	1.69	0.09	.28
.73	9.92	7.87	4.67	1.67	0.08	.27
.74	9.91	7.84	4.64	1.64	0.08	.26
.75	9.90	7.81	4.61	1.62	0.07	.25
.76	9.89	7.78	4.58	1.60	0.07	.24
.77	9.88	7.75	4.54	1.57	0.06	.23
.78	9.86	7.72	4.51	1.55	0.06	.22
.79	9.85	7.69	4.47	1.52	0.05	.21
.80	9.84	7.66	4.44	1.50	0.05	.20
.81	9.83	7.63	4.41	1.48	0.05	.19
.82	9.81	7.60	4.38	1.45	0.05	.18
.83	9.80	7.58	4.34	1.43	0.04	.17
.84	9.78	7.55	4.31	1.40	0.04	.16
.85	9.77	7.52	4.28	1.38	0.04	.15
.86	9.76	7.49	4.25	1.36	0.03	.14
.87	9.74	7.46	4.21	1.34	0.03	.13
.88	9.73	7.43	4.18	1.31	0.03	.12
.89	9.71	7.40	4.14	1.29	0.02	.11
.90	9.70	7.37	4.11	1.27	0.02	.10
.91	9.69	7.34	4.08	1.25	0.02	.09
.92	9.67	7.31	4.05	1.23	0.02	.08
.93	9.66	7.28	4.01	1.20	0.01	.07
.94	9.64	7.25	3.98	1.18	0.01	.06
.95	9.63	7.22	3.95	1.16	0.01	.05
.96	9.61	7.19	3.92	1.14	0.01	.04
.97	9.60	7.16	3.89	1.12	0.01	.03
.98	9.58	7.12	3.85	1.10	0.00	.02
.99	9.57	7.09	3.82	1.08	0.00	.01
1.00	9.55	7.06	3.79	1.06	0.00	.00
Days.	9	8	7	6	5	Days.

TABLES XIII″. - XV″.

Tables	XIII″.		XIV″.			XV″.		
Arguments	8.		9.			10.		
Days.	0	10	0	10	20	0	10	20
0.0	3.44	0.55	0.82	0.21	0.21	0.00	0.72	0.35
0.1	3.45	0.60	0.82	0.20	0.22	0.00	0.73	0.34
0.2	3.45	0.66	0.82	0.19	0.23	0.00	0.74	0.33
0.3	3.46	0.71	0.82	0.19	0.23	0.00	0.74	0.32
0.4	3.46	0.77	0.82	0.18	0.24	0.00	0.75	0.31
0.5	3.46	0.83	0.82	0.17	0.25	0.00	0.75	0.30
0.6	3.45	0.89	0.82	0.16	0.26	0.00	0.76	0.29
0.7	3.44	0.95	0.82	0.16	0.27	0.00	0.76	0.28
0.8	3.43	1.02	0.81	0.15	0.27	0.01	0.77	0.27
0.9	3.41	1.08	0.81	0.15	0.28	0.01	0.77	0.26
1.0	3.39	1.15	0.81	0.14	0.29	0.01	0.77	0.25
1.1	3.37	1.22	0.81	0.13	0.30	0.01	0.78	0.24
1.2	3.34	1.28	0.81	0.13	0.31	0.01	0.78	0.23
1.3	3.31	1.35	0.80	0.12	0.31	0.02	0.79	0.23
1.4	3.28	1.42	0.80	0.12	0.32	0.02	0.79	0.22
1.5	3.25	1.49	0.80	0.11	0.33	0.02	0.80	0.21
1.6	3.21	1.56	0.80	0.10	0.34	0.02	0.80	0.20
1.7	3.18	1.63	0.80	0.10	0.35	0.03	0.80	0.19
1.8	3.14	1.70	0.79	0.09	0.36	0.03	0.81	0.19
1.9	3.10	1.77	0.79	0.09	0.37	0.04	0.81	0.18
2.0	3.06	1.84	0.79	0.08	0.38	0.04	0.81	0.17
2.1	3.01	1.91	0.79	0.07	0.39	0.04	0.81	0.16
2.2	2.96	1.98	0.79	0.07	0.40	0.05	0.81	0.15
2.3	2.91	2.05	0.78	0.06	0.40	0.05	0.82	0.15
2.4	2.86	2.11	0.78	0.06	0.41	0.06	0.82	0.14
2.5	2.81	2.18	0.78	0.05	0.42	0.06	0.82	0.13
2.6	2.75	2.25	0.77	0.05	0.43	0.07	0.82	0.12
2.7	2.69	2.32	0.77	0.04	0.44	0.07	0.82	0.11
2.8	2.63	2.38	0.76	0.04	0.44	0.08	0.82	0.11
2.9	2.57	2.45	0.76	0.03	0.45	0.08	0.82	0.10
3.0	2.51	2.51	0.75	0.03	0.46	0.09	0.82	0.09
3.1	2.45	2.57	0.74	0.03	0.47	0.10	0.82	0.08
3.2	2.39	2.63	0.74	0.03	0.48	0.11	0.82	0.08
3.3	2.32	2.69	0.73	0.02	0.48	0.11	0.82	0.07
3.4	2.25	2.75	0.73	0.02	0.49	0.12	0.82	0.07
3.5	2.18	2.81	0.72	0.02	0.50	0.13	0.82	0.06
3.6	2.11	2.86	0.71	0.02	0.51	0.14	0.82	0.06
3.7	2.05	2.91	0.71	0.02	0.52	0.15	0.82	0.05
3.8	1.98	2.96	0.70	0.01	0.52	0.15	0.81	0.05
3.9	1.91	3.01	0.70	0.01	0.53	0.16	0.81	0.04
4.0	1.84	3.06	0.69	0.01	0.54	0.17	0.81	0.04
4.1	1.77	3.10	0.68	0.01	0.55	0.18	0.81	0.04
4.2	1.70	3.14	0.68	0.01	0.56	0.19	0.81	0.03
4.3	1.63	3.18	0.67	0.00	0.56	0.19	0.80	0.03
4.4	1.56	3.21	0.67	0.00	0.57	0.20	0.80	0.02
4.5	1.49	3.25	0.66	0.00	0.58	0.21	0.80	0.02
4.6	1.42	3.28	0.65	0.00	0.59	0.22	0.80	0.02
4.7	1.35	3.31	0.64	0.00	0.60	0.23	0.79	0.02
4.8	1.28	3.34	0.64	0.00	0.60	0.23	0.79	0.01
4.9	1.22	3.37	0.63	0.00	0.61	0.24	0.78	0.01
5.0	1.15	3.39	0.62	0.00	0.62	0.25	0.78	0.01

TABLES XIII″. - XV″.

Tables	XIII″.		XIV″.			XV″.		
Arguments	9.		9.			10.		
Days.	0	10	0	10	20	0	10	20
Days. 5.0	1″.15	3″.39	0″.62	0″.00	0″.62	0″.25	0″.78	0″.01
5.1	1.08	3.41	0.61	0.00	0.63	0.26	0.78	0.01
5.2	1.02	3.43	0.60	0.00	0.64	0.27	0.77	0.01
5.3	0.95	3.44	0.60	0.00	0.64	0.28	0.77	0.00
5.4	0.89	3.45	0.59	0.00	0.65	0.29	0.76	0.00
5.5	0.83	3.46	0.58	0.00	0.66	0.30	0.75	0.00
5.6	0.77	3.46	0.57	0.00	0.67	0.31	0.74	0.00
5.7	0.71	3.46	0.56	0.00	0.67	0.32	0.74	0.00
5.8	0.66	3.45	0.56	0.01	0.68	0.33	0.73	0.00
5.9	0.60	3.45	0.55	0.01	0.68	0.34	0.73	0.00
6.0	0.55	3.44	0.54	0.01	0.69	0.35	0.72	0.00
6.1	0.50	3.43	0.53	0.01	0.70	0.36	0.71	0.00
6.2	0.45	3.41	0.52	0.01	0.70	0.37	0.70	0.00
6.3	0.41	3.40	0.52	0.02	0.71	0.38	0.70	0.00
6.4	0.36	3.38	0.51	0.02	0.71	0.39	0.69	0.00
6.5	0.32	3.36	0.50	0.02	0.72	0.40	0.68	0.00
6.6	0.28	3.33	0.49	0.02	0.73	0.41	0.67	0.00
6.7	0.25	3.30	0.48	0.02	0.73	0.42	0.66	0.00
6.8	0.21	3.27	0.48	0.03	0.74	0.43	0.66	0.01
6.9	0.18	3.24	0.47	0.03	0.74	0.44	0.65	0.01
7.0	0.15	3.20	0.46	0.03	0.75	0.45	0.64	0.01
7.1	0.12	3.17	0.45	0.03	0.75	0.46	0.63	0.01
7.2	0.10	3.13	0.44	0.04	0.76	0.47	0.62	0.02
7.3	0.08	3.09	0.43	0.04	0.76	0.48	0.62	0.02
7.4	0.06	3.05	0.42	0.05	0.77	0.49	0.61	0.03
7.5	0.04	3.00	0.41	0.05	0.77	0.50	0.60	0.03
7.6	0.03	2.95	0.40	0.06	0.77	0.51	0.59	0.04
7.7	0.02	2.90	0.39	0.06	0.78	0.52	0.58	0.04
7.8	0.01	2.85	0.39	0.07	0.78	0.53	0.57	0.05
7.9	0.00	2.80	0.38	0.07	0.79	0.54	0.56	0.05
8.0	0.00	2.74	0.37	0.08	0.79	0.55	0.55	0.06
8.1	0.00	2.68	0.36	0.09	0.79	0.56	0.54	0.07
8.2	0.01	2.62	0.35	0.10	0.80	0.57	0.53	0.07
8.3	0.02	2.56	0.35	0.10	0.80	0.58	0.52	0.08
8.4	0.03	2.50	0.34	0.11	0.80	0.59	0.51	0.08
8.5	0.04	2.44	0.33	0.11	0.80	0.60	0.50	0.09
8.6	0.06	2.37	0.32	0.12	0.80	0.61	0.49	0.10
8.7	0.08	2.31	0.31	0.12	0.81	0.62	0.48	0.10
8.8	0.10	2.24	0.31	0.13	0.81	0.62	0.47	0.11
8.9	0.12	2.17	0.30	0.13	0.81	0.63	0.46	0.11
9.0	0.15	2.10	0.29	0.14	0.81	0.64	0.45	0.12
9.1	0.18	2.04	0.28	0.15	0.81	0.65	0.44	0.13
9.2	0.21	1.97	0.27	0.15	0.81	0.66	0.43	0.14
9.3	0.25	1.90	0.27	0.16	0.82	0.66	0.42	0.14
9.4	0.28	1.83	0.26	0.16	0.82	0.67	0.41	0.15
9.5	0.32	1.76	0.25	0.17	0.82	0.68	0.40	0.16
9.6	0.36	1.69	0.24	0.18	0.82	0.69	0.39	0.17
9.7	0.41	1.62	0.23	0.19	0.82	0.70	0.38	0.18
9.8	0.45	1.55	0.23	0.19	0.82	0.70	0.37	0.19
9.9	0.50	1.48	0.22	0.20	0.82	0.71	0.36	0.20
10.0	0.55	1.41	0.21	0.21	0.82	0.72	0.35	0.21

TABLES XVI″.-XX″.

Tables	XVI″.			XVIII″.	XIX″.			XX″.	
Arguments	11.			13.	14.			15.	
Days.	0	10	20	0	0	10	20	0	10
Days 0.0	″ 0.00	″ 0.46	″ 0.31	″ 1.68	″ 0.00	″ 0.12	″ 0.14	″ 0.03	″ 0.50
0.1	0.00	0.47	0.33	1.71	0.00	0.12	0.14	0.03	0.49
0.2	0.00	0.47	0.32	1.74	0.00	0.12	0.13	0.02	0.48
0.3	0.00	0.48	0.32	1.76	0.00	0.12	0.13	0.02	0.47
0.4	0.00	0.48	0.31	1.78	0.00	0.12	0.13	0.01	0.46
0.5	0.00	0.49	0.30	1.80	0.00	0.12	0.13	0.01	0.45
0.6	0.00	0.49	0.29	1.81	0.00	0.13	0.13	0.01	0.44
0.7	0.00	0.50	0.29	1.82	0.00	0.13	0.12	0.01	0.42
0.8	0.00	0.50	0.28	1.83	0.00	0.13	0.12	0.00	0.41
0.9	0.00	0.51	0.28	1.84	0.00	0.13	0.12	0.00	0.39
1.0	0.00	0.51	0.27	1.84	0.00	0.13	0.12	0.00	0.38
1.1	0.00	0.51	0.26	1.83	0.00	0.13	0.12	0.00	0.37
1.2	0.00	0.52	0.26	1.83	0.00	0.14	0.12	0.00	0.36
1.3	0.01	0.52	0.25	1.82	0.00	0.14	0.12	0.01	0.34
1.4	0.01	0.53	0.25	1.81	0.00	0.14	0.12	0.01	0.33
1.5	0.01	0.53	0.24	1.79	0.00	0.14	0.12	0.01	0.32
1.6	0.01	0.53	0.23	1.77	0.00	0.14	0.11	0.02	0.31
1.7	0.01	0.54	0.22	1.74	0.00	0.14	0.11	0.02	0.29
1.8	0.02	0.54	0.22	1.71	0.00	0.15	0.11	0.03	0.28
1.9	0.02	0.55	0.21	1.68	0.00	0.15	0.11	0.03	0.26
2.0	0.02	0.55	0.20	1.65	0.00	0.15	0.11	0.04	0.25
2.1	0.02	0.55	0.19	1.61	0.00	0.15	0.11	0.05	0.24
2.2	0.02	0.55	0.19	1.57	0.00	0.15	0.10	0.06	0.23
2.3	0.03	0.56	0.18	1.53	0.00	0.15	0.10	0.06	0.21
2.4	0.03	0.56	0.18	1.49	0.00	0.15	0.10	0.07	0.20
2.5	0.03	0.56	0.17	1.45	0.01	0.15	0.10	0.08	0.19
2.6	0.03	0.56	0.16	1.40	0.01	0.15	0.10	0.09	0.18
2.7	0.04	0.56	0.16	1.35	0.01	0.15	0.10	0.10	0.17
2.8	0.04	0.57	0.15	1.30	0.01	0.15	0.09	0.11	0.15
2.9	0.05	0.57	0.15	1.24	0.01	0.15	0.09	0.12	0.14
3.0	0.05	0.57	0.14	1.19	0.01	0.15	0.09	0.13	0.13
3.1	0.05	0.57	0.13	1.13	0.01	0.15	0.09	0.14	0.12
3.2	0.06	0.57	0.13	1.08	0.01	0.15	0.09	0.15	0.11
3.3	0.06	0.58	0.12	1.02	0.01	0.15	0.08	0.16	0.10
3.4	0.07	0.58	0.12	0.97	0.01	0.15	0.08	0.18	0.09
3.5	0.07	0.58	0.11	0.91	0.01	0.15	0.08	0.19	0.08
3.6	0.07	0.58	0.11	0.85	0.02	0.15	0.08	0.20	0.07
3.7	0.08	0.58	0.10	0.80	0.02	0.15	0.07	0.21	0.06
3.8	0.08	0.58	0.10	0.74	0.02	0.15	0.07	0.23	0.06
3.9	0.09	0.58	0.09	0.69	0.02	0.15	0.07	0.24	0.05
4.0	0.09	0.58	0.09	0.63	0.02	0.16	0.06	0.25	0.04
4.1	0.09	0.58	0.09	0.58	0.02	0.16	0.06	0.26	0.03
4.2	0.10	0.58	0.08	0.52	0.02	0.16	0.06	0.28	0.03
4.3	0.10	0.58	0.08	0.47	0.03	0.16	0.06	0.29	0.02
4.4	0.11	0.58	0.07	0.42	0.03	0.16	0.05	0.31	0.02
4.5	0.11	0.58	0.07	0.37	0.03	0.16	0.05	0.32	0.01
4.6	0.12	0.58	0.07	0.33	0.03	0.16	0.05	0.33	0.01
4.7	0.12	0.58	0.06	0.29	0.03	0.16	0.05	0.34	0.01
4.8	0.13	0.57	0.06	0.25	0.03	0.16	0.04	0.36	0.00
4.9	0.13	0.57	0.05	0.21	0.04	0.16	0.04	0.37	0.00
5.0	0.14	0.57	0.05	0.17	0.04	0.16	0.04	0.38	0.00

TABLES XVI''. - XX''.

Tables	XVI''.			XVIII''.	XIX''.			XX''.	
Arguments	11.			13.	14.			15.	
Days.	0	10	20	0	0	10	20	0	10
Days	"	"	"	"	"	"	"	"	"
5.0	0.14	0.57	0.05	0.17	0.04	0.16	0.04	0.38	0.00
5.1	0.15	0.57	0.05	0.14	0.04	0.16	0.04	0.39	0.00
5.2	0.15	0.57	0.04	0.11	0.04	0.16	0.04	0.41	0.00
5.3	0.16	0.56	0.04	0.09	0.04	0.16	0.04	0.42	0.00
5.4	0.16	0.56	0.03	0.07	0.04	0.16	0.04	0.44	0.01
5.5	0.17	0.56	0.03	0.05	0.05	0.16	0.04	0.45	0.01
5.6	0.18	0.56	0.03	0.03	0.05	0.16	0.03	0.46	0.01
5.7	0.18	0.56	0.03	0.02	0.05	0.16	0.03	0.47	0.01
5.8	0.19	0.55	0.02	0.01	0.05	0.16	0.03	0.48	0.02
5.9	0.19	0.55	0.02	0.01	0.05	0.16	0.03	0.49	0.02
6.0	0.20	0.55	0.02	0.00	0.05	0.16	0.03	0.50	0.02
6.1	0.21	0.55	0.02	0.01	0.05	0.16	0.03	0.51	0.03
6.2	0.22	0.54	0.02	0.01	0.06	0.16	0.03	0.52	0.04
6.3	0.22	0.54	0.01	0.02	0.06	0.16	0.02	0.53	0.04
6.4	0.23	0.53	0.01	0.03	0.06	0.16	0.02	0.54	0.05
6.5	0.24	0.53	0.01	0.05	0.06	0.16	0.02	0.55	0.06
6.6	0.25	0.53	0.01	0.07	0.06	0.16	0.02	0.56	0.07
6.7	0.25	0.52	0.01	0.09	0.06	0.16	0.02	0.57	0.08
6.8	0.26	0.52	0.01	0.11	0.07	0.16	0.01	0.57	0.08
6.9	0.26	0.51	0.00	0.14	0.07	0.16	0.01	0.58	0.09
7.0	0.27	0.51	0.00	0.17	0.07	0.16	0.01	0.59	0.10
7.1	0.28	0.51	0.00	0.20	0.07	0.16	0.01	0.59	0.11
7.2	0.28	0.50	0.00	0.24	0.07	0.16	0.01	0.60	0.12
7.3	0.29	0.50	0.00	0.28	0.07	0.16	0.01	0.60	0.13
7.4	0.29	0.49	0.00	0.32	0.07	0.16	0.01	0.61	0.14
7.5	0.30	0.49	0.00	0.37	0.08	0.16	0.01	0.61	0.15
7.6	0.31	0.48	0.00	0.42	0.08	0.16	0.01	0.61	0.16
7.7	0.32	0.48	0.00	0.47	0.08	0.16	0.01	0.61	0.17
7.8	0.32	0.47	0.00	0.52	0.08	0.16	0.01	0.62	0.19
7.9	0.33	0.47	0.00	0.58	0.08	0.16	0.01	0.62	0.20
8.0	0.34	0.46	0.00	0.63	0.08	0.16	0.01	0.62	0.21
8.1	0.35	0.45	0.00	0.69	0.08	0.16	0.01	0.62	0.22
8.2	0.35	0.45	0.00	0.74	0.09	0.16	0.01	0.62	0.24
8.3	0.36	0.44	0.01	0.80	0.09	0.16	0.01	0.61	0.25
8.4	0.36	0.44	0.01	0.85	0.09	0.16	0.01	0.61	0.27
8.5	0.37	0.43	0.01	0.91	0.09	0.16	0.00	0.61	0.28
8.6	0.38	0.42	0.01	0.97	0.09	0.15	0.00	0.61	0.29
8.7	0.38	0.42	0.01	1.02	0.09	0.15	0.00	0.60	0.31
8.8	0.39	0.41	0.02	1.08	0.10	0.15	0.00	0.60	0.32
8.9	0.39	0.41	0.02	1.13	0.10	0.15	0.00	0.59	0.34
9.0	0.40	0.40	0.02	1.19	0.10	0.15	0.00	0.59	0.35
9.1	0.41	0.39	0.02	1.24	0.10	0.15	0.00	0.58	0.36
9.2	0.41	0.39	0.03	1.30	0.10	0.15	0.00	0.57	0.37
9.3	0.42	0.38	0.03	1.35	0.10	0.15	0.00	0.57	0.39
9.4	0.42	0.38	0.03	1.40	0.11	0.15	0.00	0.56	0.40
9.5	0.43	0.37	0.03	1.45	0.11	0.15	0.00	0.55	0.41
9.6	0.44	0.36	0.04	1.50	0.11	0.14	0.00	0.54	0.42
9.7	0.44	0.36	0.04	1.54	0.11	0.14	0.00	0.53	0.43
9.8	0.45	0.35	0.04	1.56	0.11	0.14	0.00	0.52	0.45
9.9	0.45	0.35	0.05	1.62	0.12	0.14	0.00	0.51	0.46
10.0	0.46	0.34	0.05	1.65	0.12	0.14	0.00	0.50	0.47

TABLES XXI″.-XXV″.

Tables	XXI″.		XXII″.			XXIII″.	XXIV″.			XXV.″	
Arguments	16.		17.			18.	19.			20.	
Days.	0	10	0	10	20	0	0	10	20	0	10
0.0	0.18	1.15	0.12	0.03	0.05	0.70	0.00	0.09	0.01	0.22	0.04
0.1	0.14	1.20	0.12	0.03	0.05	0.70	0.00	0.09	0.04	0.22	0.04
0.2	0.11	1.25	0.12	0.03	0.05	0.70	0.00	0.09	0.04	0.22	0.05
0.3	0.09	1.30	0.12	0.03	0.05	0.69	0.00	0.09	0.03	0.22	0.05
0.4	0.07	1.34	0.12	0.03	0.05	0.69	0.00	0.09	0.03	0.22	0.06
0.5	0.06	1.39	0.12	0.03	0.06	0.69	0.00	0.10	0.03	0.22	0.06
0.6	0.05	1.43	0.12	0.02	0.06	0.68	0.00	0.10	0.03	0.22	0.06
0.7	0.04	1.46	0.12	0.02	0.06	0.67	0.00	0.10	0.03	0.22	0.07
0.8	0.05	1.49	0.12	0.02	0.06	0.66	0.00	0.10	0.03	0.22	0.07
0.9	0.06	1.51	0.12	0.02	0.06	0.65	0.00	0.10	0.02	0.22	0.08
1.0	0.07	1.53	0.12	0.02	0.06	0.64	0.00	0.10	0.02	0.22	0.08
1.1	0.09	1.54	0.12	0.02	0.06	0.63	0.00	0.10	0.02	0.22	0.09
1.2	0.11	1.55	0.12	0.02	0.06	0.62	0.00	0.10	0.02	0.22	0.09
1.3	0.14	1.56	0.12	0.02	0.07	0.60	0.00	0.10	0.02	0.22	0.10
1.4	0.17	1.55	0.12	0.02	0.07	0.59	0.00	0.10	0.02	0.22	0.10
1.5	0.21	1.54	0.12	0.02	0.07	0.57	0.00	0.10	0.02	0.22	0.11
1.6	0.25	1.53	0.11	0.01	0.07	0.55	0.00	0.10	0.01	0.22	0.11
1.7	0.30	1.51	0.11	0.01	0.07	0.53	0.00	0.10	0.01	0.22	0.12
1.8	0.35	1.48	0.11	0.01	0.08	0.51	0.00	0.10	0.01	0.21	0.12
1.9	0.40	1.45	0.11	0.01	0.08	0.49	0.00	0.10	0.01	0.21	0.13
2.0	0.45	1.42	0.11	0.01	0.08	0.47	0.00	0.10	0.01	0.21	0.13
2.1	0.51	1.38	0.11	0.01	0.08	0.45	0.00	0.10	0.01	0.21	0.13
2.2	0.56	1.33	0.11	0.01	0.08	0.43	0.00	0.10	0.01	0.21	0.14
2.3	0.62	1.28	0.11	0.01	0.08	0.40	0.00	0.10	0.01	0.20	0.14
2.4	0.68	1.23	0.11	0.01	0.08	0.38	0.00	0.10	0.01	0.20	0.15
2.5	0.74	1.18	0.11	0.01	0.09	0.36	0.00	0.10	0.00	0.20	0.15
2.6	0.81	1.12	0.11	0.00	0.09	0.34	0.00	0.10	0.00	0.20	0.16
2.7	0.87	1.06	0.11	0.00	0.09	0.32	0.00	0.10	0.00	0.19	0.16
2.8	0.93	1.00	0.11	0.00	0.09	0.29	0.00	0.10	0.00	0.19	0.17
2.9	0.98	0.94	0.11	0.00	0.09	0.27	0.00	0.10	0.00	0.18	0.17
3.0	1.03	0.88	0.11	0.00	0.09	0.25	0.00	0.10	0.00	0.18	0.18
3.1	1.08	0.82	0.11	0.00	0.09	0.23	0.00	0.10	0.00	0.17	0.18
3.2	1.13	0.75	0.11	0.00	0.09	0.21	0.00	0.10	0.00	0.17	0.19
3.3	1.18	0.69	0.11	0.00	0.09	0.19	0.00	0.10	0.00	0.16	0.19
3.4	1.22	0.62	0.11	0.00	0.09	0.17	0.00	0.10	0.00	0.16	0.20
3.5	1.26	0.56	0.11	0.00	0.10	0.15	0.01	0.10	0.00	0.15	0.20
3.6	1.29	0.50	0.10	0.00	0.10	0.13	0.01	0.10	0.00	0.15	0.20
3.7	1.32	0.44	0.10	0.00	0.10	0.11	0.01	0.10	0.00	0.14	0.20
3.8	1.35	0.39	0.10	0.00	0.10	0.10	0.01	0.10	0.00	0.14	0.21
3.9	1.37	0.34	0.10	0.00	0.10	0.08	0.01	0.10	0.00	0.13	0.21
4.0	1.39	0.29	0.10	0.00	0.10	0.07	0.01	0.10	0.00	0.13	0.21
4.1	1.40	0.25	0.10	0.00	0.10	0.05	0.01	0.10	0.00	0.13	0.21
4.2	1.40	0.21	0.10	0.00	0.10	0.04	0.01	0.10	0.00	0.12	0.21
4.3	1.41	0.17	0.10	0.00	0.10	0.03	0.01	0.10	0.00	0.12	0.22
4.4	1.41	0.13	0.10	0.00	0.10	0.02	0.01	0.10	0.00	0.11	0.22
4.5	1.40	0.10	0.10	0.00	0.10	0.01	0.01	0.10	0.00	0.11	0.22
4.6	1.39	0.08	0.09	0.00	0.11	0.01	0.02	0.10	0.00	0.10	0.22
4.7	1.37	0.07	0.09	0.00	0.11	0.01	0.02	0.10	0.00	0.10	0.22
4.8	1.34	0.06	0.09	0.00	0.11	0.00	0.02	0.10	0.00	0.09	0.22
4.9	1.31	0.05	0.09	0.00	0.11	0.00	0.02	0.10	0.00	0.09	0.22
5.0	1.28	0.05	0.09	0.00	0.11	0.00	0.02	0.10	0.00	0.08	0.22

TABLES XXI″. - XXV″.

Tables	XXI″.		XXII″.			XXIII″.	XXIV″.			XXV.″	
Arguments	16.		17.			18.	19.			20.	
Days.	0	10	0	10	20	0	0	10	20	0	10
Days. 5.0	″ 1.28	″ 0.05	″ 0.09	″ 0.00	″ 0.11	″ 0.00	″ 0.02	″ 0.10	″ 0.00	″ 0.08	″ 0.22
5.1	1.25	0.05	0.09	0.00	0.11	0.00	0.02	0.10	0.00	0.08	0.22
5.2	1.21	0.06	0.09	0.00	0.11	0.00	0.03	0.10	0.00	0.07	0.22
5.3	1.17	0.07	0.09	0.00	0.11	0.01	0.03	0.09	0.00	0.07	0.22
5.4	1.12	0.09	0.09	0.00	0.11	0.01	0.03	0.09	0.00	0.06	0.22
5.5	1.07	0.12	0.08	0.01	0.11	0.01	0.03	0.09	0.00	0.06	0.22
5.6	1.02	0.15	0.08	0.01	0.11	0.02	0.03	0.09	0.00	0.06	0.22
5.7	0.97	0.19	0.08	0.01	0.11	0.03	0.04	0.09	0.00	0.05	0.22
5.8	0.91	0.23	0.08	0.01	0.11	0.04	0.04	0.09	0.00	0.05	0.22
5.9	0.85	0.27	0.08	0.01	0.11	0.06	0.04	0.09	0.00	0.04	0.22
6.0	0.80	0.32	0.08	0.01	0.11	0.07	0.04	0.09	0.00	0.04	0.22
6.1	0.75	0.37	0.08	0.01	0.11	0.08	0.04	0.09	0.00	0.04	0.22
6.2	0.69	0.42	0.08	0.01	0.11	0.10	0.04	0.09	0.00	0.03	0.22
6.3	0.63	0.48	0.07	0.01	0.11	0.11	0.04	0.09	0.00	0.03	0.21
6.4	0.58	0.53	0.07	0.01	0.11	0.13	0.04	0.08	0.00	0.02	0.21
6.5	0.54	0.59	0.07	0.01	0.11	0.15	0.05	0.08	0.00	0.02	0.21
6.6	0.48	0.65	0.07	0.02	0.12	0.17	0.05	0.08	0.00	0.02	0.21
6.7	0.44	0.71	0.07	0.02	0.12	0.19	0.05	0.08	0.00	0.02	0.20
6.8	0.40	0.77	0.07	0.02	0.12	0.21	0.05	0.08	0.00	0.01	0.20
6.9	0.36	0.84	0.06	0.02	0.12	0.23	0.05	0.08	0.00	0.01	0.19
7.0	0.32	0.90	0.06	0.02	0.12	0.25	0.05	0.08	0.00	0.01	0.19
7.1	0.29	0.96	0.06	0.02	0.12	0.27	0.05	0.08	0.00	0.01	0.19
7.2	0.26	1.01	0.06	0.02	0.12	0.29	0.06	0.08	0.00	0.01	0.18
7.3	0.23	1.05	0.06	0.02	0.12	0.32	0.06	0.08	0.00	0.00	0.18
7.4	0.21	1.10	0.06	0.02	0.12	0.34	0.06	0.08	0.00	0.00	0.17
7.5	0.20	1.15	0.06	0.02	0.12	0.36	0.06	0.07	0.00	0.00	0.17
7.6	0.19	1.20	0.05	0.03	0.12	0.38	0.06	0.07	0.00	0.00	0.16
7.7	0.19	1.24	0.05	0.03	0.12	0.40	0.06	0.07	0.00	0.00	0.16
7.8	0.19	1.27	0.05	0.03	0.12	0.43	0.07	0.07	0.01	0.00	0.15
7.9	0.20	1.30	0.05	0.03	0.12	0.45	0.07	0.07	0.01	0.00	0.15
8.0	0.21	1.33	0.05	0.03	0.12	0.47	0.07	0.07	0.01	0.00	0.14
8.1	0.23	1.36	0.05	0.03	0.12	0.49	0.07	0.07	0.01	0.00	0.14
8.2	0.25	1.38	0.05	0.03	0.12	0.51	0.07	0.07	0.01	0.00	0.13
8.3	0.28	1.39	0.05	0.03	0.12	0.53	0.07	0.06	0.01	0.00	0.13
8.4	0.31	1.40	0.05	0.03	0.12	0.55	0.07	0.06	0.01	0.00	0.12
8.5	0.34	1.40	0.05	0.03	0.12	0.57	0.07	0.06	0.01	0.00	0.12
8.6	0.38	1.41	0.04	0.04	0.12	0.59	0.08	0.06	0.01	0.00	0.12
8.7	0.42	1.41	0.04	0.04	0.12	0.60	0.08	0.06	0.01	0.00	0.11
8.8	0.47	1.40	0.04	0.04	0.11	0.62	0.08	0.05	0.02	0.01	0.11
8.9	0.52	1.38	0.04	0.04	0.11	0.63	0.08	0.05	0.02	0.01	0.10
9.0	0.57	1.36	0.04	0.04	0.11	0.64	0.08	0.05	0.02	0.01	0.10
9.1	0.62	1.33	0.04	0.04	0.11	0.65	0.08	0.05	0.02	0.01	0.09
9.2	0.68	1.30	0.04	0.04	0.11	0.66	0.08	0.05	0.02	0.01	0.09
9.3	0.74	1.27	0.04	0.04	0.11	0.67	0.08	0.05	0.03	0.02	0.08
9.4	0.80	1.23	0.04	0.04	0.11	0.68	0.08	0.05	0.03	0.02	0.08
9.5	0.86	1.19	0.04	0.04	0.11	0.69	0.08	0.04	0.03	0.02	0.07
9.6	0.91	1.15	0.03	0.05	0.11	0.69	0.09	0.04	0.03	0.02	0.07
9.7	0.97	1.10	0.03	0.05	0.11	0.69	0.09	0.04	0.03	0.03	0.06
9.8	1.03	1.05	0.03	0.05	0.11	0.70	0.09	0.04	0.04	0.03	0.06
9.9	1.09	1.00	0.03	0.05	0.11	0.70	0.09	0.04	0.04	0.04	0.05
10.0	1.15	0.95	0.03	0.05	0.11	0.70	0.09	0.04	0.04	0.04	0.05

TABLES XXVII''. - XXXV''.

Tables	XXVII''.				XXIX''.		XXX''.	XXXI''.	XXXV''.	
Arguments	22.				24.		25.	26.	30.	
Days.	0	10	20	30	0	10	0	0	0	10
Days	"	"	"	"	"	"	"	"	"	"
0.0	0.14	0.02	0.07	0.12	0.00	0.11	0.00	0.28	0.11	0.10
0.1	0.14	0.02	0.07	0.12	0.00	0.11	0.00	0.28	0.11	0.10
0.2	0.14	0.02	0.07	0.12	0.00	0.10	0.00	0.28	0.11	0.10
0.3	0.14	0.03	0.06	0.13	0.00	0.10	0.00	0.28	0.12	0.11
0.4	0.14	0.03	0.06	0.13	0.00	0.09	0.00	0.27	0.12	0.11
0.5	0.14	0.03	0.06	0.13	0.00	0.09	0.00	0.27	0.12	0.11
0.6	0.14	0.03	0.05	0.13	0.00	0.09	0.00	0.26	0.12	0.11
0.7	0.14	0.03	0.05	0.13	0.00	0.08	0.00	0.25	0.13	0.11
0.8	0.14	0.04	0.05	0.14	0.00	0.08	0.01	0.24	0.13	0.12
0.9	0.14	0.04	0.04	0.14	0.00	0.07	0.01	0.23	0.14	0.12
1.0	0.14	0.04	0.04	0.14	0.00	0.07	0.01	0.21	0.14	0.12
1.1	0.14	0.05	0.04	0.14	0.00	0.07	0.01	0.20	0.14	0.12
1.2	0.14	0.05	0.04	0.14	0.00	0.06	0.01	0.19	0.15	0.13
1.3	0.13	0.05	0.03	0.14	0.01	0.06	0.02	0.18	0.15	0.13
1.4	0.13	0.06	0.03	0.14	0.01	0.05	0.02	0.16	0.16	0.14
1.5	0.13	0.06	0.03	0.14	0.01	0.05	0.02	0.14	0.16	0.14
1.6	0.13	0.06	0.03	0.14	0.01	0.05	0.03	0.13	0.16	0.14
1.7	0.13	0.06	0.03	0.14	0.01	0.04	0.03	0.11	0.17	0.15
1.8	0.12	0.07	0.02	0.14	0.02	0.04	0.04	0.10	0.17	0.15
1.9	0.12	0.07	0.02	0.14	0.02	0.03	0.04	0.09	0.18	0.16
2.0	0.12	0.07	0.02	0.14	0.02	0.03	0.05	0.08	0.18	0.16
2.1	0.12	0.07	0.02	0.14	0.02	0.03	0.05	0.06	0.19	0.17
2.2	0.12	0.07	0.02	0.14	0.03	0.02	0.06	0.05	0.19	0.17
2.3	0.11	0.08	0.01	0.14	0.03	0.02	0.06	0.04	0.20	0.18
2.4	0.11	0.08	0.01	0.14	0.03	0.01	0.07	0.03	0.20	0.18
2.5	0.11	0.08	0.01	0.14	0.04	0.01	0.07	0.02	0.21	0.19
2.6	0.11	0.09	0.01	0.14	0.04	0.01	0.08	0.02	0.21	0.19
2.7	0.11	0.09	0.01	0.14	0.05	0.01	0.08	0.01	0.22	0.20
2.8	0.10	0.09	0.00	0.14	0.05	0.00	0.09	0.01	0.22	0.20
2.9	0.10	0.10	0.00	0.14	0.06	0.00	0.09	0.00	0.23	0.21
3.0	0.10	0.10	0.00	0.14	0.06	0.00	0.10	0.00	0.23	0.21
3.1	0.10	0.10	0.00	0.14	0.07	0.00	0.10	0.00	0.23	0.21
3.2	0.09	0.10	0.00	0.14	0.07	0.00	0.11	0.01	0.24	0.22
3.3	0.09	0.11	0.00	0.13	0.08	0.00	0.11	0.01	0.24	0.22
3.4	0.09	0.11	0.00	0.13	0.08	0.00	0.12	0.02	0.25	0.23
3.5	0.08	0.11	0.00	0.13	0.09	0.00	0.12	0.02	0.25	0.23
3.6	0.08	0.11	0.00	0.13	0.09	0.00	0.12	0.03	0.25	0.23
3.7	0.08	0.11	0.00	0.13	0.10	0.00	0.13	0.04	0.26	0.24
3.8	0.07	0.12	0.00	0.12	0.10	0.00	0.13	0.05	0.26	0.24
3.9	0.07	0.12	0.00	0.12	0.11	0.00	0.14	0.06	0.27	0.25
4.0	0.07	0.12	0.00	0.12	0.11	0.00	0.14	0.08	0.27	0.25
4.1	0.06	0.12	0.00	0.12	0.12	0.00	0.14	0.09	0.27	0.25
4.2	0.06	0.12	0.00	0.12	0.12	0.00	0.15	0.10	0.27	0.26
4.3	0.06	0.13	0.00	0.11	0.13	0.00	0.15	0.11	0.28	0.26
4.4	0.05	0.13	0.00	0.11	0.13	0.00	0.16	0.13	0.28	0.27
4.5	0.05	0.13	0.00	0.11	0.14	0.00	0.16	0.14	0.28	0.27
4.6	0.05	0.13	0.00	0.11	0.14	0.01	0.16	0.16	0.28	0.27
4.7	0.05	0.13	0.00	0.11	0.15	0.01	0.16	0.17	0.28	0.27
4.8	0.04	0.14	0.00	0.10	0.15	0.01	0.16	0.19	0.28	0.28
4.9	0.04	0.14	0.00	0.10	0.16	0.01	0.16	0.20	0.28	0.28
5.0	0.04	0.14	0.00	0.10	0.16	0.01	0.16	0.21	0.28	0.28

TABLES XXVII″.-XXXV″.

Tables	XXVII″.				XXIX″.		XXX″.	XXXI″.	XXXV″.	
Arguments	22.				24.		25.	26.	30.	
Days.	0	10	20	30	0	10	0	0	0	10
Days. 5.0	″ 0.04	″ 0.14	″ 0.00	″ 0.10	″ 0.16	″ 0.01	″ 0.16	″ 0.21	″ 0.28	″ 0.28
5.1	0.04	0.14	0.00	0.10	0.16	0.02	0.16	0.23	0.28	0.28
5.2	0.04	0.14	0.00	0.10	0.17	0.02	0.16	0.24	0.28	0.28
5.3	0.03	0.14	0.01	0.09	0.17	0.02	0.16	0.25	0.27	0.28
5.4	0.03	0.14	0.01	0.09	0.18	0.02	0.16	0.26	0.27	0.28
5.5	0.03	0.14	0.01	0.09	0.18	0.03	0.16	0.27	0.27	0.28
5.6	0.03	0.14	0.01	0.09	0.18	0.03	0.16	0.27	0.27	0.28
5.7	0.03	0.14	0.01	0.08	0.18	0.03	0.15	0.28	0.26	0.28
5.8	0.02	0.14	0.02	0.08	0.19	0.04	0.15	0.28	0.26	0.27
5.9	0.02	0.14	0.02	0.08	0.19	0.04	0.14	0.28	0.25	0.27
6.0	0.02	0.14	0.02	0.07	0.19	0.05	0.14	0.28	0.25	0.27
6.1	0.02	0.14	0.02	0.07	0.19	0.05	0.14	0.28	0.25	0.27
6.2	0.02	0.14	0.02	0.07	0.19	0.06	0.13	0.27	0.24	0.26
6.3	0.01	0.14	0.03	0.06	0.20	0.06	0.13	0.26	0.24	0.26
6.4	0.01	0.14	0.03	0.06	0.20	0.07	0.12	0.25	0.23	0.25
6.5	0.01	0.14	0.03	0.06	0.20	0.07	0.12	0.25	0.23	0.25
6.6	0.01	0.14	0.03	0.06	0.20	0.07	0.12	0.24	0.23	0.25
6.7	0.01	0.14	0.03	0.05	0.20	0.08	0.11	0.23	0.22	0.24
6.8	0.00	0.14	0.04	0.05	0.20	0.08	0.11	0.22	0.22	0.24
6.9	0.00	0.14	0.04	0.05	0.20	0.09	0.10	0.21	0.21	0.23
7.0	0.00	0.14	0.04	0.04	0.20	0.10	0.10	0.19	0.21	0.23
7.1	0.00	0.14	0.04	0.04	0.20	0.10	0.09	0.18	0.21	0.23
7.2	0.00	0.14	0.05	0.04	0.20	0.11	0.09	0.16	0.20	0.22
7.3	0.00	0.13	0.05	0.03	0.20	0.11	0.08	0.14	0.20	0.22
7.4	0.00	0.13	0.05	0.03	0.20	0.12	0.08	0.13	0.19	0.21
7.5	0.00	0.13	0.06	0.03	0.20	0.12	0.07	0.11	0.19	0.21
7.6	0.00	0.13	0.06	0.03	0.20	0.12	0.07	0.10	0.18	0.20
7.7	0.00	0.13	0.06	0.03	0.20	0.13	0.06	0.09	0.18	0.20
7.8	0.00	0.12	0.07	0.02	0.19	0.13	0.06	0.08	0.17	0.19
7.9	0.00	0.12	0.07	0.02	0.19	0.14	0.05	0.06	0.17	0.19
8.0	0.00	0.12	0.07	0.02	0.19	0.15	0.05	0.05	0.16	0.18
8.1	0.00	0.12	0.07	0.02	0.19	0.15	0.04	0.04	0.16	0.18
8.2	0.00	0.12	0.07	0.02	0.19	0.16	0.04	0.03	0.15	0.17
8.3	0.00	0.11	0.08	0.01	0.18	0.16	0.03	0.02	0.15	0.17
8.4	0.00	0.11	0.08	0.01	0.18	0.17	0.03	0.02	0.14	0.16
8.5	0.00	0.11	0.08	0.01	0.18	0.17	0.02	0.01	0.14	0.16
8.6	0.00	0.11	0.09	0.01	0.18	0.17	0.02	0.01	0.14	0.16
8.7	0.00	0.11	0.09	0.01	0.17	0.17	0.02	0.00	0.13	0.15
8.8	0.00	0.10	0.09	0.00	0.17	0.18	0.01	0.00	0.13	0.15
8.9	0.00	0.10	0.10	0.00	0.16	0.18	0.01	0.00	0.12	0.14
9.0	0.00	0.10	0.10	0.00	0.16	0.18	0.01	0.01	0.12	0.14
9.1	0.00	0.10	0.10	0.00	0.16	0.18	0.01	0.01	0.12	0.14
9.2	0.00	0.09	0.10	0.00	0.15	0.19	0.01	0.02	0.12	0.13
9.3	0.01	0.09	0.11	0.00	0.15	0.19	0.00	0.02	0.11	0.13
9.4	0.01	0.09	0.11	0.00	0.14	0.19	0.00	0.03	0.11	0.12
9.5	0.01	0.08	0.11	0.00	0.14	0.19	0.00	0.04	0.11	0.12
9.6	0.01	0.08	0.11	0.00	0.13	0.19	0.00	0.05	0.11	0.12
9.7	0.01	0.08	0.11	0.00	0.13	0.20	0.00	0.06	0.11	0.12
9.8	0.02	0.07	0.12	0.00	0.12	0.20	0.00	0.08	0.10	0.11
9.9	0.02	0.07	0.12	0.00	0.12	0.20	0.00	0.09	0.10	0.11
10.0	0.02	0.07	0.12	0.00	0.11	0.20	0.00	0.10	0.10	0.11

TABLES VI.IV - XII.IV

TABLES	VI.IV			VII.IV				VIII.IV			XII.IV
ARGUMENTS	1.			2.				3.			7.
Days.	0	10	20	0	10	20	30	0	10	20	0
Days. 0.0	″0.24	″0.76	″0.46	″0.00	″0.01	″0.05	″0.01	″0.28	″0.08	″0.54	″0.01
0.1	0.24	0.76	0.45	0.00	0.01	0.05	0.01	0.30	0.07	0.53	0.00
0.2	0.24	0.77	0.45	0.00	0.04	0.05	0.01	0.31	0.06	0.52	0.00
0.3	0.23	0.77	0.45	0.00	0.04	0.05	0.01	0.32	0.06	0.52	0.00
0.4	0.22	0.77	0.45	0.00	0.04	0.05	0.01	0.34	0.05	0.51	0.01
0.5	0.22	0.77	0.45	0.00	0.04	0.05	0.01	0.35	0.04	0.50	0.01
0.6	0.21	0.77	0.45	0.00	0.05	0.05	0.00	0.37	0.03	0.49	0.01
0.7	0.21	0.77	0.45	0.00	0.05	0.05	0.00	0.38	0.03	0.48	0.01
0.8	0.20	0.77	0.45	0.00	0.05	0.05	0.00	0.39	0.02	0.47	0.02
0.9	0.19	0.77	0.45	0.00	0.05	0.05	0.00	0.41	0.02	0.46	0.02
1.0	0.19	0.77	0.45	0.00	0.05	0.05	0.00	0.42	0.01	0.45	0.02
1.1	0.18	0.77	0.44	0.00	0.05	0.04	0.00	0.44	0.01	0.44	0.03
1.2	0.18	0.77	0.44	0.00	0.05	0.04	0.00	0.45	0.01	0.43	0.03
1.3	0.17	0.77	0.44	0.00	0.05	0.04	0.00	0.46	0.00	0.42	0.04
1.4	0.17	0.77	0.44	0.00	0.05	0.04	0.00	0.47	0.00	0.41	0.04
1.5	0.16	0.77	0.44	0.00	0.05	0.04	0.00	0.49	0.00	0.40	0.05
1.6	0.16	0.76	0.44	0.00	0.05	0.04	0.00	0.50	0.00	0.39	0.06
1.7	0.15	0.76	0.44	0.00	0.05	0.04	0.00	0.51	0.00	0.38	0.06
1.8	0.15	0.76	0.44	0.00	0.05	0.04	0.00	0.52	0.00	0.36	0.07
1.9	0.15	0.76	0.44	0.00	0.05	0.04	0.00	0.53	0.00	0.35	0.07
2.0	0.15	0.75	0.44	0.00	0.05	0.04	0.00	0.54	0.00	0.34	0.08
2.1	0.14	0.75	0.43	0.00	0.05	0.04	0.00	0.55	0.00	0.33	0.09
2.2	0.14	0.74	0.43	0.00	0.05	0.04	0.00	0.56	0.00	0.32	0.10
2.3	0.14	0.74	0.43	0.00	0.05	0.04	0.00	0.57	0.01	0.31	0.11
2.4	0.14	0.73	0.43	0.00	0.05	0.04	0.00	0.57	0.01	0.30	0.12
2.5	0.14	0.73	0.43	0.00	0.05	0.04	0.00	0.58	0.01	0.29	0.13
2.6	0.13	0.72	0.42	0.00	0.05	0.03	0.00	0.58	0.02	0.28	0.14
2.7	0.13	0.72	0.42	0.00	0.05	0.03	0.00	0.59	0.03	0.27	0.15
2.8	0.13	0.71	0.42	0.00	0.05	0.03	0.00	0.59	0.03	0.27	0.16
2.9	0.13	0.71	0.42	0.00	0.05	0.03	0.00	0.60	0.04	0.26	0.16
3.0	0.13	0.70	0.42	0.00	0.05	0.03	0.00	0.60	0.05	0.25	0.17
3.1	0.13	0.70	0.41	0.00	0.06	0.03	0.00	0.61	0.06	0.24	0.18
3.2	0.13	0.69	0.41	0.01	0.06	0.03	0.00	0.61	0.07	0.23	0.19
3.3	0.13	0.69	0.41	0.01	0.06	0.03	0.00	0.61	0.08	0.21	0.19
3.4	0.13	0.68	0.41	0.01	0.06	0.03	0.00	0.61	0.09	0.20	0.20
3.5	0.13	0.68	0.41	0.01	0.06	0.03	0.00	0.61	0.10	0.19	0.21
3.6	0.14	0.67	0.40	0.01	0.06	0.03	0.00	0.61	0.11	0.18	0.22
3.7	0.14	0.67	0.40	0.01	0.06	0.03	0.00	0.61	0.12	0.17	0.23
3.8	0.14	0.66	0.40	0.01	0.06	0.03	0.00	0.61	0.14	0.17	0.23
3.9	0.14	0.65	0.40	0.01	0.06	0.03	0.00	0.61	0.15	0.16	0.24
4.0	0.15	0.65	0.39	0.01	0.06	0.03	0.00	0.60	0.16	0.15	0.25
4.1	0.15	0.64	0.39	0.01	0.06	0.03	0.00	0.60	0.17	0.14	0.25
4.2	0.16	0.64	0.38	0.01	0.06	0.03	0.00	0.60	0.19	0.13	0.26
4.3	0.16	0.63	0.38	0.01	0.06	0.03	0.00	0.60	0.20	0.12	0.26
4.4	0.17	0.63	0.38	0.01	0.06	0.03	0.00	0.59	0.22	0.11	0.27
4.5	0.18	0.62	0.38	0.01	0.06	0.03	0.00	0.59	0.23	0.10	0.27
4.6	0.18	0.62	0.37	0.01	0.06	0.03	0.00	0.58	0.25	0.09	0.27
4.7	0.19	0.61	0.37	0.01	0.06	0.03	0.00	0.58	0.26	0.08	0.27
4.8	0.20	0.61	0.37	0.01	0.06	0.03	0.00	0.57	0.28	0.08	0.28
4.9	0.21	0.60	0.37	0.01	0.06	0.03	0.00	0.56	0.29	0.07	0.28
5.0	0.22	0.60	0.36	0.01	0.06	0.03	0.00	0.56	0.31	0.06	0.28

TABLES VI.^{IV} - XII.^{IV}

TABLES	VI.^{IV}			VII.^{IV}				VIII.^{IV}			XII.^{IV}
ARGUMENTS	1.			2.				3.			7.
Days.	0	10	20	0	10	20	30	0	10	20	0
Days.											
5.0	0.22	0.60	0.36	0.01	0.06	0.03	0.00	0.56	0.31	0.06	0.28
5.1	0.22	0.59	0.36	0.01	0.06	0.03	0.00	0.55	0.33	0.05	0.28
5.2	0.23	0.59	0.35	0.01	0.06	0.02	0.01	0.54	0.34	0.05	0.28
5.3	0.24	0.58	0.35	0.01	0.06	0.02	0.01	0.53	0.36	0.04	0.27
5.4	0.25	0.58	0.35	0.01	0.06	0.02	0.01	0.52	0.37	0.04	0.27
5.5	0.26	0.57	0.34	0.02	0.06	0.02	0.01	0.52	0.39	0.03	0.27
5.6	0.28	0.57	0.34	0.02	0.06	0.02	0.01	0.51	0.40	0.03	0.27
5.7	0.29	0.56	0.33	0.02	0.06	0.02	0.01	0.50	0.42	0.03	0.26
5.8	0.30	0.56	0.33	0.02	0.06	0.02	0.01	0.49	0.43	0.02	0.26
5.9	0.31	0.56	0.33	0.02	0.06	0.02	0.01	0.48	0.45	0.02	0.25
6.0	0.33	0.55	0.32	0.02	0.06	0.02	0.01	0.47	0.46	0.02	0.25
6.1	0.34	0.55	0.32	0.02	0.06	0.02	0.01	0.46	0.47	0.02	0.24
6.2	0.36	0.54	0.31	0.02	0.06	0.02	0.01	0.45	0.48	0.02	0.23
6.3	0.37	0.54	0.31	0.02	0.06	0.02	0.01	0.44	0.50	0.01	0.23
6.4	0.38	0.54	0.30	0.02	0.06	0.02	0.01	0.43	0.51	0.01	0.22
6.5	0.40	0.53	0.30	0.02	0.06	0.02	0.01	0.43	0.52	0.01	0.21
6.6	0.41	0.53	0.29	0.02	0.06	0.02	0.01	0.42	0.53	0.01	0.21
6.7	0.43	0.52	0.29	0.02	0.06	0.02	0.01	0.41	0.54	0.01	0.20
6.8	0.44	0.52	0.28	0.02	0.06	0.02	0.01	0.40	0.55	0.02	0.20
6.9	0.45	0.52	0.28	0.02	0.06	0.02	0.01	0.39	0.56	0.02	0.19
7.0	0.46	0.52	0.27	0.02	0.06	0.02	0.01	0.37	0.57	0.02	0.17
7.1	0.48	0.51	0.27	0.02	0.06	0.02	0.01	0.36	0.58	0.02	0.16
7.2	0.49	0.51	0.26	0.02	0.06	0.02	0.01	0.35	0.59	0.03	0.15
7.3	0.50	0.51	0.26	0.02	0.06	0.02	0.02	0.34	0.59	0.03	0.15
7.4	0.51	0.51	0.25	0.02	0.06	0.02	0.02	0.34	0.60	0.04	0.14
7.5	0.53	0.51	0.25	0.03	0.06	0.02	0.02	0.33	0.61	0.04	0.13
7.6	0.54	0.50	0.24	0.03	0.06	0.01	0.02	0.32	0.61	0.05	0.12
7.7	0.56	0.50	0.24	0.03	0.06	0.01	0.02	0.31	0.61	0.06	0.11
7.8	0.57	0.50	0.23	0.03	0.06	0.01	0.02	0.30	0.62	0.06	0.10
7.9	0.58	0.50	0.22	0.03	0.06	0.01	0.02	0.29	0.62	0.07	0.09
8.0	0.59	0.49	0.22	0.03	0.06	0.01	0.02	0.28	0.62	0.08	0.08
8.1	0.61	0.49	0.21	0.03	0.0[illegible]	0.01	0.02	0.27	0.62	0.09	0.07
8.2	0.62	0.48	0.21	0.03	0.05	0.01	0.02	0.26	0.62	0.10	0.07
8.3	0.63	0.48	0.20	0.03	0.05	0.01	0.02	0.24	0.62	0.11	0.06
8.4	0.64	0.48	0.20	0.03	0.05	0.01	0.02	0.23	0.62	0.12	0.06
8.5	0.65	0.48	0.19	0.03	0.05	0.01	0.02	0.22	0.62	0.13	0.05
8.6	0.66	0.47	0.19	0.04	0.05	0.01	0.02	0.21	0.62	0.14	0.04
8.7	0.67	0.47	0.18	0.04	0.05	0.01	0.02	0.20	0.62	0.16	0.04
8.8	0.68	0.47	0.18	0.04	0.05	0.01	0.02	0.19	0.61	0.17	0.03
8.9	0.69	0.47	0.17	0.04	0.05	0.01	0.02	0.18	0.61	0.19	0.03
9.0	0.70	0.47	0.17	0.04	0.05	0.01	0.02	0.17	0.61	0.20	0.02
9.1	0.70	0.47	0.17	0.04	0.05	0.01	0.02	0.16	0.60	0.21	0.02
9.2	0.71	0.47	0.16	0.04	0.05	0.01	0.02	0.15	0.59	0.23	0.02
9.3	0.72	0.47	0.16	0.04	0.05	0.01	0.03	0.14	0.59	0.24	0.01
9.4	0.73	0.47	0.15	0.04	0.05	0.01	0.03	0.13	0.58	0.26	0.01
9.5	0.73	0.47	0.15	0.04	0.05	0.01	0.03	0.12	0.57	0.27	0.01
9.6	0.74	0.46	0.15	0.04	0.05	0.01	0.03	0.11	0.56	0.29	0.01
9.7	0.74	0.46	0.15	0.04	0.05	0.01	0.03	0.10	0.56	0.31	0.00
9.8	0.75	0.46	0.14	0.04	0.05	0.01	0.03	0.10	0.55	0.32	0.00
9.9	0.75	0.46	0.14	0.04	0.05	0.01	0.03	0.09	0.55	0.34	0.00
10.0	0.76	0.46	0.14	0.04	0.05	0.01	0.03	0.08	0.54	0.35	0.01

TABLES XIII.IV - XXI.IV

Tables	XIII.IV		XVIII.IV	XXI.IV			XIII.IV		XVIII.IV	XXI.IV	
Arguments	8.		13.	16.			8.		13.	16.	
Days.	0	10	0	0	10	Days.	0	10	0	0	10
0.0	0.00	0.03	0.00	0.06	0.02	5.0	0.03	0.00	0.04	0.00	0.06
0.1	0.00	0.03	0.00	0.06	0.01	5.1	0.03	0.00	0.04	0.00	0.06
0.2	0.00	0.03	0.00	0.06	0.01	5.2	0.03	0.00	0.04	0.01	0.06
0.3	0.00	0.03	0.00	0.06	0.01	5.3	0.03	0.00	0.04	0.01	0.06
0.4	0.00	0.03	0.00	0.06	0.00	5.4	0.03	0.00	0.04	0.02	0.06
0.5	0.00	0.03	0.00	0.06	0.00	5.5	0.03	0.00	0.04	0.02	0.06
0.6	0.00	0.03	0.00	0.06	0.00	5.6	0.03	0.00	0.04	0.02	0.06
0.7	0.00	0.03	0.00	0.06	0.00	5.7	0.03	0.00	0.04	0.02	0.05
0.8	0.00	0.03	0.00	0.06	0.00	5.8	0.03	0.00	0.04	0.03	0.05
0.9	0.00	0.03	0.00	0.06	0.00	5.9	0.03	0.00	0.04	0.03	0.05
1.0	0.00	0.03	0.00	0.06	0.00	6.0	0.03	0.00	0.04	0.03	0.04
1.1	0.00	0.03	0.00	0.06	0.00	6.1	0.03	0.00	0.04	0.03	0.04
1.2	0.00	0.03	0.00	0.06	0.00	6.2	0.03	0.00	0.04	0.03	0.04
1.3	0.00	0.03	0.00	0.06	0.00	6.3	0.04	0.00	0.04	0.04	0.03
1.4	0.00	0.02	0.00	0.05	0.00	6.4	0.04	0.00	0.04	0.04	0.03
1.5	0.00	0.02	0.00	0.05	0.00	6.5	0.04	0.00	0.04	0.04	0.03
1.6	0.00	0.02	0.00	0.05	0.00	6.6	0.04	0.00	0.04	0.04	0.03
1.7	0.00	0.02	0.00	0.05	0.00	6.7	0.04	0.00	0.04	0.05	0.02
1.8	0.00	0.02	0.00	0.04	0.00	6.8	0.04	0.00	0.04	0.05	0.02
1.9	0.00	0.02	0.00	0.04	0.00	6.9	0.04	0.00	0.04	0.05	0.02
2.0	0.00	0.02	0.00	0.04	0.00	7.0	0.04	0.00	0.04	0.06	0.01
2.1	0.01	0.02	0.00	0.04	0.00	7.1	0.04	0.01	0.04	0.06	0.01
2.2	0.01	0.02	0.00	0.04	0.01	7.2	0.04	0.01	0.04	0.06	0.01
2.3	0.01	0.02	0.01	0.03	0.01	7.3	0.04	0.01	0.03	0.06	0.01
2.4	0.01	0.02	0.01	0.03	0.01	7.4	0.04	0.01	0.03	0.06	0.00
2.5	0.01	0.02	0.01	0.03	0.02	7.5	0.04	0.01	0.03	0.06	0.00
2.6	0.01	0.02	0.01	0.03	0.02	7.6	0.04	0.01	0.03	0.06	0.00
2.7	0.01	0.02	0.01	0.02	0.02	7.7	0.04	0.01	0.03	0.06	0.00
2.8	0.01	0.01	0.01	0.02	0.03	7.8	0.04	0.01	0.03	0.06	0.00
2.9	0.01	0.01	0.01	0.01	0.03	7.9	0.04	0.01	0.03	0.06	0.00
3.0	0.01	0.01	0.01	0.01	0.03	8.0	0.04	0.01	0.03	0.06	0.00
3.1	0.02	0.01	0.01	0.01	0.03	8.1	0.04	0.01	0.03	0.06	0.00
3.2	0.02	0.01	0.01	0.01	0.03	8.2	0.04	0.01	0.03	0.06	0.00
3.3	0.02	0.01	0.02	0.00	0.04	8.3	0.04	0.01	0.02	0.06	0.00
3.4	0.02	0.01	0.02	0.00	0.04	8.4	0.04	0.01	0.02	0.06	0.00
3.5	0.02	0.01	0.02	0.00	0.04	8.5	0.04	0.01	0.02	0.06	0.00
3.6	0.02	0.01	0.02	0.00	0.04	8.6	0.04	0.02	0.02	0.06	0.00
3.7	0.02	0.01	0.02	0.00	0.05	8.7	0.04	0.02	0.02	0.06	0.00
3.8	0.02	0.01	0.03	0.00	0.05	8.8	0.04	0.02	0.01	0.05	0.00
3.9	0.02	0.01	0.03	0.00	0.05	8.9	0.04	0.02	0.01	0.05	0.00
4.0	0.02	0.01	0.03	0.00	0.06	9.0	0.04	0.02	0.01	0.05	0.00
4.1	0.02	0.01	0.03	0.00	0.06	9.1	0.04	0.02	0.01	0.04	0.00
4.2	0.02	0.01	0.03	0.00	0.06	9.2	0.04	0.02	0.01	0.04	0.00
4.3	0.02	0.00	0.03	0.00	0.06	9.3	0.04	0.02	0.01	0.04	0.01
4.4	0.02	0.00	0.03	0.00	0.06	9.4	0.04	0.02	0.01	0.03	0.01
4.5	0.02	0.00	0.03	0.00	0.06	9.5	0.04	0.02	0.01	0.03	0.01
4.6	0.02	0.00	0.03	0.00	0.06	9.6	0.04	0.02	0.01	0.03	0.02
4.7	0.02	0.00	0.03	0.00	0.06	9.7	0.04	0.02	0.01	0.03	0.02
4.8	0.03	0.00	0.04	0.00	0.06	9.8	0.03	0.02	0.00	0.02	0.02
4.9	0.03	0.00	0.04	0.00	0.06	9.9	0.03	0.02	0.00	0.02	0.03
5.0	0.03	0.00	0.04	0.00	0.06	10.0	0.03	0.02	0.00	0.02	0.03

LATITUDE TABLES, LXXXIII.-CIX.

ARGUMENTS, PERIODS, AND EQUATIONS.

TABLE.	ARGUMENT.	PERIOD.	EQUATION.
LXXXIII.	12′	173.31006	$846''.0 + 527''.5 \sin(2t - 2y)$.
LXXXIV.	73″	1095.1653	$25.7 - 25.7 \sin(2y - 2x)$.
LXXXV.	53	117.5394	$22.1 - 22.1 \sin(2y - 2t + z)$.
LXXXVI.	1	27.5546	$1.3 - 1.3 \sin x$.
LXXXVII.	70	329.7906	$10.3 + 10.3 \sin(2y - 2t - z)$.
LXXXVIII.	4	365.2597	$88.9 + 48.9 \sin z$.
LXXXIX.	4	365.2597	$6.63 - 1.3 \cos z$.
XC.	5′	205.8926	$4.7 + 4.7 \sin 2(x - t)$
XCI.	5	411.7852	$1.0 - 1.0 \sin(x - t)$.
XCII.	78	188.2015	$15.8 + 15.8 \sin(x + y - 2t)$.
XCIII.	73	2190.3306	$22.2 + 14.4 \sin(y - x)$.
XCIV.	12‴	346.6021	$0.6 - 0.6 \sin(y - t)$.
XCV.	79	438.3608	$0.2 - 0.2 \sin(x - y + z)$.
XCVI.	80	313.0547	$0.1 - 0.1 \sin(y + z - x)$.
XCVII.	81	124.2046	$0.7 + 0.7 \sin(y + z + x - 2t)$.
XCVIII.	82	14.8655	$1.8 + 1.8 \sin(2t - y + x)$.
XCIX.	83	39.2116	$0.7 + 0.7 \sin(2t - y - 2z)$.
C.	84	19.1434	$0.6 - 0.6 \sin(2t + y)$.
CI.	48	29.9342	$0.5 - 0.5 \sin(x + t - y)$.
CII.	85	23.5193	$0.4 + 0.4 \sin(3y - 2t)$.
CIII.	86	38.2830	$0.3 + 0.3 \sin(4t - y - 2x)$.
CIV.	87	14.2082	$0.2 - 0.2 \sin(2t + y + x)$.
CV.	88	14.6664	$0.2 + 0.2 \sin(2t + y - x)$.
CVI.	89	19.3122	$0.2 + 0.2 \sin(2t - y + 2x)$.
CVII.	90	32.2808	$0.2 - 0.2 \sin(2t - y)$.
CVIII.	77		$6.2 - 6.18 \sin 3\bar{y}$.
CIX.			$9.1 + 2.17 \cos \bar{u} - 8''.80 \sin \bar{u}$.

TABLE LXXXIII. ARGUMENT 12'.

Equation $= 846''.0 + 527''.5 \sin. (2t - 2y)$.

Days.	0		10		20		30		Days.
Days	Equation	Diff.	Equation.	Diff.	Equation.	Diff.	Equation.	Diff.	Days.
d. 0.0	322.38	0.23	333.78	0.46	411.77	1.09	546.22	1.58	d. 10.0
0.1	322.15	0.22	334.24	0.47	412.86	1.09	547.80	1.58	9.9
0.2	321.93	0.21	334.71	0.47	413.95	1.10	549.38	1.58	9.8
0.3	321.72	0.21	335.18	0.48	415.05	1.11	550.96	1.59	9.7
0.4	321.51	0.20	335.66	0.49	416.16	1.11	552.55	1.59	9.6
0.5	321.31	0.19	336.15	0.49	417.27	1.12	554.14	1.59	9.5
0.6	321.12	0.19	336.64	0.50	418.39	1.12	555.73	1.60	9.4
0.7	320.93	0.18	337.14	0.51	419.51	1.13	557.33	1.60	9.3
0.8	320.75	0.17	337.65	0.51	420.64	1.13	558.93	1.61	9.2
0.9	320.58	0.17	338.16	0.52	421.77	1.14	560.54	1.61	9.1
1.0	320.41	0.16	338.68	0.53	422.91	1.14	562.15	1.61	9.0
1.1	320.25	0.15	339.21	0.53	424.05	1.15	563.76	1.62	8.9
1.2	320.10	0.15	339.74	0.54	425.20	1.16	565.38	1.62	8.8
1.3	319.95	0.14	340.28	0.55	426.36	1.16	567.00	1.62	8.7
1.4	319.81	0.13	340.83	0.55	427.52	1.17	568.62	1.63	8.6
1.5	319.68	0.12	341.38	0.56	428.69	1.17	570.25	1.63	8.5
1.6	319.56	0.12	341.94	0.57	429.86	1.18	571.88	1.64	8.4
1.7	319.44	0.11	342.51	0.58	431.04	1.18	573.52	1.64	8.3
1.8	319.33	0.10	343.09	0.58	432.22	1.19	575.16	1.64	8.2
1.9	319.23	0.10	343.67	0.59	433.41	1.20	576.80	1.65	8.1
2.0	319.13	0.09	344.26	0.59	434.61	1.20	578.45	1.65	8.0
2.1	319.04	0.08	344.85	0.60	435.81	1.20	580.10	1.65	7.9
2.2	318.96	0.08	345.45	0.61	437.01	1.21	581.75	1.66	7.8
2.3	318.88	0.07	346.06	0.61	438.22	1.22	583.41	1.66	7.7
2.4	318.81	0.06	346.67	0.62	439.44	1.22	585.07	1.66	7.6
2.5	318.75	0.06	347.29	0.63	440.66	1.23	586.73	1.67	7.5
2.6	318.69	0.05	347.92	0.63	441.89	1.23	588.40	1.67	7.4
2.7	318.64	0.04	348.55	0.64	443.12	1.24	590.07	1.68	7.3
2.8	318.60	0.03	349.19	0.65	444.36	1.24	591.75	1.68	7.2
2.9	318.57	0.03	349.84	0.65	445.60	1.25	593.43	1.68	7.1
3.0	318.54	0.02	350.49	0.66	446.85	1.25	595.11	1.68	7.0
3.1	318.52	0.01	351.15	0.67	448.10	1.26	596.79	1.69	6.9
3.2	318.51	0.01	351.82	0.67	449.36	1.26	598.48	1.69	6.8
3.3	318.50	0.00	352.49	0.68	450.62	1.27	600.17	1.69	6.7
3.4	318.50	0.01	353.17	0.68	451.89	1.27	601.86	1.70	6.6
3.5	318.51	0.01	353.85	0.69	453.16	1.28	603.56	1.70	6.5
3.6	318.52	0.02	354.54	0.70	454.44	1.29	605.26	1.70	6.4
3.7	318.54	0.03	355.24	0.70	455.73	1.29	606.96	1.71	6.3
3.8	318.57	0.04	355.94	0.71	457.02	1.30	608.67	1.71	6.2
3.9	318.61	0.04	356.65	0.72	458.32	1.30	610.38	1.71	6.1
4.0	318.65	0.05	357.37	0.72	459.62	1.31	612.09	1.72	6.0
4.1	318.70	0.06	358.09	0.73	460.93	1.31	613.81	1.72	5.9
4.2	318.76	0.06	358.82	0.74	462.24	1.32	615.53	1.72	5.8
4.3	318.82	0.07	359.56	0.74	463.56	1.32	617.25	1.72	5.7
4.4	318.89	0.08	360.30	0.75	464.88	1.32	618.97	1.73	5.6
4.5	318.97	0.08	361.05	0.76	466.20	1.33	620.70	1.73	5.5
4.6	319.05	0.09	361.81	0.76	467.53	1.33	622.43	1.73	5.4
4.7	319.14	0.10	362.57	0.77	468.86	1.34	624.16	1.74	5.3
4.8	319.24	0.10	363.34	0.78	470.20	1.34	625.90	1.74	5.2
4.9	319.34	0.11	364.12	0.78	471.54	1.35	627.64	1.74	5.1
5.0	319.45	0.12	364.90	0.79	472.89	1.35	629.38	1.74	5.0
Days.	170		160		150		140		Days.

NOTE. — Arg. 12' = Arg. 12.

TABLE LXXXIII. ARGUMENT 12′.

Equation = 846″.0 + 527″.5 sin. $(2t - 2y)$.

Days.	0 Equation.	0 Diff.	10 Equation.	10 Diff.	20 Equation.	20 Diff.	30 Equation.	30 Diff.	Days.
d.	″		″		″		″		d
5.0	319.45	0.12	364.90	0.79	472.89	1.35	629.38	1.74	5.0
5.1	319.57	0.13	365.69	0.79	474.24	1.36	631.12	1.75	4.9
5.2	319.70	0.13	366.48	0.80	475.60	1.36	632.87	1.75	4.8
5.3	319.83	0.14	367.28	0.80	476.96	1.37	634.62	1.75	4.7
5.4	319.97	0.15	368.08	0.81	478.33	1.37	636.37	1.76	4.6
5.5	320.12	0.15	368.89	0.82	479.70	1.38	638.13	1.76	4.5
5.6	320.27	0.16	369.71	0.82	481.08	1.38	639.89	1.76	4.4
5.7	320.43	0.17	370.53	0.83	482.46	1.39	641.65	1.77	4.3
5.8	320.60	0.17	371.36	0.83	483.85	1.40	643.42	1.77	4.2
5.9	320.77	0.18	372.19	0.84	485.25	1.40	645.19	1.77	4.1
6.0	320.95	0.19	373.03	0.85	486.65	1.40	646.96	1.77	4.0
6.1	321.14	0.19	373.88	0.86	488.05	1.41	648.73	1.78	3.9
6.2	321.33	0.20	374.74	0.87	489.46	1.41	650.51	1.78	3.8
6.3	321.53	0.21	375.61	0.87	490.87	1.42	652.29	1.78	3.7
6.4	321.74	0.22	376.48	0.88	492.29	1.42	654.07	1.78	3.6
6.5	321.96	0.22	377.36	0.88	493.71	1.43	655.85	1.79	3.5
6.6	322.18	0.23	378.24	0.89	495.14	1.43	657.64	1.79	3.4
6.7	322.41	0.24	379.13	0.89	496.57	1.43	659.43	1.79	3.3
6.8	322.65	0.24	380.02	0.90	498.00	1.44	661.22	1.79	3.2
6.9	322.89	0.25	380.92	0.91	499.44	1.44	663.01	1.79	3.1
7.0	323.14	0.26	381.83	0.91	500.88	1.45	664.80	1.80	3.0
7.1	323.40	0.26	382.74	0.92	502.33	1.45	666.60	1.80	2.9
7.2	323.66	0.27	383.66	0.92	503.78	1.46	668.40	1.80	2.8
7.3	323.93	0.28	384.58	0.93	505.24	1.46	670.20	1.80	2.7
7.4	324.21	0.28	385.51	0.94	506.70	1.47	672.00	1.81	2.6
7.5	324.49	0.29	386.45	0.94	508.17	1.47	673.81	1.81	2.5
7.6	324.78	0.30	387.39	0.95	509.64	1.47	675.62	1.81	2.4
7.7	325.08	0.30	388.34	0.95	511.11	1.48	677.43	1.81	2.3
7.8	325.38	0.31	389.29	0.96	512.59	1.48	679.24	1.81	2.2
7.9	325.69	0.31	390.25	0.97	514.07	1.49	681.05	1.82	2.1
8.0	326.00	0.32	391.22	0.97	515.56	1.49	682.87	1.82	2.0
8.1	326.32	0.33	392.19	0.98	517.05	1.50	684.69	1.82	1.9
8.2	326.65	0.34	393.17	0.98	518.55	1.50	686.51	1.82	1.8
8.3	326.99	0.34	394.15	0.99	520.05	1.51	688.33	1.83	1.7
8.4	327.33	0.35	395.14	1.00	521.56	1.51	690.16	1.83	1.6
8.5	327.68	0.36	396.14	1.00	523.07	1.51	691.99	1.83	1.5
8.6	328.04	0.37	397.14	1.01	524.58	1.52	693.82	1.83	1.4
8.7	328.41	0.37	398.15	1.01	526.10	1.52	695.65	1.83	1.3
8.8	328.78	0.38	399.16	1.02	527.62	1.53	697.48	1.84	1.2
8.9	329.16	0.39	400.18	1.02	529.15	1.53	699.32	1.84	1.1
9.0	329.55	0.39	401.20	1.03	530.68	1.54	701.16	1.84	1.0
9.1	329.94	0.40	402.23	1.04	532.22	1.54	703.00	1.84	0.9
9.2	330.34	0.41	403.27	1.04	533.76	1.54	704.84	1.84	0.8
9.3	330.75	0.41	404.31	1.05	535.30	1.55	706.68	1.85	0.7
9.4	331.16	0.42	405.36	1.05	536.85	1.55	708.53	1.85	0.6
9.5	331.58	0.43	406.41	1.06	538.40	1.56	710.38	1.85	0.5
9.6	332.01	0.43	407.47	1.07	539.96	1.56	712.23	1.85	0.4
9.7	332.44	0.44	408.54	1.07	541.52	1.56	714.08	1.85	0.3
9.8	332.88	0.45	409.61	1.08	543.08	1.57	715.93	1.85	0.2
9.9	333.33	0.45	410.69	1.08	544.65	1.57	717.78	1.85	0.1
10.0	333.78	0.46	411.77	1.09	546.22	1.58	719.63	1.86	0.0
Days.	170		160		150		140		Days.

Note. — Arg. 12′ = Arg. 12.

TABLE LXXXIII. ARGUMENT 12′.

Equation = 846″.0 + 527″.5 sin. (2t — 2y).

Days.	40		50		60		70		80		Days.
Days.	Equation.	Diff.	Equation.	Diff.	Equation.	Diff.	Equation.	Diff.	Equation.	Diff.	Days.
d. 0.0	″ 719.63	1.86	″ 909.49	1.90	″ 1091.07	1.69	″ 1240.80	1.27	″ 1339.22	0.68	d. 10.0
0.1	721.49	1.86	911.39	1.90	1092.76	1.69	1242.07	1.26	1339.90	0.67	9.9
0.2	723.35	1.86	913.29	1.90	1094.45	1.69	1243.33	1.25	1340.57	0.66	9.8
0.3	725.21	1.86	915.19	1.89	1096.14	1.68	1244.58	1.25	1341.23	0.66	9.7
0.4	727.07	1.86	917.08	1.89	1097.82	1.68	1245.83	1.24	1341.89	0.65	9.6
0.5	728.93	1.87	918.97	1.89	1099.50	1.68	1247.07	1.24	1342.54	0.64	9.5
0.6	730.80	1.87	920.86	1.89	1101.18	1.67	1248.31	1.23	1343.18	0.63	9.4
0.7	732.67	1.87	922.75	1.89	1102.85	1.67	1249.54	1.23	1343.81	0.63	9.3
0.8	734.54	1.87	924.64	1.89	1104.52	1.66	1250.77	1.23	1344.44	0.62	9.2
0.9	736.41	1.87	926.53	1.89	1106.18	1.66	1252.00	1.22	1345.06	0.61	9.1
1.0	738.28	1.87	928.42	1.89	1107.84	1.66	1253.22	1.21	1345.67	0.61	9.0
1.1	740.15	1.88	930.31	1.89	1109.50	1.66	1254.43	1.21	1346.28	0.60	8.9
1.2	742.03	1.88	932.20	1.89	1111.16	1.65	1255.64	1.20	1346.88	0.60	8.8
1.3	743.91	1.88	934.09	1.88	1112.81	1.65	1256.84	1.20	1347.48	0.59	8.7
1.4	745.79	1.88	935.97	1.88	1114.46	1.64	1258.04	1.19	1348.07	0.58	8.6
1.5	747.67	1.88	937.85	1.88	1116.10	1.64	1259.23	1.19	1348.65	0.58	8.5
1.6	749.55	1.88	939.73	1.88	1117.74	1.64	1260.42	1.18	1349.23	0.57	8.4
1.7	751.43	1.88	941.61	1.88	1119.38	1.63	1261.60	1.17	1349.80	0.56	8.3
1.8	753.31	1.88	943.49	1.88	1121.01	1.63	1262.77	1.17	1350.36	0.55	8.2
1.9	755.19	1.88	945.37	1.88	1122.64	1.63	1263.94	1.16	1350.91	0.55	8.1
2.0	757.07	1.89	947.25	1.88	1124.27	1.62	1265.10	1.16	1351.46	0.54	8.0
2.1	758.96	1.89	949.13	1.87	1125.89	1.62	1266.26	1.15	1352.00	0.54	7.9
2.2	760.85	1.89	951.00	1.87	1127.51	1.62	1267.41	1.15	1352.54	0.53	7.8
2.3	762.74	1.89	952.87	1.87	1129.13	1.61	1268.56	1.14	1353.07	0.52	7.7
2.4	764.63	1.89	954.74	1.87	1130.74	1.61	1269.70	1.14	1353.59	0.52	7.6
2.5	766.52	1.89	956.61	1.87	1132.35	1.60	1270.84	1.13	1354.11	0.51	7.5
2.6	768.41	1.89	958.48	1.87	1133.95	1.60	1271.97	1.13	1354.62	0.51	7.4
2.7	770.30	1.89	960.35	1.87	1135.55	1.60	1273.10	1.12	1355.13	0.50	7.3
2.8	772.19	1.89	962.22	1.86	1137.15	1.59	1274.22	1.11	1355.63	0.49	7.2
2.9	774.08	1.90	964.08	1.86	1138.74	1.59	1275.33	1.11	1356.12	0.49	7.1
3.0	775.98	1.89	965.94	1.86	1140.33	1.58	1276.44	1.10	1356.61	0.48	7.0
3.1	777.87	1.90	967.80	1.86	1141.91	1.58	1277.54	1.10	1357.09	0.47	6.9
3.2	779.77	1.90	969.66	1.86	1143.49	1.58	1278.64	1.09	1357.56	0.46	6.8
3.3	781.67	1.90	971.52	1.86	1145.07	1.57	1279.73	1.09	1358.02	0.46	6.7
3.4	783.57	1.90	973.38	1.85	1146.64	1.57	1280.82	1.08	1358.48	0.45	6.6
3.5	785.47	1.90	975.23	1.85	1148.21	1.57	1281.90	1.07	1358.93	0.44	6.5
3.6	787.37	1.90	977.08	1.85	1149.78	1.56	1282.97	1.07	1359.37	0.44	6.4
3.7	789.27	1.90	978.93	1.85	1151.34	1.56	1284.04	1.06	1359.81	0.43	6.3
3.8	791.17	1.90	980.78	1.85	1152.90	1.55	1285.10	1.06	1360.24	0.42	6.2
3.9	793.07	1.90	982.63	1.85	1154.45	1.55	1286.16	1.05	1360.66	0.42	6.1
4.0	794.97	1.90	984.48	1.84	1156.00	1.55	1287.21	1.05	1361.08	0.41	6.0
4.1	796.87	1.90	986.32	1.84	1157.55	1.54	1288.26	1.04	1361.49	0.40	5.9
4.2	798.77	1.90	988.16	1.84	1159.09	1.54	1289.30	1.03	1361.89	0.40	5.8
4.3	800.67	1.91	990.00	1.84	1160.63	1.53	1290.33	1.03	1362.29	0.39	5.7
4.4	802.58	1.91	991.84	1.84	1162.16	1.53	1291.36	1.02	1362.68	0.38	5.6
4.5	804.49	1.91	993.68	1.84	1163.69	1.52	1292.38	1.02	1363.06	0.37	5.5
4.6	806.40	1.91	995.52	1.83	1165.21	1.52	1293.40	1.01	1363.43	0.37	5.4
4.7	808.31	1.91	997.35	1.83	1166.73	1.52	1294.41	1.00	1363.80	0.36	5.3
4.8	810.22	1.91	999.18	1.83	1168.25	1.51	1295.41	1.00	1364.16	0.35	5.2
4.9	812.13	1.91	1001.01	1.83	1169.76	1.51	1296.41	0.99	1364.51	0.35	5.1
5.0	814.04	1.91	1002.84	1.82	1171.27	1.50	1297.40	0.99	1364.86	0.34	5.0
Days.	130		120		110		100		90		Days.

NOTE. — Arg. 12′ = Arg. 12.

TABLE LXXXIII. ARGUMENT 12′.

Equation = 846″.0 + 527″.5 sin. (2*t* — 2*y*).

Days.	40		50		60		70		80		Days.
Days.	Equation.	Diff.	Equation.	Diff.	Equation.	Diff.	Equation.	Diff.	Equation.	Diff.	Days.
d	″		″		″		″		″		d
5.0	814.04	1.91	1002.84	1.82	1171.27	1.50	1297.40	0.99	1364.86	0.34	5.0
5.1	815.95	1.91	1004.66	1.82	1172.77	1.50	1298.39	0.98	1365.20	0.33	4.9
5.2	817.86	1.91	1006.48	1.82	1174.27	1.49	1299.37	0.97	1365.53	0.33	4.8
5.3	819.77	1.91	1008.30	1.82	1175.76	1.49	1300.34	0.97	1365.86	0.32	4.7
5.4	821.68	1.91	1010.12	1.82	1177.25	1.48	1301.31	0.96	1366.18	0.31	4.6
5.5	823.59	1.91	1011.94	1.82	1178.73	1.48	1302.27	0.96	1366.49	0.31	4.5
5.6	825.50	1.91	1013.76	1.81	1180.21	1.48	1303.23	0.95	1366.80	0.30	4.4
5.7	827.41	1.91	1015.57	1.81	1181.69	1.47	1304.18	0.95	1367.10	0.29	4.3
5.8	829.32	1.91	1017.38	1.81	1183.16	1.47	1305.13	0.94	1367.39	0.29	4.2
5.9	831.23	1.91	1019.19	1.81	1184.63	1.47	1306.07	0.93	1367.68	0.28	4.1
6.0	833.14	1.91	1021.00	1.80	1186.10	1.46	1307.00	0.93	1367.96	0.27	4.0
6.1	835.05	1.92	1022.80	1.80	1187.56	1.45	1307.93	0.92	1368.23	0.27	3.9
6.2	836.97	1.91	1024.60	1.80	1189.01	1.45	1308.85	0.91	1368.50	0.26	3.8
6.3	838.88	1.91	1026.40	1.80	1190.46	1.45	1309.76	0.91	1368.76	0.25	3.7
6.4	840.79	1.91	1028.20	1.79	1191.91	1.44	1310.67	0.90	1369.01	0.25	3.6
6.5	842.70	1.92	1029.99	1.79	1193.35	1.44	1311.57	0.90	1369.26	0.24	3.5
6.6	844.62	1.91	1031.78	1.79	1194.79	1.43	1312.47	0.89	1369.50	0.23	3.4
6.7	846.53	1.91	1033.57	1.79	1196.22	1.43	1313.36	0.88	1369.73	0.22	3.3
6.8	848.44	1.91	1035.36	1.78	1197.65	1.42	1314.24	0.88	1369.95	0.22	3.2
6.9	850.35	1.91	1037.14	1.78	1199.07	1.42	1315.12	0.87	1370.17	0.21	3.1
7.0	852.26	1.92	1038.92	1.78	1200.49	1.41	1315.99	0.87	1370.38	0.20	3.0
7.1	854.18	1.91	1040.70	1.78	1201.90	1.41	1316.8[illegible]	0.86	1370.58	0.20	2.9
7.2	856.09	1.91	1042.48	1.77	1203.31	1.40	1317.72	0.85	1370.78	0.19	2.8
7.3	858.00	1.91	1044.25	1.77	1204.71	1.40	1318.57	0.85	1370.97	0.18	2.7
7.4	859.91	1.91	1046.02	1.77	1206.11	1.39	1319.42	0.84	1371.15	0.18	2.6
7.5	861.82	1.92	1047.79	1.77	1207.50	1.39	1320.26	0.84	1371.33	0.17	2.5
7.6	863.74	1.91	1049.56	1.76	1208.8[illegible]	1.3[illegible]	1321.10	0.8[illegible]	1371.5[illegible]	0.1[illegible]	2.4
7.7	865.65	1.91	1051.32	1.76	1210.28	1.38	1321.93	0.82	1371.66	0.16	2.3
7.8	867.56	1.91	1053.08	1.76	1211.66	1.38	1322.75	0.81	1371.82	0.15	2.2
7.9	869.47	1.91	1054.84	1.75	1213.01	1.3[illegible]	[illegible].56	0.81	1371.97	0.14	2.1
8.0	871.38	1.91	1056.59	1.75	1214.42	1.37	1324.37	0.80	1372.11	0.14	2.0
8.1	873.29	1.91	1058.34	1.75	1215.79	1.36	1325.17	0.80	1372.25	0.13	1.9
8.2	875.20	1.91	1060.09	1.75	1217.15	1.36	1325.[illegible]7	0.79	1372.38	0.12	1.8
8.3	877.11	1.90	1061.84	1.74	1218.51	1.35	1326.76	0.78	1372.50	0.11	1.7
8.4	879.01	1.91	1063.58	1.74	1219.86	1.35	1327.54	0.78	1372.61	0.11	1.6
8.5	880.92	1.91	1065.32	1.74	1221.21	1.34	1328.32	0.77	1372.72	0.10	1.5
8.6	882.83	1.91	1067.06	1.73	1222.55	1.34	1329.09	0.76	1372.82	0.09	1.4
8.7	884.74	1.91	1068.79	1.73	1223.89	1.33	1329.85	0.76	1372.91	0.09	1.3
8.8	886.65	1.90	1070.52	1.73	1225.22	1.32	1330.61	0.75	1373.00	0.08	1.2
8.9	888.55	1.91	1072.25	1.73	1226.54	1.32	1331.36	0.74	1373.08	0.07	1.1
9.0	890.46	1.91	1073.98	1.72	1227.86	1.32	1332.10	0.74	1373.15	0.07	1.0
9.1	892.37	1.91	1075.70	1.72	1229.18	1.31	1332.84	0.73	1373.22	0.06	0.9
9.2	894.28	1.91	1077.42	1.72	1230.49	1.31	1333.57	0.73	1373.28	0.05	0.8
9.3	896.19	1.90	1079.14	1.71	1231.80	1.30	1334.30	0.72	1373.33	0.05	0.7
9.4	898.09	1.90	1080.85	1.71	1233.10	1.30	1335.02	0.71	1373.38	0.04	0.6
9.5	899.99	1.90	1082.56	1.71	1234.40	1.29	1335.73	0.71	1373.42	0.03	0.5
9.6	901.89	1.90	1084.27	1.70	1235.69	1.29	1336.44	0.70	1373.45	0.02	0.4
9.7	903.79	1.90	1085.97	1.70	1236.98	1.28	1337.14	0.70	1373.47	0.02	0.3
9.8	905.69	1.90	1087.67	1.70	1238.26	1.27	1337.84	0.69	1373.49	0.01	0.2
9.9	907.59	1.90	1089.37	1.70	1239.53	1.27	1338.53	0.69	1373.50	0.00	0.1
10.0	909.49	1.90	1091.07	1.69	1240.80	1.27	1339.22	0.68	1373.50	0.00	0.0
Days.	130		120		110		100		90		Days.

NOTE. — Arg. 12′ = Arg. 12.

TABLES LXXXIV. LXXXV.

Tables	LXXXIV.								LXXXV.			
Arguments	73″.								53.			
Days.	**0**	**100**	**200**	**300**	**400**	**500**	Days.	Days.	**0**	**10**	**20**	Days.
0	1.15″	0.95″	8.68″	21.85″	36.26″	47.29″	100	0.0	44.15″	41.82″	33.98″	10.0
01	1.11	0.99	8.79	21.99	36.40	47.37	99	0.1	44.16	41.77	33.88	9.9
02	1.07	1.03	8.90	22.14	36.53	47.45	98	0.2	44.17	41.71	33.78	9.8
03	1.03	1.07	9.01	22.28	36.67	47.53	97	0.3	44.17	41.66	33.68	9.7
04	0.99	1.11	9.12	22.43	36.80	47.61	96	0.4	44.18	41.60	33.58	9.6
05	0.95	1.16	9.23	22.58	36.93	47.68	95	0.5	44.18	41.55	33.48	9.5
06	0.91	1.20	9.34	22.72	37.06	47.76	94	0.6	44.18	41.49	33.38	9.4
07	0.87	1.25	9.46	22.87	37.19	47.83	93	0.7	44.19	41.43	33.28	9.3
08	0.83	1.29	9.57	23.01	37.32	47.91	92	0.8	44.19	41.38	33.18	9.2
09	0.79	1.34	9.69	23.16	37.45	47.98	91	0.9	44.19	41.32	33.08	9.1
10	0.76	1.39	9.81	23.31	37.58	48.05	90	1.0	44.20	41.26	32.97	9.0
11	0.72	1.44	9.92	23.45	37.71	48.12	89	1.1	44.20	41.20	32.87	8.9
12	0.69	1.49	10.04	23.60	37.84	48.19	88	1.2	44.20	41.14	32.76	8.8
13	0.65	1.54	10.15	23.75	37.97	48.26	87	1.3	44.20	41.08	32.66	8.7
14	0.62	1.59	10.27	23.90	38.10	48.33	86	1.4	44.20	41.02	32.55	8.6
15	0.59	1.64	10.39	24.05	38.23	48.40	85	1.5	44.20	40.96	32.45	8.5
16	0.56	1.69	10.51	24.19	38.36	48.47	84	1.6	44.20	40.89	32.34	8.4
17	0.53	1.74	10.63	24.34	38.49	48.54	83	1.7	44.19	40.83	32.24	8.3
18	0.50	1.80	10.75	24.48	38.62	48.61	82	1.8	44.19	40.77	32.13	8.2
19	0.47	1.85	10.87	24.63	38.75	48.68	81	1.9	44.19	40.71	32.03	8.1
20	0.44	1.91	10.99	24.78	38.87	48.74	80	2.0	44.18	40.64	31.92	8.0
21	0.41	1.97	11.11	24.92	39.00	48.81	79	2.1	44.18	40.58	31.82	7.9
22	0.39	2.02	11.23	25.07	39.13	48.87	78	2.2	44.17	40.51	31.71	7.8
23	0.36	2.08	11.35	25.22	39.25	48.94	77	2.3	44.17	40.44	31.61	7.7
24	0.34	2.14	11.47	25.37	39.38	49.00	76	2.4	44.16	40.38	31.50	7.6
25	0.32	2.20	11.60	25.52	39.50	49.06	75	2.5	44.15	40.31	31.39	7.5
26	0.29	2.26	11.72	25.66	39.63	49.12	74	2.6	44.14	40.24	31.29	7.4
27	0.27	2.32	11.84	25.81	39.75	49.18	73	2.7	44.13	40.18	31.18	7.3
28	0.25	2.38	11.97	25.96	39.87	49.24	72	2.8	44.12	40.11	31.07	7.2
29	0.23	2.44	12.09	26.11	39.99	49.30	71	2.9	44.11	40.04	30.96	7.1
30	0.21	2.50	12.22	26.26	40.11	49.35	70	3.0	44.10	39.97	30.85	7.0
31	0.19	2.56	12.35	26.40	40.24	49.41	69	3.1	44.09	39.90	30.74	6.9
32	0.17	2.63	12.47	26.55	40.36	49.47	68	3.2	44.08	39.83	30.63	6.8
33	0.16	2.69	12.60	26.70	40.48	49.52	67	3.3	44.06	39.76	30.52	6.7
34	0.14	2.76	12.73	26.85	40.60	49.58	66	3.4	44.05	39.69	30.41	6.6
35	0.13	2.83	12.86	27.00	40.72	49.63	65	3.5	44.04	39.62	30.30	6.5
36	0.11	2.90	12.98	27.14	40.84	49.69	64	3.6	44.02	39.54	30.19	6.4
37	0.10	2.97	13.11	27.29	40.96	49.74	63	3.7	44.01	39.47	30.08	6.3
38	0.09	3.04	13.24	27.44	41.08	49.79	62	3.8	43.99	39.40	29.97	6.2
39	0.08	3.11	13.37	27.58	41.20	49.84	61	3.9	43.98	39.33	29.86	6.1
40	0.07	3.18	13.50	27.73	41.31	49.89	60	4.0	43.96	39.25	29.75	6.0
41	0.06	3.25	13.63	27.88	41.43	49.94	59	4.1	43.94	39.18	29.64	5.9
42	0.05	3.32	13.76	28.03	41.55	49.99	58	4.2	43.92	39.10	29.53	5.8
43	0.04	3.39	13.89	28.17	41.66	50.04	57	4.3	43.90	39.03	29.42	5.7
44	0.03	3.46	14.02	28.32	41.78	50.09	56	4.4	43.88	38.95	29.31	5.6
45	0.02	3.54	14.15	28.47	41.89	50.13	55	4.5	43.86	38.87	29.20	5.5
46	0.01	3.61	14.28	28.62	42.01	50.18	54	4.6	43.84	38.80	29.08	5.4
47	0.01	3.69	14.41	28.76	42.12	50.22	53	4.7	43.82	38.72	28.97	5.3
48	0.01	3.76	14.55	28.91	42.24	50.27	52	4.8	43.80	38.64	28.86	5.2
49	0.00	3.84	14.68	29.05	42.35	50.31	51	4.9	43.78	38.56	28.75	5.1
50	0.00	3.92	14.82	29.20	42.46	50.35	50	5.0	43.75	38.48	28.63	5.0
Days.	**1100**	**1000**	**900**	**800**	**700**	**600**	Days.	Days.	**110**	**100**	**90**	Days.

Note. — Arg. 73″ = Arg. 76 + 321.37.

TABLES LXXXIV. LXXXV.

Tables	LXXXIV.	LXXXV.
Arguments	72″.	53.

Table LXXXIV.

Days.	0	100	200	300	400	500	Days.
50	0.00	3.92	14.82	29.20	42.46	50.35	50
51	0.00	4.00	14.95	29.35	42.57	50.39	49
52	0.00	4.08	15.08	29.49	42.68	50.43	48
53	0.00	4.16	15.22	29.64	42.79	50.47	47
54	0.00	4.24	15.35	29.78	42.90	50.51	46
55	0.00	4.32	15.49	29.93	43.01	50.55	45
56	0.00	4.40	15.62	30.08	43.12	50.59	44
57	0.00	4.48	15.76	30.22	43.23	50.63	43
58	0.01	4.57	15.89	30.37	43.34	50.66	42
59	0.01	4.65	16.03	30.51	43.45	50.70	41
60	0.02	4.74	16.17	30.65	43.55	50.73	40
61	0.03	4.83	16.30	30.80	43.66	50.76	39
62	0.04	4.91	16.44	30.94	43.76	50.79	38
63	0.05	5.00	16.58	31.09	43.87	50.82	37
64	0.06	5.09	16.72	31.23	43.97	50.85	36
65	0.07	5.18	16.86	31.37	44.07	50.88	35
66	0.08	5.27	17.00	31.52	44.18	50.91	34
67	0.09	5.36	17.14	31.66	44.28	50.94	33
68	0.10	5.45	17.28	31.81	44.38	50.97	32
69	0.11	5.54	17.42	31.95	44.48	51.00	31
70	0.13	5.63	17.56	32.09	44.58	51.02	30
71	0.14	5.72	17.70	32.24	44.68	51.05	29
72	0.16	5.81	17.84	32.38	44.78	51.07	28
73	0.18	5.91	17.98	32.53	44.88	51.10	27
74	0.20	6.00	18.12	32.67	44.98	51.12	26
75	0.22	6.10	18.26	32.81	45.07	51.14	25
76	0.24	6.19	18.40	32.95	45.17	51.16	24
77	0.26	6.29	18.54	33.09	45.27	51.18	23
78	0.28	6.38	18.68	33.23	45.36	51.20	22
79	0.30	6.48	18.82	33.37	45.46	51.22	21
80	0.32	6.58	18.97	33.51	45.55	51.23	20
81	0.34	6.68	19.11	33.65	45.65	51.25	19
82	0.37	6.78	19.25	33.79	45.74	51.26	18
83	0.39	6.88	19.39	33.93	45.83	51.28	17
84	0.42	6.98	19.53	34.07	45.92	51.29	16
85	0.45	7.08	19.68	34.21	46.01	51.30	15
86	0.47	7.18	19.82	34.35	46.10	51.32	14
87	0.50	7.28	19.96	34.49	46.19	51.33	13
88	0.53	7.39	20.11	34.63	46.28	51.34	12
89	0.56	7.49	20.25	34.77	46.37	51.35	11
90	0.59	7.60	20.40	34.90	46.45	51.36	10
91	0.62	7.70	20.54	35.04	46.54	51.37	09
92	0.65	7.81	20.68	35.18	46.63	51.38	08
93	0.69	7.91	20.83	35.31	46.71	51.38	07
94	0.72	8.02	20.97	35.45	46.80	51.39	06
95	0.76	8.13	21.12	35.58	46.88	51.39	05
96	0.79	8.24	21.26	35.72	46.97	51.39	04
97	0.83	8.35	21.41	35.86	47.05	51.40	03
98	0.87	8.46	21.55	35.99	47.13	51.40	02
99	0.91	8.57	21.70	36.13	47.21	51.40	01
100	0.95	8.68	21.85	36.26	47.29	51.40	0
Days.	1100	1000	900	800	700	600	Days.

Table LXXXV.

Days.	0	10	20	Days.
5.0	43.75	38.48	28.63	5.0
5.1	43.73	38.40	28.52	4.9
5.2	43.70	38.32	28.41	4.8
5.3	43.68	38.24	28.30	4.7
5.4	43.65	38.16	28.18	4.6
5.5	43.63	38.08	28.07	4.5
5.6	43.60	37.99	27.96	4.4
5.7	43.57	37.91	27.85	4.3
5.8	43.55	37.83	27.73	4.2
5.9	43.52	37.75	27.62	4.1
6.0	43.49	37.66	27.50	4.0
6.1	43.46	37.58	27.39	3.9
6.2	43.43	37.49	27.27	3.8
6.3	43.40	37.41	27.16	3.7
6.4	43.36	37.32	27.04	3.6
6.5	43.33	37.24	26.93	3.5
6.6	43.30	37.15	26.81	3.4
6.7	43.26	37.07	26.70	3.3
6.8	43.23	36.98	26.58	3.2
6.9	43.20	36.89	26.46	3.1
7.0	43.16	36.80	26.34	3.0
7.1	43.13	36.71	26.23	2.9
7.2	43.09	36.62	26.11	2.8
7.3	43.05	36.53	26.00	2.7
7.4	43.01	36.44	25.88	2.6
7.5	42.97	36.35	25.77	2.5
7.6	42.93	36.26	25.65	2.4
7.7	42.89	36.17	25.53	2.3
7.8	42.85	36.08	25.42	2.2
7.9	42.81	35.99	25.30	2.1
8.0	42.77	35.90	25.18	2.0
8.1	42.73	35.81	25.07	1.9
8.2	42.69	35.72	24.95	1.8
8.3	42.65	35.62	24.83	1.7
8.4	42.61	35.53	24.72	1.6
8.5	42.56	35.44	24.60	1.5
8.6	42.52	35.35	24.48	1.4
8.7	42.47	35.25	24.37	1.3
8.8	42.42	35.16	24.25	1.2
8.9	42.37	35.06	24.13	1.1
9.0	42.32	34.96	24.01	1.0
9.1	42.27	34.87	23.90	0.9
9.2	42.22	34.77	23.78	0.8
9.3	42.17	34.67	23.66	0.7
9.4	42.12	34.58	23.54	0.6
9.5	42.07	34.48	23.43	0.5
9.6	42.02	34.38	23.31	0.4
9.7	41.97	34.28	23.19	0.3
9.8	41.92	34.18	23.07	0.2
9.9	41.87	34.08	22.95	0.1
10.0	41.82	33.98	22.83	0.0
Days.	110	100	90	Days.

Note. — Arg. 73″ = Arg. 73 + 321.37.

TABLES LXXXV.-LXXXVII.

Tables	LXXXV.	LXXXVI.	LXXXVII.
Arguments	53.	1.	70.

Table LXXXV.

Days.	30	40	50	Days.
0.0	22.83	11.47	3.08	10.0
0.1	22.72	11.37	3.02	9.9
0.2	22.60	11.27	2.96	9.8
0.3	22.48	11.16	2.90	9.7
0.4	22.36	11.06	2.85	9.6
0.5	22.25	10.96	2.79	9.5
0.6	22.13	10.85	2.73	9.4
0.7	22.01	10.75	2.68	9.3
0.8	21.89	10.65	2.62	9.2
0.9	21.77	10.55	2.56	9.1
1.0	21.65	10.45	2.51	9.0
1.1	21.54	10.35	2.45	8.9
1.2	21.42	10.25	2.40	8.8
1.3	21.30	10.15	2.35	8.7
1.4	21.18	10.05	2.30	8.6
1.5	21.07	9.96	2.24	8.5
1.6	20.95	9.86	2.19	8.4
1.7	20.83	9.76	2.14	8.3
1.8	20.71	9.66	2.09	8.2
1.9	20.59	9.56	2.04	8.1
2.0	20.47	9.47	1.99	8.0
2.1	20.36	9.37	1.94	7.9
2.2	20.24	9.27	1.89	7.8
2.3	20.12	9.18	1.85	7.7
2.4	20.00	9.08	1.80	7.6
2.5	19.88	8.98	1.75	7.5
2.6	19.77	8.89	1.71	7.4
2.7	19.65	8.79	1.66	7.3
2.8	19.53	8.70	1.62	7.2
2.9	19.41	8.60	1.57	7.1
3.0	19.29	8.51	1.53	7.0
3.1	19.17	8.41	1.49	6.9
3.2	19.05	8.32	1.45	6.8
3.3	18.94	8.23	1.41	6.7
3.4	18.82	8.14	1.37	6.6
3.5	18.70	8.05	1.33	6.5
3.6	18.58	7.96	1.29	6.4
3.7	18.47	7.87	1.25	6.3
3.8	18.35	7.78	1.21	6.2
3.9	18.23	7.69	1.17	6.1
4.0	18.12	7.60	1.13	6.0
4.1	18.00	7.51	1.09	5.9
4.2	17.89	7.42	1.05	5.8
4.3	17.77	7.33	1.02	5.7
4.4	17.66	7.25	0.98	5.6
4.5	17.54	7.16	0.95	5.5
4.6	17.43	7.07	0.91	5.4
4.7	17.31	6.99	0.88	5.3
4.8	17.20	6.90	0.84	5.2
4.9	17.08	6.81	0.81	5.1
5.0	16.97	6.73	0.78	5.0
Days.	80	70	60	Days.

Table LXXXVI.

Days.	0	10	20
0.0	2.60	0.45	1.09
0.1	2.60	0.43	1.12
0.2	2.60	0.41	1.15
0.3	2.60	0.39	1.18
0.4	2.60	0.37	1.21
0.5	2.60	0.35	1.24
0.6	2.59	0.33	1.27
0.7	2.59	0.31	1.30
0.8	2.58	0.29	1.33
0.9	2.58	0.27	1.36
1.0	2.57	0.26	1.39
1.1	2.56	0.24	1.42
1.2	2.55	0.22	1.45
1.3	2.54	0.21	1.48
1.4	2.54	0.19	1.51
1.5	2.53	0.17	1.54
1.6	2.52	0.16	1.57
1.7	2.51	0.14	1.60
1.8	2.50	0.13	1.63
1.9	2.49	0.11	1.66
2.0	2.47	0.10	1.69
2.1	2.46	0.09	1.72
2.2	2.45	0.08	1.75
2.3	2.43	0.07	1.77
2.4	2.42	0.06	1.80
2.5	2.40	0.05	1.82
2.6	2.39	0.04	1.85
2.7	2.37	0.03	1.88
2.8	2.35	0.03	1.90
2.9	2.33	0.02	1.93
3.0	2.31	0.02	1.96
3.1	2.29	0.01	1.98
3.2	2.27	0.01	2.01
3.3	2.25	0.01	2.03
3.4	2.23	0.01	2.06
3.5	2.21	0.00	2.08
3.6	2.19	0.00	2.10
3.7	2.17	0.00	2.12
3.8	2.14	0.00	2.14
3.9	2.12	0.00	2.17
4.0	2.10	0.00	2.19
4.1	2.08	0.00	2.21
4.2	2.06	0.01	2.23
4.3	2.03	0.01	2.25
4.4	2.01	0.01	2.27
4.5	1.98	0.01	2.29
4.6	1.96	0.02	2.31
4.7	1.93	0.02	2.33
4.8	1.90	0.03	2.35
4.9	1.88	0.03	2.37
5.0	1.85	0.04	2.39

Table LXXXVII.

Days.	0	100	Days.
0	2.22	6.92	100
01	2.10	7.10	99
02	1.98	7.29	98
03	1.87	7.48	97
04	1.76	7.67	96
05	1.65	7.86	95
06	1.55	8.05	94
07	1.45	8.24	93
08	1.35	8.43	92
09	1.25	8.62	91
10	1.16	8.82	90
11	1.07	9.01	89
12	0.98	9.21	88
13	0.90	9.40	87
14	0.82	9.60	86
15	0.75	9.80	85
16	0.68	9.99	84
17	0.61	10.19	83
18	0.54	10.38	82
19	0.48	10.58	81
20	0.42	10.78	80
21	0.37	10.97	79
22	0.32	11.17	78
23	0.27	11.37	77
24	0.23	11.56	76
25	0.19	11.76	75
26	0.15	11.96	74
27	0.12	12.15	73
28	0.09	12.34	72
29	0.07	12.53	71
30	0.05	12.72	70
31	0.03	12.91	69
32	0.02	13.10	68
33	0.01	13.29	67
34	0.00	13.48	66
35	0.00	13.66	65
36	0.00	13.85	64
37	0.00	14.03	63
38	0.01	14.21	62
39	0.02	14.39	61
40	0.04	14.57	60
41	0.06	14.75	59
42	0.09	14.93	58
43	0.12	15.10	57
44	0.15	15.27	56
45	0.18	15.44	55
46	0.22	15.61	54
47	0.26	15.78	53
48	0.31	15.95	52
49	0.36	16.11	51
50	0.41	16.27	50
Days.	300	200	Days.

TABLES LXXXV.-LXXXVII.

Tables	LXXXV.	LXXXVI.	LXXXVII.
Arguments	53.	1.	70.

Table LXXXV. (Argument 53.)

Days.	30	40	50	Days.
5.0	16.97	6.73	0.78	5.0
5.1	16.85	6.64	0.75	4.9
5.2	16.74	6.56	0.72	4.8
5.3	16.62	6.48	0.69	4.7
5.4	16.51	6.40	0.66	4.6
5.5	16.39	6.31	0.64	4.5
5.6	16.28	6.23	0.61	4.4
5.7	16.17	6.15	0.58	4.3
5.8	16.05	6.07	0.55	4.2
5.9	15.94	5.99	0.52	4.1
6.0	15.83	5.91	0.50	4.0
6.1	15.71	5.83	0.47	3.9
6.2	15.60	5.75	0.45	3.8
6.3	15.49	5.67	0.43	3.7
6.4	15.37	5.59	0.41	3.6
6.5	15.26	5.51	0.38	3.5
6.6	15.15	5.43	0.36	3.4
6.7	15.03	5.35	0.34	3.3
6.8	14.92	5.27	0.32	3.2
6.9	14.81	5.19	0.30	3.1
7.0	14.70	5.12	0.28	3.0
7.1	14.59	5.04	0.26	2.9
7.2	14.48	4.97	0.24	2.8
7.3	14.37	4.90	0.23	2.7
7.4	14.26	4.82	0.21	2.6
7.5	14.15	4.75	0.19	2.5
7.6	14.04	4.68	0.18	2.4
7.7	13.93	4.60	0.16	2.3
7.8	13.82	4.53	0.15	2.2
7.9	13.71	4.46	0.14	2.1
8.0	13.60	4.39	0.13	2.0
8.1	13.49	4.32	0.12	1.9
8.2	13.38	4.25	0.11	1.8
8.3	13.27	4.18	0.10	1.7
8.4	13.16	4.11	0.09	1.6
8.5	13.05	4.05	0.08	1.5
8.6	12.95	3.98	0.07	1.4
8.7	12.84	3.91	0.06	1.3
8.8	12.73	3.84	0.05	1.2
8.9	12.62	3.77	0.04	1.1
9.0	12.52	3.71	0.03	1.0
9.1	12.41	3.64	0.03	0.9
9.2	12.30	3.58	0.02	0.8
9.3	12.20	3.52	0.02	0.7
9.4	12.09	3.45	0.02	0.6
9.5	11.99	3.39	0.01	0.5
9.6	11.88	3.33	0.01	0.4
9.7	11.78	3.26	0.01	0.3
9.8	11.67	3.20	0.00	0.2
9.9	11.57	3.14	0.00	0.1
10.0	11.47	3.08	0.00	0.0
Days.	80	70	60	Days.

Table LXXXVI. (Argument 1.)

Days.	0	10	20
5.0	1.85	0.04	2.39
5.1	1.82	0.05	2.40
5.2	1.80	0.06	2.42
5.3	1.77	0.07	2.43
5.4	1.75	0.08	2.45
5.5	1.72	0.09	2.46
5.6	1.69	0.10	2.47
5.7	1.66	0.11	2.49
5.8	1.63	0.13	2.50
5.9	1.60	0.14	2.51
6.0	1.57	0.16	2.52
6.1	1.54	0.17	2.53
6.2	1.51	0.19	2.54
6.3	1.48	0.21	2.54
6.4	1.45	0.22	2.55
6.5	1.42	0.24	2.56
6.6	1.39	0.26	2.57
6.7	1.36	0.27	2.58
6.8	1.33	0.29	2.58
6.9	1.30	0.31	2.59
7.0	1.27	0.33	2.59
7.1	1.24	0.35	2.60
7.2	1.21	0.37	2.60
7.3	1.18	0.39	2.60
7.4	1.15	0.41	2.60
7.5	1.12	0.43	2.60
7.6	1.09	0.45	2.60
7.7	1.06	0.47	2.60
7.8	1.04	0.50	2.60
7.9	1.01	0.52	2.60
8.0	0.98	0.54	2.60
8.1	0.95	0.57	2.60
8.2	0.92	0.59	2.59
8.3	0.90	0.62	2.59
8.4	0.87	0.64	2.58
8.5	0.84	0.67	2.58
8.6	0.81	0.70	2.57
8.7	0.78	0.73	2.56
8.8	0.76	0.76	2.55
8.9	0.73	0.78	2.54
9.0	0.70	0.81	2.54
9.1	0.67	0.84	2.53
9.2	0.64	0.87	2.52
9.3	0.62	0.90	2.51
9.4	0.59	0.92	2.50
9.5	0.57	0.95	2.48
9.6	0.54	0.98	2.47
9.7	0.52	1.01	2.46
9.8	0.50	1.04	2.44
9.9	0.47	1.06	2.43
10.0	0.45	1.09	2.42

Table LXXXVII. (Argument 70.)

Days.	0	100	Days.
50	0.41	16.27	50
51	0.47	16.43	49
52	0.53	16.59	48
53	0.59	16.74	47
54	0.66	16.89	46
55	0.73	17.04	45
56	0.80	17.19	44
57	0.88	17.33	43
58	0.96	17.47	42
59	1.05	17.61	41
60	1.14	17.75	40
61	1.23	17.88	39
62	1.33	18.01	38
63	1.43	18.14	37
64	1.53	18.27	36
65	1.63	18.39	35
66	1.74	18.51	34
67	1.85	18.63	33
68	1.96	18.75	32
69	2.07	18.86	31
70	2.19	18.96	30
71	2.31	19.07	29
72	2.44	19.17	28
73	2.57	19.27	27
74	2.70	19.36	26
75	2.83	19.45	25
76	2.97	19.54	24
77	3.11	19.62	23
78	3.25	19.70	22
79	3.39	19.78	21
80	3.54	19.86	20
81	3.69	19.93	19
82	3.84	20.00	18
83	3.99	20.06	17
84	4.15	20.12	16
85	4.31	20.18	15
86	4.47	20.23	14
87	4.63	20.28	13
88	4.80	20.33	12
89	4.97	20.37	11
90	5.14	20.41	10
91	5.31	20.45	09
92	5.48	20.48	08
93	5.65	20.51	07
94	5.83	20.53	06
95	6.01	20.55	05
96	6.19	20.57	04
97	6.37	20.58	03
98	6.55	20.59	02
99	6.73	20.60	01
100	6.92	20.60	00
Days.	300	200	Days.

TABLE LXXXVIII. ARGUMENT 4.

Equation = 88″.9 + 48″.9 sin. z.

Days.	0	10	20	30	40	50	60	70	80	90
Days.	″	″	″	″	″	″	″	″	″	″
0.0	40.00	40.67	42.76	46.22	50.93	56.76	63.55	71.08	79.14	87.48
0.1	40.00	40.68	42.78	46.26	50.98	56.82	63.62	71.15	79.22	87.56
0.2	40.00	40.70	42.81	46.30	51.03	56.88	63.69	71.23	79.30	87.64
0.3	40.00	40.71	42.84	46.34	51.09	56.95	63.76	71.31	79.38	87.73
0.4	40.00	40.73	42.87	46.38	51.14	57.01	63.83	71.39	79.46	87.81
0.5	40.00	40.74	42.90	46.42	51.19	57.07	63.91	71.47	79.55	87.89
0.6	40.00	40.76	42.93	46.46	51.25	57.14	63.98	71.55	79.63	87.98
0.7	40.00	40.77	42.96	46.50	51.30	57.20	64.05	71.63	79.71	88.06
0.8	40.00	40.79	42.99	46.54	51.36	57.27	64.12	71.71	79.79	88.15
0.9	40.00	40.80	43.02	46.58	51.41	57.33	64.19	71.79	79.87	88.23
1.0	40.00	40.82	43.05	46.63	51.47	57.40	64.27	71.87	79.96	88.32
1.1	40.00	40.83	43.08	46.67	51.52	57.46	64.34	71.94	80.04	88.40
1.2	40.00	40.85	43.11	46.71	51.57	57.53	64.41	72.02	80.12	88.48
1.3	40.00	40.86	43.14	46.75	51.63	57.59	64.48	72.10	80.20	88.57
1.4	40.00	40.88	43.17	46.80	51.68	57.66	64.56	72.18	80.29	88.65
1.5	40.01	40.90	43.20	46.84	51.73	57.72	64.63	72.26	80.37	88.74
1.6	40.01	40.91	43.23	46.88	51.79	57.79	64.70	72.34	80.45	88.82
1.7	40.01	40.93	43.26	46.93	51.84	57.85	64.78	72.42	80.54	88.90
1.8	40.01	40.94	43.29	46.97	51.90	57.92	64.85	72.50	80.62	88.99
1.9	40.01	40.96	43.32	47.01	51.95	57.98	64.92	72.58	80.70	89.07
2.0	40.02	40.98	43.35	47.06	52.01	58.05	65.00	72.66	80.79	89.16
2.1	40.02	40.99	43.38	47.10	52.06	58.11	65.07	72.73	80.87	89.24
2.2	40.02	41.01	43.41	47.14	52.12	58.18	65.14	72.81	80.95	89.33
2.3	40.02	41.03	43.44	47.19	52.17	58.24	65.22	72.89	81.03	89.41
2.4	40.03	41.04	43.47	47.23	52.23	58.31	65.29	72.97	81.12	89.50
2.5	40.03	41.06	43.50	47.27	52.28	58.37	65.36	73.05	81.20	89.58
2.6	40.03	41.08	43.53	47.32	52.34	58.44	65.44	73.13	81.28	89.67
2.7	40.04	41.09	43.56	47.36	52.40	58.51	65.51	73.21	81.37	89.75
2.8	40.04	41.11	43.59	47.41	52.45	58.57	65.59	73.29	81.45	89.84
2.9	40.04	41.13	43.62	47.45	52.51	58.64	65.66	73.37	81.53	89.92
3.0	40.05	41.15	43.66	47.50	52.57	58.71	65.74	73.45	81.62	90.01
3.1	40.05	41.16	43.69	47.54	52.62	58.77	65.81	73.53	81.70	90.10
3.2	40.06	41.18	43.72	47.59	52.68	58.84	65.88	73.61	81.78	90.18
3.3	40.06	41.20	43.75	47.63	52.74	58.90	65.96	73.69	81.86	90.27
3.4	40.07	41.22	43.78	47.68	52.79	58.97	66.03	73.77	81.95	90.35
3.5	40.07	41.24	43.82	47.72	52.85	59.04	66.10	73.85	82.03	90.43
3.6	40.08	41.26	43.85	47.77	52.91	59.10	66.18	73.93	82.11	90.52
3.7	40.08	41.28	43.88	47.82	52.96	59.17	66.25	74.01	82.20	90.60
3.8	40.09	41.30	43.91	47.86	53.02	59.24	66.33	74.09	82.28	90.69
3.9	40.09	41.32	43.94	47.91	53.08	59.30	66.40	74.17	82.36	90.77
4.0	40.10	41.34	43.98	47.96	53.14	59.37	66.48	74.25	82.45	90.85
4.1	40.10	41.36	44.01	48.00	53.19	59.44	66.55	74.33	82.53	90.94
4.2	40.11	41.38	44.04	48.05	53.25	59.50	66.63	74.41	82.61	91.02
4.3	40.11	41.40	44.08	48.09	53.31	59.57	66.70	74.49	82.70	91.11
4.4	40.12	41.42	44.11	48.14	53.37	59.64	66.78	74.57	82.78	91.19
4.5	40.12	41.44	44.14	48.18	53.43	59.71	66.85	74.65	82.86	91.28
4.6	40.13	41.46	44.18	48.23	53.48	59.77	66.93	74.73	82.95	91.36
4.7	40.14	41.48	44.21	48.28	53.54	59.84	67.00	74.81	83.03	91.44
4.8	40.14	41.50	44.25	48.32	53.60	59.91	67.08	74.89	83.12	91.53
4.9	40.15	41.52	44.28	48.37	53.66	59.98	67.15	74.97	83.20	91.61
5.0	40.16	41.54	44.32	48.42	53.72	60.05	67.23	75.06	83.29	91.69

TABLE LXXXVIII. ARGUMENT 4.

Equation = 88″.9 + 48″.9 sin. z.

Days.	0	10	20	30	40	50	60	70	80	90
Days 5.0	40.16	41.51	44.32	48.42	53.72	60.05	67.23	75.06	83.29	91.69
5.1	40.16	41.56	44.35	48.46	53.77	60.11	67.30	75.14	83.37	91.78
5.2	40.17	41.58	44.39	48.51	53.83	60.18	67.38	75.22	83.45	91.86
5.3	40.18	41.60	44.42	48.56	53.89	60.25	67.45	75.30	83.54	91.95
5.4	40.18	41.62	44.46	48.61	53.95	60.32	67.53	75.38	83.62	92.03
5.5	40.19	41.65	44.49	48.66	54.01	60.39	67.61	75.46	83.70	92.11
5.6	40.20	41.67	44.53	48.70	54.07	60.45	67.68	75.54	83.79	92.20
5.7	40.20	41.69	44.57	48.75	54.13	60.52	67.76	75.62	83.87	92.28
5.8	40.21	41.71	44.60	48.80	54.19	60.59	67.83	75.70	83.96	92.37
5.9	40.22	41.73	44.64	48.85	54.25	60.66	67.91	75.78	84.04	92.45
6.0	40.23	41.76	44.68	48.90	54.31	60.73	67.99	75.87	84.13	92.53
6.1	40.23	41.78	44.71	48.94	54.37	60.79	68.06	75.95	84.21	92.62
6.2	40.24	41.80	44.75	48.99	54.43	60.86	68.14	76.03	84.29	92.70
6.3	40.25	41.82	44.78	49.04	54.49	60.93	68.21	76.11	84.37	92.78
6.4	40.26	41.85	44.82	49.09	54.55	61.00	68.29	76.19	84.46	92.87
6.5	40.27	41.87	44.85	49.14	54.61	61.07	68.37	76.27	84.54	92.95
6.6	40.28	41.89	44.89	49.19	54.67	61.14	68.44	76.35	84.62	93.03
6.7	40.29	41.92	44.93	49.24	54.73	61.21	68.52	76.43	84.71	93.12
6.8	40.30	41.94	44.96	49.29	54.79	61.28	68.59	76.51	84.79	93.20
6.9	40.31	41.96	45.00	49.34	54.85	61.35	68.67	76.59	84.87	93.28
7.0	40.32	41.99	45.04	49.39	54.91	61.42	68.75	76.68	84.96	93.36
7.1	40.33	42.01	45.07	49.44	54.97	61.49	68.82	76.76	85.04	93.45
7.2	40.34	42.03	45.11	49.49	55.03	61.56	68.90	76.84	85.12	93.53
7.3	40.35	42.06	45.15	49.54	55.09	61.63	68.98	76.92	85.21	93.62
7.4	40.36	42.08	45.19	49.59	55.15	61.70	69.05	77.00	85.29	93.70
7.5	40.37	42.10	45.23	49.64	55.21	61.77	69.13	77.08	85.37	93.79
7.6	40.38	42.13	45.26	49.69	55.27	61.84	69.21	77.16	85.46	93.87
7.7	40.39	42.15	45.30	49.74	55.33	61.91	69.28	77.24	85.54	93.95
7.8	40.40	42.18	45.34	49.79	55.39	61.98	69.36	77.32	85.63	94.04
7.9	40.41	42.20	45.38	49.84	55.45	62.05	69.44	77.40	85.71	94.12
8.0	40.42	42.23	45.42	49.89	55.52	62.12	69.52	77.49	85.80	94.20
8.1	40.43	42.25	45.45	49.94	55.58	62.19	69.59	77.57	85.88	94.29
8.2	40.44	42.28	45.49	49.99	55.64	62.26	69.67	77.65	85.96	94.37
8.3	40.45	42.30	45.53	50.04	55.70	62.33	69.75	77.73	86.05	94.46
8.4	40.46	42.33	45.57	50.09	55.76	62.40	69.83	77.81	86.13	94.54
8.5	40.48	42.35	45.61	50.14	55.83	62.47	69.91	77.90	86.21	94.63
8.6	40.49	42.38	45.65	50.19	55.89	62.54	69.98	77.98	86.30	94.71
8.7	40.50	42.41	45.69	50.24	55.95	62.61	70.06	78.06	86.38	94.79
8.8	40.51	42.43	45.73	50.29	56.01	62.68	70.14	78.14	86.47	94.88
8.9	40.52	42.46	45.77	50.34	56.07	62.75	70.22	78.22	86.55	94.96
9.0	40.54	42.49	45.81	50.40	56.14	62.83	70.30	78.31	86.64	95.04
9.1	40.55	42.51	45.85	50.45	56.20	62.90	70.37	78.39	86.72	95.13
9.2	40.56	42.54	45.89	50.50	56.26	62.97	70.45	78.47	86.80	95.21
9.3	40.58	42.57	45.93	50.55	56.32	63.04	70.53	78.55	86.89	95.29
9.4	40.59	42.59	45.97	50.61	56.38	63.11	70.61	78.64	86.97	95.38
9.5	40.60	42.62	46.01	50.66	56.45	63.19	70.69	78.72	87.05	95.46
9.6	40.62	42.65	46.05	50.71	56.51	63.26	70.76	78.80	87.14	95.54
9.7	40.63	42.67	46.09	50.77	56.57	63.33	70.84	78.89	87.22	95.63
9.8	40.64	42.70	46.13	50.82	56.63	63.40	70.92	78.97	87.31	95.71
9.9	40.65	42.73	46.17	50.87	56.69	63.47	71.00	79.05	87.39	95.79
10.0	40.67	42.76	46.22	50.93	56.76	63.55	71.08	79.14	87.48	95.87

TABLE LXXXVIII. ARGUMENT 4.

Equation = 88″.9 + 48″.9 sin. z.

Days.	100	110	120	130	140	150	160	170	180
Days 0.0	95.87	104.05	111.79	118.85	125.02	130.13	134.02	136.58	137.73
0.1	95.96	104.13	111.87	118.92	125.08	130.18	134.06	136.60	137.74
0.2	96.04	104.21	111.94	118.99	125.14	130.22	134.09	136.62	137.74
0.3	96.12	104.29	112.02	119.05	125.19	130.27	134.12	136.64	137.75
0.4	96.21	104.37	112.09	119.12	125.25	130.31	134.15	136.66	137.75
0.5	96.29	104.45	112.17	119.19	125.31	130.36	134.18	136.67	137.75
0.6	96.37	104.53	112.24	119.25	125.36	130.40	134.22	136.69	137.76
0.7	96.46	104.61	112.31	119.32	125.42	130.45	134.25	136.71	137.76
0.8	96.54	104.69	112.39	119.38	125.47	130.49	134.28	136.73	137.76
0.9	96.62	104.77	112.46	119.45	125.53	130.54	134.31	136.75	137.76
1.0	96.70	104.85	112.53	119.51	125.58	130.58	134.34	136.76	137.77
1.1	96.79	104.93	112.61	119.58	125.64	130.63	134.37	136.78	137.77
1.2	96.87	105.01	112.68	119.64	125.69	130.67	134.40	136.80	137.77
1.3	96.95	105.09	112.75	119.71	125.75	130.71	134.43	136.82	137.77
1.4	97.04	105.17	112.83	119.77	125.80	130.76	134.46	136.83	137.77
1.5	97.12	105.25	112.90	119.84	125.86	130.80	134.49	136.85	137.78
1.6	97.20	105.33	112.97	119.90	125.91	130.84	134.52	136.87	137.78
1.7	97.29	105.41	113.05	119.97	125.97	130.89	134.55	136.88	137.78
1.8	97.37	105.49	113.12	120.03	126.02	130.93	134.58	136.90	137.78
1.9	97.45	105.57	113.19	120.10	126.08	130.97	134.61	136.92	137.78
2.0	97.53	105.64	113.26	120.16	126.13	131.01	134.64	136.93	137.79
2.1	97.62	105.72	113.34	120.23	126.19	131.06	134.67	136.95	137.79
2.2	97.70	105.80	113.41	120.29	126.24	131.10	134.70	136.96	137.79
2.3	97.78	105.88	113.48	120.36	126.30	131.14	134.73	136.98	137.79
2.4	97.87	105.96	113.56	120.42	126.35	131.18	134.76	136.99	137.79
2.5	97.95	106.04	113.63	120.49	126.41	131.22	134.79	137.01	137.80
2.6	98.03	106.12	113.70	120.55	126.46	131.27	134.82	137.02	137.80
2.7	98.12	106.20	113.78	120.62	126.51	131.31	134.85	137.04	137.80
2.8	98.20	106.28	113.85	120.68	126.57	131.35	134.88	137.05	137.80
2.9	98.28	106.36	113.92	120.74	126.62	131.39	134.91	137.07	137.80
3.0	98.36	106.43	113.99	120.80	126.67	131.43	134.93	137.08	137.80
3.1	98.45	106.51	114.07	120.87	126.73	131.48	134.96	137.10	137.80
3.2	98.53	106.59	114.14	120.93	126.78	131.52	134.99	137.11	137.80
3.3	98.61	106.67	114.21	120.99	126.83	131.56	135.02	137.12	137.80
3.4	98.69	106.75	114.28	121.06	126.89	131.60	135.05	137.14	137.80
3.5	98.77	106.82	114.35	121.12	126.94	131.64	135.07	137.15	137.80
3.6	98.86	106.90	114.43	121.18	126.99	131.68	135.10	137.16	137.79
3.7	98.94	106.98	114.50	121.25	127.05	131.72	135.13	137.18	137.79
3.8	99.02	107.06	114.57	121.31	127.10	131.76	135.16	137.19	137.79
3.9	99.10	107.14	114.64	121.37	127.15	131.80	135.19	137.20	137.79
4.0	99.18	107.21	114.71	121.43	127.20	131.84	135.21	137.21	137.79
4.1	99.27	107.29	114.79	121.50	127.26	131.88	135.24	137.23	137.78
4.2	99.35	107.37	114.86	121.56	127.31	131.92	135.27	137.24	137.78
4.3	99.43	107.45	114.93	121.62	127.36	131.96	135.29	137.25	137.78
4.4	99.51	107.53	115.00	121.69	127.41	132.00	135.32	137.27	137.78
4.5	99.59	107.60	115.07	121.75	127.46	132.04	135.35	137.28	137.78
4.6	99.68	107.68	115.14	121.81	127.52	132.08	135.37	137.29	137.77
4.7	99.76	107.76	115.21	121.88	127.57	132.12	135.40	137.31	137.77
4.8	99.84	107.84	115.28	121.94	127.62	132.16	135.42	137.32	137.77
4.9	99.92	107.92	115.35	122.00	127.67	132.20	135.45	137.33	137.77
5.0	100.00	107.99	115.42	122.06	127.72	132.24	135.47	137.34	137.77

TABLE LXXXVIII. ARGUMENT 4.

Equation = 88″.9 + 48″.9 sin. z.

Days.	100	110	120	130	140	150	160	170	180
5.0	100.00	107.99	115.42	122.06	127.72	132.21	135.47	137.34	137.77
5.1	100.09	108.07	115.49	122.13	127.78	132.28	135.50	137.36	137.76
5.2	100.17	108.15	115.56	122.19	127.83	132.32	135.52	137.37	137.76
5.3	100.25	108.23	115.63	122.25	127.88	132.36	135.55	137.38	137.76
5.4	100.33	108.30	115.70	122.31	127.93	132.40	135.57	137.39	137.76
5.5	100.41	108.38	115.77	122.37	127.98	132.43	135.60	137.40	137.75
5.6	100.50	108.46	115.84	122.43	128.03	132.47	135.62	137.41	137.75
5.7	100.58	108.53	115.91	122.49	128.08	132.51	135.65	137.42	137.75
5.8	100.66	108.61	115.98	122.55	128.13	132.55	135.67	137.43	137.74
5.9	100.74	108.69	116.05	122.61	128.18	132.59	135.70	137.44	137.74
6.0	100.82	108.76	116.12	122.67	128.23	132.62	135.72	137.45	137.73
6.1	100.91	108.84	116.19	122.73	128.28	132.66	135.75	137.46	137.72
6.2	100.99	108.92	116.26	122.79	128.33	132.70	135.77	137.47	137.72
6.3	101.07	109.00	116.33	122.85	128.38	132.74	135.80	137.48	137.71
6.4	101.15	109.07	116.40	122.91	128.43	132.77	135.82	137.49	137.71
6.5	101.23	109.15	116.47	122.97	128.48	132.81	135.85	137.50	137.70
6.6	101.31	109.23	116.54	123.03	128.53	132.85	135.87	137.51	137.70
6.7	101.39	109.30	116.61	123.09	128.58	132.88	135.89	137.52	137.69
6.8	101.47	109.38	116.68	123.15	128.63	132.92	135.92	137.53	137.69
6.9	101.55	109.46	116.75	123.21	128.68	132.96	135.94	137.54	137.68
7.0	101.63	109.53	116.81	123.27	128.72	132.99	135.96	137.54	137.68
7.1	101.72	109.61	116.88	123.33	128.77	133.03	135.99	137.55	137.67
7.2	101.80	109.69	116.95	123.39	128.82	133.07	136.01	137.56	137.67
7.3	101.88	109.76	117.02	123.45	128.87	133.10	136.03	137.57	137.66
7.4	101.96	109.84	117.09	123.51	128.92	133.14	136.05	137.58	137.65
7.5	102.04	109.92	117.16	123.57	128.96	133.18	136.07	137.58	137.65
7.6	102.12	109.99	117.23	123.63	129.01	133.21	136.10	137.59	137.64
7.7	102.20	110.07	117.30	123.69	129.06	133.25	136.12	137.60	137.64
7.8	102.28	110.14	117.37	123.75	129.11	133.28	136.14	137.61	137.63
7.9	102.36	110.22	117.44	123.81	129.16	133.32	136.16	137.62	137.63
8.0	102.44	110.29	117.50	123.87	129.21	133.35	136.18	137.62	137.62
8.1	102.53	110.37	117.57	123.93	129.25	133.39	[illegible]	137.63	137.62
8.2	102.61	110.44	117.64	123.99	129.30	133.42	136.23	137.63	137.61
8.3	102.69	110.52	117.71	124.05	129.35	133.46	136.25	137.64	137.60
8.4	102.77	110.59	117.78	124.11	129.39	133.49	136.27	137.64	137.59
8.5	102.85	110.67	117.84	124.16	129.44	133.53	136.29	137.65	137.58
8.6	102.93	110.74	117.91	124.22	129.49	133.56	136.31	137.65	137.58
8.7	103.01	110.82	117.98	124.28	129.53	133.59	136.33	137.66	137.57
8.8	103.09	110.89	118.05	124.34	129.58	133.63	136.35	137.67	137.56
8.9	103.17	110.97	118.12	124.40	129.63	133.66	136.37	137.67	137.55
9.0	103.25	111.04	118.18	124.45	129.67	133.69	136.39	137.68	137.54
9.1	103.33	111.12	118.25	124.51	129.72	133.73	136.41	137.68	137.54
9.2	103.41	111.19	118.32	124.57	129.77	133.76	136.43	137.69	137.53
9.3	103.49	111.27	118.39	124.63	129.81	133.79	136.45	137.69	137.52
9.4	103.57	111.34	118.45	124.68	129.86	133.83	136.47	137.70	137.51
9.5	103.65	111.42	118.52	124.74	129.91	133.86	136.49	137.70	137.50
9.6	103.73	111.49	118.59	124.80	129.95	133.89	136.51	137.71	137.49
9.7	103.81	111.57	118.65	124.85	130.00	133.93	136.53	137.71	137.48
9.8	103.89	111.64	118.72	124.91	130.04	133.96	136.55	137.72	137.47
9.9	103.97	111.72	118.79	124.97	130.09	133.99	136.57	137.72	137.46
10.0	104.05	111.79	118.85	125.02	130.13	134.02	136.58	137.73	137.45

TABLE LXXXVIII. ARGUMENT 4.

Equation = 88″.9 + 46″.9 sin. z.

Days.	190	200	210	220	230	240	250	260	270
Days.	″	″	″	″	″	″	″	″	″
0.0	137.45	135.72	132.62	128.23	122.67	116.12	108.76	100.82	92.53
0.1	137.44	135.70	132.59	128.18	122.61	116.05	108.69	100.74	92.45
0.2	137.43	135.67	132.55	128.13	122.55	115.98	108.61	100.66	92.37
0.3	137.42	135.65	132.51	128.08	122.49	115.91	108.53	100.58	92.28
0.4	137.41	135.62	132.47	128.03	122.43	115.84	108.46	100.50	92.20
0.5	137.40	135.60	132.43	127.98	122.37	115.77	108.38	100.41	92.11
0.6	137.39	135.57	132.40	127.93	122.31	115.70	108.30	100.33	92.03
0.7	137.38	135.55	132.36	127.88	122.25	115.63	108.23	100.25	91.95
0.8	137.37	135.52	132.32	127.83	122.19	115.56	108.15	100.17	91.86
0.9	137.36	135.50	132.28	127.78	122.13	115.49	108.07	100.09	91.78
1.0	137.34	135.47	132.24	127.72	122.06	115.42	107.99	100.00	91.69
1.1	137.33	135.45	132.20	127.67	122.00	115.35	107.92	99.92	91.61
1.2	137.32	135.42	132.16	127.62	121.94	115.28	107.84	99.84	91.53
1.3	137.31	135.40	132.12	127.57	121.88	115.21	107.76	99.76	91.44
1.4	137.29	135.37	132.08	127.52	121.81	115.14	107.68	99.68	91.36
1.5	137.28	135.35	132.04	127.46	121.75	115.07	107.60	99.59	91.28
1.6	137.27	135.32	132.00	127.41	121.69	115.00	107.53	99.51	91.19
1.7	137.25	135.29	131.96	127.36	121.62	114.93	107.45	99.43	91.11
1.8	137.24	135.27	131.92	127.31	121.56	114.86	107.37	99.35	91.02
1.9	137.23	135.24	131.88	127.26	121.50	114.79	107.29	99.27	90.94
2.0	137.21	135.21	131.84	127.20	121.43	114.71	107.21	99.18	90.85
2.1	137.20	135.19	131.80	127.15	121.37	114.64	107.14	99.10	90.77
2.2	137.19	135.16	131.76	127.10	121.31	114.57	107.06	99.02	90.69
2.3	137.18	135.13	131.72	127.05	121.25	114.50	106.98	98.94	90.60
2.4	137.16	135.10	131.68	126.99	121.18	114.43	106.90	98.86	90.52
2.5	137.15	135.07	131.64	126.94	121.12	114.35	106.82	98.77	90.43
2.6	137.14	135.05	131.60	126.89	121.06	114.28	106.75	98.69	90.35
2.7	137.12	135.02	131.56	126.83	120.99	114.21	106.67	98.61	90.27
2.8	137.11	134.99	131.52	126.78	120.93	114.14	106.59	98.53	90.18
2.9	137.10	134.96	131.48	126.73	120.87	114.07	106.51	98.45	90.10
3.0	137.08	134.93	131.43	126.67	120.80	113.99	106.43	98.36	90.01
3.1	137.07	134.91	131.39	126.62	120.74	113.92	106.36	98.28	89.92
3.2	137.05	134.88	131.35	126.57	120.68	113.85	106.28	98.20	89.84
3.3	137.04	134.85	131.31	126.51	120.62	113.78	106.20	98.12	89.75
3.4	137.02	134.82	131.27	126.46	120.55	113.70	106.12	98.03	89.67
3.5	137.01	134.79	131.22	126.41	120.49	113.63	106.04	97.95	89.58
3.6	136.99	134.76	131.18	126.35	120.42	113.56	105.96	97.87	89.50
3.7	136.98	134.73	131.14	126.30	120.36	113.48	105.88	97.78	89.41
3.8	136.96	134.70	131.10	126.24	120.29	113.41	105.80	97.70	89.33
3.9	136.95	134.67	131.06	126.19	120.23	113.34	105.72	97.62	89.24
4.0	136.93	134.64	131.01	126.13	120.16	113.26	105.64	97.53	89.16
4.1	136.92	134.61	130.97	126.08	120.10	113.19	105.57	97.45	89.07
4.2	136.90	134.58	130.93	126.02	120.03	113.12	105.49	97.37	88.99
4.3	136.88	134.55	130.89	125.97	119.97	113.05	105.41	97.29	88.90
4.4	136.87	134.52	130.84	125.91	119.90	112.97	105.33	97.20	88.82
4.5	136.85	134.49	130.80	125.86	119.84	112.90	105.25	97.12	88.74
4.6	136.83	134.46	130.76	125.80	119.77	112.83	105.17	97.04	88.65
4.7	136.82	134.43	130.71	125.75	119.71	112.75	105.09	96.95	88.57
4.8	136.80	134.40	130.67	125.69	119.64	112.68	105.01	96.87	88.48
4.9	136.78	134.37	130.63	125.64	119.58	112.61	104.93	96.79	88.40
5.0	136.76	134.34	130.58	125.58	119.51	112.53	104.85	96.70	88.32

TABLE LXXXVIII. ARGUMENT 4.

Equation $= 88''.9 + 48''.9 \sin. z.$

Days.	190	200	240	220	230	240	250	260	270
5.0	136″.76	134″.34	130″.58	125″.58	119″.51	112″.53	104″.85	96″.70	88″.32
5.1	136.75	134.31	130.54	125.53	119.45	112.46	104.77	96.62	88.23
5.2	136.73	134.28	130.49	125.47	119.38	112.39	104.69	96.54	88.15
5.3	136.71	134.25	130.45	125.42	119.32	112.31	104.61	96.46	88.06
5.4	136.69	134.22	130.40	125.36	119.25	112.24	104.53	96.37	87.98
5.5	136.67	134.18	130.36	125.31	119.19	112.17	104.45	96.29	87.89
5.6	136.66	134.15	130.31	125.25	119.12	112.09	104.37	96.21	87.81
5.7	136.64	134.12	130.27	125.19	119.05	112.02	104.29	96.12	87.73
5.8	136.62	134.09	130.22	125.14	118.99	111.94	104.21	96.04	87.64
5.9	136.60	134.06	130.18	125.08	118.92	111.87	104.13	95.96	87.56
6.0	136.58	134.02	130.13	125.02	118.85	111.79	104.05	95.87	87.48
6.1	136.57	133.99	130.09	124.97	118.79	111.72	103.97	95.79	87.39
6.2	136.55	133.96	130.04	124.91	118.72	111.64	103.89	95.71	87.31
6.3	136.53	133.93	130.00	124.85	118.65	111.57	103.81	95.63	87.22
6.4	136.51	133.89	129.95	124.80	118.59	111.49	103.73	95.54	87.14
6.5	136.49	133.86	129.91	124.74	118.52	111.42	103.65	95.46	87.05
6.6	136.47	133.83	129.86	124.68	118.45	111.34	103.57	95.38	86.97
6.7	136.45	133.79	129.81	124.63	118.39	111.27	103.49	95.29	86.89
6.8	136.43	133.76	129.77	124.57	118.32	111.19	103.41	95.21	86.80
6.9	136.41	133.73	129.72	124.51	118.25	111.12	103.33	95.13	86.72
7.0	136.39	133.69	129.67	124.45	118.18	111.04	103.25	95.04	86.64
7.1	136.37	133.66	129.63	124.40	118.12	110.97	103.17	94.96	86.55
7.2	136.35	133.63	129.58	124.34	118.05	110.89	103.09	94.88	86.47
7.3	136.33	133.59	129.53	124.28	117.98	110.82	103.01	94.79	86.38
7.4	136.31	133.56	129.49	124.22	117.91	110.74	102.93	94.71	86.30
7.5	136.29	133.53	129.44	124.16	117.84	110.67	102.85	94.63	86.21
7.6	136.27	133.49	129.39	124.11	117.78	110.59	102.77	94.54	86.13
7.7	136.25	133.46	129.35	124.05	117.71	110.52	102.69	94.46	86.05
7.8	136.23	133.42	129.30	123.99	117.64	110.44	102.61	94.37	85.96
7.9	136.21	133.39	129.25	123.93	117.57	110.37	102.53	94.29	85.88
8.0	136.18	133.35	129.20	123.87	117.50	110.29	102.44	94.20	85.80
8.1	136.16	133.32	129.16	123.81	117.44	110.22	102.36	94.12	85.71
8.2	136.14	133.28	129.11	123.75	117.37	110.14	102.28	94.04	85.63
8.3	136.12	133.25	129.06	123.69	117.30	110.07	102.20	93.95	85.54
8.4	136.10	133.21	129.01	123.63	117.23	109.99	102.12	93.87	85.46
8.5	136.07	133.18	128.96	123.57	117.16	109.92	102.04	93.79	85.37
8.6	136.05	133.14	128.92	123.51	117.09	109.84	101.96	93.70	85.29
8.7	136.03	133.10	128.87	123.45	117.02	109.76	101.88	93.62	85.21
8.8	136.01	133.07	128.82	123.39	116.95	109.69	101.80	93.53	85.12
8.9	135.99	133.03	128.77	123.33	116.88	109.61	101.72	93.45	85.04
9.0	135.96	132.99	128.72	123.27	116.81	109.53	101.63	93.36	84.96
9.1	135.94	132.96	128.68	123.21	116.75	109.46	101.55	93.28	84.87
9.2	135.92	132.92	128.63	123.15	116.68	109.38	101.47	93.20	84.79
9.3	135.89	132.88	128.58	123.09	116.61	109.30	101.39	93.12	84.71
9.4	135.87	132.85	128.53	123.03	116.54	109.23	101.31	93.03	84.62
9.5	135.85	132.81	128.48	122.97	116.47	109.15	101.23	92.95	84.54
9.6	135.82	132.77	128.43	122.91	116.40	109.07	101.15	92.87	84.46
9.7	135.80	132.74	128.38	122.85	116.33	109.00	101.07	92.78	84.37
9.8	135.77	132.70	128.33	122.79	116.26	108.92	100.99	92.70	84.29
9.9	135.75	132.66	128.28	122.73	116.19	108.84	100.91	92.62	84.21
10.0	135.72	132.62	128.23	122.67	116.12	108.76	100.82	92.53	84.13

TABLE LXXXVIII. ARGUMENT 4.

Equation = 88″.9 + 48″.9 sin z.

Days.	280	290	300	310	320	330	340	350	360
0.0	84″.13	75″.87	67″.99	60″.73	54″.31	48″.90	44″.68	41″.76	40″.23
0.1	84.04	75.78	67.91	60.66	54.25	48.85	44.64	41.73	40.22
0.2	83.96	75.70	67.83	60.59	54.19	48.80	44.60	41.71	40.21
0.3	83.87	75.62	67.76	60.52	54.13	48.75	44.57	41.69	40.20
0.4	83.79	75.54	67.68	60.45	54.07	48.70	44.53	41.67	40.20
0.5	83.70	75.46	67.61	60.39	54.01	48.66	44.49	41.65	40.19
0.6	83.62	75.38	67.53	60.32	53.95	48.61	44.46	41.62	40.18
0.7	83.54	75.30	67.45	60.25	53.89	48.56	44.42	41.60	40.18
0.8	83.45	75.22	67.38	60.18	53.83	48.51	44.39	41.58	40.17
0.9	83.37	75.14	67.30	60.11	53.77	48.46	44.35	41.56	40.16
1.0	83.29	75.06	67.23	60.05	53.72	48.42	44.32	41.54	40.16
1.1	83.20	74.97	67.15	59.98	53.66	48.37	44.28	41.52	40.15
1.2	83.12	74.89	67.08	59.91	53.60	48.32	44.25	41.50	40.14
1.3	83.03	74.81	67.00	59.84	53.54	48.28	44.21	41.48	40.14
1.4	82.95	74.73	66.93	59.77	53.48	48.23	44.18	41.46	40.13
1.5	82.86	74.65	66.85	59.71	53.43	48.18	44.14	41.44	40.12
1.6	82.78	74.57	66.78	59.64	53.37	48.14	44.11	41.42	40.12
1.7	82.70	74.49	66.70	59.57	53.31	48.09	44.08	41.40	40.11
1.8	82.61	74.41	66.63	59.50	53.25	48.05	44.04	41.38	40.11
1.9	82.53	74.33	66.55	59.44	53.19	48.00	44.01	41.36	40.10
2.0	82.45	74.25	66.48	59.37	53.14	47.96	43.98	41.34	40.10
2.1	82.36	74.17	66.40	59.30	53.08	47.91	43.94	41.32	40.09
2.2	82.28	74.09	66.33	59.24	53.02	47.86	43.91	41.30	40.09
2.3	82.20	74.01	66.25	59.17	52.96	47.82	43.88	41.28	40.08
2.4	82.11	73.93	66.18	59.10	52.91	47.77	43.85	41.26	40.08
2.5	82.03	73.85	66.10	59.04	52.85	47.72	43.82	41.24	40.07
2.6	81.95	73.77	66.03	58.97	52.79	47.68	43.78	41.22	40.07
2.7	81.86	73.69	65.96	58.90	52.74	47.63	43.75	41.20	40.06
2.8	81.78	73.61	65.88	58.84	52.68	47.59	43.72	41.18	40.06
2.9	81.70	73.53	65.81	58.77	52.62	47.54	43.69	41.16	40.05
3.0	81.62	73.45	65.74	58.71	52.57	47.50	43.66	41.15	40.05
3.1	81.53	73.37	65.66	58.64	52.51	47.45	43.62	41.13	40.04
3.2	81.45	73.29	65.59	58.57	52.45	47.41	43.59	41.11	40.04
3.3	81.37	73.21	65.51	58.51	52.40	47.36	43.56	41.09	40.04
3.4	81.28	73.13	65.44	58.44	52.34	47.32	43.53	41.08	40.03
3.5	81.20	73.05	65.36	58.37	52.28	47.27	43.50	41.06	40.03
3.6	81.12	72.97	65.29	58.31	52.23	47.23	43.47	41.04	40.03
3.7	81.03	72.89	65.22	58.24	52.17	47.19	43.44	41.03	40.02
3.8	80.95	72.81	65.14	58.18	52.12	47.14	43.41	41.01	40.02
3.9	80.87	72.73	65.07	58.11	52.06	47.10	43.38	40.99	40.02
4.0	80.79	72.66	65.00	58.05	52.01	47.06	43.35	40.98	40.02
4.1	80.70	72.58	64.92	57.98	51.95	47.01	43.32	40.96	40.01
4.2	80.62	72.50	64.85	57.92	51.90	46.97	43.29	40.94	40.01
4.3	80.54	72.42	64.78	57.85	51.84	46.93	43.26	40.93	40.01
4.4	80.45	72.34	64.70	57.79	51.79	46.88	43.23	40.91	40.01
4.5	80.37	72.26	64.63	57.72	51.73	46.84	43.20	40.90	40.01
4.6	80.29	72.18	64.56	57.66	51.68	46.80	43.17	40.88	40.00
4.7	80.20	72.10	64.48	57.59	51.63	46.75	43.14	40.86	40.00
4.8	80.12	72.02	64.41	57.53	51.57	46.71	43.11	40.85	40.00
4.9	80.04	71.94	64.34	57.46	51.52	46.67	43.08	40.83	40.00
5.0	79.96	71.87	64.27	57.40	51.47	46.63	43.05	40.82	40.00

TABLE LXXXVIII. ARGUMENT 4.

Equation = 88″.9 + 48″.9 sin z.

Days.	280	290	300	310	320	330	340	350	360
Days.									
5.0	79.96	71.87	64.27	57.40	51.47	46.63	43.05	40.82	40.00
5.1	79.87	71.79	64.19	57.33	51.41	46.58	43.02	40.80	40.00
5.2	79.79	71.71	64.12	57.27	51.36	46.54	42.99	40.79	40.00
5.3	79.71	71.63	64.05	57.20	51.30	46.50	42.96	40.77	40.00
5.4	79.63	71.55	63.98	57.14	51.25	46.46	42.93	40.76	40.00
5.5	79.55	71.47	63.91	57.07	51.19	46.42	42.90	40.74	40.00
5.6	79.46	71.39	63.83	57.01	51.14	46.38	42.87	40.73	40.00
5.7	79.38	71.31	63.76	56.95	51.09	46.34	42.84	40.71	40.00
5.8	79.30	71.23	63.69	56.88	51.03	46.30	42.81	40.70	40.00
5.9	79.22	71.15	63.62	56.82	50.98	46.26	42.78	40.69	40.00
6.0	79.14	71.08	63.55	56.76	50.93	46.22	42.76	40.67	40.00
6.1	79.05	71.00	63.47	56.69	50.87	46.17	42.73	40.65	40.00
6.2	78.97	70.92	63.40	56.63	50.82	46.13	42.70	40.64	40.00
6.3	78.89	70.84	63.33	56.57	50.77	46.09	42.67	40.63	40.00
6.4	78.80	70.76	63.26	56.51	50.71	46.05	42.65	40.62	40.00
6.5	78.72	70.69	63.19	56.45	50.66	46.01	42.62	40.60	40.00
6.6	78.64	70.61	63.11	56.38	50.61	45.97	42.59	40.59	40.00
6.7	78.55	70.53	63.04	56.32	50.55	45.93	42.57	40.58	40.00
6.8	78.47	70.45	62.97	56.26	50.50	45.89	42.54	40.56	40.01
6.9	78.39	70.37	62.90	56.20	50.45	45.85	42.51	40.55	40.01
7.0	78.31	70.30	62.83	56.14	50.40	45.81	42.49	40.54	40.01
7.1	78.22	70.22	62.75	56.07	50.34	45.77	42.46	40.52	40.01
7.2	78.14	70.14	62.68	56.01	50.29	45.73	42.43	40.51	40.01
7.3	78.06	70.06	62.61	55.95	50.24	45.69	42.41	40.50	40.02
7.4	77.98	69.98	62.54	55.89	50.19	45.65	42.38	40.49	40.02
7.5	77.90	69.91	62.47	55.83	50.14	45.61	42.35	40.48	40.02
7.6	77.81	69.83	62.40	55.76	50.09	45.57	42.33	40.46	40.02
7.7	77.73	69.75	62.33	55.70	50.04	45.53	42.30	40.45	40.03
7.8	77.65	69.67	62.26	55.64	49.99	45.49	42.28	40.44	40.03
7.9	77.57	69.59	62.19	55.58	49.94	45.45	42.25	40.43	40.03
8.0	77.49	69.52	62.12	55.52	49.89	45.42	42.23	40.42	40.04
8.1	77.40	69.44	62.05	55.45	49.84	45.38	42.20	40.41	40.04
8.2	77.32	69.36	61.98	55.39	49.79	45.34	42.18	40.40	40.04
8.3	77.24	69.28	61.91	55.33	49.74	45.30	42.15	40.39	40.05
8.4	77.16	69.21	61.84	55.27	49.69	45.26	42.13	40.38	40.05
8.5	77.08	69.13	61.77	55.21	49.64	45.23	42.10	40.37	40.06
8.6	77.00	69.05	61.70	55.15	49.59	45.19	42.08	40.36	40.06
8.7	76.92	68.98	61.63	55.09	49.54	45.15	42.06	40.35	40.07
8.8	76.84	68.90	61.56	55.03	49.49	45.11	42.03	40.34	40.07
8.9	76.76	68.82	61.49	54.97	49.44	45.07	42.01	40.33	40.08
9.0	76.68	68.75	61.42	54.91	49.39	45.04	41.99	40.32	40.08
9.1	76.59	68.67	61.35	54.85	49.34	45.00	41.96	40.31	40.09
9.2	76.51	68.59	61.28	54.79	49.29	44.96	41.94	40.30	40.09
9.3	76.43	68.52	61.21	54.73	49.24	44.93	41.92	40.29	40.10
9.4	76.35	68.44	61.14	54.67	49.19	44.89	41.89	40.28	40.10
9.5	76.27	68.37	61.07	54.61	49.14	44.85	41.87	40.27	40.11
9.6	76.19	68.29	61.00	54.55	49.09	44.82	41.85	40.26	40.11
9.7	76.11	68.21	60.93	54.49	49.04	44.78	41.82	40.25	40.12
9.8	76.03	68.14	60.86	54.43	48.99	44.75	41.80	40.24	40.12
9.9	75.95	68.06	60.79	54.37	48.94	44.71	41.78	40.23	40.13
10.0	75.87	67.99	60.73	54.31	48.90	44.68	41.76	40.23	40.14

TABLES LXXXIX. XC.

Tables	LXXXIX.				XC.				
Arguments	4.				5.				
Days.	0	100	200	300	0	100	200	300	400
0	6.61	5.34	7.00	7.81	9.11	0.46	8.75	0.88	8.26
01	6.61	5.34	7.02	7.80	9.16	0.40	8.82	0.80	8.35
02	6.59	5.34	7.04	7.79	9.20	0.34	8.89	0.72	8.44
03	6.57	5.35	7.07	7.78	9.24	0.29	8.95	0.65	8.53
04	6.54	5.35	7.09	7.77	9.27	0.24	9.01	0.58	8.61
05	6.52	5.36	7.11	7.76	9.30	0.20	9.07	0.51	8.69
06	6.50	5.37	7.13	7.75	9.33	0.16	9.12	0.45	8.76
07	6.48	5.37	7.15	7.73	9.35	0.13	9.17	0.39	8.83
08	6.46	5.38	7.17	7.72	9.37	0.10	9.21	0.33	8.90
09	6.43	5.39	7.19	7.71	9.38	0.07	9.25	0.28	8.96
10	6.41	5.40	7.21	7.69	9.39	0.05	9.28	0.23	9.02
11	6.39	5.40	7.23	7.68	9.40	0.03	9.31	0.19	9.07
12	6.37	5.41	7.25	7.67	9.40	0.02	9.33	0.15	9.12
13	6.35	5.42	7.27	7.65	9.40	0.01	9.35	0.12	9.17
14	6.32	5.42	7.29	7.64	9.39	0.00	9.37	0.09	9.21
15	6.30	5.43	7.31	7.62	9.38	0.00	9.38	0.06	9.25
16	6.28	5.41	7.33	7.61	9.36	0.01	9.39	0.04	9.28
17	6.26	5.45	7.35	7.59	9.34	0.02	9.40	0.02	9.31
18	6.21	5.46	7.37	7.58	9.31	0.03	9.40	0.01	9.33
19	6.22	5.47	7.39	7.56	9.28	0.05	9.39	0.00	9.35
20	6.20	5.48	7.41	7.55	9.25	0.07	9.38	0.00	9.37
21	6.18	5.49	7.43	7.53	9.21	0.09	9.37	0.00	9.38
22	6.16	5.50	7.15	7.52	9.17	0.12	9.35	0.01	9.39
23	6.14	5.51	7.46	7.50	9.12	0.15	9.33	0.02	9.40
24	6.12	5.52	7.18	7.49	9.07	0.19	9.30	0.03	9.40
25	6.10	5.53	7.50	7.47	9.01	0.23	9.27	0.05	9.40
26	6.08	5.54	7.51	7.45	8.95	0.28	9.24	0.07	9.39
27	6.06	5.56	7.53	7.43	8.89	0.33	9.20	0.10	9.38
28	6.01	5.57	7.55	7.41	8.82	0.39	9.16	0.13	9.36
29	6.02	5.58	7.56	7.40	8.75	0.45	9.11	0.16	9.33
30	6.00	5.60	7.58	7.38	8.68	0.51	9.06	0.20	9.30
31	5.98	5.61	7.59	7.36	8.60	0.58	9.00	0.24	9.27
32	5.96	5.62	7.61	7.34	8.52	0.65	8.94	0.29	9.24
33	5.94	5.64	7.62	7.32	8.43	0.72	8.88	0.34	9.20
34	5.92	5.65	7.64	7.30	8.34	0.80	8.81	0.40	9.16
35	5.90	5.67	7.65	7.28	8.25	0.88	8.74	0.46	9.11
36	5.88	5.68	7.66	7.26	8.15	0.97	8.67	0.52	9.06
37	5.86	5.70	7.68	7.24	8.05	1.06	8.59	0.59	9.00
38	5.85	5.71	7.69	7.22	7.95	1.15	8.51	0.66	8.94
39	5.83	5.73	7.70	7.20	7.85	1.25	8.42	0.74	8.88
40	5.81	5.75	7.72	7.18	7.74	1.35	8.33	0.82	8.81
41	5.79	5.76	7.73	7.16	7.63	1.45	8.24	0.90	8.74
42	5.77	5.78	7.74	7.14	7.52	1.56	8.14	0.99	8.66
43	5.76	5.80	7.75	7.12	7.40	1.67	8.04	1.08	8.58
44	5.74	5.81	7.76	7.10	7.28	1.78	7.94	1.17	8.50
45	5.73	5.83	7.77	7.08	7.16	1.89	7.84	1.26	8.41
46	5.71	5.85	7.78	7.06	7.04	2.01	7.73	1.36	8.32
47	5.70	5.87	7.79	7.04	6.91	2.13	7.62	1.46	8.23
48	5.68	5.89	7.80	7.02	6.78	2.25	7.51	1.57	8.13
49	5.67	5.91	7.81	7.00	6.65	2.37	7.39	1.68	8.03
50	5.65	5.93	7.82	6.98	6.52	2.49	7.27	1.79	7.93

TABLES LXXXIX. XC.

Tables	LXXXIX.				XC.				
Arguments	4.				5.				
Days.	0	100	200	300	0	100	200	300	400
Days									
50	5.65	5.93	7.82	6.98	6.52	2.49	7.27	1.79	7.93
51	5.64	5.95	7.83	6.96	6.39	2.62	7.15	1.90	7.83
52	5.62	5.97	7.84	6.94	6.26	2.75	7.03	2.02	7.72
53	5.61	5.99	7.84	6.91	6.12	2.88	6.90	2.14	7.61
54	5.59	6.01	7.85	6.89	5.98	3.01	6.77	2.26	7.49
55	5.58	6.03	7.86	6.87	5.84	3.15	6.64	2.38	7.37
56	5.57	6.05	7.87	6.85	5.70	3.29	6.51	2.51	7.25
57	5.55	6.07	7.87	6.83	5.56	3.43	6.38	2.64	7.13
58	5.54	6.09	7.88	6.80	5.42	3.57	6.25	2.77	7.01
59	5.53	6.11	7.89	6.78	5.28	3.71	6.11	2.90	6.88
60	5.51	6.13	7.89	6.76	5.14	3.85	5.97	3.03	6.75
61	5.50	6.15	7.90	6.74	4.99	3.99	5.83	3.17	6.62
62	5.49	6.17	7.91	6.72	4.85	4.13	5.69	3.30	6.49
63	5.48	6.19	7.91	6.69	4.71	4.27	5.55	3.44	6.36
64	5.47	6.21	7.91	6.67	4.56	4.41	5.41	3.58	6.23
65	5.46	6.23	7.92	6.65	4.42	4.55	5.26	3.72	6.09
66	5.45	6.25	7.92	6.62	4.28	4.70	5.12	3.86	5.95
67	5.44	6.27	7.92	6.60	4.11	4.84	4.98	4.00	5.81
68	5.44	6.30	7.92	6.58	3.99	4.98	4.84	4.14	5.67
69	5.43	6.32	7.92	6.55	3.85	5.13	4.69	4.28	5.53
70	5.42	6.34	7.93	6.53	3.71	5.27	4.55	4.42	5.39
71	5.41	6.36	7.93	6.51	3.57	5.41	4.41	4.57	5.25
72	5.40	6.38	7.93	6.49	3.43	5.55	4.26	4.71	5.11
73	5.40	6.41	7.93	6.47	3.29	5.69	4.12	4.86	4.96
74	5.39	6.43	7.93	6.44	3.15	5.83	3.98	5.01	4.82
75	5.38	6.45	7.93	6.42	3.02	5.97	3.84	5.15	4.68
76	5.38	6.47	7.93	6.40	2.89	6.11	3.70	5.29	4.53
77	5.37	6.49	7.93	6.38	2.76	6.25	3.56	5.43	4.39
78	5.37	6.52	7.93	6.36	2.63	6.39	3.42	5.57	4.25
79	5.36	6.54	7.93	6.33	2.50	6.52	3.28	5.71	4.11
80	5.36	6.56	7.92	6.31	2.37	6.65	3.14	5.85	3.96
81	5.35	6.58	7.92	6.29	2.25	6.78	3.01	5.99	3.82
82	5.35	6.60	7.92	6.27	2.13	6.91	2.88	6.13	3.68
83	5.34	6.63	7.92	6.25	2.01	7.03	2.75	6.27	3.54
84	5.34	6.65	7.92	6.23	1.89	7.15	2.62	6.40	3.40
85	5.34	6.67	7.91	6.21	1.78	7.27	2.49	6.53	3.26
86	5.34	6.69	7.91	6.19	1.67	7.39	2.36	6.66	3.12
87	5.34	6.71	7.90	6.17	1.56	7.51	2.24	6.79	2.99
88	5.33	6.74	7.90	6.15	1.46	7.62	2.12	6.92	2.86
89	5.33	6.76	7.89	6.13	1.36	7.73	2.00	7.05	2.73
90	5.33	6.78	7.88	6.11	1.26	7.84	1.88	7.17	2.60
91	5.33	6.80	7.88	6.09	1.16	7.95	1.77	7.29	2.47
92	5.33	6.82	7.87	6.07	1.07	8.05	1.66	7.41	2.34
93	5.33	6.85	7.86	6.05	0.98	8.15	1.55	7.52	2.22
94	5.33	6.87	7.86	6.03	0.89	8.25	1.45	7.63	2.10
95	5.33	6.89	7.85	6.01	0.81	8.34	1.35	7.74	1.98
96	5.33	6.91	7.84	5.99	0.73	8.43	1.25	7.85	1.87
97	5.33	6.93	7.83	5.97	0.66	8.52	1.15	7.96	1.76
98	5.34	6.96	7.82	5.95	0.59	8.60	1.06	8.06	1.65
99	5.34	6.98	7.82	5.93	0.52	8.68	0.97	8.16	1.54
100	5.34	7.00	7.81	5.91	0.46	8.75	0.88	8.26	1.44

TABLES XCI. XCII.

TABLES	XCI.	XCII.
ARGUMENTS	5.	78.

Table XCI.

Days.	0	100	200	300	400
Days. 0	1″.57	1″.85	0″.50	0″.11	1″.42
01	1.58	1.84	0.49	0.12	1.43
02	1.60	1.83	0.47	0.12	1.45
03	1.61	1.82	0.46	0.13	1.46
04	1.62	1.81	0.45	0.14	1.47
05	1.63	1.80	0.44	0.15	1.49
06	1.64	1.79	0.43	0.16	1.50
07	1.66	1.78	0.41	0.16	1.51
08	1.67	1.77	0.40	0.17	1.53
09	1.68	1.76	0.39	0.18	1.54
10	1.69	1.75	0.38	0.19	1.55
11	1.70	1.74	0.37	0.20	1.56
12	1.71	1.73	0.35	0.20	1.58
13	1.72	1.72	0.34	0.21	1.59
14	1.73	1.71	0.33	0.23	1.60
15	1.74	1.70	0.32	0.24	1.61
16	1.75	1.69	0.31	0.25	1.62
17	1.76	1.68	0.30	0.26	1.63
18	1.77	1.67	0.29	0.27	1.65
19	1.78	1.66	0.28	0.28	1.66
20	1.79	1.65	0.27	0.29	1.67
21	1.80	1.63	0.26	0.30	1.68
22	1.81	1.62	0.25	0.31	1.69
23	1.82	1.61	0.24	0.32	1.70
24	1.83	1.60	0.23	0.33	1.71
25	1.84	1.59	0.22	0.34	1.72
26	1.84	1.57	0.21	0.36	1.73
27	1.85	1.56	0.20	0.37	1.74
28	1.86	1.55	0.19	0.38	1.75
29	1.87	1.54	0.18	0.39	1.76
30	1.88	1.53	0.17	0.40	1.77
31	1.88	1.51	0.16	0.42	1.78
32	1.89	1.50	0.15	0.43	1.79
33	1.90	1.48	0.14	0.44	1.80
34	1.90	1.47	0.13	0.45	1.81
35	1.91	1.46	0.12	0.46	1.82
36	1.91	1.44	0.12	0.48	1.83
37	1.92	1.43	0.11	0.49	1.84
38	1.92	1.41	0.10	0.50	1.84
39	1.93	1.40	0.10	0.52	1.85
40	1.93	1.39	0.09	0.53	1.86
41	1.94	1.37	0.09	0.55	1.87
42	1.94	1.36	0.08	0.56	1.88
43	1.95	1.34	0.08	0.57	1.88
44	1.95	1.33	0.07	0.59	1.89
45	1.96	1.32	0.07	0.60	1.90
46	1.96	1.30	0.06	0.62	1.90
47	1.96	1.29	0.06	0.63	1.91
48	1.97	1.27	0.05	0.64	1.91
49	1.97	1.26	0.05	0.66	1.92
50	1.97	1.25	0.04	0.67	1.92

Table XCII.

Days.	0	10	20	30	40	50	Days
Days. 0.0	0″.31	0″.15	1″.72	4″.85	9″.18	14″.25	Days. 10.0
0.1	0.30	0.16	1.74	4.89	9.23	14.30	9.9
0.2	0.29	0.17	1.77	4.93	9.28	14.35	9.8
0.3	0.28	0.17	1.79	4.96	9.32	14.41	9.7
0.4	0.27	0.18	1.82	5.00	9.37	14.46	9.6
0.5	0.26	0.19	1.84	5.04	9.42	14.51	9.5
0.6	0.25	0.20	1.87	5.08	9.47	14.56	9.4
0.7	0.24	0.21	1.89	5.12	9.52	14.61	9.3
0.8	0.23	0.21	1.92	5.15	9.56	14.67	9.2
0.9	0.22	0.22	1.94	5.19	9.61	14.72	9.1
1.0	0.21	0.23	1.97	5.23	9.66	14.77	9.0
1.1	0.20	0.24	2.00	5.27	9.71	14.82	8.9
1.2	0.19	0.25	2.02	5.31	9.76	14.88	8.8
1.3	0.19	0.26	2.05	5.35	9.81	14.93	8.7
1.4	0.18	0.27	2.07	5.39	9.86	14.99	8.6
1.5	0.17	0.28	2.10	5.43	9.91	15.04	8.5
1.6	0.16	0.29	2.13	5.47	9.96	15.09	8.4
1.7	0.15	0.30	2.15	5.51	10.01	15.14	8.3
1.8	0.15	0.31	2.18	5.55	10.05	15.20	8.2
1.9	0.14	0.32	2.20	5.59	10.10	15.25	8.1
2.0	0.13	0.33	2.23	5.63	10.15	15.30	8.0
2.1	0.12	0.34	2.26	5.67	10.20	15.35	7.9
2.2	0.12	0.35	2.29	5.71	10.25	15.40	7.8
2.3	0.11	0.36	2.31	5.75	10.30	15.46	7.7
2.4	0.11	0.37	2.34	5.79	10.35	15.51	7.6
2.5	0.10	0.38	2.37	5.83	10.40	15.56	7.5
2.6	0.09	0.39	2.40	5.87	10.45	15.61	7.4
2.7	0.09	0.40	2.43	5.91	10.50	15.67	7.3
2.8	0.08	0.42	2.45	5.96	10.55	15.72	7.2
2.9	0.08	0.43	2.48	6.00	10.60	15.78	7.1
3.0	0.07	0.44	2.51	6.04	10.65	15.83	7.0
3.1	0.07	0.45	2.54	6.08	10.70	15.88	6.9
3.2	0.06	0.47	2.57	6.12	10.75	15.93	6.8
3.3	0.06	0.48	2.59	6.17	10.80	15.99	6.7
3.4	0.05	0.50	2.62	6.21	10.85	16.04	6.6
3.5	0.05	0.51	2.65	6.25	10.90	16.09	6.5
3.6	0.05	0.52	2.68	6.29	10.95	16.14	6.4
3.7	0.04	0.54	2.71	6.33	11.00	16.19	6.3
3.8	0.04	0.55	2.74	6.38	11.05	16.25	6.2
3.9	0.03	0.57	2.77	6.42	11.10	16.30	6.1
4.0	0.03	0.58	2.80	6.46	11.15	16.35	6.0
4.1	0.03	0.59	2.83	6.50	11.20	16.40	5.9
4.2	0.03	0.61	2.86	6.54	11.25	16.46	5.8
4.3	0.02	0.62	2.89	6.59	11.30	16.51	5.7
4.4	0.02	0.64	2.92	6.63	11.35	16.57	5.6
4.5	0.02	0.65	2.95	6.67	11.40	16.62	5.5
4.6	0.02	0.67	2.98	6.71	11.45	16.67	5.4
4.7	0.02	0.68	3.01	6.76	11.50	16.72	5.3
4.8	0.01	0.70	3.05	6.80	11.56	16.78	5.2
4.9	0.01	0.71	3.08	6.85	11.61	16.83	5.1
5.0	0.01	0.73	3.11	6.89	11.66	16.88	5.0
Days.	190	180	170	160	150	140	Days.

TABLES XCI. XCII.

Tables	XCI.	XCII.
Arguments	5.	78.

Table XCI. Argument 5.

Days.	0	100	200	300	400
Days	"	"	"	"	"
50	1.97	1.25	0.01	0.67	1.92
51	1.98	1.23	0.01	0.69	1.93
52	1.98	1.22	0.03	0.70	1.93
53	1.98	1.20	0.03	0.71	1.94
54	1.98	1.19	0.03	0.73	1.94
55	1.99	1.17	0.02	0.74	1.95
56	1.99	1.15	0.02	0.76	1.95
57	1.99	1.14	0.02	0.77	1.96
58	1.99	1.12	0.01	0.78	1.96
59	1.99	1.11	0.01	0.80	1.96
60	2.00	1.10	0.01	0.81	1.97
61	2.00	1.08	0.01	0.83	1.97
62	2.00	1.07	0.01	0.84	1.97
63	2.00	1.05	0.00	0.86	1.98
64	2.00	1.04	0.00	0.87	1.98
65	2.00	1.02	0.00	0.89	1.98
66	2.00	1.00	0.00	0.91	1.98
67	2.00	0.99	0.00	0.92	1.99
68	2.00	0.97	0.00	0.94	1.99
69	1.99	0.96	0.00	0.95	1.99
70	1.99	0.94	0.00	0.97	1.99
71	1.99	0.92	0.00	0.99	1.99
72	1.99	0.91	0.01	1.00	2.00
73	1.99	0.89	0.01	1.02	2.00
74	1.98	0.88	0.01	1.03	2.00
75	1.98	0.87	0.01	1.04	2.00
76	1.98	0.85	0.01	1.06	2.00
77	1.97	0.84	0.02	1.07	2.00
78	1.97	0.82	0.02	1.09	2.00
79	1.97	0.80	0.02	1.11	2.00
80	1.96	0.79	0.02	1.13	2.00
81	1.96	0.77	0.02	1.14	1.99
82	1.96	0.76	0.03	1.16	1.99
83	1.95	0.74	0.03	1.17	1.99
84	1.95	0.73	0.03	1.18	1.99
85	1.94	0.72	0.04	1.20	1.99
86	1.94	0.70	0.04	1.21	1.98
87	1.93	0.69	0.04	1.23	1.98
88	1.93	0.67	0.05	1.24	1.98
89	1.92	0.66	0.05	1.26	1.97
90	1.92	0.65	0.06	1.28	1.97
91	1.91	0.63	0.06	1.29	1.97
92	1.91	0.61	0.07	1.31	1.96
93	1.90	0.60	0.07	1.32	1.96
94	1.90	0.59	0.08	1.33	1.96
95	1.89	0.57	0.08	1.35	1.95
96	1.88	0.56	0.09	1.36	1.95
97	1.88	0.54	0.09	1.38	1.94
98	1.87	0.53	0.10	1.39	1.94
99	1.86	0.52	0.10	1.40	1.93
100	1.85	0.50	0.11	1.42	1.93

Table XCII. Argument 78.

Days.	0	10	20	30	40	50	Days
Days	"	"	"	"	"	"	Days
5.0	0.01	0.73	3.11	6.89	11.66	16.88	5.0
5.1	0.01	0.75	3.14	6.93	11.71	16.93	4.9
5.2	0.01	0.76	3.17	6.98	11.76	16.98	4.8
5.3	0.00	0.78	3.21	7.02	11.81	17.04	4.7
5.4	0.00	0.79	3.24	7.07	11.86	17.09	4.6
5.5	0.00	0.81	3.27	7.11	11.91	17.14	4.5
5.6	0.00	0.83	3.30	7.15	11.96	17.19	4.4
5.7	0.00	0.84	3.33	7.20	12.01	17.24	4.3
5.8	0.00	0.86	3.37	7.24	12.07	17.30	4.2
5.9	0.00	0.87	3.40	7.29	12.12	17.35	4.1
6.0	0.00	0.89	3.43	7.33	12.17	17.40	4.0
6.1	0.00	0.91	3.46	7.37	12.22	17.45	3.9
6.2	0.00	0.93	3.49	7.42	12.27	17.51	3.8
6.3	0.00	0.94	3.53	7.46	12.32	17.56	3.7
6.4	0.00	0.96	3.56	7.51	12.37	17.62	3.6
6.5	0.00	0.98	3.59	7.55	12.42	17.67	3.5
6.6	0.00	1.00	3.62	7.60	12.47	17.72	3.4
6.7	0.00	1.02	3.66	7.64	12.52	17.77	3.3
6.8	0.01	1.03	3.69	7.69	12.58	17.83	3.2
6.9	0.01	1.05	3.73	7.73	12.63	17.88	3.1
7.0	0.01	1.07	3.76	7.78	12.68	17.93	3.0
7.1	0.01	1.09	3.79	7.83	12.73	17.98	2.9
7.2	0.01	1.11	3.83	7.87	12.78	18.03	2.8
7.3	0.02	1.13	3.86	7.92	12.84	18.09	2.7
7.4	0.02	1.15	3.90	7.96	12.89	18.14	2.6
7.5	0.02	1.17	3.93	8.01	12.94	18.19	2.5
7.6	0.02	1.19	3.97	8.06	12.99	18.24	2.4
7.7	0.03	1.21	4.00	8.10	13.04	18.29	2.3
7.8	0.03	1.23	4.04	8.15	13.10	18.35	2.2
7.9	0.04	1.25	4.07	8.19	13.15	18.40	2.1
8.0	0.04	1.27	4.11	8.24	13.20	18.45	2.0
8.1	0.04	1.29	4.15	8.29	13.25	18.50	1.9
8.2	0.05	1.31	4.18	8.33	13.30	18.55	1.8
8.3	0.05	1.34	4.22	8.38	13.36	18.61	1.7
8.4	0.06	1.36	4.25	8.42	13.41	18.66	1.6
8.5	0.06	1.38	4.29	8.47	13.46	18.71	1.5
8.6	0.07	1.40	4.33	8.52	13.51	18.76	1.4
8.7	0.07	1.42	4.36	8.57	13.56	18.81	1.3
8.8	0.08	1.45	4.40	8.61	13.62	18.87	1.2
8.9	0.08	1.47	4.43	8.66	13.67	18.92	1.1
9.0	0.09	1.49	4.47	8.71	13.72	18.97	1.0
9.1	0.10	1.51	4.51	8.76	13.77	19.02	0.9
9.2	0.10	1.53	4.55	8.80	13.82	19.07	0.8
9.3	0.11	1.56	4.58	8.85	13.88	19.13	0.7
9.4	0.11	1.58	4.62	8.89	13.93	19.18	0.6
9.5	0.12	1.60	4.66	8.94	13.98	19.23	0.5
9.6	0.13	1.62	4.70	8.99	14.03	19.28	0.4
9.7	0.13	1.65	4.74	9.04	14.09	19.33	0.3
9.8	0.14	1.67	4.77	9.08	14.14	19.39	0.2
9.9	0.14	1.70	4.81	9.13	14.20	19.44	0.1
10.0	0.15	1.72	4.85	9.18	14.25	19.49	0.0
Days.	190	180	170	160	150	140	Days.

TABLES XCII. XCIII.

Tables	XCII.						XCIII.						
Arguments	78.						73.						

Days.	60	70	80	90	Days.	Days.	0	100	200	300	400	500	Days.
Days	"	"	"	"	Days.	Days	"	"	"	"	"	"	Days
0.0	19.49	24.31	28.21	30.73	10.0	0	7.80	8.33	10.00	12.66	16.10	20.04	100
0.1	19.54	24.35	28.24	30.75	9.9	01	7.80	8.34	10.02	12.69	16.13	20.08	99
0.2	19.59	24.40	28.27	30.76	9.8	02	7.80	8.35	10.04	12.72	16.17	20.12	98
0.3	19.64	24.44	28.31	30.78	9.7	03	7.80	8.36	10.06	12.75	16.21	20.16	97
0.4	19.69	24.49	28.34	30.79	9.6	04	7.80	8.37	10.08	12.78	16.25	20.20	96
0.5	19.74	24.53	28.37	30.81	9.5	05	7.80	8.39	10.11	12.81	16.29	20.24	95
0.6	19.79	24.57	28.40	30.83	9.4	06	7.80	8.40	10.13	12.84	16.32	20.28	94
0.7	19.84	24.62	28.43	30.84	9.3	07	7.80	8.41	10.15	12.87	16.36	20.32	93
0.8	19.90	24.66	28.47	30.86	9.2	08	7.80	8.42	10.17	12.90	16.40	20.36	92
0.9	19.95	24.71	28.50	30.87	9.1	09	7.80	8.43	10.19	12.93	16.44	20.40	91
1.0	20.00	24.75	28.53	30.89	9.0	10	7.80	8.45	10.22	12.97	16.48	20.45	90
1.1	20.05	24.79	28.56	30.91	8.9	11	7.80	8.46	10.24	13.00	16.51	20.49	89
1.2	20.10	24.84	28.59	30.92	8.8	12	7.80	8.47	10.26	13.03	16.55	20.53	88
1.3	20.15	24.88	28.62	30.94	8.7	13	7.80	8.48	10.29	13.06	16.59	20.57	87
1.4	20.20	24.93	28.65	30.95	8.6	14	7.80	8.50	10.31	13.09	16.63	20.61	86
1.5	20.25	24.97	28.68	30.97	8.5	15	7.81	8.51	10.33	13.13	16.67	20.65	85
1.6	20.30	25.01	28.71	30.98	8.4	16	7.81	8.52	10.36	13.16	16.70	20.69	84
1.7	20.35	25.05	28.74	31.00	8.3	17	7.81	8.54	10.38	13.19	16.74	20.73	83
1.8	20.40	25.10	28.77	31.01	8.2	18	7.81	8.55	10.41	13.22	16.78	20.77	82
1.9	20.45	25.14	28.80	31.03	8.1	19	7.81	8.56	10.43	13.25	16.82	20.81	81
2.0	20.50	25.18	28.83	31.04	8.0	20	7.81	8.58	10.46	13.29	16.86	20.86	80
2.1	20.55	25.22	28.86	31.05	7.9	21	7.82	8.59	10.48	13.32	16.90	20.90	79
2.2	20.60	25.26	28.89	31.07	7.8	22	7.82	8.60	10.50	13.35	16.93	20.94	78
2.3	20.65	25.31	28.92	31.08	7.7	23	7.82	8.62	10.53	13.38	16.97	20.98	77
2.4	20.70	25.35	28.95	31.10	7.6	24	7.82	8.63	10.55	13.42	17.01	21.02	76
2.5	20.75	25.39	28.98	31.11	7.5	25	7.82	8.64	10.58	13.45	17.05	21.06	75
2.6	20.80	25.43	29.01	31.12	7.4	26	7.83	8.66	10.60	13.48	17.09	21.10	74
2.7	20.85	25.47	29.04	31.13	7.3	27	7.83	8.67	10.62	13.52	17.13	21.14	73
2.8	20.90	25.52	29.06	31.15	7.2	28	7.83	8.69	10.65	13.55	17.17	21.18	72
2.9	20.95	25.56	29.09	31.16	7.1	29	7.83	8.70	10.67	13.58	17.21	21.22	71
3.0	21.00	25.60	29.12	31.17	7.0	30	7.84	8.72	10.70	13.62	17.25	21.27	70
3.1	21.05	25.64	29.15	31.18	6.9	31	7.84	8.73	10.72	13.65	17.28	21.31	69
3.2	21.10	25.68	29.18	31.19	6.8	32	7.84	8.75	10.75	13.68	17.32	21.35	68
3.3	21.15	25.73	29.20	31.21	6.7	33	7.84	8.76	10.77	13.72	17.36	21.39	67
3.4	21.20	25.77	29.23	31.22	6.6	34	7.85	8.78	10.80	13.75	17.40	21.43	66
3.5	21.25	25.81	29.26	31.23	6.5	35	7.85	8.79	10.82	13.78	17.44	21.48	65
3.6	21.30	25.85	29.29	31.24	6.4	36	7.85	8.81	10.85	13.82	17.48	21.52	64
3.7	21.35	25.89	29.32	31.25	6.3	37	7.86	8.82	10.87	13.85	17.52	21.56	63
3.8	21.40	25.93	29.34	31.26	6.2	38	7.86	8.84	10.90	13.89	17.56	21.60	62
3.9	21.45	25.97	29.37	31.27	6.1	39	7.86	8.85	10.92	13.92	17.60	21.64	61
4.0	21.50	26.01	29.40	31.28	6.0	40	7.87	8.87	10.95	13.96	17.64	21.69	60
4.1	21.55	26.05	29.43	31.29	5.9	41	7.87	8.88	10.97	13.99	17.67	21.73	59
4.2	21.60	26.09	29.45	31.30	5.8	42	7.88	8.90	11.00	14.02	17.71	21.77	58
4.3	21.64	26.13	29.48	31.31	5.7	43	7.88	8.91	11.03	14.06	17.75	21.81	57
4.4	21.69	26.17	29.50	31.32	5.6	44	7.89	8.93	11.05	14.09	17.79	21.85	56
4.5	21.74	26.21	29.53	31.33	5.5	45	7.89	8.94	11.08	14.13	17.83	21.89	55
4.6	21.79	26.25	29.56	31.34	5.4	46	7.90	8.96	11.11	14.16	17.87	21.94	54
4.7	21.84	26.29	29.58	31.35	5.3	47	7.90	8.98	11.13	14.19	17.91	21.98	53
4.8	21.89	26.33	29.61	31.36	5.2	48	7.91	9.00	11.16	14.23	17.95	22.02	52
4.9	21.94	26.37	29.63	31.37	5.1	49	7.91	9.01	11.19	14.26	17.99	22.06	51
5.0	21.99	26.41	29.66	31.38	5.0	50	7.92	9.03	11.22	14.30	18.03	22.10	50
Days.	130	120	110	100	Days.	Days.	2100	2000	1900	1800	1700	1600	Days.

TABLES XCII. XCIII.

Tables	XCII.	XCIII.
Arguments	78.	73.

Days.	60	70	80	90	Days.
5.0	21.99	26.41	29.66	31.38	5.0
5.1	22.04	26.45	29.68	31.39	4.9
5.2	22.09	26.49	29.71	31.40	4.8
5.3	22.13	26.52	29.73	31.40	4.7
5.4	22.18	26.56	29.76	31.41	4.6
5.5	22.23	26.60	29.78	31.42	4.5
5.6	22.28	26.64	29.80	31.43	4.4
5.7	22.33	26.68	29.83	31.44	4.3
5.8	22.37	26.71	29.85	31.44	4.2
5.9	22.42	26.75	29.88	31.45	4.1
6.0	22.47	26.79	29.90	31.46	4.0
6.1	22.52	26.83	29.92	31.47	3.9
6.2	22.57	26.87	29.95	31.47	3.8
6.3	22.61	26.90	29.97	31.48	3.7
6.4	22.66	26.94	30.00	31.48	3.6
6.5	22.71	26.98	30.02	31.49	3.5
6.6	22.76	27.02	30.04	31.50	3.4
6.7	22.80	27.06	30.06	31.50	3.3
6.8	22.85	27.09	30.09	31.51	3.2
6.9	22.89	27.13	30.11	31.51	3.1
7.0	22.94	27.17	30.13	31.52	3.0
7.1	22.99	27.21	30.15	31.53	2.9
7.2	23.04	27.24	30.17	31.53	2.8
7.3	23.08	27.28	30.20	31.54	2.7
7.4	23.13	27.31	30.22	31.54	2.6
7.5	23.18	27.35	30.24	31.55	2.5
7.6	23.23	27.39	30.26	31.55	2.4
7.7	23.27	27.42	30.28	31.56	2.3
7.8	23.32	27.46	30.31	31.56	2.2
7.9	23.36	27.49	30.33	31.57	2.1
8.0	23.41	27.53	30.35	31.57	2.0
8.1	23.46	27.56	30.37	31.57	1.9
8.2	23.50	27.60	30.39	31.57	1.8
8.3	23.55	27.63	30.41	31.58	1.7
8.4	23.59	27.67	30.43	31.58	1.6
8.5	23.64	27.70	30.45	31.58	1.5
8.6	23.69	27.73	30.47	31.58	1.4
8.7	23.73	27.77	30.49	31.58	1.3
8.8	23.78	27.80	30.51	31.59	1.2
8.9	23.82	27.84	30.53	31.59	1.1
9.0	23.87	27.87	30.55	31.59	1.0
9.1	23.91	27.90	30.57	31.59	0.9
9.2	23.96	27.94	30.59	31.59	0.8
9.3	24.00	27.97	30.60	31.59	0.7
9.4	24.05	28.01	30.62	31.60	0.6
9.5	24.09	28.04	30.64	31.60	0.5
9.6	24.13	28.07	30.66	31.60	0.4
9.7	24.18	28.11	30.68	31.60	0.3
9.8	24.22	28.14	30.69	31.60	0.2
9.9	24.27	28.18	30.71	31.60	0.1
10.0	24.31	28.21	30.73	31.60	0.0
Days.	130	120	110	100	Days.

Days.	0	100	200	300	400	500	Days.
50	7.92	9.03	11.22	14.30	18.03	22.10	50
51	7.92	9.04	11.24	14.33	18.07	22.14	49
52	7.93	9.06	11.27	14.37	18.11	22.18	48
53	7.93	9.08	11.30	14.40	18.15	22.22	47
54	7.94	9.09	11.32	14.44	18.19	22.26	46
55	7.94	9.11	11.35	14.47	18.23	22.31	45
56	7.95	9.13	11.38	14.51	18.27	22.35	44
57	7.96	9.14	11.40	14.54	18.31	22.39	43
58	7.96	9.16	11.43	14.58	18.35	22.43	42
59	7.97	9.18	11.46	14.61	18.39	22.47	41
60	7.98	9.20	11.49	14.65	18.43	22.51	40
61	7.98	9.22	11.51	14.68	18.47	22.55	39
62	7.99	9.23	11.54	14.72	18.51	22.60	38
63	8.00	9.25	11.57	14.75	18.55	22.64	37
64	8.00	9.27	11.60	14.79	18.59	22.68	36
65	8.01	9.29	11.63	14.82	18.63	22.72	35
66	8.02	9.31	11.65	14.86	18.67	22.76	34
67	8.02	9.33	11.68	14.89	18.71	22.81	33
68	8.03	9.35	11.71	14.93	18.75	22.85	32
69	8.04	9.37	11.74	14.96	18.79	22.89	31
70	8.05	9.39	11.77	15.00	18.83	22.93	30
71	8.05	9.40	11.80	15.03	18.87	22.97	29
72	8.06	9.42	11.82	15.07	18.91	23.01	28
73	8.07	9.44	11.85	15.10	18.95	23.06	27
74	8.08	9.46	11.88	15.14	18.99	23.10	26
75	8.09	9.48	11.91	15.17	19.03	23.14	25
76	8.09	9.50	11.94	15.21	19.07	23.18	24
77	8.10	9.52	11.97	15.25	19.11	23.22	23
78	8.11	9.54	12.00	15.28	19.15	23.26	22
79	8.12	9.56	12.03	15.32	19.19	23.30	21
80	8.13	9.58	12.06	15.36	19.23	23.34	20
81	8.14	9.60	12.09	15.39	19.27	23.38	19
82	8.15	9.62	12.12	15.43	19.31	23.43	18
83	8.16	9.64	12.15	15.47	19.35	23.47	17
84	8.17	9.66	12.18	15.50	19.39	23.51	16
85	8.18	9.68	12.21	15.54	19.43	23.55	15
86	8.19	9.70	12.24	15.58	19.47	23.59	14
87	8.20	9.72	12.27	15.61	19.51	23.63	13
88	8.21	9.74	12.30	15.65	19.55	23.67	12
89	8.22	9.76	12.33	15.69	19.59	23.71	11
90	8.23	9.78	12.36	15.73	19.64	23.75	10
91	8.24	9.80	12.39	15.76	19.68	23.80	09
92	8.25	9.82	12.42	15.80	19.72	23.84	08
93	8.26	9.84	12.45	15.84	19.76	23.88	07
94	8.27	9.86	12.48	15.87	19.80	23.92	06
95	8.28	9.89	12.51	15.91	19.84	23.96	05
96	8.29	9.91	12.54	15.95	19.88	24.00	04
97	8.30	9.93	12.57	15.98	19.92	24.04	03
98	8.31	9.95	12.60	16.02	19.96	24.08	02
99	8.32	9.97	12.63	16.06	20.00	24.12	01
100	8.33	10.00	12.66	16.10	20.04	24.16	00
Days.	2100	2000	1900	1800	1700	1600	Days.

TABLES XCIII.-XCV.

Tables	XCIII.					XCIV.		XCV.			
Arguments	73.					12‴.		79.			
Days.	**600**	**700**	**800**	**900**	**1000**	**0**	**100**	**0**	**100**	**200**	Days.
00	24.16	28.12	31.59	34.29	36.01	1.13	0.74	0.28	0.39	0.17	100
01	24.20	28.16	31.63	34.32	36.03	1.13	0.73	0.28	0.39	0.17	99
02	24.24	28.20	31.66	34.34	36.04	1.14	0.72	0.28	0.39	0.17	98
03	24.28	28.24	31.69	34.36	36.05	1.14	0.71	0.29	0.39	0.17	97
04	24.32	28.27	31.72	34.38	36.06	1.15	0.70	0.29	0.39	0.16	96
05	24.36	28.31	31.75	34.40	36.07	1.15	0.69	0.29	0.39	0.16	95
06	24.40	28.35	31.78	34.43	36.08	1.15	0.68	0.30	0.38	0.16	94
07	24.44	28.38	31.81	34.45	36.09	1.16	0.67	0.30	0.38	0.15	93
08	24.48	28.42	31.84	34.47	36.10	1.16	0.66	0.30	0.38	0.15	92
09	24.52	28.46	31.87	34.49	36.11	1.16	0.65	0.30	0.38	0.15	91
10	24.57	28.49	31.90	34.51	36.12	1.17	0.64	0.31	0.38	0.14	90
11	24.61	28.53	31.93	34.53	36.13	1.17	0.63	0.31	0.38	0.14	89
12	24.65	28.57	31.96	34.56	36.14	1.17	0.62	0.31	0.38	0.14	88
13	24.69	28.61	31.99	34.58	36.15	1.18	0.61	0.31	0.38	0.14	87
14	24.73	28.64	32.02	34.60	36.16	1.18	0.60	0.31	0.38	0.14	86
15	24.77	28.68	32.05	34.62	36.17	1.18	0.59	0.32	0.38	0.13	85
16	24.81	28.72	32.08	34.64	36.18	1.19	0.58	0.32	0.37	0.13	84
17	24.85	28.75	32.11	34.66	36.19	1.19	0.57	0.32	0.37	0.13	83
18	24.89	28.79	32.14	34.68	36.20	1.19	0.56	0.32	0.37	0.13	82
19	24.93	28.83	32.17	34.70	36.21	1.19	0.55	0.32	0.37	0.13	81
20	24.97	28.86	32.20	34.72	36.22	1.20	0.53	0.33	0.37	0.12	80
21	25.01	28.90	32.23	34.74	36.23	1.20	0.52	0.33	0.37	0.12	79
22	25.05	28.94	32.26	34.76	36.24	1.20	0.51	0.33	0.37	0.12	78
23	25.09	28.97	32.29	34.78	36.25	1.20	0.50	0.33	0.36	0.12	77
24	25.13	29.01	32.32	34.80	36.26	1.20	0.49	0.33	0.36	0.11	76
25	25.17	29.05	32.35	34.82	36.27	1.20	0.48	0.34	0.36	0.11	75
26	25.21	29.08	32.38	34.84	36.28	1.20	0.47	0.34	0.36	0.11	74
27	25.25	29.12	32.41	34.86	36.29	1.20	0.46	0.34	0.36	0.10	73
28	25.29	29.15	32.44	34.88	36.30	1.20	0.45	0.34	0.36	0.10	72
29	25.33	29.19	32.47	34.90	36.31	1.20	0.44	0.34	0.35	0.10	71
30	25.38	29.22	32.49	34.92	36.31	1.20	0.42	0.35	0.35	0.09	70
31	25.42	29.26	32.52	34.94	36.32	1.20	0.41	0.35	0.35	0.09	69
32	25.46	29.30	32.55	34.96	36.33	1.20	0.40	0.35	0.35	0.09	68
33	25.50	29.33	32.58	34.98	36.34	1.19	0.39	0.35	0.35	0.09	67
34	25.54	29.37	32.61	35.00	36.35	1.19	0.38	0.35	0.35	0.09	66
35	25.58	29.41	32.64	35.02	36.35	1.19	0.37	0.36	0.34	0.08	65
36	25.62	29.44	32.67	35.04	36.36	1.19	0.36	0.36	0.34	0.08	64
37	25.66	29.48	32.70	35.06	36.37	1.19	0.35	0.36	0.34	0.08	63
38	25.70	29.51	32.73	35.08	36.38	1.18	0.34	0.36	0.34	0.08	62
39	25.74	29.55	32.76	35.10	36.39	1.18	0.33	0.36	0.34	0.08	61
40	25.78	29.58	32.78	35.11	36.39	1.18	0.32	0.37	0.33	0.07	60
41	25.82	29.62	32.81	35.13	36.40	1.18	0.31	0.37	0.33	0.07	59
42	25.86	29.65	32.84	35.15	36.41	1.17	0.30	0.37	0.33	0.07	58
43	25.90	29.69	32.87	35.17	36.41	1.17	0.29	0.37	0.33	0.07	57
44	25.94	29.72	32.89	35.19	36.42	1.17	0.28	0.37	0.33	0.07	56
45	25.98	29.76	32.92	35.20	36.42	1.17	0.27	0.37	0.32	0.06	55
46	26.02	29.79	32.95	35.22	36.43	1.16	0.26	0.37	0.32	0.06	54
47	26.06	29.83	32.97	35.24	36.43	1.16	0.25	0.38	0.32	0.06	53
48	26.10	29.86	33.00	35.26	36.44	1.16	0.24	0.38	0.32	0.06	52
49	26.14	29.90	33.03	35.28	36.44	1.15	0.23	0.38	0.32	0.06	51
50	26.18	29.93	33.05	35.29	36.45	1.15	0.23	0.38	0.31	0.05	50
Days.	**1500**	**1400**	**1300**	**1200**	**1100**	**300**	**200**	**500**	**400**	**300**	Days.

Note. — Arg. 12‴ = Arg. 12 — 20.00.

TABLES XCIII. - XCV.

Tables	XCIII.					XCIV.		XCV.			
Arguments	73.					12‴.		79.			
Days.	600	700	800	900	1000	0	100	0	100	200	Days.
50	26.18	29.93	33.05	35.29	36.45	1.15	0.23	0.38	0.31	0.05	50
51	26.22	29.97	33.08	35.31	36.45	1.14	0.22	0.38	0.31	0.05	49
52	26.26	30.00	33.11	35.33	36.46	1.14	0.21	0.38	0.31	0.05	48
53	26.30	30.04	33.14	35.34	36.46	1.13	0.20	0.38	0.31	0.05	47
54	26.34	30.07	33.16	35.36	36.47	1.13	0.19	0.38	0.30	0.05	46
55	26.38	30.11	33.19	35.38	36.48	1.12	0.19	0.38	0.30	0.04	45
56	26.42	30.14	33.22	35.39	36.48	1.11	0.18	0.39	0.30	0.04	44
57	26.46	30.18	33.24	35.41	36.49	1.11	0.17	0.39	0.29	0.04	43
58	26.50	30.21	33.27	35.42	36.49	1.10	0.16	0.39	0.29	0.04	42
59	26.54	30.25	33.30	35.44	36.50	1.10	0.15	0.39	0.29	0.04	41
60	26.57	30.28	33.32	35.45	36.51	1.09	0.15	0.39	0.28	0.03	40
61	26.61	30.32	33.35	35.47	36.51	1.09	0.14	0.39	0.28	0.03	39
62	26.65	30.35	33.38	35.49	36.52	1.08	0.13	0.39	0.28	0.03	38
63	26.69	30.39	33.40	35.50	36.52	1.08	0.13	0.39	0.28	0.03	37
64	26.73	30.42	33.43	35.52	36.53	1.07	0.12	0.39	0.28	0.03	36
65	26.77	30.46	33.46	35.54	36.53	1.06	0.12	0.39	0.27	0.03	35
66	26.81	30.49	33.48	35.55	36.54	1.06	0.11	0.40	0.27	0.03	34
67	26.85	30.52	33.51	35.57	36.54	1.05	0.11	0.40	0.27	0.02	33
68	26.89	30.56	33.53	35.58	36.54	1.04	0.10	0.40	0.27	0.02	32
69	26.93	30.59	33.56	35.60	36.55	1.03	0.10	0.40	0.27	0.02	31
70	26.96	30.62	33.58	35.61	36.55	1.02	0.09	0.40	0.26	0.02	30
71	27.00	30.66	33.61	35.63	36.55	1.02	0.09	0.40	0.26	0.02	29
72	27.04	30.69	33.63	35.64	36.56	1.01	0.08	0.40	0.26	0.02	28
73	27.08	30.72	33.66	35.66	36.56	1.00	0.08	0.40	0.26	0.02	27
74	27.12	30.76	33.68	35.67	36.56	0.99	0.07	0.40	0.25	0.02	26
75	27.16	30.79	33.71	35.69	36.56	0.98	0.07	0.40	0.25	0.01	25
76	27.20	30.82	33.73	35.70	36.57	0.98	0.06	0.40	0.25	0.01	24
77	27.24	30.86	33.76	35.71	36.57	0.97	0.06	0.40	0.24	0.01	23
78	27.28	30.89	33.78	35.73	36.57	0.96	0.05	0.40	0.24	0.01	22
79	27.32	30.92	33.81	35.74	36.57	0.95	0.05	0.40	0.24	0.01	21
80	27.35	30.95	33.83	35.75	36.58	0.94	0.04	0.40	0.23	0.01	20
81	27.39	30.99	33.86	35.77	36.58	0.94	0.04	0.40	0.23	0.01	19
82	27.43	31.02	33.88	35.78	36.58	0.93	0.04	0.40	0.23	0.01	18
83	27.47	31.05	33.91	35.80	36.58	0.92	0.03	0.40	0.23	0.01	17
84	27.51	31.08	33.93	35.81	36.58	0.91	0.03	0.40	0.22	0.01	16
85	27.55	31.11	33.96	35.83	36.59	0.90	0.03	0.40	0.22	0.01	15
86	27.59	31.15	33.98	35.84	36.59	0.89	0.02	0.40	0.22	0.00	14
87	27.63	31.18	34.00	35.85	36.59	0.88	0.02	0.40	0.21	0.00	13
88	27.67	31.21	34.03	35.87	36.59	0.87	0.02	0.40	0.21	0.00	12
89	27.71	31.24	34.05	35.88	36.59	0.86	0.02	0.40	0.21	0.00	11
90	27.74	31.27	34.07	35.89	36.59	0.85	0.01	0.40	0.20	0.00	10
91	27.78	31.31	34.10	35.91	36.59	0.84	0.01	0.40	0.20	0.00	09
92	27.82	31.34	34.12	35.92	36.59	0.83	0.01	0.40	0.20	0.00	08
93	27.86	31.37	34.14	35.93	36.60	0.82	0.01	0.40	0.20	0.00	07
94	27.90	31.40	34.16	35.94	36.60	0.81	0.01	0.40	0.19	0.00	06
95	27.94	31.43	34.18	35.95	36.60	0.80	0.00	0.40	0.19	0.00	05
96	27.97	31.47	34.21	35.97	36.60	0.79	0.00	0.39	0.19	0.00	04
97	28.01	31.50	34.23	35.98	36.60	0.78	0.00	0.39	0.18	0.00	03
98	28.05	31.53	34.25	35.99	36.60	0.77	0.00	0.39	0.18	0.00	02
99	28.09	31.56	34.27	36.00	36.60	0.76	0.00	0.39	0.18	0.00	01
100	28.12	31.59	34.29	36.01	36.60	0.74	0.00	0.39	0.17	0.00	00
Days.	1500	1400	1300	1200	1100	300	200	500	400	300	Days.

Note. — Arg. 12‴ = Arg. 12 − 20.00.

TABLES XCVI. - CI.

Tables	XCVI.		XCVII.		XCVIII.		XCIX.		C.	CI.		
Arguments	80.		81.		82.		83.		84.	48.		
Days.	0	100	0	Days.	Days.	0	0	10	0	0	10	Days.
Days 00	″ 0.16	″ 0.14	″ 0.93	Days. 100	Days. 0.0	″ 0.96	″ 0.00	″ 0.68	″ 0.02	″ 0.75	″ 0.75	Days. 10.0
01	0.16	0.14	0.90	99	0.1	0.89	0.00	0.69	0.02	0.76	0.75	9.9
02	0.16	0.14	0.87	98	0.2	0.82	0.00	0.70	0.01	0.77	0.74	9.8
03	0.16	0.14	0.83	97	0.3	0.76	0.00	0.71	0.01	0.78	0.73	9.7
04	0.16	0.14	0.80	96	0.4	0.70	0.00	0.72	0.00	0.79	0.72	9.6
05	0.17	0.13	0.77	95	0.5	0.64	0.00	0.73	0.00	0.79	0.71	9.5
06	0.17	0.13	0.73	94	0.6	0.58	0.00	0.74	0.00	0.80	0.70	9.4
07	0.17	0.13	0.70	93	0.7	0.53	0.00	0.75	0.01	0.81	0.69	9.3
08	0.17	0.13	0.66	92	0.8	0.48	0.00	0.76	0.02	0.82	0.68	9.2
09	0.17	0.13	0.63	91	0.9	0.43	0.00	0.77	0.03	0.83	0.67	9.1
10	0.18	0.12	0.59	90	1.0	0.38	0.00	0.79	0.04	0.83	0.66	9.0
11	0.18	0.12	0.56	89	1.1	0.34	0.00	0.80	0.06	0.84	0.65	8.9
12	0.18	0.12	0.52	88	1.2	0.30	0.00	0.81	0.08	0.85	0.63	8.8
13	0.18	0.12	0.49	87	1.3	0.26	0.01	0.82	0.10	0.86	0.62	8.7
14	0.18	0.12	0.45	86	1.4	0.22	0.01	0.83	0.12	0.86	0.61	8.6
15	0.18	0.11	0.42	85	1.5	0.18	0.01	0.84	0.14	0.87	0.60	8.5
16	0.18	0.11	0.39	84	1.6	0.15	0.01	0.85	0.17	0.88	0.59	8.4
17	0.18	0.11	0.36	83	1.7	0.12	0.01	0.87	0.20	0.89	0.58	8.3
18	0.19	0.11	0.33	82	1.8	0.09	0.01	0.88	0.23	0.89	0.57	8.2
19	0.19	0.11	0.30	81	1.9	0.07	0.02	0.89	0.26	0.90	0.56	8.1
20	0.19	0.10	0.27	80	2.0	0.05	0.02	0.90	0.29	0.90	0.55	8.0
21	0.19	0.10	0.24	79	2.1	0.03	0.02	0.91	0.32	0.91	0.54	7.9
22	0.19	0.10	0.22	78	2.2	0.02	0.02	0.92	0.35	0.92	0.53	7.8
23	0.19	0.10	0.19	77	2.3	0.01	0.03	0.93	0.39	0.92	0.52	7.7
24	0.19	0.10	0.17	76	2.4	0.00	0.03	0.94	0.43	0.93	0.51	7.6
25	0.20	0.09	0.15	75	2.5	0.00	0.03	0.95	0.47	0.93	0.50	7.5
26	0.20	0.09	0.13	74	2.6	0.00	0.04	0.96	0.51	0.94	0.49	7.4
27	0.20	0.09	0.11	73	2.7	0.01	0.04	0.97	0.55	0.94	0.48	7.3
28	0.20	0.09	0.09	72	2.8	0.01	0.05	0.98	0.59	0.95	0.47	7.2
29	0.20	0.09	0.08	71	2.9	0.02	0.05	0.99	0.63	0.95	0.46	7.1
30	0.20	0.08	0.06	70	3.0	0.03	0.06	1.00	0.67	0.96	0.45	7.0
31	0.20	0.08	0.05	69	3.1	0.05	0.06	1.01	0.71	0.96	0.44	6.9
32	0.20	0.08	0.04	68	3.2	0.07	0.07	1.02	0.75	0.96	0.43	6.8
33	0.20	0.08	0.03	67	3.3	0.09	0.07	1.03	0.79	0.97	0.42	6.7
34	0.20	0.08	0.02	66	3.4	0.11	0.08	1.04	0.83	0.97	0.41	6.6
35	0.20	0.07	0.01	65	3.5	0.14	0.08	1.05	0.86	0.97	0.40	6.5
36	0.20	0.07	0.01	64	3.6	0.17	0.09	1.06	0.90	0.98	0.39	6.4
37	0.20	0.07	0.00	63	3.7	0.20	0.09	1.07	0.93	0.98	0.38	6.3
38	0.20	0.07	0.00	62	3.8	0.24	0.10	1.08	0.96	0.98	0.37	6.2
39	0.20	0.07	0.00	61	3.9	0.28	0.10	1.09	0.99	0.98	0.36	6.1
40	0.20	0.06	0.00	60	4.0	0.32	0.11	1.10	1.02	0.99	0.35	6.0
41	0.20	0.06	0.01	59	4.1	0.37	0.11	1.11	1.05	0.99	0.34	5.9
42	0.20	0.06	0.02	58	4.2	0.42	0.12	1.12	1.08	0.99	0.33	5.8
43	0.20	0.06	0.03	57	4.3	0.47	0.12	1.13	1.10	0.99	0.32	5.7
44	0.20	0.06	0.04	56	4.4	0.52	0.13	1.14	1.12	0.99	0.31	5.6
45	0.20	0.06	0.05	55	4.5	0.57	0.14	1.15	1.14	1.00	0.30	5.5
46	0.20	0.06	0.06	54	4.6	0.63	0.14	1.16	1.16	1.00	0.29	5.4
47	0.20	0.05	0.07	53	4.7	0.69	0.15	1.17	1.17	1.00	0.28	5.3
48	0.20	0.05	0.09	52	4.8	0.75	0.16	1.18	1.18	1.00	0.27	5.2
49	0.20	0.05	0.11	51	4.9	0.81	0.17	1.19	1.19	1.00	0.26	5.1
50	0.20	0.05	0.13	50	5.0	0.87	0.18	1.19	1.19	1.00	0.25	5.0
Days.	300	200	100	Days.	Days.	10	30	20	10	30	20	Days.

TABLES XCVI. - CI.

Tables	XCVI.		XCVII.		XCVIII.		XCIX.		C.	CI.		
Arguments	80.		81.		82.		83.		84.	48.		
Days.	0	100	0	Days.	Days.	0	0	10	0	0	10	Days.
50	0.20	0.05	0.13	50	5.0	0.87	0.18	1.19	1.19	1.00	0.25	5.0
51	0.20	0.05	0.15	49	5.1	0.94	0.18	1.20	1.20	1.00	0.24	4.9
52	0.20	0.05	0.17	48	5.2	1.01	0.19	1.21	1.20	1.00	0.23	4.8
53	0.20	0.04	0.20	47	5.3	1.08	0.20	1.22	1.20	1.00	0.22	4.7
54	0.20	0.04	0.22	46	5.4	1.15	0.21	1.22	1.20	1.00	0.21	4.6
55	0.20	0.04	0.25	45	5.5	1.22	0.22	1.23	1.19	0.99	0.21	4.5
56	0.19	0.04	0.28	44	5.6	1.29	0.22	1.24	1.18	0.99	0.20	4.4
57	0.19	0.04	0.30	43	5.7	1.36	0.23	1.24	1.17	0.99	0.19	4.3
58	0.19	0.04	0.33	42	5.8	1.44	0.24	1.25	1.16	0.99	0.18	4.2
59	0.19	0.03	0.36	41	5.9	1.51	0.25	1.26	1.14	0.99	0.17	4.1
60	0.19	0.03	0.39	40	6.0	1.59	0.26	1.26	1.12	0.99	0.17	4.0
61	0.19	0.03	0.42	39	6.1	1.66	0.27	1.27	1.10	0.98	0.16	3.9
62	0.19	0.03	0.46	38	6.2	1.74	0.28	1.27	1.08	0.98	0.15	3.8
63	0.19	0.03	0.49	37	6.3	1.81	0.29	1.28	1.06	0.98	0.14	3.7
64	0.19	0.03	0.53	36	6.4	1.89	0.30	1.29	1.03	0.98	0.14	3.6
65	0.19	0.03	0.56	35	6.5	1.97	0.31	1.29	1.00	0.97	0.13	3.5
66	0.19	0.03	0.60	34	6.6	2.05	0.32	1.30	0.97	0.97	0.12	3.4
67	0.19	0.02	0.63	33	6.7	2.12	0.33	1.30	0.94	0.97	0.12	3.3
68	0.19	0.02	0.67	32	6.8	2.20	0.34	1.31	0.91	0.96	0.11	3.2
69	0.19	0.02	0.70	31	6.9	2.27	0.35	1.31	0.87	0.96	0.10	3.1
70	0.19	0.02	0.74	30	7.0	2.34	0.36	1.32	0.83	0.96	0.10	3.0
71	0.18	0.02	0.77	29	7.1	2.41	0.37	1.32	0.80	0.95	0.09	2.9
72	0.18	0.02	0.81	28	7.2	2.48	0.38	1.33	0.76	0.95	0.08	2.8
73	0.18	0.02	0.84	27	7.3	2.55	0.39	1.33	0.72	0.94	0.08	2.7
74	0.18	0.02	0.88	26	7.4	2.62	0.40	1.34	0.68	0.94	0.07	2.6
75	0.18	0.02	0.91	25	7.5	2.69	0.41	1.34	0.64	0.93	0.07	2.5
76	0.18	0.01	0.94	24	7.6	2.76	0.42	1.34	0.61	0.93	0.06	2.4
77	0.18	0.01	0.98	23	7.7	2.82	0.43	1.35	0.57	0.92	0.06	2.3
78	0.17	0.01	1.01	22	7.8	2.88	0.44	1.35	0.53	0.92	0.05	2.2
79	0.17	0.01	1.04	21	7.9	2.94	0.45	1.35	0.49	0.91	0.05	2.1
80	0.17	0.01	1.07	20	8.0	3.00	0.46	1.36	0.45	0.91	0.04	2.0
81	0.17	0.01	1.10	19	8.1	3.05	0.47	1.36	0.41	0.90	0.04	1.9
82	0.17	0.01	1.13	18	8.2	3.10	0.48	1.36	0.37	0.90	0.04	1.8
83	0.17	0.01	1.16	17	8.3	3.15	0.49	1.37	0.33	0.89	0.03	1.7
84	0.17	0.01	1.18	16	8.4	3.20	0.50	1.37	0.30	0.88	0.03	1.6
85	0.17	0.00	1.21	15	8.5	3.25	0.51	1.37	0.27	0.88	0.03	1.5
86	0.16	0.00	1.23	14	8.6	3.29	0.52	1.38	0.24	0.87	0.02	1.4
87	0.16	0.00	1.25	13	8.7	3.33	0.53	1.38	0.21	0.87	0.02	1.3
88	0.16	0.00	1.27	12	8.8	3.37	0.54	1.38	0.18	0.86	0.02	1.2
89	0.16	0.00	1.29	11	8.9	3.41	0.55	1.38	0.15	0.85	0.02	1.1
90	0.16	0.00	1.31	10	9.0	3.44	0.57	1.39	0.12	0.84	0.01	1.0
91	0.16	0.00	1.33	09	9.1	3.47	0.58	1.39	0.10	0.84	0.01	0.9
92	0.15	0.00	1.34	08	9.2	3.50	0.59	1.39	0.08	0.83	0.01	0.8
93	0.15	0.00	1.36	07	9.3	3.52	0.60	1.39	0.06	0.82	0.01	0.7
94	0.15	0.00	1.37	06	9.4	3.54	0.61	1.39	0.04	0.81	0.00	0.6
95	0.15	0.00	1.38	05	9.5	3.56	0.62	1.40	0.03	0.80	0.00	0.5
96	0.15	0.00	1.39	04	9.6	3.58	0.63	1.40	0.02	0.79	0.00	0.4
97	0.14	0.00	1.39	03	9.7	3.59	0.64	1.40	0.01	0.78	0.00	0.3
98	0.14	0.00	1.40	02	9.8	3.60	0.65	1.40	0.00	0.77	0.00	0.2
99	0.14	0.00	1.40	01	9.9	3.60	0.66	1.40	0.00	0.76	0.00	0.1
100	0.14	0.00	1.40	0	10.0	3.60	0.68	1.40	0.00	0.75	0.00	0.0
Days.	300	200	100	Days.	Days.	10	30	20	10	30	20	Days.

TABLES CII.-CVII.

Tables	CII.		CIII.		CIV.	CV.	CVI.	CVII.		
Arguments	85.		86.		87.	88.	89.	90.		
Days.	0	10	0	10	0	0	0	0	10	Days.
Days 0.0	0.64″	0.04″	0.01″	0.28″	0.37″	0.12″	0.40″	0.35″	0.27″	Days 10.0
0.1	0.63	0.05	0.01	0.28	0.36	0.11	0.40	0.35	0.27	9.9
0.2	0.62	0.05	0.00	0.29	0.35	0.10	0.40	0.35	0.27	9.8
0.3	0.61	0.06	0.00	0.29	0.33	0.09	0.40	0.35	0.26	9.7
0.4	0.60	0.07	0.00	0.30	0.32	0.08	0.40	0.35	0.26	9.6
0.5	0.59	0.07	0.00	0.30	0.30	0.08	0.40	0.36	0.26	9.5
0.6	0.58	0.08	0.00	0.31	0.29	0.07	0.40	0.36	0.26	9.4
0.7	0.57	0.09	0.00	0.31	0.27	0.06	0.39	0.36	0.25	9.3
0.8	0.56	0.09	0.00	0.32	0.26	0.06	0.39	0.36	0.25	9.2
0.9	0.55	0.10	0.00	0.32	0.24	0.05	0.39	0.36	0.25	9.1
1.0	0.54	0.11	0.00	0.33	0.22	0.05	0.38	0.37	0.24	9.0
1.1	0.53	0.12	0.00	0.33	0.21	0.04	0.38	0.37	0.24	8.9
1.2	0.52	0.13	0.00	0.34	0.19	0.03	0.37	0.37	0.24	8.8
1.3	0.51	0.13	0.00	0.35	0.17	0.03	0.37	0.37	0.23	8.7
1.4	0.50	0.14	0.00	0.35	0.15	0.02	0.36	0.37	0.23	8.6
1.5	0.49	0.15	0.00	0.36	0.13	0.02	0.35	0.38	0.23	8.5
1.6	0.48	0.16	0.00	0.36	0.11	0.02	0.34	0.38	0.22	8.4
1.7	0.47	0.17	0.00	0.37	0.10	0.02	0.33	0.38	0.22	8.3
1.8	0.46	0.17	0.01	0.37	0.08	0.01	0.32	0.38	0.21	8.2
1.9	0.45	0.18	0.01	0.38	0.07	0.01	0.31	0.38	0.21	8.1
2.0	0.44	0.19	0.01	0.38	0.06	0.01	0.29	0.39	0.20	8.0
2.1	0.43	0.20	0.01	0.38	0.05	0.01	0.28	0.39	0.20	7.9
2.2	0.42	0.21	0.01	0.39	0.04	0.00	0.27	0.39	0.20	7.8
2.3	0.41	0.22	0.01	0.39	0.03	0.00	0.26	0.39	0.19	7.7
2.4	0.40	0.23	0.01	0.40	0.02	0.00	0.25	0.39	0.19	7.6
2.5	0.38	0.24	0.02	0.40	0.01	0.00	0.23	0.39	0.18	7.5
2.6	0.37	0.25	0.02	0.40	0.01	0.00	0.22	0.39	0.18	7.4
2.7	0.36	0.26	0.02	0.41	0.00	0.00	0.21	0.40	0.18	7.3
2.8	0.35	0.27	0.02	0.41	0.00	0.00	0.20	0.40	0.17	7.2
2.9	0.34	0.28	0.02	0.41	0.00	0.00	0.18	0.40	0.17	7.1
3.0	0.33	0.29	0.02	0.42	0.00	0.00	0.17	0.40	0.16	7.0
3.1	0.32	0.30	0.02	0.42	0.00	0.00	0.16	0.40	0.16	6.9
3.2	0.31	0.31	0.02	0.43	0.01	0.00	0.15	0.40	0.16	6.8
3.3	0.30	0.32	0.02	0.43	0.01	0.00	0.14	0.40	0.15	6.7
3.4	0.29	0.33	0.03	0.43	0.02	0.01	0.12	0.40	0.15	6.6
3.5	0.28	0.34	0.03	0.44	0.03	0.01	0.11	0.40	0.14	6.5
3.6	0.27	0.35	0.03	0.44	0.04	0.01	0.10	0.40	0.14	6.4
3.7	0.26	0.36	0.03	0.45	0.05	0.02	0.09	0.40	0.14	6.3
3.8	0.25	0.37	0.03	0.45	0.06	0.02	0.08	0.40	0.13	6.2
3.9	0.24	0.38	0.03	0.45	0.07	0.02	0.07	0.40	0.13	6.1
4.0	0.23	0.39	0.04	0.46	0.09	0.03	0.06	0.40	0.12	6.0
4.1	0.22	0.40	0.04	0.46	0.10	0.03	0.05	0.40	0.12	5.9
4.2	0.21	0.41	0.04	0.47	0.12	0.04	0.04	0.40	0.12	5.8
4.3	0.20	0.42	0.04	0.47	0.13	0.04	0.03	0.40	0.11	5.7
4.4	0.19	0.43	0.05	0.47	0.15	0.05	0.02	0.40	0.11	5.6
4.5	0.18	0.44	0.05	0.48	0.17	0.06	0.02	0.40	0.11	5.5
4.6	0.17	0.45	0.05	0.48	0.18	0.06	0.01	0.40	0.10	5.4
4.7	0.17	0.46	0.06	0.49	0.20	0.07	0.01	0.40	0.10	5.3
4.8	0.16	0.47	0.06	0.49	0.22	0.07	0.00	0.40	0.10	5.2
4.9	0.15	0.48	0.06	0.49	0.24	0.08	0.00	0.40	0.10	5.1
5.0	0.14	0.49	0.07	0.50	0.26	0.09	0.00	0.40	0.09	5.0
Days.	30	20	30	20	10	10	10	30	20	Days.

TABLES CII.-CVII.

Tables	CII.		CIII.		CIV.	CV.	CVI.	CVII.		
Arguments	85.		86.		87.	88.	89.	90.		
Days.	0	10	0	10	0	0	0	0	10	Days.
Days	"	"	"	"	"	"	"	"	"	Days
5.0	0.14	0.49	0.07	0.50	0.26	0.09	0.00	0.40	0.09	5.0
5.1	0.13	0.50	0.07	0.50	0.28	0.09	0.00	0.40	0.09	4.9
5.2	0.12	0.51	0.07	0.51	0.29	0.10	0.00	0.40	0.09	4.8
5.3	0.12	0.52	0.07	0.51	0.31	0.11	0.00	0.39	0.08	4.7
5.4	0.11	0.53	0.08	0.51	0.32	0.12	0.00	0.39	0.08	4.6
5.5	0.10	0.54	0.08	0.52	0.33	0.13	0.00	0.39	0.08	4.5
5.6	0.10	0.55	0.08	0.52	0.34	0.13	0.01	0.39	0.07	4.4
5.7	0.09	0.56	0.09	0.53	0.35	0.14	0.01	0.39	0.07	4.3
5.8	0.08	0.57	0.09	0.53	0.36	0.15	0.02	0.38	0.07	4.2
5.9	0.08	0.58	0.09	0.53	0.37	0.16	0.02	0.38	0.07	4.1
6.0	0.07	0.59	0.10	0.54	0.38	0.17	0.03	0.38	0.06	4.0
6.1	0.07	0.60	0.10	0.54	0.38	0.17	0.03	0.38	0.06	3.9
6.2	0.06	0.61	0.10	0.54	0.39	0.18	0.04	0.38	0.06	3.8
6.3	0.06	0.62	0.11	0.54	0.39	0.19	0.05	0.38	0.05	3.7
6.4	0.05	0.63	0.11	0.54	0.40	0.20	0.06	0.38	0.05	3.6
6.5	0.05	0.64	0.11	0.55	0.40	0.21	0.07	0.37	0.05	3.5
6.6	0.04	0.65	0.12	0.55	0.40	0.22	0.08	0.37	0.04	3.4
6.7	0.04	0.65	0.12	0.55	0.40	0.23	0.09	0.37	0.04	3.3
6.8	0.03	0.66	0.13	0.55	0.39	0.24	0.10	0.37	0.04	3.2
6.9	0.03	0.67	0.13	0.55	0.39	0.25	0.11	0.37	0.04	3.1
7.0	0.02	0.68	0.14	0.56	0.38	0.26	0.13	0.36	0.03	3.0
7.1	0.02	0.69	0.14	0.56	0.37	0.27	0.14	0.36	0.03	2.9
7.2	0.02	0.69	0.14	0.56	0.36	0.28	0.15	0.36	0.03	2.8
7.3	0.01	0.70	0.15	0.56	0.35	0.29	0.16	0.36	0.03	2.7
7.4	0.01	0.70	0.15	0.56	0.34	0.30	0.17	0.36	0.03	2.6
7.5	0.01	0.71	0.16	0.57	0.32	0.30	0.19	0.35	0.03	2.5
7.6	0.01	0.72	0.16	0.57	0.31	0.31	0.20	0.35	0.02	2.4
7.7	0.00	0.72	0.16	0.57	0.29	0.32	0.21	0.35	0.02	2.3
7.8	0.00	0.73	0.17	0.57	0.28	0.32	0.22	0.35	0.02	2.2
7.9	0.00	0.73	0.17	0.57	0.26	0.33	0.23	0.35	0.02	2.1
8.0	0.00	0.74	0.18	0.58	0.24	0.33	0.25	0.34	0.02	2.0
8.1	0.00	0.74	0.18	0.58	0.23	0.34	0.26	0.34	0.01	1.9
8.2	0.00	0.75	0.19	0.58	0.21	0.35	0.27	0.34	0.01	1.8
8.3	0.00	0.75	0.19	0.58	0.19	0.35	0.28	0.34	0.01	1.7
8.4	0.00	0.76	0.20	0.58	0.17	0.36	0.30	0.33	0.01	1.6
8.5	0.00	0.76	0.20	0.59	0.15	0.36	0.31	0.33	0.01	1.5
8.6	0.00	0.77	0.21	0.59	0.13	0.36	0.32	0.33	0.01	1.4
8.7	0.01	0.77	0.21	0.59	0.11	0.37	0.33	0.32	0.00	1.3
8.8	0.01	0.77	0.22	0.59	0.10	0.37	0.34	0.32	0.00	1.2
8.9	0.01	0.78	0.22	0.59	0.08	0.38	0.35	0.32	0.00	1.1
9.0	0.01	0.78	0.23	0.60	0.07	0.38	0.36	0.31	0.00	1.0
9.1	0.01	0.78	0.23	0.60	0.06	0.38	0.37	0.31	0.00	0.9
9.2	0.02	0.79	0.24	0.60	0.05	0.39	0.38	0.31	0.00	0.8
9.3	0.02	0.79	0.24	0.60	0.04	0.39	0.38	0.30	0.00	0.7
9.4	0.02	0.79	0.25	0.60	0.03	0.39	0.39	0.30	0.00	0.6
9.5	0.03	0.79	0.25	0.60	0.02	0.40	0.39	0.30	0.00	0.5
9.6	0.03	0.79	0.26	0.60	0.01	0.40	0.39	0.29	0.00	0.4
9.7	0.03	0.80	0.26	0.60	0.01	0.40	0.40	0.29	0.00	0.3
9.8	0.04	0.80	0.27	0.60	0.00	0.40	0.40	0.28	0.00	0.2
9.9	0.04	0.80	0.27	0.60	0.00	0.40	0.40	0.28	0.00	0.1
10.0	0.04	0.80	0.28	0.60	0.00	0.40	0.40	0.27	0.00	0.0
Days.	30	20	30	20	10	10	10	30	20	Days.

TABLE CVIII. ARGUMENT 77.

Equation = 6″.2 — 6″.18 sin. 3g.

g	210° / 90 / 330	211° / 91 / 331	212° / 92 / 332	213° / 93 / 333	214° / 94 / 334	215° / 95 / 335	216° / 96 / 336	217° / 97 / 337	218° / 98 / 338	219° / 99 / 339	g
0	12.38	12.37	12.35	12.30	12.24	12.17	12.08	11.97	11.85	11.71	3600
100	12.38	12.37	12.35	12.30	12.24	12.17	12.08	11.97	11.85	11.71	3500
200	12.38	12.37	12.35	12.30	12.24	12.17	12.08	11.97	11.85	11.71	3400
300	12.38	12.37	12.35	12.30	12.24	12.17	12.07	11.96	11.84	11.70	3300
400	12.38	12.37	12.35	12.30	12.24	12.16	12.07	11.96	11.84	11.70	3200
500	12.38	12.37	12.34	12.30	12.24	12.16	12.07	11.96	11.83	11.69	3100
600	12.38	12.37	12.34	12.30	12.23	12.16	12.06	11.95	11.83	11.69	3000
700	12.38	12.37	12.34	12.30	12.23	12.16	12.06	11.95	11.83	11.69	2900
800	12.38	12.37	12.34	12.29	12.23	12.15	12.06	11.95	11.82	11.68	2800
900	12.38	12.37	12.34	12.29	12.23	12.15	12.05	11.94	11.82	11.68	2700
1000	12.38	12.37	12.34	12.29	12.23	12.15	12.05	11.94	11.81	11.67	2600
1100	12.38	12.36	12.34	12.29	12.23	12.15	12.05	11.94	11.81	11.67	2500
1200	12.38	12.36	12.33	12.29	12.22	12.14	12.04	11.93	11.80	11.66	2400
1300	12.38	12.36	12.33	12.29	12.22	12.14	12.04	11.93	11.80	11.66	2300
1400	12.38	12.36	12.33	12.29	12.22	12.14	12.04	11.93	11.80	11.66	2200
1500	12.38	12.36	12.33	12.28	12.22	12.14	12.04	11.92	11.79	11.65	2100
1600	12.38	12.36	12.33	12.28	12.22	12.13	12.03	11.92	11.79	11.65	2000
1700	12.38	12.36	12.33	12.28	12.21	12.13	12.03	11.92	11.78	11.64	1900
1800	12.38	12.36	12.33	12.28	12.21	12.13	12.03	11.91	11.78	11.64	1800
1900	12.38	12.36	12.33	12.28	12.21	12.13	12.03	11.91	11.77	11.64	1700
2000	12.38	12.36	12.33	12.28	12.21	12.12	12.02	11.91	11.77	11.63	1600
2100	12.38	12.36	12.32	12.28	12.21	12.12	12.02	11.90	11.77	11.63	1500
2200	12.38	12.36	12.32	12.27	12.20	12.12	12.02	11.90	11.76	11.62	1400
2300	12.38	12.36	12.32	12.27	12.20	12.12	12.02	11.90	11.76	11.62	1300
2400	12.38	12.36	12.32	12.27	12.20	12.11	12.01	11.89	11.75	11.61	1200
2500	12.38	12.36	12.32	12.27	12.20	12.11	12.01	11.89	11.75	11.61	1100
2600	12.38	12.36	12.32	12.27	12.20	12.11	12.01	11.89	11.75	11.60	1000
2700	12.38	12.36	12.32	12.26	12.19	12.11	12.00	11.88	11.74	11.60	900
2800	12.38	12.36	12.32	12.26	12.19	12.10	12.00	11.88	11.74	11.59	800
2900	12.38	12.36	12.32	12.26	12.19	12.10	12.00	11.88	11.74	11.59	700
3000	12.37	12.36	12.31	12.26	12.19	12.10	11.99	11.87	11.73	11.58	600
3100	12.37	12.35	12.31	12.25	12.18	12.10	11.99	11.87	11.73	11.58	500
3200	12.37	12.35	12.31	12.25	12.18	12.09	11.99	11.87	11.73	11.57	400
3300	12.37	12.35	12.31	12.25	12.18	12.09	11.98	11.86	11.72	11.57	300
3400	12.37	12.35	12.31	12.25	12.18	12.09	11.98	11.86	11.72	11.56	200
3500	12.37	12.35	12.31	12.25	12.17	12.09	11.98	11.86	11.72	11.56	100
3600	12.37	12.35	12.30	12.24	12.17	12.08	11.97	11.85	11.71	11.55	0
g	89° / 209 / 329	88° / 208 / 328	87° / 207 / 327	86° / 206 / 326	85° / 205 / 325	84° / 204 / 324	83° / 203 / 323	82° / 202 / 322	81° / 201 / 321	80° / 200 / 320	g

TABLE CVIII. ARGUMENT 77.

Equation = 6″.2 — 6″.18 sin. 3ϑ.

ϑ	220° 100 340	221° 101 341	222° 102 342	223° 103 343	224° 104 344	225° 105 345	226° 106 346	227° 107 347	228° 108 348	229° 109 349	ϑ
0″	11.55″	11.38″	11.20″	11.00″	10.79″	10.57″	10.33″	10.09″	9.83″	9.57″	3600″
100	11.55	11.38	11.20	11.00	10.79	10.57	10.33	10.09	9.83	9.57	3500
200	11.55	11.37	11.19	10.99	10.78	10.56	10.32	10.08	9.82	9.56	3400
300	11.54	11.37	11.19	10.99	10.78	10.55	10.31	10.07	9.81	9.55	3300
400	11.53	11.36	11.18	10.98	10.77	10.55	10.31	10.06	9.80	9.54	3200
500	11.53	11.36	11.18	10.98	10.77	10.54	10.30	10.06	9.80	9.53	3100
600	11.52	11.35	11.17	10.97	10.76	10.53	10.29	10.05	9.79	9.53	3000
700	11.52	11.35	11.17	10.96	10.75	10.53	10.29	10.04	9.78	9.52	2900
800	11.51	11.34	11.16	10.96	10.75	10.52	10.28	10.03	9.78	9.51	2800
900	11.51	11.34	11.16	10.95	10.74	10.51	10.27	10.03	9.77	9.50	2700
1000	11.50	11.33	11.15	10.95	10.74	10.51	10.27	10.02	9.76	9.49	2600
1100	11.50	11.33	11.15	10.94	10.73	10.50	10.26	10.01	9.75	9.48	2500
1200	11.49	11.32	11.14	10.93	10.72	10.49	10.25	10.00	9.74	9.47	2400
1300	11.49	11.32	11.14	10.93	10.72	10.49	10.25	10.00	9.74	9.47	2300
1400	11.49	11.31	11.13	10.92	10.71	10.48	10.24	9.99	9.73	9.46	2200
1500	11.48	11.31	11.13	10.92	10.71	10.47	10.23	9.98	9.72	9.45	2100
1600	11.48	11.30	11.12	10.91	10.70	10.47	10.23	9.98	9.72	9.44	2000
1700	11.47	11.30	11.12	10.91	10.70	10.46	10.22	9.97	9.71	9.44	1900
1800	11.47	11.29	11.11	10.90	10.69	10.45	10.21	9.96	9.70	9.43	1800
1900	11.46	11.29	11.10	10.89	10.68	10.45	10.21	9.96	9.70	9.42	1700
2000	11.46	11.28	11.10	10.89	10.68	10.44	10.20	9.95	9.69	9.41	1600
2100	11.46	11.28	11.09	10.88	10.67	10.43	10.19	9.94	9.68	9.40	1500
2200	11.45	11.27	11.09	10.88	10.67	10.43	10.19	9.94	9.68	9.40	1400
2300	11.45	11.27	11.08	10.87	10.66	10.42	10.18	9.93	9.67	9.39	1300
2400	11.44	11.26	11.07	10.86	10.65	10.41	10.17	9.92	9.66	9.38	1200
2500	11.44	11.26	11.07	10.86	10.65	10.41	10.17	9.92	9.66	9.38	1100
2600	11.43	11.25	11.06	10.85	10.64	10.40	10.16	9.91	9.65	9.37	1000
2700	11.43	11.25	11.06	10.85	10.63	10.39	10.15	9.90	9.64	9.36	900
2800	11.42	11.24	11.05	10.84	10.63	10.39	10.15	9.89	9.63	9.35	800
2900	11.42	11.24	11.04	10.84	10.62	10.38	10.14	9.89	9.63	9.34	700
3000	11.41	11.23	11.04	10.83	10.61	10.37	10.13	9.88	9.62	9.34	600
3100	11.41	11.23	11.03	10.82	10.61	10.37	10.13	9.87	9.61	9.33	500
3200	11.40	11.22	11.03	10.82	10.60	10.36	10.12	9.86	9.60	9.32	400
3300	11.40	11.22	11.02	10.81	10.59	10.35	10.11	9.86	9.60	9.31	300
3400	11.39	11.21	11.02	10.81	10.59	10.35	10.11	9.85	9.59	9.31	200
3500	11.39	11.21	11.01	10.80	10.58	10.34	10.10	9.84	9.58	9.30	100
3600	11.38	11.20	11.00	10.79	10.57	10.33	10.09	9.83	9.57	9.29	0
ϑ	79° 199 319	78° 198 318	77° 197 317	76° 196 316	75° 195 315	74° 194 314	73° 193 313	72° 192 312	71° 191 311	70° 190 310	ϑ

TABLE CVIII. ARGUMENT 77.

Equation = 6″.2 — 6″.16 sin. $3g$.

g	230° 110 350	231° 111 351	232° 112 352	233° 113 353	234° 114 354	235° 115 355	236° 116 356	237° 117 357	238° 118 358	239° 119 359	g
0″	9″.29	9″.01	8″.71	8″.41	8″.11	7″.80	7″.48	7″.17	6″.85	6″.52	3600
100	9.29	9.01	8.71	8.41	8.11	7.80	7.48	7.17	6.85	6.52	3500
200	9.28	9.00	8.70	8.40	8.10	7.79	7.47	7.16	6.84	6.51	3400
300	9.27	8.99	8.69	8.39	8.09	7.78	7.46	7.15	6.83	6.50	3300
400	9.26	8.98	8.68	8.38	8.08	7.77	7.45	7.14	6.82	6.49	3200
500	9.26	8.97	8.67	8.37	8.07	7.76	7.44	7.13	6.81	6.48	3100
600	9.25	8.96	8.66	8.36	8.06	7.75	7.43	7.12	6.80	6.47	3000
700	9.24	8.96	8.66	8.36	8.06	7.74	7.42	7.11	6.79	6.47	2900
800	9.23	8.95	8.65	8.35	8.05	7.73	7.42	7.10	6.78	6.46	2800
900	9.23	8.94	8.64	8.34	8.04	7.72	7.41	7.09	6.77	6.45	2700
1000	9.22	8.93	8.63	8.33	8.03	7.71	7.40	7.08	6.76	6.44	2600
1100	9.21	8.92	8.62	8.32	8.02	7.70	7.39	7.07	6.75	6.43	2500
1200	9.20	8.91	8.61	8.31	8.01	7.69	7.38	7.06	6.74	6.42	2400
1300	9.20	8.91	8.61	8.31	8.01	7.69	7.37	7.06	6.74	6.42	2300
1400	9.19	8.90	8.60	8.30	8.00	7.68	7.36	7.05	6.73	6.41	2200
1500	9.18	8.89	8.59	8.29	7.99	7.67	7.35	7.04	6.72	6.40	2100
1600	9.17	8.88	8.58	8.28	7.98	7.66	7.35	7.03	6.71	6.39	2000
1700	9.16	8.87	8.57	8.27	7.97	7.65	7.34	7.02	6.70	6.38	1900
1800	9.15	8.86	8.56	8.26	7.96	7.64	7.33	7.01	6.69	6.37	1800
1900	9.15	8.86	8.56	8.26	7.95	7.63	7.32	7.00	6.68	6.36	1700
2000	9.14	8.85	8.55	8.25	7.94	7.63	7.31	6.99	6.67	6.35	1600
2100	9.13	8.84	8.54	8.24	7.93	7.62	7.30	6.98	6.66	6.34	1500
2200	9.12	8.83	8.53	8.23	7.92	7.61	7.29	6.97	6.65	6.33	1400
2300	9.11	8.82	8.52	8.22	7.91	7.60	7.28	6.96	6.64	6.32	1300
2400	9.10	8.81	8.51	8.21	7.90	7.59	7.27	6.95	6.63	6.31	1200
2500	9.10	8.81	8.51	8.21	7.90	7.58	7.27	6.95	6.63	6.31	1100
2600	9.09	8.80	8.50	8.20	7.89	7.57	7.26	6.94	6.62	6.30	1000
2700	9.08	8.79	8.49	8.19	7.88	7.57	7.25	6.93	6.61	6.29	900
2800	9.07	8.78	8.48	8.18	7.87	7.56	7.24	6.92	6.60	6.28	800
2900	9.07	8.77	8.47	8.17	7.86	7.55	7.23	6.91	6.59	6.27	700
3000	9.06	8.76	8.46	8.16	7.85	7.54	7.22	6.90	6.58	6.26	600
3100	9.05	8.76	8.46	8.16	7.85	7.53	7.22	6.90	6.57	6.25	500
3200	9.04	8.75	8.45	8.15	7.84	7.52	7.21	6.89	6.56	6.24	400
3300	9.04	8.74	8.44	8.14	7.83	7.51	7.20	6.88	6.55	6.23	300
3400	9.03	8.73	8.43	8.13	7.82	7.50	7.19	6.87	6.54	6.22	200
3500	9.02	8.72	8.42	8.12	7.81	7.49	7.18	6.86	6.53	6.21	100
3600	9.01	8.71	8.41	8.11	7.80	7.48	7.17	6.85	6.52	6.20	0
g	69° 189 309	68° 188 308	67° 187 307	66° 186 306	65° 185 305	64° 184 304	63° 183 303	62° 182 302	61° 181 301	60° 180 300	g

TABLE CVIII. ARGUMENT 77.

Equation = 6″.2 — 6″.18 sin $3g$.

g	240 120 0	241 121 1	242 122 2	243 123 3	244 124 4	245 125 5	246 126 6	247 127 7	248 128 8	249 129 9	g
0″	6.20″	5.88″	5.55″	5.23″	4.92″	4.60″	4.29″	3.99″	3.69″	3.39″	3600″
100	6.19	5.87	5.54	5.22	4.91	4.59	4.28	3.98	3.68	3.38	3500
200	6.18	5.86	5.53	5.21	4.90	4.58	4.27	3.97	3.67	3.37	3400
300	6.17	5.85	5.52	5.20	4.89	4.57	4.26	3.96	3.66	3.36	3300
400	6.16	5.84	5.51	5.19	4.88	4.56	4.25	3.95	3.65	3.36	3200
500	6.15	5.83	5.50	5.18	4.87	4.55	4.24	3.94	3.64	3.35	3100
600	6.14	5.82	5.50	5.18	4.86	4.55	4.24	3.94	3.64	3.34	3000
700	6.13	5.81	5.49	5.17	4.85	4.54	4.23	3.93	3.63	3.33	2900
800	6.12	5.80	5.48	5.16	4.84	4.53	4.22	3.92	3.62	3.33	2800
900	6.11	5.79	5.47	5.15	4.83	4.52	4.21	3.91	3.61	3.32	2700
1000	6.10	5.78	5.46	5.14	4.82	4.51	4.20	3.90	3.60	3.31	2600
1100	6.09	5.77	5.45	5.13	4.81	4.50	4.19	3.89	3.59	3.30	2500
1200	6.09	5.77	5.15	5.13	4.81	4.50	4.19	3.89	3.59	3.30	2400
1300	6.08	5.76	5.44	5.12	4.80	4.49	4.18	3.88	3.58	3.29	2300
1400	6.07	5.75	5.43	5.11	4.79	4.48	4.17	3.87	3.57	3.28	2200
1500	6.06	5.74	5.42	5.10	4.78	4.47	4.16	3.86	3.56	3.27	2100
1600	6.05	5.73	5.41	5.09	4.77	4.46	4.15	3.85	3.55	3.26	2000
1700	6.04	5.72	5.40	5.08	4.76	4.45	4.14	3.84	3.54	3.25	1900
1800	6.03	5.71	5.39	5.07	4.76	4.44	4.14	3.84	3.54	3.25	1800
1900	6.02	5.70	5.38	5.06	4.75	4.43	4.13	3.83	3.53	3.24	1700
2000	6.01	5.69	5.37	5.05	4.74	4.42	4.12	3.82	3.52	3.23	1600
2100	6.00	5.68	5.36	5.04	4.73	4.41	4.11	3.81	3.51	3.22	1500
2200	5.99	5.67	5.35	5.03	4.72	4.40	4.10	3.80	3.50	3.21	1400
2300	5.98	5.66	5.34	5.02	4.71	4.39	4.09	3.79	3.49	3.20	1300
2400	5.98	5.66	5.34	5.02	4.71	4.39	4.09	3.79	3.49	3.20	1200
2500	5.97	5.65	5.33	5.01	4.70	4.38	4.08	3.78	3.48	3.19	1100
2600	5.96	5.64	5.32	5.00	4.69	4.37	4.07	3.77	3.47	3.18	1000
2700	5.95	5.63	5.31	4.99	4.68	4.36	4.06	3.76	3.46	3.17	900
2800	5.94	5.62	5.30	4.98	4.67	4.35	4.05	3.75	3.45	3.17	800
2900	5.93	5.61	5.29	4.97	4.66	4.34	4.04	3.74	3.44	3.16	700
3000	5.93	5.60	5.28	4.97	4.65	4.34	4.04	3.74	3.44	3.15	600
3100	5.92	5.59	5.27	4.96	4.64	4.33	4.03	3.73	3.43	3.14	500
3200	5.91	5.58	5.26	4.95	4.63	4.32	4.02	3.72	3.42	3.14	400
3300	5.90	5.57	5.25	4.94	4.62	4.31	4.01	3.71	3.41	3.13	300
3400	5.89	5.56	5.24	4.93	4.61	4.30	4.00	3.70	3.40	3.12	200
3500	5.88	5.55	5.23	4.92	4.60	4.29	3.99	3.69	3.39	3.11	100
3600	5.88	5.55	5.23	4.92	4.60	4.29	3.99	3.69	3.39	3.11	0
g	59 179 299	58 178 298	57 177 297	56 176 296	55 175 295	54 174 294	53 173 293	52 172 292	51 171 291	50 170 290	g

TABLE CVIII. ARGUMENT 77.

Equation $= 6''.2 - 6''.18 \sin 3\bar{y}$.

$\bar{y}$	250°	251°	252°	253°	254°	255°	256°	257°	258°	259°	$\bar{y}$
	130	131	132	133	134	135	136	137	138	139	
	10	11	12	13	14	15	16	17	18	19	
0″	3.11″	2.83″	2.57″	2.31″	2.07″	1.83″	1.61″	1.40″	1.20″	1.02″	3600″
100	3.10	2.82	2.56	2.30	2.06	1.82	1.60	1.39	1.20	1.02	3500
200	3.09	2.81	2.55	2.29	2.05	1.81	1.60	1.39	1.19	1.01	3400
300	3.08	2.80	2 51	2.29	2.05	1.81	1.59	1.38	1.19	1.01	3300
400	3.08	2.80	2.54	2.28	2.04	1.80	1.59	1.38	1.18	1.00	3200
500	3.07	2.79	2.53	2.27	2.03	1.79	1.58	1.37	1.18	1.00	3100
600	3.06	2.78	2.52	2.27	2.03	1.79	1.58	1.37	1.17	0.99	3000
700	3.05	2.77	2.51	2.26	2.02	1.78	1.57	1.36	1.17	0.99	2900
800	3.05	2.77	2.51	2.25	2.01	1.77	1.56	1.35	1.16	0.98	2800
900	3.04	2.76	2 50	2.25	2.01	1.77	1.56	1.35	1.16	0.98	2700
1000	3.03	2.75	2.49	2.24	2.00	1.76	1.55	1.34	1.15	0.97	2600
1100	3.02	2.74	2.48	2.23	1.99	1.75	1.55	1.34	1.15	0.97	2500
1200	3.02	2.74	2.48	2.23	1.99	1.75	1.54	1.33	1.14	0.96	2400
1300	3.01	2.73	2.47	2.22	1.98	1.74	1.53	1.32	1.14	0.96	2300
1400	3.00	2.72	2.46	2.21	1.97	1.73	1.53	1.32	1.13	0.95	2200
1500	2.99	2.72	2.46	2.21	1.97	1.73	1.52	1.31	1.13	0.95	2100
1600	2.99	2.71	2.45	2.20	1.96	1.72	1.52	1.31	1.12	0.94	2000
1700	2.98	2.70	2.44	2.19	1.95	1.72	1.51	1.30	1.12	0.94	1900
1800	2.97	2.70	2.44	2.19	1.95	1.71	1.51	1.30	1.11	0.93	1800
1900	2.96	2.69	2.43	2.18	1.94	1.71	1.50	1.29	1.11	0.93	1700
2000	2.96	2.68	2.42	2.17	1.93	1.70	1.49	1.28	1.10	0.93	1600
2100	2.95	2.68	2.42	2.17	1.93	1.70	1.49	1.28	1.10	0.92	1500
2200	2.94	2.67	2.41	2.16	1.92	1.69	1.48	1.27	1.09	0.92	1400
2300	2.93	2.66	2.40	2.15	1.91	1.69	1.48	1.27	1.09	0.91	1300
2400	2.93	2.66	2.40	2.15	1.91	1.68	1.47	1.26	1.08	0.91	1200
2500	2.92	2.65	2.39	2.14	1.90	1.67	1.46	1.26	1.08	0.91	1100
2600	2.91	2.64	2.38	2.13	1.89	1.67	1.46	1.25	1.07	0.90	1000
2700	2.90	2.63	2.37	2.13	1.89	1.66	1.45	1.25	1.07	0.90	900
2800	2.89	2.63	2.37	2.12	1.88	1.66	1.45	1.24	1.06	0.89	800
2900	2.88	2.62	2.36	2.11	1.87	1.65	1.44	1.24	1.06	0.89	700
3000	2.88	2.61	2.35	2.11	1.87	1.64	1.43	1.23	1.05	0.88	600
3100	2.87	2.60	2.34	2.10	1.86	1.64	1.43	1.23	1.05	0.88	500
3200	2.86	2.60	2.34	2.09	1.85	1.63	1.42	1.22	1.04	0.87	400
3300	2.85	2.59	2.33	2.09	1.85	1.63	1.42	1.22	1.04	0.87	300
3400	2.84	2.58	2.32	2.08	1.84	1.62	1.41	1.21	1.03	0.86	200
3500	2.83	2.57	2.31	2.07	1.83	1.62	1.41	1.21	1.03	0.86	100
3600	2.83	2.57	2.31	2.07	1.83	1.61	1.40	1.20	1.02	0.85	0
	49°	48°	47°	46°	45°	44°	43°	42°	41°	40°	
$\bar{y}$	169	168	167	166	165	164	163	162	161	160	$\bar{y}$
	289	288	287	286	285	284	283	282	281	280	

TABLE CVIII. ARGUMENT 77.

Equation = 6″.2 — 6″.16 sin 3ϑ.

ϑ	260°	261°	262°	263°	264°	265°	266°	267°	268°	269°	ϑ
	140	141	142	143	144	145	146	147	148	149	
	20	21	22	23	24	25	26	27	28	29	
0″	0″.85	0″.69	0″.55	0″.43	0″.32	0″.23	0″.16	0″.10	0″.05	0″.03	3600″
100	0.85	0.69	0.55	0.43	0.32	0.23	0.16	0.10	0.05	0.03	3500
200	0.84	0.68	0.55	0.43	0.32	0.23	0.16	0.10	0.05	0.03	3400
300	0.84	0.68	0.54	0.42	0.32	0.23	0.16	0.10	0.05	0.03	3300
400	0.83	0.68	0.54	0.42	0.31	0.22	0.15	0.09	0.05	0.03	3200
500	0.83	0.67	0.54	0.42	0.31	0.22	0.15	0.09	0.05	0.03	3100
600	0.82	0.67	0.53	0.41	0.31	0.22	0.15	0.09	0.05	0.03	3000
700	0.82	0.67	0.53	0.41	0.31	0.22	0.15	0.09	0.05	0.03	2900
800	0.81	0.66	0.53	0.41	0.30	0.21	0.14	0.09	0.05	0.03	2800
900	0.81	0.66	0.52	0.40	0.30	0.21	0.14	0.09	0.05	0.02	2700
1000	0.80	0.66	0.52	0.40	0.30	0.21	0.14	0.09	0.01	0.02	2600
1100	0.80	0.65	0.52	0.40	0.30	0.21	0.14	0.09	0.04	0.02	2500
1200	0.79	0.65	0.51	0.39	0.29	0.20	0.13	0.08	0.04	0.02	2400
1300	0.79	0.65	0.51	0.39	0.29	0.20	0.13	0.08	0.04	0.02	2300
1400	0.78	0.64	0.51	0.39	0.29	0.20	0.13	0.08	0.04	0.02	2200
1500	0.78	0.64	0.50	0.39	0.29	0.20	0.13	0.08	0.04	0.02	2100
1600	0.77	0.63	0.50	0.38	0.28	0.20	0.13	0.08	0.01	0.02	2000
1700	0.77	0.63	0.50	0.38	0.28	0.20	0.13	0.08	0.01	0.02	1900
1800	0.77	0.63	0.49	0.38	0.28	0.19	0.12	0.08	0.01	0.02	1800
1900	0.76	0.62	0.49	0.38	0.28	0.19	0.12	0.08	0.04	0.02	1700
2000	0.76	0.62	0.49	0.37	0.27	0.19	0.12	0.07	0.04	0.02	1600
2100	0.75	0.61	0.48	0.37	0.27	0.19	0.12	0.07	0.01	0.02	1500
2200	0.75	0.61	0.48	0.37	0.27	0.19	0.12	0.07	0.01	0.02	1400
2300	0.74	0.61	0.48	0.37	0.27	0.19	0.12	0.07	0.04	0.02	1300
2400	0.74	0.60	0.47	0.36	0.26	0.18	0.11	0.07	0.04	0.02	1200
2500	0.73	0.60	0.47	0.36	0.26	0.18	0.11	0.07	0.04	0.02	1100
2600	0.73	0.59	0.47	0.36	0.26	0.18	0.11	0.07	0.04	0.02	1000
2700	0.73	0.59	0.46	0.35	0.26	0.18	0.11	0.07	0.01	0.02	900
2800	0.72	0.59	0.46	0.35	0.25	0.18	0.11	0.06	0.03	0.02	800
2900	0.72	0.58	0.46	0.35	0.25	0.18	0.11	0.06	0.03	0.02	700
3000	0.71	0.58	0.45	0.34	0.25	0.17	0.11	0.06	0.03	0.02	600
3100	0.71	0.57	0.45	0.34	0.25	0.17	0.11	0.06	0.03	0.02	500
3200	0.70	0.57	0.45	0.34	0.24	0.17	0.11	0.06	0.03	0.02	400
3300	0.70	0.56	0.44	0.33	0.24	0.17	0.10	0.06	0.03	0.02	300
3400	0.70	0.56	0.44	0.33	0.24	0.17	0.10	0.06	0.03	0.02	200
3500	0.69	0.56	0.44	0.33	0.24	0.17	0.10	0.06	0.03	0.02	100
3600	0.69	0.55	0.43	0.32	0.23	0.16	0.10	0.05	0.03	0.02	0
	39°	38°	37°	36°	35°	34°	33°	32°	31°	30°	
ϑ	159	158	157	156	155	154	153	152	151	150	ϑ
	279	278	277	276	275	274	273	272	271	270	

TABLE CIX.

Equation $= 9''.1 + 2''.17 \cos. \bar{u} - 8''.80 \sin. \bar{u}$.

	0°	10°	20°	30°	40°	50°	60°	70°	80°
0° 0″	11″.27	9″.71	8″.13	6″.58	5″.10	3″.75	2″.56	1″.57	0″.81
0 1000	11.23	9.67	8.09	6.54	5.06	3.71	2.53	1.54	0.79
0 2000	11.19	9.62	8.05	6.50	5.02	3.68	2.50	1.52	0.77
0 3000	11.14	9.58	8.00	6.45	4.98	3.61	2.47	1.50	0.75
1 400	11.10	9.54	7.96	6.41	4.94	3.61	2.44	1.47	0.73
1 1400	11.06	9.49	7.91	6.37	4.90	3.57	2.42	1.45	0.72
1 2400	11.02	9.45	7.87	6.33	4.86	3.54	2.39	1.43	0.70
1 3400	10.97	9.41	7.82	6.28	4.82	3.51	2.36	1.40	0.68
2 800	10.93	9.36	7.78	6.24	4.78	3.47	2.33	1.38	0.66
2 1800	10.89	9.32	7.73	6.20	4.75	3.44	2.30	1.36	0.65
2 2800	10.85	9.28	7.69	6.16	4.71	3.40	2.27	1.33	0.63
3 200	10.80	9.23	7.65	6.12	4.67	3.37	2.24	1.31	0.62
3 1200	10.76	9.19	7.61	6.07	4.63	3.34	2.21	1.29	0.60
3 2200	10.72	9.15	7.56	6.03	4.59	3.30	2.18	1.27	0.59
3 3200	10.67	9.10	7.52	5.99	4.56	3.27	2.15	1.24	0.57
4 600	10.63	9.06	7.48	5.95	4.52	3.24	2.13	1.22	0.56
4 1600	10.58	9.01	7.44	5.91	4.48	3.20	2.10	1.20	0.55
4 2600	10.54	8.97	7.39	5.87	4.44	3.17	2.07	1.18	0.53
5 0	10.49	8.92	7.35	5.83	4.41	3.14	2.04	1.16	0.52
5 1000	10.45	8.88	7.31	5.78	4.37	3.10	2.01	1.14	0.50
5 2000	10.40	8.84	7.27	5.74	4.33	3.07	1.98	1.12	0.49
5 3000	10.36	8.79	7.22	5.70	4.30	3.04	1.96	1.10	0.48
6 400	10.32	8.75	7.18	5.66	4.26	3.00	1.93	1.08	0.46
6 1400	10.27	8.70	7.14	5.62	4.22	2.97	1.90	1.06	0.45
6 2400	10.23	8.66	7.09	5.58	4.19	2.94	1.88	1.04	0.44
6 3400	10.18	8.61	7.05	5.54	4.15	2.90	1.85	1.02	0.42
7 800	10.14	8.57	7.01	5.50	4.11	2.87	1.82	1.00	0.41
7 1800	10.10	8.52	6.96	5.46	4.08	2.84	1.80	0.98	0.40
7 2800	10.06	8.48	6.92	5.42	4.01	2.81	1.77	0.96	0.39
8 200	10.01	8.44	6.88	5.38	4.00	2.77	1.74	0.94	0.37
8 1200	9.97	8.39	6.83	5.34	3.97	2.74	1.72	0.92	0.36
8 2200	9.93	8.35	6.79	5.30	3.93	2.71	1.69	0.90	0.35
8 3200	9.88	8.31	6.75	5.26	3.89	2.68	1.67	0.88	0.34
9 600	9.84	8.26	6.71	5.22	3.86	2.65	1.64	0.86	0.33
9 1600	9.79	8.22	6.67	5.18	3.82	2.62	1.62	0.84	0.32
9 2600	9.75	8.18	6.62	5.14	3.78	2.59	1.59	0.82	0.31
10 0	9.71	8.13	6.58	5.10	3.75	2.56	1.57	0.81	0.30

TABLE CIX.

Equation = $9''.1 + 2''.17 \cos. \omega - 8''.80 \sin. \omega$.

	90°	100°	110°	120°	130°	140°	150°	160°	170°
0° 0″	0″.30	0″.05	0″.09	0″.39	0″.97	1″.79	2″.82	4″.05	5″.44
0 1000	0.29	0.05	0.09	0.40	0.98	1.81	2.85	4.08	5.48
0 2000	0.28	0.05	0.10	0.41	1.00	1.84	2.88	4.12	5.52
0 3000	0.27	0.05	0.10	0.43	1.02	1.86	2.92	4.16	5.56
1 400	0.26	0.05	0.11	0.44	1.04	1.89	2.95	4.19	5.60
1 1400	0.25	0.04	0.11	0.46	1.06	1.91	2.98	4.23	5.64
1 2400	0.24	0.04	0.12	0.47	1.08	1.94	3.02	4.27	5.68
1 3400	0.23	0.04	0.12	0.49	1.10	1.96	3.05	4.30	5.72
2 800	0.22	0.04	0.13	0.50	1.12	1.99	3.08	4.34	5.76
2 1800	0.21	0.04	0.14	0.52	1.14	2.02	3.12	4.38	5.80
2 2800	0.20	0.04	0.14	0.53	1.16	2.04	3.15	4.41	5.84
3 200	0.19	0.04	0.15	0.54	1.18	2.07	3.18	4.45	5.88
3 1200	0.18	0.04	0.16	0.56	1.21	2.10	3.22	4.49	5.92
3 2200	0.17	0.04	0.16	0.57	1.23	2.13	3.25	4.53	5.96
3 3200	0.17	0.04	0.17	0.59	1.25	2.15	3.28	4.57	6.00
4 600	0.16	0.04	0.18	0.60	1.28	2.18	3.32	4.60	6.04
4 1600	0.15	0.04	0.18	0.62	1.30	2.21	3.35	4.64	6.08
4 2600	0.14	0.04	0.19	0.63	1.32	2.24	3.38	4.68	6.12
5 0	0.14	0.04	0.20	0.65	1.35	2.27	3.42	4.72	6.17
5 1000	0.13	0.04	0.20	0.66	1.37	2.30	3.45	4.76	6.21
5 2000	0.13	0.04	0.21	0.68	1.39	2.33	3.48	4.80	6.25
5 3000	0.12	0.05	0.22	0.70	1.42	2.36	3.52	4.84	6.29
6 400	0.12	0.05	0.23	0.71	1.44	2.39	3.55	4.88	6.33
6 1400	0.11	0.05	0.24	0.73	1.47	2.42	3.59	4.92	6.38
6 2400	0.11	0.05	0.25	0.75	1.49	2.45	3.62	4.96	6.42
6 3400	0.10	0.05	0.26	0.76	1.52	2.48	3.66	5.00	6.46
7 800	0.10	0.06	0.27	0.78	1.54	2.51	3.69	5.04	6.50
7 1800	0.09	0.06	0.28	0.80	1.57	2.54	3.73	5.08	6.55
7 2800	0.09	0.06	0.29	0.82	1.59	2.57	3.76	5.12	6.59
8 200	0.08	0.06	0.30	0.83	1.61	2.60	3.80	5.16	6.63
8 1200	0.08	0.07	0.31	0.85	1.64	2.63	3.83	5.20	6.67
8 2200	0.07	0.07	0.32	0.87	1.66	2.66	3.87	5.24	6.72
8 3200	0.07	0.07	0.34	0.89	1.69	2.69	3.90	5.28	6.76
9 600	0.06	0.08	0.35	0.91	1.71	2.72	3.94	5.32	6.80
9 1600	0.06	0.08	0.36	0.93	1.74	2.75	3.97	5.36	6.84
9 2600	0.06	0.08	0.37	0.95	1.76	2.78	4.01	5.40	6.88
10 0	0.05	0.09	0.39	0.97	1.79	2.82	4.05	5.44	6.93

TABLE CIX.

Equation $= 9''.1 + 2''.17 \cos. \bar{u} - 8''.80 \sin. \bar{u}$.

	180	190	200	210	220	230	240	250	260
0° 0''	6''.93	8''.49	10''.07	11''.62	13''.10	14''.45	15''.64	16''.63	17''.39
0 1000	6.97	8.53	10.11	11.67	13.14	14.49	15.67	16.66	17.41
0 2000	7.01	8.57	10.15	11.71	13.18	14.52	15.70	16.68	17.43
0 3000	7.05	8.62	10.20	11.75	13.22	14.56	15.73	16.70	17.45
1 400	7.09	8.66	10.24	11.79	13.26	14.59	15.76	16.73	17.47
1 1400	7.14	8.70	10.29	11.84	13.30	14.63	15.79	16.75	17.48
1 2400	7.18	8.75	10.33	11.88	13.34	14.66	15.82	16.77	17.50
1 3400	7.22	8.79	10.38	11.92	13.38	14.70	15.85	16.80	17.52
2 800	7.26	8.83	10.42	11.96	13.42	14.73	15.88	16.82	17.54
2 1800	7.31	8.88	10.47	12.00	13.45	14.76	15.90	16.84	17.55
2 2800	7.35	8.92	10.52	12.05	13.49	14.80	15.93	16.87	17.57
3 200	7.39	8.96	10.56	12.09	13.53	14.83	15.96	16.89	17.58
3 1200	7.44	9.01	10.60	12.13	13.57	14.86	15.99	16.91	17.60
3 2200	7.48	9.05	10.64	12.17	13.61	14.90	16.02	16.93	17.61
3 3200	7.53	9.10	10.68	12.21	13.64	14.93	16.05	16.96	17.63
4 600	7.57	9.14	10.73	12.25	13.68	14.96	16.08	16.98	17.64
4 1600	7.62	9.19	10.77	12.29	13.72	15.00	16.11	17.00	17.66
4 2600	7.66	9.23	10.81	12.33	13.76	15.03	16.14	17.02	17.67
5 0	7.71	9.28	10.85	12.37	13.79	15.06	16.16	17.04	17.68
5 1000	7.75	9.32	10.90	12.41	13.83	15.10	16.19	17.06	17.70
5 2000	7.79	9.37	10.94	12.46	13.87	15.13	16.22	17.08	17.71
5 3000	7.84	9.41	10.98	12.50	13.90	15.16	16.24	17.10	17.72
6 400	7.88	9.46	11.03	12.54	13.94	15.20	16.27	17.12	17.74
6 1400	7.92	9.50	11.07	12.58	13.98	15.23	16.30	17.14	17.75
6 2400	7.97	9.55	11.11	12.62	14.01	15.26	16.32	17.16	17.76
6 3400	8.01	9.59	11.16	12.66	14.05	15.30	16.35	17.18	17.78
7 800	8.05	9.64	11.20	12.70	14.09	15.33	16.38	17.20	17.79
7 1800	8.10	9.68	11.24	12.74	14.12	15.36	16.40	17.22	17.80
7 2800	8.14	9.72	11.29	12.78	14.16	15.39	16.43	17.24	17.81
8 200	8.18	9.77	11.33	12.82	14.20	15.43	16.46	17.26	17.83
8 1200	8.23	9.81	11.37	12.86	14.23	15.46	16.48	17.28	17.84
8 2200	8.27	9.85	11.41	12.90	14.27	15.49	16.51	17.30	17.85
8 3200	8.31	9.90	11.46	12.94	14.31	15.52	16.53	17.32	17.86
9 600	8.36	9.94	11.50	12.98	14.34	15.55	16.56	17.34	17.87
9 1600	8.40	9.99	11.54	13.02	14.38	15.58	16.58	17.36	17.88
9 2600	8.44	10.03	11.58	13.06	14.42	15.61	16.61	17.38	17.89
10 0	8.49	10.07	11.62	13.10	14.45	15.64	16.63	17.39	17.90

TABLE CIX.

Equation = 9″.1 + 2″.17 cos ů — 6″.80 sin ů.

	270°	280°	290°	300°	310°	320°	330°	340°	350°
0° 0″	17.90″	18.15″	18.11″	17.81″	17.23″	16.41″	15.38″	14.15″	12.76″
0 1000	17.91	18.15	18.11	17.80	17.22	16.39	15.35	14.12	12.72
0 2000	17.92	18.15	18.10	17.79	17.20	16.36	15.32	14.08	12.68
0 3000	17.93	18.15	18.10	17.77	17.18	16.34	15.28	14.04	12.64
1 400	17.94	18.16	18.09	17.76	17.16	16.31	15.25	14.01	12.60
1 1400	17.95	18.16	18.09	17.74	17.14	16.29	15.22	13.97	12.56
1 2400	17.96	18.16	18.08	17.73	17.12	16.26	15.18	13.93	12.52
1 3400	17.97	18.16	18.08	17.71	17.10	16.24	15.15	13.90	12.48
2 800	17.98	18.16	18.07	17.70	17.08	16.21	15.12	13.86	12.44
2 1800	17.99	18.16	18.06	17.68	17.06	16.18	15.08	13.82	12.40
2 2800	18.00	18.16	18.06	17.67	17.04	16.16	15.05	13.79	12.36
3 200	18.01	18.16	18.05	17.66	17.02	16.13	15.02	13.75	12.32
3 1200	18.02	18.16	18.04	17.64	16.99	16.10	14.98	13.71	12.28
3 2200	18.03	18.16	18.04	17.63	16.97	16.07	14.95	13.67	12.24
3 3200	18.03	18.16	18.03	17.61	16.95	16.05	14.92	13.64	12.20
4 600	18.04	18.16	18.02	17.60	16.92	16.02	14.88	13.60	12.16
4 1600	18.05	18.16	18.02	17.58	16.90	15.99	14.85	13.56	12.12
4 2600	18.06	18.16	18.01	17.56	16.88	15.96	14.82	13.52	12.08
5 0	18.06	18.16	18.00	17.55	16.85	15.93	14.78	13.48	12.03
5 1000	18.07	18.16	18.00	17.54	16.83	15.90	14.75	13.44	11.99
5 2000	18.08	18.16	17.99	17.52	16.81	15.87	14.72	13.40	11.95
5 3000	18.08	18.15	17.98	17.50	16.78	15.84	14.68	13.36	11.91
6 400	18.09	18.15	17.97	17.49	16.76	15.81	14.65	13.32	11.87
6 1400	18.09	18.15	17.96	17.47	16.73	15.78	14.61	13.28	11.82
6 2400	18.10	18.15	17.95	17.45	16.71	15.75	14.58	13.24	11.78
6 3400	18.10	18.15	17.94	17.44	16.68	15.72	14.54	13.20	11.74
7 800	18.11	18.14	17.93	17.42	16.66	15.69	14.51	13.16	11.70
7 1800	18.11	18.14	17.92	17.40	16.63	15.66	14.47	13.12	11.65
7 2800	18.12	18.14	17.91	17.38	16.61	15.63	14.44	13.08	11.61
8 200	18.12	18.14	17.90	17.37	16.59	15.60	14.40	13.04	11.57
8 1200	18.13	18.13	17.89	17.35	16.56	15.57	14.37	13.00	11.53
8 2200	18.13	18.13	17.88	17.33	16.54	15.54	14.33	12.96	11.49
8 3200	18.14	18.13	17.86	17.31	16.51	15.51	14.30	12.92	11.44
9 600	18.14	18.12	17.85	17.29	16.49	15.48	14.26	12.88	11.40
9 1600	18.14	18.12	17.84	17.27	16.46	15.45	14.23	12.84	11.36
9 2600	18.15	18.12	17.83	17.25	16.44	15.42	14.19	12.80	11.32
10 0	18.15	18.11	17.81	17.23	16.41	15.38	14.15	12.76	11.27

TABLE CX. ARGUMENT 1″.

Equation $= 348''.0 + 186''.8 \cos x + 10''.3 \cos 2x + 0''.6 \cos 3x$.

Days.	0		1		2		3		4		5	
Days.	Equation.	Diff.	Equation.	Diff.	Equation.	Diff.	Equation.	Diff.	Equation.	Diff.	Equation.	Diff.
d .00	338.66″	.42	297.79″	.39	260.80″	.35	229.06″	.29	203.60″	.22	185.13″	.15
.01	338.24	.42	297.40	.39	260.45	.35	228.77	.29	203.38	.22	184.98	.14
.02	337.82	.42	297.01	.39	260.10	.34	228.48	.29	203.16	.22	184.84	.15
.03	337.40	.43	296.62	.39	259.76	.35	228.19	.28	202.94	.22	184.69	.14
.04	336.97	.42	296.23	.39	259.41	.34	227.91	.29	202.72	.22	184.55	.15
.05	336.55	.42	295.84	.39	259.07	.34	227.62	.28	202.50	.22	184.40	.14
.06	336.13	.42	295.45	.39	258.73	.34	227.34	.28	202.28	.21	184.26	.14
.07	335.71	.42	295.06	.39	258.39	.34	227.06	.28	202.07	.22	184.12	.14
.08	335.29	.42	294.67	.39	258.05	.34	226.78	.28	201.85	.21	183.98	.14
.09	334.87	.42	294.28	.39	257.71	.34	226.50	.28	201.64	.21	183.84	.14
.10	334.45	.42	293.89	.39	257.37	.34	226.22	.28	201.43	.22	183.70	.14
.11	334.03	.42	293.50	.39	257.03	.34	225.94	.28	201.21	.21	183.56	.14
.12	333.61	.42	293.11	.39	256.69	.34	225.66	.28	201.00	.21	183.42	.14
.13	333.19	.42	292.72	.39	256.35	.34	225.38	.28	200.79	.21	183.28	.14
.14	332.77	.42	292.33	.39	256.01	.34	225.10	.28	200.58	.21	183.14	.14
.15	332.35	.41	291.94	.38	255.67	.34	224.82	.28	200.37	.21	183.00	.13
.16	331.94	.42	291.56	.38	255.33	.34	224.54	.27	200.16	.21	182.87	.14
.17	331.52	.42	291.18	.38	254.99	.33	224.27	.28	199.95	.21	182.73	.13
.18	331.10	.42	290.80	.38	254.66	.34	223.99	.27	199.74	.21	182.60	.14
.19	330.68	.42	290.42	.38	254.32	.33	223.72	.28	199.53	.21	182.46	.13
.20	330.26	.42	290.04	.39	253.99	.34	223.44	.27	199.32	.21	182.33	.13
.21	329.84	.42	289.65	.39	253.65	.33	223.17	.28	199.11	.20	182.20	.13
.22	329.42	.42	289.26	.38	253.32	.34	222.89	.27	198.91	.21	182.07	.13
.23	329.00	.42	288.88	.38	252.98	.33	222.62	.28	198.70	.20	181.94	.13
.24	328.58	.42	288.50	.38	252.65	.33	222.34	.27	198.50	.21	181.81	.13
.25	328.16	.42	288.12	.38	252.32	.33	222.07	.27	198.29	.20	181.68	.13
.26	327.74	.42	287.74	.38	251.99	.33	221.80	.27	198.09	.20	181.55	.13
.27	327.32	.41	287.36	.38	251.66	.33	221.53	.27	197.89	.20	181.42	.13
.28	326.91	.41	286.98	.38	251.33	.33	221.26	.27	197.69	.20	181.29	.12
.29	326.50	.41	286.60	.38	251.00	.33	220.99	.27	197.49	.20	181.17	.13
.30	326.09	.41	286.22	.38	250.67	.33	220.72	.27	197.29	.20	181.04	.12
.31	325.68	.41	285.84	.38	250.34	.33	220.45	.27	197.09	.20	180.92	.13
.32	325.27	.42	285.46	.38	250.01	.33	220.18	.27	196.89	.20	180.79	.12
.33	324.85	.42	285.08	.38	249.68	.33	219.91	.27	196.69	.20	180.67	.13
.34	324.43	.41	284.70	.38	249.35	.33	219.64	.27	196.49	.20	180.54	.12
.35	324.02	.41	284.32	.38	249.02	.32	219.37	.26	196.29	.19	180.42	.12
.36	323.61	.41	283.94	.38	248.70	.32	219.11	.27	196.10	.19	180.30	.12
.37	323.20	.41	283.56	.37	248.38	.32	218.84	.26	195.91	.19	180.18	.12
.38	322.79	.42	283.19	.37	248.06	.32	218.58	.26	195.72	.19	180.06	.12
.39	322.37	.42	282.82	.37	247.74	.32	218.32	.25	195.53	.19	179.94	.12
.40	321.95	.42	282.45	.38	247.42	.33	218.07	.26	195.34	.19	179.82	.12
.41	321.53	.41	282.07	.38	247.09	.32	217.81	.26	195.15	.19	179.70	.11
.42	321.12	.41	281.69	.38	246.77	.32	217.55	.26	194.96	.19	179.59	.12
.43	320.71	.41	281.31	.37	246.45	.32	217.29	.26	194.77	.19	179.47	.11
.44	320.30	.41	280.94	.37	246.13	.32	217.03	.26	194.58	.19	179.36	.12
.45	319.89	.41	280.57	.37	245.81	.32	216.77	.26	194.39	.19	179.24	.11
.46	319.48	.41	280.20	.37	245.49	.32	216.51	.26	194.20	.18	179.13	.11
.47	319.07	.41	279.83	.37	245.17	.32	216.25	.25	194.02	.19	179.02	.11
.48	318.66	.41	279.46	.37	244.85	.32	216.00	.26	193.83	.18	178.91	.11
.49	318.25	.41	279.09	.37	244.53	.33	215.74	.25	193.65	.19	178.80	.11
.50	317.84	.41	278.72	.37	244.20	.32	215.49	.25	193.46	.18	178.69	.11

NOTE. — Arg. 1″ = Arg. 1 + 13.77728.

TABLES

OF THE MOON'S PARALLAX.

These tables are constructed from the formulæ given by Walker in a report to Prof. Bache, the Superintendent of the Coast Survey, and printed on the 114th page of the *Coast Survey Report for* 1848; and from those given by Adams on the 263d page of Volume XIII. of the *Proceedings of the Royal Astronomical Society* of London. They are substituted for the Tables given in the first edition of this work.

The following is the formula for the Horizontal Parallax.

EQUATION. sin Moon's Equatorial Horizontal Parallax =	AUTHORITY.	ARGUMENT.	TABLE.	EQUATION. Moon's Equatorial Horizontal Parallax, continued.	AUTHORITY.	ARGUMENT.	TABLE.
3000″.	A.			− 0″.31 cos [120]	W. & A.	15	121
+ 345.58	A.		110	+ 0.39	A.		122
+ 186.51 cos [1]	A.	1	110	− 0.11 cos [11]	A.	16	122
+ 10.17 cos [2]	A.		110	+ 0.28 cos [22]	A.		122
+ 0.63 cos [3]	A.		110	+ 0.14	A.		123
+ 0.04 cos [4]	A.		110	+ 0.14 cos [110]	A.	17	123
+ 29.44	A.		111	+ 0.22	A.		124
− 0.95 cos [10]	A.	3	111	+ 0.22 cos [1′21]	A.	18	124
+ 28.23 cos [20]	A.		111	+ 0.12	A.		125
+ 0.26 cos [40]	A.		111	− 0.12 cos [2′3]	A.	19	125
+ 34.67	A.		112	+ 0.12	A.		126
+ 34.30 cos [21′]	A.	2	112	+ 0.12 cos [1′02]	A.	20	126
+ 0.37 cos [42′]	A.		112	+ 0.09	A.		127
+ 1.45	W. & A.		113	+ 0.09 cos [2′20]	A.	22	127
+ 1.45 cos [1′21′]	W. & A.	6	113	+ 0.09	A.		128
+ 3.09	A.		114	− 0.09 cos [202′1]	A.	23	128
+ 3.09 cos [21]	A.	7	114	+ 0.10	A.		129
+ 1.92	W. & A.		115	− 0.10 cos [102]	A.	24	129
+ 1.92 cos [1′20]	W. & A.	8	115	+ 0.05	A.		130
+ 1.16	A.		116	+ 0.05 cos [2′21′]	A.	28	130
+ 1.16 cos [1′01]	A.	9	116	+ 0.06	A.		131
+ 0.95	A.		117	+ 0.06 cos [1′41′]	A.	30	131
− 0.95 cos [101]	A.	10	117	+ 0.40	A.		132
+ 0.71	A.		118	− 0.40 cos [100]	A.	4	132
− 0.71 cos [2001′]	A.	11	118	+ 0.32	A.		133
+ 0.60	A.		119	+ 0.01 cos [1′1]	A.	5	133
+ 0.60 cos [41′]	A.	13	119	− 0.31 cos [2′2]	A.		133
+ 0.23	W. & A.		120	+ 0.11	A.		134
− 0.23 cos [121′]	W. & A.	14	120	− 0.11 cos [202′0]	A.	12′	134
+ 0.31	W. & A.		121				

Table 135 contains the excess of the Moon's Horizontal Parallax above its sine.

TABLE CX. ARGUMENT 1.

Equation $= 345''.58 + 186''.51 \cos x + 10''.17 \cos 2x + 0''.63 \cos 3x + 0''.04 \cos 4x$.

Days.	0		1		2		3		4		5	
Days.	Equation.	Diff.	Equation.	Diff.	Equation.	Diff.	Equation.	Diff.	Equation.	Diff.	Equation.	Diff.
d.	"		"		"		"		"		"	
.00	334.49	.42	377.26	.43	419.92	.42	459.95	.38	494.63	.31	521.36	22
.01	334.91	.42	377.69	.43	420.34	.42	460.33	.37	494.94	.31	521.58	.21
.02	335.33	.43	378.12	.43	420.76	.41	460.70	.38	495.25	.31	521.79	.22
.03	335.76	.42	378.55	.43	421.17	.42	461.08	.37	495.56	.31	522.01	.21
.04	336.18	.42	378.98	.43	421.59	.42	461.45	.38	495.87	.31	522.22	.22
.05	336.60	.42	379.41	.43	422.01	.42	461.83	.37	496.18	.30	522.44	.21
.06	337.02	.42	379.84	.43	422.43	.41	462.20	.38	496.48	.31	522.65	.21
.07	337.44	.43	380.27	.44	422.84	.42	462.58	.37	496.79	.30	522.86	.22
.08	337.87	.42	380.71	.43	423.26	.41	462.95	.38	497.09	.31	523.08	.21
.09	338.29	.42	381.14	.43	423.67	.42	463.33	.37	497.40	.30	523.29	.21
.10	338.71	.42	381.57	.43	424.09	.41	463.70	.37	497.70	.30	523.50	.21
.11	339.13	.42	382.00	.43	424.50	.42	464.07	.37	498.00	.30	523.71	.20
.12	339.55	.43	382.43	.43	424.92	.41	464.44	.38	498.30	.31	523.91	.21
.13	339.98	.42	382.86	.43	425.33	.42	464.82	.37	498.61	.30	524.12	.20
.14	340.40	.42	383.29	.43	425.75	.41	465.19	.37	498.91	.30	524.32	.21
.15	340.82	.42	383.72	.43	426.16	.41	465.56	.37	499.21	.30	524.53	.20
.16	341.24	.43	384.15	.43	426.57	.41	465.93	.37	499.51	.29	524.73	.20
.17	341.67	.42	384.58	.43	426.98	.42	466.30	.36	499.80	.30	524.93	.21
.18	342.09	.43	385.01	.43	427.40	.41	466.66	.37	500.10	.29	525.14	.20
.19	342.52	.42	385.44	.43	427.81	.41	467.03	.37	500.39	.30	525.34	.20
.20	342.94	.42	385.87	.43	428.22	.41	467.40	.36	500.69	.29	525.54	.20
.21	343.36	.43	386.30	.43	428.63	.41	467.76	.37	500.98	.30	525.74	.19
.22	343.79	.42	386.73	.41	429.04	.42	468.13	.36	501.28	.29	525.93	.20
.23	344.21	.43	387.17	.43	429.46	.41	468.49	.37	501.57	.30	526.13	.19
.24	344.64	.42	387.60	.43	429.87	.41	468.86	.36	501.87	.29	526.32	.20
.25	345.06	.43	388.03	.43	430.28	.41	469.22	.36	502.16	.29	526.52	.19
.26	345.49	.42	388.46	.43	430.69	.41	469.58	.37	502.45	.29	526.71	.19
.27	345.91	.43	388.89	.43	431.10	.41	469.95	.36	502.74	.28	526.90	.19
.28	346.34	.42	389.32	.43	431.51	.41	470.31	.37	503.02	.29	527.09	.19
.29	346.76	.43	389.75	.43	431.92	.41	470.68	.36	503.31	.29	527.28	.19
.30	347.19	.43	390.18	.43	432.33	.41	471.04	.36	503.60	.28	527.47	.19
.31	347.62	.43	390.61	.43	432.74	.41	471.40	.36	503.88	.28	527.66	.18
.32	348.05	.42	391.04	.42	433.15	.40	471.76	.35	504.16	.29	527.84	.19
.33	348.47	.43	391.46	.43	433.55	.41	472.11	.36	504.45	.28	528.03	.18
.34	348.90	.43	391.89	.43	433.96	.41	472.47	.36	504.73	.28	528.21	.19
.35	349.33	.43	392.32	.43	434.37	.41	472.83	.36	505.01	.28	528.40	.18
.36	349.76	.42	392.75	.43	434.78	.40	473.19	.35	505.29	.28	528.58	.18
.37	350.18	.43	393.18	.43	435.18	.41	473.54	.36	505.57	.28	528.76	.18
.38	350.61	.42	393.61	.43	435.59	.40	473.90	.35	505.85	.28	528.94	.18
.39	351.03	.43	394.04	.43	435.99	.41	474.25	.36	506.13	.28	529.12	.18
.40	351.46	.43	394.47	.43	436.40	.40	474.61	.35	506.41	.28	529.30	.17
.41	351.89	.43	394.90	.43	436.80	.41	474.96	.36	506.69	.27	529.47	.18
.42	352.32	.42	395.33	.42	437.21	.40	475.32	.35	506.96	.28	529.65	.17
.43	352.74	.43	395.75	.43	437.61	.41	475.67	.36	507.24	.27	529.82	.18
.44	353.17	.43	396.18	.43	438.02	.40	476.03	.35	507.51	.28	530.00	.17
.45	353.60	.43	396.61	.43	438.42	.40	476.38	.35	507.79	.27	530.17	.17
.46	354.03	.43	397.04	.43	438.82	.40	476.73	.35	508.06	.27	530.34	.17
.47	354.46	.42	397.47	.42	439.22	.41	477.08	.35	508.33	.27	530.51	.17
.48	354.88	.43	397.89	.43	439.63	.40	477.43	.35	508.60	.27	530.68	.17
.49	355.31	.43	398.32	.43	440.03	.40	477.78	.35	508.87	.27	530.85	.17
.50	355.74	.43	398.75	.43	440.43	.40	478.13	.35	509.14	.27	531.02	.16

TABLE CX. ARGUMENT 1.

Equation $= 345''.58 + 186''.51 \cos x + 10''.17 \cos 2x + 0''.63 \cos 3x + 0''.04 \cos 4x.$

Days.	0		1		2		3		4		5	
Days.	Equation.	Diff.	Equation.	Diff.	Equation.	Diff.	Equation.	Diff.	Equation.	Diff.	Equation.	Diff.
d.	"		"		"		"		"		"	
.50	355.74	.43	398.75	.43	440.43	.40	478.13	.35	509.14	.27	531.02	.16
.51	356.17	.43	399.18	.43	440.83	.40	478.48	.34	509.41	.26	531.18	.17
.52	356.60	.43	399.61	.42	441.23	.40	478.82	.35	509.67	.27	531.35	.16
.53	357.03	.43	400.03	.43	441.63	.40	479.17	.34	509.94	.26	531.51	.17
.54	357.46	.43	400.46	.43	442.03	.40	479.51	.35	510.20	.27	531.68	.16
.55	357.89	.43	400.89	.43	442.43	.40	479.86	.34	510.47	.26	531.84	.16
.56	358.32	.43	401.32	.42	442.83	.40	480.20	.34	510.73	.26	532.00	.16
.57	358.75	.42	401.74	.43	443.23	.40	480.54	.35	510.99	.27	532.16	.15
.58	359.17	.43	402.17	.42	443.63	.40	480.89	.34	511.26	.26	532.31	.16
.59	359.60	.43	402.59	.43	444.03	.40	481.23	.34	511.52	.26	532.47	.16
.60	360.03	.43	403.02	.43	444.43	.40	481.57	.34	511.78	.26	532.63	.15
.61	360.46	.43	403.45	.42	444.83	.39	481.91	.34	512.04	.25	532.78	.15
.62	360.89	.43	403.87	.43	445.22	.40	482.25	.34	512.29	.26	532.93	.16
.63	361.32	.43	404.30	.42	445.62	.39	482.59	.34	512.55	.25	533.09	.15
.64	361.75	.43	404.72	.43	446.01	.40	482.93	.34	512.80	.26	533.24	.15
.65	362.18	.43	405.15	.43	446.41	.39	483.27	.34	513.06	.25	533.39	.15
.66	362.61	.43	405.58	.42	446.80	.40	483.61	.33	513.31	.25	533.54	.14
.67	363.04	.43	406.00	.43	447.20	.39	483.94	.34	513.56	.26	533.68	.15
.68	363.47	.43	406.43	.42	447.59	.40	484.28	.33	513.82	.25	533.83	.14
.69	363.90	.43	406.85	.43	447.99	.39	484.61	.34	514.07	.25	533.97	.15
.70	364.33	.43	407.28	.42	448.38	.39	484.95	.33	514.32	.25	534.12	.14
.71	364.76	.43	407.70	.43	448.77	.39	485.28	.33	514.57	.24	534.26	.14
.72	365.19	.43	408.13	.42	449.16	.40	485.61	.34	514.81	.25	534.40	.14
.73	365.62	.43	408.55	.43	449.56	.39	485.95	.33	515.06	.24	534.54	.14
.74	366.05	.43	408.98	.42	449.95	.39	486.28	.33	515.30	.25	534.68	.14
.75	366.48	.43	409.40	.42	450.34	.39	486.61	.33	515.55	.24	534.82	.14
.76	366.91	.43	409.82	.42	450.73	.39	486.94	.33	515.79	.24	534.96	.13
.77	367.34	.44	410.24	.43	451.12	.38	487.27	.32	516.03	.25	535.09	.14
.78	367.78	.43	410.67	.42	451.50	.39	487.59	.33	516.28	.24	535.23	.13
.79	368.21	.43	411.09	.42	451.89	.39	487.92	.33	516.52	.24	535.36	.14
.80	368.64	.43	411.51	.42	452.28	.39	488.25	.33	516.76	.24	535.50	.13
.81	369.07	.43	411.93	.42	452.67	.39	488.58	.32	517.00	.24	535.63	.13
.82	369.50	.43	412.35	.43	453.06	.38	488.90	.33	517.24	.23	535.76	.13
.83	369.93	.43	412.78	.42	453.44	.39	489.23	.32	517.47	.24	535.89	.13
.84	370.36	.43	413.20	.42	453.83	.39	489.55	.33	517.71	.24	536.02	.13
.85	370.79	.43	413.62	.42	454.22	.38	489.88	.32	517.95	.23	536.15	.12
.86	371.22	.43	414.04	.42	454.60	.39	490.20	.32	518.18	.23	536.27	.13
.87	371.65	.44	414.46	.43	454.99	.38	490.52	.32	518.41	.24	536.40	.12
.88	372.09	.43	414.89	.42	455.37	.39	490.84	.32	518.65	.23	536.52	.13
.89	372.52	.43	415.31	.42	455.76	.38	491.16	.32	518.88	.23	536.65	.12
.90	372.95	.43	415.73	.42	456.14	.38	491.48	.32	519.11	.23	536.77	.12
.91	373.38	.43	416.15	.42	456.52	.38	491.80	.32	519.34	.23	536.89	.12
.92	373.81	.43	416.57	.42	456.90	.39	492.12	.32	519.57	.22	537.01	.11
.93	374.24	.43	416.99	.42	457.29	.38	492.43	.32	519.79	.23	537.12	.12
.94	374.67	.43	417.41	.42	457.67	.38	492.75	.32	520.02	.23	537.24	.12
.95	375.10	.43	417.83	.42	458.05	.38	493.07	.31	520.25	.22	537.36	.11
.96	375.53	.43	418.25	.42	458.43	.38	493.38	.31	520.47	.22	537.47	.11
.97	375.96	.44	418.67	.41	458.81	.38	493.69	.32	520.69	.23	537.58	.12
.98	376.40	.43	419.08	.42	459.19	.38	494.01	.31	520.92	.22	537.70	.11
.99	376.83	.43	419.50	.42	459.57	.38	494.32	.31	521.14	.22	537.81	.11
1.00	377.26	.43	419.92	.43	459.95	.38	494.63	.31	521.36	.22	537.92	.11

TABLE CX. ARGUMENT 1.

Equation = 345".58 + 186".51 cos x + 10".17 cos $2x$ + 0".63 cos $3x$ + 0".04 cos $4x$.

Days.	6		7		8		9		10		11	
Days.	Equation.	Diff.	Equation.	Diff.	Equation.	Diff.	Equation.	Diff.	Equation.	Diff.	Equation.	Diff.
d. .00	537.92"	.11	542.88"	.01	535.80"	.13	517.31"	.24	488.99"	.33	453.16"	.39
.01	538.03	.10	542.87	.01	535.67	.13	517.07	.24	488.66	.32	452.77	.39
.02	538.13	.11	542.86	.02	535.54	.14	516.83	.24	488.34	.33	452.38	.38
.03	538.24	.10	542.84	.01	535.40	.13	516.59	.24	488.01	.32	452.00	.39
.04	538.34	.11	542.83	.01	535.27	.13	516.35	.24	487.69	.33	451.61	.39
.05	538.45	.10	542.82	.02	535.14	.14	516.11	.25	487.36	.33	451.22	.39
.06	538.55	.10	542.80	.02	535.00	.14	515.86	.24	487.03	.33	450.83	.39
.07	538.65	.10	542.78	.02	534.86	.13	515.62	.25	486.70	.33	450.44	.39
.08	538.75	.10	542.76	.02	534.73	.14	515.37	.24	486.37	.33	450.05	.39
.09	538.85	.10	542.74	.02	534.59	.14	515.13	.25	486.04	.33	449.66	.39
.10	538.95	.09	542.72	.03	534.45	.14	514.88	.25	485.71	.33	449.27	.39
.11	539.04	.10	542.69	.02	534.31	.15	514.63	.25	485.38	.34	448.88	.40
.12	539.14	.09	542.67	.03	534.16	.14	514.38	.25	485.04	.33	448.48	.39
.13	539.23	.10	542.64	.02	534.02	.15	514.13	.25	484.71	.34	448.09	.40
.14	539.33	.09	542.62	.03	533.87	.14	513.88	.25	484.37	.33	447.69	.39
.15	539.42	.09	542.59	.03	533.73	.15	513.63	.25	484.04	.34	447.30	.39
.16	539.51	.09	542.56	.03	533.58	.15	513.38	.26	483.70	.34	446.91	.40
.17	539.60	.09	542.53	.04	533.43	.15	513.12	.25	483.36	.33	446.51	.39
.18	539.69	.09	542.49	.03	533.28	.15	512.87	.26	483.03	.34	446.12	.40
.19	539.78	.09	542.46	.03	533.13	.15	512.61	.25	482.69	.34	445.72	.39
.20	539.87	.08	542.43	.04	532.98	.16	512.36	.26	482.35	.34	445.33	.40
.21	539.95	.08	542.39	.04	532.82	.15	512.10	.26	482.01	.34	444.93	.40
.22	540.03	.09	542.35	.03	532.67	.16	511.84	.25	481.67	.35	444.53	.39
.23	540.12	.08	542.32	.04	532.51	.15	511.59	.26	481.32	.34	444.14	.40
.24	540.20	.08	542.28	.04	532.36	.16	511.33	.26	480.98	.34	443.74	.40
.25	540.28	.08	542.24	.04	532.20	.16	511.07	.26	480.64	.34	443.34	.40
.26	540.36	.08	542.20	.05	532.04	.16	510.81	.27	480.30	.35	442.94	.40
.27	540.44	.07	542.15	.04	531.88	.17	510.54	.26	479.95	.34	442.54	.40
.28	540.51	.08	542.11	.05	531.71	.16	510.28	.27	479.61	.35	442.14	.40
.29	540.59	.08	542.06	.04	531.55	.16	510.01	.26	479.26	.34	441.74	.40
.30	540.67	.07	542.02	.05	531.39	.17	509.75	.27	478.92	.35	441.34	.40
.31	540.74	.07	541.97	.05	531.22	.16	509.48	.27	478.57	.35	440.94	.40
.32	540.81	.07	541.92	.05	531.06	.17	509.21	.26	478.22	.34	440.54	.40
.33	540.88	.07	541.87	.05	530.89	.16	508.95	.27	477.88	.35	440.14	.40
.34	540.95	.07	541.82	.05	530.73	.17	508.68	.27	477.53	.35	439.74	.40
.35	541.02	.07	541.77	.06	530.56	.17	508.41	.27	477.18	.35	439.34	.40
.36	541.09	.06	541.71	.05	530.39	.17	508.14	.28	476.83	.35	438.94	.41
.37	541.15	.07	541.66	.06	530.22	.18	507.86	.27	476.48	.36	438.53	.40
.38	541.22	.06	541.60	.05	530.04	.17	507.59	.28	476.12	.35	438.13	.41
.39	541.28	.07	541.55	.06	529.87	.17	507.31	.27	475.77	.35	437.72	.40
.40	541.35	.06	541.49	.06	529.70	.18	507.04	.28	475.42	.35	437.32	.41
.41	541.41	.06	541.43	.06	529.52	.18	506.76	.27	475.07	.36	436.91	.40
.42	541.47	.05	541.37	.07	529.34	.17	506.49	.28	474.71	.35	436.51	.41
.43	541.52	.06	541.30	.06	529.17	.18	506.21	.27	474.36	.36	436.10	.40
.44	541.58	.06	541.24	.06	528.99	.18	505.94	.28	474.00	.35	435.70	.41
.45	541.64	.05	541.18	.07	528.81	.18	505.66	.28	473.65	.36	435.29	.41
.46	541.69	.05	541.11	.07	528.63	.18	505.38	.28	473.29	.36	434.88	.40
.47	541.74	.06	541.04	.07	528.45	.19	505.10	.29	472.93	.35	434.48	.41
.48	541.80	.05	540.97	.07	528.26	.18	504.81	.28	472.58	.36	434.07	.40
.49	541.85	.05	540.90	.07	528.08	.18	504.53	.28	472.22	.36	433.67	.41
.50	541.90	.05	540.83	.07	527.90	.19	504.25	.29	471.86	.36	433.26	.41

TABLE CX. ARGUMENT 1.

$$\text{Equation} = 345''.58 + 186''.51 \cos x + 10''.17 \cos 2x + 0''.63 \cos 3x + 0''.04 \cos 4x.$$

Days.	6		7		8		9		10		11	
Days.	Equation.	Diff.	Equation.	Diff.	Equation.	Diff.	Equation.	Diff.	Equation.	Diff.	Equation.	Diff.
d.	″		″		″		″		″		″	
.50	541.90	.05	540.83	.07	527.90	.19	504.25	.29	471.86	.36	433.26	.41
.51	541.95	.05	540.76	.08	527.71	.19	503.96	.29	471.50	.36	432.85	.41
.52	542.00	.04	540.68	.07	527.52	.18	503.67	.28	471.14	.37	432.44	.41
.53	542.04	.05	540.61	.08	527.34	.19	503.39	.29	470.77	.36	432.03	.41
.54	542.09	.05	540.53	.07	527.15	.19	503.10	.29	470.41	.36	431.62	.41
.55	542.14	.04	540.46	.08	526.96	.19	502.81	.29	470.05	.36	431.21	.41
.56	542.18	.04	540.38	.08	526.77	.20	502.52	.29	469.69	.37	430.80	.41
.57	542.22	.04	540.30	.08	526.57	.19	502.23	.29	469.32	.36	430.39	.41
.58	542.26	.04	540.22	.08	526.38	.20	501.94	.29	468.96	.37	429.98	.41
.59	542.30	.04	540.14	.08	526.18	.19	501.65	.29	468.59	.36	429.57	.41
.60	542.34	.04	540.06	.09	525.99	.20	501.36	.29	468.23	.37	429.16	.41
.61	542.38	.03	539.97	.08	525.79	.20	501.07	.30	467.86	.37	428.75	.41
.62	542.41	.04	539.89	.09	525.59	.20	500.77	.29	467.49	.36	428.34	.42
.63	542.45	.03	539.80	.08	525.39	.20	500.48	.30	467.13	.37	427.92	.41
.64	542.48	.04	539.72	.09	525.19	.20	500.18	.29	466.76	.37	427.51	.41
.65	542.52	.03	539.63	.09	524.99	.20	499.89	.30	466.39	.37	427.10	.41
.66	542.55	.03	539.54	.09	524.79	.21	499.59	.30	466.02	.37	426.69	.42
.67	542.58	.02	539.45	.10	524.58	.20	499.29	.30	465.65	.37	426.27	.41
.68	542.60	.03	539.35	.09	524.38	.21	498.99	.30	465.28	.37	425.86	.42
.69	542.63	.03	539.26	.09	524.17	.20	498.69	.30	464.91	.37	425.44	.41
.70	542.66	.02	539.17	.10	523.97	.21	498.39	.30	464.54	.37	425.03	.42
.71	542.68	.02	539.07	.10	523.76	.21	498.09	.30	464.17	.37	424.61	.41
.72	542.70	.03	538.97	.09	523.55	.21	497.79	.31	463.80	.38	424.20	.42
.73	542.73	.02	538.88	.10	523.34	.21	497.48	.30	463.42	.37	423.78	.41
.74	542.75	.02	538.78	.10	523.13	.21	497.18	.30	463.05	.37	423.37	.42
.75	542.77	.02	538.68	.10	522.92	.21	496.88	.31	462.68	.38	422.95	.42
.76	542.79	.02	538.58	.11	522.71	.22	496.57	.31	462.30	.37	422.53	.41
.77	542.81	.01	538.47	.10	522.49	.21	496.26	.30	461.93	.38	422.12	.42
.78	542.82	.02	538.37	.11	522.28	.22	495.96	.31	461.55	.37	421.70	.41
.79	542.84	.02	538.26	.10	522.06	.21	495.65	.31	461.18	.38	421.29	.42
.80	542.86	.01	538.16	.11	521.85	.22	495.34	.31	460.80	.38	420.87	.42
.81	542.87	.01	538.05	.11	521.63	.22	495.03	.31	460.42	.38	420.45	.42
.82	542.88	.01	537.94	.10	521.41	.22	494.72	.32	460.04	.37	420.03	.41
.83	542.89	.01	537.84	.11	521.19	.22	494.40	.31	459.67	.38	419.62	.42
.84	542.90	.01	537.73	.11	520.97	.22	494.09	.31	459.29	.38	419.20	.42
.85	542.91	.00	537.62	.12	520.75	.22	493.78	.31	458.91	.38	418.78	.42
.86	542.91	.01	537.50	.11	520.53	.23	493.47	.32	458.53	.38	418.36	.42
.87	542.92	.00	537.39	.12	520.30	.22	493.15	.31	458.15	.38	417.94	.42
.88	542.92	.01	537.27	.11	520.08	.23	492.84	.32	457.77	.38	417.52	.42
.89	542.93	.00	537.16	.12	519.85	.22	492.52	.31	457.39	.38	417.10	.42
.90	542.93	.00	537.04	.12	519.63	.23	492.21	.32	457.01	.38	416.68	.42
.91	542.93	.00	536.92	.12	519.40	.23	491.89	.32	456.63	.39	416.26	.42
.92	542.93	.01	536.80	.13	519.17	.23	491.57	.32	456.24	.38	415.84	.42
.93	542.92	.00	536.67	.12	518.94	.23	491.25	.32	455.86	.39	415.42	.42
.94	542.92	.00	536.55	.12	518.71	.23	490.93	.32	455.47	.38	415.00	.42
.95	542.92	.01	536.43	.13	518.48	.23	490.61	.32	455.09	.39	414.58	.42
.96	542.91	.01	536.30	.12	518.25	.24	490.29	.33	454.70	.38	414.16	.42
.97	542.90	.00	536.18	.13	518.01	.23	489.96	.32	454.32	.39	413.74	.43
.98	542.90	.01	536.05	.12	517.78	.24	489.64	.33	453.93	.38	413.31	.42
.99	542.89	.01	535.93	.13	517.54	.23	489.31	.32	453.55	.39	412.89	.42
1.00	542.88	.01	535.80	.13	517.31	.24	488.99	.33	453.16	.39	412.47	.42

TABLE CX. ARGUMENT 1.

Equation $= 345''.58 + 186''.51 \cos x + 10''.17 \cos 2x + 0''.63 \cos 3x + 0''.04 \cos 4x$.

Days.	12		13		14		15		16		17	
Days.	Equation.	Diff.	Equation.	Diff.	Equation.	Diff.	Equation.	Diff.	Equation.	Diff.	Equation.	Diff.
d. .00	412.47″	.42	369.62″	.43	327.07″	.42	286.97″	.38	251.04″	.33	220.59″	.27
.01	412.05	.42	369.19	.43	326.65	.41	286.59	.38	250.71	.33	220.32	.27
.02	411.63	.43	368.76	.44	326.24	.42	286.21	.38	250.38	.34	220.05	.28
.03	411.20	.42	368.32	.43	325.82	.41	285.83	.38	250.04	.33	219.77	.27
.04	410.78	.42	367.89	.43	325.41	.42	285.45	.38	249.71	.33	219.50	.27
.05	410.36	.42	367.46	.43	324.99	.41	285.07	.38	249.38	.33	219.23	.27
.06	409.94	.43	367.03	.43	324.58	.42	284.69	.38	249.05	.33	218.96	.27
.07	409.51	.42	366.60	.43	324.16	.41	284.31	.38	248.72	.33	218.69	.26
.08	409.09	.43	366.17	.43	323.75	.42	283.93	.38	248.39	.33	218.43	.27
.09	408.66	.42	365.74	.43	323.33	.41	283.55	.38	248.06	.33	218.16	.27
.10	408.24	.42	365.31	.43	322.92	.41	283.17	.38	247.73	.33	217.89	.27
.11	407.82	.43	364.88	.43	322.51	.41	282.79	.38	247.40	.33	217.62	.26
.12	407.39	.42	364.45	.43	322.10	.42	282.41	.37	247.07	.32	217.36	.27
.13	406.97	.43	364.02	.43	321.68	.41	282.04	.38	246.75	.33	217.09	.26
.14	406.54	.42	363.59	.43	321.27	.41	281.66	.38	246.42	.33	216.83	.27
.15	406.12	.43	363.16	.43	320.86	.41	281.28	.37	246.09	.32	216.56	.26
.16	405.69	.42	362.73	.43	320.45	.41	280.91	.38	245.77	.33	216.30	.26
.17	405.27	.43	362.30	.43	320.04	.42	280.53	.37	245.44	.32	216.04	.27
.18	404.84	.42	361.87	.43	319.62	.41	280.16	.38	245.12	.33	215.77	.26
.19	404.42	.43	361.44	.43	319.21	.41	279.78	.37	244.79	.32	215.51	.26
.20	403.99	.43	361.01	.43	318.80	.41	279.41	.37	244.47	.32	215.25	.26
.21	403.56	.42	360.58	.43	318.39	.41	279.04	.38	244.15	.32	214.99	.26
.22	403.14	.43	360.15	.43	317.98	.42	278.66	.37	243.83	.32	214.73	.25
.23	402.71	.42	359.72	.43	317.56	.41	278.29	.38	243.51	.32	214.48	.26
.24	402.29	.43	359.29	.43	317.15	.41	277.91	.37	243.19	.32	214.22	.26
.25	401.86	.43	358.86	.43	316.74	.41	277.54	.37	242.87	.32	213.96	.26
.26	401.43	.43	358.43	.43	316.33	.41	277.17	.37	242.55	.32	213.70	.25
.27	401.00	.42	358.00	.42	315.92	.40	276.80	.37	242.23	.32	213.45	.26
.28	400.58	.43	357.58	.43	315.52	.41	276.43	.37	241.91	.32	213.19	.25
.29	400.15	.43	357.15	.43	315.11	.41	276.06	.37	241.59	.32	212.94	.26
.30	399.72	.43	356.72	.43	314.70	.41	275.69	.37	241.27	.31	212.68	.25
.31	399.29	.42	356.29	.43	314.29	.41	275.32	.37	240.96	.32	212.43	.25
.32	398.87	.43	355.86	.42	313.88	.40	274.95	.36	240.64	.31	212.18	.26
.33	398.44	.42	355.44	.43	313.48	.41	274.59	.37	240.33	.32	211.92	.25
.34	398.02	.43	355.01	.43	313.07	.41	274.22	.37	240.01	.31	211.67	.25
.35	397.59	.43	354.58	.43	312.66	.41	273.85	.37	239.70	.31	211.42	.25
.36	397.16	.43	354.15	.43	312.25	.40	273.48	.36	239.39	.31	211.17	.25
.37	396.73	.43	353.72	.42	311.85	.41	273.12	.37	239.08	.32	210.92	.24
.38	396.30	.43	353.30	.43	311.44	.40	272.75	.36	238.76	.31	210.68	.25
.39	395.87	.43	352.87	.43	311.04	.41	272.39	.37	238.45	.31	210.43	.25
.40	395.44	.43	352.44	.43	310.63	.40	272.02	.36	238.14	.31	210.18	.25
.41	395.01	.43	352.01	.43	310.23	.41	271.66	.37	237.83	.31	209.93	.24
.42	394.58	.42	351.58	.42	309.82	.40	271.29	.36	237.52	.31	209.69	.25
.43	394.16	.43	351.16	.43	309.42	.41	270.93	.37	237.21	.31	209.44	.24
.44	393.73	.43	350.73	.43	309.01	.40	270.56	.36	236.90	.31	209.20	.25
.45	393.30	.43	350.30	.43	308.61	.40	270.20	.36	236.59	.31	208.95	.24
.46	392.87	.43	349.87	.42	308.21	.40	269.84	.36	236.28	.30	208.71	.24
.47	392.44	.43	349.45	.43	307.81	.41	269.48	.37	235.98	.31	208.47	.25
.48	392.01	.43	349.02	.42	307.40	.40	269.11	.36	235.67	.30	208.22	.24
.49	391.58	.43	348.60	.43	307.00	.40	268.75	.36	235.37	.31	207.98	.24
.50	391.15	.43	348.17	.43	306.60	.40	268.39	.36	235.06	.30	207.74	.24

TABLE CX. ARGUMENT 1.

Equation $= 345''.58 + 186''.51 \cos x + 10''.17 \cos 2x + 0''.63 \cos 3x + 0''.04 \cos 4x$.

Days.	12		13		14		15		16		17	
Days.	Equation.	Diff.	Equation.	Diff.	Equation.	Diff.	Equation.	Diff.	Equation.	Diff.	Equation.	Diff.
d	"		"		"		"		"		"	
.50	391.15	.43	348.17	.43	306.60	.40	268.39	.36	235.06	.30	207.74	.24
.51	390.72	.43	347.74	.42	306.20	.40	268.03	.36	234.76	.31	207.50	.24
.52	390.29	.43	347.32	.43	305.80	.40	267.67	.35	234.45	.30	207.26	.23
.53	389.86	.43	346.89	.42	305.40	.40	267.32	.36	234.15	.31	207.03	.24
.54	389.43	.43	346.47	.43	305.00	.40	266.96	.36	233.84	.30	206.79	.24
.55	389.00	.43	346.04	.43	304.60	.40	266.60	.36	233.54	.30	206.55	.23
.56	388.57	.43	345.61	.42	304.20	.40	266.24	.35	233.24	.30	206.32	.24
.57	388.14	.43	345.19	.43	303.80	.40	265.89	.36	232.94	.30	206.08	.23
.58	387.71	.43	344.76	.42	303.40	.40	265.53	.35	232.64	.30	205.85	.24
.59	387.28	.43	344.34	.43	303.00	.40	265.18	.36	232.34	.30	205.61	.23
.60	386.85	.43	343.91	.42	302.60	.40	264.82	.35	232.04	.30	205.38	.23
.61	386.42	.43	343.49	.43	302.20	.40	264.47	.36	231.74	.30	205.15	.23
.62	385.99	.43	343.06	.42	301.80	.39	264.11	.35	231.44	.29	204.92	.24
.63	385.56	.43	342.64	.43	301.41	.40	263.76	.36	231.15	.30	204.68	.23
.64	385.13	.43	342.21	.42	301.01	.40	263.40	.35	230.85	.30	204.45	.23
.65	384.70	.43	341.79	.42	300.61	.39	263.05	.35	230.55	.29	204.22	.23
.66	384.27	.43	341.37	.43	300.22	.40	262.70	.35	230.26	.30	203.99	.23
.67	383.84	.43	340.94	.42	299.82	.39	262.35	.35	229.96	.29	203.76	.22
.68	383.41	.43	340.52	.43	299.43	.40	262.00	.35	229.67	.30	203.54	.23
.69	382.98	.43	340.09	.42	299.03	.39	261.65	.35	229.37	.29	203.31	.23
.70	382.55	.43	339.67	.42	298.64	.39	261.30	.35	229.08	.29	203.08	.22
.71	382.12	.43	339.25	.42	298.25	.40	260.95	.35	228.79	.29	202.86	.23
.72	381.69	.44	338.83	.43	297.85	.39	260.60	.35	228.50	.29	202.63	.22
.73	381.25	.43	338.40	.42	297.46	.40	260.25	.35	228.21	.29	202.41	.23
.74	380.82	.43	337.98	.42	297.06	.39	259.90	.35	227.92	.29	202.18	.22
.75	380.39	.43	337.56	.42	296.67	.39	259.55	.35	227.63	.29	201.96	.22
.76	379.96	.43	337.14	.42	296.28	.39	259.20	.34	227.34	.29	201.74	.22
.77	379.53	.43	336.72	.43	295.89	.40	258.86	.35	227.05	.28	201.52	.23
.78	379.10	.43	336.29	.42	295.49	.39	258.51	.34	226.77	.29	201.29	.22
.79	378.67	.43	335.87	.42	295.10	.39	258.17	.35	226.48	.29	201.07	.22
.80	378.24	.43	335.45	.42	294.71	.39	257.82	.34	226.19	.29	200.85	.22
.81	377.81	.43	335.03	.42	294.32	.39	257.48	.34	225.90	.28	200.63	.22
.82	377.38	.44	334.61	.42	293.93	.39	257.14	.35	225.62	.29	200.41	.21
.83	376.94	.43	334.19	.42	293.54	.39	256.79	.34	225.33	.28	200.20	.22
.84	376.51	.43	333.77	.42	293.15	.39	256.45	.34	225.05	.29	199.98	.22
.85	376.08	.43	333.35	.42	292.76	.39	256.11	.34	224.76	.28	199.76	.21
.86	375.65	.43	332.93	.42	292.37	.39	255.77	.34	224.48	.28	199.55	.22
.87	375.22	.43	332.51	.42	291.98	.38	255.43	.35	224.20	.28	199.33	.21
.88	374.79	.43	332.09	.42	291.60	.39	255.08	.34	223.92	.28	199.12	.22
.89	374.36	.43	331.67	.42	291.21	.39	254.74	.34	223.64	.28	198.90	.21
.90	373.93	.43	331.25	.42	290.82	.39	254.40	.34	223.36	.28	198.69	.21
.91	373.50	.43	330.83	.42	290.43	.38	254.06	.34	223.08	.28	198.48	.21
.92	373.07	.44	330.41	.41	290.05	.39	253.72	.33	222.80	.28	198.27	.21
.93	372.63	.43	330.00	.42	289.66	.38	253.39	.34	222.52	.28	198.06	.21
.94	372.20	.43	329.58	.42	289.28	.39	253.05	.34	222.24	.28	197.85	.21
.95	371.77	.43	329.16	.42	288.89	.38	252.71	.33	221.96	.27	197.64	.21
.96	371.34	.43	328.74	.42	288.51	.39	252.38	.34	221.69	.28	197.43	.20
.97	370.91	.43	328.32	.41	288.12	.38	252.04	.33	221.41	.27	197.23	.21
.98	370.48	.43	327.91	.42	287.74	.39	251.71	.34	221.14	.28	197.02	.20
.99	370.05	.43	327.49	.42	287.35	.38	251.37	.33	220.86	.27	196.82	.21
1.00	369.62	.43	327.07	.42	286.97	.38	251.04	.33	220.59	.27	196.61	.20

TABLE CX. ARGUMENT 1.

Equation = $345''.56 + 186''.51 \cos x + 10''.17 \cos 2x + 0''.63 \cos 3x + 0''.04 \cos 4x$.

	18		19		20		21		22	
Days.	Equation.	Diff.	Equation.	Diff.	Equation.	Diff.	Equation.	Diff.	Equation.	Diff.
d. .00	196.61	.20	179.77	.13	170.51	.05	169.03	.03	175.37	.10
.01	196.41	.21	179.64	.13	170.46	.05	169.06	.02	175.47	.11
.02	196.20	.20	179.51	.12	170.41	.06	169.08	.03	175.58	.10
.03	196.00	.21	179.39	.13	170.35	.05	169.11	.02	175.68	.11
.04	195.79	.20	179.26	.13	170.30	.05	169.13	.03	175.79	.10
.05	195.59	.20	179.13	.13	170.25	.05	169.16	.03	175.89	.11
.06	195.39	.20	179.00	.12	170.20	.05	169.19	.03	176.00	.11
.07	195.19	.20	178.88	.13	170.15	.04	169.22	.03	176.11	.10
.08	194.99	.20	178.75	.12	170.11	.05	169.25	.03	176.21	.11
.09	194.79	.20	178.63	.13	170.06	.05	169.28	.03	176.32	.11
.10	194.59	.20	178.50	.12	170.01	.04	169.31	.03	176.43	.11
.11	194.39	.19	178.38	.12	169.97	.05	169.34	.04	176.54	.11
.12	194.20	.20	178.26	.13	169.92	.04	169.38	.03	176.65	.12
.13	194.00	.19	178.13	.12	169.88	.05	169.41	.04	176.77	.11
.14	193.81	.20	178.01	.12	169.83	.04	169.45	.03	176.88	.11
.15	193.61	.19	177.89	.12	169.79	.04	169.48	.04	176.99	.12
.16	193.42	.19	177.77	.12	169.75	.04	169.52	.04	177.11	.11
.17	193.23	.20	177.65	.11	169.71	.04	169.56	.03	177.22	.12
.18	193.03	.19	177.54	.12	169.67	.04	169.59	.04	177.34	.11
.19	192.84	.19	177.42	.12	169.63	.04	169.63	.04	177.45	.12
.20	192.65	.19	177.30	.11	169.59	.04	169.67	.04	177.57	.12
.21	192.46	.19	177.19	.11	169.55	.04	169.71	.04	177.69	.12
.22	192.27	.18	177.08	.12	169.51	.03	169.75	.05	177.81	.12
.23	192.09	.19	176.96	.11	169.48	.04	169.80	.04	177.93	.12
.24	191.90	.19	176.85	.11	169.44	.04	169.84	.04	178.05	.12
.25	191.71	.18	176.74	.11	169.40	.03	169.88	.05	178.17	.12
.26	191.53	.19	176.63	.11	169.37	.03	169.93	.04	178.29	.12
.27	191.34	.18	176.52	.11	169.34	.04	169.97	.05	178.41	.13
.28	191.16	.19	176.41	.11	169.30	.03	170.02	.04	178.54	.12
.29	190.97	.18	176.30	.11	169.27	.03	170.06	.05	178.66	.12
.30	190.79	.18	176.19	.11	169.24	.03	170.11	.05	178.78	.13
.31	190.61	.18	176.08	.11	169.21	.03	170.16	.05	178.91	.13
.32	190.43	.19	175.97	.10	169.18	.02	170.21	.05	179.04	.12
.33	190.24	.18	175.87	.11	169.16	.03	170.26	.05	179.16	.13
.34	190.06	.18	175.76	.11	169.13	.03	170.31	.05	179.29	.13
.35	189.88	.18	175.65	.10	169.10	.02	170.36	.05	179.42	.13
.36	189.70	.18	175.55	.10	169.08	.03	170.41	.06	179.55	.13
.37	189.52	.17	175.45	.11	169.05	.02	170.47	.05	179.68	.13
.38	189.35	.18	175.34	.10	169.03	.03	170.52	.06	179.81	.13
.39	189.17	.18	175.24	.10	169.00	.02	170.58	.05	179.94	.13
.40	188.99	.17	175.14	.10	168.98	.02	170.63	.06	180.07	.13
.41	188.82	.18	175.04	.10	168.96	.02	170.69	.06	180.20	.14
.42	188.64	.17	174.94	.09	168.94	.03	170.75	.05	180.34	.13
.43	188.47	.18	174.85	.10	168.91	.02	170.80	.06	180.47	.14
.44	188.29	.17	174.75	.10	168.89	.02	170.86	.06	180.61	.13
.45	188.12	.17	174.65	.09	168.87	.02	170.92	.06	180.74	.14
.46	187.95	.17	174.56	.10	168.85	.01	170.98	.06	180.88	.14
.47	187.78	.17	174.46	.09	168.84	.02	171.04	.07	181.02	.14
.48	187.61	.17	174.37	.10	168.82	.01	171.11	.06	181.16	.14
.49	187.44	.17	174.27	.09	168.81	.02	171.17	.06	181.30	.14
.50	187.27	.17	174.18	.09	168.79	.01	171.23	.06	181.44	.14

TABLE CX. ARGUMENT 1.

Equation = $345''.58 + 186''.51 \cos x + 10''.17 \cos 2x + 0''.63 \cos 3x + 0''.04 \cos 4x$.

	18		19		20		21		22	
Days.	Equation.	Diff.	Equation.	Diff.	Equation.	Diff.	Equation.	Diff.	Equation.	Diff.
d.	"		"		"		"			
.50	187.27	.17	174.18	.09	168.79	.01	171.23	.06	181.44	.14
.51	187.10	.17	174.09	.09	168.78	.01	171.29	.07	181.58	.14
.52	186.93	.16	174.00	.10	168.77	.02	171.36	.06	181.72	.15
.53	186.77	.17	173.90	.09	168.75	.01	171.42	.07	181.87	.14
.54	186.60	.17	173.81	.09	168.74	.01	171.49	.06	182.01	.14
.55	186.43	.16	173.72	.09	168.73	.01	171.55	.07	182.15	.15
.56	186.27	.16	173.63	.08	168.72	.01	171.62	.07	182.30	.14
.57	186.11	.17	173.55	.09	168.71	.01	171.69	.07	182.44	.15
.58	185.94	.16	173.46	.08	168.70	.01	171.76	.07	182.59	.14
.59	185.78	.16	173.38	.09	168.69	.01	171.83	.07	182.73	.15
.60	185.62	.16	173.29	.08	168.68	.01	171.90	.07	182.88	.15
.61	185.46	.16	173.21	.09	168.67	.00	171.97	.07	183.03	.15
.62	185.30	.16	173.12	.08	168.67	.01	172.04	.08	183.18	.15
.63	185.14	.16	173.04	.09	168.66	.00	172.12	.07	183.33	.15
.64	184.98	.16	172.95	.08	168.66	.01	172.19	.07	183.48	.15
.65	184.82	.16	172.87	.08	168.65	.00	172.26	.08	183.63	.15
.66	184.66	.15	172.79	.08	168.65	.01	172.34	.08	183.78	.16
.67	184.51	.16	172.71	.07	168.64	.00	172.42	.07	183.94	.15
.68	184.35	.15	172.64	.08	168.64	.00	172.49	.08	184.09	.16
.69	184.20	.16	172.56	.08	168.64	.01	172.57	.08	184.25	.15
.70	184.04	.15	172.48	.08	168.65	.00	172.65	.08	184.40	.16
.71	183.89	.15	172.40	.07	168.65	.00	172.73	.08	184.56	.15
.72	183.74	.16	172.33	.08	168.65	.01	172.81	.09	184.71	.16
.73	183.58	.15	172.25	.07	168.66	.00	172.90	.08	184.87	.15
.74	183.43	.15	172.18	.08	168.66	.00	172.98	.08	185.02	.16
.75	183.28	.15	172.10	.07	168.66	.01	173.06	.08	185.18	.16
.76	183.13	.14	172.03	.07	168.67	.01	173.14	.09	185.34	.16
.77	182.99	.15	171.96	.08	168.68	.00	173.23	.08	185.50	.17
.78	182.84	.14	171.88	.07	168.68	.01	173.31	.09	185.67	.16
.79	182.70	.15	171.81	.07	168.69	.01	173.40	.08	185.83	.16
.80	182.55	.15	171.74	.07	168.70	.01	173.48	.09	185.99	.16
.81	182.40	.14	171.67	.07	168.71	.01	173.57	.09	186.15	.17
.82	182.26	.15	171.60	.06	168.72	.01	173.66	.09	186.32	.16
.83	182.11	.14	171.54	.07	168.73	.01	173.75	.09	186.48	.17
.84	181.97	.15	171.47	.07	168.74	.01	173.84	.09	186.65	.16
.85	181.82	.14	171.40	.06	168.75	.01	173.93	.09	186.81	.17
.86	181.68	.14	171.34	.06	168.76	.02	174.02	.09	186.98	.17
.87	181.54	.14	171.28	.07	168.78	.01	174.11	.10	187.15	.16
.88	181.40	.14	171.21	.06	168.79	.02	174.21	.09	187.31	.17
.89	181.26	.14	171.15	.06	168.81	.01	174.30	.09	187.48	.17
.90	181.12	.14	171.09	.06	168.82	.02	174.39	.10	187.65	.17
.91	180.98	.13	171.03	.06	168.84	.02	174.49	.09	187.82	.17
.92	180.85	.14	170.97	.06	168.86	.02	174.58	.10	187.99	.18
.93	180.71	.13	170.91	.06	168.88	.02	174.68	.09	188.17	.17
.94	180.58	.14	170.85	.06	168.90	.02	174.77	.10	188.34	.17
.95	180.44	.13	170.79	.06	168.92	.02	174.87	.10	188.51	.18
.96	180.31	.14	170.73	.05	168.94	.02	174.97	.10	188.69	.17
.97	180.17	.13	170.68	.06	168.96	.03	175.07	.10	188.86	.18
.98	180.04	.14	170.62	.05	168.99	.02	175.17	.10	189.04	.17
.99	179.90	.13	170.57	.06	169.01	.02	175.27	.10	189.21	.18
1.00	179.77	.13	170.51	.05	169.03	.03	175.37	.10	189.39	.18

TABLE CX. ARGUMENT 1.

Equation $= 345''.58 + 186''.51 \cos x + 10''.17 \cos 2x + 0''.63 \cos 3x + 0''.04 \cos 4x$.

	23		24		25		26		27	
Days.	Equation.	Diff.	Equation.	Diff.	Equation.	Diff.	Equation.	Diff.	Equation.	Diff.
d.	″		″		″		″		″	
.00	189.39	.18	210.74	.25	238.84	.31	272.85	.37	311.56	.41
.01	189.57	.18	210.99	.25	239.15	.32	273.22	.36	311.97	.40
.02	189.75	.18	211.24	.25	239.47	.31	273.58	.37	312.37	.41
.03	189.93	.18	211.49	.25	239.78	.32	273.95	.36	312.78	.40
.04	190.11	.18	211.74	.25	240.10	.31	274.31	.37	313.18	.41
.05	190.29	.18	211.99	.25	240.41	.32	274.68	.37	313.59	.41
.06	190.47	.18	212.24	.26	240.73	.32	275.05	.37	314.00	.41
.07	190.65	.19	212.50	.25	241.05	.31	275.42	.37	314.41	.40
.08	190.84	.18	212.75	.26	241.36	.32	275.79	.37	314.81	.41
.09	191.02	.18	213.01	.25	241.68	.32	276.16	.37	315.22	.41
.10	191.20	.19	213.26	.26	242.00	.32	276.53	.37	315.63	.41
.11	191.39	.19	213.52	.25	242.32	.32	276.90	.37	316.04	.41
.12	191.58	.18	213.77	.26	242.64	.31	277.27	.38	316.45	.41
.13	191.76	.19	214.03	.25	242.95	.32	277.65	.37	316.86	.41
.14	191.95	.19	214.28	.26	243.27	.32	278.02	.37	317.27	.41
.15	192.14	.19	214.54	.26	243.59	.32	278.39	.37	317.68	.41
.16	192.33	.19	214.80	.26	243.91	.33	278.76	.38	318.09	.41
.17	192.52	.19	215.06	.26	244.24	.32	279.14	.37	318.50	.41
.18	192.71	.19	215.32	.26	244.56	.33	279.51	.38	318.91	.41
.19	192.90	.19	215.58	.26	244.89	.32	279.89	.37	319.32	.41
.20	193.09	.19	215.84	.26	245.21	.32	280.26	.38	319.73	.41
.21	193.28	.20	216.10	.27	245.53	.33	280.64	.37	320.14	.42
.22	193.48	.19	216.37	.26	245.86	.32	281.01	.38	320.56	.41
.23	193.67	.20	216.63	.27	246.18	.33	281.39	.37	320.97	.42
.24	193.87	.19	216.90	.26	246.51	.32	281.76	.38	321.39	.41
.25	194.06	.20	217.16	.27	246.83	.33	282.14	.38	321.80	.41
.26	194.26	.20	217.43	.27	247.16	.33	282.52	.38	322.21	.42
.27	194.46	.19	217.70	.26	247.49	.32	282.90	.37	322.63	.41
.28	194.65	.20	217.96	.27	247.81	.33	283.27	.38	323.04	.42
.29	194.85	.20	218.23	.27	248.14	.33	283.65	.38	323.46	.41
.30	195.05	.20	218.50	.27	248.47	.33	284.03	.38	323.87	.41
.31	195.25	.20	218.77	.27	248.80	.33	284.41	.38	324.28	.42
.32	195.45	.20	219.04	.27	249.13	.34	284.79	.38	324.70	.41
.33	195.65	.20	219.31	.27	249.47	.33	285.17	.38	325.11	.42
.34	195.85	.20	219.58	.27	249.80	.33	285.55	.38	325.53	.41
.35	196.05	.21	219.85	.27	250.13	.33	285.93	.38	325.94	.42
.36	196.26	.20	220.12	.27	250.46	.34	286.31	.39	326.36	.41
.37	196.46	.21	220.39	.28	250.80	.33	286.70	.38	326.77	.42
.38	196.67	.20	220.67	.27	251.13	.34	287.08	.39	327.19	.41
.39	196.87	.21	220.94	.27	251.47	.33	287.47	.38	327.60	.42
.40	197.08	.21	221.21	.28	251.80	.34	287.85	.38	328.02	.42
.41	197.29	.21	221.49	.28	252.14	.33	288.23	.39	328.44	.42
.42	197.50	.20	221.77	.27	252.47	.34	288.62	.38	328.86	.41
.43	197.70	.21	222.04	.28	252.81	.33	289.00	.39	329.27	.42
.44	197.91	.21	222.32	.28	253.14	.34	289.39	.38	329.69	.42
.45	198.12	.21	222.60	.28	253.48	.34	289.77	.39	330.11	.42
.46	198.33	.21	222.88	.28	253.82	.34	290.16	.38	330.53	.42
.47	198.54	.22	223.16	.28	254.16	.33	290.54	.39	330.95	.41
.48	198.76	.21	223.44	.28	254.49	.34	290.93	.38	331.36	.42
.49	198.97	.21	223.72	.28	254.83	.34	291.31	.39	331.78	.42
.50	199.18	.22	224.00	.28	255.17	.34	291.70	.39	332.20	.42

TABLE CX. ARGUMENT 1.

Equation $= 345''.58 + 186''.51 \cos x + 10''.17 \cos 2x + 0''.63 \cos 3x + 0''.04 \cos 4x$.

	23		24		25		26		27	
Days.	Equation	Diff	Equation.	Diff.	Equation.	Diff.	Equation	Diff.	Equation.	Diff
d.	"		"		"		"		"	
.50	199.18	.22	224.00	.28	255.17	.34	291.70	.39	332.20	.42
.51	199.40	.21	224.28	.28	255.51	.34	292.09	.39	332.62	.42
.52	199.61	.22	224.56	.29	255.85	.35	292.48	.39	333.04	.42
.53	199.83	.21	224.85	.28	256.20	.34	292.87	.39	333.46	.42
.54	200.04	.22	225.13	.28	256.54	.34	293.26	.39	333.88	.42
.55	200.26	.22	225.41	.29	256.88	.35	293.65	.39	334.30	.42
.56	200.48	.22	225.70	.28	257.23	.34	294.04	.39	334.72	.42
.57	200.70	.21	225.98	.29	257.57	.35	294.43	.39	335.14	.43
.58	200.91	.22	226.27	.28	257.92	.34	294.82	.39	335.57	.42
.59	201.13	.22	226.55	.29	258.26	.35	295.21	.39	335.99	.42
.60	201.35	.22	226.84	.29	258.61	.35	295.60	.39	336.41	.42
.61	201.57	.22	227.13	.29	258.96	.34	295.99	.39	336.83	.42
.62	201.79	.23	227.42	.29	259.30	.35	296.38	.40	337.25	.43
.63	202.02	.22	227.71	.29	259.65	.34	296.78	.39	337.68	.42
.64	202.24	.22	228.00	.29	259.99	.35	297.17	.39	338.10	.42
.65	202.46	.23	228.29	.29	260.34	.35	297.56	.39	338.52	.42
.66	202.69	.22	228.58	.29	260.69	.35	297.95	.40	338.94	.42
.67	202.91	.23	228.87	.30	261.04	.35	298.35	.39	339.36	.43
.68	203.14	.22	229.17	.29	261.39	.35	298.74	.40	339.79	.42
.69	203.36	.23	229.46	.29	261.74	.35	299.14	.39	340.21	.42
.70	203.59	.23	229.75	.30	262.09	.35	299.53	.40	340.63	.42
.71	203.82	.23	230.05	.29	262.44	.35	299.93	.40	341.05	.43
.72	204.05	.23	230.34	.30	262.79	.36	300.33	.39	341.48	.42
.73	204.28	.23	230.64	.29	263.15	.35	300.72	.40	341.90	.43
.74	204.51	.23	230.93	.30	263.50	.35	301.12	.40	342.33	.42
.75	204.74	.23	231.23	.30	263.85	.36	301.52	.40	342.75	.42
.76	204.97	.24	231.53	.30	264.21	.35	301.92	.40	343.17	.43
.77	205.21	.23	231.83	.29	264.56	.36	302.32	.39	343.60	.42
.78	205.44	.24	232.12	.30	264.92	.35	302.71	.40	344.02	.43
.79	205.68	.23	232.42	.30	265.27	.36	303.11	.40	344.45	.42
.80	205.91	.24	232.72	.30	265.63	.36	303.51	.40	344.87	.43
.81	206.15	.23	233.02	.30	265.99	.35	303.91	.40	345.30	.42
.82	206.38	.24	233.32	.31	266.34	.36	304.31	.40	345.72	.43
.83	206.62	.23	233.63	.30	266.70	.35	304.71	.40	346.15	.42
.84	206.85	.24	233.93	.30	267.05	.36	305.11	.40	346.57	.43
.85	207.09	.24	234.23	.30	267.41	.36	305.51	.40	347.00	.43
.86	207.33	.24	234.53	.31	267.77	.36	305.91	.40	347.43	.42
.87	207.57	.24	234.84	.30	268.13	.36	306.31	.41	347.85	.43
.88	207.81	.24	235.14	.31	268.49	.36	306.72	.40	348.28	.42
.89	208.05	.24	235.45	.30	268.85	.36	307.12	.40	348.70	.43
.90	208.29	.24	235.75	.31	269.21	.36	307.52	.40	349.13	.43
.91	208.53	.25	236.06	.31	269.57	.36	307.92	.40	349.56	.43
.92	208.78	.24	236.37	.30	269.93	.37	308.32	.41	349.99	.42
.93	209.02	.25	236.67	.31	270.30	.36	308.73	.40	350.41	.43
.94	209.27	.24	236.98	.31	270.66	.36	309.13	.40	350.84	.43
.95	209.51	.25	237.29	.31	271.02	.37	309.53	.41	351.27	.43
.96	209.76	.24	237.60	.31	271.39	.36	309.94	.40	351.70	.43
.97	210.00	.25	237.91	.31	271.75	.37	310.31	.41	352.13	.42
.98	210.25	.24	238.22	.31	272.12	.36	310.75	.40	352.55	.43
.99	210.49	.25	238.53	.31	272.48	.37	311.15	.41	352.98	.43
1.00	210.74	.25	238.84	.31	272.85	.37	311.56	.41	353.41	.43

TABLE CXI. ARGUMENT 3.

Equation $= 29''.44 - 0''.95 \cos t + 28''.23 \cos 2t + 0''.26 \cos 4t$.

Days.	0	1	2	3	4	5	6	7	8	9	Days.
Days. .00	58.73	57″.33	50″.97	40″.89	28″.95	17″.33	8″.01	2″.53	1″.75	5″.78	Days. 1.00
.01	58.74	57.29	50.89	40.78	28.83	17.22	7.93	2.50	1.77	5.84	.99
.02	58.75	57.25	50.80	40.66	28.71	17.11	7.86	2.47	1.79	5.91	.98
.03	58.77	57.20	50.72	40.55	28.58	17.01	7.78	2.43	1.80	5.97	.97
.04	58.78	57.16	50.63	40.43	28.46	16.90	7.71	2.40	1.82	6.04	.96
.05	58.79	57.12	50.55	40.32	28.34	16.79	7.63	2.37	1.84	6.10	.95
.06	58.80	57.08	50.46	40.20	28.22	16.68	7.56	2.34	1.86	6.17	.94
.07	58.81	57.03	50.37	40.09	28.10	16.58	7.49	2.31	1.88	6.23	.93
.08	58.81	56.99	50.29	39.97	27.98	16.47	7.41	2.29	1.90	6.30	.92
.09	58.82	56.94	50.20	39.86	27.86	16.37	7.34	2.26	1.92	6.36	.91
.10	58.83	56.90	50.11	39.74	27.74	16.26	7.27	2.23	1.94	6.43	.90
.11	58.84	56.85	50.02	39.62	27.62	16.16	7.20	2.20	1.96	6.50	.89
.12	58.84	56.81	49.93	39.51	27.50	16.05	7.13	2.18	1.99	6.57	.88
.13	58.85	56.76	49.84	39.39	27.37	15.95	7.05	2.15	2.01	6.63	.87
.14	58.85	56.72	49.75	39.28	27.25	15.84	6.98	2.13	2.04	6.70	.86
.15	58.86	56.67	49.66	39.16	27.13	15.74	6.91	2.10	2.06	6.77	.85
.16	58.86	56.62	49.57	39.04	27.01	15.64	6.84	2.08	2.08	6.84	.84
.17	58.87	56.57	49.48	38.93	26.89	15.53	6.77	2.05	2.11	6.91	.83
.18	58.87	56.53	49.38	38.81	26.77	15.43	6.71	2.03	2.13	6.98	.82
.19	58.88	56.48	49.29	38.70	26.65	15.32	6.64	2.00	2.16	7.05	.81
.20	58.88	56.43	49.20	38.58	26.53	15.22	6.57	1.98	2.18	7.12	.80
.21	58.88	56.38	49.11	38.46	26.41	15.12	6.50	1.96	2.21	7.19	.79
.22	58.88	56.33	49.02	38.34	26.29	15.02	6.43	1.94	2.24	7.26	.78
.23	58.88	56.27	48.92	38.23	26.17	14.91	6.37	1.92	2.26	7.34	.77
.24	58.88	56.22	48.83	38.11	26.05	14.81	6.30	1.90	2.29	7.41	.76
.25	58.88	56.17	48.74	37.99	25.93	14.71	6.23	1.88	2.32	7.48	.75
.26	58.88	56.12	48.64	37.87	25.81	14.61	6.17	1.86	2.35	7.55	.74
.27	58.88	56.06	48.55	37.75	25.69	14.51	6.10	1.84	2.38	7.63	.73
.28	58.87	56.01	48.45	37.64	25.57	14.41	6.04	1.82	2.41	7.70	.72
.29	58.87	55.95	48.36	37.52	25.45	14.31	5.97	1.80	2.44	7.78	.71
.30	58.87	55.90	48.26	37.40	25.33	14.21	5.91	1.78	2.47	7.85	.70
.31	58.87	55.84	48.16	37.28	25.21	14.11	5.85	1.76	2.50	7.93	.69
.32	58.86	55.79	48.07	37.16	25.09	14.01	5.78	1.75	2.53	8.00	.68
.33	58.86	55.73	47.97	37.05	24.97	13.91	5.72	1.73	2.57	8.08	.67
.34	58.85	55.68	47.88	36.93	24.85	13.81	5.65	1.72	2.60	8.15	.66
.35	58.85	55.62	47.78	36.81	24.73	13.71	5.59	1.70	2.63	8.23	.65
.36	58.84	55.56	47.68	36.69	24.61	13.61	5.53	1.69	2.67	8.31	.64
.37	58.83	55.50	47.58	36.57	24.50	13.51	5.47	1.67	2.70	8.38	.63
.38	58.83	55.45	47.49	36.45	24.38	13.42	5.41	1.66	2.74	8.46	.62
.39	58.82	55.39	47.39	36.33	24.27	13.32	5.35	1.64	2.77	8.53	.61
.40	58.81	55.33	47.29	36.21	24.15	13.22	5.29	1.63	2.81	8.61	.60
.41	58.80	55.27	47.19	36.08	24.03	13.12	5.23	1.62	2.85	8.69	.59
.42	58.79	55.21	47.09	35.97	23.91	13.03	5.17	1.61	2.88	8.77	.58
.43	58.78	55.15	46.99	35.86	23.80	12.93	5.12	1.59	2.92	8.85	.57
.44	58.77	55.09	46.89	35.74	23.68	12.84	5.06	1.58	2.95	8.93	.56
.45	58.76	55.03	46.79	35.62	23.56	12.74	5.00	1.57	2.99	9.01	.55
.46	58.75	54.97	46.69	35.50	23.44	12.65	4.94	1.56	3.03	9.09	.54
.47	58.73	54.90	46.59	35.38	23.32	12.55	4.89	1.55	3.07	9.17	.53
.48	58.72	54.84	46.49	35.26	23.21	12.46	4.83	1.55	3.11	9.26	.52
.49	58.70	54.77	46.39	35.14	23.09	12.36	4.78	1.54	3.15	9.34	.51
.50	58.69	54.71	46.29	35.02	22.97	12.27	4.72	1.53	3.19	9.42	.50
Days.	29	28	27	26	25	24	23	22	21	20	Days.

TABLE CXI. ARGUMENT 3.

Equation = $29''.44 - 0''.95 \cos t + 28''.23 \cos 2t + 0''.26 \cos 4t$.

Days.	0	1	2	3	4	5	6	7	8	9	Days.
.50	58.69	54.71	46.29	35.02	22.97	12.27	4.72	1.53	3.19	9.42	.50
.51	58.67	54.65	46.19	34.90	22.85	12.18	4.67	1.52	3.23	9.50	.49
.52	58.66	54.58	46.09	34.78	22.73	12.08	4.61	1.52	3.27	9.58	.48
.53	58.64	54.52	45.98	34.65	22.62	11.99	4.56	1.51	3.32	9.67	.47
.54	58.63	54.45	45.88	34.53	22.50	11.89	4.50	1.51	3.36	9.75	.46
.55	58.61	54.39	45.78	34.41	22.38	11.80	4.45	1.50	3.40	9.83	.45
.56	58.59	54.32	45.68	34.29	22.26	11.71	4.40	1.50	3.44	9.92	.44
.57	58.57	54.25	45.57	34.17	22.15	11.62	4.35	1.49	3.49	10.00	.43
.58	58.56	54.19	45.47	34.05	22.03	11.53	4.29	1.49	3.53	10.09	.42
.59	58.54	54.12	45.36	33.93	21.92	11.44	4.24	1.48	3.58	10.17	.41
.60	58.52	54.05	45.26	33.81	21.80	11.35	4.19	1.48	3.62	10.26	.40
.61	58.50	53.98	45.15	33.69	21.69	11.26	4.14	1.48	3.67	10.35	.39
.62	58.48	53.91	45.05	33.57	21.57	11.17	4.09	1.48	3.71	10.43	.38
.63	58.46	53.84	44.94	33.44	21.46	11.08	4.04	1.47	3.76	10.52	.37
.64	58.44	53.77	44.84	33.32	21.34	10.99	3.99	1.47	3.80	10.60	.36
.65	58.42	53.70	44.73	33.20	21.23	10.90	3.94	1.47	3.85	10.69	.35
.66	58.40	53.63	44.62	33.08	21.12	10.81	3.89	1.47	3.90	10.78	.34
.67	58.37	53.56	44.52	32.96	21.00	10.72	3.84	1.47	3.95	10.87	.33
.68	58.35	53.48	44.41	32.84	20.89	10.64	3.80	1.47	3.99	10.95	.32
.69	58.32	53.41	44.31	32.72	20.77	10.55	3.75	1.47	4.04	11.04	.31
.70	58.30	53.34	44.20	32.60	20.66	10.46	3.70	1.47	4.09	11.13	.30
.71	58.27	53.27	44.09	32.48	20.55	10.37	3.65	1.47	4.14	11.22	.29
.72	58.25	53.19	43.98	32.36	20.43	10.29	3.61	1.48	4.19	11.31	.28
.73	58.22	53.12	43.88	32.23	20.32	10.20	3.57	1.48	4.25	11.40	.27
.74	58.20	53.04	43.77	32.11	20.20	10.12	3.52	1.49	4.30	11.49	.26
.75	58.17	52.97	43.66	31.99	20.09	10.03	3.48	1.49	4.35	11.58	.25
.76	58.14	52.89	43.55	31.87	19.98	9.95	3.44	1.50	4.40	11.67	.24
.77	58.11	52.82	43.44	31.75	19.87	9.86	3.39	1.50	4.45	11.76	.23
.78	58.09	52.74	43.34	31.62	19.75	9.78	3.35	1.51	4.51	11.86	.22
.79	58.06	52.67	43.23	31.50	19.64	9.69	3.30	1.51	4.56	11.95	.21
.80	58.03	52.59	43.12	31.38	19.53	9.61	3.26	1.52	4.61	12.04	.20
.81	58.00	52.51	43.01	31.26	19.42	9.53	3.22	1.53	4.67	12.13	.19
.82	57.97	52.43	42.90	31.14	19.31	9.44	3.18	1.54	4.72	12.22	.18
.83	57.93	52.36	42.79	31.02	19.19	9.35	3.14	1.54	4.78	12.32	.17
.84	57.90	52.28	42.68	30.90	19.08	9.27	3.10	1.55	4.83	12.41	.16
.85	57.87	52.20	42.57	30.78	18.97	9.19	3.06	1.56	4.89	12.50	.15
.86	57.84	52.12	42.46	30.66	18.86	9.11	3.02	1.57	4.95	12.59	.14
.87	57.80	52.04	42.35	30.54	18.75	9.03	2.98	1.58	5.00	12.69	.13
.88	57.77	51.96	42.23	30.41	18.64	8.95	2.95	1.59	5.06	12.78	.12
.89	57.73	51.88	42.12	30.29	18.53	8.87	2.91	1.60	5.11	12.88	.11
.90	57.70	51.80	42.01	30.17	18.42	8.79	2.87	1.61	5.17	12.97	.10
.91	57.66	51.72	41.90	30.05	18.31	8.71	2.84	1.62	5.23	13.07	.09
.92	57.63	51.64	41.79	29.93	18.20	8.63	2.80	1.63	5.29	13.16	.08
.93	57.59	51.55	41.67	29.80	18.09	8.55	2.77	1.65	5.35	13.26	.07
.94	57.56	51.47	41.56	29.68	17.98	8.47	2.73	1.66	5.41	13.35	.06
.95	57.52	51.39	41.45	29.56	17.87	8.39	2.70	1.67	5.47	13.45	.05
.96	57.48	51.31	41.34	29.44	17.76	8.31	2.67	1.69	5.53	13.55	.04
.97	57.44	51.22	41.23	29.32	17.65	8.24	2.63	1.70	5.59	13.65	.03
.98	57.41	51.14	41.11	29.19	17.55	8.16	2.60	1.72	5.66	13.74	.02
.99	57.37	51.05	41.00	29.07	17.44	8.09	2.56	1.73	5.72	13.84	.01
1.00	57.33	50.97	40.89	28.95	17.33	8.00	2.53	1.75	5.78	13.94	.00
Days.	29	28	27	26	25	24	23	22	21	20	Days.

TABLES CXI. CXII.

Tables	CXI.	CXII.
Arguments	3.	2.

Table CXI. Argument 3.

Days.	10	11	12	13	14	Days.
Days. .00	13.94	24.87	36.65	47.15	54.39	Days. 1.00
.01	14.04	24.99	36.76	47.24	54.44	.99
.02	14.14	25.10	36.88	47.33	54.49	.98
.03	14.24	25.22	36.99	47.42	54.54	.97
.04	14.34	25.33	37.11	47.51	54.59	.96
.05	14.44	25.45	37.22	47.60	54.64	.95
.06	14.54	25.57	37.33	47.69	54.69	.94
.07	14.64	25.69	37.45	47.78	54.74	.93
.08	14.74	25.80	37.56	47.86	54.78	.92
.09	14.84	25.92	37.68	47.95	54.83	.91
.10	14.94	26.04	37.79	48.04	54.88	.90
.11	15.04	26.16	37.90	48.13	54.92	.89
.12	15.14	26.28	38.02	48.22	54.97	.88
.13	15.25	26.39	38.13	48.30	55.01	.87
.14	15.35	26.51	38.25	48.39	55.06	.86
.15	15.45	26.63	38.36	48.48	55.10	.85
.16	15.55	26.75	38.47	48.57	55.14	.84
.17	15.65	26.87	38.58	48.65	55.18	.83
.18	15.76	26.98	38.70	48.74	55.23	.82
.19	15.86	27.10	38.81	48.82	55.27	.81
.20	15.96	27.22	38.92	48.91	55.31	.80
.21	16.06	27.34	39.03	48.99	55.35	.79
.22	16.17	27.46	39.14	49.08	55.39	.78
.23	16.27	27.57	39.26	49.16	55.43	.77
.24	16.38	27.69	39.37	49.25	55.47	.76
.25	16.48	27.81	39.48	49.33	55.51	.75
.26	16.59	27.93	39.59	49.41	55.55	.74
.27	16.69	28.05	39.70	49.49	55.59	.73
.28	16.80	28.16	39.81	49.58	55.62	.72
.29	16.90	28.28	39.92	49.66	55.66	.71
.30	17.01	28.40	40.03	49.74	55.70	.70
.31	17.12	28.52	40.14	49.82	55.74	.69
.32	17.22	28.64	40.25	49.90	55.77	.68
.33	17.33	28.76	40.36	49.98	55.81	.67
.34	17.43	28.88	40.47	50.06	55.84	.66
.35	17.54	29.00	40.58	50.14	55.88	.65
.36	17.65	29.12	40.69	50.22	55.91	.64
.37	17.76	29.24	40.80	50.30	55.94	.63
.38	17.86	29.35	40.90	50.37	55.98	.62
.39	17.97	29.47	41.01	50.45	56.01	.61
.40	18.08	29.59	41.12	50.53	56.04	.60
.41	18.19	29.71	41.23	50.61	56.07	.59
.42	18.30	29.83	41.34	50.68	56.10	.58
.43	18.40	29.95	41.44	50.76	56.13	.57
.44	18.51	30.07	41.55	50.83	56.16	.56
.45	18.62	30.19	41.66	50.91	56.19	.55
.46	18.73	30.31	41.77	50.98	56.22	.54
.47	18.84	30.43	41.87	51.06	56.24	.53
.48	18.95	30.54	41.98	51.13	56.26	.52
.49	19.06	30.66	42.08	51.21	56.29	.51
.50	19.17	30.78	42.19	51.28	56.32	.50
Days.	19	18	17	16	15	Days.

Table CXII. Argument 2.

Days.	0		10	
Days.	Equation.	Diff.	Equation.	Diff.
0.0	33.66	.68	66.71	.27
0.1	34.34	.68	66.44	.28
0.2	35.02	.68	66.16	.30
0.3	35.70	.67	65.86	.31
0.4	36.37	.68	65.55	.32
0.5	37.05	.68	65.23	.33
0.6	37.73	.67	64.90	.35
0.7	38.40	.68	64.55	.35
0.8	39.08	.68	64.20	.37
0.9	39.76	.67	63.83	.39
1.0	40.43	.67	63.44	.39
1.1	41.10	.67	63.05	.40
1.2	41.77	.66	62.65	.41
1.3	42.43	.67	62.24	.43
1.4	43.10	.66	61.81	.44
1.5	43.76	.66	61.37	.44
1.6	44.42	.65	60.93	.46
1.7	45.07	.65	60.47	.47
1.8	45.72	.65	60.00	.47
1.9	46.37	.64	59.53	.49
2.0	47.01	.64	59.04	.50
2.1	47.65	.63	58.54	.50
2.2	48.28	.63	58.04	.52
2.3	48.91	.62	57.52	.52
2.4	49.53	.62	57.00	.53
2.5	50.15	.61	56.47	.54
2.6	50.76	.60	55.93	.55
2.7	51.36	.60	55.38	.56
2.8	51.96	.60	54.82	.56
2.9	52.56	.58	54.26	.57
3.0	53.14	.58	53.69	.58
3.1	53.72	.57	53.11	.59
3.2	54.29	.57	52.52	.59
3.3	54.86	.55	51.93	.60
3.4	55.41	.55	51.33	.61
3.5	55.96	.54	50.72	.61
3.6	56.50	.53	50.11	.62
3.7	57.03	.52	49.49	.62
3.8	57.55	.52	48.87	.63
3.9	58.07	.50	48.24	.63
4.0	58.57	.50	47.61	.64
4.1	59.07	.48	46.97	.64
4.2	59.55	.48	46.33	.65
4.3	60.03	.47	45.68	.65
4.4	60.50	.45	45.03	.65
4.5	60.95	.45	44.38	.66
4.6	61.40	.43	43.72	.66
4.7	61.83	.43	43.06	.67
4.8	62.26	.41	42.39	.66
4.9	62.67	.41	41.73	.67
5.0	63.08	.39	41.06	.67

Tables	CXI.	CXII.
Arguments	2.	2.

Days.	10	11	12	13	14	Days.
Days.						Days.
.50	19.17	30.78	42.19	51.28	56.32	.50
.51	19.28	30.90	42.30	51.35	56.35	.49
.52	19.39	31.02	42.40	51.42	56.37	.48
.53	19.50	31.13	42.51	51.50	56.40	.47
.54	19.61	31.25	42.61	51.57	56.42	.46
.55	19.72	31.37	42.72	51.64	56.45	.45
.56	19.83	31.49	42.82	51.71	56.47	.44
.57	19.94	31.61	42.93	51.78	56.49	.43
.58	20.05	31.73	43.03	51.85	56.52	.42
.59	20.16	31.85	43.14	51.92	56.54	.41
.60	20.27	31.97	43.24	51.99	56.56	.40
.61	20.38	32.09	43.34	52.06	56.58	.39
.62	20.50	32.21	43.44	52.13	56.60	.38
.63	20.61	32.32	43.55	52.19	56.62	.37
.64	20.73	32.44	43.65	52.26	56.64	.36
.65	20.84	32.56	43.75	52.33	56.66	.35
.66	20.95	32.68	43.85	52.40	56.68	.34
.67	21.06	32.80	43.95	52.46	56.69	.33
.68	21.18	32.91	44.06	52.53	56.71	.32
.69	21.29	33.03	44.16	52.59	56.72	.31
.70	21.40	33.15	44.26	52.66	56.74	.30
.71	21.51	33.27	44.36	52.72	56.75	.29
.72	21.63	33.39	44.46	52.79	56.77	.28
.73	21.74	33.50	44.56	52.85	56.78	.27
.74	21.86	33.62	44.66	52.92	56.80	.26
.75	21.97	33.74	44.76	52.98	56.81	.25
.76	22.09	33.86	44.86	53.04	56.82	.24
.77	22.20	33.97	44.96	53.10	56.83	.23
.78	22.32	34.09	45.05	53.16	56.85	.22
.79	22.43	34.20	45.15	53.22	56.86	.21
.80	22.55	34.32	45.25	53.28	56.87	.20
.81	22.66	34.44	45.35	53.34	56.88	.19
.82	22.78	34.56	45.45	53.40	56.89	.18
.83	22.89	34.67	45.54	53.46	56.90	.17
.84	23.01	34.79	45.64	53.52	56.91	.16
.85	23.12	34.91	45.74	53.58	56.92	.15
.86	23.24	35.03	45.84	53.64	56.93	.14
.87	23.35	35.14	45.93	53.69	56.93	.13
.88	23.47	35.26	46.03	53.75	56.94	.12
.89	23.58	35.37	46.12	53.80	56.94	.11
.90	23.70	35.49	46.22	53.86	56.95	.10
.91	23.82	35.61	46.31	53.91	56.95	.09
.92	23.93	35.72	46.41	53.97	56.96	.08
.93	24.05	35.84	46.50	54.02	56.96	.07
.94	24.16	35.95	46.60	54.08	56.97	.06
.95	24.28	36.07	46.69	54.13	56.97	.05
.96	24.40	36.19	46.78	54.18	56.97	.04
.97	24.52	36.30	46.87	54.23	56.97	.03
.98	24.63	36.42	46.97	54.29	56.98	.02
.99	24.75	36.53	47.06	54.34	56.98	.01
1.00	24.87	36.65	47.15	54.39	56.98	.00
Days.	19	18	17	16	15	Days.

Days.	0		10	
Days.	Equation.	Diff.	Equation.	Diff.
5.0	63.08	.39	41.06	.67
5.1	63.47	.38	40.39	.68
5.2	63.85	.37	39.71	.67
5.3	64.22	.35	39.04	.68
5.4	64.57	.35	38.36	.67
5.5	64.92	.33	37.69	.68
5.6	65.25	.32	37.01	.68
5.7	65.57	.31	36.33	.67
5.8	65.88	.30	35.66	.68
5.9	66.18	.28	34.98	.68
6.0	66.46	.27	34.30	.68
6.1	66.73	.27	33.62	.67
6.2	67.00	.23	32.95	.68
6.3	67.23	.23	32.27	.67
6.4	67.46	.22	31.60	.68
6.5	67.68	.21	30.92	.67
6.6	67.89	.19	30.25	.66
6.7	68.08	.18	29.59	.67
6.8	68.26	.16	28.92	.66
6.9	68.42	.16	28.26	.66
7.0	68.58	.14	27.60	.66
7.1	68.72	.12	26.94	.65
7.2	68.84	.11	26.29	.66
7.3	68.95	.10	25.63	.64
7.4	69.05	.08	24.99	.64
7.5	69.13	.07	24.35	.64
7.6	69.20	.06	23.71	.63
7.7	69.26	.04	23.08	.63
7.8	69.30	.03	22.45	.62
7.9	69.33	.01	21.83	.62
8.0	69.34	.00	21.21	.61
8.1	69.34	.02	20.60	.61
8.2	69.32	.03	19.99	.60
8.3	69.29	.04	19.39	.59
8.4	69.25	.05	18.80	.59
8.5	69.20	.08	18.21	.58
8.6	69.12	.08	17.63	.58
8.7	69.04	.10	17.05	.56
8.8	68.94	.11	16.49	.56
8.9	68.83	.12	15.93	.55
9.0	68.71	.14	15.38	.55
9.1	68.57	.15	14.83	.53
9.2	68.42	.17	14.30	.53
9.3	68.25	.18	13.77	.52
9.4	68.07	.19	13.25	.52
9.5	67.88	.21	12.73	.50
9.6	67.67	.22	12.23	.49
9.7	67.45	.23	11.74	.49
9.8	67.22	.25	11.25	.48
9.9	66.97	.26	10.77	.46
10.0	66.71	.27	10.31	.46

Tables	CXII.				CXIII.				CXIV.	CXV.	
Arguments	2.				6.				7.	8.	

Days.	20		30	
Days.	Equation.	D.ff.	Equation.	Diff.
0.0	10.31	.46	21.79	.62
0.1	9.85	.45	22.41	.63
0.2	9.40	.44	23.04	.63
0.3	8.96	.42	23.67	.64
0.4	8.54	.42	24.31	.64
0.5	8.12	.41	24.95	.65
0.6	7.71	.40	25.60	.65
0.7	7.31	.38	26.25	.65
0.8	6.93	.38	26.90	.66
0.9	6.55	.36	27.56	.66
1.0	6.19	.36	28.22	.66
1.1	5.83	.34	28.88	.67
1.2	5.49	.33	29.55	.67
1.3	5.16	.32	30.22	.67
1.4	4.84	.31	30.89	.67
1.5	4.53	.30	31.56	.67
1.6	4.23	.29	32.23	.68
1.7	3.94	.27	32.91	.67
1.8	3.67	.27	33.58	.68
1.9	3.40	.25	34.26	.68
2.0	3.15	.23	34.94	.67
2.1	2.92	.23	35.61	.68
2.2	2.69	.22	36.29	.68
2.3	2.47	.20	36.97	.68
2.4	2.27	.19	37.65	.67
2.5	2.08	.17	38.32	.68
2.6	1.91	.17	39.00	.67
2.7	1.74	.15	39.67	.68
2.8	1.59	.14	40.35	.67
2.9	1.45	.13	41.02	.67
3.0	1.32	.12	41.69	.66
3.1	1.20	.10	42.35	.67
3.2	1.10	.09	43.02	.66
3.3	1.01	.07	43.68	.66
3.4	0.94	.07	44.34	.65
3.5	0.87	.05	44.99	.65
3.6	0.82	.04	45.64	.65
3.7	0.78	.02	46.29	.64
3.8	0.76	.02	46.93	.64
3.9	0.74	.00	47.57	.63
4.0	0.74	.01	48.20	.63
4.1	0.75	.03	48.83	.63
4.2	0.78	.04	49.46	.61
4.3	0.82	.05	50.07	.61
4.4	0.87	.06	50.68	.61
4.5	0.93	.08	51.29	.60
4.6	1.01	.09	51.89	.59
4.7	1.10	.10	52.48	.58
4.8	1.20	.11	53.06	.58
4.9	1.31	.13	53.64	.57
5.0	1.44	.14	54.21	.57

Days.	0	10	20	30	0	0	10
0.0	1.30	2.89	0.94	0.25	2.70	1.68	0.52
0.1	1.33	2.88	0.92	0.27	2.91	1.76	0.47
0.2	1.35	2.88	0.89	0.28	3.11	1.84	0.42
0.3	1.38	2.87	0.87	0.30	3.31	1.91	0.38
0.4	1.40	2.87	0.84	0.31	3.51	1.99	0.33
0.5	1.43	2.86	0.82	0.33	3.71	2.07	0.28
0.6	1.46	2.85	0.80	0.35	3.90	2.15	0.24
0.7	1.48	2.85	0.77	0.36	4.09	2.23	0.21
0.8	1.51	2.84	0.75	0.38	4.28	2.30	0.17
0.9	1.53	2.81	0.72	0.39	4.47	2.38	0.14
1.0	1.56	2.83	0.70	0.41	4.65	2.46	0.11
1.1	1.59	2.82	0.68	0.43	4.82	2.53	0.09
1.2	1.61	2.81	0.66	0.45	4.98	2.61	0.07
1.3	1.64	2.81	0.64	0.47	5.14	2.68	0.05
1.4	1.66	2.80	0.62	0.49	5.28	2.75	0.03
1.5	1.69	2.79	0.60	0.51	5.42	2.82	0.02
1.6	1.72	2.78	0.58	0.53	5.55	2.89	0.01
1.7	1.74	2.77	0.56	0.55	5.67	2.95	0.01
1.8	1.77	2.75	0.53	0.57	5.77	3.02	0.00
1.9	1.79	2.74	0.51	0.59	5.87	3.08	0.00
2.0	1.82	2.73	0.49	0.61	5.95	3.14	0.00
2.1	1.84	2.72	0.47	0.63	6.02	3.20	0.01
2.2	1.87	2.70	0.45	0.65	6.08	3.26	0.02
2.3	1.89	2.69	0.44	0.68	6.12	3.31	0.03
2.4	1.92	2.67	0.42	0.70	6.15	3.37	0.05
2.5	1.94	2.66	0.40	0.72	6.17	3.42	0.07
2.6	1.96	2.65	0.38	0.74	6.18	3.46	0.10
2.7	1.99	2.63	0.36	0.77	6.17	3.51	0.12
2.8	2.01	2.62	0.35	0.79	6.15	3.55	0.15
2.9	2.04	2.60	0.33	0.82	6.12	3.59	0.18
3.0	2.06	2.59	0.31	0.84	6.07	3.63	0.21
3.1	2.08	2.57	0.30	0.86	6.01	3.66	0.25
3.2	2.11	2.55	0.28	0.89	5.94	3.69	0.29
3.3	2.13	2.54	0.27	0.91	5.86	3.72	0.33
3.4	2.16	2.52	0.25	0.94	5.76	3.74	0.38
3.5	2.18	2.50	0.24	0.96	5.66	3.77	0.42
3.6	2.20	2.48	0.23	0.98	5.54	3.79	0.47
3.7	2.22	2.46	0.21	1.01	5.41	3.81	0.53
3.8	2.24	2.44	0.20	1.03	5.27	3.82	0.59
3.9	2.26	2.42	0.18	1.06	5.12	3.83	0.64
4.0	2.28	2.40	0.17	1.08	4.97	3.84	0.70
4.1	2.30	2.38	0.16	1.11	4.80	3.84	0.76
4.2	2.32	2.36	0.15	1.13	4.63	3.84	0.82
4.3	2.35	2.34	0.13	1.16	4.45	3.83	0.89
4.4	2.37	2.32	0.12	1.18	4.27	3.83	0.95
4.5	2.39	2.30	0.11	1.21	4.08	3.82	1.02
4.6	2.41	2.28	0.10	1.24	3.89	3.81	1.09
4.7	2.43	2.26	0.09	1.26	3.69	3.79	1.16
4.8	2.45	2.24	0.09	1.29	3.49	3.77	1.24
4.9	2.47	2.22	0.08	1.31	3.29	3.75	1.31
5.0	2.49	2.20	0.07	1.34	3.09	3.73	1.38

TABLES CXII.-CXV.

Tables	CXII.				CXIII.				CXIV.	CXV.	
Arguments	2.				6.				7.	8.	

Days.	20		30	
Days.	Equation.	Diff.	Equation.	Diff.
5.0	1.41	.14	51.21	.57
5.1	1.58	.15	54.78	.56
5.2	1.73	.17	55.34	.55
5.3	1.90	.17	55.89	.54
5.4	2.07	.19	56.43	.54
5.5	2.26	.20	56.97	.53
5.6	2.46	.22	57.50	.52
5.7	2.68	.22	58.02	.50
5.8	2.90	.24	58.52	.49
5.9	3.14	.25	59.01	.48
6.0	3.39	.26	59.49	.47
6.1	3.65	.27	59.96	.47
6.2	3.92	.29	60.43	.46
6.3	4.21	.30	60.89	.45
6.4	4.51	.31	61.34	.44
6.5	4.82	.32	61.78	.43
6.6	5.14	.33	62.21	.42
6.7	5.47	.34	62.63	.41
6.8	5.81	.35	63.04	.39
6.9	6.16	.37	63.43	.38
7.0	6.53	.37	63.81	.37
7.1	6.90	.39	64.18	.36
7.2	7.29	.40	64.54	.34
7.3	7.69	.40	64.88	.33
7.4	8.09	.42	65.21	.32
7.5	8.51	.43	65.53	.31
7.6	8.94	.44	65.84	.30
7.7	9.38	.44	66.14	.29
7.8	9.82	.46	66.43	.28
7.9	10.28	.47	66.71	.26
8.0	10.75	.47	66.97	.25
8.1	11.22	.49	67.22	.23
8.2	11.71	.49	67.45	.22
8.3	12.20	.50	67.67	.20
8.4	12.70	.51	67.87	.19
8.5	13.21	.52	68.06	.18
8.6	13.73	.53	68.24	.17
8.7	14.26	.54	68.41	.15
8.8	14.80	.54	68.56	.14
8.9	15.34	.56	68.70	.13
9.0	15.90	.56	68.83	.11
9.1	16.46	.56	68.94	.10
9.2	17.02	.58	69.04	.08
9.3	17.60	.58	69.12	.07
9.4	18.18	.58	69.19	.06
9.5	18.76	.60	69.25	.04
9.6	19.36	.60	69.29	.03
9.7	19.96	.60	69.32	.02
9.8	20.56	.61	69.34	.00
9.9	21.17	.62	69.34	.00
10.0	21.79	.62	69.34	.00

Days.	0	10	20	30	0	0	10
5.0	2.49	2.20	0.07	1.34	3.09	3.73	1.38
5.1	2.51	2.18	0.06	1.37	2.89	3.70	1.46
5.2	2.52	2.15	0.06	1.39	2.69	3.67	1.54
5.3	2.54	2.13	0.05	1.42	2.49	3.63	1.61
5.4	2.55	2.10	0.05	1.44	2.29	3.60	1.69
5.5	2.57	2.08	0.04	1.47	2.10	3.56	1.77
5.6	2.58	2.06	0.03	1.50	1.91	3.51	1.85
5.7	2.60	2.03	0.03	1.52	1.73	3.47	1.93
5.8	2.62	2.01	0.02	1.55	1.55	3.42	2.00
5.9	2.63	1.98	0.02	1.57	1.38	3.37	2.08
6.0	2.65	1.96	0.01	1.60	1.21	3.32	2.16
6.1	2.66	1.94	0.01	1.63	1.06	3.26	2.24
6.2	2.68	1.91	0.01	1.65	0.91	3.20	2.31
6.3	2.69	1.89	0.00	1.68	0.77	3.14	2.39
6.4	2.71	1.86	0.00	1.70	0.64	3.08	2.46
6.5	2.72	1.84	0.00	1.73	0.52	3.02	2.54
6.6	2.73	1.81	0.00	1.76	0.42	2.95	2.61
6.7	2.74	1.79	0.00	1.78	0.32	2.89	2.69
6.8	2.76	1.76	0.00	1.81	0.24	2.82	2.76
6.9	2.77	1.74	0.00	1.83	0.17	2.75	2.83
7.0	2.78	1.71	0.00	1.86	0.11	2.68	2.90
7.1	2.79	1.68	0.00	1.88	0.06	2.61	2.96
7.2	2.80	1.66	0.00	1.91	0.03	2.53	3.03
7.3	2.81	1.63	0.01	1.93	0.01	2.46	3.09
7.4	2.82	1.61	0.01	1.96	0.00	2.38	3.15
7.5	2.83	1.58	0.01	1.98	0.01	2.31	3.21
7.6	2.84	1.55	0.02	2.01	0.03	2.23	3.26
7.7	2.84	1.53	0.02	2.03	0.06	2.15	3.32
7.8	2.85	1.50	0.03	2.05	0.10	2.08	3.37
7.9	2.85	1.48	0.03	2.08	0.16	2.00	3.42
8.0	2.86	1.45	0.04	2.10	0.23	1.92	3.47
8.1	2.86	1.42	0.05	2.12	0.31	1.84	3.51
8.2	2.87	1.40	0.05	2.14	0.41	1.76	3.55
8.3	2.87	1.37	0.06	2.17	0.51	1.69	3.59
8.4	2.88	1.35	0.06	2.19	0.63	1.61	3.63
8.5	2.88	1.32	0.07	2.21	0.76	1.53	3.67
8.6	2.88	1.29	0.08	2.23	0.90	1.45	3.70
8.7	2.89	1.27	0.09	2.25	1.04	1.38	3.73
8.8	2.89	1.24	0.10	2.27	1.20	1.30	3.75
8.9	2.90	1.22	0.11	2.29	1.36	1.23	3.77
9.0	2.90	1.19	0.12	2.31	1.53	1.16	3.79
9.1	2.90	1.16	0.13	2.33	1.71	1.09	3.81
9.2	2.90	1.14	0.14	2.35	1.90	1.02	3.82
9.3	2.90	1.11	0.15	2.37	2.09	0.96	3.83
9.4	2.90	1.09	0.17	2.39	2.28	0.89	3.84
9.5	2.90	1.06	0.18	2.41	2.47	0.82	3.84
9.6	2.90	1.04	0.19	2.43	2.67	0.76	3.84
9.7	2.90	1.01	0.21	2.45	2.87	0.70	3.83
9.8	2.89	0.99	0.22	2.47	3.07	0.64	3.83
9.9	2.89	0.96	0.24	2.49	3.27	0.58	3.82
10.0	2.89	0.94	0.25	2.51	3.48	0.52	3.81

TABLES CXVI. - CXIX.

TABLES	CXVI.			CXVII.			CXVIII.			CXIX.
ARGUMENTS	9.			10.			11.			13.
Days.	0	10	20	0	10	20	0	10	20	0
Days. 0.0	1.14″	2.16″	0.15″	0.99″	0.31″	1.89″	0.80″	0.14″	1.41″	0.26″
0.1	1.16	2.15	0.14	0.97	0.33	1.89	0.78	0.15	1.41	0.29
0.2	1.19	2.14	0.13	0.95	0.35	1.88	0.77	0.16	1.41	0.32
0.3	1.21	2.12	0.12	0.92	0.36	1.88	0.75	0.17	1.42	0.35
0.4	1.24	2.11	0.11	0.90	0.38	1.87	0.74	0.18	1.42	0.39
0.5	1.26	2.10	0.10	0.88	0.40	1.87	0.72	0.19	1.42	0.42
0.6	1.28	2.08	0.09	0.86	0.42	1.86	0.70	0.20	1.42	0.46
0.7	1.31	2.07	0.08	0.83	0.44	1.85	0.69	0.21	1.42	0.50
0.8	1.33	2.06	0.07	0.81	0.46	1.85	0.67	0.23	1.42	0.54
0.9	1.36	2.04	0.06	0.78	0.48	1.84	0.66	0.24	1.42	0.58
1.0	1.38	2.03	0.05	0.76	0.50	1.83	0.64	0.25	1.42	0.62
1.1	1.40	2.01	0.04	0.74	0.52	1.82	0.62	0.26	1.42	0.66
1.2	1.43	1.99	0.04	0.72	0.54	1.81	0.61	0.28	1.42	0.70
1.3	1.45	1.98	0.03	0.69	0.57	1.80	0.59	0.29	1.41	0.73
1.4	1.48	1.96	0.03	0.67	0.59	1.79	0.58	0.31	1.41	0.77
1.5	1.50	1.94	0.02	0.65	0.61	1.78	0.56	0.32	1.41	0.80
1.6	1.52	1.92	0.02	0.63	0.63	1.77	0.54	0.33	1.41	0.83
1.7	1.54	1.90	0.02	0.61	0.65	1.75	0.53	0.35	1.40	0.87
1.8	1.57	1.89	0.01	0.58	0.68	1.74	0.51	0.36	1.40	0.90
1.9	1.59	1.87	0.01	0.56	0.70	1.72	0.50	0.38	1.39	0.93
2.0	1.61	1.85	0.01	0.54	0.72	1.71	0.48	0.39	1.39	0.96
2.1	1.63	1.83	0.00	0.52	0.74	1.70	0.46	0.41	1.38	0.99
2.2	1.65	1.81	0.00	0.50	0.76	1.68	0.45	0.42	1.38	1.02
2.3	1.68	1.78	0.00	0.48	0.79	1.67	0.43	0.44	1.37	1.04
2.4	1.70	1.76	0.00	0.46	0.81	1.65	0.42	0.45	1.37	1.07
2.5	1.72	1.74	0.00	0.44	0.83	1.64	0.40	0.47	1.36	1.09
2.6	1.74	1.72	0.00	0.42	0.85	1.62	0.39	0.49	1.35	1.11
2.7	1.76	1.70	0.00	0.40	0.88	1.60	0.37	0.50	1.34	1.12
2.8	1.79	1.67	0.01	0.39	0.90	1.59	0.36	0.52	1.34	1.14
2.9	1.81	1.65	0.01	0.37	0.93	1.57	0.34	0.53	1.33	1.15
3.0	1.83	1.63	0.01	0.35	0.95	1.55	0.33	0.55	1.32	1.17
3.1	1.85	1.61	0.01	0.33	0.97	1.53	0.32	0.57	1.31	1.18
3.2	1.87	1.59	0.02	0.31	1.00	1.51	0.30	0.58	1.30	1.19
3.3	1.88	1.56	0.02	0.30	1.02	1.50	0.29	0.60	1.29	1.19
3.4	1.90	1.54	0.03	0.28	1.05	1.48	0.27	0.61	1.28	1.20
3.5	1.92	1.52	0.03	0.26	1.07	1.46	0.26	0.63	1.27	1.20
3.6	1.94	1.50	0.04	0.25	1.09	1.44	0.25	0.65	1.26	1.20
3.7	1.96	1.47	0.04	0.23	1.11	1.42	0.24	0.66	1.25	1.19
3.8	1.97	1.45	0.05	0.22	1.14	1.40	0.22	0.68	1.24	1.19
3.9	1.99	1.42	0.05	0.20	1.16	1.38	0.21	0.69	1.23	1.18
4.0	2.01	1.40	0.06	0.19	1.18	1.36	0.20	0.71	1.22	1.17
4.1	2.03	1.38	0.07	0.18	1.20	1.34	0.19	0.73	1.21	1.15
4.2	2.04	1.35	0.08	0.16	1.22	1.32	0.18	0.74	1.20	1.14
4.3	2.06	1.33	0.09	0.15	1.25	1.29	0.17	0.76	1.18	1.12
4.4	2.07	1.30	0.10	0.13	1.27	1.27	0.16	0.77	1.17	1.10
4.5	2.09	1.28	0.11	0.12	1.29	1.25	0.15	0.79	1.16	1.08
4.6	2.10	1.26	0.12	0.11	1.31	1.23	0.14	0.81	1.15	1.05
4.7	2.12	1.23	0.13	0.10	1.33	1.21	0.13	0.82	1.13	1.03
4.8	2.13	1.21	0.14	0.09	1.36	1.18	0.12	0.84	1.12	1.00
4.9	2.15	1.18	0.15	0.08	1.38	1.16	0.11	0.85	1.10	0.98
5.0	2.16	1.16	0.16	0.07	1.40	1.14	0.10	0.87	1.09	0.95

TABLES CXVI.-CXIX.

Tables	CXVI.			CXVII.			CXVIII.			CXIX.
Arguments	9.			10.			11.			12.
Days.	0	10	20	0	10	20	0	10	20	0
5.0	2.16	1.16	0.16	0.07	1.40	1.14	0.10	0.87	1.09	0.95
5.1	2.17	1.14	0.17	0.06	1.42	1.12	0.09	0.89	1.08	0.92
5.2	2.18	1.11	0.19	0.05	1.44	1.09	0.08	0.90	1.06	0.88
5.3	2.19	1.09	0.20	0.05	1.46	1.07	0.08	0.92	1.05	0.85
5.4	2.20	1.06	0.22	0.04	1.48	1.04	0.07	0.93	1.03	0.81
5.5	2.21	1.04	0.23	0.03	1.50	1.02	0.06	0.95	1.02	0.78
5.6	2.22	1.02	0.25	0.03	1.52	1.00	0.05	0.97	1.00	0.74
5.7	2.23	0.99	0.26	0.02	1.54	0.98	0.05	0.98	0.99	0.71
5.8	2.24	0.97	0.28	0.02	1.55	0.95	0.04	1.00	0.97	0.67
5.9	2.25	0.94	0.29	0.01	1.57	0.93	0.04	1.01	0.96	0.64
6.0	2.26	0.92	0.31	0.01	1.59	0.91	0.03	1.03	0.94	0.60
6.1	2.27	0.90	0.33	0.01	1.61	0.89	0.03	1.04	0.92	0.56
6.2	2.27	0.87	0.35	0.01	1.62	0.86	0.02	1.06	0.91	0.53
6.3	2.28	0.85	0.36	0.00	1.64	0.84	0.02	1.07	0.89	0.49
6.4	2.28	0.82	0.38	0.00	1.65	0.81	0.01	1.09	0.88	0.46
6.5	2.29	0.80	0.40	0.00	1.67	0.79	0.01	1.10	0.86	0.42
6.6	2.29	0.78	0.42	0.00	1.68	0.77	0.01	1.11	0.84	0.39
6.7	2.30	0.76	0.44	0.00	1.70	0.75	0.01	1.13	0.83	0.35
6.8	2.30	0.73	0.45	0.00	1.71	0.72	0.00	1.14	0.81	0.32
6.9	2.31	0.71	0.47	0.00	1.73	0.70	0.00	1.16	0.80	0.28
7.0	2.31	0.69	0.49	0.01	1.74	0.68	0.00	1.17	0.78	0.25
7.1	2.31	0.67	0.51	0.01	1.75	0.66	0.00	1.18	0.76	0.22
7.2	2.31	0.65	0.53	0.01	1.76	0.64	0.00	1.19	0.75	0.20
7.3	2.32	0.62	0.56	0.01	1.78	0.61	0.00	1.21	0.73	0.17
7.4	2.32	0.60	0.58	0.02	1.79	0.59	0.00	1.22	0.72	0.15
7.5	2.32	0.58	0.60	0.02	1.80	0.57	0.00	1.23	0.70	0.12
7.6	2.32	0.56	0.62	0.03	1.81	0.55	0.00	1.24	0.68	0.10
7.7	2.32	0.54	0.64	0.04	1.82	0.53	0.00	1.25	0.67	0.08
7.8	2.32	0.51	0.67	0.04	1.82	0.51	0.01	1.26	0.65	0.07
7.9	2.32	0.49	0.69	0.05	1.83	0.49	0.01	1.27	0.64	0.05
8.0	2.31	0.47	0.71	0.06	1.84	0.47	0.01	1.28	0.62	0.04
8.1	2.31	0.45	0.73	0.07	1.85	0.45	0.01	1.29	0.60	0.03
8.2	2.31	0.43	0.75	0.08	1.86	0.43	0.02	1.30	0.59	0.02
8.3	2.30	0.42	0.78	0.08	1.86	0.41	0.02	1.31	0.57	0.01
8.4	2.30	0.40	0.80	0.09	1.87	0.39	0.03	1.32	0.56	0.01
8.5	2.30	0.38	0.82	0.10	1.88	0.37	0.03	1.33	0.54	0.00
8.6	2.29	0.36	0.84	0.11	1.88	0.35	0.04	1.34	0.52	0.00
8.7	2.29	0.34	0.87	0.12	1.88	0.33	0.04	1.34	0.50	0.01
8.8	2.28	0.33	0.89	0.14	1.89	0.32	0.05	1.35	0.49	0.01
8.9	2.28	0.31	0.92	0.15	1.89	0.30	0.05	1.35	0.48	0.02
9.0	2.27	0.29	0.94	0.16	1.89	0.28	0.06	1.36	0.46	0.03
9.1	2.26	0.27	0.96	0.17	1.90	0.26	0.07	1.37	0.44	0.04
9.2	2.25	0.26	0.99	0.19	1.90	0.25	0.07	1.37	0.43	0.06
9.3	2.24	0.24	1.01	0.20	1.90	0.23	0.08	1.38	0.41	0.07
9.4	2.23	0.23	1.04	0.22	1.90	0.22	0.08	1.38	0.40	0.09
9.5	2.22	0.22	1.06	0.23	1.90	0.20	0.09	1.39	0.38	0.11
9.6	2.21	0.20	1.08	0.25	1.90	0.19	0.10	1.39	0.37	0.13
9.7	2.20	0.19	1.11	0.26	1.90	0.18	0.11	1.40	0.35	0.16
9.8	2.18	0.17	1.13	0.28	1.89	0.16	0.12	1.40	0.34	0.18
9.9	2.17	0.16	1.16	0.29	1.89	0.15	0.13	1.41	0.32	0.21
10.0	2.16	0.15	1.18	0.31	1.89	0.14	0.14	1.41	0.31	0.24

TABLES CXX.-CXXIV.

Tables	CXX.			CXXI.		CXXII.		CXXIII.			CXXIV.
Arguments	14.			15.		16.		17.			18.
Days.	0	10	20	0	10	0	10	0	10	20	0
Days 0.0	″ 0.25	″ 0.03	″ 0.43	″ 0.43	″ 0.55	″ 0.64	″ 0.15	″ 0.13	″ 0.25	″ 0.00	″ 0.21
0.1	0.25	0.03	0.43	0.42	0.56	0.61	0.17	0.13	0.25	0.00	0.23
0.2	0.24	0.03	0.43	0.40	0.57	0.59	0.18	0.14	0.25	0.00	0.24
0.3	0.24	0.04	0.44	0.39	0.57	0.56	0.20	0.14	0.24	0.00	0.25
0.4	0.23	0.04	0.44	0.37	0.58	0.54	0.22	0.15	0.24	0.00	0.27
0.5	0.23	[illegible]	0.44	0.36	0.59	0.51	0.24	0.15	0.24	0.00	0.28
0.6	0.22	[illegible]	0.44	0.35	0.59	0.48	0.26	0.15	0.24	0.00	0.29
0.7	0.22	0.05	0.44	0.34	0.60	0.45	0.29	0.15	0.24	0.00	0.30
0.8	0.21	0.05	0.45	0.32	0.60	0.43	0.31	0.16	0.23	0.00	0.32
0.9	0.21	0.06	0.45	0.31	0.61	0.40	0.34	0.16	0.23	0.00	0.33
1.0	0.20	0.06	0.45	0.30	0.61	0.37	0.37	0.16	0.23	0.00	0.34
1.1	0.19	0.06	0.45	0.29	0.61	0.34	0.40	0.17	0.23	0.00	0.35
1.2	0.19	0.06	0.45	0.27	0.61	0.31	0.43	0.17	0.23	0.00	0.36
1.3	0.18	0.07	0.46	0.26	0.62	0.29	0.45	0.17	0.22	0.00	0.37
1.4	0.18	0.07	0.46	0.24	0.62	0.26	0.48	0.18	0.22	0.00	0.38
1.5	0.17	0.07	0.46	0.23	0.62	0.24	0.51	0.18	0.22	0.00	0.39
1.6	0.17	0.07	0.46	0.22	0.62	0.22	0.54	0.18	0.21	0.00	0.40
1.7	0.16	0.08	0.46	0.21	0.62	0.20	0.56	0.18	0.21	0.00	0.41
1.8	0.16	0.08	0.46	0.19	0.61	0.18	0.59	0.19	0.21	0.00	0.42
1.9	0.15	0.09	0.46	0.18	0.61	0.17	0.61	0.19	0.20	0.00	0.42
2.0	0.15	0.09	0.46	0.17	0.61	0.15	0.64	0.19	0.20	0.00	0.43
2.1	0.15	0.09	0.46	0.16	0.61	0.14	0.66	0.20	0.20	0.01	0.43
2.2	0.14	0.10	0.46	0.15	0.60	0.13	0.68	0.20	0.20	0.01	0.43
2.3	0.14	0.10	0.46	0.13	0.60	0.12	0.70	0.20	0.19	0.01	0.44
2.4	0.13	0.11	0.46	0.12	0.59	0.12	0.71	0.21	0.19	0.01	0.44
2.5	0.13	0.11	0.46	0.11	0.59	0.11	0.73	0.21	0.19	0.01	0.44
2.6	0.13	0.11	0.46	0.10	0.58	0.11	0.74	0.21	0.18	0.01	0.44
2.7	0.12	0.12	0.46	0.09	0.58	0.11	0.75	0.21	0.18	0.01	0.44
2.8	0.12	0.12	0.46	0.08	0.57	0.11	0.76	0.22	0.18	0.02	0.43
2.9	0.11	0.13	0.46	0.07	0.57	0.12	0.77	0.22	0.17	0.02	0.43
3.0	0.11	0.13	0.46	0.06	0.56	0.12	0.78	0.22	0.17	0.02	0.43
3.1	0.11	0.14	0.46	0.05	0.55	0.13	0.78	0.22	0.17	0.02	0.42
3.2	0.10	0.14	0.46	0.05	0.54	0.14	0.78	0.22	0.17	0.02	0.42
3.3	0.10	0.15	0.45	0.04	0.53	0.15	0.77	0.23	0.16	0.02	0.41
3.4	0.09	0.15	0.45	0.04	0.52	0.16	0.77	0.23	0.16	0.02	0.41
3.5	0.09	0.16	0.45	0.03	0.51	0.17	0.76	0.23	0.16	0.03	0.40
3.6	0.09	0.16	0.45	0.03	0.50	0.19	0.75	0.23	0.15	0.03	0.39
3.7	0.08	0.17	0.45	0.02	0.49	0.21	0.74	0.23	0.15	0.03	0.38
3.8	0.08	0.17	0.44	0.02	0.48	0.23	0.72	0.24	0.15	0.03	0.37
3.9	0.07	0.18	0.44	0.01	0.47	0.25	0.71	0.24	0.14	0.03	0.36
4.0	0.07	0.18	0.44	0.01	0.46	0.27	0.69	0.24	0.14	0 04	0.35
4.1	0.07	0.19	0.44	0.01	0.45	0.29	0.67	0.24	0.14	0.04	0.34
4.2	0.06	0.19	0.44	0.01	0.43	0.31	0.65	0.24	0.13	0.04	0.33
4.3	0.06	0.20	0.43	0.00	0.42	0.33	0.63	0.24	0.13	0.04	0.32
4.4	0.05	0.20	0.43	0.00	0.40	0.35	0.60	0.25	0.13	0 05	0.30
4.5	0.05	0.21	0.43	0.00	0.39	0.37	0.58	0.25	0.12	0.05	0.29
4.6	0.05	0.21	0.43	0.00	0.38	0.39	0.55	0.25	0.12	0.05	0.28
4.7	0.05	0.22	0.43	0.00	0.36	0.41	0.53	0.25	0.12	0.05	0.26
4.8	0.04	0.22	0.42	0.01	0.35	0.43	0.50	0.26	0.11	0.06	0.25
4.9	0.04	0.23	0.42	0.01	0.33	0.45	0.47	0.26	0.11	0.06	0.23
5.0	0.04	0.23	0.42	0.01	0.32	0.47	0.44	0.26	0.11	0.06	0.22

TABLES CXX.–CXXIV.

Tables	CXX.			CXXI.		CXXII.		CXXIII.			CXXIV.
Arguments	14.			15.		16.		17.			18.
Days.	0	10	20	0	10	0	10	0	10	20	0
Days 5.0	″0.04	″0.23	″0.42	″0.01	″0.32	″0.47	″0.44	0.26	″0.11	″0.06	″0.22
5.1	0.04	0.23	0.42	0.01	0.31	0.48	0.41	0.26	0.10	0.06	0.21
5.2	0.04	0.24	0.42	0.02	0.29	0.50	0.39	0.26	0.10	0.06	0.19
5.3	0.03	0.24	0.41	0.02	0.28	0.51	0.36	0.27	0.10	0.07	0.18
5.4	0.03	0.25	0.41	0.03	0.26	0.52	0.34	0.27	0.09	0.07	0.16
5.5	0.03	0.25	0.41	0.03	0.25	0.53	0.31	0.27	0.09	0.07	0.15
5.6	0.03	0.26	0.41	0.04	0.24	0.54	0.29	0.27	0.09	0.07	0.14
5.7	0.03	0.26	0.40	0.05	0.23	0.55	0.26	0.27	[illegible]	0.08	0.13
5.8	0.02	0.27	0.40	0.05	0.21	0.55	0.24	0.27	0.08	0.08	0.11
5.9	0.02	0.27	0.39	0.06	0.20	0.56	0.21	0.27	0.08	0.08	0.10
6.0	0.02	0.28	0.39	0.07	0.19	0.56	0.19	0.28	0.08	0.09	0.09
6.1	0.02	0.28	0.39	0.08	0.18	0.56	0.17	0.28	0.08	0.09	0.08
6.2	0.02	0.29	0.38	0.09	0.17	0.55	0.16	0.28	0.07	0.09	0.07
6.3	0.01	0.29	0.38	0.10	0.15	0.55	0.14	0.28	0.07	0.10	0.06
6.4	0.01	0.30	0.37	0.11	0.14	0.54	0.13	0.28	0.07	0.10	0.05
6.5	0.01	0.30	0.37	0.12	0.13	0.53	0.12	0.28	0.06	0.10	0.04
6.6	0.01	0.31	0.37	0.13	0.12	0.52	0.12	0.28	0.06	0.11	0.03
6.7	0.01	0.31	0.36	0.14	0.11	0.51	0.11	0.28	0.06	0.11	0.03
6.8	0.00	0.32	0.36	0.16	0.10	0.50	0.11	0.28	0.05	0.11	0.02
6.9	0.00	0.32	0.35	0.17	0.09	0.48	0.11	0.28	0.05	0.12	0.02
7.0	0.00	0.33	0.35	0.18	0.08	0.47	0.11	0.28	0.05	0.12	0.01
7.1	0.00	0.33	0.35	0.19	0.07	0.45	0.12	0.28	0.05	0.12	0.01
7.2	0.00	0.34	0.34	0.20	0.06	0.43	0.12	0.28	0.05	0.13	0.01
7.3	0.00	0.34	0.34	0.22	0.05	0.41	0.13	0.28	0.04	0.13	0.00
7.4	0.00	0.35	0.33	0.23	0.05	0.39	0.13	0.28	0.04	0.13	0.00
7.5	0.00	0.35	0.33	0.24	0.04	0.37	0.14	0.28	0.04	0.14	0.00
7.6	0.00	0.35	0.33	0.25	0.03	0.35	0.15	0.28	0.04	0.14	0.00
7.7	0.00	0.36	0.32	0.27	0.03	0.33	0.16	0.28	0.04	0.14	0.00
7.8	0.00	0.36	0.32	0.28	0.02	0.31	0.18	0.28	0.03	0.15	0.01
7.9	0.00	0.37	0.31	0.30	0.02	0.29	0.20	0.28	0.03	0.15	0.01
8.0	0.00	0.37	0.31	0.31	0.01	0.27	0.22	0.28	0.03	0.15	0.01
8.1	0.00	0.37	0.31	0.32	0.01	0.25	0.24	0.28	0.03	0.16	0.02
8.2	0.00	0.38	0.30	0.34	0.01	0.23	0.26	0.28	0.03	0.16	0.02
8.3	0.00	0.38	0.30	0.35	0.00	0.21	0.28	0.27	0.02	0.16	0.03
8.4	0.00	0.39	0.29	0.37	0.00	0.19	0.30	0.27	0.02	0.17	0.04
8.5	0.00	0.39	0.29	0.38	0.00	0.17	0.32	0.27	0.02	0.17	0.05
8.6	0.00	0.39	0.29	0.39	0.00	0.16	0.34	0.27	0.02	0.17	0.06
8.7	0.00	0.39	0.28	0.40	0.00	0.15	0.36	0.27	0.02	0.17	0.07
8.8	0.01	0.40	0.28	0.42	0.00	0.14	0.38	0.27	0.01	0.18	0.08
8.9	0.01	0.40	0.27	0.43	0.00	0.13	0.40	0.27	0.01	0.18	0.09
9.0	0.01	0.40	0.27	0.44	0.00	0.12	0.42	0.27	0.01	0.18	0.10
9.1	0.01	0.40	0.26	0.45	0.01	0.12	0.44	0.27	0.01	0.19	0.11
9.2	0.01	0.41	0.25	0.46	0.01	0.11	0.46	0.27	0.01	0.19	0.12
9.3	0.02	0.41	0.25	0.48	0.01	0.11	0.48	0.26	0.01	0.19	0.14
9.4	0.02	0.42	0.24	0.49	0.02	0.11	0.49	0.26	0.01	0.20	0.15
9.5	0.02	0.42	0.23	0.50	0.02	0.11	0.51	0.26	0.01	0.20	0.16
9.6	0.02	0.42	0.23	0.51	0.03	0.12	0.52	0.26	0.01	0.20	0.17
9.7	0.02	0.42	0.22	0.52	0.03	0.12	0.53	0.26	0.01	0.20	0.19
9.8	0.03	0.43	0.22	0.53	0.04	0.13	0.54	0.25	0.00	0.21	0.20
9.9	0.03	0.43	0.21	0.54	0.04	0.14	0.54	0.25	0.00	0.21	0.22
10.0	0.03	0.43	0.21	0.55	0.05	0.15	0.55	0.25	0.00	0.21	0.23

TABLES CXXV.-CXXVIII.

Tables	CXXV.			CXXVI.		CXXVII.				CXXVIII.		
Arguments	**19.**			**20.**		**22.**				**23.**		
Days.	**0**	**10**	**20**	**0**	**10**	**0**	**10**	**20**	**30**	**0**	**10**	**20**
Days. 0.0	″ 0.15	″ 0.04	″ 0.24	″ 0.08	″ 0.03	″ 0.09	″ 0.03	″ 0.18	″ 0.02	″ 0.09	″ 0.05	″ 0.17
0.1	0.15	0.04	0.24	0.08	0.03	0.09	0.02	0.18	0.03	0.09	0.05	0.17
0.2	0.14	0.04	0.24	0.09	0.02	0.10	0.02	0.18	0.03	0.09	0.05	0.17
0.3	0.14	0.05	0.23	0.09	0.02	0.10	0.02	0.18	0.03	0.08	0.05	0.16
0.4	0.13	0.05	0.23	0.10	0.01	0.11	0.02	0.18	0.04	0.08	0.05	0.16
0.5	0.13	0.05	0.23	0.10	0.01	0.11	0.01	0.18	0.04	0.08	0.06	0.16
0.6	0.13	0.05	0.23	0.11	0.01	0.11	0.01	0.18	0.04	0.08	0.06	0.16
0.7	0.12	0.05	0.23	0.11	0.01	0.12	0.01	0.18	0.04	0.08	0.06	0.16
0.8	0.12	0.06	0.23	0.12	0.00	0.12	0.01	0.17	0.05	0.07	0.07	0.15
0.9	0.11	0.06	0.23	0.12	0.00	0.13	0.01	0.17	0.05	0.07	0.07	0.15
1.0	0.11	0.06	0.23	0.13	0.00	0.13	0.01	0.17	0.05	0.07	0.07	0.15
1.1	0.11	0.07	0.23	0.13	0.00	0.13	0.01	0.17	0.05	0.07	0.07	0.15
1.2	0.11	0.07	0.23	0.14	0.00	0.13	0.01	0.17	0.06	0.07	0.07	0.15
1.3	0.10	0.07	0.22	0.14	0.00	0.14	0.00	0.17	0.06	0.06	0.08	0.14
1.4	0.10	0.08	0.22	0.15	0.00	0.14	0.00	0.17	0.06	0.06	0.08	0.14
1.5	0.10	0.08	0.22	0.15	0.00	0.14	0.00	0.17	0.07	0.06	0.08	0.14
1.6	0.10	0.08	0.22	0.16	0.00	0.14	0.00	0.16	0.07	0.06	0.08	0.14
1.7	0.09	0.08	0.22	0.16	0.00	0.14	0.00	0.16	0.08	0.06	0.08	0.14
1.8	0.09	0.09	0.21	0.17	0.00	0.15	0.00	0.16	0.08	0.05	0.09	0.13
1.9	0.08	0.09	0.21	0.17	0.00	0.15	0.00	0.16	0.09	0.05	0.09	0.13
2.0	0.08	0.09	0.21	0.18	0.00	0.15	0.00	0.15	0.09	0.05	0.09	0.13
2.1	0.08	0.09	0.21	0.18	0.00	0.16	0.00	0.15	0.09	0.05	0.09	0.13
2.2	0.08	0.09	0.21	0.19	0.00	0.16	0.00	0.15	0.10	0.05	0.09	0.13
2.3	0.07	0.10	0.20	0.19	0.01	0.16	0.00	0.15	0.10	0.04	0.10	0.12
2.4	0.07	0.10	0.20	0.20	0.01	0.16	0.00	0.14	0.11	0.04	0.10	0.12
2.5	0.07	0.10	0.20	0.20	0.01	0.17	0.00	0.14	0.11	0.04	0.10	0.12
2.6	0.07	0.11	0.19	0.20	0.01	0.17	0.00	0.14	0.11	0.04	0.10	0.12
2.7	0.07	0.11	0.19	0.21	0.01	0.17	0.00	0.14	0.11	0.04	0.10	0.12
2.8	0.06	0.11	0.19	0.21	0.02	0.17	0.01	0.13	0.12	0.03	0.11	0.11
2.9	0.06	0.12	0.18	0.22	0.02	0.17	0.01	0.13	0.12	0.03	0.11	0.11
3.0	0.06	0.12	0.18	0.22	0.02	0.17	0.01	0.13	0.12	0.03	0.11	0.11
3.1	0.06	0.12	0.18	0.22	0.03	0.17	0.01	0.12	0.12	0.03	0.11	0.11
3.2	0.05	0.13	0.18	0.22	0.03	0.17	0.01	0.12	0.13	0.03	0.12	0.11
3.3	0.05	0.13	0.17	0.23	0.03	0.18	0.02	0.12	0.13	0.02	0.12	0.10
3.4	0.04	0.13	0.17	0.23	0.04	0.18	0.02	0.11	0.13	0.02	0.12	0.10
3.5	0.04	0.14	0.17	0.23	0.04	0.18	0.02	0.11	0.14	0.02	0.12	0.10
3.6	0.04	0.14	0.17	0.23	0.04	0.18	0.02	0.11	0.14	0.02	0.13	0.10
3.7	0.04	0.14	0.17	0.23	0.05	0.18	0.02	0.10	0.14	0.02	0.13	0.10
3.8	0.03	0.15	0.16	0.24	0.05	0.18	0.03	0.10	0.14	0.01	0.13	0.09
3.9	0.03	0.15	0.16	0.24	0.06	0.18	0.03	0.09	0.15	0.01	0.13	0.09
4.0	0.03	0.15	0.16	0.24	0.06	0.18	0.03	0.09	0.15	0.01	0.13	0.09
4.1	0.03	0.15	0.15	0.24	0.07	0.18	0.03	0.09	0.15	0.01	0.14	0.09
4.2	0.03	0.15	0.15	0.24	0.07	0.18	0.03	0.08	0.15	0.01	0.14	0.08
4.3	0.02	0.16	0.15	0.24	0.08	0.18	0.04	0.08	0.15	0.01	0.14	0.08
4.4	0.02	0.16	0.14	0.24	0.08	0.18	0.04	0.07	0.16	0.01	0.14	0.08
4.5	0.02	0.16	0.14	0.24	0.09	0.18	0.04	0.07	0.16	0.01	0.14	0.08
4.6	0.02	0.17	0.14	0.24	0.09	0.18	0.04	0.07	0.16	0.01	0.15	0.07
4.7	0.02	0.17	0.13	0.24	0.10	0.18	0.05	0.07	0.16	0.01	0.15	0.07
4.8	0.01	0.17	0.13	0.24	0.10	0.17	0.05	0.06	0.16	0.00	0.15	0.07
4.9	0.01	0.18	0.13	0.24	0.11	0.17	0.05	0.06	0.17	0.00	0.15	0.06
5.0	0.01	0.18	0.12	0.24	0.11	0.17	0.06	0.06	0.17	0.00	0.15	0.06

TABLES CXXV.-CXXVIII.

Tables	CXXV.			CXXVI.		CXXVII.				CXXVIII.		
Arguments	19.			20.		22.				23.		
Days.	0	10	20	0	10	0	10	20	30	0	10	20
5.0	0.01	0.18	0.12	0.24	0.11	0.17	0.06	0.06	0.17	0.00	0.15	0.06
5.1	0.01	0.18	0.12	0.24	0.12	0.17	0.06	0.06	0.17	0.00	0.15	0.06
5.2	0.01	0.18	0.12	0.24	0.12	0.17	0.06	0.05	0.17	0.00	0.16	0.06
5.3	0.01	0.19	0.11	0.23	0.13	0.16	0.07	0.05	0.17	0.00	0.16	0.05
5.4	0.01	0.19	0.11	0.23	0.13	0.16	0.07	0.05	0.18	0.00	0.16	0.05
5.5	0.01	0.19	0.11	0.23	0.14	0.16	0.07	0.04	0.18	0.00	0.16	0.05
5.6	0.01	0.19	0.10	0.23	0.14	0.16	0.08	0.04	0.18	0.00	0.16	0.05
5.7	0.01	0.19	0.10	0.22	0.15	0.16	0.08	0.04	0.18	0.00	0.16	0.05
5.8	0.00	0.20	0.10	0.22	0.15	0.15	0.08	0.04	0.18	0.00	0.17	0.04
5.9	0.00	0.20	0.09	0.21	0.16	0.15	0.09	0.03	0.18	0.00	0.17	0.04
6.0	0.00	0.20	0.09	0.21	0.16	0.15	0.09	0.03	0.18	0.00	0.17	0.04
6.1	0.00	0.20	0.09	0.21	0.17	0.15	0.09	0.03	0.18	0.00	0.17	0.04
6.2	0.00	0.20	0.09	0.20	0.17	0.15	0.10	0.03	0.18	0.00	0.17	0.04
6.3	0.00	0.21	0.08	0.20	0.18	0.14	0.10	0.03	0.18	0.00	0.17	0.03
6.4	0.00	0.21	0.08	0.19	0.18	0.14	0.10	0.02	0.18	0.00	0.17	0.03
6.5	0.00	0.21	0.08	0.19	0.19	0.14	0.11	0.02	0.18	0.00	0.17	0.03
6.6	0.00	0.21	0.07	0.19	0.19	0.14	0.11	0.02	0.18	0.00	0.17	0.03
6.7	0.00	0.21	0.07	0.18	0.20	0.13	0.11	0.02	0.18	0.00	0.17	0.03
6.8	0.00	0.21	0.07	0.18	0.20	0.13	0.12	0.02	0.17	0.00	0.18	0.03
6.9	0.00	0.22	0.06	0.17	0.21	0.13	0.12	0.01	0.17	0.00	0.18	0.02
7.0	0.00	0.22	0.06	0.17	0.21	0.12	0.12	0.01	0.17	0.00	0.18	0.02
7.1	0.00	0.22	0.06	0.17	0.21	0.12	0.13	0.01	0.17	0.00	0.18	0.02
7.2	0.00	0.22	0.06	0.16	0.21	0.12	0.13	0.01	0.17	0.00	0.18	0.02
7.3	0.00	0.22	0.05	0.16	0.22	0.11	0.13	0.01	0.17	0.01	0.18	0.02
7.4	0.00	0.22	0.05	0.15	0.22	0.11	0.14	0.00	0.17	0.01	0.18	0.02
7.5	0.00	0.23	0.05	0.15	0.22	0.11	0.14	0.00	0.17	0.01	0.18	0.02
7.6	0.00	0.23	0.05	0.14	0.22	0.11	0.14	0.00	0.16	0.01	0.18	0.02
7.7	0.00	0.23	0.05	0.14	0.22	0.10	0.14	0.00	0.16	0.01	0.18	0.02
7.8	0.01	0.23	0.04	0.13	0.23	0.10	0.15	0.00	0.16	0.01	0.18	0.01
7.9	0.01	0.23	0.04	0.13	0.23	0.10	0.15	0.00	0.16	0.01	0.18	0.01
8.0	0.01	0.23	0.04	0.12	0.23	0.09	0.15	0.00	0.16	0.01	0.18	0.01
8.1	0.01	0.23	0.04	0.11	0.23	0.09	0.15	0.00	0.15	0.01	0.18	0.01
8.2	0.01	0.23	0.04	0.11	0.23	0.08	0.16	0.00	0.15	0.01	0.18	0.01
8.3	0.01	0.24	0.03	0.10	0.23	0.08	0.16	0.00	0.15	0.02	0.18	0.01
8.4	0.01	0.24	0.03	0.10	0.24	0.07	0.16	0.00	0.14	0.02	0.18	0.01
8.5	0.01	0.24	0.03	0.09	0.24	0.07	0.16	0.00	0.14	0.02	0.18	0.01
8.6	0.01	0.24	0.03	0.09	0.24	0.07	0.17	0.00	0.14	0.02	0.18	0.01
8.7	0.01	0.24	0.03	0.08	0.24	0.06	0.17	0.00	0.14	0.02	0.18	0.01
8.8	0.02	0.24	0.02	0.08	0.24	0.06	0.17	0.01	0.13	0.02	0.18	0.00
8.9	0.02	0.24	0.02	0.07	0.24	0.05	0.17	0.01	0.13	0.03	0.18	0.00
9.0	0.02	0.24	0.02	0.07	0.24	0.05	0.17	0.01	0.13	0.03	0.18	0.00
9.1	0.02	0.24	0.02	0.07	0.24	0.05	0.17	0.01	0.13	0.03	0.18	0.00
9.2	0.02	0.24	0.02	0.06	0.23	0.05	0.17	0.01	0.12	0.03	0.18	0.00
9.3	0.03	0.24	0.01	0.06	0.23	0.04	0.18	0.01	0.12	0.03	0.17	0.00
9.4	0.03	0.24	0.01	0.05	0.23	0.04	0.18	0.01	0.12	0.03	0.17	0.00
9.5	0.03	0.24	0.01	0.05	0.23	0.04	0.18	0.01	0.11	0.04	0.17	0.00
9.6	0.03	0.24	0.01	0.05	0.23	0.04	0.18	0.01	0.11	0.04	0.17	0.00
9.7	0.03	0.24	0.01	0.04	0.23	0.04	0.18	0.02	0.10	0.04	0.17	0.00
9.8	0.04	0.24	0.01	0.04	0.22	0.03	0.18	0.02	0.10	0.04	0.17	0.00
9.9	0.04	0.24	0.01	0.03	0.22	0.03	0.18	0.02	0.09	0.04	0.17	0.00
10.0	0.04	0.24	0.01	0.03	0.22	0.03	0.18	0.02	0.09	0.05	0.17	0.00

TABLES CXXIX.-CXXXII.

Tables	CXXIX.		CXXX.				CXXXI.			CXXXII.			
Arguments	24.		28.				30.			4.			
Days.	0	10	0	10	20	30	0	10	Days.	0	100	200	300
0.0	0.12	0.20	0.04	0.10	0.05	0.00	0.05	0.06	0	0.41	0.01	0.51	0.76
0.1	0.11	0.20	0.04	0.10	0.05	0.00	0.05	0.06	1	0.40	0.01	0.52	0.76
0.2	0.11	0.20	0.04	0.10	0.05	0.00	0.04	0.05	2	0.40	0.01	0.52	0.75
0.3	0.10	0.20	0.05	0.10	0.05	0.00	0.04	0.05	3	0.39	0.01	0.53	0.75
0.4	0.10	0.20	0.05	0.10	0.05	0.00	0.03	0.05	4	0.38	0.01	0.54	0.75
0.5	0.09	0.20	0.05	0.10	0.05	0.00	0.03	0.04	5	0.37	0.01	0.54	0.74
0.6	0.09	0.20	0.05	0.10	0.05	0.00	0.03	0.04	6	0.37	0.02	0.55	0.74
0.7	0.08	0.20	0.05	0.10	0.05	0.00	0.03	0.04	7	0.36	0.02	0.56	0.73
0.8	0.08	0.20	0.05	0.10	0.04	0.00	0.02	0.03	8	0.35	0.02	0.56	0.73
0.9	0.07	0.20	0.05	0.10	0.04	0.00	0.02	0.03	9	0.35	0.02	0.57	0.73
1.0	0.07	0.20	0.05	0.10	0.04	0.00	0.02	0.03	10	0.34	0.02	0.58	0.72
1.1	0.07	0.19	0.05	0.10	0.04	0.00	0.02	0.03	11	0.33	0.03	0.58	0.72
1.2	0.06	0.19	0.05	0.10	0.04	0.00	0.02	0.02	12	0.33	0.03	0.59	0.71
1.3	0.06	0.19	0.06	0.10	0.04	0.00	0.01	0.02	13	0.32	0.03	0.60	0.71
1.4	0.05	0.19	0.06	0.10	0.04	0.00	0.01	0.02	14	0.31	0.03	0.60	0.70
1.5	0.05	0.19	0.06	0.10	0.04	0.00	0.01	0.01	15	0.30	0.04	0.61	0.70
1.6	0.05	0.18	0.06	0.10	0.04	0.00	0.01	0.01	16	0.30	0.04	0.61	0.69
1.7	0.04	0.18	0.06	0.10	0.04	0.00	0.01	0.01	17	0.29	0.04	0.62	0.69
1.8	0.04	0.18	0.06	0.10	0.03	0.00	0.00	0.01	18	0.28	0.05	0.62	0.68
1.9	0.03	0.17	0.06	0.10	0.03	0.00	0.00	0.00	19	0.28	0.05	0.63	0.68
2.0	0.03	0.17	0.06	0.10	0.03	0.00	0.00	0.00	20	0.27	0.05	0.63	0.67
2.1	0.03	0.17	0.06	0.10	0.03	0.00	0.00	0.00	21	0.26	0.06	0.64	0.67
2.2	0.03	0.16	0.06	0.10	0.03	0.00	0.00	0.00	22	0.26	0.06	0.64	0.66
2.3	0.02	0.16	0.06	0.10	0.03	0.00	0.00	0.00	23	0.25	0.06	0.65	0.66
2.4	0.02	0.15	0.06	0.10	0.03	0.00	0.00	0.00	24	0.25	0.07	0.65	0.65
2.5	0.02	0.15	0.06	0.10	0.03	0.00	0.00	0.00	25	0.24	0.07	0.66	0.65
2.6	0.01	0.15	0.06	0.10	0.03	0.00	0.00	0.00	26	0.24	0.07	0.66	0.64
2.7	0.01	0.14	0.06	0.10	0.03	0.01	0.00	0.00	27	0.23	0.08	0.67	0.64
2.8	0.01	0.14	0.07	0.10	0.03	0.01	0.01	0.00	28	0.23	0.08	0.67	0.63
2.9	0.01	0.13	0.07	0.10	0.03	0.01	0.01	0.00	29	0.22	0.09	0.68	0.63
3.0	0.01	0.13	0.07	0.10	0.03	0.01	0.01	0.00	30	0.21	0.09	0.68	0.62
3.1	0.00	0.13	0.07	0.10	0.03	0.01	0.01	0.00	31	0.21	0.09	0.69	0.62
3.2	0.00	0.12	0.07	0.09	0.03	0.01	0.01	0.00	32	0.20	0.10	0.69	0.61
3.3	0.00	0.12	0.07	0.09	0.02	0.01	0.02	0.01	33	0.19	0.10	0.70	0.61
3.4	0.00	0.11	0.07	0.09	0.02	0.01	0.02	0.01	34	0.19	0.11	0.70	0.60
3.5	0.00	0.11	0.07	0.09	0.02	0.01	0.02	0.01	35	0.18	0.11	0.71	0.60
3.6	0.00	0.10	0.07	0.09	0.02	0.01	0.02	0.01	36	0.18	0.12	0.71	0.59
3.7	0.00	0.10	0.07	0.09	0.02	0.01	0.02	0.01	37	0.17	0.12	0.72	0.59
3.8	0.00	0.09	0.07	0.09	0.02	0.01	0.03	0.02	38	0.17	0.13	0.72	0.58
3.9	0.00	0.09	0.07	0.09	0.02	0.01	0.03	0.02	39	0.16	0.13	0.73	0.58
4.0	0.00	0.08	0.07	0.09	0.02	0.01	0.03	0.02	40	0.16	0.14	0.73	0.57
4.1	0.00	0.08	0.08	0.09	0.02	0.01	0.03	0.02	41	0.15	0.14	0.73	0.56
4.2	0.00	0.07	0.08	0.09	0.02	0.01	0.04	0.03	42	0.15	0.15	0.74	0.56
4.3	0.01	0.07	0.08	0.09	0.02	0.01	0.04	0.03	43	0.14	0.15	0.74	0.55
4.4	0.01	0.06	0.08	0.09	0.02	0.02	0.05	0.03	44	0.14	0.16	0.74	0.54
4.5	0.01	0.06	0.08	0.09	0.02	0.02	0.05	0.04	45	0.13	0.16	0.75	0.54
4.6	0.01	0.06	0.08	0.09	0.02	0.02	0.05	0.04	46	0.13	0.17	0.75	0.53
4.7	0.01	0.05	0.08	0.09	0.02	0.02	0.06	0.04	47	0.12	0.17	0.75	0.52
4.8	0.02	0.05	0.08	0.09	0.01	0.02	0.06	0.05	48	0.12	0.18	0.76	0.52
4.9	0.02	0.04	0.08	0.09	0.01	0.02	0.07	0.05	49	0.11	0.18	0.76	0.51
5.0	0.02	0.04	0.08	0.09	0.01	0.02	0.07	0.05	50	0.11	0.19	0.76	0.50

TABLES CXXIX.-CXXXII.

Tables	CXXIX.		CXXX.				CXXXI.			CXXXII.			
Arguments	24.		28.				30.			4.			
Days.	0	10	0	10	20	30	0	10	Days.	0	100	200	300
5.0	0.02	0.04	0.08	0.09	0.01	0.02	0.07	0.05	50	0.11	0.19	0.76	0.50
5.1	0.03	0.04	0.08	0.09	0.01	0.02	0.07	0.05	51	0.10	0.19	0.77	0.50
5.2	0.03	0.03	0.08	0.09	0.01	0.02	0.07	0.06	52	0.10	0.20	0.77	0.49
5.3	0.03	0.03	0.08	0.08	0.01	0.02	0.08	0.06	53	0.09	0.20	0.77	0.48
5.4	0.03	0.02	0.08	0.08	0.01	0.02	0.08	0.07	54	0.09	0.21	0.77	0.48
5.5	0.04	0.02	0.08	0.08	0.01	0.02	0.08	0.07	55	0.08	0.22	0.77	0.47
5.6	0.04	0.02	0.09	0.08	0.01	0.02	0.08	0.07	56	0.08	0.22	0.78	0.46
5.7	0.04	0.02	0.09	0.08	0.01	0.02	0.09	0.08	57	0.07	0.23	0.78	0.46
5.8	0.05	0.01	0.09	0.08	0.01	0.02	0.09	0.08	58	0.07	0.24	0.78	0.45
5.9	0.05	0.01	0.09	0.08	0.01	0.02	0.10	0.09	59	0.07	0.24	0.78	0.44
6.0	0.06	0.01	0.09	0.08	0.01	0.03	0.10	0.09	60	0.06	0.25	0.78	0.44
6.1	0.06	0.01	0.09	0.08	0.01	0.03	0.10	0.09	61	0.06	0.26	0.79	0.43
6.2	0.06	0.01	0.09	0.08	0.01	0.03	0.10	0.09	62	0.05	0.26	0.79	0.42
6.3	0.07	0.00	0.09	0.08	0.01	0.03	0.11	0.10	63	0.05	0.27	0.79	0.42
6.4	0.07	0.00	0.09	0.08	0.01	0.03	0.11	0.10	64	0.05	0.27	0.79	0.41
6.5	0.08	0.00	0.09	0.08	0.01	0.03	0.11	0.10	65	0.04	0.28	0.79	0.40
6.6	0.08	0.00	0.09	0.08	0.01	0.03	0.11	0.10	66	0.04	0.28	0.79	0.40
6.7	0.09	0.00	0.09	0.08	0.01	0.03	0.11	0.10	67	0.04	0.29	0.79	0.39
6.8	0.09	0.00	0.09	0.07	0.00	0.03	0.12	0.11	68	0.03	0.30	0.80	0.38
6.9	0.10	0.00	0.09	0.07	0.00	0.03	0.12	0.11	69	0.03	0.30	0.80	0.37
7.0	0.10	0.00	0.09	0.07	0.00	0.03	0.12	0.11	70	0.03	0.31	0.80	0.37
7.1	0.10	0.00	0.09	0.07	0.00	0.03	0.12	0.11	71	0.02	0.31	0.80	0.36
7.2	0.11	0.00	0.09	0.07	0.00	0.03	0.12	0.11	72	0.02	0.32	0.80	0.35
7.3	0.11	0.00	0.09	0.07	0.00	0.04	0.12	0.12	73	0.02	0.33	0.80	0.35
7.4	0.12	0.00	0.09	0.07	0.00	0.04	0.12	0.12	74	0.02	0.33	0.80	0.34
7.5	0.12	0.00	0.09	0.07	0.00	0.04	0.12	0.12	75	0.02	0.34	0.80	0.33
7.6	0.12	0.00	0.09	0.07	0.00	0.04	0.12	0.12	76	0.01	0.34	0.80	0.33
7.7	0.13	0.00	0.09	0.07	0.00	0.04	0.12	0.12	77	0.01	0.35	0.80	0.32
7.8	0.13	0.01	0.10	0.07	0.00	0.04	0.12	0.12	78	0.01	0.36	0.80	0.31
7.9	0.14	0.01	0.10	0.07	0.00	0.04	0.12	0.12	79	0.01	0.36	0.80	0.31
8.0	0.14	0.01	0.10	0.07	0.00	0.04	0.12	0.12	80	0.01	0.37	0.80	0.30
8.1	0.15	0.01	0.10	0.07	0.00	0.04	0.12	0.12	81	0.01	0.38	0.80	0.29
8.2	0.15	0.01	0.10	0.07	0.00	0.04	0.11	0.12	82	0.00	0.39	0.80	0.29
8.3	0.16	0.02	0.10	0.06	0.00	0.04	0.11	0.12	83	0.00	0.40	0.80	0.28
8.4	0.16	0.02	0.10	0.06	0.00	0.04	0.11	0.11	84	0.00	0.40	0.80	0.28
8.5	0.16	0.02	0.10	0.06	0.00	0.04	0.11	0.11	85	0.00	0.41	0.80	0.27
8.6	0.17	0.03	0.10	0.06	0.00	0.04	0.11	0.11	86	0.00	0.42	0.79	0.26
8.7	0.17	0.03	0.10	0.06	0.00	0.04	0.10	0.11	87	0.00	0.42	0.79	0.26
8.8	0.18	0.03	0.10	0.06	0.00	0.05	0.10	0.11	88	0.00	0.43	0.79	0.25
8.9	0.18	0.04	0.10	0.06	0.00	0.05	0.10	0.10	89	0.00	0.44	0.79	0.24
9.0	0.18	0.04	0.10	0.06	0.00	0.05	0.09	0.10	90	0.00	0.44	0.79	0.24
9.1	0.18	0.04	0.10	0.06	0.00	0.05	0.09	0.10	91	0.00	0.45	0.78	0.23
9.2	0.19	0.05	0.10	0.06	0.00	0.05	0.09	0.10	92	0.00	0.46	0.78	0.22
9.3	0.19	0.05	0.10	0.05	0.00	0.05	0.08	0.09	93	0.00	0.47	0.78	0.22
9.4	0.19	0.06	0.10	0.05	0.00	0.05	0.08	0.09	94	0.00	0.47	0.78	0.21
9.5	0.19	0.06	0.10	0.05	0.00	0.05	0.08	0.09	95	0.00	0.48	0.77	0.20
9.6	0.19	0.06	0.10	0.05	0.00	0.05	0.08	0.09	96	0.00	0.48	0.77	0.20
9.7	0.19	0.07	0.10	0.05	0.00	0.05	0.07	0.08	97	0.00	0.49	0.77	0.19
9.8	0.20	0.07	0.10	0.05	0.00	0.06	0.07	0.08	98	0.01	0.50	0.77	0.18
9.9	0.20	0.08	0.10	0.05	0.00	0.06	0.06	0.07	99	0.01	0.50	0.76	0.18
10.0	0.20	0.09	0.10	0.05	0.00	0.06	0.06	0.07	100	0.01	0.51	0.76	0.17

TABLES CXXXIII.-CXXXV.

Tables	CXXXIII.					CXXXIV.		CXXXV.	
Arguments	5.					12'.		Hor. Par.	
Days.	0	100	200	300	400	0	100	Hor. Par.	0″
Days	″	″	″	″	″	″	″	Seconds.	″
0	0.20	0.46	0.17	0.50	0.11	0.10	0.07	3000	0.11
1	0.21	0.45	0.18	0.49	0.12	0.10	0.07	3010	0.11
2	0.22	0.44	0.19	0.48	0.13	0.11	0.06	3020	0.11
3	0.23	0.43	0.20	0.47	0.14	0.11	0.06	3030	0.11
4	0.24	0.43	0.20	0.46	0.14	0.12	0.05	3040	0.11
5	0.25	0.42	0.21	0.45	0.15	0.12	0.05	3050	0.11
6	0.26	0.41	0.22	0.44	0.16	0.12	0.05	3060	0.11
7	0.27	0.40	0.23	0.43	0.17	0.12	0.05	3070	0.11
8	0.28	0.39	0.24	0.42	0.18	0.13	0.04	3080	0.11
9	0.29	0.38	0.25	0.42	0.18	0.13	0.04	3090	0.12
10	0.30	0.37	0.26	0.41	0.19	0.13	0.04	3100	0.12
11	0.31	0.36	0.27	0.40	0.20	0.14	0.03	3110	0.12
12	0.32	0.35	0.28	0.39	0.21	0.14	0.03	3120	0.12
13	0.33	0.34	0.29	0.38	0.22	0.14	0.03	3130	0.12
14	0.34	0.33	0.29	0.37	0.22	0.15	0.02	3140	0.12
15	0.35	0.32	0.30	0.36	0.23	0.15	0.02	3150	0.12
16	0.36	0.31	0.31	0.35	0.24	0.15	0.02	3160	0.12
17	0.37	0.30	0.32	0.34	0.25	0.16	0.02	3170	0.12
18	0.38	0.29	0.33	0.33	0.26	0.16	0.01	3180	0.12
19	0.38	0.29	0.34	0.33	0.27	0.17	0.01	3190	0.13
20	0.39	0.28	0.35	0.32	0.28	0.17	0.01	3200	0.13
21	0.40	0.27	0.36	0.31	0.29	0.17	0.01	3210	0.13
22	0.41	0.26	0.37	0.30	0.30	0.18	0.01	3220	0.13
23	0.42	0.25	0.38	0.29	0.31	0.18	0.00	3230	0.13
24	0.43	0.24	0.39	0.28	0.31	0.19	0.00	3240	0.13
25	0.44	0.23	0.40	0.27	0.32	0.19	0.00	3250	0.14
26	0.45	0.22	0.41	0.26	0.33	0.19	0.00	3260	0.14
27	0.46	0.21	0.42	0.25	0.34	0.19	0.00	3270	0.14
28	0.47	0.20	0.43	0.24	0.35	0.20	0.00	3280	0.14
29	0.47	0.20	0.43	0.23	0.36	0.20	0.00	3290	0.14
30	0.48	0.19	0.44	0.22	0.37	0.20	0.00	3300	0.14
31	0.49	0.18	0.45	0.21	0.38	0.20	0.00	3310	0.14
32	0.50	0.17	0.46	0.20	0.39	0.20	0.00	3320	0.14
33	0.51	0.16	0.47	0.19	0.40	0.20	0.00	3330	0.15
34	0.51	0.16	0.47	0.19	0.41	0.21	0.00	3340	0.15
35	0.52	0.15	0.48	0.18	0.42	0.21	0.00	3350	0.15
36	0.53	0.14	0.49	0.17	0.43	0.21	0.00	3360	0.15
37	0.54	0.13	0.50	0.16	0.44	0.21	0.00	3370	0.15
38	0.54	0.13	0.51	0.15	0.45	0.21	0.00	3380	0.15
39	0.55	0.12	0.51	0.15	0.45	0.21	0.00	3390	0.15
40	0.55	0.12	0.52	0.14	0.46	0.21	0.01	3400	0.15
41	0.56	0.11	0.53	0.13	0.47	0.21	0.01	3410	0.15
42	0.57	0.10	0.54	0.12	0.48	0.21	0.01	3420	0.16
43	0.57	0.10	0.54	0.11	0.49	0.22	0.01	3430	0.16
44	0.58	0.09	0.55	0.11	0.49	0.22	0.01	3440	0.16
45	0.58	0.09	0.55	0.10	0.50	0.22	0.01	3450	0.16
46	0.59	0.08	0.56	0.09	0.51	0.22	0.01	3460	0.16
47	0.59	0.07	0.57	0.09	0.52	0.22	0.01	3470	0.16
48	0.60	0.07	0.57	0.08	0.53	0.22	0.02	3480	0.17
49	0.60	0.06	0.58	0.08	0.53	0.22	0.02	3490	0.17
50	0.61	0.06	0.58	0.07	0.54	0.22	0.02	3500	0.17

TABLES CXXXIII. - CXXXV.

Tables	CXXXIII.					CXXXIV.		CXXXV.	
Arguments	5.					12'.		Hor. Par.	
Days.	0	100	200	300	400	0	100	Hor. Par.	0''
Days	''	''	''	''	''			Seconds.	''
50	0.61	0.06	0.58	0.07	0.54	0.22	0.02	3500	0.17
51	0.61	0.05	0.59	0.07	0.55	0.22	0.02	3510	0.17
52	0.61	0.05	0.59	0.06	0.56	0.22	0.02	3520	0.17
53	0.61	0.04	0.60	0.05	0.56	0.22	0.03	3530	0.17
54	0.62	0.04	0.60	0.05	0.57	0.22	0.03	3540	0.17
55	0.62	0.03	0.61	0.04	0.57	0.22	0.03	3550	0.17
56	0.62	0.03	0.61	0.03	0.58	0.22	0.03	3560	0.18
57	0.62	0.03	0.61	0.03	0.58	0.22	0.04	3570	0.18
58	0.62	0.03	0.61	0.03	0.59	0.21	0.04	3580	0.18
59	0.63	0.02	0.62	0.02	0.59	0.21	0.04	3590	0.18
60	0.63	0.02	0.62	0.02	0.60	0.21	0.05	3600	0.18
61	0.63	0.02	0.62	0.02	0.60	0.21	0.05	3610	0.18
62	0.63	0.02	0.62	0.02	0.60	0.20	0.05	3620	0.18
63	0.63	0.02	0.62	0.02	0.61	0.20	0.06	3630	0.19
64	0.63	0.02	0.63	0.01	0.61	0.20	0.06	3640	0.19
65	0.63	0.02	0.63	0.01	0.62	0.19	0.06	3650	0.19
66	0.63	0.02	0.63	0.01	0.62	0.19	0.06	3660	0.19
67	0.63	0.02	0.63	0.00	0.62	0.19	0.07	3670	0.19
68	0.63	0.02	0.63	0.00	0.62	0.18	0.07	3680	0.19
69	0.62	0.02	0.63	0.00	0.63	0.18	0.08	3690	0.20
70	0.62	0.02	0.63	0.00	0.63	0.18	0.08	3700	0.20
71	0.62	0.02	0.63	0.00	0.63	0.18	0.08	3710	0.20
72	0.62	0.02	0.63	0.00	0.63	0.17	0.09	3720	0.20
73	0.62	0.02	0.63	0.00	0.63	0.17	0.09	3730	0.20
74	0.61	0.03	0.62	0.00	0.63	0.17	0.10	3740	0.20
75	0.61	0.03	0.62	0.00	0.63	0.17	0.10	3750	0.21
76	0.61	0.03	0.62	0.00	0.63	0.16	0.10	3760	0.21
77	0.61	0.03	0.62	0.00	0.63	0.16	0.11	3770	0.21
78	0.60	0.04	0.62	0.00	0.63	0.16	0.11	3780	0.21
79	0.60	0.04	0.61	0.01	0.63	0.15	0.12	3790	0.21
80	0.59	0.05	0.61	0.01	0.63	0.15	0.12	3800	0.21
81	0.59	0.05	0.61	0.01	0.62	0.15	0.12	3810	0.21
82	0.58	0.06	0.61	0.01	0.62	0.14	0.13	3820	0.22
83	0.58	0.06	0.60	0.02	0.62	0.14	0.13	3830	0.22
84	0.57	0.07	0.60	0.02	0.62	0.13	0.14	3840	0.22
85	0.57	0.07	0.59	0.03	0.61	0.13	0.14	3850	0.22
86	0.56	0.08	0.59	0.03	0.61	0.13	0.14	3860	0.22
87	0.55	0.09	0.58	0.03	0.61	0.12	0.15	3870	0.23
88	0.55	0.09	0.58	0.04	0.61	0.12	0.15	3880	0.23
89	0.54	0.10	0.57	0.04	0.60	0.11	0.16	3890	0.23
90	0.54	0.10	0.57	0.05	0.60	0.11	0.16	3900	0.23
91	0.53	0.11	0.56	0.05	0.60	0.11	0.16	3910	0.23
92	0.52	0.12	0.55	0.06	0.59	0.10	0.17	3920	0.23
93	0.51	0.12	0.55	0.06	0.59	0.10	0.17	3930	0.24
94	0.51	0.13	0.54	0.07	0.58	0.09	0.18	3940	0.24
95	0.50	0.13	0.54	0.07	0.58	0.09	0.18	3950	0.24
96	0.49	0.14	0.53	0.08	0.57	0.09	0.18	3960	0.24
97	0.48	0.15	0.52	0.09	0.56	0.08	0.18	3970	0.24
98	0.47	0.16	0.51	0.10	0.56	0.08	0.19	3980	0.25
99	0.47	0.17	0.51	0.10	0.55	0.07	0.19	3990	0.25
100	0.46	0.17	0.50	0.11	0.55	0.07	0.19	4000	0.25

TABLE CXXXVI.

Argument, *Horizontal Parallax.*

Hor. Par.	0″	1000″	2000″	3000″	Hor. Par.	0″	1000″	2000″	3000″
0″	0.00″	272.27″	544.55″	816.82″	500″	136.14″	408.41″	680.69″	952.96″
10	2.72	274.99	547.27	819.54	510	138.86	411.13	683.41	955.68
20	5.45	277.72	549.99	822.27	520	141.58	413.85	686.13	958.40
30	8.17	280.44	552.72	824.99	530	144.31	416.58	688.86	961.13
40	10.89	283.17	555.44	827.72	540	147.03	419.30	691.58	963.85
50	13.61	285.89	558.16	830.44	550	149.75	422.02	694.30	966.57
60	16.33	288.61	560.88	833.16	560	152.47	424.74	697.02	969.29
70	19.06	291.33	563.60	835.88	570	155.19	427.47	699.74	972.02
80	21.78	294.06	566.33	838.61	580	157.92	430.19	702.47	974.74
90	24.51	296.78	569.05	841.33	590	160.64	432.92	705.19	977.47
100	27.23	299.50	571.77	844.05	600	163.36	435.64	707.91	980.19
110	29.95	302.22	574.49	846.77	610	166.08	438.36	710.63	982.91
120	32.67	304.95	577.22	849.49	620	168.81	441.08	713.36	985.63
130	35.40	307.67	579.94	852.22	630	171.53	443.81	716.08	988.36
140	38.12	310.40	582.67	854.94	640	174.26	446.53	718.81	991.08
150	40.84	313.12	585.39	857.66	650	176.98	449.25	721.53	993.80
160	43.56	315.84	588.11	860.38	660	179.70	451.97	724.25	996.52
170	46.28	318.56	590.83	863.11	670	182.42	454.70	726.97	999.24
180	49.01	321.29	593.56	865.83	680	185.15	457.42	729.70	1001.97
190	51.73	324.01	596.28	868.56	690	187.87	460.15	732.42	1004.69
200	54.45	326.73	599.00	871.28	700	190.59	462.87	735.14	1007.41
210	57.17	329.45	601.72	874.00	710	193.31	465.59	737.86	1010.13
220	59.90	332.17	604.45	876.72	720	196.04	468.31	740.58	1012.86
230	62.62	334.90	607.17	879.45	730	198.76	471.04	743.31	1015.58
240	65.35	337.62	609.90	882.17	740	201.49	473.76	746.03	1018.31
250	68.07	340.34	612.62	884.89	750	204.21	476.48	748.75	1021.03
260	70.79	343.06	615.34	887.61	760	206.93	479.20	751.47	1023.75
270	73.51	345.79	618.06	890.33	770	209.65	481.92	754.20	1026.47
280	76.24	348.51	620.79	893.06	780	212.38	484.65	756.92	1029.20
290	78.96	351.24	623.51	895.78	790	215.10	487.37	759.65	1031.92
300	81.68	353.96	626.23	898.50	800	217.82	490.09	762.37	1034.64
310	84.40	356.68	628.95	901.22	810	220.54	492.81	765.09	1037.36
320	87.13	359.40	631.67	903.95	820	223.26	495.54	767.81	1040.08
330	89.85	362.13	634.40	906.67	830	225.99	498.26	770.54	1042.81
340	92.58	364.85	637.12	909.40	840	228.71	500.99	773.26	1045.53
350	95.30	367.57	639.84	912.12	850	231.43	503.71	775.98	1048.25
360	98.02	370.29	642.56	914.84	860	234.15	506.43	778.70	1050.97
370	100.74	373.01	645.29	917.56	870	236.88	509.15	781.42	1053.70
380	103.47	375.74	648.01	920.29	880	239.60	511.88	784.15	1056.42
390	106.19	378.46	650.74	923.01	890	242.33	514.60	786.87	1059.15
400	108.91	381.18	653.46	925.73	900	245.05	517.32	789.59	1061.87
410	111.63	383.90	656.18	928.45	910	247.77	520.04	792.31	1064.59
420	114.35	386.63	658.90	931.18	920	250.49	522.76	795.04	1067.31
430	117.08	389.35	661.63	933.90	930	253.22	525.49	797.76	1070.04
440	119.80	392.08	664.35	936.63	940	255.94	528.21	800.49	1072.76
450	122.52	394.80	667.07	939.35	950	258.66	530.93	803.21	1075.48
460	125.24	397.52	669.79	942.07	960	261.38	533.65	805.93	1078.20
470	127.97	400.24	672.52	944.79	970	264.10	536.38	808.65	1080.93
480	130.69	402.97	675.24	947.52	980	266.83	539.10	811.38	1083.65
490	133.42	405.69	677.97	950.24	990	269.55	541.83	814.10	1086.38
500	136.14	408.41	680.69	952.96	1000	272.27	544.55	816.82	1089.10

The Latitude $= A \sin \bar{y} + B \cos \bar{y} + C$, as given on page 12, Introduction, may be put in the form,

$$\text{Latitude} = 18500'' \sin \bar{y} + A' \sin \bar{y} + B \cos \bar{y} + C,$$

in which A' only differs from A in that the constant 900″ is subtracted instead of the constant 17600″ added. Compute 18500″ sin $\bar{y}$ for each noon and midnight from Table CXXXVII.; compute $A' \sin \bar{y} + B \cos \bar{y} + C$ for each noon, using five place decimals, and interpolate to midnight. In Table CXXXVII., the columns headed P. P. are the proportional parts for the seconds of $\bar{y}$.

TABLE CXXXVII. ARGUMENT 77.

Equation = 18500″ sin $\bar{y}$.

$\bar{y}$	180°− 0°+ Equation.	P. P.	181°− 1°+ Equation.	P. P.	182°− 2°+ Equation.	P. P.	183°− 3°+ Equation.	P. P.	184°− 4°+ Equation.	P. P.	185°− 5°+ Equation.	P. P.	$\bar{y}$
0′	0° 0′ 0.00″	0.00″	0° 5′ 22.87″	0.00″	0 10′ 45.64″	0.00″	0 16′ 8.22″	0.00″	0 21′ 30.49″	0.00″	0° 26′ 52.38″	0.00″	60′
1	0 5.38	0.09	5 28.25	0.09	10 51.02	0.09	16 13.60	0.09	21 35.86	0.09	26 57.74	0.09	59
2	0 10.76	0.18	5 33.63	0.18	10 56.40	0.18	16 18.98	0.18	21 41.23	0.18	27 3.10	0.18	58
3	0 16.14	0.27	5 39.01	0.27	11 1.78	0.27	16 24.35	0.27	21 46.60	0.27	27 8.46	0.27	57
4	0 21.52	0.36	5 44.39	0.36	11 7.16	0.36	16 29.72	0.36	21 51.97	0.36	27 13.82	0.36	56
5	0 26.90	0.45	5 49.77	0.45	11 12.53	0.45	16 35.09	0.45	21 57.33	0.45	27 19.18	0.45	55
6	0 0 32.28	0.54	0 5 55.15	0.54	0 11 17.91	0.54	0 16 40.47	0.54	0 22 2.70	0.53	0 27 24.54	0.53	54
7	0 37.66	0.63	6 0.53	0.63	11 23.29	0.63	16 45.85	0.63	22 8.07	0.62	27 29.90	0.62	53
8	0 43.04	0.72	6 5.91	0.72	11 28.67	0.72	16 51.22	0.72	22 13.44	0.71	27 35.26	0.71	52
9	0 48.42	0.81	6 11.29	0.81	11 34.05	0.81	16 56.59	0.81	22 18.80	0.80	27 40.62	0.80	51
10	0 53.81	0.90	6 16.67	0.90	11 39.42	0.90	17 1.96	0.90	22 24.17	0.89	27 45.98	0.89	50
11	0 0 59.19	0.99	0 6 22.05	0.99	0 11 44.80	0.99	0 17 7.34	0.99	0 22 29.54	0.98	0 27 51.34	0.98	49
12	1 4.57	1.08	6 27.43	1.08	11 50.18	1.08	17 12.71	1.08	22 34.91	1.07	27 56.70	1.07	48
13	1 9.95	1.17	6 32.81	1.17	11 55.56	1.17	17 18.08	1.17	22 40.28	1.16	28 2.06	1.16	47
14	1 15.33	1.26	6 38.19	1.26	12 00.94	1.26	17 23.45	1.26	22 45.64	1.25	28 7.42	1.25	46
15	1 20.72	1.35	6 43.57	1.35	12 6.31	1.35	17 28.82	1.35	22 51.00	1.34	28 12.78	1.34	45
16	0 1 26.10	1.44	0 6 48.95	1.44	0 12 11.69	1.44	0 17 34.20	1.44	0 22 56.37	1.43	0 28 18.14	1.42	44
17	1 31.48	1.53	6 54.33	1.53	12 17.07	1.53	17 39.57	1.53	23 1.74	1.52	28 23.50	1.51	43
18	1 36.86	1.62	6 59.71	1.62	12 22.45	1.62	17 44.94	1.62	23 7.11	1.61	28 28.86	1.60	42
19	1 42.24	1.71	7 5.09	1.72	12 27.83	1.71	17 50.31	1.71	23 12.47	1.70	28 34.22	1.69	41
20	1 47.63	1.79	7 10.47	1.79	12 33.20	1.79	17 55.68	1.79	23 17.83	1.79	28 39.57	1.78	40
21	0 1 53.01	1.88	0 7 15.85	1.88	0 12 38.58	1.88	0 18 1.06	1.88	0 23 23.20	1.88	0 28 44.93	1.87	39
22	1 58.39	1.97	7 21.23	1.97	12 43.96	1.97	18 6.43	1.97	23 28.57	1.97	28 50.29	1.96	38
23	2 3.77	2.06	7 26.61	2.06	12 49.31	2.06	18 11.80	2.06	23 33.94	2.06	28 55.65	2.05	37
24	2 9.15	2.15	7 31.99	2.15	12 54.71	2.15	18 17.17	2.15	23 39.30	2.15	29 1.01	2.14	36
25	2 14.54	2.24	7 37.37	2.24	13 0.08	2.24	18 22.54	2.24	23 44.66	2.24	29 6.36	2.23	35
26	0 2 19.92	2.33	0 7 42.75	2.33	0 13 5.46	2.33	0 18 27.92	2.33	0 23 50.03	2.32	0 29 11.72	2.32	34
27	2 25.30	2.42	7 48.13	2.42	13 10.84	2.42	18 33.29	2.42	23 55.40	2.41	29 17.08	2.41	33
28	2 30.68	2.51	7 53.51	2.51	13 16.22	2.51	18 38.66	2.51	24 0.77	2.50	29 22.44	2.50	32
29	2 36.06	2.60	7 58.89	2.60	13 21.59	2.60	18 44.03	2.60	24 6.13	2.59	29 27.80	2.59	31
30	2 41.44	2.69	8 4.27	2.69	13 26.96	2.69	18 49.40	2.69	24 11.49	2.68	29 33.15	2.68	30
31	0 2 46.82	2.78	0 8 9.65	2.78	0 13 32.34	2.78	0 18 54.78	2.78	0 24 16.86	2.77	0 29 38.51	2.77	29
32	2 52.21	2.87	8 15.03	2.87	13 37.72	2.87	19 0.15	2.87	24 22.23	2.86	29 43.87	2.86	28
33	2 57.58	2.96	8 20.41	2.96	13 43.10	2.96	19 5.52	2.96	24 27.60	2.95	29 49.23	2.95	27
34	3 2.97	3.05	8 25.79	3.05	13 48.47	3.05	19 10.89	3.05	24 32.96	3.04	29 54.58	3.04	26
35	3 8.36	3.14	8 31.17	3.14	13 53.84	3.14	19 16.26	3.14	24 38.32	3.13	29 59.93	3.13	25
36	0 3 13.74	3.23	0 8 36.55	3.23	0 13 59.22	3.23	0 19 21.63	3.23	0 24 43.69	3.22	0 30 5.29	3.21	24
37	3 19.12	3.32	8 41.93	3.32	14 4.60	3.32	19 27.00	3.32	24 49.06	3.31	30 10.65	3.30	23
38	3 24.50	3.41	8 47.31	3.41	14 9.98	3.41	19 32.37	3.41	24 54.42	3.40	30 16.01	3.39	22
39	3 29.88	3.50	8 52.69	3.50	14 15.35	3.50	19 37.74	3.50	24 59.78	3.49	30 21.36	3.48	21
40	3 35.26	3.59	8 58.07	3.59	14 20.72	3.59	19 43.11	3.58	25 5.14	3.58	30 26.71	3.57	20
41	0 3 40.64	3.68	0 9 3.45	3.68	0 14 26.10	3.68	0 19 48.48	3.67	0 25 10.51	3.67	0 30 32.07	3.66	19
42	3 46.02	3.77	9 8.83	3.77	14 31.48	3.77	19 53.85	3.76	25 15.87	3.76	30 37.43	3.75	18
43	3 51.40	3.86	9 14.21	3.86	14 36.86	3.86	19 59.22	3.85	25 21.23	3.85	30 42.78	3.84	17
44	3 56.78	3.95	9 19.59	3.95	14 42.23	3.95	20 4.59	3.94	25 26.59	3.94	30 48.13	3.93	16
45	4 2.16	4.04	9 24.96	4.04	14 47.60	4.04	20 9.96	4.03	25 31.95	4.03	30 53.48	4.02	15
46	0 4 7.54	4.13	0 9 30.34	4.13	0 14 52.98	4.13	0 20 15.33	4.12	0 25 37.32	4.11	0 30 58.84	4.10	14
47	4 12.92	4.22	9 35.72	4.22	14 58.36	4.22	20 20.70	4.21	25 42.68	4.20	31 4.20	4.19	13
48	4 18.30	4.31	9 41.10	4.31	15 3.74	4.31	20 26.07	4.30	25 48.04	4.29	31 9.55	4.28	12
49	4 23.68	4.40	9 46.48	4.40	15 9.11	4.40	20 31.44	4.39	25 53.40	4.38	31 14.90	4.37	11
50	4 29.07	4.48	9 51.86	4.48	15 14.48	4.48	20 36.81	4.47	25 58.76	4.47	31 20.25	4.46	10
51	0 4 34.45	4.57	0 9 57.24	4.57	0 15 19.86	4.57	0 20 42.18	4.56	0 26 4.13	4.56	0 31 25.61	4.55	9
52	4 39.83	4.66	10 2.62	4.66	15 25.24	4.66	20 47.55	4.65	26 9.49	4.65	31 30.97	4.64	8
53	4 45.21	4.75	10 8.00	4.75	15 30.61	4.75	20 52.92	4.74	26 14.85	4.74	31 36.32	4.73	7
54	4 50.59	4.84	10 13.38	4.84	15 35.98	4.84	20 58.29	4.83	26 20.21	4.83	31 41.67	4.82	6
55	4 55.97	4.93	10 18.75	4.93	15 41.35	4.93	21 3.65	4.92	26 25.57	4.92	31 47.02	4.91	5
56	0 5 1.35	5.02	0 10 24.13	5.02	0 15 46.73	5.02	0 21 9.02	5.01	0 26 30.94	5.01	0 31 52.38	5.00	4
57	5 6.73	5.11	10 29.51	5.11	15 52.11	5.11	21 14.39	5.10	26 36.30	5.10	31 57.73	5.09	3
58	5 12.11	5.20	10 34.89	5.20	15 57.48	5.20	21 19.76	5.19	26 41.66	5.19	32 3.08	5.18	2
59	5 17.49	5.29	10 40.27	5.29	16 2.85	5.29	21 25.13	5.28	26 47.02	5.28	32 8.43	5.27	1
60	5 22.87	5.38	10 45.64	5.38	16 8.22	5.38	21 30.49	5.37	26 52.38	5.37	32 13.78	5.36	0
$\bar{y}$	179°+ 359°−		178°+ 358°−		177°+ 357°−		176°+ 356°−		175°+ 355°−		174°+ 354°−		$\bar{y}$

TABLE CXXXVII. ARGUMENT 77.

Equation = 18500″ sin $\bar{y}$.

$\bar{y}$	186°– 6°+ Equation.	P. P.	187°– 7°+ Equation.	P. P.	188°– 8°+ Equation.	P. P.	189°– 9°+ Equation.	P. P.	190°– 10°+ Equation.	P. P.	191°– 11°+ Equation.	P. P.	y
0′	0° 32′ 13.78″	0.00″	0° 37′ 34.58″	0.00″	0° 42′ 54.70″	0.00″	0° 48′ 14.03″	0.00″	0 53 32.49	0.00	0 58 49.97	0.00	60′
1	32 19.14	0.09	37 39.92	0.09	43 0.03	0.09	48 19.35	0.09	53 37.79	0.09	58 55.26	0.09	59
2	32 24.49	0.18	37 45.26	0.18	43 5.36	0.18	48 24.67	0.18	53 43.09	0.18	59 0.54	0.18	58
3	32 29.84	0.27	37 50.60	0.27	43 10.69	0.27	48 29.98	0.27	53 48.39	0.27	59 5.82	0.27	57
4	32 35.19	0.36	37 55.94	0.36	43 16.02	0.36	48 35.29	0.36	53 53.69	0.36	59 11.10	0.36	56
5	32 40.54	0.45	38 1.28	0.45	43 21.34	0.45	48 40.60	0.45	53 58.99	0.45	59 16.38	0.44	55
6	0 32 45.89	0.53	0 38 6.62	0.53	0 43 26.67	0.53	0 48 45.92	0.53	0 54 4.29	0.53	0 59 21.66	0.53	54
7	32 51.24	0.62	38 11.96	0.62	43 32.00	0.62	48 51.24	0.62	54 9.59	0.62	59 26.94	0.62	53
8	32 56.59	0.71	38 17.30	0.71	43 37.33	0.71	48 56.55	0.71	54 14.89	0.71	59 32.22	0.71	52
9	33 1.94	0.80	38 22.64	0.80	43 42.66	0.80	49 1.86	0.80	54 20.19	0.80	59 37.50	0.79	51
10	33 7.29	0.89	38 27.98	0.89	43 47.98	0.89	49 7.17	0.89	54 25.48	0.88	59 42.78	0.88	50
11	0 33 12.64	0.98	0 38 33.32	0.98	0 43 53.31	0.98	0 49 12.49	0.98	0 54 30.78	0.97	0 59 48.06	0.97	49
12	33 17.99	1.07	38 38.66	1.07	43 58.64	1.07	49 17.80	1.07	54 36.08	1.06	59 53.34	1.06	48
13	33 23.34	1.16	38 44.00	1.16	44 3.97	1.16	49 23.11	1.16	54 41.38	1.15	59 58.62	1.15	47
14	33 28.69	1.25	38 49.34	1.25	44 9.29	1.25	49 28.42	1.25	54 46.67	1.24	1 0 3.90	1.23	46
15	33 34.04	1.34	38 54.68	1.34	44 14.61	1.34	49 33.73	1.34	54 51.96	1.33	0 9.18	1.32	45
16	0 33 39.39	1.42	0 39 0.02	1.42	0 44 19.94	1.42	0 49 39.04	1.42	0 54 57.26	1.41	1 0 14.46	1.41	44
17	33 44.74	1.51	39 5.36	1.51	44 25.27	1.51	49 44.35	1.51	55 2.56	1.50	0 19.74	1.50	43
18	33 50.09	1.60	39 10.70	1.60	44 30.60	1.60	49 49.66	1.60	55 7.85	1.59	0 25.02	1.58	42
19	33 55.44	1.69	39 16.04	1.69	44 35.92	1.69	49 54.97	1.69	[illegible]	[illegible]	[illegible]	1.67	41
20	34 0.78	1.78	39 21.37	1.78	44 41.21	1.77	50 0.28	1.77	55 18.43	1.76	0 35.56	1.76	40
21	0 34 6.23	1.87	0 39 26.71	1.87	0 44 46.57	1.86	0 50 5.59	1.86	0 55 23.73	1.85	1 0 40.84	1.85	39
22	34 11.58	1.96	39 32.05	1.96	44 51.90	1.95	50 10.90	1.95	55 29.03	1.94	0 46.12	1.94	38
23	34 16.93	2.05	39 37.39	2.05	44 57.22	2.04	50 16.21	2.04	55 34.32	2.03	[illegible]	[illegible]	37
24	34 22.28	2.14	39 42.72	2.14	45 2.54	2.13	50 21.52	2.13	55 39.61	2.12	[illegible]	2.11	36
25	34 27.52	2.23	39 48.05	2.23	45 7.86	2.22	50 26.82	2.22	[illegible]	[illegible]	[illegible]	[illegible]	35
26	0 34 32.88	2.31	0 39 53.39	2.31	0 45 13.19	2.30	0 50 32.11	2.30	0 55 50.20	2.29	[illegible]	[illegible]	34
27	34 38.23	2.40	39 58.73	2.40	45 18.51	2.39	50 37.45	2.39	55 55.49	2.38	[illegible]	[illegible]	33
28	34 43.58	2.49	40 4.07	2.49	45 23.83	2.48	50 42.76	2.48	56 0.78	2.47	1 17.77	2.46	32
29	34 48.92	2.58	40 9.40	2.58	45 29.15	2.57	50 48.07	2.57	56 6.07	2.56	1 23.04	2.55	31
30	34 54.26	2.67	40 14.73	2.67	45 34.47	2.66	50 53.37	2.66	56 11.36	2.65	1 28.31	2.64	30
31	0 34 59.61	2.76	0 40 20.07	2.76	0 45 39.80	2.75	0 50 58.68	2.75	0 56 16.65	2.74	1 [illegible]	2.73	29
32	35 4.96	2.85	40 25.41	2.85	45 45.12	2.84	51 3.99	2.84	56 21.94	2.83	1 38.86	[illegible]	28
33	35 10.31	2.94	40 30.75	2.94	45 50.44	2.93	51 9.30	2.93	56 27.23	2.92	1 44.13	2.91	27
34	35 15.65	3.03	40 36.08	3.03	45 55.76	3.02	51 14.61	3.02	56 32.52	3.01	1 49.40	3.00	26
35	35 20.99	3.12	40 41.41	3.12	46 1.08	3.11	51 19.91	3.11	56 37.81	3.10	1 54.67	3.09	25
36	0 35 26.34	3.21	0 40 46.75	3.20	0 46 6.40	3.19	0 51 25.22	3.19	0 56 43.10	3.19	1 1 59.95	3.18	24
37	35 31.69	3.30	40 52.09	3.29	46 11.72	3.28	51 30.53	3.27	56 48.39	3.27	2 5.22	3.27	23
38	35 37.04	3.39	40 57.42	3.38	46 17.04	3.37	51 35.84	3.36	56 53.68	3.36	2 10.49	3.35	22
39	35 42.38	3.48	41 2.75	3.47	46 22.36	3.46	51 41.14	3.45	56 58.97	[illegible]	2 15.76	3.43	21
40	35 47.72	3.57	41 8.08	3.56	46 27.68	3.55	51 46.41	3.54	57 4.26	3.53	2 21.03	3.51	20
41	0 35 53.07	3.66	0 41 13.42	3.65	0 46 33.00	3.64	0 51 51.75	3.63	0 57 9.55	3.62	1 2 26.30	3.60	19
42	35 58.42	3.75	41 18.75	3.74	46 38.32	3.73	51 57.06	3.72	57 14.84	3.71	2 31.57	3.69	18
43	36 3.76	3.84	41 24.08	3.83	46 43.64	3.82	52 2.36	3.81	57 20.13	3.79	2 36.84	3.57	17
44	36 9.10	3.93	41 29.41	3.92	46 48.96	3.91	52 7.66	3.90	57 25.42	3.88	2 42.11	3.66	16
45	36 14.44	4.02	41 34.74	4.01	46 54.28	4.00	52 12.96	3.99	57 30.70	3.97	2 47.38	3.95	15
46	0 36 19.79	4.10	0 41 40.08	4.09	0 46 59.60	4.08	0 52 18.27	4.07	0 57 35.91	4.05	1 2 52.65	4.04	14
47	36 25.14	4.19	41 45.41	4.18	47 4.92	4.17	52 23.58	4.16	57 41.28	4.14	2 57.92	4.13	13
48	36 30.48	4.28	41 50.74	4.27	47 10.24	4.26	52 28.88	4.25	57 46.57	4.23	3 3.19	4.22	12
49	36 35.82	4.37	41 56.07	4.36	47 15.56	4.35	52 34.18	4.34	57 51.85	4.32	3 8.46	4.31	11
50	36 41.16	4.46	42 1.40	4.45	47 20.87	4.43	52 39.48	4.43	57 57.13	4.41	3 13.72	4.39	10
51	0 36 46.51	4.55	0 42 6.73	4.54	0 47 26.19	4.52	0 52 44.79	4.52	0 58 2.42	4.50	1 3 18.99	4.48	9
52	36 51.85	4.64	42 12.06	4.63	47 31.51	4.61	52 50.09	4.61	58 7.71	4.59	3 24.26	4.57	8
53	36 57.19	4.73	42 17.39	4.72	47 36.83	4.70	52 55.39	4.70	58 12.99	4.68	3 29.53	4.65	7
54	37 2.53	4.82	42 22.72	4.81	47 42.14	4.79	53 0.60	4.79	58 18.27	4.77	3 34.79	4.74	6
55	37 7.87	4.91	42 28.05	4.90	47 47.45	4.88	53 5.99	4.88	58 23.55	4.85	3 40.05	4.83	5
56	0 37 13.22	4.99	0 42 33.38	4.98	0 47 52.77	4.96	0 53 11.29	4.96	0 58 28.84	4.94	1 3 45.32	4.92	4
57	37 18.56	5.08	42 38.71	5.07	47 58.09	5.05	53 16.59	5.05	58 34.13	5.03	3 50.59	5.01	3
58	37 23.90	5.17	42 44.04	5.16	48 3.41	5.14	53 21.89	5.14	58 39.41	5.12	3 55.85	5.09	2
59	37 29.24	5.26	42 49.37	5.25	48 8.72	5.23	53 27.19	5.23	58 44.69	5.20	4 1.11	5.18	1
60	37 34.58	5.35	42 54.70	5.34	48 14.03	5.32	53 32.49	5.31	58 49.97	5.29	4 6.37	5.27	0
$\bar{y}$	173°+ 353°–		172°+ 352°–		171°+ 351°–		170°+ 350°–		169°+ 349°–		168°+ 348°–		$\bar{y}$

TABLE CXXXVII. ARGUMENT 77.

Equation = 18500″ sin $\bar{y}$.

$\bar{y}$	192°— 12°+ Equation.	P. P.	193°— 13°+ Equation.	P. P.	194°— 14°+ Equation.	P. P.	195°— 15°+ Equation.	P. P.	196°— 16°+ Equation.	P. P.	197°— 17°+ Equation.	P. P.	$\bar{y}$
0′	1° 4′ 6.37″	0.00″	1° 9′ 21.59″	0.00″	1° 14′ 35.55″	0.00″	1° 19′ 48.17″	0.00″	1° 24′ 59.29″	0.00″	1° 30′ 8.88″	0.00″	60′
1	4 11.64	0.09	9 26.84	0.09	14 40.77	0.09	19 53.37	0.09	25 4.47	0.09	30 14.03	0.09	59
2	4 16.90	0.18	9 32.08	0.18	14 45.99	0.18	19 58.57	0.18	25 9.64	0.18	30 19.18	0.18	58
3	4 22.16	0.27	9 37.32	0.26	14 51.21	0.26	20 3.77	0.26	25 14.81	0.26	30 24.32	0.26	57
4	4 27.42	0.36	9 42.56	0.35	14 56.43	0.34	20 8.97	0.35	25 19.98	0.34	30 29.46	0.34	56
5	4 32.68	0.44	9 47.80	0.43	15 1.65	0.43	20 14.16	0.43	25 25.15	0.43	30 34.60	0.43	55
6	1 4 37.95	0.53	1 9 53.04	0.52	1 15 6.87	0.51	1 20 19.36	0.52	1 25 30.32	0.52	1 30 39.75	0.52	54
7	4 43.21	0.62	9 58.28	0.61	15 12.09	0.60	20 24.56	0.61	25 35.49	0.61	30 44.89	0.60	53
8	4 48.47	0.71	10 3.52	0.69	15 17.31	0.69	20 29.75	0.69	25 40.66	0.69	30 50.03	0.69	52
9	4 53.73	0.79	10 8.76	0.78	15 22.53	0.78	20 34.94	0.77	25 45.83	0.77	30 55.17	0.78	51
10	4 58.99	0.88	10 14.00	0.87	15 27.74	0.87	20 40.13	0.86	25 51.00	0.86	31 0.31	0.86	50
11	1 5 4.25	0.97	1 10 19.24	0.96	1 15 32.96	0.96	1 20 45.32	0.94	1 25 56.17	0.94	1 31 5.45	0.94	49
12	5 9.51	1.06	10 24.48	1.05	15 38.18	1.04	20 50.51	1.03	26 1.34	1.03	31 10.59	1.02	48
13	5 14.77	1.15	10 29.72	1.14	15 43.40	1.13	20 55.70	1.12	26 6.51	1.12	31 15.73	1.11	47
14	5 20.03	1.33	10 34.96	1.22	15 48.62	1.22	21 0.89	1.21	26 11.67	1.20	31 20.87	1.20	46
15	5 25.29	1.32	10 40.20	1.31	15 53.83	1.30	21 6.08	1.30	26 16.83	1.29	31 26.01	1.28	45
16	1 5 30.55	1.41	1 10 45.44	1.40	1 15 59.05	1.38	1 21 11.27	1.39	1 26 22.00	1.38	1 31 31.15	1.36	44
17	5 35.81	1.49	10 50.68	1.48	16 4.27	1.47	21 16.46	1.47	26 27.17	1.46	31 36.29	1.45	43
18	5 41.07	1.58	10 55.92	1.57	16 9.49	1.56	21 21.65	1.56	26 32.33	1.54	31 41.45	1.53	42
19	5 46.33	1.67	11 1.16	1.66	16 14.70	1.65	21 26.84	1.65	26 37.49	1.63	31 46.57	1.62	41
20	5 51.58	1.75	11 6.39	1.74	16 19.91	1.74	21 32.03	1.73	26 42.65	1.72	31 51.70	1.71	40
21	1 5 56.84	1.84	1 11 11.63	1.82	1 16 25.13	1.82	1 21 37.22	1.82	1 26 47.82	1.80	1 31 56.84	1.80	39
22	6 2.10	1.93	11 16.87	1.91	16 30.35	1.90	21 42.41	1.90	26 52.99	1.89	32 1.98	1.88	38
23	6 7.36	2.01	11 22.11	2.00	16 35.56	1.99	21 47.60	1.99	26 58.15	1.97	32 7.12	1.97	37
24	6 12.61	2.10	11 27.34	2.09	16 40.77	2.07	21 52.79	2.07	27 3.31	2.06	32 12.25	2.06	36
25	6 17.86	2.19	11 32.57	2.18	16 45.98	2.16	21 57.97	2.16	27 8.47	2.15	32 17.38	2.14	35
26	1 6 23.12	2.28	1 11 37.81	2.26	1 16 51.19	2.25	1 22 3.16	2.24	1 27 13.64	2.23	1 32 22.52	2.23	34
27	6 28.38	2.37	11 43.04	2.35	16 56.40	2.33	22 8.45	2.33	27 18.80	2.32	32 27.66	2.32	33
28	6 33.63	2.45	11 48.27	2.44	17 1.61	2.42	22 13.54	2.42	27 23.96	2.41	32 32.80	2.40	32
29	6 38.88	2.54	11 53.50	2.53	17 6.82	2.51	22 18.73	2.51	27 29.12	2.49	32 37.93	2.49	31
30	6 44.13	2.63	11 58.73	2.62	17 12.03	2.60	22 23.91	2.59	27 34.28	2.58	32 43.06	2.57	30
31	1 6 49.39	2.72	1 12 3.96	2.70	1 17 17.24	2.68	1 22 29.10	2.67	1 27 39.44	2.66	1 32 48.20	2.66	29
32	6 54.64	2.81	12 9.19	2.79	17 22.45	2.77	22 34.29	2.76	27 44.60	2.75	32 53.33	2.75	28
33	6 59.89	2.90	12 14.42	2.87	17 27.66	2.86	22 39.48	2.84	27 49.76	2.84	32 58.46	2.81	27
34	7 5.14	2.99	12 19.65	2.96	17 32.87	2.95	22 44.66	2.93	27 54.92	2.92	33 3.59	2.92	26
35	7 10.39	3.07	12 24.88	3.04	17 38.07	3.03	22 49.84	3.02	28 0.07	3.01	33 8.72	3.00	25
36	1 7 15.65	3.16	1 12 30.11	3.13	1 17 43.28	3.12	1 22 55.03	3.11	1 28 5.23	3.09	1 33 13.85	3.09	24
37	7 20.90	3.25	12 35.34	3.22	17 48.49	3.21	23 0.21	3.19	28 10.39	3.18	33 18.98	3.18	23
38	7 26.15	3.33	12 40.57	3.31	17 53.70	3.30	23 5.39	3.28	28 15.55	3.26	33 24.11	3.26	22
39	7 31.40	3.41	12 45.80	3.40	17 58.91	3.39	23 10.57	3.36	28 20.70	3.35	33 29.24	3.34	21
40	7 36.65	3.50	12 51.03	3.49	18 4.11	3.47	23 15.75	3.45	28 25.85	3.44	33 34.36	3.42	20
41	1 7 41.90	3.59	1 12 56.26	3.58	1 18 9.32	3.55	1 23 20.93	3.54	1 28 31.01	3.52	1 33 39.49	3.50	19
42	7 47.15	3.68	13 1.49	3.66	18 14.53	3.64	23 26.11	3.62	28 36.17	3.61	33 44.62	3.59	18
43	7 52.40	3.76	13 6.72	3.75	18 19.74	3.73	23 31.29	3.69	28 41.33	3.70	33 49.75	3.68	17
44	7 57.65	3.85	13 11.95	3.84	18 24.94	3.82	23 36.47	3.77	28 46.48	3.78	33 54.87	3.77	16
45	8 2.90	3.94	13 17.18	3.92	18 30.14	3.91	23 41.65	3.86	28 51.63	3.87	33 59.99	3.85	15
46	1 8 8.15	4.03	1 13 22.41	4.01	1 18 35.35	3.99	1 23 46.83	3.95	1 28 56.79	3.96	1 34 5.12	3.94	14
47	8 13.40	4.11	13 27.64	4.09	18 40.56	4.08	23 52.01	4.04	29 1.95	4.04	34 10.25	4.02	13
48	8 18.65	4.20	13 32.87	4.18	18 45.76	4.17	23 57.19	4.13	29 7.10	4.13	34 15.37	4.11	12
49	8 23.90	4.29	13 38.10	4.27	18 50.96	4.26	24 2.37	4.22	29 12.25	4.21	34 20.49	4.20	11
50	8 29.14	4.38	13 43.32	4.36	18 56.16	4.34	24 7.54	4.31	29 17.40	4.30	34 25.61	4.28	10
51	1 8 34.39	4.47	1 13 48.55	4.44	1 19 1.37	4.42	1 24 12.72	4.40	1 29 22.55	4.39	1 34 30.74	4.37	9
52	8 39.64	4.55	13 53.78	4.53	19 6.57	4.51	24 17.90	4.49	29 27.70	4.48	34 35.86	4.45	8
53	8 44.89	4.64	13 59.00	4.62	19 11.77	4.60	24 23.08	4.58	29 32.85	4.57	34 40.98	4.54	7
54	8 50.13	4.73	14 4.22	4.70	19 16.97	4.68	24 28.25	4.66	29 38.00	4.66	34 46.10	4.62	6
55	8 55.37	4.81	14 9.44	4.78	19 22.17	4.77	24 33.42	4.75	29 43.15	4.74	34 51.22	4.71	5
56	1 9 0.62	4.90	1 14 14.67	4.87	1 19 27.37	4.86	1 24 38.60	4.84	1 29 48.30	4.83	1 34 56.34	4.79	4
57	9 5.87	4.98	14 19.89	4.96	19 32.57	4.94	24 43.78	4.92	29 53.45	4.92	35 1.46	4.88	3
58	9 11.11	5.07	14 24.11	5.05	19 37.77	5.03	24 48.95	5.00	29 58.60	5.00	35 6.58	4.97	2
59	9 16.35	5.16	14 29.33	5.14	19 42.97	5.12	24 54.12	5.09	30 3.74	5.08	35 11.70	5.05	1
60	9 21.59	5.25	14 35.55	5.23	19 48.16	5.21	24 59.29	5.18	30 8.88	5.16	35 16.81	5.13	0
$\bar{y}$	167°+ 347°—		166°+ 346°—		165°+ 345°—		164°+ 344°—		163°+ 343°—		162°+ 342°—		$\bar{y}$

TABLE CXXXVII. ARGUMENT 77.

Equation = 18500″ sin $\bar{y}$.

$\bar{y}$	198°− 18°+ Equation.	P. P.	199°− 19°+ Equation.	P. P.	200°− 20°+ Equation.	P. P.	201°− 21°+ Equation.	P. P.	202°− 22°+ Equation.	P. P.	203°− 23°+ Equation.	P. P.	$\bar{y}$
0′	1° 35′ 16″.81	0″.00	1° 40′ 24″.01	0″.00	1° 45′ 27″.38	0″.00	1° 50′ 29″.81	0″.00	1° 55′ 30″.21	0″.00	2° 0′ 28″.52	0″.00	60′
1	35 21.93	0.09	40 29.10	0.08	45 32.44	0.08	50 34.84	0.08	55 35.21	0.08	0 33.48	0.08	59
2	35 27.05	0.18	40 33.19	0.17	45 37.00	0.17	50 39.86	0.16	55 40.20	0.16	0 38.44	0.16	58
3	35 32.17	0.25	40 38.28	0.26	45 42.55	0.25	50 44.88	0.25	55 45.19	0.25	0 43.39	0.24	57
4	35 37.28	0.34	40 43.36	0.34	45 47.60	0.33	50 49.90	0.34	55 50.18	0.33	0 48.34	0.32	56
5	35 42.39	0.43	40 48.44	0.42	45 52.65	0.42	50 54.92	0.42	55 55.17	0.41	0 53.29	0.41	55
6	1 35 47.51	0.52	1 40 53.53	0.50	1 45 57.71	0.50	1 50 59.94	0.50	1 56 0.15	0.48	2 0 58.24	0.49	54
7	35 52.63	0.61	40 58.62	0.59	46 2.76	0.59	51 4.96	0.59	56 5.13	0.56	1 3.19	0.58	53
8	35 57.74	0.69	41 3.70	0.67	46 7.81	0.67	51 9.98	0.67	56 10.11	0.65	1 8.14	0.66	52
9	36 2.85	0.77	41 8.78	0.76	46 12.86	0.75	51 15.00	0.75	56 15.10	0.74	1 13.09	0.74	51
10	36 7.96	0.85	41 13.86	0.85	46 17.91	0.84	51 20.02	0.83	56 20.09	0.83	1 18.04	0.82	50
11	1 36 13.08	0.94	1 41 18.95	0.93	1 46 22.97	0.92	1 51 25.04	0.92	1 56 25.07	0.91	2 1 22.99	0.90	49
12	36 18.19	1.02	41 24.04	1.02	46 28.02	1.00	51 30.06	1.00	56 30.05	1.00	1 27.94	0.98	48
13	36 23.30	1.11	41 29.12	1.10	46 33.07	1.09	51 35.08	1.08	56 35.03	1.08	1 32.88	1.07	47
14	36 28.41	1.19	41 34.20	1.19	46 38.12	1.17	51 40.10	1.16	56 40.01	1.16	1 37.82	1.15	46
15	36 33.52	1.27	41 39.28	1.27	46 43.17	1.26	51 45.11	1.25	56 45.00	1.24	1 42.76	1.23	45
16	1 36 38.63	1.35	1 41 44.36	1.35	1 46 48.22	1.34	1 51 50.13	1.34	1 56 49.98	1.32	2 1 47.71	1.31	44
17	36 43.74	1.44	41 49.44	1.44	46 53.27	1.43	51 55.15	1.42	56 54.96	1.41	1 52.65	1.39	43
18	36 48.85	1.52	41 54.52	1.52	46 58.32	1.51	52 0.16	1.50	56 59.93	1.50	1 57.59	1.48	42
19	36 53.96	1.61	41 59.60	1.60	47 3.37	1.60	52 5.17	1.59	57 4.91	1.58	2 2.53	1.56	41
20	36 59.07	1.70	42 4.68	1.69	47 8.41	1.68	52 10.18	1.67	57 9.89	1.66	2 7.47	1.65	40
21	1 37 4.18	1.79	1 42 9.76	1.78	1 47 13.46	1.76	1 52 15.20	1.76	1 57 14.87	1.74	2 2 12.41	1.73	39
22	37 9.29	1.87	42 14.84	1.86	47 18.51	1.85	52 20.21	1.84	57 19.85	1.82	2 17.35	1.81	38
23	37 14.41	1.96	42 19.92	1.95	47 23.55	1.93	52 25.22	1.92	57 24.83	1.90	2 22.29	1.90	37
24	37 19.51	2.05	42 24.99	2.04	47 28.59	2.01	52 30.23	2.01	57 29.80	1.98	2 27.21	1.98	36
25	37 24.61	2.14	42 30.06	2.13	47 33.63	2.10	52 35.24	2.10	57 34.77	2.06	2 32.17	2.06	35
26	1 37 29.72	2.23	1 42 35.14	2.21	1 47 38.68	2.18	1 52 40.25	2.18	1 57 39.75	2.15	2 2 37.11	2.14	34
27	37 34.83	2.31	42 40.22	2.29	47 43.72	2.26	52 45.26	2.26	57 44.73	2.23	2 42.05	2.22	33
28	37 39.93	2.39	42 45.29	2.37	47 48.76	2.35	52 50.27	2.34	57 49.70	2.31	2 46.99	2.30	32
29	37 45.03	2.47	42 50.36	2.45	47 53.80	2.44	52 55.27	2.42	57 54.67	2.40	2 51.93	2.38	31
30	37 50.13	2.55	42 55.43	2.53	47 58.84	2.52	53 0.28	2.50	57 59.64	2.48	2 56.86	2.47	30
31	1 37 55.24	2.64	1 43 0.51	2.61	1 48 3.88	2.60	1 53 5.29	2.58	1 58 4.62	2.56	2 3 1.80	2.55	29
32	38 0.34	2.72	43 5.58	2.70	48 8.92	2.69	53 10.29	2.66	58 9.59	2.65	3 6.74	2.63	28
33	38 5.44	2.81	43 10.65	2.78	48 13.96	2.77	53 15.30	2.74	58 14.56	2.74	3 11.67	2.71	27
34	38 10.54	2.90	43 15.72	2.87	48 19.00	2.86	53 20.30	2.82	58 19.53	2.82	3 16.60	2.79	26
35	38 15.64	2.98	43 20.79	2.95	48 24.04	2.94	53 25.30	2.91	58 24.50	2.90	3 21.53	2.88	25
36	1 38 20.74	3.07	1 43 25.86	3.04	1 48 29.08	3.03	1 53 30.30	3.00	1 58 29.47	2.98	2 3 26.46	2.96	24
37	38 25.84	3.15	43 30.93	3.12	48 34.12	3.11	53 35.30	3.09	58 34.44	3.06	3 31.39	3.04	23
38	38 30.94	3.24	43 36.00	3.21	48 39.16	3.20	53 40.31	3.18	58 39.41	3.14	3 36.32	3.12	22
39	38 36.04	3.32	43 41.07	3.30	48 44.19	3.28	53 45.31	3.26	58 44.38	3.22	3 41.25	3.20	21
40	38 41.14	3.40	43 46.14	3.38	48 49.22	3.36	53 50.31	3.34	58 49.34	3.31	3 46.18	3.29	20
41	1 38 46.24	3.49	1 43 51.21	3.46	1 48 54.26	3.44	1 53 55.31	3.42	1 58 54.31	3.40	2 3 51.11	3.37	19
42	38 51.34	3.58	43 56.28	3.55	48 59.30	3.53	54 0.31	3.50	58 59.27	3.48	3 56.04	3.46	18
43	38 56.44	3.66	44 1.35	3.63	49 4.33	3.61	54 5.31	3.59	59 4.23	3.56	4 0.97	3.54	17
44	39 1.54	3.75	44 6.41	3.72	49 9.36	3.70	54 10.31	3.67	59 9.19	3.65	4 5.90	3.62	16
45	39 6.63	3.84	44 11.47	3.80	49 14.39	3.78	54 15.31	3.75	59 14.15	3.73	4 10.82	3.70	15
46	1 39 11.73	3.92	1 44 16.54	3.89	1 49 19.42	3.87	1 54 20.31	3.84	1 59 19.11	3.82	2 4 15.75	3.79	14
47	39 16.83	4. 0	44 21.61	3.96	49 24.45	3.95	54 25.31	3.91	59 24.07	3.90	4 20.68	3.87	13
48	39 21.92	4. 8	44 26.67	4.04	49 29.48	4.03	54 30.30	4.00	59 29.03	3.98	4 25.60	3.95	12
49	39 27.01	4.16	44 31.73	4.13	49 34.51	4.11	54 35.30	4.09	59 33.99	4.06	4 30.52	4.03	11
50	39 32.11	4.25	44 36.79	4.22	49 39.54	4.20	54 40.30	4.18	59 38.95	4.14	4 35.44	4.12	10
51	1 39 37.21	4.33	1 44 41.85	4.30	1 49 44.57	4.28	1 54 45.30	4.26	1 59 43.91	4.22	2 4 40.36	4.21	9
52	39 42.30	4.42	44 46.91	4.39	49 49.60	4.37	54 50.29	4.35	59 48.87	4.31	4 45.28	4.29	8
53	39 47.39	4.50	44 51.97	4.47	49 54.63	4.45	54 55.29	4.44	59 53.83	4.39	4 50.20	4.37	7
54	39 52.48	4.59	44 57.03	4.56	49 59.66	4.54	55 0.28	4.52	59 58.79	4.47	4 55.12	4.45	6
55	39 57.57	4.68	45 2.09	4.64	50 4.68	4.62	55 5.27	4.60	2 0 3.74	4.55	5 0.04	4.53	5
56	1 40 2.66	4.76	1 45 7.15	4.73	1 50 9.71	4.70	1 55 10.26	4.68	2 0 8.70	4.64	2 5 4.96	4.61	4
57	40 7.75	4.85	45 12.21	4.82	50 14.74	4.79	55 15.25	4.76	0 13.66	4.72	5 9.88	4.69	3
58	40 12.84	4.93	45 17.27	4.90	50 19.77	4.87	55 20.24	4.85	0 18.62	4.80	5 14.80	4.77	2
59	40 17.93	5.02	45 22.33	4.99	50 24.79	4.96	55 25.24	4.93	0 23.57	4.88	5 19.72	4.85	1
60	40 23.01	5.10	45 27.38	5.07	50 29.81	5.04	55 30.23	5.01	0 28.52	4.97	5 24.63	4.94	0
$\bar{y}$	161°+ 341°−		160°+ 340°−		159°+ 339°−		158°+ 338°−		157°+ 337°−		156°+ 336°−		$\bar{y}$

TABLE CXXXVII. ARGUMENT 77.

Equation = 18500″ sin ȳ.

ȳ	204°— 24°+		205°— 25°+		206°— 26°+		207°— 27°+		208°— 28°+		209°— 29°+		ȳ
	Equation.	P.P.	Equation.	P.P.	Equation.	P.P.	Equation.	P.P.	Equation.	P.P.	Equation.	P.P.	
0′	2° 5′ 24.63″	0.00″	2° 10′ 18.46″	0.00″	2° 15′ 9.87″	0.00″	2° 19′ 58.82″	0.00″	2° 24′ 45.22″	0.00″	2° 29′ 28.98″	0.00″	60′
1	5 29.55	0.08	10 23.34	0.08	15 14.71	0.08	20 3.62	0.08	24 49.97	0.08	29 33.69	0.08	59
2	5 34.47	0.16	10 28.21	0.16	15 19.55	0.16	20 8.42	0.16	24 54.72	0.16	29 38.40	0.16	58
3	5 39.38	0.24	10 33.08	0.24	15 24.38	0.24	20 13.21	0.24	24 59.47	0.24	29 43.11	0.24	57
4	5 44.29	0.32	10 37.95	0.32	15 29.21	0.32	20 18.00	0.32	25 4.22	0.32	29 47.81	0.32	56
5	5 49.20	0.41	10 42.82	0.40	15 34.04	0.40	20 22.79	0.40	25 8.97	0.40	29 52.51	0.40	55
6	2 5 54.12	0.49	2 10 47.70	0.48	2 15 38.88	0.48	2 20 27.58	0.48	2 25 13.72	0.48	2 29 57.22	0.48	54
7	5 59.03	0.57	10 52.57	0.56	15 43.71	0.56	20 32.37	0.56	25 18.47	0.56	30 1.92	0.56	53
8	6 3.94	0.65	10 57.44	0.64	15 48.54	0.64	20 37.16	0.64	25 23.22	0.64	30 6.62	0.64	52
9	6 8.85	0.73	11 2.31	0.72	15 53.37	0.72	20 41.95	0.72	25 27.97	0.72	30 11.32	0.71	51
10	6 13.76	0.82	11 7.18	0.81	15 58.20	0.80	20 46.74	0.80	25 32.72	0.79	30 16.02	0.78	50
11	2 6 18.67	0.90	2 11 12.05	0.89	2 16 3.03	0.88	2 20 51.53	0.88	2 25 37.46	0.87	2 30 20.72	0.86	49
12	6 23.58	0.98	11 16.92	0.97	16 7.86	0.96	20 56.32	0.96	25 42.20	0.95	30 25.42	0.94	48
13	6 28.49	1.06	11 21.79	1.05	16 12.69	1.04	21 1.10	1.04	25 46.94	1.03	30 30.11	1.02	47
14	6 33.40	1.14	11 26.66	1.13	16 17.52	1.12	21 5.88	1.12	25 51.68	1.11	30 34.80	1.10	46
15	6 38.30	1.22	11 31.53	1.21	16 22.34	1.20	21 10.66	1.20	25 56.41	1.19	30 39.50	1.18	45
16	2 6 43.21	1.30	2 11 36.40	1.29	2 16 27.17	1.28	2 21 15.45	1.28	2 26 1.15	1.27	2 30 44.20	1.26	44
17	6 48.12	1.38	11 41.27	1.37	16 32.00	1.36	21 20.23	1.36	26 5.89	1.35	30 48.90	1.34	43
18	6 53.02	1.46	11 46.14	1.45	16 36.82	1.44	21 25.01	1.44	26 10.63	1.43	30 53.90	1.42	42
19	6 57.92	1.54	11 51.00	1.53	16 41.64	1.52	21 29.79	1.52	26 15.37	1.51	30 58.28	1.49	41
20	7 2.82	1.63	11 55.86	1.62	16 46.46	1.61	21 34.57	1.59	26 20.10	1.58	31 2.97	1.56	40
21	2 7 7.73	1.71	2 12 0.73	1.70	2 16 51.29	1.69	2 21 39.35	1.67	2 26 24.84	1.66	2 31 7.66	1.64	39
22	7 12.63	1.80	12 5.59	1.78	16 56.11	1.77	21 44.13	1.75	26 29.58	1.74	31 12.35	1.72	38
23	7 17.53	1.88	12 10.45	1.86	17 0.93	1.85	21 48.91	1.83	26 34.31	1.82	31 17.04	1.80	37
24	7 22.43	1.96	12 15.31	1.94	17 5.75	1.93	21 53.69	1.91	26 39.04	1.90	31 21.73	1.88	36
25	7 27.34	2.04	12 20.17	2.02	17 10.57	2.01	21 58.46	1.99	26 43.77	1.98	31 26.42	1.96	35
26	2 7 32.23	2.12	2 12 25.03	2.10	2 17 15.39	2.09	2 22 3.24	2.07	2 26 48.51	2.06	2 31 31.11	2.04	34
27	7 37.13	2.20	12 29.89	2.18	17 20.21	2.17	22 8.02	2.15	26 53.24	2.14	31 35.80	2.12	33
28	7 42.03	2.28	12 34.75	2.26	17 25.03	2.25	22 12.79	2.23	26 57.97	2.22	31 40.48	2.20	32
29	7 46.93	2.36	12 39.61	2.34	17 29.85	2.33	22 17.57	2.31	27 2.70	2.29	31 45.16	2.27	31
30	7 51.83	2.45	12 44.46	2.42	17 34.66	2.41	22 22.34	2.39	27 7.43	2.36	31 49.84	2.34	30
31	2 7 56.73	2.53	2 12 49.32	2.50	2 17 39.48	2.49	2 22 27.11	2.47	2 27 12.16	2.44	2 31 54.53	2.42	29
32	8 1.63	2.62	12 54.18	2.58	17 44.30	2.57	22 31.88	2.55	27 16.89	2.52	31 59.21	2.50	28
33	8 6.53	2.70	12 59.03	2.66	17 49.11	2.65	22 36.65	2.63	27 21.62	2.60	32 3.89	2.58	27
34	8 11.42	2.78	13 3.88	2.74	17 53.92	2.73	22 41.42	2.71	27 26.35	2.68	32 8.57	2.66	26
35	8 16.31	2.87	13 8.73	2.82	17 58.73	2.81	22 46.20	2.79	27 31.07	2.76	32 13.25	2.74	25
36	2 8 21.21	2.95	2 13 13.59	2.90	2 18 3.54	2.89	2 22 50.97	2.87	2 27 35.80	2.84	2 32 17.93	2.82	24
37	8 26.10	3.03	13 18.44	2.98	18 8.35	2.97	22 55.74	2.95	27 40.53	2.92	32 22.61	2.90	23
38	8 30.99	3.11	13 23.29	3.06	18 13.16	3.05	23 0.51	3.03	27 45.25	3.00	32 27.29	2.98	22
39	8 35.88	3.19	13 28.14	3.14	18 17.97	3.13	23 5.28	3.11	27 50.97	3.08	32 31.96	3.05	21
40	8 40.77	3.26	13 32.99	3.23	18 22.78	3.21	23 10.04	3.19	27 54.69	3.15	32 36.64	3.12	20
41	2 8 45.66	3.34	2 13 37.84	3.31	2 18 27.59	3.29	2 23 14.81	3.27	2 27 59.41	3.23	2 32 41.32	3.20	19
42	8 50.55	3.42	13 42.69	3.39	18 32.40	3.37	23 19.58	3.35	28 4.13	3.31	32 46.60	3.28	18
43	8 55.44	3.50	13 47.54	3.47	18 37.21	3.45	23 24.34	3.43	28 8.85	3.39	32 50.67	3.36	17
44	9 0.33	3.58	13 52.39	3.55	18 42.02	3.53	23 29.10	3.51	28 13.57	3.47	32 55.34	3.44	16
45	9 5.21	3.66	13 57.24	3.63	18 46.82	3.61	23 33.86	3.59	28 18.29	3.55	33 0.01	3.52	15
46	2 9 10.10	3.75	2 14 2.09	3.71	2 18 51.63	3.69	2 23 38.63	3.67	2 28 23.01	3.63	2 33 4.68	3.60	14
47	9 14.99	3.83	14 6.94	3.79	18 56.44	3.77	23 43.39	3.75	28 27.73	3.71	33 9.35	3.68	13
48	9 19.87	3.91	14 11.79	3.87	19 1.24	3.85	23 48.15	3.83	28 32.45	3.79	33 14.02	3.76	12
49	9 24.75	4.00	14 16.63	3.95	19 6.04	3.94	23 52.91	3.91	28 37.16	3.87	33 18.69	3.83	11
50	9 29.63	4.08	14 21.47	4.04	19 10.84	4.01	23 57.67	3.99	28 41.87	3.74	33 23.36	3.90	10
51	2 9 34.52	4.16	2 14 26.32	4.12	2 19 15.64	4.09	2 24 2.43	4.07	2 28 46.59	4.02	2 33 28.03	3.98	9
52	9 39.40	4.24	14 31.16	4.20	19 20.44	4.17	24 7.19	4.15	28 51.30	4.10	33 32.70	4.06	8
53	9 44.28	4.32	14 36.00	4.28	19 25.24	4.25	24 11.95	4.23	28 56.01	4.18	33 37.37	4.14	7
54	9 49.16	4.40	14 40.84	4.36	19 30.04	4.33	24 16.71	4.31	29 0.72	4.26	33 42.04	4.22	6
55	9 54.05	4.49	14 45.68	4.44	19 34.84	4.41	24 20.46	4.39	29 5.43	4.34	33 46.69	4.30	5
56	2 9 58.94	4.57	2 14 50.52	4.52	2 19 39.64	4.49	2 24 26.22	4.47	2 29 10.14	4.42	2 33 51.36	4.38	4
57	10 3.82	4.65	14 55.36	4.60	19 44.44	4.57	24 30.97	4.55	29 14.85	4.50	33 56.02	4.46	3
58	10 8.70	4.73	15 0.20	4.68	19 49.24	4.65	24 35.72	4.63	29 19.56	4.58	34 0.68	4.54	2
59	10 13.58	4.81	15 5.04	4.74	19 54.03	4.73	24 40.47	4.71	29 23.27	4.66	34 5.34	4.61	1
60	10 18.46	4.90	15 9.87	4.85	19 58.82	4.82	24 45.22	4.78	29 28.98	4.73	34 10.00	4.68	0
ȳ	155°+ 335°—		154°+ 334°—		153°+ 333°—		152°+ 332°—		151°+ 331°—		150°+ 330°—		ȳ

TABLE CXXXVII. ARGUMENT 77.

Equation = 18500″ sin $\bar{y}$.

$\bar{y}$	210°− 30°+ Equation.	P. P.	211°− 31°+ Equation.	P. P.	212°− 32°+ Equation.	P. P.	213°− 33°+ Equation.	P. P.	214°− 34°+ Equation.	P. P.	215°− 35°+ Equation.	P. P.	$\bar{y}$
0	2 34 10.00	0.00	2 38 48.20	0.00	2 43 23.51	0.00	2 47 55.82	0.00	2 52 25.07	0.00	2 56 51.16	0.00	60
1	34 14.66	0.08	38 52.82	0.08	43 28.08	0.08	48 0.34	0.08	52 29.53	0.08	56 55.57	0.08	59
2	34 19.32	0.16	38 57.43	0.16	43 32.64	0.16	48 4.86	0.16	52 33.99	0.16	56 59.98	0.16	58
3	34 23.98	0.24	39 2.04	0.24	43 37.20	0.24	48 9.37	0.24	52 38.45	0.24	57 4.39	0.24	57
4	34 28.64	0.32	39 6.65	0.32	43 41.76	0.32	48 13.88	0.32	52 42.91	0.32	57 8.79	0.31	56
5	34 33.29	0.40	39 11.26	0.40	43 46.32	0.40	48 18.39	0.40	52 47.37	0.39	57 13.19	0.38	55
6	2 34 37.95	0.48	2 39 15.87	0.48	2 43 50.88	0.48	2 48 22.90	0.47	2 52 51.83	0.46	2 57 17.60	0.45	54
7	34 42.61	0.56	39 20.48	0.56	43 55.44	0.55	48 27.41	0.54	52 56.29	0.53	57 22.00	0.52	53
8	34 47.26	0.63	39 25.09	0.63	44 0.00	0.62	48 31.92	0.61	53 0.74	0.60	57 26.40	0.59	52
9	34 51.91	0.70	39 29.69	0.70	44 4.56	0.69	48 36.43	0.68	53 5.19	0.67	57 30.80	0.66	51
10	34 56.56	0.77	39 34.29	0.77	44 9.11	0.76	48 40.93	0.75	53 9.64	0.74	57 35.20	0.73	50
11	2 35 1.22	0.85	2 39 38.90	0.85	2 44 13.67	0.84	2 48 45.43	0.83	2 53 14.09	0.82	2 57 39.60	0.81	49
12	35 5.87	0.93	39 43.50	0.93	44 18.22	0.92	48 49.93	0.91	53 18.54	0.90	57 44.00	0.89	48
13	35 10.52	1.01	39 48.10	1.01	44 22.77	1.00	48 54.43	0.99	53 22.99	0.98	57 48.40	0.97	47
14	35 15.17	1.09	39 52.70	1.09	44 27.32	1.08	48 58.93	1.07	53 27.44	1.06	57 52.79	1.04	46
15	35 19.82	1.17	39 57.30	1.17	44 31.87	1.16	49 3.43	1.15	53 31.89	1.13	57 57.18	1.11	45
16	2 35 24.47	1.25	2 40 1.90	1.25	2 44 36.42	1.23	2 49 7.93	1.22	2 53 36.34	1.20	2 58 1.58	1.18	44
17	35 29.12	1.33	40 6.50	1.32	44 40.97	1.30	49 12.43	1.29	53 40.79	1.27	58 5.97	1.25	43
18	35 33.77	1.41	40 11.10	1.39	44 45.52	1.37	49 16.93	1.36	53 45.24	1.34	58 10.36	1.32	42
19	35 38.42	1.48	40 15.70	1.46	44 50.07	1.44	49 21.42	1.44	53 49.68	1.41	58 14.75	1.39	41
20	35 43.06	1.55	40 20.29	1.53	44 54.61	1.50	49 25.91	1.50	53 54.12	1.48	58 19.14	1.46	40
21	2 35 47.71	1.63	2 40 24.89	1.61	2 44 59.16	1.59	2 49 30.41	1.58	2 53 58.57	1.56	2 58 23.53	1.54	39
22	35 52.35	1.71	40 29.49	1.69	45 3.71	1.67	49 34.91	1.66	54 3.01	1.64	58 27.92	1.62	38
23	35 56.99	1.79	40 34.08	1.77	45 8.25	1.75	49 39.40	1.74	54 7.45	1.72	58 32.31	1.70	37
24	36 1.63	1.87	40 38.67	1.85	45 12.79	1.83	49 43.89	1.82	54 11.89	1.80	58 36.70	1.77	36
25	36 6.27	1.95	40 43.26	1.93	45 17.33	1.91	49 48.38	1.89	54 16.33	1.87	58 41.08	1.84	35
26	2 36 10.91	2.03	2 40 47.86	2.01	2 45 21.88	1.99	2 49 52.88	1.96	2 54 20.77	1.94	2 58 45.47	1.91	34
27	36 15.55	2.11	40 52.45	2.09	45 26.42	2.06	49 57.37	2.03	54 25.21	2.01	58 49.86	1.98	33
28	36 20.19	2.18	40 57.04	2.16	45 30.96	2.13	50 1.86	2.10	54 29.65	2.08	58 54.24	2.05	32
29	36 24.83	2.25	41 1.63	2.23	45 35.50	2.20	50 6.35	2.17	54 34.08	2.15	58 58.62	2.12	31
30	36 29.46	2.32	41 6.22	2.30	45 40.04	2.27	50 10.84	2.24	54 38.51	2.22	59 3.00	2.19	30
31	2 36 34.10	2.40	2 41 10.81	2.38	2 45 44.58	2.35	2 50 15.33	2.32	2 54 42.95	2.30	2 59 7.38	2.27	29
32	36 38.74	2.48	41 15.40	2.46	45 49.12	2.43	50 19.82	2.40	54 47.38	2.38	59 11.76	2.35	28
33	36 43.37	2.56	41 19.99	2.54	45 53.66	2.51	50 24.31	2.48	54 51.81	2.46	59 16.14	2.43	27
34	36 48.00	2.64	41 24.58	2.62	45 58.20	2.59	50 28.79	2.56	54 56.24	2.54	59 20.52	2.50	26
35	36 52.63	2.72	41 29.16	2.70	46 2.73	2.67	50 33.27	2.64	55 0.67	2.61	59 24.90	2.57	25
36	2 36 57.26	2.80	2 41 33.75	2.78	2 46 7.27	2.74	2 50 37.75	2.71	2 55 5.10	2.68	2 59 29.28	2.64	24
37	37 1.89	2.88	41 38.33	2.85	46 11.80	2.81	50 42.23	2.78	55 9.53	2.75	59 33.66	2.71	23
38	37 6.52	2.96	41 42.91	2.92	46 16.33	2.88	50 46.71	2.85	55 13.96	2.82	59 38.03	2.78	22
39	37 11.15	3.02	41 47.49	2.98	46 20.86	2.95	50 51.19	2.92	55 18.39	2.89	59 42.40	2.85	21
40	37 15.78	3.09	41 52.07	3.06	46 25.39	3.02	50 55.67	2.99	55 22.81	2.96	59 46.77	2.92	20
41	2 37 20.41	3.17	2 41 56.65	3.14	2 46 29.92	3.10	2 51 0.15	3.07	2 55 27.24	3.04	2 59 51.14	3.00	19
42	37 25.04	3.24	42 1.23	3.22	46 32.45	3.18	51 4.63	3.15	55 31.67	3.12	59 55.51	3.08	18
43	37 29.67	3.33	42 5.81	3.30	46 38.98	3.26	51 9.11	3.23	55 36.10	3.20	59 59.88	3.16	17
44	37 34.30	3.41	42 10.39	3.38	46 43.50	3.34	51 13.58	3.31	55 40.52	3.28	3 0 4.25	3.23	16
45	37 38.92	3.49	42 14.96	3.46	46 48.02	3.42	51 18.05	3.39	55 44.94	3.36	0 8.61	3.30	15
46	2 37 43.54	3.57	2 42 19.54	3.54	2 46 52.55	3.49	2 51 22.53	3.46	2 55 49.36	3.43	3 0 12.98	3.37	14
47	37 48.16	3.65	42 24.12	3.62	46 57.07	3.56	51 27.00	3.53	55 53.78	3.50	0 17.35	3.44	13
48	37 52.78	3.73	42 28.69	3.69	47 1.59	3.63	51 31.47	3.60	55 58.20	3.57	0 21.71	3.51	12
49	37 57.40	3.80	42 33.26	3.76	47 6.11	3.70	51 35.94	3.67	56 2.62	3.64	0 26.07	3.58	11
50	38 2.03	3.87	42 37.83	3.83	47 10.63	3.77	51 40.41	3.74	56 7.04	3.70	0 30.43	3.65	10
51	2 38 6.65	3.95	2 42 42.40	3.91	2 47 15.15	3.85	2 51 44.88	3.82	2 56 11.46	3.78	3 0 34.79	3.73	9
52	38 11.27	4.03	42 46.97	3.99	47 19.67	3.93	51 49.35	3.90	56 15.88	3.86	0 39.15	3.81	8
53	38 15.89	4.11	42 51.54	4.07	47 24.19	4.01	51 53.82	3.98	56 20.29	3.94	0 43.51	3.89	7
54	38 20.51	4.19	42 56.11	4.15	47 28.71	4.09	51 58.29	4.06	56 24.70	4.02	0 47.87	3.96	6
55	38 25.12	4.27	43 0.68	4.23	47 33.23	4.17	52 2.75	4.14	56 29.11	4.09	0 52.23	4.03	5
56	2 38 29.74	4.35	2 43 5.25	4.31	2 47 37.75	4.25	2 52 7.22	4.21	2 56 33.52	4.16	3 0 56.59	4.10	4
57	38 34.36	4.43	43 9.82	4.38	47 42.27	4.32	52 11.69	4.28	56 37.93	4.23	1 0.96	4.17	3
58	38 38.98	4.50	43 14.39	4.45	47 46.79	4.39	52 16.15	4.35	56 42.34	4.30	1 5.32	4.24	2
59	38 43.59	4.57	43 18.95	4.52	47 51.31	4.46	52 20.61	4.42	56 46.75	4.37	1 9.68	4.31	1
60	38 48.20	4.64	43 23.51	4.59	47 55.82	4.53	52 25.07	4.49	56 51.16	4.44	1 14.04	4.38	0
$\bar{y}$	149°+ 329°−		148°+ 328°−		147°+ 327°−		146°+ 326°−		145°+ 325°−		144°+ 324°−		$\bar{y}$

TABLE CXXXVII. ARGUMENT 77.

Equation = 18500″ sin $\bar{y}$.

$\bar{y}$	216°— 36°+ Equation.	P. P.	217°— 37°+ Equation.	P. P.	218°— 38°+ Equation.	P. P.	219°— 39°+ Equation.	P. P.	220°— 40°+ Equation.	P. P.	221°— 41°+ Equation.	P. P.	$\bar{y}$
0′	3° 1′ 14.02″	0.00″	3° 5′ 33.58″	0.00″	3° 9′ 49.76″	0.00″	3° 14′ 2.42″	0.00″	3° 18′ 11.57″	0.00″	3° 22′ 17.09″	0.00″	60′
1	1 18.38	0.08	5 37.88	0.08	9 53.98	0.07	14 6.60	0.07	18 15.69	0.07	22 21.15	0.07	59
2	1 22.74	0.16	5 42.18	0.15	9 58.22	0.14	14 10.78	0.14	18 19.81	0.14	22 25.21	0.13	58
3	1 27.09	0.23	5 46.48	0.22	10 2.46	0.21	14 14.96	0.21	18 23.93	0.21	22 29.27	0.20	57
4	1 31.44	0.30	5 50.77	0.29	10 6.70	0.28	14 19.14	0.28	18 28.05	0.28	22 33.33	0.26	56
5	1 35.79	0.37	5 55.06	0.36	10 10.94	0.35	14 23.32	0.35	18 32.17	0.34	22 37.38	0.33	55
6	3 1 40.14	0.44	3 5 59.35	0.43	3 10 15.18	0.42	3 14 27.50	0.42	3 18 36.29	0.41	3 22 41.44	0.39	54
7	1 44.49	0.51	6 3.64	0.50	10 19.42	0.49	14 31.68	0.49	18 40.41	0.48	22 45.50	0.46	53
8	1 48.84	0.58	6 7.93	0.57	10 23.65	0.56	14 35.85	0.56	18 44.52	0.54	22 49.55	0.53	52
9	1 53.19	0.65	6 12.22	0.64	10 27.88	0.63	14 40.02	0.63	18 48.63	0.61	22 53.60	0.60	51
10	1 57.53	0.72	6 16.51	0.71	10 32.11	0.70	14 44.19	0.69	18 52.74	0.68	22 57.65	0.67	50
11	3 2 1.87	0.80	3 6 20.80	0.79	3 10 36.34	0.77	3 14 48.36	0.76	3 18 56.85	0.75	3 23 1.70	0.74	49
12	2 6.21	0.88	6 25.09	0.86	10 40.57	0.84	14 52.53	0.83	19 0.96	0.82	23 5.75	0.81	48
13	2 10.55	0.95	6 29.38	0.93	10 44.80	0.91	14 56.70	0.90	19 5.07	0.90	23 9.80	0.88	47
14	2 14.89	1.02	6 33.66	1.00	10 49.02	0.98	15 0.87	0.97	19 9.18	0.97	23 13.85	0.94	46
15	2 19.23	1.09	6 37.94	1.07	10 53.24	1.05	15 5.04	1.04	19 13.29	1.04	23 17.90	1.01	45
16	3 2 23.57	1.16	3 6 42.23	1.14	3 10 57.47	1.12	3 15 9.21	1.11	3 19 17.40	1.11	3 23 21.95	1.08	44
17	2 27.91	1.23	6 46.51	1.21	11 1.69	1.19	15 13.38	1.18	19 21.51	1.18	23 26.00	1.14	43
18	2 32.25	1.30	6 50.79	1.28	11 5.91	1.26	15 17.55	1.25	19 25.62	1.24	23 30.04	1.21	42
19	2 36.58	1.37	6 55.07	1.35	11 10.13	1.33	15 21.71	1.32	19 29.72	1.30	23 34.08	1.28	41
20	2 40.91	1.44	6 59.35	1.42	11 14.35	1.40	15 25.87	1.38	19 33.82	1.36	23 38.12	1.34	40
21	3 2 45.25	1.52	3 7 3.63	1.50	3 11 18.57	1.48	3 15 30.03	1.45	3 19 37.92	1.43	3 23 42.16	1.41	39
22	2 49.59	1.60	7 7.91	1.58	11 22.79	1.55	15 34.19	1.52	19 42.02	1.50	23 46.20	1.48	38
23	2 53.92	1.68	7 12.19	1.65	11 27.01	1.62	15 38.35	1.59	19 46.12	1.57	23 50.24	1.55	37
24	2 58.25	1.75	7 16.46	1.72	11 31.23	1.69	15 42.51	1.66	19 50.22	1.64	23 54.28	1.62	36
25	3 2.58	1.82	7 20.73	1.79	11 35.44	1.76	15 46.67	1.73	19 54.32	1.71	23 58.31	1.69	35
26	3 3 6.91	1.89	3 7 25.01	1.86	3 11 39.66	1.83	3 15 50.83	1.80	3 19 58.42	1.78	3 24 2.35	1.75	34
27	3 11.24	1.96	7 29.28	1.93	11 43.88	1.90	15 54.99	1.87	20 2.52	1.85	24 6.38	1.82	33
28	3 15.57	2.03	7 33.55	2.00	11 48.10	1.97	15 59.14	1.94	20 6.61	1.92	24 10.41	1.88	32
29	3 19.90	2.10	7 37.82	2.07	11 52.31	2.04	16 3.29	2.01	20 10.70	1.99	24 14.44	1.95	31
30	3 24.23	2.17	7 42.09	2.14	11 56.52	2.11	16 7.44	2.08	20 14.79	2.05	24 18.47	2.02	30
31	3 3 28.56	2.25	3 7 46.36	2.22	3 12 0.73	2.18	3 16 11.59	2.15	3 20 18.88	2.12	3 24 22.50	2.09	29
32	3 32.89	2.33	7 50.63	2.29	12 4.94	2.25	16 15.74	2.22	20 22.97	2.19	24 26.53	2.16	28
33	3 37.21	2.40	7 54.90	2.36	12 9.15	2.32	16 19.89	2.29	20 27.06	2.25	24 30.56	2.23	27
34	3 41.53	2.47	7 59.17	2.43	12 13.36	2.39	16 24.04	2.36	20 31.15	2.32	24 34.58	2.30	26
35	3 45.85	2.54	8 3.44	2.50	12 17.57	2.46	16 28.19	2.43	20 35.23	2.39	24 38.60	2.36	25
36	3 3 50.17	2.61	3 8 7.71	2.57	3 12 21.78	2.53	3 16 32.34	2.50	3 20 39.32	2.46	3 24 42.63	2.42	24
37	3 54.49	2.68	8 11.97	2.64	12 25.99	2.60	16 36.49	2.57	20 43.41	2.53	24 46.65	2.49	23
38	3 58.81	2.75	8 16.23	2.71	12 30.19	2.67	16 40.64	2.64	20 47.49	2.59	24 50.67	2.56	22
39	4 3.13	2.82	8 20.49	2.78	12 34.39	2.74	16 44.78	2.71	20 51.57	2.66	24 54.69	2.63	21
40	4 7.44	2.89	8 24.75	2.85	12 38.59	2.81	16 48.92	2.77	20 55.65	2.73	24 58.71	2.69	20
41	3 4 11.76	2.97	3 8 29.01	2.93	3 12 42.79	2.88	3 16 53.06	2.84	3 20 59.74	2.80	3 25 2.73	2.76	19
42	4 16.08	3.05	8 33.27	3.00	12 46.99	2.95	16 57.20	2.91	21 3.82	2.87	25 6.75	2.82	18
43	4 20.39	3.12	8 37.52	3.07	12 51.19	3.02	17 1.34	2.98	21 7.90	2.95	25 10.77	2.89	17
44	4 24.70	3.19	8 41.77	3.14	12 55.39	3.09	17 5.48	3.05	21 11.98	3.01	25 14.79	2.96	16
45	4 29.01	3.26	8 46.02	3.21	12 59.58	3.16	17 9.62	3.12	21 16.06	3.08	25 18.81	3.03	15
46	3 4 33.32	3.33	3 8 50.27	3.28	3 13 3.78	3.23	3 17 13.76	3.19	3 21 20.14	3.15	3 25 22.83	3.10	14
47	4 37.63	3.40	8 54.52	3.35	13 7.98	3.30	17 17.90	3.26	21 24.22	3.22	25 26.85	3.16	13
48	4 41.94	3.47	8 58.77	3.42	13 12.17	3.37	17 22.03	3.33	21 28.30	3.29	25 30.86	3.23	12
49	4 46.25	3.54	9 3.02	3.49	13 16.36	3.44	17 26.16	3.40	21 32.37	3.35	25 34.87	3.30	11
50	4 50.56	3.61	9 7.27	3.56	13 20.55	3.51	17 30.29	3.46	21 36.44	3.41	25 38.88	3.36	10
51	3 4 54.87	3.69	3 9 11.52	3.64	3 13 24.74	3.58	3 17 34.42	3.53	3 21 40.51	3.48	3 25 42.89	3.43	9
52	4 59.18	3.77	9 15.77	3.71	13 28.93	3.65	17 38.55	3.60	21 44.58	3.55	25 46.90	3.50	8
53	5 3.48	3.84	9 20.02	3.78	13 33.12	3.72	17 42.68	3.63	21 48.65	3.62	25 50.91	3.57	7
54	5 7.78	3.91	9 24.27	3.85	13 37.31	3.79	17 46.81	3.74	21 52.72	3.69	25 54.92	3.64	6
55	5 12.08	3.98	9 28.51	3.92	13 41.50	3.86	17 50.94	3.81	21 56.78	3.75	25 58.92	3.70	5
56	3 5 16.38	4.05	3 9 32.76	3.99	3 13 45.69	3.93	3 17 55.07	3.88	3 22 0.85	3.82	3 26 2.92	3.77	4
57	5 20.68	4.12	9 37.01	4.06	13 49.88	4.00	17 59.20	3.95	22 4.91	3.89	26 6.92	3.84	3
58	5 24.98	4.19	9 41.26	4.13	13 54.06	4.07	18 3.33	4.02	22 8.97	3.95	26 10.92	3.90	2
59	5 29.28	4.26	9 45.50	4.20	13 58.24	4.14	18 7.45	4.09	22 13.03	4.02	26 14.92	3.96	1
60	5 33.58	4.33	9 49.74	4.27	14 2.42	4.21	18 11.57	4.15	22 17.09	4.09	26 18.92	4.03	0
$\bar{y}$	143°+ 323°—		142°+ 322°—		141°+ 321°—		140°+ 320°—		139°+ 319°—		138°+ 318°—		$\bar{y}$

TABLE CXXXVII. ARGUMENT 77.

Equation = 18500″ sin $\bar{y}$.

$\bar{y}$	222°− 42°+ Equation.	P. P.	223°− 43°+ Equation.	P. P.	224°− 44°+ Equation.	P. P.	225°− 45°+ Equation.	P. P.	226°− 46°+ Equation.	P. P.	227°− 47°+ Equation.	P. P.	$\bar{y}$
′	° ′ ″	″	° ′ ″	″	° ′ ″	″	° ′ ″	″	° ′ ″	″	° ′ ″	″	′
0	3 26 18.92	0.00	3 30 16.97	0.00	3 34 11.18	0.00	3 38 1.47	0.00	3 41 47.78	0.00	3 45 30.04	0.00	60
1	26 22.92	0.07	30 20.91	0.06	34 15.05	0.06	38 5.27	0.06	41 51.52	0.06	45 33.71	0.06	59
2	26 26.91	0.13	30 24.85	0.12	34 18.92	0.12	38 9.07	0.12	41 55.26	0.12	45 37.38	0.12	58
3	26 30.91	0.20	30 28.78	0.18	34 22.79	0.18	38 12.87	0.18	41 59.00	0.18	45 41.05	0.18	57
4	26 34.90	0.26	30 32.71	0.24	34 26.66	0.24	38 16.67	0.24	42 2.73	0.24	45 44.72	0.24	56
5	26 38.90	0.33	30 36.64	0.30	34 30.55	0.30	38 20.47	0.30	42 6.46	0.30	45 48.38	0.30	55
6	3 26 42.89	0.40	3 30 40.57	0.37	3 34 34.39	0.37	3 38 24.27	0.37	3 42 10.19	0.36	3 45 52.05	0.36	54
7	26 46.88	0.47	30 44.50	0.44	34 38.25	0.44	38 28.07	0.42	42 13.92	0.42	45 55.71	0.42	53
8	26 50.87	0.54	30 48.43	0.51	34 42.11	0.51	38 31.87	0.49	42 17.65	0.48	45 59.37	0.48	52
9	26 54.86	0.60	30 52.36	0.58	34 45.97	0.57	38 35.66	0.56	42 21.38	0.55	46 3.03	0.54	51
10	26 58.86	0.66	30 56.28	0.65	34 49.83	0.64	38 39.45	0.63	42 25.11	0.62	46 6.69	0.61	50
11	3 27 2.85	0.73	3 31 0.21	0.72	3 34 53.69	0.71	3 38 43.25	0.70	3 42 28.84	0.68	3 46 10.35	0.67	49
12	27 6.84	0.79	31 4.13	0.79	34 57.55	0.78	38 47.05	0.76	42 32.57	0.74	46 14.01	0.73	48
13	27 10.83	0.86	31 8.05	0.86	35 1.41	0.84	38 50.84	0.83	42 36.29	0.80	46 17.66	0.79	47
14	27 14.81	0.92	31 11.97	0.92	35 5.27	0.91	38 54.63	0.90	42 40.01	0.86	46 21.31	0.85	46
15	27 18.79	0.98	31 15.89	0.99	35 9.12	0.98	38 58.42	0.96	42 43.73	0.93	46 24.96	0.91	45
16	3 27 22.78	1.05	3 31 19.81	1.06	3 35 12.98	1.04	3 39 2.21	1.02	3 42 47.45	0.99	3 46 28.61	0.97	44
17	27 26.76	1.12	31 23.73	1.12	35 16.83	1.10	39 6.00	1.08	42 51.17	1.05	46 32.26	1.03	43
18	27 30.74	1.19	31 27.65	1.18	35 20.68	1.16	39 9.79	1.14	42 54.89	1.11	46 35.91	1.09	42
19	27 34.72	1.26	31 31.57	1.24	35 24.53	1.22	39 13.58	1.20	42 58.61	1.17	46 39.56	1.15	41
20	27 38.70	1.33	31 35.49	1.30	35 28.38	1.28	39 17.36	1.26	43 2.32	1.24	46 43.20	1.21	40
21	3 27 42.68	1.40	3 31 39.40	1.37	3 35 32.23	1.35	3 39 21.11	1.32	3 43 6.04	1.30	3 46 46.85	1.27	39
22	27 46.66	1.47	31 43.31	1.44	35 36.08	1.41	39 24.92	1.38	43 9.75	1.36	46 50.50	1.33	38
23	27 50.61	1.53	31 47.22	1.50	35 39.93	1.47	39 28.70	1.45	43 13.46	1.42	46 54.14	1.39	37
24	27 54.61	1.60	31 51.13	1.57	35 43.77	1.51	39 32.48	1.51	43 17.17	1.48	46 57.78	1.45	36
25	27 58.58	1.67	31 55.01	1.61	35 47.61	1.60	39 36.26	1.57	43 20.88	1.54	47 1.42	1.51	35
26	3 28 2.55	1.73	3 31 58.95	1.71	3 35 51.46	1.66	3 39 40.04	1.64	3 43 24.59	1.60	3 47 5.06	1.57	34
27	28 6.52	1.80	32 2.86	1.77	35 55.30	1.73	39 43.82	1.70	43 28.30	1.66	47 8.70	1.63	33
28	28 10.49	1.86	32 6.76	1.81	35 59.14	1.80	39 47.59	1.76	43 32.01	1.72	47 12.34	1.69	32
29	28 14.46	1.92	32 10.66	1.90	36 2.98	1.86	39 51.36	1.82	43 35.72	1.78	47 15.98	1.75	31
30	28 18.42	1.99	32 14.56	1.96	36 6.82	1.92	39 55.13	1.88	43 39.42	1.85	47 19.62	1.82	30
31	3 28 22.39	2.06	3 32 18.46	2.03	3 36 10.66	1.98	3 39 58.90	1.94	3 43 43.12	1.91	3 47 23.24	1.88	29
32	28 26.36	2.12	32 22.36	2.10	36 14.50	2.05	40 2.67	2.00	43 46.82	1.97	47 26.89	1.94	28
33	28 30.32	2.19	32 26.26	2.17	36 18.34	2.11	40 6.00	2.07	43 50.57	2.03	47 30.53	2.00	27
34	28 34.28	2.25	32 30.16	2.21	36 22.17	2.18	40 10.21	2.13	43 54.23	2.09	47 34.16	2.06	26
35	28 38.24	2.32	32 34.06	2.30	36 26.00	2.25	40 13.97	2.19	43 57.93	2.15	47 37.79	2.12	25
36	3 28 42.20	2.39	3 32 37.96	2.37	3 36 29.83	2.31	3 40 17.74	2.25	3 44 1.63	2.21	3 47 41.42	2.18	24
37	28 46.16	2.46	32 41.86	2.43	36 33.66	2.38	40 21.51	2.32	44 0.00	2.28	47 45.05	2.24	23
38	28 50.12	2.53	32 45.76	2.49	36 37.49	2.44	40 25.27	2.38	44 9.03	2.34	47 48.68	2.30	22
39	28 54.08	2.59	32 49.65	2.55	36 41.32	2.51	40 29.03	2.44	44 12.72	2.41	47 52.31	2.36	21
40	28 58.04	2.65	32 53.54	2.60	36 45.15	2.56	40 32.79	2.51	44 16.40	2.47	47 55.93	2.42	20
41	3 29 2.00	2.72	3 32 57.43	2.68	3 36 48.98	2.63	3 40 36.55	2.58	3 44 20.09	2.53	3 47 59.56	2.48	19
42	29 5.96	2.79	33 1.32	2.74	36 52.81	2.70	40 40.31	2.64	44 23.78	2.59	48 3.18	2.54	18
43	29 9.91	2.86	33 5.21	2.81	36 56.63	2.76	40 44.07	2.70	44 27.47	2.66	48 6.80	2.60	17
44	29 13.86	2.92	33 9.10	2.87	37 0.45	2.83	40 47.83	2.77	44 31.16	2.72	48 10.42	2.66	16
45	29 17.81	2.99	33 12.99	2.94	37 4.27	2.90	40 51.58	2.83	44 34.85	2.78	48 14.04	2.72	15
46	3 29 21.76	3.06	3 33 16.88	3.00	3 37 8.09	2.96	3 40 55.34	2.90	3 44 38.54	2.84	3 48 17.66	2.78	14
47	29 25.71	3.13	33 20.77	3.06	37 11.91	3.02	40 59.09	2.96	44 42.22	2.90	48 21.28	2.84	13
48	29 29.66	3.19	33 24.65	3.13	37 15.73	3.08	41 2.84	3.02	44 45.90	2.96	48 24.90	2.90	12
49	20 33.61	3.25	33 28.53	3.19	37 19.55	3.14	41 6.59	3.08	44 49.58	3.02	48 28.51	2.96	11
50	29 37.55	3.32	33 32.41	3.26	37 23.37	3.20	41 10.34	3.14	44 53.26	3.09	48 32.12	3.03	10
51	3 29 41.50	3.39	3 33 36.29	3.33	3 37 27.19	3.27	3 41 14.09	3.20	3 44 56.95	3.15	3 48 35.73	3.09	9
52	29 45.45	3.46	33 40.17	3.40	37 31.01	3.33	41 17.84	3.27	45 0.63	3.21	48 39.34	3.15	8
53	29 49.39	3.53	33 44.05	3.46	37 34.82	3.40	41 21.59	3.33	45 4.31	3.27	48 42.95	3.21	7
54	29 53.33	3.60	33 47.93	3.52	37 38.63	3.46	41 25.33	3.39	45 7.99	3.34	48 46.56	3.27	6
55	29 57.27	3.66	33 51.81	3.59	37 42.44	3.53	41 29.07	3.45	45 11.67	3.40	48 50.17	3.33	5
56	3 30 1.21	3.73	3 33 55.69	3.65	3 37 46.25	3.59	3 41 32.82	3.52	3 45 15.34	3.46	3 48 53.78	3.39	4
57	30 5.15	3.80	33 59.57	3.72	37 50.06	3.65	41 36.56	3.58	45 19.01	3.52	48 57.38	3.45	3
58	30 9.09	3.86	34 3.44	3.79	37 53.87	3.72	41 40.30	3.64	45 22.68	3.58	49 0.98	3.51	2
59	30 13.03	3.92	34 7.31	3.85	37 57.67	3.78	41 44.04	3.71	45 26.36	3.64	49 4.58	3.57	1
60	30 16.97	3.98	34 11.18	3.90	38 1.47	3.84	41 47.78	3.77	45 30.04	3.71	49 8.18	3.63	0
$\bar{y}$	137°+ 317°−		136°+ 316°−		135°+ 315°−		134°+ 314°−		133°+ 313°−		132°+ 312°−		$\bar{y}$

TABLE CXXXVII. ARGUMENT 77.

Equation = 18500″ sin $\bar{y}$.

$\bar{y}$	228°– 48°+		229°– 49°+		230°– 50°+		231°– 51°+		232°– 52°+		233°– 53°+		$\bar{y}$
	Equation.	P. P.	Equation.	P. P.	Equation.	P. P.	Equation.	P. P.	Equation.	P. P.	Equation.	P. P.	
′	° ′ ″	″	° ′ ″	″	° ′ ″	″	° ′ ″	″	° ′ ″	″	° ′ ″	″	′
0	3 49 8.18	0.00	3 52 42.13	0.00	3 56 11.82	0.00	3 59 37.20	0.00	4 2 58.18	0.00	4 6 14.76	0.00	60
1	49 11.78	0.06	52 45.66	0.06	56 15.28	0.06	59 40.59	0.06	3 1.49	0.06	6 18.00	0.06	59
2	49 15.38	0.12	52 49.19	0.12	56 18.74	0.12	59 43.98	0.12	3 4.80	0.11	6 21.24	0.11	58
3	49 18.98	0.18	52 52.72	0.18	56 22.20	0.18	59 47.36	0.18	3 8.11	0.17	6 24.48	0.16	57
4	49 22.57	0.24	52 56.24	0.24	56 25.65	0.24	59 50.74	0.24	3 11.42	0.22	6 27.71	0.22	56
5	49 26.16	0.30	52 59.76	0.30	56 29.10	0.29	59 54.12	0.29	3 14.73	0.28	6 30.94	0.27	55
6	3 49 29.76	0.36	3 53 3.29	0.36	3 56 32.55	0.35	3 59 57.50	0.34	4 3 18.04	0.33	4 6 34.17	0.32	54
7	49 33.35	0.42	53 6.81	0.42	56 36.00	0.40	4 0 0.88	0.40	3 21.35	0.39	6 37.40	0.38	53
8	49 36.94	0.48	53 10.33	0.48	56 39.45	0.46	0 4.26	0.46	3 24.65	0.45	6 40.63	0.43	52
9	49 40.53	0.54	53 13.85	0.53	56 42.90	0.52	0 7.64	0.51	3 27.95	0.50	6 43.86	0.48	51
10	49 44.12	0.59	53 17.37	0.58	56 46.35	0.57	0 11.01	0.56	3 31.25	0.55	6 47.09	0.53	50
11	3 49 47.71	0.65	3 53 20.89	0.64	3 56 49.80	0.63	4 0 14.38	0.62	4 3 34.56	0.61	4 6 50.32	0.59	49
12	49 51.30	0.71	53 24.41	0.70	56 53.25	0.68	0 17.75	0.67	3 37.86	0.66	6 53.54	0.64	48
13	49 54.89	0.77	53 27.93	0.76	56 56.69	0.74	0 21.12	0.73	3 41.16	0.72	6 56.76	0.70	47
14	49 58.48	0.83	53 31.44	0.82	57 0.13	0.80	0 24.49	0.78	3 44.46	0.78	6 59.98	0.76	46
15	50 2.06	0.89	53 34.95	0.88	57 3.57	0.85	0 27.86	0.84	3 47.76	0.83	7 3.20	0.81	45
16	3 50 5.65	0.95	3 53 38.46	0.94	3 57 7.01	0.91	4 0 31.23	0.89	4 3 51.06	0.89	4 7 6.42	0.86	44
17	50 9.23	1.01	53 41.97	1.00	57 10.45	0.96	0 34.60	0.95	3 54.36	0.94	7 9.64	0.91	43
18	50 12.81	1.07	53 45.48	1.06	57 13.89	1.02	0 37.97	1.00	3 57.65	0.99	7 12.86	0.97	42
19	50 16.39	1.13	53 48.99	1.12	57 17.33	1.08	0 41.33	1.05	4 0.94	1.04	7 16.07	1.02	41
20	50 19.97	1.19	53 52.50	1.17	57 20.76	1.14	0 44.69	1.11	4 4.23	1.09	7 19.28	1.07	40
21	3 50 23.55	1.25	3 53 56.01	1.23	3 57 24.20	1.20	4 0 48.05	1.17	4 4 7.52	1.14	4 7 22.49	1.13	39
22	50 27.13	1.31	53 59.52	1.29	57 27.63	1.26	0 51.41	1.22	4 10.81	1.20	7 25.70	1.18	38
23	50 30.70	1.37	54 3.02	1.35	57 31.06	1.31	0 54.77	1.28	4 14.10	1.26	7 28.91	1.23	37
24	50 34.27	1.43	54 6.52	1.41	57 34.49	1.37	0 58.13	1.32	4 17.38	1.31	7 32.12	1.28	36
25	50 37.84	1.49	54 10.02	1.47	57 37.92	1.42	1 1.49	1.38	4 20.66	1.37	7 35.33	1.34	35
26	3 50 41.41	1.55	3 54 13.52	1.52	3 57 41.35	1.48	4 1 4.85	1.43	4 4 23.94	1.42	4 7 38.54	1.40	34
27	50 44.98	1.61	54 17.02	1.58	57 44.78	1.54	1 8.21	1.49	4 27.22	1.48	7 41.75	1.45	33
28	50 48.55	1.67	54 20.52	1.64	57 48.21	1.60	1 11.55	1.55	4 30.50	1.54	7 44.96	1.50	32
29	50 52.12	1.73	54 24.02	1.69	57 51.63	1.66	1 14.90	1.61	4 33.78	1.59	7 48.16	1.55	31
30	50 55.68	1.78	54 27.51	1.75	57 55.05	1.72	1 18.25	1.67	4 37.05	1.64	7 51.36	1.60	30
31	3 50 59.25	1.84	3 54 31.01	1.81	3 57 58.47	1.78	4 1 21.60	1.73	4 4 40.33	1.70	4 7 54.56	1.66	29
32	51 2.81	1.90	54 34.50	1.87	58 1.89	1.83	1 24.95	1.79	4 43.60	1.75	7 57.76	1.72	28
33	51 6.37	1.96	54 37.99	1.93	58 5.31	1.89	1 28.30	1.85	4 46.87	1.81	8 0.96	1.78	27
34	51 9.93	2.02	54 41.48	1.99	58 8.73	1.95	1 31.64	1.91	4 50.14	1.87	8 4.16	1.84	26
35	51 13.49	2.08	54 44.97	2.05	58 12.15	2.01	1 34.98	1.97	4 53.41	1.92	8 7.36	1.89	25
36	3 51 17.05	2.14	3 54 48.46	2.10	3 58 15.57	2.06	4 1 38.32	2.03	4 4 56.68	1.98	4 8 10.56	1.94	24
37	51 20.61	2.20	54 51.95	2.16	58 18.99	2.12	1 41.66	2.08	4 59.95	2.03	8 13.75	1.99	23
38	51 24.17	2.26	54 55.44	2.22	58 22.40	2.18	1 44.99	2.13	5 3.22	2.09	8 16.94	2.04	22
39	51 27.73	2.32	54 58.92	2.27	58 25.81	2.24	1 48.32	2.18	5 6.48	2.14	8 20.13	2.09	21
40	51 31.28	2.37	55 2.40	2.33	58 29.22	2.29	1 51.68	2.23	5 9.74	2.19	8 23.32	2.13	20
41	3 51 34.84	2.43	3 55 5.88	2.39	3 58 32.63	2.35	4 1 55.02	2.29	4 5 13.00	2.25	4 8 26.51	2.19	19
42	51 38.39	2.49	55 9.36	2.45	58 36.04	2.41	1 58.36	2.34	5 16.26	2.30	8 29.69	2.25	18
43	51 41.94	2.55	55 12.84	2.51	58 39.45	2.47	2 1.69	2.40	5 19.52	2.36	8 32.87	2.31	17
44	51 45.49	2.61	55 16.32	2.56	58 42.86	2.52	2 5.02	2.46	5 22.78	2.41	8 36.05	2.36	16
45	51 49.04	2.67	55 19.80	2.62	58 46.26	2.58	2 8.35	2.51	5 26.04	2.46	8 39.23	2.42	15
46	3 51 52.59	2.73	3 55 23.28	2.68	3 58 49.67	2.64	4 2 11.68	2.57	4 5 29.30	2.52	4 8 42.41	2.46	14
47	51 56.14	2.79	55 26.76	2.74	58 53.07	2.70	2 15.01	2.63	5 32.56	2.58	8 45.59	2.51	13
48	51 59.69	2.85	55 30.23	2.80	58 56.47	2.76	2 18.34	2.69	5 35.81	2.64	8 48.77	2.56	12
49	52 3.23	2.91	55 33.70	2.86	58 59.87	2.81	2 21.67	2.74	5 39.06	2.69	8 51.95	2.61	11
50	52 6.77	2.97	55 37.17	2.92	59 3.27	2.86	2 24.99	2.79	5 42.31	2.74	8 55.11	2.66	10
51	3 52 10.31	3.03	3 55 40.64	2.98	3 59 6.67	2.92	4 2 28.32	2.85	4 5 45.56	2.79	4 8 58.29	2.72	9
52	52 13.85	3.09	55 44.11	3.04	59 10.07	2.98	2 31.64	2.90	5 48.81	2.84	9 1.46	2.77	8
53	52 17.39	3.15	55 47.58	3.10	59 13.47	3.04	2 34.96	2.96	5 52.06	2.89	9 4.63	2.82	7
54	52 20.93	3.21	55 51.05	3.16	59 16.86	3.09	2 38.28	3.02	5 55.31	2.95	9 7.80	2.87	6
55	52 24.47	3.27	55 54.51	3.22	59 20.25	3.15	2 41.60	3.08	5 58.55	3.01	9 10.97	2.93	5
56	3 52 28.01	3.33	3 55 57.98	3.28	3 59 23.64	3.21	4 2 44.92	3.14	4 6 1.80	3.07	4 9 14.14	2.98	4
57	52 31.54	3.39	56 1.44	3.33	59 27.03	3.27	2 48.24	3.19	6 5.04	3.12	9 17.32	3.04	3
58	52 35.07	3.45	56 4.90	3.38	59 30.42	3.32	2 51.57	3.24	6 8.28	3.18	9 20.49	3.10	2
59	52 38.60	3.51	56 8.36	3.44	59 33.81	3.38	2 54.87	3.29	6 11.52	3.23	9 23.66	3.15	1
60	52 42.13	3.56	56 11.82	3.50	59 37.20	3.43	2 58.18	3.34	6 14.76	3.29	9 26.82	3.20	0
$\bar{y}$	131°+ 311°–		130°+ 310°–		129°+ 309°–		128°+ 308°–		127°+ 307°–		126°+ 306°–		$\bar{y}$

TABLE CXXXVII. ARGUMENT 77.

Equation = 18500″ sin $\bar{y}$.

$\bar{y}$	234°– 54°+		235°– 55°+		236°– 56°+		237°– 57°+		238°– 58°+		239°– 59°+		$\bar{y}$
	Equation.	P. P.	Equation.	P. P.	Equation.	P. P.	Equation.	P. P.	Equation.	P. P.	Equation.	P. P.	
0′	4° 9′ 26.82″	0.00″	4° 12′ 34.31″	0.00″	4° 15′ 37.19″	0.00″	4° 18′ 35.40″	0.00″	4° 21′ 28.89″	0.00″	4° 24′ 17.59″	0.00″	60′
1	9 29.99	0.05	12 37.40	0.05	15 40.20	0.05	18 38.33	0.05	21 31.74	0.05	24 20.36	0.05	59
2	9 33.15	0.10	12 40.49	0.10	15 43.21	0.10	18 41.26	0.10	21 34.59	0.10	24 23.13	0.10	58
3	9 36.31	0.15	12 43.57	0.15	15 46.22	0.15	18 44.19	0.15	21 37.44	0.15	24 25.90	0.14	57
4	9 39.47	0.20	12 46.65	0.20	15 49.22	0.20	18 47.12	0.20	21 40.29	0.20	24 28.67	0.18	56
5	9 42.63	0.26	12 49.73	0.25	15 52.22	0.25	18 50.04	0.25	21 43.14	0.25	24 31.43	0.23	55
6	4 9 45.79	0.31	4 12 52.81	0.30	4 15 55.22	0.30	4 18 52.97	0.30	4 21 45.99	0.29	4 24 34.19	0.28	54
7	9 48.95	0.36	12 55.89	0.35	15 58.22	0.35	18 55.89	0.35	21 48.83	0.34	24 36.96	0.32	53
8	9 52.10	0.42	12 58.97	0.40	16 1.22	0.40	18 58.81	0.40	21 51.67	0.38	24 39.72	0.37	52
9	9 55.25	0.47	13 2.04	0.45	16 4.22	0.45	19 1.73	0.44	21 54.51	0.42	24 42.48	0.42	51
10	9 58.40	0.52	13 5.11	0.51	16 7.22	0.50	19 4.65	0.48	21 57.35	0.47	24 45.24	0.46	50
11	4 10 1.55	0.58	4 13 8.18	0.56	4 16 10.22	0.55	4 19 7.57	0.53	4 22 0.19	0.52	4 24 48.00	0.51	49
12	10 4.70	0.63	13 11.25	0.61	16 13.21	0.60	19 10.49	0.57	22 3.03	0.57	24 50.76	0.56	48
13	10 7.84	0.69	13 14.32	0.66	16 16.20	0.65	19 13.40	0.62	22 5.86	0.61	24 53.52	0.60	47
14	10 10.98	0.74	13 17.39	0.71	16 19.19	0.70	19 16.31	0.67	22 8.69	0.66	24 56.27	0.65	46
15	10 14.12	0.79	13 20.46	0.76	16 22.18	0.75	19 19.22	0.72	22 11.52	0.70	24 59.02	0.70	45
16	4 10 17.26	0.84	4 13 23.53	0.81	4 16 25.17	0.80	4 19 22.13	0.77	4 22 14.35	0.75	4 25 1.78	0.74	44
17	10 20.40	0.89	13 26.60	0.86	16 28.16	0.85	19 25.04	0.82	22 17.18	0.80	25 4.53	0.79	43
18	10 23.54	0.94	13 29.66	0.91	16 31.15	0.90	19 27.95	0.87	22 20.01	0.85	25 7.28	0.83	42
19	10 26.68	0.99	13 32.72	0.96	16 34.13	0.95	19 30.86	0.92	22 22.84	0.89	25 10.02	0.87	41
20	10 29.82	1.04	13 35.78	1.02	16 37.11	0.99	19 33.76	0.96	22 25.66	0.94	25 12.76	0.91	40
21	4 10 32.96	1.10	4 13 38.84	1.07	4 16 40.09	1.04	4 19 36.67	1.01	4 22 28.48	0.99	4 25 15.51	0.96	39
22	10 36.10	1.15	13 41.90	1.12	16 43.07	1.09	19 39.57	1.05	22 31.31	1.03	25 18.25	1.00	38
23	10 39.23	1.20	13 44.96	1.17	16 46.05	1.14	19 42.47	1.10	22 34.13	1.08	25 20.99	1.05	37
24	10 42.36	1.26	13 48.02	1.22	16 49.03	1.19	19 45.37	1.15	22 36.95	1.12	25 23.73	1.10	36
25	10 45.49	1.31	13 51.07	1.27	16 52.01	1.24	19 48.27	1.20	22 39.77	1.17	25 26.47	1.14	35
26	4 10 48.62	1.36	4 13 54.13	1.32	4 16 54.99	1.29	4 19 51.17	1.25	4 22 42.59	1.22	4 25 29.21	1.19	34
27	10 51.75	1.41	13 57.18	1.37	16 57.97	1.34	19 54.07	1.30	22 45.41	1.26	25 31.95	1.23	33
28	10 54.88	1.46	14 0.23	1.42	17 0.94	1.39	19 56.96	1.35	22 48.23	1.30	25 34.68	1.28	32
29	10 58.01	1.51	14 3.28	1.47	17 3.91	1.44	19 59.85	1.40	22 51.04	1.35	25 37.41	1.33	31
30	11 1.14	1.57	14 6.33	1.52	17 6.88	1.49	20 2.74	1.45	22 53.85	1.40	25 40.14	1.37	30
31	4 11 4.27	1.62	4 14 9.38	1.57	4 17 9.85	1.54	4 20 5.63	1.50	4 22 56.66	1.45	4 25 42.87	1.42	29
32	11 7.40	1.67	14 12.43	1.62	17 12.82	1.59	20 8.52	1.55	22 59.47	1.50	25 45.60	1.46	28
33	11 10.52	1.71	14 15.48	1.67	17 15.79	1.64	20 11.41	1.60	23 2.28	1.55	25 48.33	1.51	27
34	11 13.64	1.80	14 18.52	1.72	17 18.76	1.69	20 14.30	1.65	23 5.09	1.59	25 51.06	1.56	26
35	11 16.76	1.85	14 21.56	1.77	17 21.72	1.74	20 17.18	1.70	23 7.89	1.64	25 53.78	1.60	25
36	4 11 19.88	1.90	4 14 24.60	1.82	4 17 24.68	1.79	4 20 20.07	1.74	4 23 10.70	1.68	4 25 56.51	1.65	24
37	11 23.00	1.95	14 27.64	1.87	17 27.64	1.84	20 22.95	1.79	23 13.50	1.73	25 59.23	1.69	23
38	11 26.12	2.00	14 30.68	1.92	17 30.60	1.89	20 25.83	1.84	23 16.30	1.77	26 1.95	1.74	22
39	11 29.23	2.05	14 33.72	1.97	17 33.56	1.94	20 28.71	1.88	23 19.10	1.82	26 4.67	1.78	21
40	11 32.34	2.09	14 36.75	2.03	17 36.52	1.98	20 31.59	1.93	23 21.90	1.87	26 7.39	1.82	20
41	4 11 35.45	2.14	4 14 39.79	2.08	4 17 39.48	2.03	4 20 34.47	1.98	4 23 24.70	1.92	4 26 10.11	1.87	19
42	11 38.56	2.19	14 42.82	2.13	17 42.44	2.08	20 37.35	2.02	23 27.50	1.96	26 12.83	1.91	18
43	11 41.67	2.25	14 45.85	2.18	17 45.39	2.13	20 40.22	2.07	23 30.29	2.01	26 15.54	1.96	17
44	11 44.77	2.30	14 48.88	2.23	17 48.34	2.18	20 43.09	2.12	23 33.08	2.06	26 18.25	2.00	16
45	11 47.88	2.35	14 51.91	2.28	17 51.29	2.23	20 45.96	2.17	23 35.87	2.10	26 20.96	2.05	15
46	4 11 50.98	2.40	4 14 54.95	2.33	4 17 54.24	2.28	4 20 48.83	2.21	4 23 38.66	2.15	4 26 23.67	2.09	14
47	11 54.08	2.45	14 57.97	2.38	17 57.19	2.33	20 51.70	2.26	23 41.45	2.20	26 26.38	2.14	13
48	11 57.18	2.50	15 1.00	2.43	18 0.14	2.38	20 54.57	2.31	23 44.24	2.24	26 29.09	2.18	12
49	12 0.28	2.55	15 4.02	2.48	18 3.09	2.43	20 57.44	2.36	23 47.03	2.29	26 31.79	2.23	11
50	12 3.38	2.61	15 7.04	2.54	18 6.03	2.48	21 0.30	2.41	23 49.81	2.34	26 34.49	2.28	10
51	4 12 6.48	2.67	4 15 10.06	2.59	4 18 8.97	2.53	4 21 3.17	2.46	4 23 52.60	2.39	4 26 37.19	2.32	9
52	12 9.58	2.72	15 13.08	2.64	18 11.91	2.58	21 6.04	2.50	23 55.38	2.43	26 39.89	2.36	8
53	12 12.68	2.77	15 16.10	2.69	18 14.85	2.63	21 8.90	2.55	23 58.16	2.48	26 42.59	2.41	7
54	12 15.77	2.82	15 19.12	2.74	18 17.79	2.68	21 11.75	2.60	24 0.94	2.52	26 45.29	2.46	6
55	12 18.86	2.87	15 22.13	2.79	18 20.73	2.73	21 14.61	2.65	24 3.72	2.56	26 47.99	2.50	5
56	4 12 21.95	2.92	4 15 25.15	2.84	4 18 23.67	2.78	4 21 17.47	2.70	4 24 6.50	2.61	4 26 50.69	2.54	4
57	12 25.04	2.97	15 28.16	2.89	18 26.61	2.83	21 20.33	2.75	24 9.28	2.66	26 53.39	2.59	3
58	12 28.13	3.02	15 31.17	2.94	18 29.54	2.88	21 23.19	2.80	24 12.05	2.71	26 56.09	2.63	2
59	12 31.22	3.08	15 34.18	2.99	18 32.47	2.93	21 26.04	2.85	24 14.82	2.76	26 58.78	2.68	1
60	12 34.31	3.13	15 37.19	3.05	18 35.40	2.98	21 28.89	2.89	24 17.59	2.81	27 1.48	2.73	0
$\bar{y}$	125°+ 305°–		124°+ 304°–		123°+ 303°–		122°+ 302°–		121°+ 301°–		120°+ 300°–		$\bar{y}$

TABLE CXXXVII. ARGUMENT 77.

Equation = 18500″ sin $\bar{y}$.

$\bar{y}$	240°— 60°+ Equation.	P. P.	241°— 61°+ Equation.	P. P.	242°— 62°+ Equation.	P. P.	243°— 63°+ Equation.	P. P.	244°— 64°+ Equation.	P. P.	245°— 65°+ Equation.	P. P.	$\bar{y}$
0′	4° 27′ 1″.47	0″.00	4° 29′ 40″.47	0″.00	4° 32′ 14″.53	0″.00	4° 34′ 43″.62	0″.00	4° 37′ 7″.68	0″.00	4° 39′ 26″.69	0″.00	60′
1	27 4.16	0.05	29 43.08	0.05	32 17.06	0.04	34 46.06	0.04	37 10.04	0.04	39 28.96	0.04	59
2	27 6.85	0.10	29 45.69	0.10	32 19.59	0.08	34 48.50	0.08	37 12.39	0.08	39 31.23	0.08	58
3	27 9.54	0.14	29 48.30	0.14	32 22.11	0.12	34 50.94	0.12	37 14.74	0.12	39 33.50	0.12	57
4	27 12.23	0.18	29 50.90	0.19	32 24.63	0.16	34 53.38	0.16	37 17.09	0.16	39 35.77	0.16	56
5	27 14.91	0.22	29 53.50	0.23	32 27.15	0.20	34 55.81	0.20	37 19.45	0.20	39 38.04	0.20	55
6	4 27 17.59	0.27	4 29 56.10	0.27	4 32 29.67	0.25	4 34 58.25	0.24	4 37 21.80	0.24	4 39 40.31	0.23	54
7	27 20.27	0.32	29 58.70	0.31	32 32.19	0.29	35 0.68	0.28	37 24.15	0.28	39 42.58	0.27	53
8	27 22.95	0.36	30 1.30	0.35	32 34.71	0.33	35 3.11	0.32	37 26.50	0.32	39 44.84	0.30	52
9	27 25.63	0.40	30 3.90	0.39	32 37.22	0.38	35 5.54	0.36	37 28.85	0.36	39 47.10	0.34	51
10	27 28.31	0.44	30 6.49	0.43	32 39.73	0.42	35 7.97	0.40	37 31.19	0.39	39 49.36	0.37	50
11	4 27 30.99	0.48	4 30 9.09	0.47	4 32 42.24	0.46	4 35 10.40	0.44	4 37 33.54	0.43	4 39 51.62	0.40	49
12	27 33.67	0.52	30 11.68	0.52	32 44.75	0.50	35 12.83	0.48	37 35.89	0.47	39 53.88	0.44	48
13	27 36.34	0.57	30 14.27	0.56	32 47.26	0.54	35 15.26	0.52	37 38.23	0.51	39 56.14	0.48	47
14	27 39.01	0.61	30 16.86	0.60	32 49.77	0.58	35 17.69	0.56	37 40.57	0.55	39 58.40	0.52	46
15	27 41.68	0.66	30 19.45	0.64	32 52.27	0.62	35 20.11	0.60	37 42.91	0.59	40 0.65	0.55	45
16	4 27 44.35	0.70	4 30 22.04	0.68	4 32 54.78	0.66	4 35 22.53	0.64	4 37 45.25	0.63	4 40 2.90	0.59	44
17	27 47.02	0.74	30 24.63	0.73	32 57.28	0.70	35 24.95	0.68	37 47.59	0.67	40 5.15	0.63	43
18	27 49.69	0.78	30 27.22	0.78	32 59.78	0.74	35 27.37	0.72	37 49.93	0.70	40 7.40	0.66	42
19	27 52.35	0.83	30 29.80	0.82	33 2.28	0.78	35 29.79	0.76	37 52.26	0.74	40 9.65	0.70	41
20	27 55.01	0.88	30 32.38	0.86	33 4.78	0.83	35 32.21	0.80	37 54.59	0.77	40 11.90	0.74	40
21	4 27 57.67	0.93	4 30 34.96	0.90	4 33 7.28	0.87	4 35 34.63	0.84	4 37 56.92	0.81	4 40 14.15	0.78	39
22	28 0.33	0.97	30 37.54	0.94	33 9.78	0.91	35 37.04	0.88	37 59.25	0.85	40 16.39	0.82	38
23	28 2.99	1.02	30 40.12	0.98	33 12.28	0.96	35 39.45	0.92	38 1.58	0.89	40 18.63	0.85	37
24	28 5.65	1.07	30 42.70	1.03	33 14.77	1.00	35 41.86	0.96	38 3.91	0.93	40 20.87	0.89	36
25	28 8.31	1.11	30 45.27	1.08	33 17.26	1.04	35 44.27	1.00	38 6.23	0.97	40 23.11	0.93	35
26	4 28 10.97	1.16	4 30 47.84	1.12	4 33 19.75	1.08	4 35 46.68	1.04	4 38 8.55	1.01	4 40 25.35	0.97	34
27	28 13.63	1.20	30 50.41	1.16	33 22.24	1.12	35 49.08	1.08	38 10.87	1.05	40 27.59	1.00	33
28	28 16.28	1.25	30 52.98	1.20	33 24.73	1.16	35 51.48	1.12	38 13.19	1.09	40 29.83	1.04	32
29	28 18.93	1.29	30 55.55	1.24	33 27.22	1.20	35 53.88	1.16	38 15.51	1.13	40 31.06	1.08	31
30	28 21.58	1.33	30 58.12	1.28	33 29.70	1.25	35 56.28	1.20	38 17.82	1.16	40 34.29	1.12	30
31	4 28 24.23	1.38	4 31 0.69	1.32	4 33 32.19	1.31	4 35 58.68	1.24	4 38 20.11	1.20	4 40 36.62	1.16	29
32	28 26.88	1.43	31 3.26	1.36	33 34.67	1.35	36 1.08	1.28	38 22.45	1.24	40 38.85	1.20	28
33	28 29.53	1.47	31 5.82	1.41	33 37.15	1.39	36 3.48	1.32	38 24.76	1.28	40 41.08	1.23	27
34	28 32.18	1.52	31 8.38	1.46	33 39.63	1.43	36 5.87	1.36	38 27.07	1.32	40 43.31	1.27	26
35	28 34.82	1.56	31 10.94	1.50	33 42.11	1.47	36 8.26	1.40	38 29.38	1.36	40 45.43	1.30	25
36	4 28 37.46	1.61	4 31 13.50	1.54	4 33 44.59	1.51	4 36 10.65	1.44	4 38 31.69	1.40	4 40 47.65	1.34	24
37	28 40.10	1.65	31 16.06	1.58	33 47.07	1.55	36 13.04	1.48	38 34.00	1.44	40 49.87	1.38	23
38	28 42.74	1.69	31 18.62	1.63	33 49.54	1.59	36 15.43	1.52	38 36.31	1.48	40 52.09	1.41	22
39	28 45.38	1.73	31 21.18	1.67	33 51.81	1.63	36 17.82	1.56	38 38.61	1.52	40 54.31	1.45	21
40	28 48.02	1.77	31 23.73	1.71	33 54.48	1.66	36 20.21	1.60	38 40.91	1.55	40 56.53	1.49	20
41	4 28 50.66	1.82	4 31 26.28	1.75	4 33 56.95	1.70	4 36 22.60	1.64	4 38 43.21	1.59	4 40 58.75	1.53	19
42	28 53.29	1.86	31 28.83	1.80	33 59.42	1.74	36 24.99	1.68	38 45.51	1.63	41 0.97	1.57	18
43	28 55.92	1.91	31 31.38	1.84	34 1.89	1.78	36 27.38	1.72	38 47.81	1.67	41 3.18	1.60	17
44	28 58.55	1.96	31 33.93	1.88	34 4.36	1.82	36 29.76	1.76	38 50.11	1.71	41 5.39	1.64	16
45	29 1.18	2.00	31 36.48	1.92	34 6.82	1.86	36 32.14	1.80	38 52.41	1.75	41 7.60	1.68	15
46	4 29 3.81	2.04	4 31 39.03	1.96	4 34 9.29	1.92	4 36 34.52	1.84	4 38 54.71	1.79	4 41 9.81	1.71	14
47	29 6.44	2.09	31 41.58	2.01	34 11.75	1.96	36 36.90	1.88	38 57.01	1.83	41 12.02	1.74	13
48	29 9.07	2.13	31 44.12	2.06	34 14.21	2.00	36 39.28	1.92	38 59.30	1.87	41 14.23	1.78	12
49	29 11.69	2.17	31 46.66	2.10	34 16.67	2.04	36 41.66	1.96	39 1.59	1.91	41 16.43	1.82	11
50	29 14.31	2.21	31 49.20	2.14	34 19.13	2.08	36 44.03	2.00	39 3.88	1.94	41 18.63	1.86	10
51	4 29 16.93	2.26	4 31 51.74	2.19	4 34 21.59	2.12	4 36 46.40	2.04	4 39 6.17	1.98	4 41 20.83	1.90	9
52	29 19.55	2.30	31 54.28	2.23	34 24.05	2.16	36 48.77	2.08	39 8.46	2.02	41 23.03	1.93	8
53	29 22.13	2.35	31 56.82	2.27	34 26.50	2.20	36 51.14	2.12	39 10.75	2.06	41 25.23	1.97	7
54	29 24.79	2.39	31 59.35	2.32	34 28.95	2.24	36 53.51	2.16	39 13.03	2.10	41 27.43	2.00	6
55	29 27.41	2.44	32 1.88	2.36	34 31.40	2.28	36 55.88	2.20	39 15.31	2.14	41 29.63	2.04	5
56	4 29 30.03	2.48	4 32 4.41	2.40	4 34 33.85	2.32	4 36 58.24	2.24	4 39 17.59	2.18	4 41 31.83	2.08	4
57	29 32.64	2.54	32 6.94	2.44	34 36.30	2.36	37 0.60	2.28	39 19.87	2.22	41 34.02	2.12	3
58	29 35.25	2.58	32 9.43	2.48	34 38.74	2.40	37 2.96	2.32	39 22.15	2.25	41 36.21	2.16	2
59	29 37.86	2.62	32 12.00	2.52	34 41.18	2.44	37 5.32	2.36	39 24.42	2.29	41 38.40	2.20	1
60	29 40.47	2.65	32 14.53	2.57	34 43.62	2.49	37 7.68	2.40	39 26.69	2.32	41 40.59	2.23	0
$\bar{y}$	119°+ 299°—		118°+ 298°—		117°+ 297°—		116°+ 296°—		115°+ 295°—		114°+ 294°—		$\bar{y}$

TABLE CXXXVII. ARGUMENT 77.

Equation = 18500″ sin $\bar{y}$.

$\bar{y}$	246°– 66°+ Equation.	P. P.	247°– 67°+ Equation.	P. P.	248°– 68°+ Equation.	P. P.	249°– 69°+ Equation.	P. P.	250°– 70°+ Equation.	P. P.	251°– 71°+ Equation.	P. P.	$\bar{y}$
0′	4° 41′ 40″.59	0″.00	4° 43′ 49″.34	0″.00	4° 45′ 52″.90	0″.00	4° 47′ 51″.23	0″.00	4° 49′ 44″.31	0″.00	4° 51′ 32″.10	0″.00	60′
1	41 42.78	0.04	43 51.44	0.04	45 54.92	0.03	47 53.16	0.03	49 46.15	0.03	51 33.85	0.03	59
2	41 44.97	0.08	43 53.54	0.08	45 56.93	0.06	47 55.09	0.06	49 47.99	0.06	51 35.60	0.06	58
3	41 47.16	0.12	43 55.64	0.11	45 58.94	0.09	47 57.01	0.09	49 49.83	0.09	51 37.35	0.09	57
4	41 49.31	0.15	43 57.74	0.14	46 0.95	0.12	47 58.93	0.12	49 51.66	0.12	51 39.09	0.12	56
5	41 51.52	0.18	43 59.84	0.18	46 2.96	0.16	48 0.85	0.15	49 53.49	0.15	51 40.83	0.15	55
6	4 41 53.70	0.22	4 44 1.94	0.21	4 46 4.97	0.20	4 48 2.77	0.18	4 49 55.32	0.18	4 51 42.57	0.18	54
7	41 55.88	0.25	44 4.03	0.25	46 6.98	0.23	48 4.69	0.21	49 57.15	0.21	51 44.31	0.21	53
8	41 58.06	0.29	44 6.12	0.28	46 8.98	0.26	48 6.61	0.24	49 58.98	0.24	51 46.05	0.24	52
9	42 0.24	0.33	44 8.21	0.31	46 10.98	0.30	48 8.53	0.28	50 0.81	0.27	51 47.79	0.27	51
10	42 2.41	0.36	44 10.30	0.34	46 12.98	0.33	48 10.44	0.32	50 2.64	0.30	51 49.53	0.29	50
11	4 42 4.58	0.40	4 44 12.39	0.38	4 46 14.98	0.36	4 48 12.35	0.35	4 50 4.47	0.33	4 51 51.27	0.32	49
12	42 6.75	0.43	44 14.48	0.41	46 16.98	0.40	48 14.26	0.38	50 6.30	0.36	51 53.00	0.35	48
13	42 8.92	0.47	44 16.56	0.44	46 18.98	0.43	48 16.17	0.41	50 8.12	0.39	51 54.73	0.38	47
14	42 11.09	0.51	44 18.64	0.48	46 20.97	0.47	48 18.08	0.44	50 9.94	0.42	51 56.46	0.41	46
15	42 13.26	0.54	44 20.72	0.52	46 22.97	0.50	48 19.99	0.47	50 11.76	0.45	51 58.19	0.43	45
16	4 42 15.43	0.58	4 44 22.80	0.55	4 46 24.97	0.53	4 48 21.90	0.50	4 50 13.58	0.48	4 51 59.92	0.46	44
17	42 17.60	0.62	44 24.88	0.58	46 26.96	0.56	48 23.81	0.53	50 15.40	0.51	52 1.65	0.49	43
18	42 19.76	0.66	44 26.96	0.61	46 28.95	0.60	48 25.71	0.56	50 17.22	0.54	52 3.38	0.51	42
19	42 21.92	0.69	44 29.04	0.65	46 30.94	0.63	48 27.61	0.60	50 19.03	0.57	52 5.10	0.54	41
20	42 24.08	0.72	44 31.11	0.69	46 32.93	0.66	48 29.51	0.63	50 20.84	0.60	52 6.82	0.57	40
21	4 42 26.24	0.76	4 44 33.18	0.72	4 46 34.92	0.69	4 48 31.41	0.66	4 50 22.65	0.63	4 52 8.54	0.60	39
22	42 28.40	0.80	44 35.25	0.76	46 36.91	0.73	48 33.31	0.69	50 24.46	0.66	52 10.26	0.63	38
23	42 30.56	0.83	44 37.32	0.80	46 38.89	0.76	48 35.21	0.72	50 26.27	0.69	52 11.98	0.66	37
24	42 32.71	0.86	44 39.39	0.83	46 40.87	0.80	48 37.10	0.75	50 28.08	0.72	52 13.70	0.69	36
25	42 34.86	0.90	44 41.46	0.87	46 42.85	0.83	48 38.99	0.78	50 29.88	0.75	52 15.42	0.72	35
26	4 42 37.01	0.94	4 44 43.53	0.90	4 46 44.83	0.86	4 48 40.88	0.81	4 50 31.68	0.78	4 52 17.14	0.75	34
27	42 39.16	0.98	44 45.59	0.94	46 46.81	0.90	48 42.77	0.84	50 33.48	0.81	52 18.86	0.78	33
28	42 41.31	1.01	44 47.65	0.97	46 48.78	0.93	48 44.66	0.88	50 35.28	0.84	52 20.57	0.81	32
29	42 43.46	1.04	44 49.71	1.00	46 50.75	0.96	48 46.54	0.91	50 37.08	0.87	52 22.28	0.84	31
30	42 45.61	1.07	44 51.77	1.03	46 52.72	0.99	48 48.42	0.95	50 38.87	0.90	52 23.99	0.86	30
31	4 42 47.76	1.11	4 44 53.83	1.07	4 46 54.69	1.02	4 48 50.30	0.98	4 50 40.67	0.93	4 52 25.70	0.89	29
32	42 49.91	1.14	44 55.89	1.11	46 56.66	1.06	48 52.18	1.01	50 42.46	0.96	52 27.41	0.92	28
33	42 52.05	1.18	44 57.95	1.14	46 58.63	1.09	48 54.06	1.04	50 44.25	0.99	52 29.12	0.94	27
34	42 54.19	1.22	45 0.00	1.17	47 0.60	1.12	48 55.94	1.07	50 46.04	1.02	52 30.82	0.97	26
35	42 56.33	1.25	45 2.05	1.20	47 2.56	1.16	48 57.82	1.10	50 47.83	1.05	52 32.52	1.00	25
36	4 42 58.47	1.28	4 45 4.10	1.24	4 47 4.53	1.19	4 48 59.70	1.13	4 50 49.62	1.08	4 52 34.22	1.03	24
37	43 0.61	1.32	45 6.15	1.27	47 6.49	1.23	49 1.58	1.16	50 51.41	1.11	52 35.92	1.06	23
38	43 2.75	1.36	45 8.20	1.30	47 8.45	1.26	49 3.45	1.19	50 53.20	1.14	52 37.62	1.09	22
39	43 4.88	1.39	45 10.25	1.33	47 10.41	1.29	49 5.32	1.22	50 54.98	1.17	52 39.31	1.11	21
40	43 7.01	1.43	45 12.29	1.37	47 12.37	1.32	49 7.19	1.26	50 56.76	1.20	52 41.00	1.14	20
41	4 43 9.14	1.47	4 45 14.34	1.40	4 47 14.33	1.35	4 49 9.06	1.29	4 50 58.54	1.23	4 52 42.69	1.17	19
42	43 11.27	1.50	45 16.38	1.44	47 16.29	1.38	49 10.93	1.32	51 0.32	1.26	52 44.38	1.20	18
43	43 13.40	1.54	45 18.42	1.47	47 18.24	1.42	49 12.80	1.35	51 2.10	1.29	52 46.07	1.22	17
44	43 15.52	1.57	45 20.46	1.50	47 20.19	1.45	49 14.67	1.38	51 3.87	1.32	52 47.75	1.25	16
45	43 17.64	1.60	45 22.50	1.54	47 22.14	1.48	49 16.53	1.41	51 5.05	1.35	52 49.93	1.28	15
46	4 43 19.76	1.63	4 45 24.54	1.58	4 47 24.09	1.51	4 49 18.39	1.44	4 51 7.43	1.38	4 52 51.11	1.31	14
47	43 21.88	1.67	45 26.58	1.61	47 26.04	1.54	49 20.25	1.47	51 9.20	1.41	52 52.79	1.34	13
48	43 24.00	1.71	45 28.61	1.64	47 27.99	1.57	49 22.11	1.50	51 10.97	1.44	52 54.47	1.37	12
49	43 26.12	1.75	45 30.64	1.68	47 29.93	1.61	49 23.97	1.53	51 12.74	1.47	52 56.15	1.40	11
50	43 28.24	1.79	45 32.67	1.72	47 31.87	1.65	49 25.83	1.57	51 14.51	1.50	52 57.83	1.43	10
51	4 43 30.36	1.83	4 45 34.70	1.76	4 47 33.81	1.68	4 49 27.69	1.60	4 51 16.28	1.53	4 52 59.51	1.46	9
52	43 32.48	1.86	45 36.73	1.79	47 35.75	1.72	49 29.54	1.63	51 18.05	1.56	53 1.19	1.49	8
53	43 34.59	1.90	45 38.76	1.83	47 37.69	1.75	49 31.39	1.66	51 19.81	1.59	53 2.86	1.52	7
54	43 36.70	1.94	45 40.78	1.86	47 39.63	1.79	49 33.24	1.69	51 21.57	1.62	53 4.53	1.54	6
55	43 38.81	1.98	45 42.80	1.90	47 41.57	1.81	49 35.09	1.72	51 23.33	1.65	53 6.20	1.57	5
56	4 43 40.92	2.01	4 45 44.82	1.93	4 47 43.51	1.84	4 49 36.94	1.75	4 51 25.09	1.68	4 53 7.87	1.60	4
57	43 43.03	2.05	45 46.84	1.97	47 45.44	1.87	49 38.79	1.78	51 26.85	1.71	53 9.54	1.63	3
58	43 45.14	2.09	45 48.86	2.00	47 47.37	1.90	49 40.63	1.82	51 28.60	1.74	53 11.21	1.66	2
59	43 47.24	2.12	45 50.88	2.03	47 49.30	1.94	49 42.47	1.85	51 30.35	1.77	53 12.88	1.69	1
60	43 49.34	2.15	45 52.90	2.06	47 51.23	1.98	49 44.31	1.89	51 32.10	1.80	53 14.54	1.71	0
$\bar{y}$	113°+ 293°–		112°+ 292°–		111°+ 291°–		110°+ 290°–		109°+ 289°–		108°+ 288°–		$\bar{y}$

TABLE CXXXVII. ARGUMENT 77.

Equation = 18500″ sin $\bar{y}$.

$\bar{y}$	252°– 72°+ Equation.	P. P.	253°– 73°+ Equation.	P. P.	254°– 74°+ Equation.	P. P.	255°– 75°+ Equation.	P. P.	256°– 76°+ Equation.	P. P.	257°– 77°+ Equation.	P. P.	$\bar{y}$
0′	4° 53′ 14.54″	0.00″	4 54 51.64	0.00	4 56 23.35	0.00	4 57 49.63	0.00	4 59 10.47	0.00	5 0 25.84	0.00	60′
1	53 16.20	0.03	54 53.21	0.03	56 24.83	0.03	57 51.02	0.02	59 11.77	0.02	0 27.04	0.02	59
2	53 17.86	0.06	54 54.78	0.06	56 26.31	0.06	57 52.41	0.04	59 13.07	0.04	0 28.25	0.04	58
3	53 19.52	0.09	54 56.35	0.09	56 27.79	0.08	57 53.80	0.07	59 14.37	0.06	0 29.46	0.06	57
4	53 21.18	0.12	54 57.92	0.12	56 29.27	0.11	57 55.19	0.09	59 15.67	0.08	0 30.67	0.08	56
5	53 22.84	0.15	54 59.49	0.14	56 30.75	0.13	57 56.58	0.12	59 16.96	0.10	0 31.88	0.10	55
6	4 53 24.50	0.17	4 55 1.06	0.17	4 56 32.23	0.15	4 57 57.97	0.14	4 59 18.25	0.12	5 0 33.08	0.12	54
7	53 26.15	0.20	55 2.62	0.20	56 33.70	0.18	57 59.35	0.17	59 19.54	0.14	0 34.28	0.14	53
8	53 27.80	0.23	55 4.18	0.22	56 35.17	0.20	58 0.73	0.19	59 20.83	0.17	0 35.48	0.16	52
9	53 29.45	0.25	55 5.74	0.24	56 36.64	0.22	58 2.11	0.21	59 22.12	0.19	0 36.68	0.18	51
10	53 31.10	0.27	55 7.30	0.26	56 38.11	0.24	58 3.49	0.23	59 23.41	0.21	0 37.88	0.19	50
11	4 53 32.75	0.30	4 55 8.86	0.28	4 56 39.58	0.27	4 58 4.87	0.25	4 59 24.70	0.23	5 0 39.07	0.21	49
12	53 34.40	0.32	55 10.42	0.31	56 41.05	0.30	58 6.24	0.27	59 25.99	0.25	0 40.26	0.23	48
13	53 36.04	0.35	55 11.97	0.34	56 42.51	0.32	58 7.61	0.30	59 27.27	0.27	0 41.45	0.25	47
14	53 37.68	0.37	55 13.52	0.36	56 43.97	0.34	58 8.98	0.32	59 28.55	0.29	0 42.64	0.27	46
15	53 39.32	0.40	55 15.07	0.38	56 45.43	0.36	58 10.35	0.34	59 29.83	0.32	0 43.83	0.29	45
16	4 53 40.96	0.43	4 55 16.62	0.41	4 56 46.89	0.39	4 58 11.72	0.37	4 59 31.11	0.34	5 0 45.02	0.31	44
17	53 42.60	0.45	55 18.17	0.43	56 48.35	0.41	58 13.09	0.39	59 32.39	0.36	0 46.21	0.33	43
18	53 44.24	0.48	55 19.72	0.45	56 49.81	0.43	58 14.45	0.41	59 33.67	0.38	0 47.39	0.35	42
19	53 45.88	0.51	55 21.26	0.48	56 51.27	0.45	58 15.81	0.43	59 34.94	0.40	0 48.57	0.37	41
20	53 47.51	0.54	55 22.80	0.51	56 52.72	0.48	58 17.17	0.45	59 36.21	0.42	0 49.75	0.39	40
21	4 53 49.14	0.57	4 55 24.34	0.54	4 56 54.17	0.51	4 58 18.53	0.47	4 59 37.48	0.44	5 0 50.93	0.41	39
22	53 50.77	0.60	55 25.88	0.57	56 55.62	0.53	58 19.89	0.49	59 38.75	0.46	0 52.11	0.43	38
23	53 52.40	0.63	55 27.42	0.60	56 57.07	0.56	58 21.25	0.52	59 40.02	0.48	0 53.29	0.45	37
24	53 54.03	0.65	55 28.96	0.62	56 58.52	0.58	58 22.61	0.54	59 41.29	0.50	0 54.46	0.47	36
25	53 55.66	0.68	55 30.50	0.65	56 59.97	0.60	58 23.96	0.56	59 42.55	0.53	0 55.63	0.49	35
26	4 53 57.28	0.71	4 55 32.04	0.68	4 57 1.41	0.63	4 58 25.32	0.59	4 59 43.81	0.55	5 0 56.80	0.51	34
27	53 58.90	0.73	55 33.58	0.70	57 2.85	0.65	58 26.68	0.61	59 45.07	0.57	0 57.97	0.53	33
28	54 0.52	0.76	55 35.11	0.72	57 4.29	0.68	58 28.03	0.63	59 46.33	0.59	0 59.14	0.55	32
29	54 2.14	0.79	55 36.64	0.74	57 5.73	0.70	58 29.38	0.65	59 47.59	0.61	1 0.31	0.57	31
30	54 3.76	0.81	55 38.17	0.76	57 7.17	0.72	58 30.73	0.68	59 48.84	0.63	1 1.47	0.58	30
31	4 54 5.38	0.84	4 55 39.70	0.79	4 57 8.61	0.75	4 58 32.08	0.70	4 59 50.17	0.65	5 1 2.63	0.60	29
32	54 7.00	0.86	55 41.23	0.81	57 10.05	0.77	58 33.43	0.72	59 51.35	0.67	1 3.79	0.62	28
33	54 8.61	0.89	55 42.76	0.84	57 11.48	0.80	58 34.78	0.75	59 52.60	0.69	1 4.95	0.64	27
34	54 10.22	0.92	55 44.28	0.86	57 12.91	0.82	58 36.12	0.77	59 53.85	0.72	1 6.11	0.66	26
35	54 11.83	0.95	55 45.80	0.88	57 14.34	0.84	58 37.46	0.79	59 55.10	0.74	1 7.27	0.68	25
36	4 54 13.44	0.97	4 55 47.32	0.91	4 57 15.77	0.87	4 58 38.80	0.81	4 59 56.35	0.76	5 1 8.43	0.70	24
37	54 15.05	1.00	55 48.84	0.94	57 17.20	0.90	58 40.14	0.83	59 57.60	0.78	1 9.59	0.72	23
38	54 16.66	1.03	55 50.36	1.97	57 18.63	0.92	58 41.48	0.85	59 58.84	0.80	1 10.74	0.74	22
39	54 18.27	1.06	55 51.88	1.00	57 20.05	0.94	58 42.81	0.87	5 0 0.08	0.82	1 11.89	0.76	21
40	54 19.87	1.08	55 53.39	1.02	57 21.47	0.96	58 44.14	0.90	0 1.32	0.84	1 13.04	0.77	20
41	4 54 21.47	1.11	4 55 54.90	1.05	4 57 22.89	0.98	4 58 45.47	0.92	5 0 2.56	0.86	5 1 14.19	0.79	19
42	54 23.07	1.14	55 56.41	1.07	57 24.31	1.00	58 46.80	0.94	0 3.80	0.88	1 15.34	0.81	18
43	54 24.67	1.16	55 57.92	1.10	57 25.73	1.03	58 48.13	0.97	0 5.04	0.90	1 16.49	0.83	17
44	54 26.27	1.19	55 59.43	1.12	57 27.15	1.05	58 49.45	0.99	0 6.28	0.93	1 17.63	0.85	16
45	54 27.87	1.21	56 0.93	1.14	57 28.57	1.08	58 50.77	1.01	0 7.51	0.95	1 18.77	0.87	15
46	4 54 29.47	1.24	4 56 1.44	1.17	4 57 29.98	1.10	4 58 52.09	1.04	5 0 8.74	0.97	5 1 19.91	0.89	14
47	54 31.06	1.27	56 3.94	1.19	57 31.39	1.12	58 53.41	1.06	0 9.97	0.99	1 21.05	0.91	13
48	54 32.65	1.29	56 5.44	1.22	57 32.80	1.15	58 54.73	1.09	0 11.20	1.01	1 22.19	0.93	12
49	54 34.24	1.32	56 6.94	1.25	57 33.21	1.18	58 56.05	1.11	0 12.43	1.03	1 23.33	0.95	11
50	54 35.83	1.35	56 8.44	1.28	57 35.62	1.20	58 57.37	1.13	0 13.66	1.05	1 24.46	0.97	10
51	4 54 37.42	1.38	4 56 9.94	1.31	4 57 37.03	1.23	4 58 58.69	1.15	5 0 14.89	1.07	5 1 25.59	0.99	9
52	54 39.01	1.40	56 11.44	1.33	57 38.44	1.25	59 0.01	1.17	0 16.11	1.09	1 26.72	1.01	8
53	54 40.59	1.43	56 12.93	1.36	57 39.84	1.28	59 1.32	1.20	0 17.33	1.11	1 27.85	1.03	7
54	54 42.17	1.46	56 14.42	1.38	57 41.24	1.30	59 2.63	1.22	0 18.55	1.14	1 28.98	1.05	6
55	54 43.75	1.48	56 15.91	1.40	57 42.64	1.32	59 3.94	1.24	0 19.77	1.16	1 30.11	1.07	5
56	4 54 45.33	1.51	4 56 17.40	1.43	4 57 44.04	1.34	4 59 5.25	1.27	5 0 20.99	1.18	5 1 31.24	1.09	4
57	54 46.91	1.53	56 18.89	1.45	57 45.44	1.36	59 6.56	1.29	0 22.21	1.20	1 32.37	1.11	3
58	54 48.49	1.56	56 20.38	1.48	57 46.84	1.38	59 7.87	1.31	0 23.42	1.22	1 33.49	1.13	2
59	54 50.07	1.59	56 21.87	1.51	57 48.24	1.40	59 9.17	1.33	0 24.63	1.24	1 34.61	1.15	1
60	54 51.64	1.62	56 23.35	1.53	57 49.63	1.44	59 10.47	1.35	0 25.84	1.26	1 35.73	1.16	0
$\bar{y}$	107°+ 287°–		106°+ 286°–		105°+ 285°–		104°+ 284°–		103°+ 283°–		102°+ 282°–		$\bar{y}$

TABLE CXXXVII. ARGUMENT 77.

Equation = 18500″ sin $\bar{y}$.

$\bar{y}$	258°– 78°+		259°– 79°+		260°– 80°+		261°– 81°+		262°– 82°+		263°– 83°+		$\bar{y}$
	Equation.	P. P.	Equation.	P. P.	Equation.	P. P.	Equation.	P. P.	Equation.	P. P.	Equation.	P. P.	
0′	5° 1′ 35.73″	0.00″	5° 2′ 40.10″	0.00″	5° 3′ 38.94″	0.00″	5° 4′ 32.23″	0.00″	5° 5′ 19.96″	0.00″	5° 6′ 2.10″	0.00″	60′
1	1 36.85	0.02	2 41.13	0.02	3 39.87	0.01	4 33.07	0.01	5 20.71	0.01	6 2.75	0.01	59
2	1 37.97	0.04	2 42.16	0.04	3 40.80	0.03	4 33.91	0.02	5 21.46	0.02	6 3.40	0.02	58
3	1 39.09	0.05	2 43.18	0.05	3 41.73	0.04	4 34.75	0.04	5 22.21	0.03	6 4.05	0.03	57
4	1 40.20	0.07	2 44.20	0.07	3 42.66	0.06	4 35.58	0.05	5 22.95	0.04	6 4.70	0.04	56
5	1 41.31	0.09	2 45.22	0.08	3 43.59	0.07	4 36.41	0.06	5 23.69	0.05	6 5.35	0.05	55
6	5 1 42.42	0.11	5 2 46.24	0.10	5 3 44.52	0.09	5 4 37.24	0.07	5 5 24.43	0.07	5 6 6.00	0.06	54
7	1 43.53	0.12	2 47.26	0.12	3 45.44	0.10	4 38.00	0.09	5 25.17	0.08	6 6.64	0.07	53
8	1 44.64	0.14	2 48.28	0.13	3 46.36	0.11	4 38.90	0.10	5 25.90	0.09	6 7.29	0.08	52
9	1 45.75	0.16	2 49.29	0.14	3 47.28	0.13	4 39.73	0.11	5 26.63	0.10	6 7.93	0.09	51
10	1 46.85	0.18	2 50.30	0.16	3 48.20	0.15	4 40.56	0.13	5 27.36	0.12	6 8.57	0.10	50
11	5 1 47.95	0.20	5 2 51.31	0.17	5 3 49.12	0.16	5 4 41.39	0.14	5 5 28.10	0.13	5 6 9.21	0.11	49
12	1 49.05	0.22	2 52.32	0.18	3 50.04	0.18	4 42.22	0.15	5 28.83	0.14	6 9.85	0.12	48
13	1 50.15	0.24	2 53.33	0.20	3 50.96	0.19	4 43.04	0.17	5 29.56	0.15	6 10.49	0.13	47
14	1 51.25	0.25	2 54.33	0.22	3 51.87	0.21	4 43.86	0.18	5 30.29	0.16	6 11.13	0.14	46
15	1 52.34	0.27	2 55.33	0.23	3 52.78	0.22	4 44.68	0.19	5 31.02	0.18	6 11.76	0.15	45
16	5 1 53.43	0.28	5 2 56.33	0.25	5 3 53.69	0.24	5 4 45.50	0.21	5 5 31.74	0.19	5 6 12.39	0.16	44
17	1 54.52	0.30	2 57.33	0.27	3 54.60	0.25	4 46.32	0.22	5 32.46	0.20	6 13.02	0.17	43
18	1 55.61	0.32	2 58.33	0.28	3 55.51	0.26	4 47.14	0.23	5 33.18	0.21	6 13.65	0.18	42
19	1 56.70	0.34	2 59.33	0.30	3 56.42	0.28	4 47.95	0.25	5 33.90	0.22	6 14.28	0.19	41
20	1 57.79	0.36	3 0.33	0.33	3 57.32	0.29	4 48.76	0.26	5 34.62	0.23	6 14.90	0.20	40
21	5 1 58.88	0.38	5 3 1.33	0.35	5 3 58.22	0.31	5 4 49.57	0.27	5 5 35.34	0.24	5 6 15.53	0.21	39
22	1 59.97	0.40	3 2.32	0.37	3 59.12	0.32	4 50.38	0.29	5 36.06	0.25	6 16.15	0.22	38
23	2 1.05	0.41	3 3.31	0.38	4 0.02	0.33	4 51.19	0.30	5 36.77	0.26	6 16.77	0.23	37
24	2 2.13	0.43	3 4.30	0.40	4 0.92	0.35	4 51.99	0.31	5 37.48	0.27	6 17.39	0.24	36
25	2 3.21	0.45	3 5.29	0.41	4 1.82	0.36	4 52.79	0.33	5 38.19	0.29	6 18.01	0.25	35
26	5 2 4.29	0.46	5 3 6.28	0.43	5 4 2.72	0.37	5 4 53.59	0.34	5 5 38.90	0.30	5 6 18.63	0.26	34
27	2 5.37	0.48	3 7.27	0.44	4 3.61	0.39	4 54.39	0.35	5 39.61	0.31	6 19.24	0.27	33
28	2 6.45	0.50	3 8.25	0.46	4 4.50	0.40	4 55.19	0.37	5 40.32	0.32	6 19.85	0.28	32
29	2 7.53	0.52	3 9.23	0.47	4 5.39	0.42	4 55.99	0.38	5 41.03	0.33	6 20.46	0.29	31
30	2 8.61	0.54	3 10.21	0.48	4 6.28	0.44	4 56.79	0.40	5 41.73	0.35	6 21.07	0.30	30
31	5 2 9.68	0.56	5 3 11.19	0.51	5 4 7.17	0.45	5 4 57.59	0.41	5 5 42.43	0.36	5 6 21.68	0.31	29
32	2 10.75	0.58	3 12.17	0.52	4 8.06	0.47	4 58.38	0.42	5 43.13	0.37	6 22.29	0.32	28
33	2 11.82	0.59	3 13.15	0.54	4 8.94	0.49	4 59.17	0.44	5 43.83	0.38	6 22.89	0.33	27
34	2 12.89	0.61	3 14.12	0.55	4 9.82	0.50	4 59.96	0.45	5 44.53	0.40	6 23.49	0.34	26
35	2 13.96	0.63	3 15.09	0.57	4 10.70	0.51	5 0.75	0.46	5 45.22	0.41	6 24.09	0.35	25
36	5 2 15.03	0.65	5 3 16.06	0.59	5 4 11.58	0.53	5 5 1.54	0.48	5 5 45.91	0.42	5 6 24.69	0.36	24
37	2 16.09	0.66	3 17.03	0.60	4 12.46	0.54	5 2.33	0.49	5 46.60	0.44	6 25.29	0.37	23
38	2 17.15	0.68	3 18.00	0.62	4 13.34	0.56	5 3.11	0.50	5 47.29	0.45	6 25.89	0.38	22
39	2 18.21	0.70	3 18.97	0.64	4 14.21	0.57	5 3.89	0.51	5 47.98	0.46	6 26.49	0.39	21
40	2 19.27	0.72	3 19.94	0.65	4 15.08	0.59	5 4.67	0.53	5 48.67	0.47	6 27.08	0.40	20
41	5 2 20.33	0.74	5 3 20.91	0.66	5 4 15.95	0.60	5 5 5.45	0.54	5 5 49.36	0.48	5 6 27.67	0.41	19
42	2 21.38	0.76	3 21.87	0.68	4 16.82	0.62	5 6.23	0.55	5 50.05	0.49	6 28.26	0.42	18
43	2 22.43	0.77	3 22.83	0.70	4 17.69	0.63	5 7.01	0.57	5 50.73	0.50	6 28.85	0.43	17
44	2 23.48	0.79	3 23.79	0.71	4 18.56	0.65	5 7.78	0.58	5 51.41	0.52	6 29.44	0.44	16
45	2 24.53	0.81	3 24.75	0.73	4 19.43	0.66	5 8.55	0.59	5 52.09	0.53	6 30.03	0.45	15
46	5 2 25.58	0.83	5 3 25.76	0.74	5 4 20.30	0.68	5 5 9.32	0.61	5 5 52.77	0.54	5 6 30.62	0.46	14
47	2 26.63	0.84	3 26.67	0.76	4 21.16	0.69	5 10.09	0.62	5 53.45	0.55	6 31.20	0.47	13
48	2 27.67	0.86	3 27.62	0.78	4 22.02	0.70	5 10.86	0.63	5 54.13	0.57	6 31.78	0.48	12
49	2 28.71	0.88	3 28.57	0.80	4 22.88	0.72	5 11.63	0.65	5 54.80	0.58	6 32.36	0.49	11
50	2 29.75	0.90	3 29.52	0.82	4 23.74	0.74	5 12.39	0.66	5 55.47	0.59	6 32.94	0.50	10
51	5 2 30.79	0.92	5 3 30.47	0.84	5 4 24.60	0.75	5 5 13.15	0.68	5 5 56.13	0.60	5 6 33.52	0.51	9
52	2 31.83	0.94	3 31.42	0.86	4 25.46	0.77	5 13.91	0.69	5 56.80	0.61	6 34.10	0.52	8
53	2 32.87	0.95	3 32.37	0.87	4 26.31	0.78	5 14.67	0.70	5 57.47	0.62	6 34.67	0.53	7
54	2 33.91	0.97	3 33.31	0.89	4 27.16	0.80	5 15.43	0.72	5 58.14	0.63	6 35.24	0.54	6
55	2 34.94	0.99	3 34.25	0.91	4 28.01	0.81	5 16.19	0.73	5 58.81	0.64	6 35.81	0.55	5
56	5 2 35.98	1.01	5 3 35.19	0.92	5 4 28.86	0.83	5 5 16.95	0.75	5 5 59.47	0.65	5 6 36.38	0.56	4
57	2 37.01	1.02	3 36.13	0.93	4 29.71	0.84	5 17.71	0.76	6 0.13	0.66	6 36.95	0.57	3
58	2 38.04	1.04	3 37.07	0.95	4 30.55	0.86	5 18.46	0.77	6 0.79	0.67	6 37.52	0.58	2
59	2 39.07	1.06	3 38.01	0.96	4 31.39	0.87	5 19.21	0.78	6 1.45	0.68	6 38.09	0.59	1
60	2 40.10	1.08	3 38.94	0.98	4 32.23	0.88	5 19.96	0.79	6 2.10	0.70	6 38.65	0.60	0
$\bar{y}$	101°+ 281°–		100°+ 280°–		99°+ 279°–		98°+ 278°–		97°+ 277°–		96°+ 276°–		$\bar{y}$

TABLE CXXXVII. ARGUMENT 77.

Equation = 18500″ sin $\bar{y}$.

$\bar{y}$	264°– 84°+ Equation.	P.P.	265°– 85°+ Equation.	P.P.	266°– 86°+ Equation.	P.P.	267°– 87°+ Equation.	P.P.	268°– 88°+ Equation.	P.P.	269°– 89°+ Equation.	P.P.	$\bar{y}$
0′	5° 6′ 38.65″	0.00″	5° 7′ 9.60″	0.00″	5° 7′ 34.92″	0.00″	5° 7′ 54.65″	0.00″	5° 8′ 8.73″	0.00″	5° 8′ 17.18″	0.00″	60′
1	6 39.21	0.01	7 10.07	0.01	7 35.29	0.01	7 54.93	0.00	8 8.92	0.00	8 17.27	0.00	59
2	6 39.77	0.02	7 10.54	0.01	7 35.66	0.01	7 55.21	0.01	8 9.11	0.00	8 17.36	0.00	58
3	6 40.33	0.03	7 11.01	0.02	7 36.03	0.02	7 55.49	0.01	8 9.29	0.01	8 17.45	0.00	57
4	6 40.89	0.04	7 11.47	0.02	7 36.40	0.02	7 55.76	0.01	8 9.47	0.01	8 17.54	0.00	56
5	6 41.45	0.05	7 11.93	0.03	7 36.77	0.03	7 56.03	0.02	8 9.65	0.01	8 17.62	0.00	55
6	5 6 42.01	0.06	5 7 12.39	0.04	5 7 37.14	0.03	5 7 56.30	0.02	5 8 9.83	0.01	5 8 17.71	0.01	54
7	6 42.56	0.07	7 12.85	0.04	7 37.51	0.04	7 56.57	0.03	8 10.01	0.02	8 17.79	0.01	53
8	6 43.11	0.08	7 13.31	0.05	7 37.87	0.04	7 56.84	0.03	8 10.19	0.02	8 17.87	0.01	52
9	6 43.66	0.09	7 13.77	0.06	7 38.23	0.05	7 57.11	0.04	8 10.36	0.02	8 17.95	0.01	51
10	6 44.21	0.09	7 14.22	0.07	7 38.59	0.05	7 57.38	0.04	8 10.53	0.02	8 18.03	0.01	50
11	5 6 44.76	0.10	5 7 14.67	0.07	5 7 38.95	0.06	5 7 57.65	0.04	5 8 10.70	0.02	5 8 18.11	0.01	49
12	6 45.30	0.11	7 15.12	0.08	7 39.31	0.06	7 57.92	0.04	8 10.87	0.02	8 18.19	0.01	48
13	6 45.84	0.12	7 15.57	0.09	7 39.67	0.07	7 58.18	0.05	8 11.04	0.02	8 18.27	0.02	47
14	6 46.38	0.12	7 16.02	0.10	7 40.03	0.07	7 58.44	0.05	8 11.21	0.03	8 18.34	0.02	46
15	6 46.92	0.13	7 16.46	0.10	7 40.38	0.08	7 58.70	0.05	8 11.37	0.03	8 18.41	0.02	45
16	5 6 47.46	0.14	5 7 16.91	0.11	5 7 40.73	0.08	5 7 58.96	0.06	5 8 11.53	0.03	5 8 18.48	0.02	44
17	6 48.00	0.15	7 17.35	0.12	7 41.08	0.09	7 59.22	0.06	8 11.69	0.03	8 18.55	0.02	43
18	6 48.54	0.16	7 17.79	0.12	7 41.43	0.09	7 59.48	0.06	8 11.85	0.04	8 18.62	0.02	42
19	6 49.07	0.17	7 18.23	0.13	7 41.78	0.10	7 59.73	0.07	8 12.01	0.04	8 18.69	0.02	41
20	6 49.60	0.17	7 18.67	0.14	7 42.13	0.11	7 59.98	0.07	8 12.17	0.05	8 18.75	0.02	40
21	5 6 50.13	0.18	5 7 19.11	0.15	5 7 42.47	0.12	5 8 0.23	0.07	5 8 12.33	0.05	5 8 18.81	0.02	39
22	6 50.66	0.19	7 19.55	0.15	7 42.81	0.12	8 0.48	0.07	8 12.48	0.05	8 18.87	0.02	38
23	6 51.19	0.20	7 19.98	0.16	7 43.15	0.13	8 0.73	0.08	8 12.63	0.05	8 18.93	0.02	37
24	6 51.72	0.20	7 20.41	0.17	7 43.49	0.13	8 0.97	0.08	8 12.78	0.06	8 18.99	0.02	36
25	6 52.24	0.21	7 20.84	0.18	7 43.83	0.14	8 1.21	0.08	8 12.93	0.06	8 19.04	0.02	35
26	5 6 52.76	0.22	5 7 21.27	0.18	5 7 44.17	0.14	5 8 1.45	0.09	5 8 13.08	0.06	5 8 19.09	0.02	34
27	6 53.28	0.23	7 21.70	0.19	7 44.50	0.15	8 1.69	0.09	8 13.23	0.06	8 19.14	0.02	33
28	6 53.80	0.24	7 22.12	0.20	7 44.83	0.15	8 1.93	0.10	8 13.38	0.07	8 19.19	0.02	32
29	6 54.32	0.25	7 22.54	0.20	7 45.16	0.16	8 2.17	0.10	8 13.52	0.07	8 19.24	0.02	31
30	6 54.83	0.26	7 22.96	0.21	7 45.49	0.16	8 2.40	0.11	8 13.66	0.07	8 19.29	0.02	30
31	5 6 55.34	0.27	5 7 23.38	0.22	5 7 45.82	0.16	5 8 2.63	0.11	5 8 13.80	0.07	5 8 19.34	0.02	29
32	6 55.85	0.28	7 23.80	0.23	7 46.15	0.17	8 2.86	0.12	8 13.94	0.08	8 19.38	0.02	28
33	6 56.36	0.29	7 24.22	0.23	7 46.47	0.17	8 3.09	0.12	8 14.08	0.08	8 19.42	0.02	27
34	6 56.87	0.29	7 24.63	0.24	7 46.79	0.18	8 3.32	0.13	8 14.22	0.08	8 19.46	0.03	26
35	6 57.38	0.30	7 25.04	0.25	7 47.11	0.18	8 3.55	0.13	8 14.35	0.09	8 19.50	0.03	25
36	5 6 57.89	0.31	5 7 25.45	0.25	5 7 47.43	0.19	5 8 3.78	0.13	5 8 14.48	0.09	5 8 19.54	0.03	24
37	6 58.40	0.32	7 25.86	0.26	7 47.75	0.19	8 4.00	0.14	8 14.61	0.09	8 19.58	0.03	23
38	6 58.90	0.33	7 26.27	0.27	7 48.07	0.20	8 4.22	0.14	8 14.74	0.09	8 19.62	0.03	22
39	6 59.40	0.33	7 26.68	0.27	7 48.39	0.20	8 4.44	0.14	8 14.87	0.09	8 19.65	0.03	21
40	6 59.90	0.34	7 27.09	0.28	7 48.70	0.21	8 4.66	0.15	8 15.00	0.09	8 19.68	0.03	20
41	5 7 0.40	0.35	5 7 27.50	0.28	5 7 49.01	0.21	5 8 4.88	0.15	5 8 15.12	0.09	5 8 19.71	0.03	19
42	7 0.90	0.36	7 27.91	0.29	7 49.32	0.22	8 5.10	0.15	8 15.24	0.10	8 19.74	0.03	18
43	7 1.40	0.37	7 28.31	0.30	7 49.63	0.22	8 5.32	0.15	8 15.36	0.10	8 19.77	0.03	17
44	7 1.90	0.38	7 28.71	0.31	7 49.94	0.23	8 5.53	0.16	8 15.48	0.10	8 19.80	0.04	16
45	7 2.39	0.38	7 29.11	0.31	7 50.25	0.23	8 5.74	0.16	8 15.60	0.11	8 19.82	0.04	15
46	5 7 2.88	0.39	5 7 29.51	0.32	5 7 50.56	0.24	5 8 5.95	0.17	5 8 15.72	0.11	5 8 19.84	0.04	14
47	7 3.37	0.40	7 29.91	0.33	7 50.86	0.24	8 6.16	0.17	8 15.84	0.11	8 19.86	0.04	13
48	7 3.86	0.41	7 30.31	0.34	7 51.16	0.25	8 6.37	0.17	8 15.95	0.11	8 19.88	0.04	12
49	7 4.35	0.42	7 30.70	0.34	7 51.46	0.26	8 6.58	0.18	8 16.06	0.12	8 19.90	0.04	11
50	7 4.84	0.43	7 31.09	0.35	7 51.76	0.27	8 6.78	0.18	8 16.17	0.12	8 19.92	0.04	10
51	5 7 5.32	0.44	5 7 31.48	0.36	5 7 52.06	0.27	5 8 6.98	0.18	5 8 16.28	0.12	5 8 19.94	0.04	9
52	7 5.80	0.45	7 31.87	0.36	7 52.36	0.28	8 7.18	0.19	8 16.39	0.12	8 19.95	0.04	8
53	7 6.28	0.46	7 32.26	0.37	7 52.65	0.28	8 7.38	0.19	8 16.50	0.13	8 19.96	0.04	7
54	7 6.76	0.47	7 32.65	0.38	7 52.94	0.29	8 7.58	0.19	8 16.60	0.13	8 19.97	0.05	6
55	7 7.24	0.47	7 33.03	0.39	7 53.23	0.29	8 7.78	0.20	8 16.70	0.13	8 19.98	0.05	5
56	5 7 7.72	0.48	5 7 33.41	0.39	5 7 53.52	0.30	5 8 7.97	0.20	5 8 16.80	0.14	5 8 19.98	0.05	4
57	7 8.19	0.49	7 33.79	0.40	7 53.81	0.30	8 8.16	0.21	8 16.90	0.14	8 19.99	0.05	3
58	7 8.66	0.50	7 34.17	0.41	7 54.09	0.31	8 8.35	0.21	8 17.00	0.14	8 19.99	0.05	2
59	7 9.13	0.51	7 34.55	0.41	7 54.37	0.31	8 8.54	0.22	8 17.09	0.14	8 20.00	0.05	1
60	7 9.60	0.51	7 34.92	0.42	7 54.65	0.32	8 8.73	0.22	8 17.18	0.14	8 20.00	0.05	0
$\bar{y}$	95°+ 275°–		94°+ 274°–		93°+ 273°–		92°+ 272°–		91°+ 271°–		90°+ 270°–		$\bar{y}$

www.ingramcontent.com/pod-product-compliance
Lightning Source LLC
Chambersburg PA
CBHW021405210326
41599CB00011B/1013

* 9 7 8 3 3 3 7 3 9 6 2 2 0 *